ESSENTIAL PUBLIC HEALTH

Series Editor: R

Essentials of Environmental Health

THIRD EDITION

Robert H. Friis, PhD
Emeritus Professor and Chair,
Department of Health Science - California State University, Long Beach

JONES & BARTLETT LEARNING

World Headquarters
Jones & Bartlett Learning
5 Wall Street
Burlington, MA 01803
978-443-5000
info@jblearning.com
www.jblearning.com

Jones & Bartlett Learning books and products are available through most bookstores and online booksellers. To contact Jones & Bartlett Learning directly, call 800-832-0034, fax 978-443-8000, or visit our website, www.jblearning.com.

Substantial discounts on bulk quantities of Jones & Bartlett Learning publications are available to corporations, professional associations, and other qualified organizations. For details and specific discount information, contact the special sales department at Jones & Bartlett Learning via the above contact information or send an email to specialsales@jblearning.com.

13693-7

Production Credits
VP, Product Management: David D. Cella
Director of Product Management: Michael Brown
Product Specialist: Carter McAlister
Production Manager: Carolyn Rogers Pershouse
Production Editor: Kelly Sylvester
Senior Marketing Manager: Sophie Fleck Teague
Marketing Communications Manager: Katie Hennessy
Production Services Manager: Colleen Lamy
Manufacturing and Inventory Control Supervisor: Amy Bacus
Composition: codeMantra U.S. LLC
Cover Design: Kristin E. Parker
Director of Rights & Media: Joanna Gallant
Rights & Media Specialist: Wes DeShano
Media Development Editor: Shannon Sheehan
Cover Image (Title Page, Part Opener, Chapter Opener): © Jean-Luc Rivard/EyeEm/Getty Images
Printing and Binding: LSC Communications
Cover Printing: LSC Communications

Library of Congress Cataloging-in-Publication Data
Names: Friis, Robert H., author.
Title: Essentials of environmental health / Robert H. Friis.
Description: Third edition. | Burlington, MA: Jones & Bartlett Learning, [2019] | Includes bibliographical references and index.
Identifiers: LCCN 2017058190 | ISBN 9781284123975 (paperback)
Subjects: LCSH: Environmental health.
Classification: LCC RA565 .F75 2019 | DDC 362.1—dc23
LC record available at https://lccn.loc.gov/2017058190

6048

Printed in the United States of America
22 21 20 19 18 10 9 8 7 6 5 4 3 2 1

To C.A.F.

Contents

Prologue

Dr. Robert Friis' *Essentials of Environmental Health* was the first text in the *Essential Public Health* Series. In the decade since the first edition was published, the importance of environmental health has been brought to the world's attention by crises and challenges ranging from oil spills to climate change, and from lead in water to pandemic influenza to Zika.

The third edition of *Essentials of Environmental Health* builds upon the previous editions. It keeps up with the changes in environmental health to produce a state-of-the-art introductory text. The third edition has been updated extensively to reflect the rapidly changing context of the environmental health field. Many new charts and figures have been added. Examples of new topics in this edition are the following:

- *Healthy People 2020* environmental objectives
- Expanded coverage of climate change
- Current status of population growth
- Mushroom poisoning
- Cap-and-trade policies in California
- Zika virus infection
- Lead contamination of water supply in Flint, Michigan
- Colony collapse disorder among honey bees
- Fukushima Daiichi Nuclear Power Plant
- Changing distribution of unintentional injury deaths

Dr. Friis also has produced a comprehensive set of ancillary materials accessible in the Navigate 2 Advantage package that accompanies this text. These resources provide the most recent information to broaden students' knowledge of environmental health. They also provide faculty with resources to challenge students and deepen their understanding of environmental health.

The *Essential Public Health* Series now includes over 20 books covering the full range of introductory texts. Dr. Friis has also authored *Epidemiology 101* as part of the series. *Epidemiology 101* is a core undergraduate public health text and a key component of the "101" approach to undergraduate public health education that also includes *Public Health 101* and *Global Health 101*. *Essentials of Environmental Health* has rapidly become a core text as well, key to a comprehensive curriculum in public health.

Dr. Friis brings to all his writing in environmental health and epidemiology a lifetime commitment to teaching, a personal connection to students as they begin their study of public health, and an impressive ability to clearly present complex subjects. I know that you will enjoy and benefit from this text.

Richard Riegelman, MD, MPH, PhD
Series Editor—Essential Public Health

© Jean-Luc Rivard/EyeEm/Getty Images

Preface

When you follow the media, dramatic pictures and accounts of chemical spills, industrial fires, and other environmental disasters are not unusual. If you read the newspaper, surf the Internet, watch television, or merely observe what is happening around you, you will become aware of the environmental threats that are occurring on a worldwide scale. Since the publication of the first edition of *Essentials of Environmental Health* in 2007, our awareness of how threats to the environment endanger the survival of all life on the planet has continued to grow.

Within the past decade, public debate about global warming—a controversial topic—has escalated. Some worldwide climate changes such as extreme weather events have been attributed to global warming. The devastation caused by the August 2005 Hurricane Katrina in New Orleans was particularly moving. Other examples of extreme climate variations are increases in desertification in some parts of the world; at the same time, other areas such as Pakistan and China have experienced periodic massive flooding, for example, during the summer of 2010. In 2017, hurricanes devastated Puerto Rico and other Caribbean islands as well as Texas and Florida on the United States mainland.

As a consequence of the warming of the earth's temperature, mountain glaciers, masses of ice in Greenland, and glaciers and sea ice in the Arctic and Antarctic regions have been melting. Everyone should be concerned about the potential impact of these dramatic trends and incidents on the global physical environment and the health and survival of the world's population. At the same time, a hopeful attitude is warranted because of the progress that has been made in protecting environmental quality.

I have always been interested in the environment, especially the impact of rapid growth in California, my home state. During my lifetime, I have witnessed the conversion of pastoral northern California farmlands and orchards into densely populated urban zones with consequential environmental degradation. Although destruction of Brazilian rainforests is a focus of environmental advocates, wide swaths of the redwood forests of northern California and the wooded areas of the Pacific Northwest also are impacted by deforestation and are in need of protection.

In addition to urbanization in my home state, rapid population growth has also been a global phenomenon. In my travels over the past decades, I have observed this phenomenon in Mexico and some parts of Europe. One example is the transformation of small cities in Baja, California into booming metropolises. Another example is the residential and commercial development that now fill the open, snow-covered fields that I traipsed through as a student in western Germany on the way to language classes. Despite urban growth, German cities have adopted exemplary programs for urban planning and recycling solid wastes. Other northern European countries, such as Sweden, have implemented development policies that preserve open space and encourage citizens to use public transportation. Member states of the European Union have been leaders in the development of wind and solar power, alternative fuels such as biodiesel, and energy-saving technologies.

Increasing population and development pressures affect southern Europe as well. As a result of water overdraft, some European cities along the Mediterranean have had salt water intrude into their public water supplies. When I first traveled to Europe as a student, I was impressed with the fact that people used bottled water instead of tap water, and later I began to understand why—most of the time, the tap water simply was not as palatable as bottled water. Nowadays, however, it is also common for people in the United States to tote around bottles of "prestige water," perhaps because of their fear of contaminants and microbes that may be present in the tap water. (In most cases, this fear is unwarranted because the quality of municipal tap water is highly regulated.)

Back in California, we now know that because of the introduction of toxic chemicals and pesticides, sewage contamination, and overfishing, our once abundant fisheries are declining and fishing must be limited. For example, it was once possible to consume fresh abalone, but now this delicious crustacean is almost unavailable from the wild due to excessive human predation. Seafood lovers are advised not to consume certain species of fish, which may be contaminated with mercury and other harmful substances.

During the spring and summer of 2010, a catastrophe at the *Deepwater Horizon* oil rig off the Louisiana coast caused the largest marine oil spill up to that time—about 5 million barrels. Aside from the immediate impact on wildlife and the fishing and tourism industries, the long-term environmental effects will require years to be ascertained. This catastrophe has called into question the safety of offshore oil drilling operations and focused attention on the adverse environmental consequences of oil extraction in general.

Still another concern is the impact of environmental factors on the safety of the global food supply. During late summer 2010, millions of eggs produced in the midwestern United States and sold widely by supermarket chains were found to be contaminated with *Salmonella* bacteria. Inspection of the poultry farms revealed that chickens were raised in despicable environmental conditions, which could contribute to the proliferation of *Salmonella* contamination. Within the past decade, other foodborne disease outbreaks in the United States have been associated with tomatoes (*Salmonella*) as well as lettuce and ground beef (*Escherichia coli*). Noteworthy is the increasing internationalization of the world's food supply—adulterated and unsafe foodstuffs produced in one country can be shipped globally and threaten all of humanity. An example is the melamine-contaminated baby formula manufactured in China.

Since 2010, environmental issues have continued to command our attention. Global political conflicts and war have been an impetus for migration to the European Union and North America. Migration impacts the environment and taxes the resources of host countries. Another continuing issue is the lack of clean water in many countries. Unsafe water provides an avenue for the transmission of cholera, amebiasis, and other deadly waterborne scourges. Often, many of the same countries that do not have reliable water supplies also lack sanitary toilet facilities.

An ongoing environmental topic involves the safety of nuclear power plants. In 2011, an earthquake and tsunami caused a meltdown of the Fukushima Daiichi Nuclear Reactor in northern Japan and release of radioactive materials into the surrounding area. This event followed the earlier meltdowns of the reactors at Three Mile Island in Pennsylvania (1979) and Chernobyl, Ukraine (1986).

The picture is not entirely bleak, however; much progress has been made in informing the public about environmental health hazards and introducing regulations and procedures for the control of these hazards. Notable are the use of unleaded fuels and catalytic converters in automobiles to control air pollution and lead contamination. Not only have important environmental protections such as these been implemented in the United States, they also have been adopted by European countries and elsewhere. Global recognition of the adverse impacts of climate change has led the world community, especially countries that are heavy users of fossil fuels, to formulate and adopt protocols that aim to reduce emissions of greenhouse gases.

Another example of a successful environmental policy that is now enforced in many parts of the world is the regulation of smoking in alcohol-serving establishments. This policy originated as California's Smokefree Bars Law, which controls occupational exposure to environmental tobacco smoke and also protects bar and restaurant patrons. At national and international levels, regulatory agencies conduct research and develop and enforce laws that control environmental health hazards. Although these efforts hold much promise for maintaining environmental quality, it must be kept in mind that policy and regulation are strongly influenced by the political process. Often economic considerations such as the need to maintain jobs and the prospect of increased taxation must be balanced against environmental protections.

This text addresses the important topics and methodological approaches in the environmental health field in order to provide a diversity of learners (from the beginning student to the experienced health professional) with an overview of the field. I believe that knowledge of environmental health issues can lead to an appreciation of humankind's connection with the earth and the precarious balance between human activities and environmental resources.

Robert H. Friis, PhD

Acknowledgments

My colleagues and students were extremely helpful in providing comments and background information necessary to complete this project. For the first edition of the text, I wish to thank the following former students: Ibtisam Khoury, Lillian Camacho, Sheetal Monga, Manohar Sukumar, and Heidi Burkey from California State University, Long Beach (CSULB); and Nada Hamade from the University of California, Irvine. Students helped with literature searches, reviewed written text materials, and provided feedback. I also acknowledge the contributions of Dr. Robert Phalen and Dr. Yee-lean Lee, University of California, Irvine; Dr. Jan Semenza, European Center for Disease Prevention and Control; Dr. Glenn Paulson, School of Public Health, University of Medicine and Dentistry of New Jersey; and Dr. Michelle Saint-Germain and Dr. Javier Lopez-Zetina, CSULB. These professional colleagues carefully reviewed chapters that were relevant to their areas of expertise.

For the second edition, I express my gratitude to the following students who helped with literature searches and review of the text: Sarah Long, Paula Griego, Jane Kil, and Dexter Dizon. For the third edition, I thank CSULB graduate student Nutan Kafle for her assistance with literature searches.

Also, I would like to thank anonymous reviewers whose insightful comments greatly enhanced the quality of this text. Mike Brown, Director of Product Management for Jones & Bartlett Learning, provided continued encouragement and motivation for completion of the project; Jones & Bartlett Learning staff offered much expertise. Finally, my wife, Carol Friis, was involved extensively with this project for all three editions—she provided detailed editorial comments, verified the accuracy of the references, and helped with many other aspects of the project. Without her support and assistance, completion of the text would not have been possible.

About the Author

Robert H. Friis, PhD, is Professor and Chair, emeritus, of the Department of Health Science at California State University, Long Beach (CSULB), and former Director of the CSULB-Veterans Affairs Medical Center, Long Beach, Joint Studies Institute. He is a past president of the Southern California Public Health Association. He serves or has served on the advisory boards of several health-related organizations, including the California Health Interview Survey. He previously retired from the University of California, Irvine, where he was an Associate Clinical Professor in the Department of Medicine, Department of Neurology, and School of Social Ecology. Dr. Friis is an epidemiologist by training and profession and has an enduring fascination with environmental health.

As a health department epidemiologist, he led investigations into environmental health problems such as chemical spills, air pollution, and foodborne illness. He has taught courses on epidemiology, environmental health, and statistics at universities in New York City and southern California. The topics of his research, publications, and presentations include tobacco use, mental health, chronic disease, disability, minority health, and psychosocial epidemiology.

Dr. Friis has been principal investigator or coinvestigator on grants and contracts from the University of California's Tobacco-Related Disease Research Program, the National Institutes of Health, and other agencies. This funding has supported investigations into topics such as tobacco control policies, geriatric health, depression in Hispanic populations, and infectious disease transmission in nursing homes. His academic interests have led him to conduct research in Mexico City and European countries. He has been a visiting professor at the Center for Nutrition and Toxicology, Karolinska Institute, Stockholm, Sweden; the Max Planck Institute, Munich, Germany; and the Medizinische Fakultät Carl Gustav Carus of the Dresden Technical University, Dresden, Germany. He reviews articles frequently for scientific journals and is on the international editorial board of *Public Health* and the *Journal of Public Health*. Dr. Friis is a fellow of the Royal Society of Public Health, lifetime member of the governing council of the Southern California Public Health Association, member of the Society for Epidemiologic Research, and member of the American Public Health Association. His awards include a postdoctoral fellowship for study at the Institute for Social Research, University of Michigan, and the Achievement Award for Scholarly and Creative Activity from California State University, Long Beach.

He is author/coauthor of the following texts published by Jones & Bartlett Learning:

- *Epidemiology for Public Health Practice*, with Thomas A. Sellers (editions one through five)
- *Essentials of Environmental Health* (editions one and two)
- *Epidemiology 101* (editions one and two)
- *Occupational Health and Safety for the 21st Century*

He is also the author/coauthor of texts on biostatistics and community/public health and is the editor of the *Praeger Handbook of Environmental Health*. Dr. Friis has helped to create online courses for several universities and other organizations.

Introduction

The purpose of this text is to inform the reader about the key areas of environmental health and instill awareness about the crucial role of the environment in the health of the planet and all living creatures. Organized according to three major domains—background, environmental disease agents, and applications—the text begins with background material and "tools of the trade" (environmental epidemiology, environmental toxicology, and environmental policy and regulation). The text then covers specific agents (e.g., microbial agents, toxic metals, pesticides, and ionizing and nonionizing radiation) of environmentally related diseases. Finally, applications and domains of environmental health are addressed (water and air quality, food safety, waste disposal, occupational health, and injuries).

This work is intended for graduate and undergraduate students who take environmental health courses in a variety of settings. Often, these courses are offered by schools of public health and health science departments. The text can also be used in online courses and instruction in intensive courses offered in a nontraditional format. Taking a nontechnical approach, the text should be accessible and interesting to students who have not had a great deal of previous introductory background, especially in the sciences. Nevertheless, the text should appeal to more advanced students as well. In order to generate interest in the subject matter, the author has included many examples and illustrations of environmental health issues. Text boxes throughout provide detailed information on selected topics. Other instructional aids are a list of learning objectives at the beginning of each chapter and study questions and exercises at the end.

A summary of the content of each chapter follows. Part I: Background of the Field, includes Chapters 1 through 4. Chapter 1 illustrates the role of environmental health in contemporary society, presents examples from the history of environmental health, and delimits the scope of the environmental health field; in addition, career opportunities are featured. Chapter 2 covers environmental epidemiology, one of the fundamental disciplines used in the study of environmental health. The subject of Chapter 3 is environmental toxicology, which, along with environmental epidemiology, is one of the key disciplines of the environmental health field. Chapter 4 focuses on environmental policy and regulation. Sometimes this content is placed near the end of environmental health texts. The author has elected to move it closer to the beginning because an appreciation of policy issues is crucial to the understanding of specific domains of environmental health. In addition, Chapter 4 contains extensive coverage of major environmental regulatory agencies and major US environmental health laws. It is important to cover these topics early in the text because references to agencies and laws are made in the chapters that follow. The reader may want to use this material for reference and consult it later while reading the remaining chapters.

The next group of chapters, Part II: Agents of Environmental Disease, Chapters 5 through 8, covers agents of environmental disease. The respective topics are zoonotic and vector-borne diseases, toxic metals and elements, pesticides and other organic chemicals, and ionizing and nonionizing radiation. Part III: Applications of Environmental Health, Chapters 9 through 14 deals with applications of environmental health: water quality, air quality, food safety, solid and liquid wastes, occupational health, and injuries.

Other components of the text are a glossary of key definitions and a list of abbreviations. For additional information and learning aids that reinforce the didactic content of the text, the author recommends that readers access the *Navigate 2 Advantage Access* online component. Some instructors enhance the learning experience by conducting field visits to environmentally relevant sites in the community (e.g., the municipal water plant). Other means to support the course are the use of videos and online resources.

Abbreviations

Term	Definition
APHIS	Animal and Plant Health Inspection Service (of USDA)
AQI	air quality index
ATSDR	Agency for Toxic Substances and Disease Registry
BHA	butylated hydroxyl anisole
BHT	butylated hydroxytoluene
BLL	blood lead level
BLS	Bureau of Labor Statistics
BRI	building-related illness
BSE	bovine spongiform encephalopathy
CAA	Clean Air Act
CDC	Centers for Disease Control and Prevention
CERCLA	Comprehensive Environmental Response, Compensation, and Liability Act, 1980
CFR	case fatality rate
CFSAN	Center for Food Safety and Applied Nutrition
CHD	coronary heart disease
Ci	Curie
CNS	central nervous system
CO	carbon monoxide
COHb	carboxyhemoglobin
COPD	chronic obstructive pulmonary disease
Cr (VI)	hexavalent chromium
CTS	carpal tunnel syndrome
CWP	coal workers' pneumoconiosis
dB	decibel
DBPs	disinfection byproducts (of water)
DDT	dichlorodiphenyltrichloroethane
DES	diethylstilbestrol
DHF/DSS	dengue hemorrhagic fever/dengue shock syndrome
DHHS	Department of Health and Human Services
DNA	deoxyribonucleic acid
EBCLIS	EMF and Breast Cancer on Long Island Study
EDTA	ethylenediaminetetraacetic acid
EEA	European Environment Agency
EIA	environmental impact assessment
ELF	extremely low frequency radiation
EMF	electromagnetic field
EPA	U.S. Environmental Protection Agency
ETS	environmental tobacco smoke
EU	European Union
FAO	Food and Agricultural Organization (of United Nations)
FDA	Food and Drug Administration
FEMA	Federal Emergency Management Agency
FIFRA	Federal Insecticide, Fungicide, and Rodenticide Act, 1996
FoodNet	Foodborne Diseases Active Surveillance Network (of CDC)
FSIS	Food Safety and Inspection Service (of USDA)
FWS	Fish and Wildlife Service
GM	genetically modified
GRAS	generally recognized as safe
HACCP	hazard analysis of critical control points
HAV	hepatitis A virus
HBV	hepatitis B virus
HCV	hepatitis C virus
HIA	health impact assessment
HIV/AIDS	human immunodeficiency virus/ acquired immunodeficiency syndrome
HPS	hantavirus pulmonary syndrome
HSEES	Hazardous Substances Emergency Events Surveillance (of ATSDR)
HUS	hemolytic uremic syndrome
IARC	International Agency for Research on Cancer
LD_{50}	lethal dose 50
MIC	methyl isocyanate
MM	malignant melanoma
MSD	musculoskeletal disorder
MSG	monosodium glutamate
MSHA	Mine Safety and Health Administration
MSW	municipal solid waste
NAAQS	National Ambient Air Quality Standards
NCHS	National Center for Health Statistics
NEPA	National Environmental Policy Act, 1969

NIEHS	National Institute of Environmental Health Sciences
NIH	National Institutes of Health
NIOSH	National Institute for Occupational Safety and Health
NMSC	non-melanoma skin cancer
NO_x	nitrogen oxides
NORA	National Occupational Research Agenda
NPL	National Priorities List
NRDC	Natural Resources Defense Council
NTOF	National Traumatic Occupational Fatalities Surveillance System
NTP	National Toxicology Program
O_3	ozone
OAQPS	Office of Air Quality Planning and Standards (of EPA)
OECD	European Organization for Economic Cooperation and Development
OPs	organophosphates
OR	odds ratio
OSCAR	Osteoporosis with Cadmium as a Risk Factor [study]
OSHA	Occupational Safety and Health Administration
PAHs	polycyclic aromatic hydrocarbons
PAYT	pay-as-you-throw
PCBs	polychlorinated biphenyls
PEP	post-exposure prophylaxis
PM	particulate matter
POPs	persistent organic pollutants
ppb	parts per billion
PPCPs	pharmaceutical and personal care products
PPE	personal protective equipment
ppm	parts per million
PSP	paralytic shellfish poison
RCRA	Resource Conservation and Recovery Act, 1976
RDD	radiological dispersal device
RF	radio frequency
RMSF	Rocky Mountain spotted fever
RR	relative risk
RVF	Rift Valley fever
SARA	Superfund Amendments and Reauthorization Act
SBS	sick building syndrome
SCE	sister chromatid exchange
SDWD	Safe Drinking Water Act
SI	System International
SMR	standardized mortality ratio
SO_2	sulfur dioxide
SPL	sound pressure level
TCDD	2,3,7,8-tetrachlorodibenzo-p-dioxin
TCE	trichloroethylene
TFR	total fertility rate
TLV	threshold limit value
TSCA	Toxic Substances Control Act, 1976
UNEP	United Nations Environment Programme
USDA	U.S. Department of Agriculture
USGS	United States Geological Survey
USPHS	U.S. Public Health Service
UVR	ultraviolet radiation
vCJD	Creutzfeldt-Jakob disease, new variant
VHF	viral hemorrhagic fever
VOCs	volatile organic compounds
VSP	Vessel Sanitation Program (operated by CDC)
WHO	World Health Organization
WNV	West Nile virus

© Jean-Luc Rivard/EyeEm/Getty Images

PART I

Background of the Field

© Jean-Luc Rivard/EyeEm/Getty Images

CHAPTER 1

Introduction: The Environment At Risk

LEARNING OBJECTIVES

By the end of this chapter the reader will be able to:

- Describe how environmental health problems influence our lives.
- Discuss the potential impacts of population growth upon the environment.
- State a definition of the term *environmental health*.
- List at least five major events in the history of environmental health.
- Summarize employment opportunities in the environmental health field.

Introduction

This chapter will illustrate how the environment impacts the health of people and survival of every living being on the planet. You will learn about key terms used in environmental health and the scope of the field. The focus will be on distinguishing features of the field and the basic concepts, which are essential to this discipline. For example, one of these concepts is the relationship between world population growth and the environment. Another concept relates to historically significant environmental events and how they influenced the topics that are of current importance to the environmental health field. An additional topic involves employment classifications, career roles, and opportunities for environmental health workers. The chapter will conclude with an overview of the textbook: the roles of **environmental epidemiology** and **toxicology**, policy aspects of environmental health, examples of environmentally related agents and diseases, and specific content areas of environmental health such as air quality, water quality, food safety, and waste disposal.

Progress and Challenges in Protecting Our Environment

Although much progress has been made in protecting our environment, many lingering challenges confront humanity. Maintaining environmental quality is a pressing task for the 21st century. Often achievements in environment quality are limited primarily to the developed world, which has the financial wherewithal to address environmental health.

Improvement in environmental quality is an official goal of the US government, as articulated in *Healthy People 2020*. This goal (number 8, Environmental Health) is formatted as follows: "Promote health for all through a healthy environment."[1] A list of environmental objectives is shown in **TABLE 1.1**.

TABLE 1.1 Objectives for *Healthy People 2020*—Environmental Health Goal: Promote Health for All through a Healthy Environment

Objective
Outdoor Air Quality
EH-1 Reduce the number of days the Air Quality Index (AQI) exceeds 100, weighted by population and AQI.
EH-2 Increase use of alternative modes of transportation for work.
EH-3 Reduce air toxic emissions to decrease the risk of adverse health effects caused by mobile, area, and major sources of airborne toxics.
Water Quality
EH-4 Increase the proportion of persons served by community water systems who receive a supply of drinking water that meets the regulations of the Safe Drinking Water Act.
EH-5 Reduce waterborne disease outbreaks arising from water intended for drinking among persons served by community water systems.
EH-6 Reduce per capita domestic water withdrawals with respect to use and conservation.
EH-7 Increase the proportion of days that beaches are open and safe for swimming.
Toxics and Waste
EH-8 Reduce blood lead levels in children.
EH-9 Minimize the risks to human health and the environment posed by hazardous sites.
EH-10 Reduce pesticide exposures that result in visits to a health care facility.
EH-11 Reduce the amount of toxic pollutants released into the environment.
EH-12 Increase recycling of municipal solid waste.
Healthy Homes and Healthy Communities
EH-13 Reduce indoor allergen levels.
EH-14 Increase the proportion of homes with an operating radon mitigation system for persons living in homes at risk for radon exposure.
EH-15 Increase the proportion of new single-family homes (SFH) constructed with radon-reducing features, especially in high-radon-potential areas.
EH-16 Increase the proportion of the Nation's elementary, middle, and high schools that have official school policies and engage in practices that promote a healthy and safe physical school environment.
EH-17 (Developmental) Increase the proportion of persons living in pre-1978 housing that has been tested for the presence of lead-based paint or related hazards.

EH-18 Reduce the number of U.S. homes that are found to have lead-based paint or related hazards.
EH-19 Reduce the proportion of occupied housing units that have moderate or severe physical problems.
Infrastructure and Surveillance
EH-20 Reduce exposure to selected environmental chemicals in the population, as measured by blood and urine concentrations of the substances or their metabolites.
EH-21 Improve quality, utility, awareness, and use of existing information systems for environmental health.
EH-22 Increase the number of States, Territories, Tribes, and the District of Columbia that monitor diseases or conditions that can be caused by exposure to environmental hazards.
EH-23 Reduce the number of public schools located within 150 meters of major highways in the United States.
Global Environmental Health
EH-24 Reduce the global burden of disease due to poor water quality, sanitation, and insufficient hygiene.

Modified from US Department of Health and Human Services. Office of Disease Prevention and Health Promotion. *Healthy People 2020*: Environmental Health. Available at: https://www.healthypeople.gov/2020/topics-objectives/topic/environmental-health/objectives. Accessed January 17, 2017.

According to *Healthy People 2020*:

> Humans interact with the environment constantly. These interactions affect quality of life, years of healthy life lived, and health disparities.... Maintaining a healthy environment is central to increasing quality of life and years of healthy life. Globally, nearly 25 percent of all deaths and the total disease burden can be attributed to environmental factors. Environmental factors are diverse and far reaching.... Poor environmental quality has its greatest impact on people whose health status is already at risk. Therefore, environmental health must address the societal and environmental factors that increase the likelihood of exposure and disease.[1]

Protecting the environment means creating a world in which the air is safe to breathe, the water is safe to drink, the land is arable and free from toxins, wastes are managed effectively, infectious diseases are kept at bay, and natural areas are preserved. **FIGURE 1.1** illustrates a beautifully maintained natural area in the United States. Crucial environmental dimensions also include the impacts of disasters, the built environment, and availability of nutritious foods.[1]

The requirements of a growing world population need to be balanced against the demands for environmental preservation. Although developed countries such as the United States have made substantial progress in clearing the air and reducing air pollution, significant challenges to the environment and human health remain. For example, among the current and persistent threats to the environment in the United States are the following: trash that fouls our beaches, hazardous wastes (including radioactive wastes) leaching from disposal sites, continuing episodes of air pollution, exposures to toxic chemicals, destruction of the land through deforestation, and global warming.

The hallmarks of environmental degradation are not difficult to find: Warning signs posted on beaches advise bathers not to enter ocean water that is unsafe because of sewage contamination. In some areas of the United States, drinking water is threatened by toxic chemicals

FIGURE 1.1 A natural ecosystem in the United States. Maintaining environmental quality is a pressing task for the 21st century.

that are leaching from disposal sites. Too many factories continue to belch thick, black smoke or emit unseen pollutants. Avoidance of air pollution, which at best insults our aesthetic senses and at worst endangers our health, is often impossible. Society's appetite for lumber and new housing to satisfy the burgeoning population has resulted in clear-cutting of forests and destruction of wildlife habitats in order to accommodate new habitations for humans. Continued use of fossil fuels contributes to poor air quality and climate change.

Pollution and population growth, often associated with adverse economic circumstances, are closely connected with environmental health. In his classic article, the late Professor Warren Winkelstein wrote that "the three P's—pollution, population, and poverty—are principal determinants of health worldwide. . . ."[2(p932)] The three P's are interrelated: Population growth is associated with poverty, and both poverty and population growth are associated with pollution.

An example of the first "P" is pollution from combustion of fossil fuels (e.g., petroleum and coal), which disperses greenhouse gases along with other pollutants into the atmosphere. This process is believed to be a cause of global warming that in turn may have wide-ranging adverse effects. One such effect is to advance the range of disease-carrying insects, bringing them into new geographic areas; for example, mosquito-borne diseases such as the West Nile virus and dengue fever may appear in areas that previously were free from these conditions. (Refer to the chapter on zoonotic and vector-borne diseases for more information.) The second "P" is population, which is growing exponentially in many parts of the world, especially the less-developed areas, and may result in a worldwide population of up to 10 to 12 billion people during the 21st century; the presence of so many people may exceed the carrying capacity (defined later in the chapter) of the earth by a factor of two. The third "P," poverty, is linked to population growth; poverty is one of the well-recognized determinants of adverse health outcomes.

A recent environmentally related adverse health outcome may be attributed, at least in part, to one of the three P's: population growth (which is associated with urban crowding). As a result of known and unknown environmental and other factors, threats to the human population periodically arise from infectious disease agents. (This topic is discussed in the chapter on zoonotic and vector-borne diseases.) For example, influenza viruses threaten the world's population from time to time. Environmental factors that are likely to advance the spread of influenza viruses include intensive animal husbandry practices needed to supply food to the world's growing population. These practices create extremely crowded conditions among food animals coupled with their close residential proximity to humans. Often such farm animals are treated with antibiotics that contribute to the proliferation of antibiotic resistant strains of bacteria.

Several years ago, public health officials became concerned about the possible occurrence of a human pandemic of avian influenza, caused by the avian influenza A (H5N1) virus. Large outbreaks of avian influenza occurred on poultry farms in Asia. Apparently, some transmission of the virus from birds to humans also occurred. The disease (called bird flu) produces a severe human illness that has a high fatality rate. Health officials were concerned that the virus might mutate, enabling human-to-human transmission; if human-to-human transmission of the virus erupted, a pandemic might result. Contributing to the possible epidemic transmission of influenza (and other communicable diseases) is the ability of human beings to travel rapidly from one area of the globe to another.

An example of a global outbreak of influenza was the pandemic caused by swine flu (H1N1 influenza). In 2009, swine flu spread through North America to other parts of the globe. The World Health Organization (WHO) declared a pandemic of influenza was underway.

Another example of a condition that threatens the human population as well as all life on earth is global climate change. **EXHIBIT 1.1** presents a case study of global climate change. More information on this topic appears in the chapter on air quality. Refer to **FIGURE 1.2** for ways climate change threatens your health.

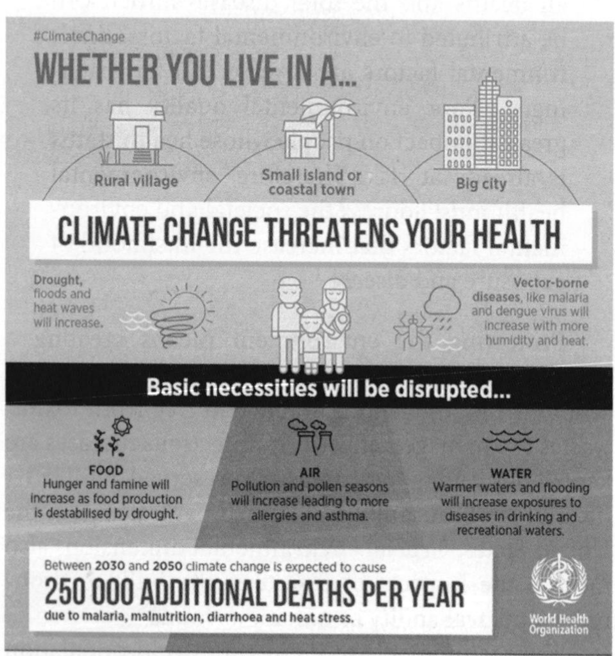

FIGURE 1.2 Climate change threatens your health.
Reproduced from World Health Organization. Climate change and human health. Available at: http://www.who.int/globalchange/mediacentre/events/climate-health-conference/climatechange-infographic2.jpg?ua=1. Accessed March 22, 2017.

EXHIBIT 1.1

Case Study: Climate Change: the Most Pressing Environmental Health Issue for the 21st Century

Two leading health bodies, a US governmental agency and a global health organization, have made the following declarations:

Climate change endangers human health, affecting all sectors of society, both domestically and globally. The environmental consequences of climate change, both those already observed and those that are anticipated, such as sea-level rise, changes in precipitation resulting in flooding and drought, heat waves, more intense hurricanes and storms, and degraded air quality, will affect human health both directly and indirectly. Addressing the effects of climate change on human health is especially challenging because both the surrounding environment and the decisions that people make influence health. For example, increases in the frequency and severity of regional heat waves—likely outcomes of climate change—have the potential to harm a lot of people.

—National Institute of Environmental Health Sciences

*****

- "Climate change affects the social and environmental determinants of health—clean air, safe drinking water, sufficient food and secure shelter.
- Between 2030 and 2050, climate change is expected to cause approximately 250 000 additional deaths per year, from malnutrition, malaria, diarrhea, and heat stress.
- The direct damage costs to health (i.e., excluding costs in health-determining sectors such as agriculture and water and sanitation), is estimated to be between US$ 2–4 billion/year by 2030.
- Areas with weak health infrastructure—mostly in developing countries—will be the least able to cope without assistance to prepare and respond.
- Reducing emissions of greenhouse gases through better transport, food, and energy-use choices can result in improved health, particularly through reduced air pollution."

—World Health Organization

Modified and reproduced from Portier CJ, Thigpen TK, Carter SR, et al. *A Human Perspective on Climate Change*: A Report Outlining the Research Needs on the Human Health Effects of Climate Change. Research Triangle Park, NC: Environmental Health Perspectives/National Institute of Environmental Health Sciences; 2010; and World Health Organization. Climate change and health. June 2016. Available at: http://www.who.int/mediacentre/factsheets/fs266/en/. Accessed March 22, 2017

Significance of the Environment for Human Health

The environment is intimately connected with human health, illness, and mortality. For example, although figures on the role of environmental factors in global mortality vary considerably, the environment is undoubtedly a salient influence on human deaths. Some estimates from the 1990s placed the toll of the world's deaths caused by environmental factors at around 40%.[3] World Health Organization data from 2012 suggested that almost one-quarter of global deaths result from an unhealthy environment. Exposures to potentially hazardous agents such as microbes, toxic chemicals and metals, pesticides, and ionizing radiation account for many of the forms of environmentally associated morbidity (acute and chronic conditions, allergic responses, and disability) and mortality that occur in today's world. These environmentally related determinants are believed to be important for the development of chronic diseases such as cancer, although most chronic diseases are thought to be the result of complex interactions between environmental and genetic factors.[4] All human beings are affected in some way by exposure to environmental hazards associated with lifestyle: at work, at home, during recreation, or while traveling on the expressway. **TABLE 1.2** provides examples of the scope of disease burden associated with exposure to environmental hazards.

Vulnerable Subgroups of the Population

The elderly, persons with disabilities and chronic diseases, pregnant women, and children are more likely to be affected by environmental hazards than are members of the general population. With respect to age, research from WHO underscored the differential effects of environmental influences across the life course.[5] Age groups most likely to be impacted are children younger than 5 years, children as old as 10 years to a lesser degree, and older adults from 50 to 75 years.

TABLE 1.2 The Scope of Environmental Health Problems in the World and the United States

- The World Health Organization (WHO) estimated that in 2012, approximately 12.6 million deaths across the globe (23% of all deaths) were linked to environmental sources.[a]
- The US Environmental Protection Agency reported that in 2015, industry released 881 million pounds (400 million kilograms) of toxic chemicals into the air and water (690 million pounds [313 million kilograms] into the air and 191 million pounds [87 million kilograms] into the water). On the positive side, a declining trend in the release of these chemicals occurred between 2005 and 2015.[b]
- Elevated blood levels of lead continue to be an important problem in the United States, with children living in at least 4 million households that expose them to excessive amounts of lead.[c]
- The number of people with asthma in the United States increased to 8% of the population in 2009; environmental factors such as tobacco smoke and air pollution are asthma triggers.[d]
- "Strong evidence exists that industrial chemicals widely disseminated in the environment are important contributors to… the global, silent pandemic of neurodevelopmental toxicity."[e]
- "Using air quality standards established by WHO, experts have estimated that 1.3 billion of the world's urban inhabitants breathe air that exceeds these quality standards."[f]
- Environmental factors are thought to contribute significantly to various forms of cancer, including cervical cancer, prostate cancer, and breast cancer.

Data from:

[a]Prüss-Üstün A, Wolf J, Corvalan CF, et al. Preventing disease through healthy environments. Geneva, Switzerland: World Health Organization; 2016.

[b]US Environmental Protection Agency. Introduction to the 2015 TRI national analysis. Available at: https://www.epa.gov/sites/production/files/2017-01/documents/tri_na_2015_complete_english.pdf. Accessed June 29, 2017.

[c]Centers for Disease Control and Prevention. Lead home page. Available at: https://www.cdc.gov/nceh/lead. Accessed June 30, 2017.

[d]Centers for Disease Control and Prevention. Vital signs. Asthma in the US. Available at: https:// www.cdc.gov/vitalsigns/asthma/index.html. Accessed June 30, 2017.

[e]Grandjean P, Landrigan PJ. Neurobehavioural effects of developmental toxicity. *Lancet Neurol*. 2014;13(3):330-338.

[f]Butterfield PG. Upstream reflections on environmental health: An abbreviated history and framework for action. *Advances in Nursing Science*. 2002;25:34.

Children represent an especially vulnerable group with respect to exposure to hazardous materials, including pesticides and toxic chemicals. Their immune systems and detoxifying organs are still developing and are not fully capable of responding to environmental toxins.[6] Children may be exposed more often than adults to toxic chemicals in the ambient outdoor air and in the soil because they spend more time outside.[7,8]

Environmental Health and the Developing World

Residents of developing countries suffer far more from problems associated with environmental degradation than do those who live in developed countries; this observation holds true despite the fact that developed countries are highly industrialized and disseminate vast quantities of pollutants into the environment from industrial processes and motor vehicles. In comparison with developing countries, wealthy nations provide better access to medical care and are better able to finance pollution controls.

In the developing world, the pursuit of natural resources has caused widespread deforestation of tropical rain forests and destruction of wildlife habitat. Although these two issues have been the focus of much publicity, less widely publicized environmental hazards such as water contamination, air pollution, unsanitary food, and crowding take a steep toll in both morbidity and mortality in developing countries.[9]

One region of the world that at present confronts serious environmental threats is Asia. Many of the countries in this region are experiencing declines in the amount of forest land, unintentional conversion of arable land to desert, and rising levels of pollution. In order to meet the demands of the rapidly increasing populations of South Asia, rural farmers clear forests and cultivate land that erodes easily and eventually becomes useless for agriculture.[10] **Runoff** from the land contributes to water pollution. The world's most populous country, China, faces many challenging environmental problems including water shortages in the northwest; severe air pollution in major cities, such as Beijing; and increasing desertification.[11]

Environmental Risk Transition

The term **environmental risk transition** has been used to characterize changes in environmental risks that happen as a consequence of economic development in the less-developed regions of the world. Environmental risk transition is characterized by the following circumstances:

In the poorest societies, household risks caused by poor food, air, and water quality tend to dominate. The major risks existing in developing countries today are of this type—diarrhea is attributable to poor water/sanitation/hygiene, acute respiratory diseases to poor housing and indoor air pollution from poor quality household fuels, and malaria to poor housing quality, although all are of course influenced by other factors as well (malnutrition in particular). . . . As these problems are brought under control, a new set tends to be created at the regional and global level through long-term and long-range pollutants, such as acid rain precursors, ozone-depleting chemicals, and greenhouse gases.[12(p38)]

Population and the Environment

Currently increasing at a geometric rate, the human population threatens to overwhelm available resources; some areas of the world face periodic food scarcity and famine. A number of factors have contributed to population growth, including increases in fertility and reductions in mortality. One of the consequences of population growth has been to encourage the conversion of large rural and forested areas of the earth into cities. Urbanization is linked to numerous adverse implications for the health of populations, including increasing rates of morbidity and mortality. Refer to the following text box, which discusses the consequences of continued population growth.

HOMO SAPIENS—A SUICIDAL SPECIES?

Largely as a result of human action, profound changes are occurring in our environment. . . . The basic cause of almost all of these problems is the world's large and growing human population, which consumes so much energy and produces such large quantities of toxic wastes. . . . Environmental changes, if accompanied by economic and political instability, could lead to the collapse of organized health services. In an era of scarcities of food, water, and other resources, and of a threat to survival, priorities should be reassessed.[13(pp121,123)]

Population Growth Trends

The human population has grown exponentially over the past 200 years and reached 6 billion in June 1999.[14] By 2017, this number reached 7.5 billion. The current trend is for world population growth to continue at a high rate, as noted in the following passage:

> Every day we share the earth and its resources with 250,000 more people than the day before. Every year there are another 90 million mouths to feed. That is the equivalent of adding a city the size of Philadelphia to the world population every week; a Los Angeles every two weeks; a Mexico every year; and a United States and Canada every three years.[15(p30)]

FIGURE 1.3 characterizes this burgeoning growth for a single year—2002. During that year the world population increased by 2⅓ persons per second, or 141 persons per minute. This annual growth rate would be equivalent to a Boeing 737 jetliner carrying a new group of 141 passengers each minute.

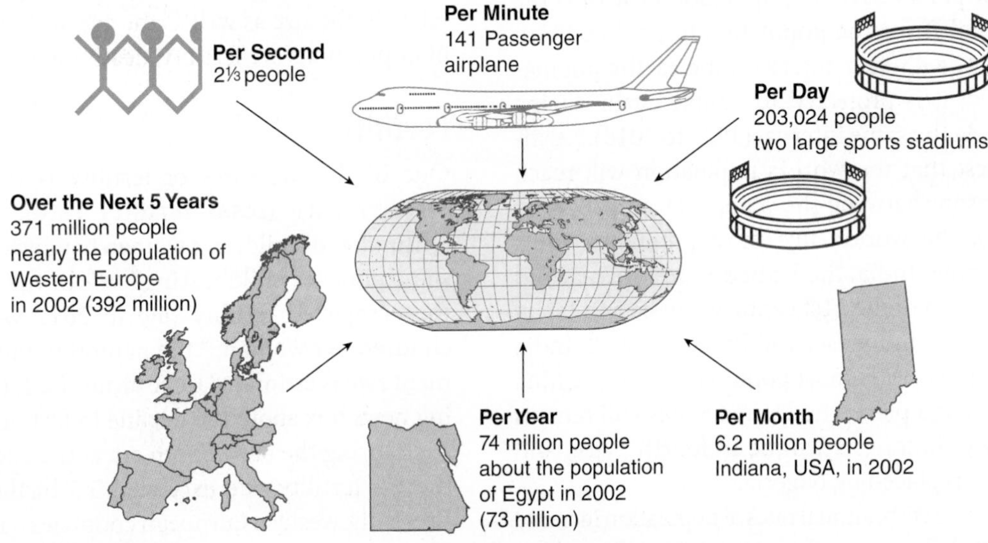

FIGURE 1.3 Net additions to the world: 2002. In 2002, the world gained 2⅓ people per second.
Modified from US Census Bureau. International Population Reports WP/02. *Global Population Profile: 2002*. Washington, DC: US Government Printing Office; 2004:14.

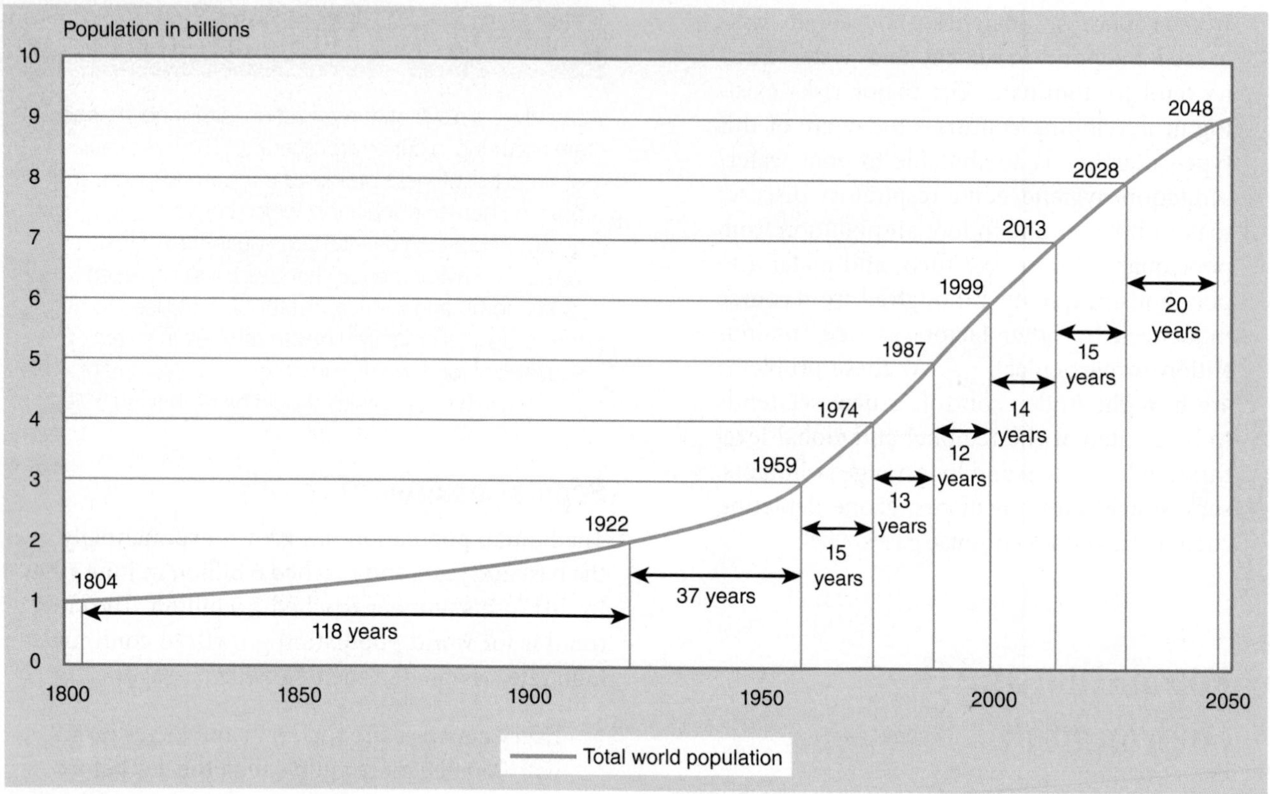

FIGURE 1.4 Time to successive billions in world population: 1800 to 2050.
Reproduced from US Census Bureau, International Population Reports WP/02. *Global Population Profile: 2002*. Washington, DC: US Government Printing Office; 2004:11.

From the origin of the species *Homo sapiens* (assumed to be about 250,000 years ago) to CE 1800, the population of the world grew by 1 billion individuals.[16] From 1800 to 1922 (122 years), the population added another 1 billion persons. Since 1922, the population has increased at a phenomenal rate: Another billion persons were added after 37 years, 15 years, and 13 years, respectively. Only 12 years elapsed before an additional billion persons were added between 1987 and 1999. See **FIGURE 1.4**.

Another perspective on population growth is the time that it takes for the population to double. From 1931 to 1974 (a 43-year interval), the earth's population doubled; it is projected to double again during approximately the same interval (1974 to 2018).[17] Estimates suggest that the world's population will reach 8 billion persons between the years 2018 and 2028.

In 1950, the world's five most populous countries were China, India, the United States, Russia, and Japan; at the turn of the 21st century, Russia and Japan were replaced by Indonesia and Brazil. In 2050, India will become the world's most populous country; China will fall to second place, the United States will remain in third place, Indonesia will be in fourth place, and Brazil will be replaced by Nigeria.

Around the 1960s, annual rates of population increase topped out at slightly more than 2% (an 81 million absolute increase annually since the 1980s).[18] Demographers project that the human population eventually may stabilize at a size—about 10 billion persons—that is about three quarters larger than it was around 2000.

Population Dynamics

The term **population dynamics** refers to the ever-changing interrelationships among the set of variables that influence the demographic makeup of populations as well as the variables that influence the growth and decline of population sizes. Among the factors that relate to the size as well as the age and sex composition of populations are fertility, death rates, and migration.

Fertility

One of the measures of fertility is the **completed fertility rate (total fertility rate)**, which is the "[n]umber of children a woman has given birth to when she completes childbearing."[19(p2)] In the United States, the completed fertility rate in 2012 was around 2.0 children per woman;[19] the natural population replacement rate is estimated to be around 2.1. (See the breaking news box about the decline in fertility).

During the baby boom era at the end of the 1950s, the US fertility rate exceeded 3.5 births per woman. Presently, western European countries have low fertility rates; also, the rates are declining in most regions of the developing world.[20] The United States, Canada, Japan, South Korea, Thailand, China, and many countries in

BREAKING NEWS:

"The US fertility rate fell [in 2016] to the lowest point since record keeping started more than a century ago, according to statistics released by the Centers for Disease Control and Prevention." (CNN, August 11, 2016). The general fertility rate (total number of births per 1,000 women aged 15–14) declined to 62.0 in the fourth quarter of 2016. In comparison, the general fertility rate was 118.0 in 1960.

Europe are at or below the replacement rate for fertility. Despite the declines in fertility rates in some Asian countries to levels approaching replacement rates, the populations in these countries will continue to increase because of births among the large cohort of persons of childbearing age who were born when fertility rates were high. Those individuals form a substantial proportion of the population in China and other rapidly industrializing countries. However, the overall trend is for the world's population to age and be composed of increasing numbers of older individuals.

In comparison with the developed world, the fertility rates are considerably higher (at about 4.0 births per woman) in many Asian countries, Latin American countries, and African countries. In the future, their relatively higher fertility rates will enable these regions of the developing world to claim the largest population sizes. (See **FIGURE 1.5**.)

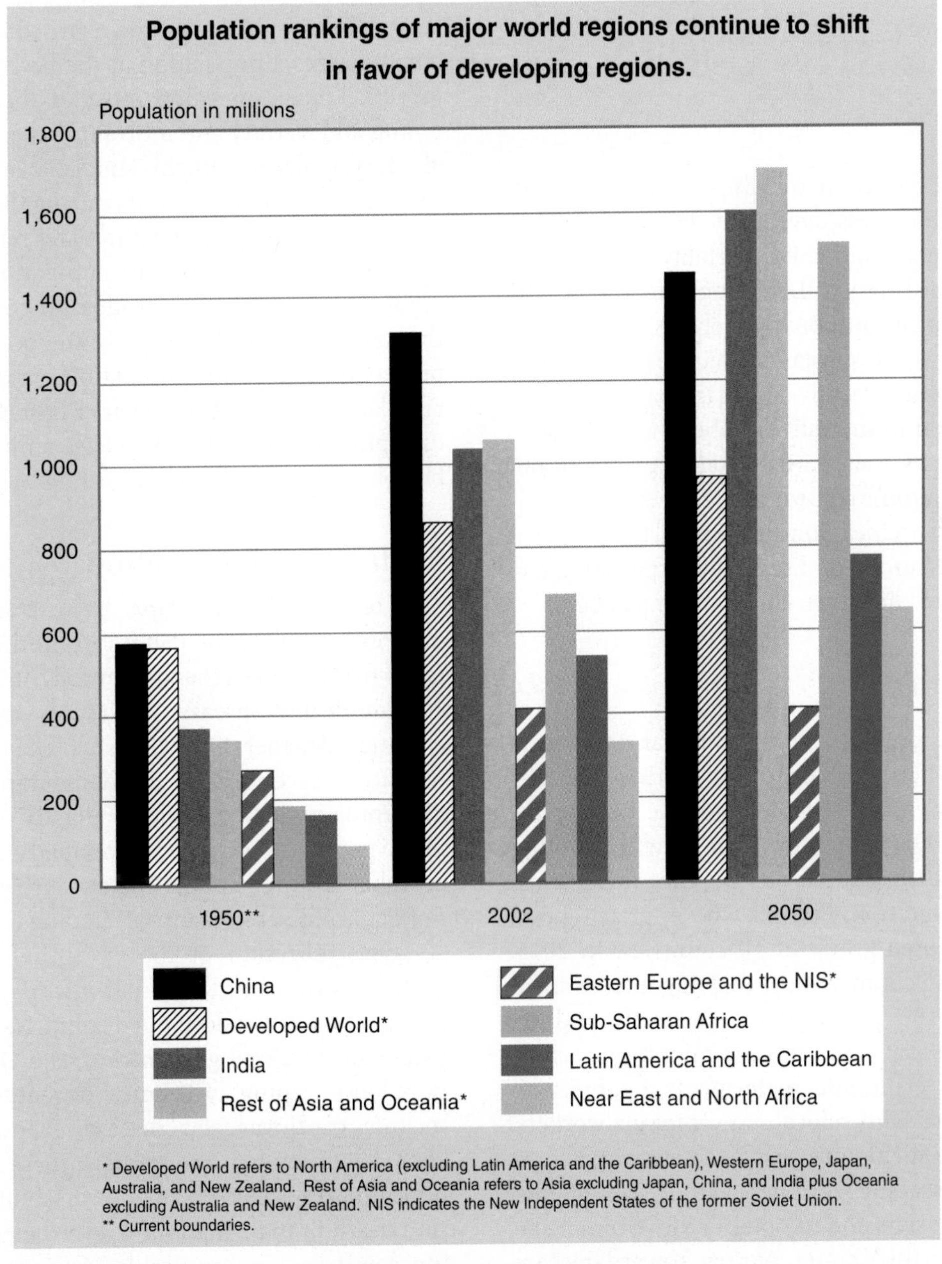

FIGURE 1.5 Regional distribution of global population: 1950, 2002, and 2050.

Reproduced from US Census Bureau. International Population Reports WP/02. *Global Population Profile: 2002*. Washington, DC: US Government Printing Office; 2004:12.

MORTALITY TERMS

Burden of disease is defined as "[t]he impact of disease in a population. An approach to the analysis of health problems, including loss of healthy years of life." One measure of the burden of disease is disability-adjusted life years (DALYs).

Life expectancy (expectation of life) refers to "[t]he average number of years an individual is expected to live if current mortality rates continue to apply." *Life expectancy at birth* is the "[a]verage number of years a newborn baby can be expected to live if current mortality trends continue."

Disability-adjusted life years (DALYs) refer to "adjustment of life expectancy to allow for long-term disability as estimated from official statistics." "A DALY lost is a measure of the burden of disease on a defined population."

Porta M, ed. *A Dictionary of Epidemiology*. 6th ed. New York, NY: Oxford University Press; 2014.

Mortality

Mortality has declined markedly over time in both industrialized and less-developed countries. Adult mortality and infant and child mortality have demonstrated downward trends. Declining mortality in the developed world began approximately 200 years ago; in the developing world, substantial declines in mortality have occurred more recently during the past 50 years or so. The reduction in mortality has been accomplished through measures that have included public health improvements, famine control, and increased availability of drugs and vaccines. Some additional terms related to mortality are burden of disease, life expectancy, and disability-adjusted life years. (Refer to the text box.)

Migration

Migration has continued to feed global population growth; more than 1 billion of the world's residents are migrants. Census estimates indicate that by the year 2050, the US population will grow by another 100 million and that about one-third of this growth will be from migration. Persons who migrate tend to cluster in a limited group of 10 countries. In 2015, the three leading countries for receiving international migrants were the United States, Germany, and the Russian Federation.[21] Reasons for migration include the search for economic betterment; a large proportion of those who relocate are migrant workers. Forced migration (forcible displacement of persons) is a means of escaping from persecution for religious and political reasons and to obtain relief from unstable conditions in one's home country. Toward the conclusion of 2015, more than 65 million persons were displaced. Many were refugees from Syria, Afghanistan, and Somalia.[21]

Demographic Transition

Demographic transition is the alteration over time in a population's fertility, mortality, and makeup.[14] (Note that demographic transition theory does not include the effects of migration upon the age and sex composition of a population.) According to the demographic transition theory, developed societies have progressed through three stages that have affected their age and sex distributions.

The three phases can be demonstrated by hypothetical population pyramids, which are graphs that show the distribution of a population according to age and sex. Examples of the population pyramids at stages 1 through 3 are shown in **FIGURE 1.6**. Stage 1 characterizes a population at the first stage of demographic transition when most of the population is young and fertility and mortality rates are high; overall, the population remains small. Stage 2 shows a drop in mortality rates that occurs during the demographic transition; at this stage fertility rates remain high, and there is a rapid increase in population, particularly among the younger age groups. In comparison with the narrow triangular shape of the population distribution at stage 1, this population pyramid also is triangular in shape but has a wider base. Stage 3 reflects dropping fertility rates that cause a more even distribution of the population according to age and sex.

Epidemiologic Transition

The term **epidemiologic transition** is used to describe a shift in the pattern of morbidity and mortality from causes related primarily to infectious and communicable diseases to causes associated with chronic, degenerative diseases. The epidemiologic transition accompanies the demographic transition. The epidemiologic transition already has taken place in the populations of most developed countries (a process that required approximately one century) but has not occurred yet in many developing countries.

One reflection of the epidemiologic transition is the growing burden of chronic, degenerative diseases, especially in developed countries and to a lesser extent in developing countries, as a consequence of population aging. Chronic, degenerative diseases include cardiovascular diseases, cancer, neuropsychiatric conditions, and injuries; these conditions are becoming the major causes of disability and premature death in many nations. Nevertheless, in developing countries communicable and infectious diseases remain the leading causes of morbidity and mortality.

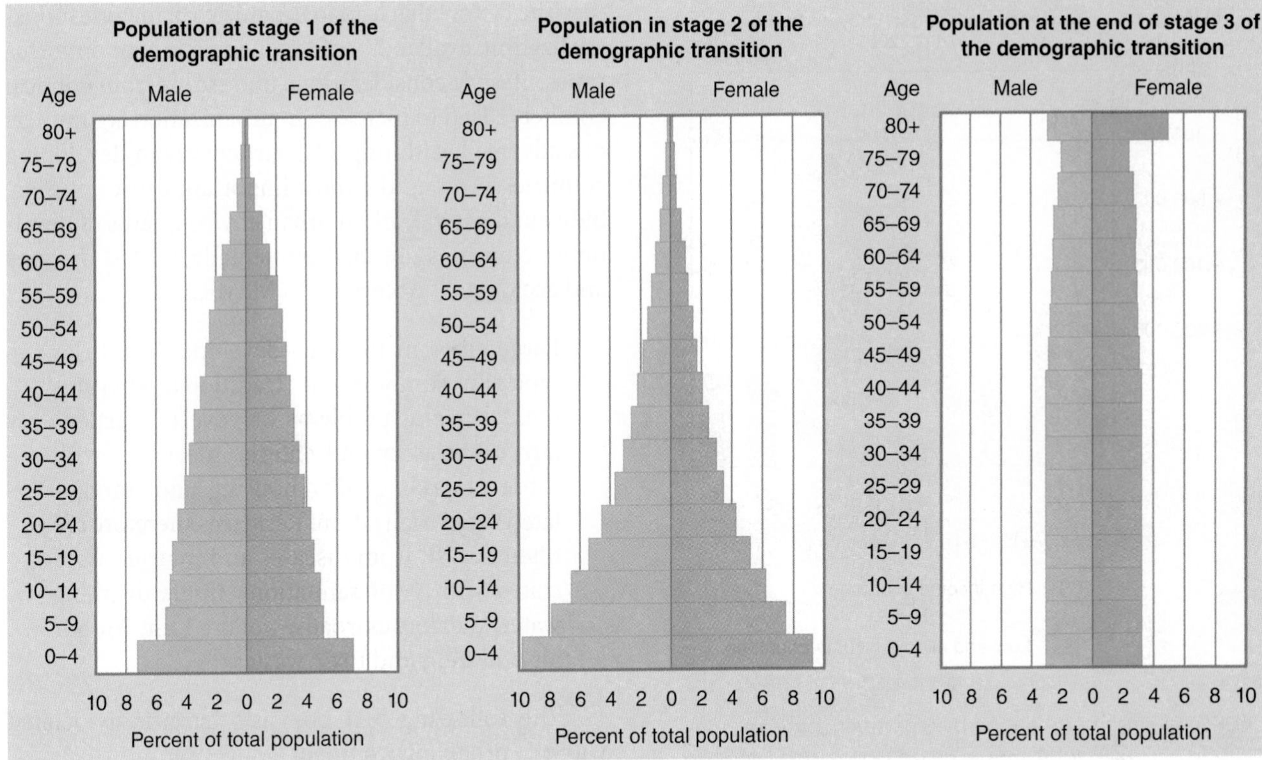

FIGURE 1.6 The demographic transition in three stages of age and sex composition: stage 1 (left), stage 2 (middle), and stage 3 (right).
Reproduced from Kinsella K, He W. US Census Bureau. International Population Reports P95/09-1. *An Aging World: 2008*. Washington, DC: US Government Printing Office; 2009:20.

Consequences of Population Increases

Rapid growth of the world's population contributes to the deterioration of the environment through widespread depletion of natural resources and by causing the levels of air, water, and other forms of pollution to increase. Also, the resources available per person decrease as the total number of individuals on the planet continues to increase geometrically. Consequently, population growth is a determinant of the number of persons who live in poverty. In already crowded regions, an even larger population means that the size of most people's living spaces must decline and population density must increase. Population density and associated urban crowding are dimensions of environmental degradation associated with increases in the spread of infectious and communicable diseases.

Unless significant technological innovations can be introduced, merely feeding the world's hungry population will become problematic. Many developing countries, where population growth rates are among the highest in the world, are reaching the limit of their abilities to provide for the economic and social needs of their citizens. The United Nations Secretariat states that:

> ...excessive population pressure in specific geographical areas can pose serious ecological hazards, including soil erosion, desertification, dwindling supplies of firewood, deforestation and the degradation of sources of fresh water. Often the link between population pressure and those types of environmental stress is the growth in the relative and absolute number of persons living in poverty. The result is marginalization of small-scale farmers and pressure on larger numbers to migrate from distressed areas. In many cases the result is also the prevalence of environmentally related diseases.[22(p32)]

The effects of rapid growth of the world's population include:

- Urbanization
- Overtaxing carrying capacity (defined later in this chapter)
- Food insecurity
- Loss of biodiversity

Urbanization and the Environment

The past two centuries have seen a rapid increase in the number of cities over the entire globe.[23] The proportion of urban residents has increased from about 5% in 1800 to 50% in 2000 and is expected to reach about 66% by 2030.[22] **FIGURE 1.7** illustrates the growth of the world's urban population between 1975 and 2015 for low- and middle-income countries in comparison with high-income countries. Although the

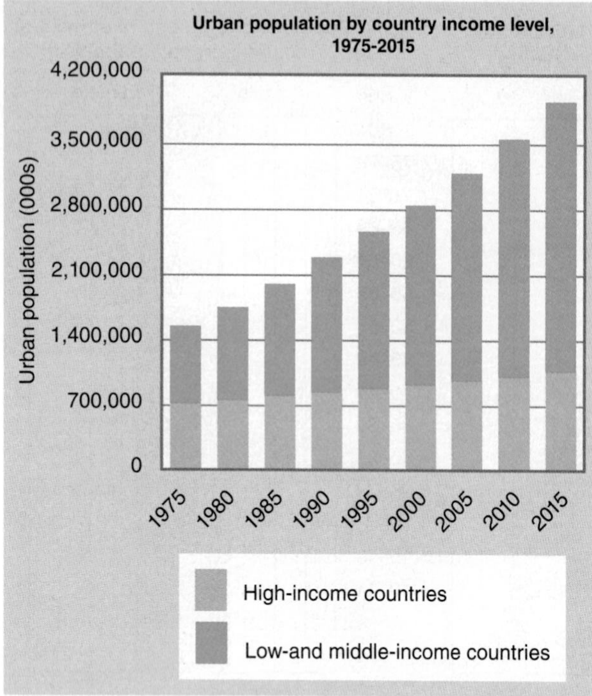

FIGURE 1.7 Urban population by country income level, 1975-2015 (1.84% total annual urban population growth between 2015 and 2020).

Reproduced from World Health Organization. *Urban population growth*. Available at: http://who.int/gho/urban_health/situation_trends/urban_population_growth/en/index1.html. Accessed July 2, 2017.

proportion of urban residents has increased in both categories, a relatively larger growth in the number of urban residents is projected to occur among low-income countries. This trend is apparent in the figure and is forecast to continue into the future. On average, the total urban population will grow by 1.84% annually between 2015 and 2020. (See Figure 1.7.)

The factors that lead to urbanization include industrialization, availability of food, employment opportunities, lifestyle considerations, and escape from political conflict.[23] Tied to increases in urbanization are numerous adverse health impacts, particularly in developing countries. Among the most important causes of morbidity and early death in urban environments of developing countries are environmentally related diseases and accidents.[24] According to McMichael:

> Large cities in the least developed countries typically combine the traditional environmental health problems of poverty, particularly respiratory and enteric infections, with those of poor quality housing and unregulated industrialization. Residents therefore are often at **risk** from diseases and injuries associated with poor sanitation, unsafe drinking water, dangerous roads, polluted air, indoor air pollution and toxic wastes.[23(p1119)]

The following text box lists hazards associated with the urban environment.

Megacities

The term **megacity** denotes an urbanized area that has 10 million or more inhabitants; in 2016, there were 31 megacities that contained slightly more than 6.8% of the world's population.[25] Examples of megacities and their respective 2016 populations (in millions) are Tokyo (38.1), Shanghai (24.5), Mumbai (formerly known as Bombay; 21.4), Sao Paulo (21.3), and Mexico City (21.2). The two megacities in the United States are New York (18.6) and Los Angeles (12.3).

HAZARDS TO HEALTH WITHIN THE URBAN ENVIRONMENT

1. Biological pathogens or pollutants within the human environment that impair human health—including pathogenic agents and their vectors (and reservoirs); for instance, the many pathogenic microorganisms in human excreta, airborne pathogens (for instance, those responsible for acute respiratory infections and tuberculosis), and disease vectors such as malaria-carrying (*Anopheline*) mosquitoes
2. Chemical pollutants within the human environment, including those added to the environment by human activities (e.g., industrial wastes) and chemical agents present in the environment independent of human activities
3. The availability, cost, and quality of natural resources on which human health depends—for instance, food, water, and fuel
4. Physical hazards (e.g., high risks of flooding in houses and settlements built on floodplains or of mud slides or landslides for houses on slopes)
5. Aspects of the built environment with negative consequences on physical or psychosocial health (e.g., overcrowding, inadequate protection against noise, inadequate provision of infrastructure, services, and common areas)
6. Natural resource degradation (e.g., of soil and water quality) caused by wastes from city-based producers or consumers that impacts on the health/livelihoods of some urban dwellers
7. National/global environmental degradation with more indirect but long-term influences on human health

Modified and reproduced with permission from Satterthwaite D. The impact on health of urban environments. *Environ Urban*. 1993;5:88.

FIGURE 1.8 Street scene in a crowded megacity (Mexico City).

Megacities have major influences upon the environment in a number of ways (e.g., demands for energy, potable water, construction materials, food, sewage processing, and solid waste disposal). **FIGURE 1.8** shows street life in a crowded megacity.

Carrying Capacity

Carrying capacity is "[t]he maximum number of individuals that can be supported sustainably by a given environment."[26] Both human and nonhuman populations may be threatened with disastrous consequences when available resources are exhausted. "Like a bacterial colony in a culture medium, we are susceptible to depletion of nutriments and to poisoning by our own waste products."[15(p123)]

Animal Populations

In the animal kingdom, the carrying capacity of an environment governs population size. In nature, the factors of food availability, reproductive behavior, and infectious diseases tend to keep animal populations in check. An example of an animal population kept in check by food availability follows: The U.S. Coast Guard shipped 29 reindeer to St. Matthew Island in the Bering Sea during the World War II era.[26] The deer were intended as a source of meat for personnel on the island; however, no deer were ever culled and all 29 remained when the war ended. An abundance of deer fodder was available on the island. By the early 1960s, the original deer population had swelled to 6,000 animals. Soon afterward, as a result of overgrazing and depletion of food sources for the deer, the population—having declined to fewer than 50 animals in 1966—faced extinction.

In a given area, the growth of animal populations appears to be sequenced according to the following characteristic patterns:

"Logistic growth, responding to immediate negative feedback, as carrying capacity is approached

Domed or capped growth, responding to deferred negative feedback but necessitating a period of excess mortality

Irruptive growth, with a chaotic post-crash pattern."[18(p978)]

Human Populations

The factors that lead to the crash of animal populations are similar to those that could threaten the survival of the human race. Human life is not possible without adequate food, breathable air, and safe water. Agricultural land must continue to be arable. There needs to be a diversity of plant and animal species. If these components of the human life support system are disrupted by overpopulation of the planet, the species *Homo sapiens* could suffer a population crash. This outcome would be in line with Malthusian predictions.

In 1798, Thomas Malthus authored *First Essay on Population*, which theorized that the human population had the potential to grow exponentially.[16] According to this scenario, the population could outstrip available resources. Malthus suggested that "positive checks" for excessive population growth rates were epidemics of disease, starvation, and population reduction through warfare. The growth of the population could be constrained also through "preventive checks" such as not allowing people to marry.

Endangerment of the human population through ecological damage is not far fetched: Previous history has recorded incidents of decimation and collapse of civilizations that were associated with disruption of the environment. It is believed that approximately 5,000 years ago, Mesopotamia, a renowned ancient civilization, declined as a result of agricultural practices that caused soil erosion, buildup of salt in the soil, and the filling of irrigation channels with silt.[27] During medieval times, crowded cities of Europe were devastated by plague and other infectious diseases. In the interval between the 13th and 16th centuries, global temperatures declined by approximately 1°C (1.8°F), contributing to the decimation of societies that were located in the far north (e.g., Viking settlements in Greenland).

Food Insecurity and Famine

The term **food insecurity** refers to a situation in which supplies of wholesome foods are uncertain or may have limited availability. Food insecurity and famine may occur when the carrying capacity in a particular geographic area is exceeded. An illustration of the effect of exceeding the carrying capacity in a local geographic area is the occurrence of a local subsistence crisis, which follows when the ability of land and available water to produce food are overtaxed.[18] In theory, low nutritional levels that accompany local subsistence crises may cause population mortality to increase so that mortality is brought into balance with fertility, stabilizing the population size. Periodically, food insecurity is a reality in some developing regions. For example, food insecurity endangers as much as one-third of Africa, and the prognosis for increasing the food supply in some African countries is poor.[27]

Loss of Biodiversity

The word *biodiversity* is formed from the combination of *biological* and *diversity*. An adequate definition of biodiversity is not readily available. Nevertheless, the term **biodiversity** generally refers to the different types and variability of animal and plant species and ecosystems in which they live.[28] With respect to a particular geographic area, biodiversity involves diversity in the genes of a population of a given species, diversity in the number of species, and diversity in habitats. Biodiversity is considered to be an essential dimension of human health.[29]

The dramatic human population growth during the past few decades and concomitant increases in urbanization and industrialization have caused the physical environment to be degraded substantially; one of the consequences of unchecked population growth is hypothesized to be accelerated loss of biodiversity. Human activities are thought to be related to the spread of harmful insect vectors, extinctions of species, and loss of flora; some of these plants and trees could be the source of valuable commodities such as new pharmaceuticals. Ultimately, loss of biodiversity may pose a danger to food production as a result of the growth in numbers of invasive species and the eradication of helpful plants and insects. An example of the loss of biodiversity is the destruction of tropical rain forests that has culminated in the extinction of some flowering plants that may have had future medical value.[29]

Definitions Used in the Environmental Health Field

The Environment

The term **environment** refers to "the complex of physical, chemical, and biotic factors (as climate, soil, and living things) that act upon an organism or an ecological community and ultimately determine its form and survival."[30] This definition pertains to the physical environment. Examples of physical environmental factors (as noted previously) that affect human health include toxic chemicals, metallic compounds, ionizing and nonionizing radiation, and physical and mechanical energy. These factors will be discussed in more detail later in the text.

The term *environment* captures the notion of factors that are external to the individual, as opposed to internal factors such as genetic makeup. In contrast to the physical environment, described in the foregoing definition, the **social environment** encompasses influences upon the individual that arise from societal and cultural factors. Among the major determinants of health are the environment (physical and social), personal lifestyle factors, constitutional factors such as heredity and human biology, and healthcare systems dimensions such as access to and quality of medical care and methods for organization of healthcare systems.[31] A model that describes these aspects of health is the **ecological model**, which proposes that the determinants of health (environmental, biological, and behavioral) interact and are interlinked over the life course of individuals. (Refer to **FIGURE 1.9**.) From the model it may be inferred that the environment is one component of many interacting dimensions that affect the health of populations.

Ecological System (Ecosystem)

Ecosystems are one of the important dimensions of life in the biosphere. All life on earth survives in the biosphere, which consists of the atmosphere and the earth's surface and oceans. The biosphere covers a narrow range from about 6 miles (9.6 kilometers) above the earth to the surface of the earth to the deepest ocean trenches, some of which are 36,000 feet deep (about 11,000 meters). One of the crucial aspects of the earth's biosphere is energy flow; the ultimate source of energy for all living beings on earth is the sun. Energy flows from the sun in the form of electromagnetic radiation (e.g., ultraviolet radiation, infrared radiation, and visible light). Only a small percentage of the energy produced by the sun impinges upon the earth. Plants absorb some portions of the sun's electromagnetic radiation and convert them into nutrients and oxygen via the process known as photosynthesis. This energy is then transferred to other life forms through the food chain, for example, via herbivores that eat the plants themselves or carnivores that eat other animals.

"An ecosystem is a dynamic complex of plant, animal, and microorganism communities and the nonliving environment interacting as a functional

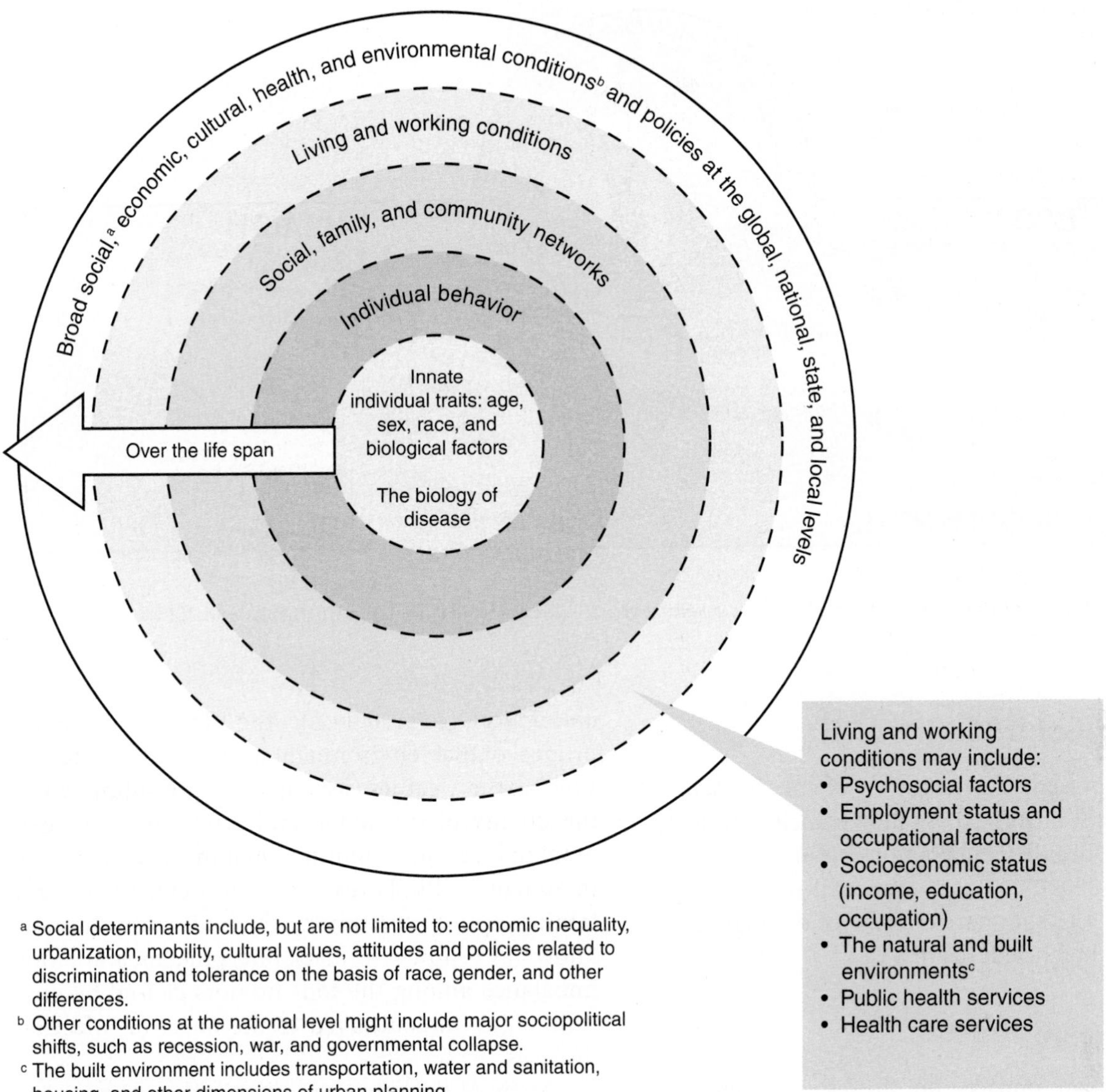

FIGURE 1.9 The ecological model of population health.

Modified and reproduced with permission from *Who Will Keep the Public Healthy? Educating Public Health Professionals for the 21st Century*, © 2003 by the National Academy of Sciences, courtesy of the National Academies Press, Washington, D.C., p. 33; and from Dahlgren G, Whitehead M. *Policies and Strategies to Promote Social Equity in Health*. Stockholm, Sweden: Institute for Futures Studies; 1991.

unit. Humans are an integral part of ecosystems. Ecosystems vary enormously in size: a temporary pond in a tree hollow and an ocean basin can both be ecosystems."[32(p3)] The interconnected components of an **ecosystem** are in a steady state; disrupting one of the components can disrupt the entire ecosystem. **FIGURE 1.10** suggests that the health of the ecosystem is associated with the health of human beings as well as that of domestic animals and wildlife.

Survival of the human population depends upon ecosystems, which aid in supplying clean air and water as part of the earth's life support system.[33] Ecosystems are being degraded with increasing rapidity because of human environmental impacts such as urbanization and deforestation. Degradation of ecosystems poses environmental dangers such as loss of the oxygen-producing capacity of plants and loss of biodiversity.

Environmental Health

The field of **environmental health** has a broad focus and includes a number of subspecializations. For example, occupational health often is regarded as a topic that is closely allied with environmental health and is a subset of broader environmental health concerns. Consequently, in view of its broad reach, the term *environmental health* does not have a single definition, nor is it easy to define. According to the World Health Organization:

> Environmental health addresses all the physical, chemical, and biological factors external to a person, and all the related factors impacting behaviours. It encompasses the assessment and control of those environmental factors that can potentially affect health. It is targeted towards preventing disease and creating health-supportive environments.[34]

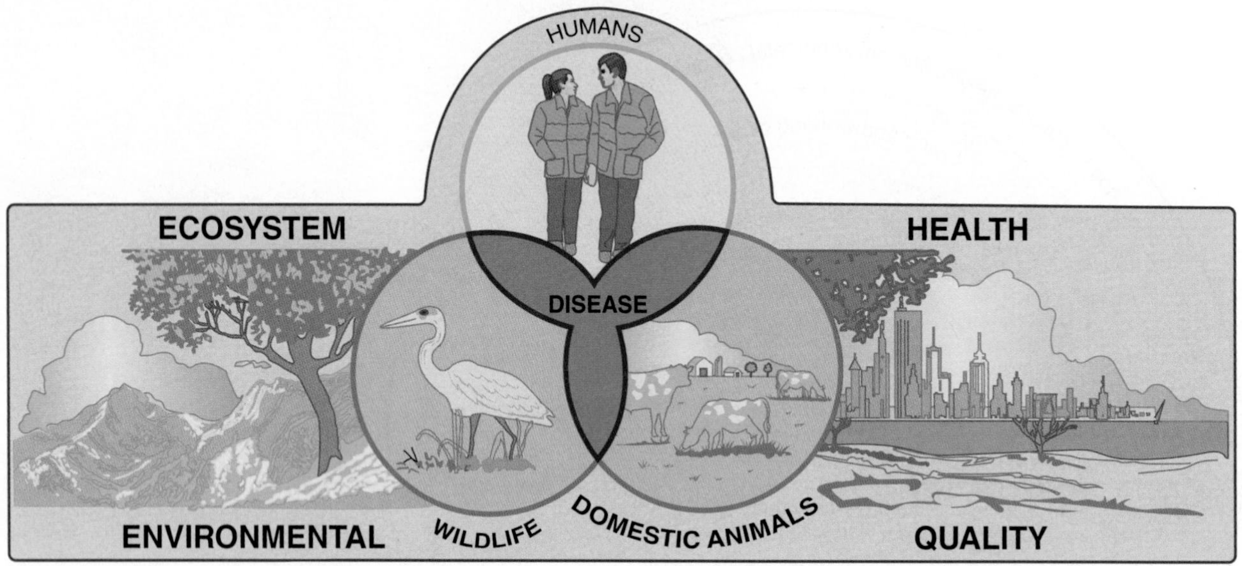

FIGURE 1.10 Ecosystem health.
Reproduced from Friend M. *Disease Emergence and Reemergence. The Wildlife-Human Connection*. Reston, VA: US Geological Survey, Circular 1285, p. 131. Illustration by John M. Evans.

Historical Background

This section presents a brief review of environmental health history, categorized as follows: ancient history, occupational health (contributions from about 1500 to the mid-1800s), and environmental history post-1800. **FIGURE 1.11** summarizes some of the highlights in environmental health history.

Ancient History

Negative human impacts on the environment are thought to have begun many thousands of years ago. One of the initial targets of human activity was forests, which were cut down for use as timber and burned to clear land for agriculture and human settlements. Deforestation subsequently led to soil erosion that caused rivers and bays to be fouled with silt.

The observations, insights, and writings of the ancient Greeks are noteworthy for the history of environmental health.[35] Around the 5th century BCE, the ancient Greek philosophers had developed the concept of the relationship between environmental factors and human health; instead of advocating for the workings of supernatural factors and the belief that magic potions would have curative powers, their philosophical position linked the influence of environment to disease.

Hippocrates, who lived between 460 and 370 BCE, often is referred to as "the father of medicine." (See **FIGURE 1.12**.) Hippocrates emphasized the role of the environment as an influence on people's health and health status in his work titled *On Airs, Waters, and Places* (ca 400 BCE). The Greek philosopher proposed that environmental and climatic factors such as the weather, seasons, and prevailing winds; the quality of air, water, and food; and one's geographic location were influential in causing changes in human health. He espoused the doctrine of maintaining equilibrium among the body's four humors, known as yellow bile, black bile, phlegm, and blood; imbalance among the four humors caused by environmental influences led to the onset of infectious diseases.

Many of the principles identified by Hippocrates regarding the impact of the environment on human health and disease remain credible despite the great increases in medical knowledge that have occurred since Hippocrates' time.[35] For example, now it is known that polluted water is associated with many types of waterborne infections (e.g., cholera and cryptosporidiosis [discussed later in this text]). Consistent with the belief that air is a factor in diseases is the origin of the term *malaria* (bad air), a disease that is carried by airborne mosquitoes that dwell in standing pools of water.

For many years, people have known about the harmful effects of heavy metals.[36] Hippocrates identified the toxic properties of lead.[37] The toxic properties of sulfur and zinc were pointed out by the Roman scholar Pliny the Elder (CE 29–79) during the 1st century CE; Pliny invented a mask constructed from the bladder of an animal for protection against dusts and metal fumes. During the 2nd century CE, the renowned Greek physician Galen (CE 129–200) outlined the pathological aspects of lead toxicity and

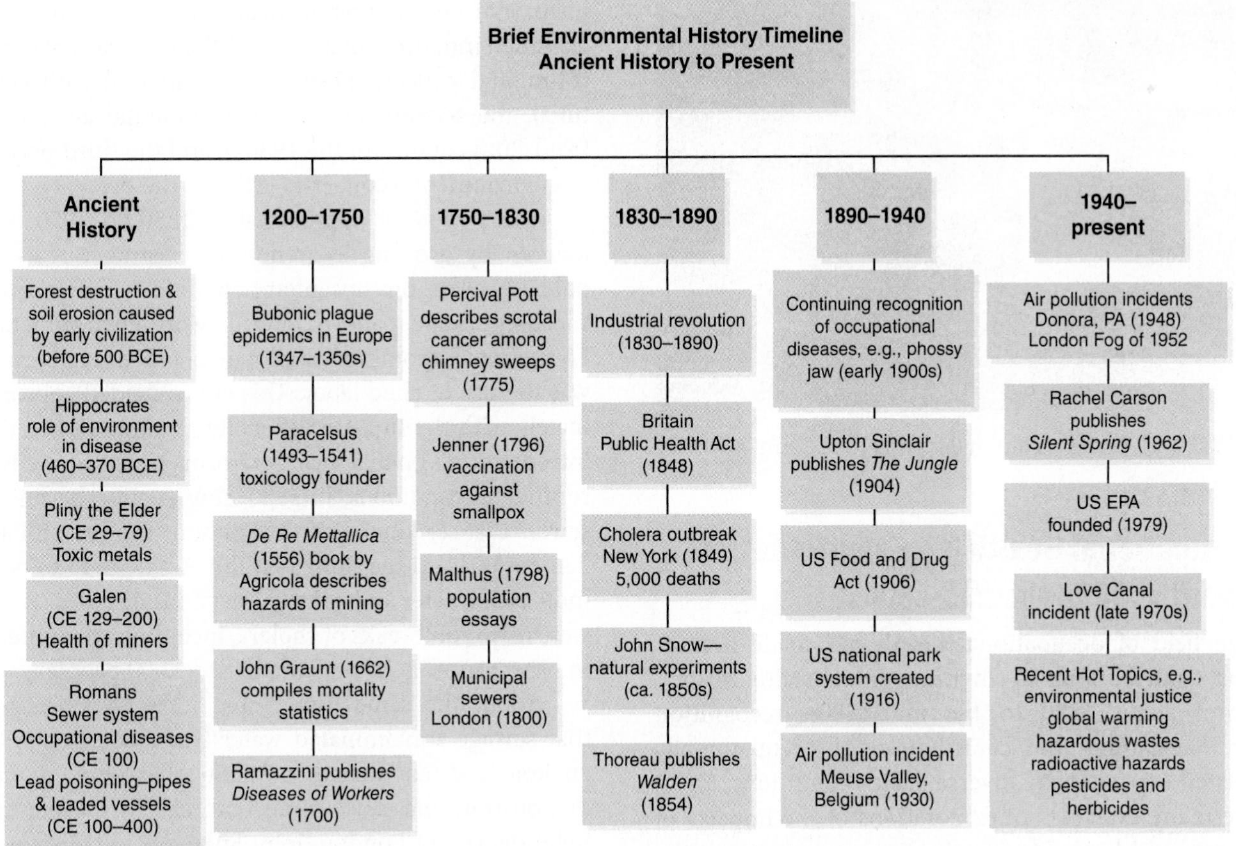
Brief Environmental History Timeline
Ancient History to Present
Ancient History
Forest destruction & soil erosion caused by early civilization (before 500 BCE)
Hippocrates role of environment in disease (460–370 BCE)
Pliny the Elder (CE 29–79) Toxic metals
Galen (CE 129–200) Health of miners
Romans Sewer system Occupational diseases (CE 100) Lead poisoning–pipes & leaded vessels (CE 100–400)
1200–1750
Bubonic plague epidemics in Europe (1347–1350s)
Paracelsus (1493–1541) toxicology founder
De Re Mettallica (1556) book by Agricola describes hazards of mining
John Graunt (1662) compiles mortality statistics
Ramazzini publishes Diseases of Workers (1700)
1750–1830
Percival Pott describes scrotal cancer among chimney sweeps (1775)
Jenner (1796) vaccination against smallpox
Malthus (1798) population essays
Municipal sewers London (1800)
1830–1890
Industrial revolution (1830–1890)
Britain Public Health Act (1848)
Cholera outbreak New York (1849) 5,000 deaths
John Snow— natural experiments (ca. 1850s)
Thoreau publishes Walden (1854)
1890–1940
Continuing recognition of occupational diseases, e.g., phossy jaw (early 1900s)
Upton Sinclair publishes The Jungle (1904)
US Food and Drug Act (1906)
US national park system created (1916)
Air pollution incident Meuse Valley, Belgium (1930)
1940–present
Air pollution incidents Donora, PA (1948) London Fog of 1952
Rachel Carson publishes Silent Spring (1962)
US EPA founded (1979)
Love Canal incident (late 1970s)
Recent Hot Topics, e.g., environmental justice global warming hazardous wastes radioactive hazards pesticides and herbicides

FIGURE 1.11 Brief environmental history time line: ancient history to the present.

FIGURE 1.12 Hippocrates.

Courtesy of US National Library of Medicine, National Institutes of Health. History of medicine division. Available at: http://ihm.nlm.nih.gov/luna/servlet/view/search?q=B014553. Accessed July 1, 2017.

suggested that mists from acids could endanger the health of copper miners.[37]

The ancient Romans developed the first infrastructure for maintaining public health. Among their innovations were systems for the transport of water and sewage, heating devices for water and for rooms, and communal baths.[38,39] Beginning about 500 BCE, the Etruscans constructed a sewer called the Cloaca Maxima in Rome. As the city grew, a system of aqueducts that supplied fresh water and a web of sewers called cloacae were installed.

The Romans brought many of these innovations to their settlements all over Europe. Roman aqueducts and baths can been seen today in many parts of Europe. An example of the Roman baths at Baden-Baden, Germany is shown in **FIGURE 1.13**. The Romans used lead pipes to supply the homes of the affluent, who probably suffered from chronic lead poisoning. After the decline of their empire (possibly due, in part, to chronic lead poisoning), many of the hygiene-related contributions of the Romans were forgotten; for several centuries, the European world endured the abhorrent sanitary conditions of the medieval era, with its periodic outbreaks of epidemics of plague, cholera, and other pestilence.

FIGURE 1.13 Roman baths at Baden-Baden, Germany.

Occupational Health (Contributions from About 1500 to the Mid-1800s)

The field of occupational health has made numerous contributions to environmental health history. From about 1500 to the mid-1800s, recognition grew regarding the contribution of occupationally related exposures to adverse health conditions. There are many examples of explorations of the impacts of unsafe and hazardous working environments on the health of workers, especially the effects of exposures to toxic metals and hazards that occurred among miners. Among the historically important figures in occupational health were Paracelsus (1493–1541), Agricola (1494–1555), Bernardino Ramazzini (1633–1714), Percival Pott (1714–1788), and Alice Hamilton (1869–1970). See the chapter on environmental toxicology for information on Paracelsus, the chapter on environmental epidemiology for information about Percival Pott, and the chapter on occupational health for a discussion of Agricola, Ramazzini, and Hamilton; the chapter on occupational health also provides information on other historically important individuals in the field of occupational health.

Although his contributions were not limited specifically to occupational health, John Graunt (one of the early compilers of vital statistics data) published *Natural and Political Observations Made upon the Bills of Mortality* in 1662. Sometimes Graunt is referred to as the Columbus of statistics because his book made a fundamental contribution by attempting to demonstrate the quantitative characteristics of birth and death data.

Environmental History Post-1800

Just before the commencement of the 1800s, Jenner (in 1796) devised a method for vaccinating against smallpox; in 1798, Malthus wrote his well-known essays on population, mentioned earlier in this chapter. The history of environmental health since 1800 may be classified into three major eras:[40] the first wave of environmental concern (19th century to mid-20th century), the second wave of environmental awareness (mid-20th century to the 1980s), and the third period of environmental concern (1980s to the present).

The period of approximately 1850 to 1950 was marked by growing awareness of existing threats to public health from unsanitary conditions, detrimental social conditions, and hazardous work environments. For example, a common employment practice in Europe was the use of child labor. This era coincided approximately with the Industrial Revolution and marked the introduction of public health reforms to improve environmental conditions. In 1800, construction began on sewers that served the city of London. The British Parliament enacted the Public Health Act in 1848 to promote clean water and control infectious diseases. There were major outbreaks of cholera, including an outbreak in New York City in 1849 that killed 5,000 people.

About the same time, John Snow hypothesized that sewage-contaminated water was associated with cholera and conducted a "natural experiment" to demonstrate the cause of an outbreak in the present Soho district of London. John Snow (1813–1858) was an English anesthesiologist who innovated several of the key epidemiologic methods that remain valid and in use today. In Snow's time, the mechanism for the causation of infectious diseases was largely unknown. The Dutchman Anton van Leeuwenhoek (1632–1723) had used the microscope to observe microorganisms (bacteria and yeast). However, the connection between microorganisms and disease had not yet been ascertained. During Snow's time, one of the explanations for infectious diseases such as cholera was the miasmic theory of disease, which alleged that illnesses were caused by clouds of noxious matter. Snow rejected the miasmatic theory and showed the connection between polluted water and cholera. Because of his pioneering work, Snow is regarded as an icon in the history of public health and continues to be influential during the 21st century. (More information about Snow's work is provided in the chapter on environmental epidemiology.)

In the United States, Lemuel Shattuck published the 1850 *Report on the Sanitary Conditions of Massachusetts*. Shattuck argued for the creation of a state health department and local health boards. Among other issues, Shattuck's report dealt in detail with the topic of environmental sanitation and its connection with health. Although not implemented by the state legislature at the time, Shattuck's proposed recommendations were extremely farsighted and innovative and became a major influence in the development of subsequent public health practice. Ultimately, years later, many of Shattuck's proposals were adopted by public health departments and are in use today.

Several additional developments marked the late 19th century and early 20th century. Henry David Thoreau published *Walden* in 1854; this book extolled the virtues of a simple life in a beautiful natural environment. Thoreau suggested that nature enriched the lives of human beings and therefore should be respected.

Beginning in 1900, Walter Reed, who was a US Army medical officer, investigated the causes of yellow fever, which was a scourge of US troops in the Caribbean. He showed that yellow fever was a mosquito-borne affliction. Following this discovery, Major William Gorgas was dispatched to Havana, Cuba, where he implemented a highly successful mosquito eradication program. This action led to a drastic reduction in yellow fever cases in Havana; later Gorgas conducted a mosquito control program in the Panama Canal Zone, making possible the construction of the canal.

In *The Jungle* (1906), Upton Sinclair described the deplorable conditions in the meat processing industry in Chicago. Sinclair's exposé was instrumental in passage of the first Food and Drug Act that was instituted in the United States in 1906. (Some of the US laws regarding food safety are described in the chapter on food safety.) One other development that reflected the public's concern for the environment was creation of the US National Park System in 1916.

During the second wave of environmental concern, defined approximately from the middle of the 1950s to the 1980s, environmental issues continued to come to the forefront. The period witnessed the occurrence of several noteworthy air pollution incidents, including the fatal 1930 incident in the Meuse Valley, Belgium; an air pollution episode that caused numerous deaths in Donora, Pennsylvania, in 1948; and the deadly London fog of 1952.[41] (More information on these incidents is presented in the chapter on air quality.)

Awareness increased regarding the potential health hazards of toxic chemicals. In the United States, efforts were made to protect ecologically sensitive areas from toxic hazards and from overdevelopment. Additional legislation in the United States modified food and drug laws designed to regulate toxins and the use of additives in food. Rachel Carson published *Silent Spring* in 1962, which highlighted the potential dangers of pesticides. In 1970, the Environmental Protection Agency (EPA) was founded to address environmental concerns at the federal government level (more information on this topic is presented in the chapter on environmental policy and regulation).

The topic of toxic waste disposal also was the focus of much attention during the 1970s. For example, when residents discovered that their homes had been constructed on a former toxic waste site referred to as the Love Canal, they became alarmed about possible adverse health effects that might be linked to the waste site. Love Canal became a cause célèbre for environmental activists. (This topic is covered in more detail in the chapter on solid and liquid waste.)

The most recent period in environmental history (the third wave of environmental concern—1980s to the present) has been marked by high population growth rates, industrialization, and urbanization. Specific concerns have continued regarding the effects of toxic chemicals in the environment.

A new topic has been the emission of greenhouse gases and their possible contribution to global warming. **TABLE 1.3** presents a compilation of some of the contemporary issues that are relevant to environmental health. Although this list is not exhaustive, it identifies several of the major "hot topics" in the environmental health field.

The topics shown in Table 1.3 will be covered in this text. However, let us select four of the issues—global climate change, pesticides and herbicides, air quality, and war and terrorism—and consider them briefly. For example, one issue that commands our attention (and that has generated extensive coverage in the media) is the prospect of global climate change including global warming (and production of greenhouse gases). Refer back to Exhibit 1.1. Among the outcomes believed to be associated with global warming are changes in the distribution of insect vectors that can carry diseases such as malaria and the West Nile virus. Elsewhere in this text, in the chapter on air quality, global warming—its hypothesized causes, extent, and effects—will be considered in more detail.

The impact of toxic pesticides and toxic chemicals is a major issue for environmental health. For example, toxic materials have been introduced into the drinking water supplies of some communities. In November 2005, an explosion at a factory in northeastern China caused about 100 tons (about 91 metric tons) of benzene and other hazardous chemicals to be released into the Songhua River. This incident led Chinese officials to shut off the water taps in Harbin because of potential contamination of the water supply in that city. A noteworthy incident in the United States was contamination of municipal water in Flint, Michigan in 2015, when the EPA detected lead contamination in the water used by households there. The events in China and Michigan, as well as many other similar occurrences in which toxic chemicals

TABLE 1.3 Examples of Hot Topics in Environmental Health

Air quality	Nuclear power
Conservation	Oceans
Endangered species/Wildlife impacts	Pesticides and herbicides
Energy resources	Pollution
Environmental justice	Radioactive waste
Environmental protection	Recycling
Forests	Solid waste
Global warming/Global climate change	War and terrorism
Greenhouse gases	Water resources
Hazardous wastes	Wetlands
Land use	

have intruded into the public water supply, raise the issue of what can be done to prevent and abate such hazards.

A related issue concerns the runoff of rainwater that overtaxes sewage processing facilities and results in pollution of public beaches and groundwater. Carelessly discarded solvents and other toxic chemicals pose dangers to aquifers; the author will provide more information on toxic pesticides and toxic chemicals in the chapter on pesticides and other organic chemicals.

Still another issue is the impact of air quality on human health, including the role of air pollution in causing cancer and lung diseases as well as aggravating chronic conditions such as heart disease. Some regions face a continuing and growing threat to the environment from air pollution. Several US cities, such as those located in the Los Angeles basin of southern California, face occasional episodes of significant air pollution. Fortunately, air quality has shown improvement in southern California and elsewhere in the United States during the past few decades. In contrast, many cities in the rapidly industrializing nations of the developing world are experiencing declines in air quality due to the increasing use of fossil fuels.

Finally, war and terrorism can have devastating impacts upon the environment. Some of these potential impacts include the destruction of fauna and flora, exposure of the population to hazardous radiation from spent munitions, and water pollution caused by the manufacture of nuclear weapons. (Refer to the chapter on ionizing and nonionizing radiation.) Recently, health officials and the public have been concerned about threats to the environment from the intentional release of infectious biological agents such as the agent that causes anthrax. (Refer to the chapter on zoonotic and vector-borne diseases.)

Careers in the Environmental Health Field

The field of environmental health provides numerous career roles and possible occupations. Private industry, government units, universities, and private research organizations employ environmental health workers in diverse functions. **TABLE 1.4** gives a detailed, although not exhaustive, list of occupations that have a connection with environmental health. Following is a description of some of these occupations.

Hygienist

In the work environment, professional industrial hygienists are responsible for control of hazards that

TABLE 1.4 Professions Involved in Environmental Health

Academics, lecturers, teachers, teacher trainers	Ergonomists
Agriculturists	Fire safety officers
Agronomists	Food inspectors
Architects	Food safety specialists
Bacteriologists	Geneticists
Biochemists	Geographers
Chemical process engineers	Geologists
Civil engineers	Health promotion experts
Climatologists	Hydrogeologists
Communications experts	Hydrologists
Disaster preparedness specialists	Hygienists
Ecologists	Information scientists
Economists	Laboratory assistants/technicians
Engineering specialists (with postgraduate qualifications)	Marine scientists
Entomologists	Materials technologists
Environmental biologists	Medical specialists (with postgraduate qualifications in the public health area)
Environmental chemists	Meteorologists
Environmental engineers	Microbiologists
Environmental health administrators	Noise inspectors
Environmental health educators	Nuclear safety managers
Environmental health managers	Nutritionists
Environmental health officers	Occupational health nurses
Environmental health planners	Occupational health physicians
Environmental health technicians	Occupational hygienists
Environmental lawyers	Physicists
Epidemiologists (with medical degree)	Political scientists
Epidemiologists (without medical degree)	Pollution inspectors

(continues)

TABLE 1.4 Professions Involved in Environmental Health *(continued)*

Psychologists	Social scientists
Public health nurses	Social workers
Public health physicians	Soil scientists
Public health veterinarians	Statisticians
Public relations experts	Technical assistants
Risk assessors	Toxicologists
Rural and urban planners	Transport planners/managers
Safety inspectors	Water quality inspectors
Sanitary engineers	Zoologists
Sanitary officers	

Reproduced with permission from Fitzpatrick M, Bonnefoy X. *Environmental Health Services in Europe 4: Guidance on the Development of Educational and Training Curricula*. Copenhagen, Denmark: World Health Organization Regional Office for Europe; 1999:12. WHO Regional Publications, European Series, No. 84.

may affect the workers as well as hazards that may impact the community. They are involved with the design and installation of control systems for hazards in the occupational and environmental setting. They require training in the epidemiologic and biologic aspects of environmental hazards and also in toxicology. Industrial hygienists work closely with engineers who design and maintain industrial processes.[42]

Toxicologist

As a general description of the field, it may be said that toxicology concerns the effects of poisons. Among the many subspecializations in the field of toxicology are medical, veterinary, forensic, and environmental applications. The field of environmental toxicology specializes in the effects of toxic chemicals upon the environment and living creatures such as human beings and wildlife. Occupational and industrial toxicologists investigate the effects of chemicals found in the workplace upon the health of workers. Toxicologists are employed in academia as professors and researchers, by government agencies, by hospitals, and in various private industry settings. (See **FIGURE 1.14**.)

FIGURE 1.14 Researchers in a toxicology laboratory.
Courtesy of Dr. Arezoo Campbell, Department of Community and Environmental Medicine, University of California, Irvine.

Environmental Health Inspector

Public health departments provide many job opportunities for environmental health workers. Environmental health inspectors, who work mainly for state and local governments, are responsible for monitoring and enforcing government regulations for environmental quality. This employment category includes pollution inspectors, noise inspectors, and water quality inspectors. Such personnel help to monitor the treatment and disposal of sewage, refuse, and garbage. They also may visit toxic waste dumps, factories, and other sources of pollution in order to collect air, water, and waste samples for testing. They may be sent out to follow up on complaints from the community, attempt to determine the sources and nature of environmental pollution, and provide necessary background data for enforcement actions.

Food Inspector/Food Safety Specialist

Food inspectors and food safety specialists are involved with the cleanliness and safety of foods and beverages consumed by the public. They inspect restaurants, dairies, food processing plants, and other food preparation venues in order to control biological hazards from sources such as *Escherichia coli*, *Salmonella*, and other foodborne agents. Their purview of responsibility also may extend to hospitals and other institutional settings. They may be responsible for examining the methods for handling, processing, and serving of food so that these procedures are in compliance with sanitation rules and regulations. **FIGURE 1.15** demonstrates a food safety inspector examining a batch of food.

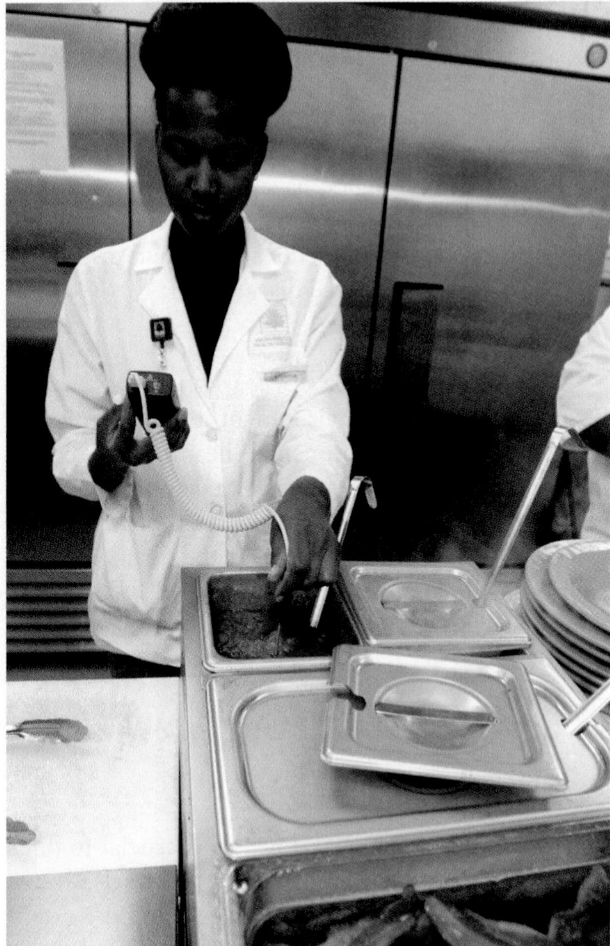

FIGURE 1.15 Public health inspector checking a kitchen's sanitary conditions.
Courtesy of CDC/Amanda Mills

Vector Control Specialist

Vector control specialists (not listed in Table 1.4) are responsible for the enforcement of various public health laws, sanitary codes, and other regulations related to the spread of disease by vectors. Examples of vectors are insects such as mosquitoes and flies, rodents, and other animals and arthropods that carry disease organisms. Vector control specialists are involved with controlling rabies, mosquito-borne encephalitis, tick-borne diseases, and zoonotic diseases. At the local government level, vector control specialists may be responsible for conducting community education programs, monitoring animal bites, collecting specimens for testing, and developing other procedures for control of diseases carried by vectors. **FIGURE 1.16** illustrates a vector control specialist at work.

FIGURE 1.16 Vector control specialist spraying insecticidal dust for control of plague, 1993.
Courtesy of CDC

Researcher/Research Analyst

In universities and research units, individuals who have specialized training in environmental health conduct basic research on the risks associated with exposures to certain specific hazards and conduct statistical analyses of the impact of such exposures on human populations. Although this category is not listed in Table 1.4, examples of employment titles found in research settings are laboratory scientist and technician, epidemiologist, and statistician.

Occupational Health Physician/ Occupational Health Nurse

These professionals are involved with the prevention and treatment of occupationally related illnesses and injuries. They investigate hazards in the work environment and develop procedures for their abatement. They also conduct health education programs for the prevention of work-related diseases.

Environmental Lawyer

Beyond the research environment, there is an active field known as environmental health law. Closely linked with environmental health law are environmental policies. Specialists in this field provide input to government agencies, assist in the formulation of environmental policies, and may be involved in litigation concerning environmental health problems.

Conclusion and Overview of the Text

The study of environmental health is crucial to one's understanding of the hazards and potential adverse effects posed by environmental agents and the extent to which environmental factors play a role in human disease. This foundation is essential for being an effective advocate for preventing environmentally caused diseases and for more advanced study of environmental health issues. This chapter has provided an introduction to the environmental health field: definitions, terminology and concepts, historical background, and career opportunities. The field of environmental health draws heavily upon **epidemiology** and toxicology.[43] As a result, environmental health makes abundant use of terminology from these disciplines. Epidemiologic aspects of environmental health include such well-known methods as case ascertainment, continuous surveillance of hazards, and development of tools for evaluation of intervention programs. Epidemiologic studies have been crucial for delineating the health effects of exposure to pollutants at the population level. Chapter 2 will introduce the topic of environmental epidemiology.

One of the contributions of toxicology to the environmental health field is in assessing **dose–response relationships**, which describe the responses of organisms to exposures to toxic substances. A dose–response relationship can be illustrated by an S-shaped curve known as a dose–response curve.[44] An additional concern of toxicology is exposure assessment, which involves procedures for determining what levels of specific hazards (e.g., a toxic chemical such as DDT) produce symptoms, disease, or other adverse effects. Chapter 3 provides more information on toxicology. Chapter 4 provides a discussion of environmental policy making and environmental health, including regulatory agencies and specific environmental laws that have been enacted in the United States. Chapters 5 through 9 cover agents such as microbes, toxic metals, radiation, and organic chemicals; these agents pose risks of disease from environmental factors. Chapter 5 focuses on environmental aspects of zoonotic and vector-borne diseases. Chapter 6 describes hazards that arise from toxic metals and elements (e.g., arsenic and chromium). Chapter 7 reveals hazards that arise from toxic chemicals such as pesticides, herbicides, and organophosphates. Chapter 8 focuses on ionizing radiation (for example, radiation emitted during the Chernobyl nuclear power plant accident and from X-ray machines used in body scans) and nonionizing radiation (for example, radio frequency radiation from cell phones and ultraviolet radiation from the sun). Chapter 9 covers water quality and pollution. Chapter 10 focuses on air quality and the impact of climate change. Chapter 11 covers the broad topic of food safety, including foodborne diseases caused by microbial agents and how additives and chemicals affect the food supply. Chapter 12 covers solid waste disposal and sewage processing. Chapter 13 presents information on occupational health. Finally, Chapter 14 demonstrates the role of environmental factors in unintentional injuries, a leading cause of mortality.

Study Questions and Exercises

1. Define the following terms:
   a. Population dynamics
   b. Completed fertility rate (total fertility rate)
   c. Environmental health
   d. Environmental risk transition
   e. Environment, physical and social
   f. Demographic transition
   g. Epidemiologic transition
2. Discuss three of the six major objectives (Outdoor Air Quality) for goal number 8, environmental health, as described in *Healthy People 2020*. What steps can be taken in your community to accomplish the objectives you have selected?
3. Describe the three P's that are principal determinants of health worldwide. Discuss how the three P's could be considered interrelated characteristics. Can you think of other

consequences of the three P's that are not discussed in the text?

4. This chapter discussed how environmental exposures affect our daily lives.
   a. Illustrate the types of environmental health problems that you have in your own community.
   b. Review and summarize at least three articles on environmental health that you have seen in the media, e.g., on the web or in a newspaper, during the past week. Are any of these articles relevant to the community where you live?
   c. Why is maintenance of environmental quality important for human health?
5. Discuss the role of population growth in human health. How might recent outbreaks of diseases such as the bird flu or pandemic H1N1 be linked to population growth? In addition to population growth, what other environmental factors could lead to pandemics such as those associated with influenza viruses?
6. Describe the types of environmental health problems that prevail in the developing world. Give at least three examples.
7. Demonstrate population growth trends over the past two centuries. What is the likelihood that current exponential population growth rates will continue? Take a stand for or against population growth at present high levels.
8. Describe variables that affect the size of a population. What countries or regions of the world are projected to experience stable or declining population sizes? What areas are expected to have the greatest increases in population size?
9. Define and discuss the following terms:
   a. Ecological model of population health
   b. Ecosystem
10. Summarize the contributions of the early Greeks to environmental health. How do Hippocrates' explanations of disease etiology compare with current beliefs about the role of the environment in human illness?
11. List and discuss five of today's most pressing environmental health issues ("hot topics"). In addition to the material presented in the text, you also may use your own ideas.
12. Explain why environmental health is an important field of employment. List specific employment roles in environmental health. Consult the World Wide Web for employment listings (e.g., published by state, local, and federal government agencies) and summarize the requirements and functions of three job titles.

For Further Reading

The Population Bomb, Paul R. Ehrlich, 1968
Walden, Henry David Thoreau, 1854
The Jungle, Upton Sinclair, 1906.

References

1. HealthyPeople.gov. 2020 Topics & Objectives: Environmental Health. *Healthy People 2020.* Available at: https://www.healthypeople.gov/2020/topics-objectives/topic/environmental-health. Accessed January 17, 2017.
2. Winkelstein W Jr. Determinants of worldwide health. *Am J Public Health.* 1992;82:931–932.
3. National Institute of Environmental Health Sciences. Forum: killer environment. *Environ Health Perspect.* 1999;107:A62–A63.
4. Butterfield PG. Upstream reflections on environmental health: an abbreviated history and framework for action. *Adv Nurs Sci.* 2002;25(1):32–49.
5. Prüss-Üstün A, Wolf J, Corvalán CF, et al. Preventing disease through healthy environments: a global assessment of the burden of disease from environmental risks. Geneva, Switzerland: World Health Organization; 2016.
6. World Health Organization. WHO to highlight specific health risks to children from environmental dangers. Available at: http://www.who.int/mediacentre/news/notes/note04/en/. Accessed June 27, 2017.
7. Carlson JE. Children's Environmental Health Network, Emeryville, California. *Environ Health Perspect.* 1998;106(Suppl 3):785–786.
8. Goldman LR, Koduru S. Chemicals in the environment and developmental toxicity to children: a public health and policy perspective. *Environ Health Perspect.* 2000;108(Suppl 3):443–448.
9. Chukwuma C Sr. Environmental health concepts and issues—a viewpoint. *Intern J Environ Studies.* 2001;58:631–644.
10. Mellor JW. The intertwining of environmental problems and poverty. *Environ.* 1988;30:8–30.
11. Camp SL. Population pressure, poverty and the environment. *Integration.* June 1992;24–27.
12. Smith KR. Environment and health: issues for the new U.S. administration. *Environ.* 2001;43:34–41.
13. Last JM. *Homo sapiens*—a suicidal species? *World Health Forum.* 1991;12:121–126.
14. US Census Bureau. *Global Population Profile: 2002.* International Population Reports WP/02. Washington, DC: US Government Printing Office; 2004.
15. Hinrichsen D. The decisive decade: what we can do about population. *Amicus J.* 1990;12(1):30–33.
16. Raleigh VS. World population and health in transition. *BMJ.* 1999;319:981–984.
17. Brown LR. Feeding six billion. *World Watch.* September/October 1989;32–40.
18. McMichael AJ, Powles JW. Human numbers, environment, sustainability, and health. *BMJ.* 1999;319:977–980.
19. Monte LM, Ellis RR. Fertility of women in the United States: 2012. *Current Population Reports,* P20-575. Washington, DC: US Census Bureau; 2014.
20. Gelbard A, Haub C. Population "explosion" not over for half the world. *Popul Today.* 1998;26:1–2.
21. International Organization for Migration. 2015 Global Migration Trends. Factsheet. Available at: https://publications.iom.int/system/files/global_migration_trends_2015_factsheet.pdf. Accessed July 6, 2017.

22. United Nations Environment Programme. Population and the environment. *Popul Bull UN*. 1987;(21–22):32–44.
23. McMichael AJ. The urban environment and health in a world of increasing globalization: issues for developing countries. *Bull World Health Organ*. 2000;78:1117–1126.
24. Satterthwaite D. The impact on health of urban environments. *Environ Urban*. 1993;5:87–111.
25. United Nations, Department of Economic and Social Affairs, Population Division (2016). The World's Cities in 2016—Data Booklet (ST/ESA/SER.A/392).
26. PopulationMatters.Carryingcapacity;2011.Availableat:https://populationmatters.org/wp-content/uploads/D20Carrying capacity.pdf Accessed June 27, 2017.
27. McMichael AJ. Ecological disruption and human health: the next great challenge to public health [editorial]. *Aust J Public Health*. 1992;16:3–5.
28. Congress of the United States. Office of Technology Assessment. Technologies to maintain biological diversity. OTA Brief Report; March 1987. Available at: http://ota.fas.org/reports/RB-Technologies%20To%20Maintain%20Biological%20Diversity.pdf. Accessed June 28, 2017.
29. Marwick C. Scientists stress biodiversity-human health links. *JAMA*. 1995;273:1246.
30. Merriam-Webster Online dictionary. Environment. Available at: https://www.merriam-webster.com/dictionary/environment. Accessed June 28, 2017.
31. The National Academies, Institute of Medicine. *Who Will Keep the Public Healthy? Educating Public Health Professionals for the 21st Century*. Washington, DC: The National Academies Press; 2003.
32. Millennium Ecosystem Assessment. *Ecosystems and Human Well-Being: A Framework for Assessment*. Washington, DC: Island Press; 2003.
33. World Health Organization. Damage to ecosystems poses growing threat to human health [press release]. March 29, 2005. Available at: http://www.who.int/globalchange/publications/masynthesis/en/. Accessed June 27, 2017.
34. World Health Organization. Health topics: environmental health. Available at: http://www.searo.who.int/topics/environmental_health/en/. Accessed June 28, 2017.
35. Franco DA, Williams CE. "Airs, Waters, Places" and other Hippocratic writings: inferences for control of foodborne and waterborne disease. *J Environ Health*. 2000;62(10):9–14.
36. Jårup L. Hazards of heavy metal contamination. *British Medical Journal*. 2003;68:167-182.
37. US Department of Labor, Occupational Safety and Health Administration Office of Training and Education. Industrial hygiene. Available at: https://www.osha.gov/dte/library/industrial_hygiene/industrial_hygiene.pdf. Accessed June 28, 2017.
38. Environmentalhistory.org. Environmental history timeline. Classical 1000 BCE–500 CE. Ancient Rome. Available at: http://environmentalhistory.org/ancient/classical-1000-bce-500-ce/. Accessed June 27, 2017.
39. History on the net.com. The Romans—public health. Available at: http://www.historyonthenet.com/the-romans-public-health/. Accessed June 27, 2017.
40. Yassi A, Kjellström T, de Kok T, et al. *Basic Environmental Health*. New York, NY: Oxford University Press; 2001.
41. Bell ML, Davis DL. Reassessment of the lethal London fog of 1952: novel indicators of acute and chronic consequences of acute exposure to air pollution. *Environ Health Perspect*. 2001;109(Suppl 3):389–394.
42. Sherwood RJ. Cause and control: education and training of professional industrial hygienists for 2020. *Am Ind Hyg Assoc J*. 1992;53:398–403.
43. Goldstein BD. Environmental risks and public health. *Ann NY Acad Sci*. 2001;933:112–118.
44. Friis RH, Sellers TA. *Epidemiology for Public Health Practice*. 5th ed. Burlington, MA: Jones & Bartlett Learning; 2014.

© Jean-Luc Rivard/EyeEm/Getty Images

CHAPTER 2
Environmental Epidemiology

LEARNING OBJECTIVES

By the end of this chapter the reader will be able to:

- Define the term environmental epidemiology.
- Describe three major historical events in environmental epidemiology.
- Provide examples of epidemiologic tools used in environmental health.
- Identify types of associations found between environmental hazards and health outcomes.
- Compare study designs used in environmental epidemiology.

Introduction

In this chapter you will learn that epidemiology is one of the fundamental disciplines used in the study of environmental health. For example, by using epidemiology it may be possible to connect environmental hazards such as air pollution or toxic chemicals with cancer and other adverse health outcomes. The launching point for our discussion will be a definition of the term *environmental epidemiology*. You will acquire information about the scope of this discipline and be able to define several of the special quantitative measures used to study the occurrence of environmental health problems in populations.

Next, we will trace the key historical developments in environmental epidemiology. Some of these historical benchmarks include concerns of the ancient Greeks about diseases caused by the environment, the observations of Sir Percival Pott on scrotal cancer among chimney sweeps in England (including Pott's clever public health recommendation for prevention of scrotal cancer), the work of John Snow on cholera, and later work on the role of toxic substances in the etiology of cancer. Closely linked to quantitative measures used by environmental epidemiology are the major study designs described in this chapter. See **FIGURE 2.1**.

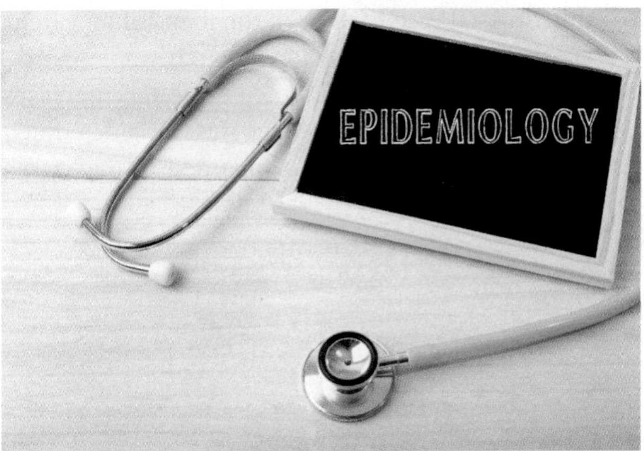

FIGURE 2.1 Epidemiology and environmental health.
© shahreen/Shutterstock

Research Topics for Environmental Epidemiology

Environmental epidemiology is a complex field that in some cases provides keen insights into environmentally caused diseases and in others provides unclear results that must be followed up by other types of studies. A special concern of the discipline is causality—whether research findings represent cause-and-effect associations. In order to explore such associations, the field employs experimental, quasi-experimental, and observational study designs. The latter (i.e., observational designs) include cross-sectional, ecologic, case-control, and cohort studies. More information on causality and epidemiologic study designs is provided later in this chapter.

Epidemiology is one of the research fields that seeks answers to crucial environmental questions, such as those that pertain to the domains of air pollution, chemicals, climate change, and water pollution.[1] Refer to the infographic presented in **FIGURE 2.2**.

- Air pollution continues to be a global public health issue. Associated with increasing urbanization of developing regions of the world are increasing levels of air pollution. Epidemiologic research has helped to identify adverse health effects of air pollution among vulnerable groups such as children and the elderly. A related concern pertains to the health impacts of exposure to environmental tobacco smoke.
- Potentially toxic chemicals such as pesticides, asbestos, lead, and mercury have been implicated in cancer, adverse reproductive outcomes, nervous system impacts, and numerous other health outcomes. These have been the focus of an extensive body of epidemiologic research.
- Scientists have documented gradual increases in global temperatures over past decades. Such changes have been accompanied by extreme climatic events, for example, high heat disasters in cities and flooding in coastal areas. Epidemiologic investigations will help to document adverse outcomes linked to global warming and inform policy decisions in response to climate change.
- Potable water has become increasingly scarce in the arid regions of the globe, depriving many of the world's inhabitants of safe water. In some parts of the United States, water supplies have also become compromised. An example is the intrusion of lead into the public drinking water supply from aging pipelines, as seen in Flint, Michigan in 2105. Pollution from urban runoff harms the nation's beaches and adversely affects marine life and seafood. Epidemiologic investigations have been instrumental in identifying the adverse health effects associated with such water pollution.
- As the sophistication of methodology in genetics, data analysis, and other cutting-edge disciplines has grown, the capacity of epidemiologic research to explore a panoply of intriguing issues has been enhanced. Some of these involve possible environmental concomitants of neurologic conditions such as Alzheimer's disease, interaction of environmental exposures with our genetic makeup, and exposure to radiation from nuclear power plants.

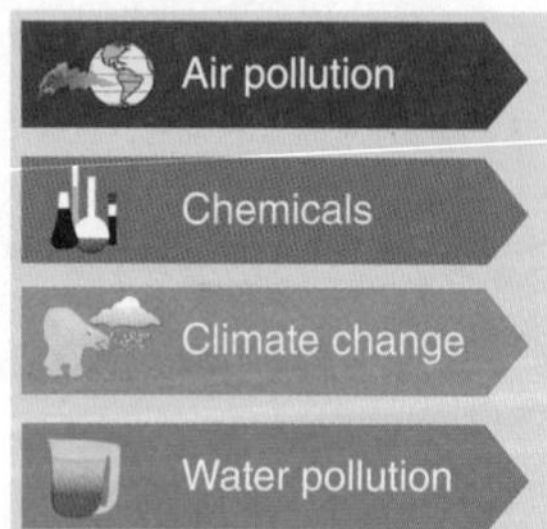

FIGURE 2.2 Environmental topics for epidemiology.
Reproduced from National Institutes of Health, National Institute of Environmental Health Sciences. What is environmental health? Available at: https://kidsenvirohealth.nlm.nih.gov. Accessed April 3, 2017.

In summary, epidemiology is the method of choice to address issues such as the foregoing ones. Refer to the following text box for a further discussion of epidemiology and environmental health.

Definition of Environmental Epidemiology

Epidemiology is concerned with the study of the distribution and determinants of health and diseases, morbidity, injuries, disability, and mortality in populations.[2] Epidemiologic studies are applied to the control of health problems in populations. Epidemiology is one of the core disciplines used to examine the associations between environmental hazards and health outcomes. The term *environmental epidemiology* refers to the study of diseases and health conditions (occurring in the population) that are linked to environmental factors.[3,4] The exposures, which most of the time are outside the control of the individual, usually may be considered involuntary and stem from ambient and occupational environments.[5] According to this conception of environmental

EPIDEMIOLOGY'S UNIQUE CONTRIBUTION TO ENVIRONMENTAL HEALTH

Epidemiology makes a special contribution to environmental health through its focus on entire populations and by its use of descriptive and analytic methodologies. Clinical observations help to identify and diagnose individual patients who are afflicted by environmental hazards. However, this information may not be sufficient to discern how adverse health effects of environmental exposures are distributed in the community.

The chain of research often begins when local public health departments receive complaints of diseases such as asthma from people living in the departments' jurisdictions. When these individual case reports are collated, epidemiologists may be able to develop hypotheses regarding how these outcomes are related to environmental factors. Epidemiologists may then decide to conduct a broader investigation of the entire community in order to delineate specific groups of persons who are being affected as well as their location in relation to the presence of hazardous environmental exposures.

The findings of epidemiologic research can aid in controlling environmental exposures and developing health policies for protecting the public. For example, epidemiologic studies of air pollution conducted during the 20th century showed that people exposed to high levels of air pollution experienced increased mortality in comparison with those whose air was less polluted. More recent findings suggested a relationship between living in proximity to a heavily traveled motorway and adverse health effects connected with emissions from passing vehicles.

In response to findings such as these, government agencies and stakeholders introduced regulations to clean the air. These measures included reductions of "smokestack" emissions from factories and adoption of air quality standards for automobile exhaust. Without epidemiologic research, the ties between air pollution and specific adverse health effects might not have been obvious. Epidemiology has also uncovered connections between other forms of environmental pollution such as exposures to toxic chemicals and adverse health outcomes. For these reasons, epidemiology makes a unique contribution to the study of environmental health issues.

epidemiology, standard epidemiologic methods are used to study the association between environmental factors (exposures) and health outcomes. Examples of topics studied include air and water pollution, the occupational environment with its possible use of physical and chemical agents, and the psychosocial environment.[6]

As noted previously, for an environmentally associated health outcome to be considered a topic of environmental epidemiology, exposure factors must lie outside the individual's immediate control. Hazards associated with smoking can be explored as an exposure dimension that is either under or not under the control of the individual. As an example of the former, studies of the health effects of smoking among individuals who smoke would not be a usual concern of environmental epidemiology. However, exposure of populations to secondhand cigarette smoke would be a concern because nonsmokers and vulnerable groups such as children cannot control whether they are exposed to environmental tobacco smoke.

Thus, traditionally, environmental epidemiology has tended to focus on health effects linked to degradation of the air we breathe, the water we drink, and the food we eat.[7] With the advances achieved during the 20th century in environmental sanitation and control of disease-causing biological organisms, attention to chemical and physical impacts upon the environment has increased. Some of the agents and environmental factors being focused on are lead toxicity, particulates from diesel exhaust, and exposures to pesticides and halogenated compounds (a compound that is a combination of a halogen such as chlorine or iodine and one or more elements). Halogenated compounds include **polychlorinated biphenyls (PCBs)**, which through biological processes can become increasingly concentrated in foodstuffs, can pose hazards as potential carcinogens, and can impact the reproductive system. More recent concerns of environmental health include the reemerging infectious diseases (see the chapter on zoonotic and vector-borne diseases) and the effects of climate changes due to global warming.

Although the relationship between environmental exposures and their unknown hazards remains a concern of environmental epidemiology, the field has evolved to include a broader approach: identification of previously unrecognized exposures to known hazardous agents and the quantification of such risks, estimation of the amount of exposures that individuals have to environmental hazards, assessment of risks associated with exposures (discussed in the chapter on environmental toxicology), and evaluation of procedures to prevent exposures.[4] Similarly, in the related field of occupational health, the goals of epidemiologic research encompass the description of exposure–response gradients, discovery of how occupational hazards may cause harmful effects, characterization of vulnerable workers, and input into programs for the prevention of occupationally related diseases.[8]

This discussion regarding the definition of environmental epidemiology leads to the issue of the types of work performed by epidemiologists. **FIGURES 2.3** and **2.4** highlight two of several of the diverse settings where epidemiologists work. As indicated by Figure 2.3, a cadre of epidemiologists is situated primarily in a laboratory where they research disease agents such as the Zika virus. Others—as demonstrated by the team in Figure 2.4—analyze research data that may reveal patterns and associations with respect to the occurrence of adverse and other health outcomes.

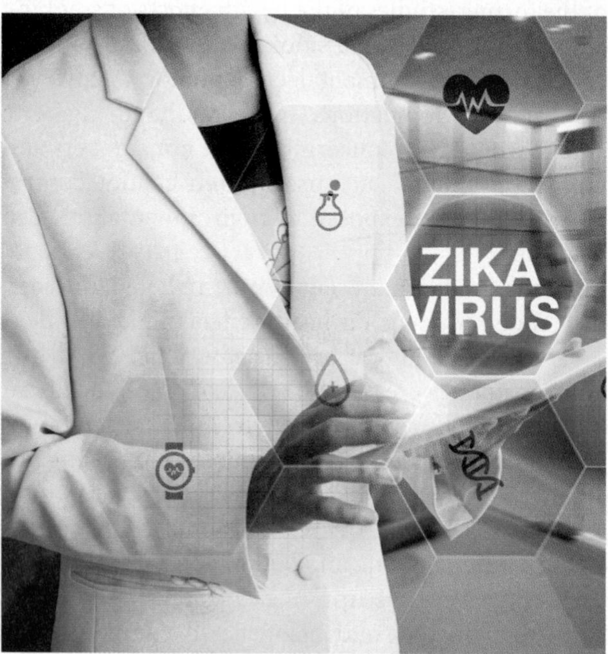

FIGURE 2.3 Exploring the Zika virus epidemic.

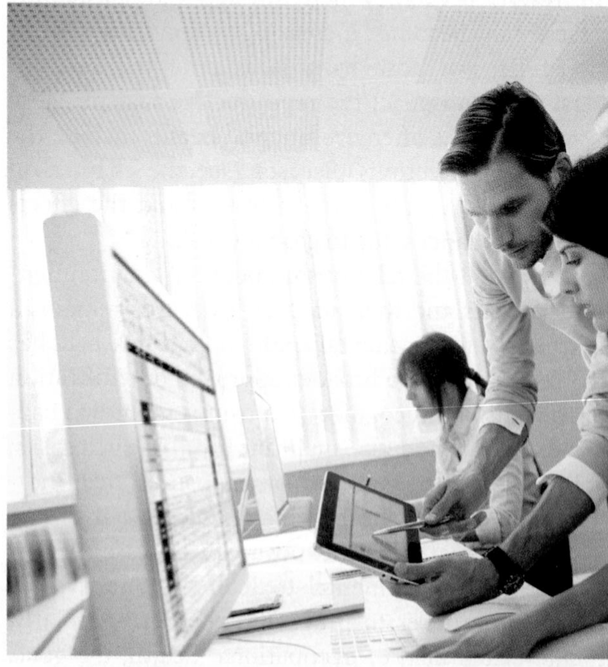

FIGURE 2.4 Epidemiologic data analysis.

Contributions of Epidemiology to Environmental Health

Epidemiology aids the environmental health field through:

- Concern with populations
- Use of observational data
- Methodology for study designs
- Descriptive and analytic studies

Epidemiology is important to the study of environmental health problems because (1) many exposures and health effects associated with the environment occur at the population level; (2) the epidemiologic methods of natural experiments and observational techniques are appropriate; (3) the study designs used in epidemiologic research can be applied directly to the study of environmental health issues; and (4) epidemiology aids in the development of hypotheses and the study of causal relationships.

Concern with Populations

In contrast with clinical medicine's traditional focus on the individual, a unique characteristic of epidemiology is that it studies the entire population and hence is sometimes called population medicine. Refer to **FIGURE 2.5**. For example, epidemiologic studies of lung disease may examine the occurrence of lung cancer mortality across counties or among regional geographic subdivisions known as census tracts. Investigators may want to determine whether lung cancer mortality is higher in areas with higher concentrations of "smokestack" industries in comparison with areas that have lower levels of air pollution or are relatively free from air pollution. The alternative approach of the clinician would be to concentrate on the diagnosis and treatment of lung cancer among specific individuals.

Use of Observational Data

In examining the occurrence of health and disease in human populations, researchers often are prohibited

FIGURE 2.5 Epidemiology's population focus.

from using experimental methods because of ethical issues such as potential dangers to subjects. Studies of the population's health present a challenge that is partially met by epidemiology because epidemiology is primarily an observational science that takes advantage of naturally occurring situations in order to study the occurrence of disease.

Methodology for Study Designs

In the realm of environmental health, epidemiologic research generally aims to portray the frequency of disease occurrence in the population or to link disease outcomes to specific exposures.[9] In order to research environmentally caused disease in the population, the field of environmental epidemiology uses characteristic study designs: cross-sectional, ecologic, case-control, and cohort. For example, these methods are useful in and linked closely to the field of risk assessment (discussed in the chapter on environmental toxicology). Professor A.H. Smith writes, "The epidemiologic input to environmental risk assessment involves the interpretation of epidemiological studies and their application to estimating the potential health risks to populations from known or estimated environmental exposures."[10(p124)]

Two Classes of Epidemiologic Studies: Descriptive and Analytic

The term **descriptive epidemiology** refers to the depiction of the occurrence of disease in populations according to classification by person, place, and time variables. Examples of person variables are demographic characteristics such as sex, age, and race/ethnicity. Place variables denote geographic locations including a specific country or countries, areas within countries, and regions where localized patterns of disease may occur. Illustrations of time variables are a decade, a year, a month, a week, or a day. Descriptive studies, regarded as a fundamental approach by epidemiologists, aim to delineate the patterns and manner in which disease occurs in populations.[11]

An example of a pattern derived from descriptive studies is disease clustering, which refers to "... a closely grouped series of events or cases of a disease or other health-related phenomena with well-defined distribution patterns in relation to time or place or both. The term is normally used to describe aggregation of relatively uncommon events or diseases (e.g., leukemia, multiple sclerosis)."[12]

Clustering may suggest common exposure of the population to an environmental hazard; it also may be purely spurious—due to the operation of chance. One cause of spurious clustering is called the Texas Sharpshooter Effect, discussed in the text box.

In the field of occupational health—which in many respects is emblematic of the general field of environmental health—"*descriptive studies* provide information for setting priorities, identifying hazards, and formulating hypotheses for new occupational risk."[13(p944)] A historical example (discussed later in this chapter) is William Farr's work showing that Cornwall metal miners had higher mortality from all causes than the general population.[13]

Analytic epidemiology examines causal (etiologic) hypotheses regarding the association between exposures and health conditions. "Etiologic studies are planned examinations of causality and the natural history of disease. These studies have required increasingly sophisticated analytic methods as the importance of low-level exposures is explored and greater refinement in exposure-effect relationships is sought."[13(p945)] The field of analytic epidemiology proposes and evaluates causal models that employ both outcome variables and exposure variables.

The exposure variables in epidemiologic research include contact with toxic substances, potential carcinogens, or air pollution. In other cases, exposure may be to biological agents or to forms of energy such as ionizing and nonionizing radiation, noise, and

THE TEXAS SHARPSHOOTER EFFECT

A traveler passing through a small town in Texas noted a remarkable display of sharpshooting. On almost every barn he passed there was a target with a single bullet hole that uncannily passed through the center of the bull's-eye. He was so intrigued by this that he stopped at a nearby gas station to ask about the sharpshooter. With a chuckle, the attendant told him that the shooting was the work of Old Joe. Old Joe would first shoot at the side of a barn and then paint targets centered over his bullet holes so that each shot appeared to pass through the center of the target. . . . In a random distribution of cases of cancer over a geographic area, some cases will appear to occur very close together just on the basis of random variation. The occurrence of a group of cases of a disease close together in time and place at the time of their diagnosis is called a cluster.

Reproduced from Grufferman S. Methodologic approaches to studying environmental factors in childhood cancer. *Environ Health Perspect*. 1998;106(suppl 3):882.

extremes of temperature. For an environmental epidemiologic research study to be valid, the level of exposure in a population must be assessed validly.

The outcome variable in epidemiologic studies is usually a specific disease, cause of mortality, or health condition. Accurate clinical assessments of an outcome such as lung cancer are vitally important to the quality of epidemiologic research.

One approach of analytic epidemiology is to take advantage of naturally occurring situations or events in order to test causal hypotheses. These naturally occurring events are referred to as **natural experiments**, defined as "naturally occurring circumstances in which subsets of the population have different levels of exposure to a hypothesized causal factor in a situation resembling an actual experiment."[12] An example is the work of John Snow, discussed later in this chapter. Many past or ongoing natural experiments are relevant to environmental epidemiology. For example, in some regions of the United States, health legislation prohibits smoking in public areas in order to prevent exposure to secondhand smoke. At the same time, this activity may be considered a natural experiment that impacts human health and that can be studied by environmental epidemiologists.

Measures of Disease Frequency Used in Epidemiology

A number of quantitative terms, useful in environmental epidemiology, have been developed to characterize the occurrence of disease, morbidity, and mortality in populations. Particularly noteworthy are the two terms **prevalence** and **incidence**, which can be stated as frequencies or raw numbers of cases. In order to make comparisons among populations that differ in size, statisticians divide the number of cases by the population size. Several examples follow.

The term *prevalence* refers to the number of existing cases of or deaths from a disease or health condition in a population at some designated time. More specifically, **point prevalence** refers to all cases of or deaths from a disease or health condition that exist at a particular point in time relative to a specific population from which the cases are derived. Prevalence measures are used to describe the scope and distribution of health outcomes in the population. By revealing a snapshot of disease occurrence in the population, prevalence data contribute to the accomplishment of two of the primary functions of descriptive epidemiology: to assess variations in the occurrence of disease in populations and to aid in the development of etiologic hypotheses.

Comparisons among populations that differ in size cannot be accomplished directly by using frequency or prevalence data. In order to make such comparisons, prevalence (usually referring to point prevalence) may be expressed as a proportion formed by dividing the number of cases that occur in a population by the size of the population in which the cases occur.

$$\text{Point prevalence} = \frac{\text{Number of persons ill}}{\text{Total number in the group}} \text{ at a point in time}$$

The term *incidence* refers to the occurrence of new disease or mortality within a defined period of observation (e.g., a week, month, year, or other time period) in a specified population. Those members of the population who are capable of developing the disease or condition being studied are known as the **population at risk**.

The **incidence rate** denotes "[t]he RATE at which new events occur in a population."[12] The new events may be cases of disease or some other outcome of interest. Statistically speaking, the incidence rate is a rate because of the specification of a time period during which the new cases occur. (Several variations of incidence rates exist, but a discussion of all of them is beyond the scope of this chapter.) The incidence rate used most often in public health is formed by dividing the number of new cases that occur during a time period by the average number of individuals in the population at risk.

$$\text{Incidence rate} = \frac{\text{Number of new cases over a time period}}{\text{Average population at risk during the same time period}} \times \text{multiplier (e.g., 100,000)}$$

Incidence measures are central to the study of causal mechanisms with regard to how exposures affect health outcomes. Incidence measures are used to describe the risks associated with certain exposures; they can be used to estimate in a population "the probability of someone in that population developing the disease during a specified period, conditional on not dying first from another disease."[9(p23)]

One additional measure covered in this section is known as the case fatality rate (CFR). (Note that the chapter on zoonotic and vector-borne diseases will refer to the CFR.) The CFR, which provides a measure of the lethality of a disease, is defined as the number of deaths due to a specific disease within a specified time period divided by the number of cases of that disease during the same time period multiplied by 100. The formula is expressed as follows:

$$\text{CFR}(\%) = \frac{\begin{array}{c}\text{Number of deaths}\\ \text{due to disease "X"}\end{array}}{\begin{array}{c}\text{Number of cases}\\ \text{of disease "X"}\end{array}} \times \begin{array}{c}\text{100 during a}\\ \text{time period}\end{array}$$

The numerator and denominator refer to the same time period. For example, suppose that 45 cases of hantavirus infection occurred in a western US state during a year of interest. Of these cases, 22 were fatal. The CFR would be:

$$\text{CFR}(\%) = \frac{22}{45} \times 100 = 48.9\%$$

Brief History of Environmental Epidemiology

Hippocrates

Environmental epidemiology has a long history that dates back 2,000 or more years.[14] For example, in about 400 BC the ancient Greek authority Hippocrates expounded on the role of environmental factors such as water quality and the air in causing diseases.[14] He produced the well-known book *On Airs, Waters, and Places.* Experts in the field confirm that these writings form the historical cornerstone of environmental epidemiology. Hippocrates' work and the writings of many of the ancients did not delineate specific known agents involved in the causality of health problems, but referred more generically to air, water, and food. In this respect, early epidemiology shares with contemporary epidemiology the frequent lack of complete knowledge of the specific agents of environmentally associated diseases.

Sir Percival Pott

Sir Percival Pott, a London surgeon, was significant to the history of environmental epidemiology because he is thought to be the first individual to describe an environmental cause of cancer. (See **FIGURE 2.6**.)

In 1775, Pott made the astute observation that chimney sweeps had a high incidence of scrotal cancer (in comparison with male workers in other occupations). He argued that chimney sweeps were prone to this malady as a consequence of their contact with soot.[15] (See **FIGURE 2.7**.)

In a book entitled *Chirurgical Observations Relative to the Cataract, the Polypus of the Nose, the Cancer of the Scrotum, the Different Kinds of Ruptures, and the Mortification of the Toes and Feet*, published in London in 1775, Pott developed a chapter called

PERCIVALL POTT Esq.r

Engraved by Heath, from a Picture of Sir Joshua Reynolds.

Publish'd March 10th 1790, by J. Johnson, St Paul's Church Yard.

FIGURE 2.6 Percival Pott, F.R.S., 1714–1788.

FIGURE 2.7 A chimney sweep.

"A Short Treatise of the Chimney Sweeper's Cancer." This brief work of only 725 words is noteworthy because

> … it provided the first clear description of an environmental cause of cancer, suggested a way to prevent the disease, and led indirectly to the synthesis of the first known pure carcinogen and the isolation of the first carcinogenic chemical to be obtained from a natural product. No wonder therefore that Pott's observation has come to be regarded as the foundation stone on w[h]ich the knowledge of cancer prevention has been built![15(p521)]

In Pott's own words,

> [E]very body . . . is acquainted with the disorders to which painters, plummers, glaziers, and the workers in white lead are liable; but there is a disease as peculiar to a certain set of people which has not, at least to my knowledge, been publickly noteced; I mean the chimney-sweepers' cancer. . . . The fate of these people seems singularly hard; in their early infancy, they are most frequently treated with great brutality, and almost starved with cold and hunger; they are thrust up narrow, and sometimes hot chimnies, where they are bruised, burned, and almost suffocated; and when they get to puberty, become peculiary [*sic*] liable to a noisome, painful and fatal disease. Of this last circumstance there is not the least doubt though perhaps it may not have been sufficiently attended to, to make it generally known. Other people have cancers of the same part; and so have others besides lead-workers, the Poictou colic, and the consequent paralysis; but it is nevertheless a disease to which they are particularly liable; and so are chimney-sweepers to the cancer of the scrotum and testicles. The disease, in these people . . . seems to derive its origin from a lodgment of soot in the rugae of the scrotum.[15(pp521–522)]

Following his conclusions about the relationship between scrotal cancer and chimney sweeping, Pott established an occupational hygiene control measure—the recommendation that chimney sweeps bathe once a week.

John Snow

During the mid-1800s, English anesthesiologist John Snow (see **FIGURE 2.8**) linked a cholera outbreak in London to contaminated water from the Thames River. His methodology for investigating the cholera outbreak of

FIGURE 2.8 John Snow.

JOHN SNOW'S INVESTIGATION OF A CHOLERA OUTBREAK IN LONDON, CIRCA 1849

A section of London, designated the Broad Street neighborhood (now part of the Soho district), became the focus of Snow's detective work. Two water companies, the Lambeth Company and the Southwark and Vauxhall Company, provided water in such a manner that adjacent houses could receive water from two different sources. One of the companies, the Lambeth Company, relocated its water sources to a section of the Thames River that was less contaminated. During a later cholera outbreak in 1854, Snow observed that a higher proportion of residents who used the water from the Southwark and Vauxhall Company developed cholera than did residents who used water from the Lambeth Company. Snow's efforts to show a correspondence between changes in the water supply and occurrence of cholera became known as a natural experiment.

Here is Snow's graphic description of the cholera outbreak that occurred in 1849.

The most terrible outbreak of cholera which ever occurred in this kingdom, is probably that which took place in Broad Street, Golden Square, and the adjoining streets, a few weeks ago.... The mortality in this limited area probably equals any that was ever caused in this country, even by the plague; and it was much more sudden, as the greater number of cases terminated in a few hours.... Many houses were closed altogether, owing to the death of the proprietors; and, in a great number of instances, the tradesmen who remained had sent away their families: so that in less than six days from the commencement of the outbreak, the most afflicted streets were deserted by more than three-quarters of their inhabitants.[16(p38)]

Snow's pioneering approach illustrated the use of both descriptive and analytic epidemiology. One of his first activities was to plot the cholera deaths in relation to a pump that he hypothesized was the cause of the cholera outbreak. Each death was shown on the map (**FIGURE 2.9**) as a short line. An arrow in the figure points to the location of the Broad Street pump. "As soon as I became acquainted with the situation and the extent of this irruption of cholera, I suspected some contamination of the water of the much-frequented street-pump in Broad Street, near the end of Cambridge Street.... On proceeding to the spot, I found that nearly all the deaths had taken place within a short distance of the pump."[16(pp38–39)] The handle of the pump was later removed—a public health measure to control the outbreak. In Snow's time, many European cities took water for domestic use directly from rivers, which often were contaminated with microorganisms.

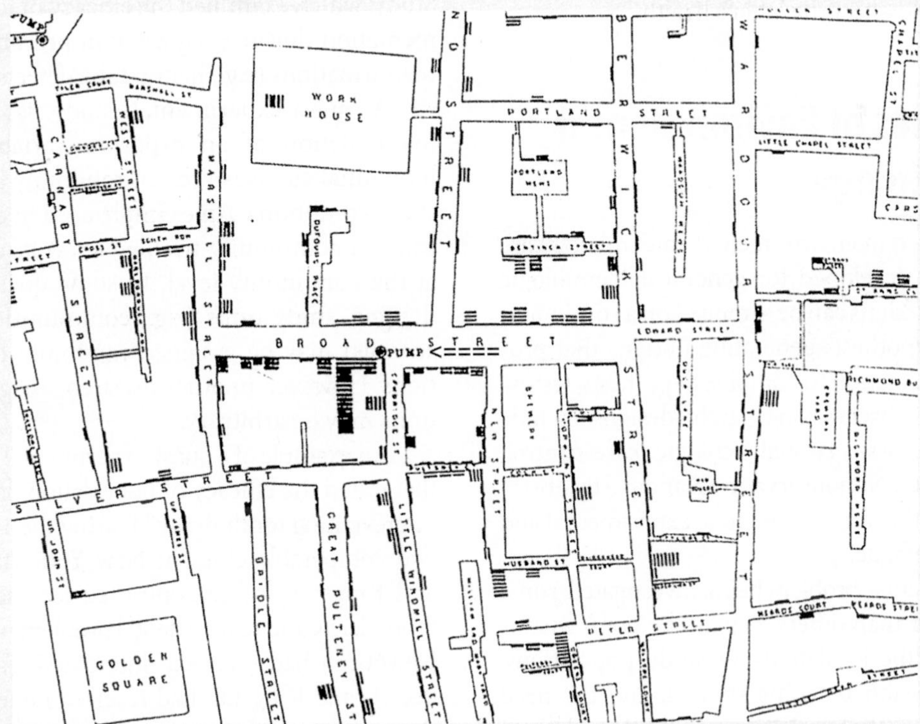

FIGURE 2.9 Map of cholera cases in the Broad Street area. Each case is indicated by a short line.

Modified from Snow J. *Snow on Cholera*. Cambridge, MA: Harvard University Press; 1965.

1849 was known as a "natural experiment," a methodology used currently in the study of environmental health problems. Refer to the text box for more information.

In addition to utilizing the method of natural experiment, John Snow provided expert witness testimony on behalf of industry with respect to environmental exposures to potential disease agents.[17] Snow attempted to extrapolate from the health effects of exposures to high doses of environmental substances what the effects of low doses would be. On January 23, 1855, the Nuisances Removal and Diseases Prevention Amendments bill was introduced in the British Parliament. This bill was a reform of Victorian public health legislation that followed the 1854 cholera outbreak described in the foregoing paragraph.[17] The intent of the bill was to control release into the atmosphere of fumes from operations such as gas works, silk-boiling works, and bone-boiling factories. Snow contended that these odors were not a disease hazard in the community.[18] The thesis of Snow's argument was that deleterious health effects from the low levels of exposure experienced in the community were unlikely, given the knowledge about higher-level exposures among those who worked in the factories. Snow argued that the workers in the factories were not suffering any ill health effects or dying from the exposure. Therefore, it was unlikely that the much lower exposures experienced by the members of the larger community would affect the latter's health.

Strategies of Environmental Epidemiology

Study designs used in environmental epidemiology are similar to those developed for general epidemiologic research. Study designs can be arranged on a continuum ranging from hypothesis-generating designs that provide limited information to complex hypothesis-testing designs.[8] Purely observational study designs include case series and cross-sectional, ecologic, case-control, and cohort studies. Nonobservational and partly observational designs that are used include experimental and quasi-experimental designs.

For the particular problem being investigated, some designs are better than others, depending upon what is to be achieved, the availability of study populations, the disease or health outcome studied, and the need to uncover disease etiology.[8] Examples of the subset of observational designs that are used for hypothesis generation include cross-sectional studies, case series, some types of ecologic correlations, and proportionate mortality comparisons.[8] The subset of observational designs that are used for hypothesis testing includes cohort and case-control studies. One of the distinguishing characteristics of study designs is whether they involve the individual or group as the unit of analysis. With the exception of ecologic studies, all the designs presented in this chapter use the individual as the unit of analysis.

Experimental Studies

Consider the use of experimental studies in environmental health research; in epidemiology, experimental studies are implemented as intervention studies. An **intervention study** is "[a]n investigation involving intentional change in some aspect of the status of the subjects, e.g., introduction of a preventive or therapeutic regimen or an intervention designed to test a hypothesized relationship. . . ."[12]

Two intervention study designs are clinical trials (randomized controlled trials) and quasi-experiments (community trials). A simple illustration of the former is a classic experimental design in which there is manipulation of an exposure variable and random assignment of subjects to either a treatment group or a control group. The exposure variable might be a new drug or other regimen. Some uses of randomized controlled trials are to test the efficacy of new medications, medical regimens, and vaccines. Among the many examples of a clinical trial is the Medical Research Council Vitamin Study, which examined the efficacy of folic acid supplementation during pregnancy in preventing congenital malformations (e.g., neural tube defects).[2]

A quasi-experimental study is one in which manipulation of an exposure variable occurs, but individual subjects are not randomly allocated to the study conditions. An example used in epidemiology is called a community trial, which tests an intervention at the community level. In some quasi-experimental designs, study units (e.g., communities, counties, or schools) may be assigned randomly to study conditions. However, in other research, assignment of study units may be arbitrary.

An example of a quasi-experimental study was a trial that tested the efficacy of fluoridation of drinking water in preventing tooth decay.[9] During the 1940s and 1950s, two comparable cities in New York State—Newburgh and Kingston—were contrasted for the occurrence of tooth decay and related dental problems among children. Newburgh had received fluoridated water for about a decade and Kingston had received none. In Newburgh, the frequency of dental problems decreased by about one half and increased slightly in Kingston.[9] In this quasi-experimental design, individual subjects were not randomized to the study conditions.

For several reasons, the use of experimental methods in environmental epidemiology is difficult to achieve; consequently, observational methods are usually more feasible to implement. Rothman points out:

> Randomized assignment of individuals into groups with different environmental exposures generally is impractical, if not unethical; community intervention trials for environmental exposures have been conducted, although seldom (if ever) with random assignment. Furthermore, the benefits of randomization are heavily diluted when the number of randomly assigned units is small, as when communities rather than individuals are randomized. Thus, environmental epidemiology consists nearly exclusively of non-experimental epidemiology. Ideally, such studies use individuals as the unit of measurement; but often environmental data are available only for groups of individuals, and investigators turn to so-called ecologic studies to learn what they can.[6(p20)]

Consequently, in order to study the effects of environmental exposures when dealing with human populations, researchers must use observational methods, and, in fact, the majority of research on health outcomes associated with the environment uses observational methods.[19]

FIGURE 2.10 Advertisement promoting diethylstilbestrol (DES).

Available at: https://desinfo411.wordpress.com/tag/massachusetts-general-hospital/. Accessed April 3, 2017

Case Series

A **case series study** is one in which information about patients who share a disease in common is gathered over time. Although this type of study is among the weakest for making causal assertions, a case series can be useful for developing hypotheses for further study. Usually information from a case series study is considered to be preliminary and a starting point for more complex investigations. However, some astute clinicians have used information from series of cases to make important observations. An example comes from the work of Herbst and Scully, who were the first to describe the association between exposure to diethylstilbestrol (DES) during mothers' pregnancies and risk of clear-cell cervicovaginal cancer among six female adolescents and young adults.[20] (Refer to **FIGURE 2.10** for an advertisement for diethylstilbestrol [DES, des*PLEX*®], which was administered to "prevent abortion, miscarriage and premature labor.")

Cross-Sectional Studies

A **cross-sectional study** is defined as one

> ...that examines the relationship between diseases (or other health outcomes) and other variables of interest as they exist in a defined population at one particular time. The presence or absence of disease and the presence or absence of the other variables... are determined in each member of the study population or in a representative sample at one particular time.[12]

Thus, a cross-sectional study is a type of prevalence study in which the distribution of disease and exposure are determined, although it is not imperative for the study to include both exposure and disease. A cross-sectional study may focus only on the latter.[2] Cross-sectional designs make a one-time assessment of the prevalence of disease in a sample that in most situations has been sampled randomly from the parent population of interest.[9] Cross-sectional studies may be used to formulate hypotheses that can be followed up in analytic studies.

Here is an example of a cross-sectional study: As part of an asthma reduction program conducted in Passaic, New Jersey during the 1998 through 1999 school year, investigators conducted a survey of a community in which all third graders were targeted.[21] The study children and their parents were given self-report symptom questionnaires. A total of 976 children and 818 parents returned the questionnaire. A respiratory therapist collected spirometry (lung function)

TABLE 2.1 Population Distribution and Percentage of Physician-Interpreted Abnormal Spirometry Readings (n = 455)

Race/Ethnicity	Population Distribution	Abnormal Spirometry (%)
Dominican	22.6	28.2
Mexican	19.8	10.0
Puerto Rican	19.3	8.0
Mixed other Hispanic	9.7	4.5
Peruvian	4.0	11.1
Colombian	2.6	8.3
Black	11.2	39.2
White	4.0	27.8
Asian	3.7	47.1
Mixed non-Hispanic	3.1	14.3

Modified and reproduced with permission from Freeman NCG, Schneider D, and McGarvey P. School-based screening for asthma in third-grade urban children: the Passaic asthma reduction effort survey. *Am J Public Health*. 2002;92:45.

readings from 615 children (approximately 58% of the target population). The study demonstrated that about half the children experienced self-reported asthma-related symptoms. However, because self-reports were not associated closely with the results of the spirometry tests, the investigators concluded that the self-reported data from children were not good predictors of asthma risk. From the spirometry results, about 22% of the children had abnormal results, with significant differences occurring by race and ethnicity. More abnormal evaluations were found for blacks and Asians in comparison with other groups. **TABLE 2.1** reports the results of the spirometry evaluation.

Ecologic Studies

Ecologic studies are different from most other types of epidemiologic research in regard to the unit of analysis. An **ecologic study** (also called an ecological study) is "a study in which the units of analysis are populations or groups of people rather than individuals."[12] For example, the occurrence of an outcome of interest (e.g., a disease, mortality, health effect) might be assessed over different geographic areas—states, census tracts, or counties. To illustrate, one could study "the relationship between the distribution of income and mortality rates in states or provinces."[12] The assumption is made that outcome rates would be comparable in exposed and nonexposed groups if the exposure did not take place in the exposed group. In the foregoing example, if the outcome were mortality from cancer, researchers might hypothesize that persons living in lower-income areas have greater exposure to environmental carcinogens than those who live in higher-income areas, producing differences in cancer mortality.

Ecologic analyses have been used to correlate air pollution with adverse health effects such as mortality. Instead of correlating individual exposure to air pollution with mortality, the researcher measures the association between average exposure to air pollution within a census tract and the average mortality in that census tract. Other types of geographic subdivisions besides census tracts may be used as well. This type of study attempts to demonstrate that mortality is higher in more polluted census tracts than in less polluted census tracts.

A major problem of the ecologic technique for the study of air pollution (and for virtually all ecologic studies) stems from uncontrolled factors. Examples relevant to air pollution include individual levels of smoking and smoking habits, occupational exposure to respiratory hazards and air pollution, differences in social class and other demographic factors, genetic background, and length of residence in the area.[8] Nonetheless, ecologic studies may open the next generation of investigations; the interesting observations gathered in ecologic studies may provide the impetus for more carefully designed studies. The next wave of studies that build on ecologic studies then may attempt to take advantage of more rigorous analytic study designs.

Ecologic studies have examined the association between water quality and both stroke and coronary diseases. A group of studies has demonstrated that hardness of the domestic water supply is associated inversely with risk of cerebrovascular mortality and cardiovascular diseases. However, a Japanese investigation did not support a relationship between water hardness and cerebrovascular diseases. In the latter ecologic study, the unit of analysis was municipalities (population subdivisions in Japan that consisted of from 6,000 to 3 million inhabitants). In analyzing the 1995 death rates from strokes in relationship to the values of water hardness, the researchers did not find statistically significant associations across municipalities.[22]

Other ecologic studies have examined the possible association between use of agricultural pesticides and childhood cancer incidence. For example, a total of 7,143 incident cases of invasive cancer diagnosed among children younger than age 15 were reported to the California Cancer Registry during the years 1988 to 1994. (Note that a registry is a centralized database for collection of information about a disease.) In this ecologic study, the unit of analysis was census blocks, with average annual pesticide exposure estimated per square mile. The study showed no overall association between pesticide exposure determined by this method and childhood cancer incidence rates. However, a significant increase of childhood leukemia rates was linked to census block groups that had the highest use of one form of pesticide, called propargite.[23]

Case-Control Studies

In a **case-control study**, subjects who participate in the study are defined on the basis of the presence or absence of an outcome of interest. The cases are those who have the outcome or disease of interest, and the controls are those who do not. In a case-control study, cases and controls generally are matched according to criteria such as sex, age, race, or other variables. Exposure to a factor is determined retrospectively, meaning that exposure has already occurred in the past. One method to determine past exposure is for the investigator to interview cases and controls regarding their exposure history. An advantage of case-control studies is that they can examine many potential exposures. For example, subjects may be queried about one or more exposures that they may have had in the past; in some variations of this approach, it may be possible to conduct direct measurements of the environment for various types of exposures. A disadvantage of case-control studies is that, in most circumstances, they can examine only one or a few outcomes.[8]

Researchers have a variety of sources available for the selection of cases and controls. For example, they may use patients from hospitals, specialized clinics, or medical practices. Sometimes, advertisements in media solicit cases. Cases may be selected from disease registries such as cancer registries. Controls can be either healthy persons or those affected by a disease that is etiologically unrelated to the outcome of interest. For example, investigators may identify as controls patients from hospitals or clinics; however, these control patients must not have been affected by the outcome of interest. In other studies, controls may be friends or relatives of the cases or be from the community.

The measure of association between exposure and outcome used in case-control studies is known as the **odds ratio (OR)**. A particular form of OR, the exposure-odds ratio, refers to "the ratio of the odds in favor of exposure among the cases [A/C] to the odds in favor of exposure among non cases [the controls, B/D]."[12] **TABLE 2.2** illustrates the method for labeling cells in a case-control study. This table is called a 2 × 2 table.

The OR is defined as $\frac{A/C}{B/D}$, which can be expressed as $\frac{AD}{BC}$.

TABLE 2.2 Table for a Case-Control Study

		Disease Status—Outcome of Interest	
		Yes (Cases)	**No (Controls)**
Exposure Status	Yes	A	B
	No	C	D
	Total	A + C	B + D

An odds ratio of more than 1 suggests a positive association between the exposure and disease or other outcome (provided that the results are statistically significant—a concept that will not be discussed here).

Calculation example: Suppose we have the following data from a case-control study: A = 9, B = 4, C = 95, D = 88. The OR is calculated as follows:

$$OR = \frac{AD}{BC} = \frac{(9)(88)}{(4)(95)} = 2.08$$

In this sample calculation, the OR is greater than 1, suggesting that the odds of the disease are higher among the exposed persons than among the nonexposed persons.

Case-control studies are very common in environmental epidemiologic research. For example, environmental health researchers have been concerned about the possible health effects of exposure to electromagnetic fields (EMFs). A case-control study among female residents of Long Island, New York examined the possible association between exposure to EMFs and breast cancer.[24] Eligible subjects were those who were younger than age 75 years and who had lived in the study area for 15 years or longer. Cases (n = 576) consisted of women diagnosed with in situ or invasive breast cancer. Controls (n = 585) were selected from the same community by random digit-dialing procedures. Several types of measurement of EMFs were taken in the subjects' homes and by mapping overhead power lines. The investigators reported that the odds ratio between EMF exposure and breast cancer was not statistically significantly different from 1; thus, the results suggested that there was no association between breast cancer and residential EMF exposure.

In comparison with cross-sectional study designs, case-control studies may provide more complete exposure data, especially when the exposure information is collected from the friends and relatives of cases who died of a particular cause. Nevertheless, some unmeasured exposure variables as well as methodological biases (a term discussed later in this chapter) may remain in case-control studies. For example, in studies of health and air pollution, exposure levels are difficult to quantify precisely. Also, it may be difficult to measure unknown and unobserved factors, including smoking habits and occupational exposures to air pollution, which affect the lungs.[8]

Cohort Studies

A **cohort study** design classifies subjects according to their exposure to a factor of interest and then observes them over time to document the occurrence of new cases (incidence) of disease or other health events. Cohort studies are a type of longitudinal design, meaning that subjects are followed over an extended period of time. Using cohort studies, epidemiologists are able to evaluate many different outcomes (causes of death) but few exposures.[8]

Cohort studies may be either prospective or retrospective. At the inception of a prospective cohort study, participating individuals must be certified as being free from the outcome of interest. As these individuals are followed into the future, the occurrence of new cases of the disease is noted. A retrospective cohort study (historical cohort study) is "conducted by reconstructing data about persons at a time or times in the past. This method uses existing records about the health or other relevant aspects of a population as it was at some time in the past and determines the current (or subsequent) status of members of this population with respect to the condition of interest."[12] An example of a retrospective cohort study would be one that examined mortality among an occupational cohort such as shipyard workers who were employed at a specific naval yard during a defined time interval (e.g., during World War II).

The measure of association used in cohort studies is called **relative risk (RR)**, the ratio of the incidence rate of a disease or health outcome in an exposed group to the incidence rate of the disease or condition in a nonexposed group. As noted previously, an incidence rate may be interpreted as the risk of occurrence of an outcome that is associated with a particular exposure. The RR provides a ratio of two risks—the risk associated with an exposure in comparison with the risk associated with nonexposure.

Mathematically, the term *relative risk* is defined as A/(A + B) (the rate [incidence] of the disease or condition in the exposed group) divided by C/(C + D) (the rate [incidence] of the disease or condition in the nonexposed group). A 2 × 2 table for the elements used in the calculation of a relative risk is shown in **TABLE 2.3**.

TABLE 2.3 Table for a Cohort Study

		Disease Status		
		Yes	No	Total
Exposure Status	Yes	A	B	A + B
	No	C	D	C + D

$$RR = \frac{A/(A+B)}{C/(C+D)}$$

Calculation example: Suppose that we are researching whether exposure to solvents is associated with risk of liver cancer. From a cohort study of industrial workers, we find that three persons who worked with solvents developed liver cancer (cell A of Table 2.3) and 104 did not (cell B). Two cases of liver cancer occurred among nonexposed workers (cell C) in the same type of industry. The remaining 601 nonexposed workers (cell D) did not develop liver cancer. The RR is:

$$RR = \frac{3/(3+104)}{2/(2+601)} = 8.45$$

We may interpret relative risk in a manner that is similar to that of the odds ratio. For example, a relative risk greater than 1 (and statistically significant) indicates that the risk of disease is greater in the exposed group than in the nonexposed group. In other words, there is a positive association between exposure and the outcome under study. In the calculation example, the risk of developing liver cancer is eight times greater among workers who were exposed to solvents than among those who were not exposed to solvents.

Sometimes a relative risk calculation yields a value that is less than 1. If the relative risk is less than 1 (and statistically significant), the risk is lower among the exposed group. This level of risk (i.e., less than 1) sometimes is called a protective effect.

Accurate disease verification is necessary to optimize measures of relative risk; disease misclassification affects estimates of relative risk. The type of disease and method of diagnosis affect accuracy of diagnosis.[8] To illustrate, death certificates are used frequently as a source of information about the diagnosis of a disease. Information from death certificates regarding cancer as the underlying cause of death is believed to be more accurate than the information for other diagnoses such as those for nonmalignant conditions. Nevertheless, the accuracy of diagnoses of cancer as a cause of death varies according to the particular form of cancer.

Cohort studies are applied widely in environmental health. For example, they have been used to examine the effects of environmental and work-related exposures to potentially toxic agents. One concern of cohort studies has been exposure of female workers to occupationally related reproductive hazards and adverse pregnancy outcomes.[25]

A second example is an Australian study that examined the health impacts of occupational exposure to insecticides.[26] The investigators selected a cohort of 1,999 outdoor workers known to be employed as field officers or laboratory staff for the New South Wales Board of Tick Control between 1935 and 1996. Only male subjects were selected for the study. A control cohort consisted of 1,984 men who worked as outdoor field officers at any time since 1935. Occupational monitoring programs demonstrated that members of the exposure cohort had worked with pesticides, including DDT. The investigators carefully evaluated exposure status and health outcomes such as mortality from various chronic diseases and cancer. They reported an association between exposure to pesticides and adverse health effects, particularly for asthma, diabetes, and some forms of cancer including pancreatic cancer.

Study Endpoints Used in Environmental Epidemiologic Research

In evaluating the health effects of occupational exposures to toxic agents, researchers may study various endpoints, including measures derived from self-report questionnaires, results of direct physical examinations, and mortality experience in a population. The endpoints also may be keyed to any of a number of stages in the natural progression of disease (e.g., presymptomatic, symptomatic, or permanent dysfunction).[27]

In some studies, self-reported symptom rates are used as a measure of the effects of low-level chemical exposure. Occupational health investigators can design and administer self-report questionnaires inexpensively. Self-reports to questionnaires, however, may not always be reliable, and although they correlate often with clinical diagnoses they also may differ markedly.[6]

Physiologic or clinical examinations are other means to evaluate adverse health effects. For example, in a study of respiratory diseases, pulmonary function tests, such as forced expiratory volume, may be an appropriate indicator. Although clinical examinations may provide "harder" evidence of health effects than self-reports, such examinations may be expensive or impractical to conduct in the case of workers who have left employment.

In other studies, mortality is the outcome of interest; research on mortality frequently uses a retrospective cohort study design.[7] Mortality experience in an employment cohort can be compared with the expected mortality in the general population (national, regional, state, or county) by using the standardized mortality ratio (SMR), which is defined as "the ratio of the number of deaths observed in the study group or population to the number that would be expected if the study population had the same specific rates as the standard population. Often multiplied by 100."[12] Typically the SMR is denoted by a percentage; when the percentage is greater than 100%, the SMR in the study population is elevated above that found in the comparison population. Conversely, when the SMR is less than 100%, the mortality experience in the study population is lower than that of the comparison population.

One also can contrast the mortality experience of exposed workers with the mortality rate of nonexposed workers in the same industry. For example, production workers might be compared with drivers or office workers. Another option is to identify a second industry or occupation that is comparable in terms of skill level, educational requirements, or geographic location but in which the exposure of interest is not present.

The use of mortality as a study endpoint has several advantages, including the fact that it may be relevant to agents that have a subtle effect over a long time period. Although any fatal chronic disease may be investigated, mortality from cancer often is studied as an outcome variable in occupational exposures. According to Monson, "Cancer specifically tends to be a fatal illness; its presence is usually indicated on the death certificate. Also, cancer is a fairly specific disease and is less subject to random misclassification than, say, one of the cardiovascular diseases."[28(p106)]

Causality in Epidemiologic Studies

One of the fundamental models of causality used in epidemiologic studies is the **epidemiologic triangle**, which includes three major factors: agent, host, and environment. Although this model has been applied to the field of infectious disease epidemiology, it also provides a framework for organizing the causality of other types of environmental problems. Refer to **FIGURE 2.11** for an illustration.

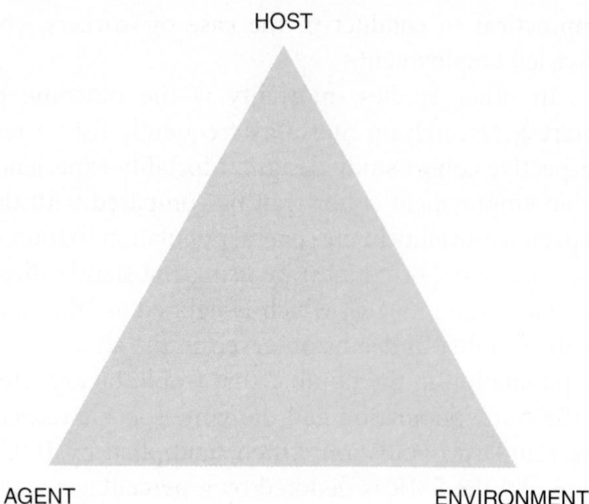

FIGURE 2.11 The epidemiologic triangle.

Reproduced from Friis RH, Sellers TA. *Epidemiology for Public Health Practice*. 5th ed. Burlington, MA: Jones and Bartlett Learning; 2014:493.

The term *environment* is defined as the domain in which disease-causing agents may exist, survive, or originate; it consists of "[a]ll that which is external to the individual human host."[12] The **host** is "[a] person or other living animal, including birds and arthropods, that affords subsistence or lodgment to an INFECTIOUS AGENT under natural conditions."[12] A human host is a person who is afflicted with a disease; or, from the epidemiologic perspective, the term host denotes an affected group or population. An **agent** (of disease) refers to

> ... a factor (e.g., a microorganism, chemical substance, form of radiation, mechanical, behavioral, social agent or process) whose presence, excessive presence, or (in deficiency diseases) relative absence is essential for the occurrence of a disease. A disease may have a single agent, a number of independent alternative agents (at least one of which must be present), or a complex of two or more factors whose combined presence is essential for or contributes to the development of the disease or other outcome.[12]

In environmental health, agent factors can include (but are not limited to) particulate matter from pollution, toxic chemicals and pesticides, and microbes. Examples of agent factors covered in this text are:

- Microbial agents responsible for zoonotic diseases
- Microbial agents linked to foodborne illness
- Toxic chemicals including pesticides
- Toxic metals
- Airborne particulates and gases
- Radiation: ionizing and nonionizing

These agents are relevant to many of the environmental problems discussed in this text, including hazardous waste disposal, zoonotic illnesses, foodborne illnesses, unintentional injuries, occupational illnesses, and adverse health outcomes associated with water and air pollution.

Criteria of Causality

The epidemiologic triangle provides a framework for viewing hypothesized relationships among agent, host, and environmental factors in causation of disease. One of the central concerns of environmental epidemiology is to be able to assert that a causal association exists between an agent factor and a disease in the host. Hill pointed out that in the realm of occupational health, extreme conditions in the physical environment or exposure to known toxic chemicals should be invariably injurious.[29] More commonly the situation occurs in which weaker associations have been observed between certain aspects of the environment and the occurrence of health events. An example would be the development of lung diseases among persons exposed to dusts (e.g., miners who work in dusty, unventilated mines). Hill raised the question of how one moves from such an observed association to the verdict of causation (e.g., exposure to coal dust *causes* coal miner's pneumoconiosis). A second example is the perplexing question of the extent to which studies reveal a causal association between a specific environmental exposure and a particular form of cancer.[20]

Hill proposed a situation in which there is a clear association between two variables and in which statistical tests have suggested that this association is not due to chance. For example, data have revealed that smoking is associated with lung cancer in humans and that chance can be ruled out as being responsible for this observed association. The 1964 US government report *Smoking and Health* stated that "the evaluation of a causal association does not depend solely upon evidence from a probabilistic statement derived from statistics, but is a matter of judgment that depends upon several criteria."[30] Similarly, Hill listed nine causal criteria that need to be taken into account in the assessment of a causal association between factor A and disease B. For the purposes of this text, we will consider seven of the criteria, which are included in **TABLE 2.4**.

TABLE 2.4 Hill's Criteria of Causality

Strength	Biological gradient
Consistency	Plausibility
Specificity	Coherence
Temporality	

Strength

Strong associations give support to a causal relationship between factor and disease. Hill provided the example of the very large increase in scrotal cancer (by a factor of 200 times) among chimney sweeps in comparison with workers who were not exposed occupationally to tars and mineral oils. Another example arises from the steeply elevated lung cancer mortality rates among heavy cigarette smokers in comparison with nonsmokers (20 to 30 times higher). Hill also cautioned that we should not be too ready to dismiss causal associations when the strength of the association is small, because there are many examples of causal relationships that are characterized by weak associations.[29] One example would be exposure to an infectious agent such as meningococcus that produces relatively few clinical cases of meningococcal meningitis.

Consistency

According to Hill, a consistent association is one that has been observed repeatedly "by different persons, in different places, circumstances and times. . . ."[29(p296)] An example of consistency comes from research on the relationship between smoking and lung cancer, a relationship that was found repeatedly in many retrospective and prospective studies.

Specificity

A specific association is one that is constrained to a particular disease–exposure relationship. In a specific association, a given disease results from a given exposure and not from other types of exposures. Hill gave the example of an association that "is limited to specific workers and to particular sites and types of disease and there is no association between the work and other modes of dying. . . ."[29(p297)] Returning to the smoking–lung cancer example, one may argue that the association is not specific, because "the death rate among smokers is higher than the death rate of non-smokers from many causes of death. . . ."[29(p297)] Nevertheless, Hill argued that one-to-one causation is unusual, because many diseases have more than one causal factor.

Temporality

This criterion specifies that we must observe the cause before the effect; Hill stated that we cannot put the cart before the horse. For example, if we assert that air pollution causes lung cancer, we first must exclude persons who have lung cancer from our study; then we must follow those who are exposed to air pollution to determine whether lung cancer develops.

Biological Gradient

A biological gradient also is known as a dose–response curve (discussed in the chapter on environmental toxicology), which shows a linear trend in the association between exposure and disease. An example arises from the linear association between the number of cigarettes smoked and the lung cancer death rate.

Plausibility

This criterion states that an association must be biologically plausible from the standpoint of contemporary biological knowledge. The association between exposure to tars and oils and the development of scrotal cancer is plausible in view of current knowledge about carcinogenesis. However, this knowledge was not available when Pott made his observations during the 18th century.

Coherence

This criterion suggests that "the cause-and-effect interpretation of our data should not seriously conflict with the generally known facts of the natural history and biology of the disease. . . ."[29(p298)] Examples related to cigarette smoking and lung cancer come from the rise in the number of lung cancer deaths associated with an increase in smoking, as well as lung cancer mortality differences between men (who smoke more and have higher lung cancer mortality rates) and women (who smoke less and have lower rates).

Bias in Environmental Epidemiologic Studies

Epidemiologic studies may be impacted by **bias**, which is defined as the "[s]ystematic deviation of results or inferences from truth, or processes leading to such deviation. An error in the conception and design of a study—or in the collection, analysis, interpretation, reporting, publication, or review of data—leading to results or conclusions that are systematically (as opposed to randomly) different from truth."[12] There are many types of bias; particularly important for environmental epidemiology are those that impact study procedures. Examples of such bias are related to how the study was designed, the method of data collection, interpretation and review of findings, and procedures used in data analysis. For example, in measurements of exposures and outcomes, faulty measurement devices may introduce biases into study designs.

A complete discussion of all the kinds of bias is beyond the scope of the text; however, we will consider two types of bias, *recall bias* and *selection bias.* The former is particularly relevant to case-control studies. Recall bias refers to the fact that cases may remember an exposure more clearly than controls.[19] The consequence of recall bias is to reduce the reliability of exposure information gathered from control groups. Selection bias is defined as "bias in the estimated association or effect of an exposure on the outcome that arises from the procedures used to select individuals into the study or the analysis."[12] The effect of selection bias may be to cause systematic differences in characteristics between participants and nonparticipants in research. An example of selection bias is the **healthy worker effect**, which may reduce the validity of exposure data when employed persons are chosen as research subjects in studies of occupational health. Monson states that the healthy worker effect refers to the "observation that employed populations tend to have a lower mortality experience than the general population."[28(p114)] The healthy worker effect may have an impact on occupational mortality studies in several ways. People whose life expectancy is shortened by disease are less likely to be employed than healthy persons. One consequence of this phenomenon would be a reduced (or attenuated) measure of effect for an exposure that increases morbidity or mortality; that is, because the general population includes both employed and unemployed individuals, the mortality rate of that population may be somewhat elevated in comparison with a population in which everyone is healthy enough to work. As a result, any excess mortality associated with a given occupational exposure is more difficult to detect when the healthy worker effect is operative. The healthy worker effect is likely to be stronger for nonmalignant causes of mortality, which usually produce worker attrition during an earlier career phase, than for malignant causes of mortality, which typically have longer latency periods and occur later in life. In addition, healthier workers may have greater total exposure to occupational hazards than those who leave the workforce at an earlier age because of illness.

Another example of study bias is **confounding**, which denotes "the distortion of a measure of the effect of an exposure on an outcome due to the association of the exposure with other factors that influence the occurrence of the outcome."[12] See **FIGURE 2.12**. Confounding factors are associated with disease risk and produce a different distribution of outcomes in the exposure groups than in the comparison groups. The existence of confounding factors that occur in the exposed group may lead to invalid conclusions from a study.

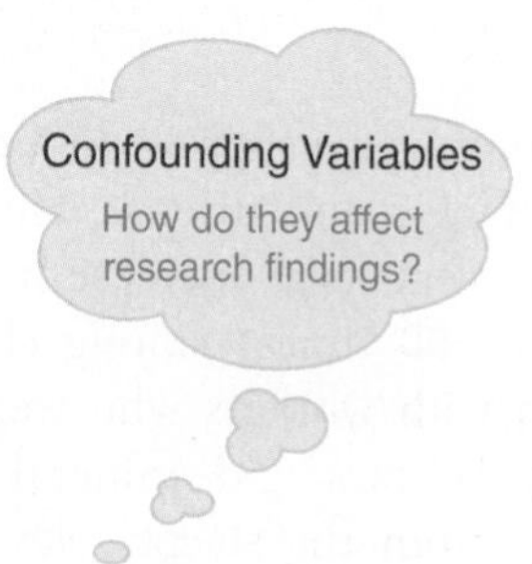

FIGURE 2.12 Confounding variables.

An instance of confounding arises from the possible association between exposure of workers to occupational dusts and development of lung cancer. One of the types of dust encountered in the workplace is silica (e.g., from sand used in sandblasting). In a retrospective cohort study, one might compare the workers' mortality rates for lung cancer with those of the general population (by using SMRs). Suppose we find that the SMR for lung cancer of workers exposed to silica is greater than 100% (i.e., exceeds the rate of the nonexposed population). One conclusion is that the workers have a higher risk of lung cancer than the nonexposed population. However, the issue of confounding also should be considered: Employees exposed to silica are usually blue-collar workers who, as a rule, have higher smoking rates than the general population (that might be used as a comparison population). When smoking rates are taken into account, the strength of the association between silica exposure and lung cancer is reduced—suggesting that smoking is a confounder that needs to be considered in the association.[31]

Limitations and Deficiencies of Environmental Epidemiology

According to Buffler, the three major requirements for the successful epidemiologic investigation of environmental exposures are: "(1) direct and accurate estimates of the exposures experienced by individual members of the study population, (2) direct and accurate determination of the disease status of individual members of the study population, and (3) appropriate statistical summarization and analysis of the individual data pertaining to disease and exposure."[14(p131)] To the extent that these requirements are not met, limitations are introduced into epidemiologic studies. Other limiting

TABLE 2.5 Limitations Faced by Epidemiologists in Studying Relationships between Exposure and Disease Outcomes in Relation to Community Environmental Pollution

Limitations in detecting disease
1. Long and variable latency periods between exposure and disease diagnosis
2. Etiologic nonspecificity of disease clinical features
3. Small population size coupled with low disease frequency
4. Observer bias in reporting illness occurrence
Limitations in measuring exposure
1. Dependence on indirect, surrogate estimates of exposure and dose (distance from pollution site, etc.)
2. Uncertainty regarding pathways of exposure
3. Probable low-dose levels in most settings
4. Frequent inability to develop useful dose–response data

Reproduced with permission from Heath CW Jr. Uses of epidemiologic information in pollution episode management. *Arch Environ Health*. 1988;43:76.

factors include the long latency periods and infrequent occurrence that characterize many environmentally associated diseases. (Refer to **TABLE 2.5**.)

Long Latency Periods

A consideration that limits one's ability to derive causal inferences from epidemiologic studies is the long **latency period** phenomenon.[9] The term *latency period* refers to the time interval between initial exposure to a disease-causing agent (e.g., environmental risk factor or exposure) and the appearance of a disease or its manifestations in the host.[12] Note that the occurrence of disease can be conceptualized in a number of different ways depending on the measure used, such as screening tests and observation of clinical signs and symptoms. Environmentally caused diseases, for example, some forms of cancer, have latency periods that span many years. These long latency periods reduce the epidemiologist's ability to ascertain definitively the outcomes of exposure.[9] Examples are asbestos-related diseases, which in many cases do not appear until many years after initial exposure.[32]

Low Incidence and Prevalence

Another limiting factor of studies concerns the infrequent occurrence of certain diseases that are the target of environmental epidemiologic studies.[9] An example is the occurrence of childhood cancers, which have been examined in relation to environmental factors such as toxic chemicals.[11] The incidence of cancers among children is much lower than that of adults—17.0 per 100,000 among children aged 0 to 14 years and 436.6 per 100,000 for all ages (2014 data). When diseases are uncommon, one's ability to make precise estimates of exposure–disease associations is reduced. The researcher also may be dependent upon less powerful research designs—descriptive and case-control studies.

Difficulties in Exposure Assessment

Several authorities have stressed the requirement for accurate assessment of exposures in epidemiologic studies of environmental health. Rothman points out that "Atop the list of methodologic problems is the problem of exposure assessment, a problem that extends through all of epidemiologic research but is a towering obstacle in environmental epidemiology."[6(p19)] Gardner points out that "Epidemiological methods of investigation are incomplete without good quality exposure data to parallel information on health. The need for monitoring environmental and biological exposure is paramount to the successful interpretation of results and implementation of any required preventive programs."[33(p108)]

For high levels of exposures to toxic agents that produce clear and immediate effects, causation is clear cut.[34] Examples are the release of toxic gases in Bhopal, India, in 1984, and the 1986 Chernobyl nuclear reactor disaster in the former Soviet Union. Moreover, although earlier generations of studies led to the control of intense environmental exposures that were strongly correlated with disease outcomes, the focus of contemporary research has shifted. Modern studies examine low levels of exposure that potentially are associated with low levels of risk.[14]

Low-level environmental exposures challenge epidemiologic researchers who, when dealing with them, have difficulty applying standard laboratory methods used to determine exposure levels. Consequently, researchers are unable to establish definitively whether exposure to a particular agent has occurred. In the ambient environment, not only may several exposures be mixed, but also the levels of exposures may be uncertain.[34]

Examples of exposure measurements used in environmental epidemiology include the following: samples of toxic fumes in a manufacturing plant, ozone readings in the community, and distances of housing tracts from high-tension power lines that emit electromagnetic radiation. All of these measures are prone to error because they are indirect measures of exposure and do not provide direct information on the amount of exposure that an individual may actually receive.[6] As noted, a common method for approximating exposure is the use of proxies (substitutes for direct measures). An example of such a measure is the previously noted distance of a housing tract from the source of an environmental hazard. These proxy or surrogate measures are usually too diffuse to establish exposure definitively.

Nonspecific Effects

A specific health outcome is one that is usually associated with a particular exposure, and only that exposure. When an outcome is nonspecific, it can be associated with several or many different environmental exposures. The majority of diseases and conditions thought to be related to environmental exposures are influenced by many factors.[32] Consequently, any particular environmental exposure probably will not be associated with a specific outcome. Further complicating the picture of exposure determination is the fact that we are exposed to hundreds of chemicals in the environment; these chemicals often are mixed, clouding our knowledge of the level of exposure that took place. Exposures to any of these chemicals could produce outcomes that are similar to one another.

Summary of Characteristics, Weaknesses, and Strengths of Environmental Epidemiology

Several of the key characteristics, weaknesses, and strengths of environmental epidemiology are shown in **TABLE 2.6**. One strength is the ability to deal with "real world" problems, for example community exposure to environmental contaminants; a second strength is the possibility of examining complex problems that involve multiple variables (e.g., exposure, demographic, and outcome variables); a third is the capability to impel environmental action, even though the level of exposure and etiologic mechanisms of health effects have not been ascertained definitively.

The weaknesses include the fact that exposure levels in environmental epidemiology studies are difficult to measure precisely; also, there may be many uncontrolled variables that can bias the results.[35] In the words of Grandjean, "The quality of environmental epidemiology research can be considered from two perspectives, one representing methodological issues, the other dealing with the usefulness of the work. These two views are connected, because a study of superior quality is likely to be of greater validity and therefore more useful. Still, an imperfect study can be of great relevance, and epidemiologists must therefore tackle the challenging balance between being an advocate for particular policies and being a skeptical ivory-tower scientist."[36(p158)] Nevertheless, these weaknesses do not negate the fact that environmental epidemiology has made, and will continue to make, important contributions to the environmental health field.

Conclusion

Environmental epidemiology is one of the fields that research fundamental questions regarding the role of environmental exposures in human health. The discipline traces its history from the time of Hippocrates and from early studies of occupational cancer during the late 18th century. Also historically significant were Snow's investigations of cholera during the mid-19th century. Epidemiology, with its emphasis on observation as well as focus on populations, contributes important methodological tools—particularly with respect to study design.

Descriptive and analytic designs are the two major classes of epidemiologic studies. In order to infer causality among observed associations between environmental exposures and adverse health outcomes demonstrated by epidemiologic studies, one must

TABLE 2.6 Characteristics, Weaknesses, and Strengths of Environmental Epidemiology

Characteristics	Weaknesses	Strengths
Deals with nondisease effects.	Sample size is insufficient.	Engages the real world.
Involves numerous variables.	Important variables "uncontrolled."	Unique perspective on disease/health.
Tends to be community specific.	Exposure estimation is invalid.	Basis for action despite ignorance of mechanism.

Reproduced with permission from Goldsmith JR. Keynote address: improving the prospects for environmental epidemiology. *Arch Environ Health*. 1988;43:71.

apply the criteria of causality. Biases such as confounding can impact epidemiologic research. Although environmental epidemiology has yielded noteworthy insights, one needs to keep in mind the limitations of the discipline. For more information about epidemiology, consult Friis and Sellers[2] or one of the other introductory texts that is available.

Study Questions and Exercises

1. Define the following terms:
   a. Epidemiology
   b. Environmental epidemiology
   c. Descriptive epidemiology
   d. Natural experiments
   e. Prevalence
   f. Incidence
   g. Case fatality rate
   h. Odds ratio
   i. Relative risk
2. What is meant by a cause in environmental epidemiology? Apply Hill's criteria of causality to an example of an association between a specific environmental exposure and health outcome.
3. Explain the reason why studies of the health effects of smoking among individuals who smoke would not be a concern of environmental epidemiology. Explain the reason why exposure to secondhand cigarette smoke is a concern of this discipline.
4. Define the following terms and discuss how each affects the validity of epidemiologic study designs:
   a. Bias
   b. Confounding
   c. Latency period
   d. Exposure assessment
5. List the reasons why epidemiology is important to research studies of environmental health. What are some of the important limitations of the epidemiologic approach with respect to the study of environmental health problems?
6. Explain why epidemiology sometimes is called "population medicine." State how epidemiology contrasts with clinical medicine.
7. Explain the difference between descriptive and analytic epidemiology. Give examples of how both types of study design are utilized in the field of environmental health.
8. What does early epidemiology (e.g., Hippocrates) share in common with contemporary epidemiology in terms of examining the causality of health problems?
9. Describe the importance of the contributions of Sir Percival Pott to environmental health, particularly in the area of cancer prevention.
10. Explain the work of John Snow using the methodology of the natural experiment.
11. Name the study designs that are used for hypothesis testing and those that are used for generating hypotheses.
12. Explain why most studies conducted in the field of environmental epidemiology are nonexperimental.
13. Explain how ecologic analysis is used to study the health effects of air pollution. Give examples of uncontrolled factors that may affect ecologic study results.
14. Explain why cross-sectional studies are defined as prevalence studies. Give an example of a cross-sectional study.
15. Explain why cohort studies are an improvement over case-control studies with respect to measurement of exposure data.

For Further Reading

Snow on Cholera, John Snow, 1855.

References

1. U.S. National Library of Medicine, National Institutes of Health, National Institute of Environmental Health Sciences. What is environmental health? Available at: https://kidsenvirohealth.nlm.nih.gov. Accessed April 3, 2017.
2. Friis RH, Sellers TA. *Epidemiology for Public Health Practice*. 5th ed. Burlington, MA: Jones & Bartlett Learning; 2014.
3. Pekkanen J, Pearce N. Environmental epidemiology: challenges and opportunities. *Environ Health Perspect*. 2001;109:1–5.
4. Terracini B. Environmental epidemiology: a historical perspective. In: Elliott P, Cuzick J, English D, Stern R, eds. *Geographical and Environmental Epidemiology: Methods for Small-Area Studies*. New York, NY: Oxford University Press; 1992.
5. Acquavella JF, Friedlander BR, Ireland BK. Interpretation of low to moderate relative risks in environmental epidemiologic studies. *Ann Rev Public Health*. 1994;15:179–201.
6. Rothman KJ. Methodologic frontiers in environmental epidemiology. *Environ Health Perspect*. 1993;101 (suppl 4):19–21.
7. Hertz-Picciotto I, Brunekreef B. Environmental epidemiology: where we've been and where we're going. *Epidemiol*. 2001;12:479–481.
8. Blair A, Hayes RB, Stewart PA, Zahm SH. Occupational epidemiologic study design and application. *Occup Med*. 1996;11:403–419.
9. Morgenstern H, Thomas D. Principles of study design in environmental epidemiology. *Environ Health Perspect*. 1993;101 (Suppl 4):23–38.
10. Smith AH. Epidemiologic input to environmental risk assessment. *Arch Environ Health*. 1988;43:124–127.
11. Grufferman S. Methodologic approaches to studying environmental factors in childhood cancer. *Environ Health Perspect*. 1998;106(Suppl 3): 881–886.

12. Porta M, ed. *A Dictionary of Epidemiology*. 6th ed. New York, NY: Oxford University Press; 2014.
13. Wegman DH. The potential impact of epidemiology on the prevention of occupational disease. *Am J Public Health*. 1992;82:944–954.
14. Buffler PA. Epidemiology needs and perspectives in environmental epidemiology. *Arch Environ Health*. 1988;43: 130–132.
15. Doll R. Pott and the path to prevention. *Arch Geschwulstforsch*. 1975;45:521–531.
16. Snow J. *Snow on Cholera*. Cambridge, MA: Harvard University Press; 1965.
17. Lilienfeld DE. John Snow: the first hired gun? *Am J Epidemiol*. 2000;152:4–9.
18. Sandler DP. John Snow and modern-day environmental epidemiology. *Am J Epidemiol*. 2000;152:1–3.
19. Prentice RL, Thomas D. Methodologic research needs in environmental epidemiology: data analysis. *Environ Health Perspect*. 1993;101(Suppl 4):39–48.
20. DeBaun MR, Gurney JG. Environmental exposure and cancer in children. A conceptual framework for the pediatrician. *Pediatr Clin North Am*. 2001;48:1215–1221.
21. Freeman NCG, Schneider D, McGarvey P. School-based screening for asthma in third-grade urban children: the Passaic asthma reduction effort survey. *Am J Public Health*. 2002;92:45–46.
22. Miyake Y, Iki M. Ecologic study of water hardness and cerebrovascular mortality in Japan. *Arch Environ Health*. 2003;58:163–166.
23. Reynolds P, Von Behren J, Gunier RB, et al. Childhood cancer and agricultural pesticide use: an ecologic study in California. *Environ Health Perspect*. 2002;110:319–324.
24. Schoenfeld ER, O'Leary ES, Henderson K, et al. Electromagnetic fields and breast cancer on Long Island: a case-control study. *Am J Epidemiol*. 2003;158:47–58.
25. Taskinen HK. Epidemiological studies in monitoring reproductive effects. *Environ Health Perspect*. 1993;101(Suppl 3): 279–283.
26. Beard J, Sladden T, Morgan G, et al. Health impacts of pesticide exposure in a cohort of outdoor workers. *Environ Health Perspect*. 2003;111: 724–730.
27. Neutra R, Goldman L, Smith D, et al. Study, endpoints, goals, and prioritization for a program in hazardous chemical epidemiology. *Arch Environ Health*. 1988;43(2):94–99.
28. Monson RR. *Occupational Epidemiology*. Boca Raton, FL: CRC Press; 1990.
29. Hill AB. The environment and disease: association or causation? *Proc R Soc Med*. 1965;58:295–300.
30. US Department of Health, Education, and Welfare, Public Health Service, Centers for Disease Control. *Smoking and Health. Report of the Advisory Committee to the Surgeon General of the Public Health Service*. PHS Publication No. 1103. Washington, DC: US Government Printing Office; 1964.
31. Steenland K, Greenland S. Monte Carlo sensitivity analysis and Bayesian analysis of smoking as an unmeasured confounder in a study of silica and lung cancer. *Am J Epidemiol*. 2004;160:384–392.
32. Grandjean P. Epidemiology of environmental hazards. *Public Health Rev*. 1993–1994;21:255–262.
33. Gardner MJ. Epidemiological studies of environmental exposure and specific diseases. *Arch Environ Health*. 1988;43:102–108.
34. Health CW, Jr. Uses of epidemiologic information in pollution episode management. *Arch Environ Health*. 1988;43:75–80.
35. Goldsmith JR. Keynote address: improving the prospects for environmental epidemiology. *Arch Environ Health*. 1988;43: 69–74.
36. Grandjean P. Seven deadly sins of environmental epidemiology and the virtues of precaution. *Epidemiology*. 2008;19:158–162.

© Jean-Luc Rivard/EyeEm/Getty Images

CHAPTER 3
Environmental Toxicology

LEARNING OBJECTIVES

By the end of this chapter the reader will be able to:

- Illustrate two applications of toxicology in environmental health.
- Define three important terms used in the field of toxicology.
- State five factors that affect responses to a toxic chemical.
- Discuss the steps involved in risk assessment.
- Describe methods for human exposure assessment.

Introduction

Did you know that toxicology—a.k.a. the science of poisons—helps scientists learn about the hazards associated with potentially harmful environmental chemicals? Toxicology aids in determining the safety of chemicals and their associated risks. You will learn why toxicology is a crucial discipline for environmental health research. This chapter will define terms used in toxicology, describe its relationships with other scientific disciplines, and provide an overview of the discipline as it applies to environmental health. Examples of how toxicology helps to delineate factors related to people's responses to toxic chemicals will be provided. An additional topic will be the links between toxicology and risk assessment. From this information you will be better equipped to explore the environmental health applications covered in the remainder of the text.

Toxicology, a Cornerstone of Environmental Health

First, let us explore why toxicology is important to the environmental health field and the methods it uses to illuminate environmental health problems. The National Toxicology Program (NTP) emphasizes the central role of toxicology in identifying the potential hazards of the numerous chemicals in use in the United States. The NTP states:

> More than 80,000 chemicals are registered for use in the United States. Each year, an estimated 2,000 new ones are introduced for use in such everyday items as foods, personal care products, prescription drugs, household cleaners, and lawn care products. We do not know the effects of many of these chemicals

on our health, yet we may be exposed to them while manufacturing, distributing, using, and disposing of them or when they become pollutants in our air, water, or soil. Relatively few chemicals are thought to pose a significant risk to human health. However, safeguarding public health depends on identifying both what the effects of these chemicals are and at what levels of exposure they may become hazardous to humans—that is, understanding their toxicology.[1]

Toxicology overlaps other disciplines, including physiology, pharmacology, and pathology, and to some extent epidemiology, chemistry, and statistics. A basic assumption in toxicology is that "[a]ll substances are poisons; there is none that is not a poison. The right dose differentiates a poison from a remedy."[2(p2)] We can think of toxicology and epidemiology as working hand in hand, the latter concerned with the occurrence and etiology of disease, and the former studying dose–response relationships and mechanisms of action in order to better understand the adverse health effects linked to chemicals. The coupling of toxicology and epidemiology with genomic methods (employed in the study of genes) enables scientists to assess risks associated with toxic substances.[3]

Toxicology contributes to the armamentarium of tools that are crucial to the description and characterization of environmental chemicals and the responses of living organisms to these chemicals. Related to the functions of toxicology are assessment of exposure, risk, and hazards. **Exposure** is defined as "[p]roximity and/or contact with a source of a disease agent in such a manner that effective transmission of the agent or harmful effects of the agent may occur."[4] The terms **risk** and **hazard** are defined later in the chapter. In the following textbox, noted epidemiologist Douglas L. Weed describes how epidemiology and toxicology are compatible in delving into the complex nature of disease causation.[5]

EPIDEMIOLOGY, TOXICOLOGY, AND DISEASE CAUSATION

Toxicologists and epidemiologists often work together to assess the available scientific evidence relating to potential environmental disease-causing hazards.... The assessment of available evidence lays the foundation for the problem of complexity: Relevant evidence arises from toxicologic and epidemiological investigations, and reflects the acquisition of knowledge from many levels of scientific understanding: molecular, cellular, tissue, organ systems, complete organisms (man and mouse), relationships between individuals, and on to social and political processes that may impact human health.... Consider, as a representative example, the evidence involved in considering whether electrical and magnetic fields are causes of cancer, reproductive and developmental disabilities, and neurobiologic dysfunction (e.g., learning and behavioral disabilities).... Studies range along a continuum, starting at the level of atoms, simple molecules, larger molecules such as DNA including adducts and repair mechanisms, proteins and their synthesis, intracellular environments (e.g., calcium levels), cell-signaling pathways and other extracellular phenomena, tissues (e.g., cell cultures, bones, nerves, polyps, and tumors), and on to the studies in intact individuals (e.g., mice) wherein toxicologists study tumor incidence in rodents and observational studies wherein epidemiologists study the relationship between field exposures and the incidence of diseases and disorders.

Modified from Weed DL. Environmental epidemiology basics and proof of cause-effect. *Toxicology*. 2002;181–182:399–401.

Description of Toxicology

The field of toxicology has been in existence for many centuries, although in its early history the field may have operated under a rubric that is different from its current name. Also, modern toxicology has evolved a number of subspecializations.

Definition of Toxicology

According to its traditional definition, toxicology is the science of poisons. Refer to **FIGURE 3.1**. A more complete definition is "[t]he study of the adverse effects of chemicals on living organisms."[6(p6)] The science of toxicology is not confined to the study of humans, but can be applied

FIGURE 3.1 Toxicology: the science of poisons.

to other species and organisms as well. Through in vivo (in living organisms) and in vitro (in "glass," e.g., cell culture) studies, toxicologists examine such health effects of chemical exposures as carcinogenesis (production of cancer), and damage to internal organs, to the developing fetus, and to the reproductive system. Several types of cancer have been tied to chemical exposure, including carcinoma of the lung, breast, and prostate gland, and some forms of skin cancer and leukemia.

History of Toxicology

Toxicology may be regarded as a field that, in some respects, has a venerable history, yet has a short history in the sense that much of the information regarding toxicology has been acquired in the last few decades. Since early history, human beings have been mystified by the nature of chemicals; they also became fascinated with the use of poisons. Ancient civilizations, early cave dwellers, and isolated primitive tribes were aware of the existence of toxic plants and animal venoms. In ancient Greece and Rome, poisons were used for suicides and executions, and to accomplish political aims. Socrates, who lived from 470 to 399 BCE, was executed by poison, as was Theophrastus (370–286 BCE).

FIGURE 3.2 Paracelsus (1493–1541).

Around the fourth century BCE, poisonings (especially by use of arsenic) grew more frequent in the Roman Empire; Nero was said to have used arsenic to poison Claudius in order to serve political ambitions.

Poisonings also were employed frequently during the Middle Ages to do away with rivals, including spouses and politicians. Beginning about the 17th century and onward, physicians and others developed a gradually increasing awareness of the toxic effects of exposure to industrial metals and chemicals that were then in common use.

Paracelsus (born Phillippus Theophrastus Aureolus Bombastus von Hohenheim) is considered to be one of the founders of modern toxicology.[7] His image is shown in **FIGURE 3.2**. Active during the time of da Vinci and Copernicus, Paracelsus contributed to the discipline during the early 16th century. Among his contributions were several important concepts, including the dose–response relationship, which refers to the observation that the effects of a poison are related to the strength of its dose, and the notion of target organ specificity of chemicals.

Terminology Used in the Field of Toxicology

One aspect of toxicology is the examination of the mechanisms by which chemicals produce toxic effects on living organisms and their tissues. Virtually all known chemicals (even sodium chloride—common table salt) have the capacity to produce toxic effects, such as injury or death, depending upon the amount ingested. This wide spectrum of doses is related to a particular chemical in question, for example, from botulinum toxin (highly toxic in tiny doses) to ethyl alcohol (toxic in much larger doses than botulinum toxin).[6] Measures of lethality (discussed later in this chapter) may not describe fully a chemical's spectrum of toxicity; chemicals that have low acute toxicity may have other effects such as carcinogenicity or teratogenicity. This section introduces a number of terms that are used to describe toxic materials. It should be noted that some of the terms tend to overlap conceptually and are not always differentiated clearly in the literature.

Mathieu Orfila was another pioneer in the field of toxicology. In the 1800s, he authored a number of significant works, among them *Trait des poisons* (1813). This work described in great detail various types of poisons and their bodily effects, a development that contributed to the foundations of forensic toxicology.

Fields within Toxicology

What type of work does a toxicologist perform and what are the specializations within toxicology? A **toxicologist** is a scientist who has received extensive training in order to investigate in living organisms "the adverse effects of chemicals . . . (including their cellular, biochemical, and molecular mechanisms of action) and assess the probability of their occurrence."[6(p6)] The field of toxicology comprises several key areas, including regulatory matters (*regulatory toxicology*), medico-legal issues (*forensic toxicology*), and clinical manifestations of disease related to toxic substances (*clinical toxicology*).[6] In addition, the specializations known as *environmental toxicology*, *reproductive toxicology*, and *developmental toxicology* are particularly relevant to environmental health problems.

The field of **environmental toxicology** is defined as "...the study of the impacts of pollutants upon the structure and function of ecological systems."[8] For example, environmental toxicology examines how environmental exposures to chemical pollutants may present risks to biological organisms, particularly animals, birds, and fish. The field of **ecotoxicology** is concerned with the effects of pollutants on ecosystems. (Refer to the textbox, What Is Ecotoxicology?)

In several respects, the mission of environmental toxicology overlaps those of reproductive and developmental toxicology. *Reproductive toxicology* examines the association between environmental chemicals and adverse effects upon the reproductive system.[9] Exposure to chemicals may arise from a number of environmental sources including the workplace, the home, and the medications and foods that we consume. On the list of hazardous chemicals and other agents that are suspected of impacting human reproduction negatively are pesticides, drugs, heavy metals, and hormones.[10]

WHAT IS ECOTOXICOLOGY?

The field of *ecotoxicology* is "...the branch of Toxicology concerned with the study of toxic effects, caused by natural or synthetic pollutants, to the constituents of ecosystems, animal (including human), vegetable and microbial, in an integral context." This important subfield of environmental toxicology was proposed in 1969 at a scientific meeting in Stockholm, Sweden. Broadly speaking, ecotoxicology investigates dispersion of pollutants into the physical environment, their impact upon biological chains such as food chains, and their toxic effects within ecosystems.

Data from Truhaut R. Ecotoxicology: principles and perspectives. *Ecotoxicol Environ Saf.* 1977;1(2):151-173.

Developmental toxicology researches the effects of natural and man-made chemicals (some classified as teratogens—substances that cause birth defects) on prenatal development. Other chemicals that can produce developmental toxicity are included in the broad class known as **xenobiotics**, which are "[c]hemical substances that are foreign to the biological system. They include naturally occurring compounds, drugs, environmental agents, carcinogens, [and] insecticides. . . ."[11] Some specific examples of xenobiotics are antibiotics, dioxins, and polychlorinated biphenyls (PCBs).

Poison

The term **poison** is "defined as any agent capable of producing a deleterious response in a biological system."[6(p6)] Examples of deleterious responses are death or serious impairment of biological functioning. Some definitions of a poison affirm that it is an agent that produces immediate effects such as lethality or sickness even when present in small doses.[12]

Toxic Agent

The term **toxic agent** refers very generally to a material or factor that can be harmful to biological systems.[12] Examples are physical energy (e.g., heat and ionizing and nonionizing radiation), substances derived from biological sources (e.g., black widow spider venom), and almost all chemicals.

Toxicity

Toxicity is defined as "the degree to which something is poisonous."[13] Toxicity is related to a material's physical and chemical properties. Some chemicals have low innate toxicity (e.g., ethyl alcohol, sodium chloride); others have high toxicity (e.g., dioxins and botulinum toxin formed by the bacteria that cause botulism). Substances that have low toxicity must be ingested in large amounts in order for them to have toxic effects; the converse is true of chemicals that have high toxicity. For example, ingestion of large amounts of water (which has low toxicity) is necessary in order to produce water intoxication. In contrast, injection with a small amount of highly toxic insect venom such as spider venom may be sufficient to cause severe damage to the body or even death. The venom from the bite of a black widow spider shown in **FIGURE 3.3** can cause serious illness, but usually does not result in death.

Toxic Substance

A material that has toxic properties is called a **toxic substance.**[12] This substance can be a single toxic chemical (e.g., arsenic, lead) or a mixture of toxic chemicals

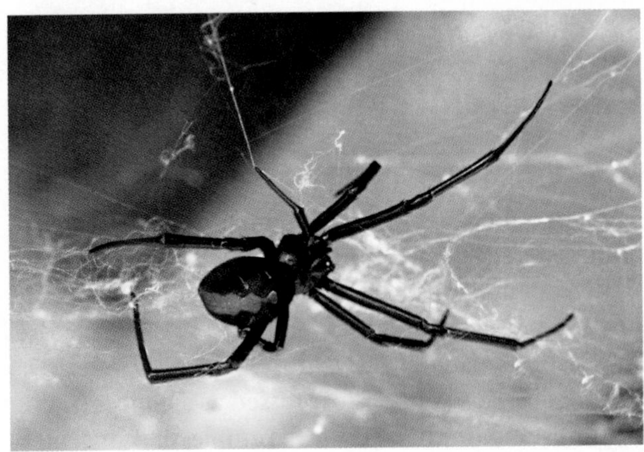

FIGURE 3.3 Black widow spider.

(e.g., gasoline). Toxic substances also can be categorized as organic toxins and inorganic toxins. Organic toxins denote substances originally taken from living organisms; such chemicals are carbon containing, frequently made up of large molecules, and capable of being synthesized. Inorganic toxins refer to specific chemicals, such as minerals, that have not been extracted from living organisms and do not follow the structure of a toxin derived from a living organism; such toxins generally are made up of small molecules.

Toxicant

Toxicants denote toxic substances that are manmade or result from human (anthropogenic) activity.[6] The toxic effect may occur directly or indirectly. The chemical potassium cyanide is directly toxic to the body; in contrast, methanol, a form of alcohol (wood alcohol) is indirectly toxic. Through the action of the liver, methanol is converted to formaldehyde and then to formic acid, a toxic metabolite.[14]

Toxin

A **toxin** usually refers to a toxic substance made by living organisms including reptiles, insects, plants, and microorganisms. To give one illustration, certain bacteria produce toxins that may act directly on the nervous or gastrointestinal system to produce symptoms of toxic effects. Foodborne botulism caused by the bacterium *Clostridium botulinum* is an environmental hazard associated with improperly canned foods and other unsafe practices in food preparation. Toxin production by microorganisms differs from disease causation in that the former involves actual invasion and multiplication of microorganisms and consequent organ and cell damage. *Systemic toxins* are those that affect the entire body or multiple organ systems; *target organ toxins* affect specific parts of the body.

FIGURE 3.4 Poisonous mushrooms.

Other toxins originate from plants. Some mushrooms (e.g., *Amanita phalloides*, "death cap") are highly poisonous. (See **FIGURE 3.4**.) There are many examples of plants that are toxic:

- Poison hemlock
- Foxglove
- Poison oak/poison ivy
- Rhubarb, especially the leaves, which have high levels of oxalates
- Some houseplants such as dieffenbachia

Refer to the following textbox for an example of poisonings caused by nicotine, a highly toxic substance derived from the tobacco plant, and by mushrooms.

CASE STUDY: CLASSIC EPISODES OF POISONINGS FROM PLANT-ORIGIN TOXINS

- **Nicotine poisoning from ground beef:** In Michigan during early 2003, some supermarket customers fell ill after consuming ground beef. The customers reported nausea, vomiting, dizziness, and—in one case—an irregular heartbeat. Samples of ground beef submitted by the supermarket were tested at a regional medical center and determined to have high levels of nicotine, which is a toxin derived from tobacco. Epidemiologic investigators interviewed the victims in order to examine the range of symptoms, their patterns, and their consistency with clinical presentations. A total of 92 persons among the interviewees demonstrated illnesses consistent with nicotine poisoning. A legal investigation led to the arrest of a store employee who was accused of putting a commercial insecticide that contained nicotine into 200 pounds of meat that was later purchased by unsuspecting customers.

- **Fatalities caused by ingesting "death caps":** The Centers for Disease Control and Prevention (CDC) have highlighted the dangers associated with popular interest in collecting mushrooms in the wild. Aficionados of wild mushrooms should observe the utmost caution in making certain that the fungi are safe for consumption; in some instances, one has extreme difficulty is distinguishing nontoxic from poisonous species. One of the species associated with poisonings is called *Amanita phalloides*. These mushrooms are also called "death cap" mushrooms for good reason; they cause about 90% of deaths from mushroom poisoning worldwide. Death caps, which can be confused with nontoxic species, do not have a distinct smell or taste. Cooking does not destroy their toxins. Consumption of these highly toxic mushrooms can result in severe hepatitis and even liver failure. Unfortunately, there is no antidote for this poison. (See Figure 3.4 for an illustration of poisonous mushrooms, including *A. phalloides*.)
- **1981: Mushroom poisoning among Laotian refugees:** During December 1981, a small group of Laotian refugees gathered mushrooms in Sonoma County in northern California. In Laos, a customary practice is to test the safety of mushrooms by boiling them with rice. If the rice turns red, the mushrooms are believed to be poisonous. As this was not the case in the Sonoma County incident, the refugees assumed the mushrooms were safe and consumed them. All of the seven mushroom hunters developed gastrointestinal distress; three required intensive care treatment. All recovered within a week.
- **1997:** The CDC reported two deaths in 1997 associated with eating *A. phalloides* mushrooms in northern California.
- **2016:** The California Poison Control System probed 14 poisoning cases suspected of being caused by eating wild *A. phalloides* mushrooms. The poisonings, which occurred in December in five northern California counties, necessitated three liver transplants; a child sustained permanent neurologic impairment. The remaining individuals recovered.

Data from Centers for Disease Control and Prevention. Nicotine poisoning after ingestion of contaminated ground beef—Michigan, 2003. *MMWR*. 2003;52:413–415; Centers for Disease Control and Prevention. Mushroom poisoning among Laotian Refugees—1981. *MMWR*. 1982;31:287-288; Centers for Disease Control and Prevention. Amanita phalloides mushroom poisoning—Northern California, January 1997. *MMWR*. 1997;46:489-492; Centers for Disease Control and Prevention. *Amanita phalloides* Mushroom Poisonings—Northern California, December 2016. *MMWR*. 2017;66:549-553.

The Concept of a Dose and Related Terms

Dose

The term **dose** refers to "the amount of a substance administered at one time."[15] There are several ways of describing a dose such as exposure dose, absorbed dose, administered dose, total dose, external dose, internal dose, and effective dose. (Refer to **TABLE 3.1**.) In practice, dose often is expressed as a concentration of a substance in the body, for example, the concentration per milliliter (ml) of blood.

Toxicologists take into account the total dose, how often each individual dose occurs, and the time period during which the dosing occurs in order to describe the effects of a dose. When a dose is fractionated (broken up over a period of time), the effects may be different from those that transpire when a dose is administered all at one time. For example, poison that is fatal in a single, concentrated dose may no longer be fatal when the same dose is broken down into small units and given over time. Another consideration in the lethality or other effects of a dose relates to the body size of the subject. Young children, who have small body sizes, are more affected by a specific dose than are large adults who are given the same dose. For this reason, environmental exotoxins (those originating from external sources such as lead and mercury) at a given concentration may present a greater hazard to children than to adults.

Dose–Response Curves

The dose–response curve is a type of graph that is used to describe the effect of exposure to a chemical or toxic substance upon an organism such as an experimental animal. Toxicologists have devised two types of dose–response curves: one for the responses of an individual to a chemical and one for a population. For both types of curves, the dose is indicated along the x axis, and the response is shown along the y axis.

A typical dose–response curve for an individual exposed to ethyl alcohol is shown in **FIGURE 3.5**. The curve, which assumes a one-time exposure, shows a graded and increasing response as the dose of alcohol increases. The spectrum of effects can range from no effect when the dose is zero to death when the dose is increased to a toxic level.

With respect to the population, a dose–response relationship is "[t]he relationship of observed responses or outcomes in a population to varying levels of a beneficial or harmful agent."[4] The response could be measured as the percentage of exposed animals showing

TABLE 3.1 Ways to Describe a Dose

Term	Definition
Exposure dose	The amount of a xenobiotic encountered in the environment.[a]
External dose	A dose that results from contact with environmental sources, e.g., environmental contamination.
Absorbed dose	The actual amount of the exposed dose that enters the body.[a]
Internal dose	A synonym for absorbed dose.[a] Various interpretations: The absorbed dose can reflect short-term exposure, previous day exposure, or long-term exposure.[b]
Administered dose	The quantity of a substance that is administered usually orally or by injection.[a]
Effective dose	Indicates the effectiveness of a substance. Normally, effective dose refers to a beneficial effect such as relief of pain. It may also stand for a harmful effect such as paralysis.[a]
Total dose	The sum of all individual doses.[a]

[a]National Library of Medicine. *ToxTutor*. Dose. Available at: https://toxtutor.nlm.nih.gov/02-001.html. Accessed March 10, 2017.
[b]Klaasen CD, Watkins JB III, eds. *Casarett & Doull's Essentials of Toxicology*. 3rd ed. New York, NY: McGraw-Hill, 2015.

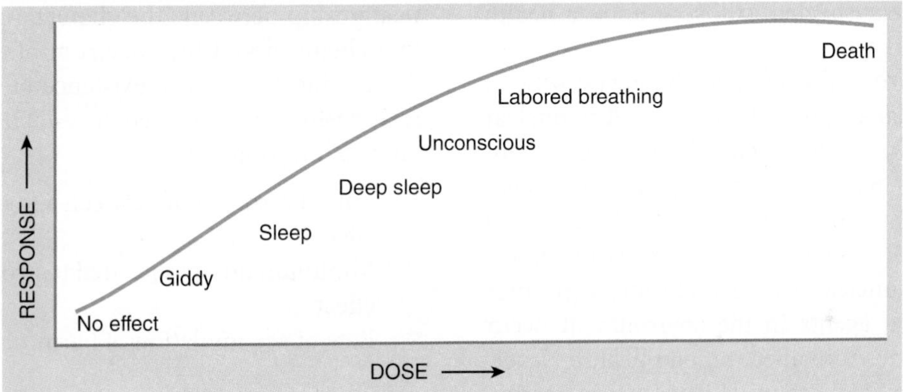

FIGURE 3.5 Individual dose–response curve.

Reproduced with permission from Marczewski AE, Kamrin M. *Toxicology for the Citizen*. 2nd ed. East Lansing, Mich: Michigan State University, Center for Integrative Toxicology; 1991:5

a particular effect, or it could reflect the effect in an individual subject.

The population dose–response curve, which has a sigmoid shape (S-shaped), is also a cumulative percentage response curve. (Refer to **FIGURE 3.6**.) This type of dose–response curve (i.e., population based) also is called a quantal curve.[6] The estimates of LD_{10}, LD_{50}, and LD_{90} appear in the figure.

FIGURE 3.7 illustrates the threshold of a dose–response curve. At the beginning of the curve, there is a flat portion suggesting that at low levels an increase in dosage produces no effect; this is known as the subthreshold phase. After the threshold is reached, the curve rises steeply and then progresses to a linear phase, where an increase in response is proportional to the increase in dose. When the maximal response is reached, the curve flattens out. A dose–response relationship is one of the indicators used to assess a

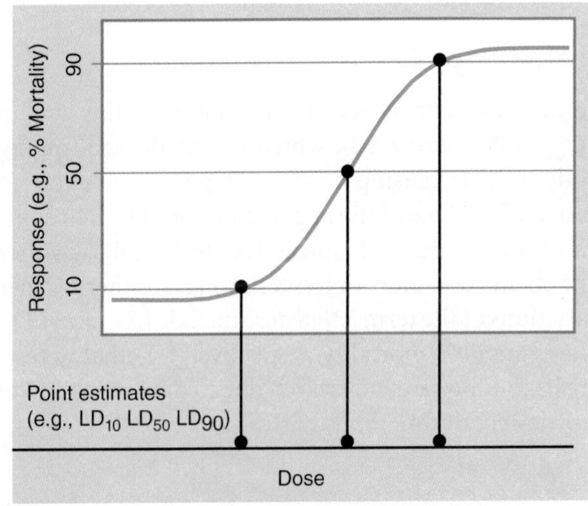

FIGURE 3.6 A population dose–response curve.

Modified from Guidelines for Ecological Risk Assessment. US Environmental Protection Agency, Risk Assessment Forum. Washington, DC, EPA/630/R095/002F; 1998:81.

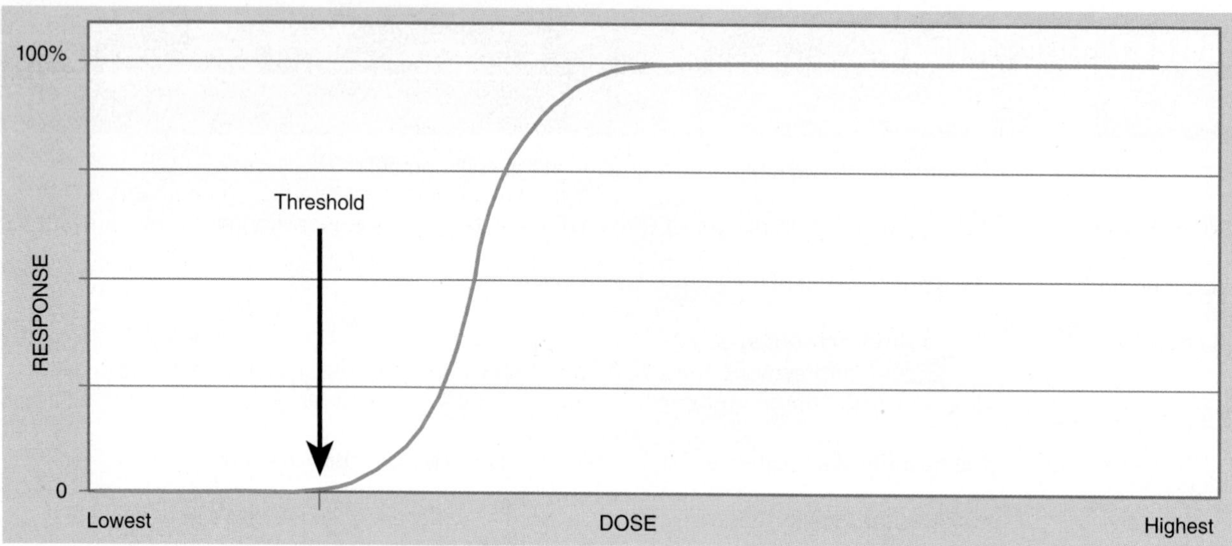

FIGURE 3.7 The threshold of a dose–response curve.

Image from *Chemicals, the Environment, & You: Exploration in Science and Human Health*. Copyright © 2000 by BSCS & Videodiscovery. All rights reserved. Used with permission. Reproduced from National Institutes of Health, National Institute of Environmental Health Sciences. *Chemicals, the Environment and You. Teachers Guide*. Available at: https://science.education.nih.gov/supplements/webversions/Chemicals/guide/lesson3-1.html. Accessed July 6, 2017.

causal effect of a suspected exposure upon a health outcome.

The term **threshold** refers to the lowest dose at which a particular response may occur. It is unclear whether exposure to toxic chemicals at low (subthreshold) levels may produce any health response. Although society's concern about the health effects of exposure to environmental pollutants has increased, it still remains unclear whether chronic exposures to toxic chemical agents in the environment occur at high enough doses to affect human health. Nevertheless, environmental scientists have had increasing concerns over the long-term effects of low-level exposures to toxic substances, exposures such as those that may take place in the workplace.

Lethal Dose 50 (LD_{50})

To describe toxic effects, toxicologists use the symbol **LD_{50} (lethal dose 50)**, which is "the dosage (mg/kg body weight) causing death in 50 percent of exposed animals."[6(p7)] One of the applications of LD_{50} is to compare the toxicities of chemicals (i.e., to describe whether one chemical is more or less toxic than another). Other variations of the term *lethal dose* include LD_{10} and LD_{90} (10% and 90% mortality, respectively). Lethality tests are becoming rare in research due to the availability of less destructive methods of study.

Dose–Response Relationships

One of the most basic toxicologic concepts, a dose–response relationship, refers to a type of correlative relationship between "the characteristics of exposure to a chemical and the spectrum of effects caused by the chemical."[16(p61)] The existence of a dose–response relationship may be used to establish the following kinds of information[15]:

- Causal association between a toxin and biological effects
- Minimum dosages needed to produce a biological effect
- Rate of accumulation of harmful effects

Factors that Affect Responses to a Toxic Chemical

Several factors determine whether an individual will respond to exposure to a toxic chemical and the type of response that occurs.[17] The nature of a toxic effect depends on the innate toxicity of the chemical, whether it is in sufficient concentration, and how it impinges upon a somatic location as a consequence of the route and site of exposure. In addition, exposure needs to take place for a sufficient time duration and frequency: The effects of a one-time exposure of short duration are usually different from those of an exposure that happens on several occasions over a long period of time. Other factors are related to a person's sensitivity to a chemical, for example, one's ability to metabolize the chemical (through biological processes that convert the chemical into either less harmful or more harmful substances). The concentration and toxicity of the chemical are affected by the following:

- Route of entry into the body
- Received dose of the chemical
    - Duration of exposure
    - Interactions that transpire among multiple chemicals
    - Individual sensitivity

Route and Site of Entry

The most frequent sites of exposure to environmental (i.e., xenobiotic) chemicals include the gastrointestinal tract, the respiratory system, and the skin. (Refer to **TABLE 3.2**.) With respect to route of exposure, chemicals may enter the body intentionally (as in the example of medical interventions) or unintentionally (through injuries or environmental exposures). Illustrations of the routes of exposure corresponding to the sites of entry are ingestion (e.g., consumption of contaminated food or drink), injections into the bloodstream, contact with the surface of the skin (topical mode), and inhalation. With respect to inhalation, when we breathe contaminated air (for instance, at work), potentially dangerous particles and gases may enter our bodies and be absorbed into the bloodstream by the alveoli of the lungs without producing any warning symptoms. These chemicals later may cause serious lung diseases, including lung cancer. In some cases the absorbed chemicals may affect other organs such as the brain.

As mentioned previously, other routes of exposure include skin contact, eye contact, and ingestion. The skin forms a protective barrier that can be compromised by dehydration, physical damage, caustic chemicals, and organic solvents; the result is that chemicals may bypass the barrier and enter the bloodstream. The eyes are particularly vulnerable to chemical exposure, especially when chemicals splash into the eye during an accident. Finally, employees who work with toxic chemicals may unwittingly ingest such chemicals when they smoke or eat on the job; in addition, workers unintentionally may transport them home on their clothing and skin.

How quickly a chemical produces acute and other effects depends on the site and route of exposure.[6] Among possible sites and routes, contact with the skin generally produces the slowest response, and direct injection into the bloodstream typically yields the fastest and strongest effects. (Refer to Table 3.2.)

TABLE 3.2 Ranking of the Relative Speed of Effect According to the Site and Route of Exposure

Site	Route	Relative Speed of Effect[a]
Bloodstream	Intravenous	++++++++
Lungs	Inhalation	+++++++
Other	Intraperitoneal	++++++
Other	Subcutaneous	+++++
Other	Intramuscular	++++
Other	Intradermal	+++
Gastrointestinal	Ingestion (oral route)	++
Skin	Dermal	+

[a] The greater the number of plus signs (+), the faster the effect; e.g., when chemicals enter the bloodstream directly they usually have the fastest effect.

Data from Eaton DL, Gilbert SG. Principles of toxicology. In: Klaasen CD, Watkins JB III, eds. *Casarett & Doull's Essentials of Toxicology*. 3rd ed. New York, NY: McGraw-Hill, 2015:9.

Length and Duration of Exposure

Toxicologists describe the duration of exposure in a range from acute to chronic. Relevant terms are defined in the following box.

The duration of exposure is important from the toxicologic point of view because the effects of exposure to many chemicals may vary depending on whether the exposure is minimal and takes place on a single occasion for a short time period or is heavy and long-standing. Some workers (for example, agricultural workers) receive exposures to pesticides in greater concentrations and over much longer time periods than does the general population. Acute exposures may produce acute effects that disappear rapidly, although this outcome is not invariable. In comparison with short-term exposures, chronic exposures may allow for the buildup of effects over time with the resulting accumulation of biological damage. However, such accumulation may not occur if the chemical is excreted before the next exposure happens.

Effects of Chemical Mixtures

In the "real world" of environmental exposures, most exposures are to multiple chemicals rather than to a

LIST OF TERMS THAT DESCRIBE DURATION OF EXPOSURE

Acute—Usually a single exposure for less than 24 hours
Subacute—Repeated exposure for 1 month or less
Subchronic—Repeated exposure for 1 to 3 months
Chronic—Repeated exposure for more than 3 months

Reproduced from Eaton DL, Gilbert SG. Principles of toxicology. In: Klaasen CD, Watkins JB III, eds. *Casarett & Doull's Essentials of Toxicology*. 3rd ed. New York, NY: McGraw-Hill, 2015;9.

single chemical. By way of example, the ambient air in a factory may contain a soup of chemicals that impinge upon the production workers. Toxicologists and others have observed that when chemicals mix, they may produce surprising effects—sometimes their combined effects may be greater and sometimes less than expected. When an organism is exposed to two or more chemicals, their combined effects may be additive, synergistic, antagonistic, or coalitive; some chemicals may cause potentiation. **Additive** means that the combination of two chemicals produces an effect that is equal to their individual effects added together. The term **synergism** (from the Greek word *synergos*—working together) indicates that the combined effect of exposures to two or more chemicals is greater than the sum of their individual effects.

There are many examples of synergism in the field of environmental health. One is the interaction between asbestos and smoking in causing lung cancer. The prominent medical researcher Irvine Selikoff and associates demonstrated that lung cancer mortality risk among asbestos insulation workers was much higher among those who smoked in comparison with those who did not smoke.[18] A second example pertains to the synergistic effects of solvents (e.g., n-hexane and methyl ethyl ketone [MEK]). These chemicals, when acting simultaneously on the nervous system, produce a much greater combined effect than would be expected from their simple additive individual effects.[19]

Another type of interaction between two chemicals is called **potentiation**, which happens when one chemical that is not toxic causes another chemical to become more toxic.[6] For example, isopropanol by itself is not toxic to the liver but has the capacity to increase the liver toxicity of carbon tetrachloride when exposure of the two chemicals occurs together. In a **coalitive interaction**, several agents that have no known toxic effects interact to produce a toxic effect.[20]

The term **antagonism** means that "two chemicals administered together interfere with each other's actions or one interferes with the action of the other. . . ."[6(p8)] In illustration, supplemental vitamins may reduce the effects of needed prescription medicines such as antibiotics; an example is the antagonistic relationship between calcium and tetracycline.

Individual Responses to Toxic Exposures

The responses of individuals to toxic substances can vary greatly, ranging from no apparent response to severe responses. Responses may vary according to age, sex, race, and health status. Other influences include the person's genetic background, use of medications, consumption of alcohol, and pregnancy status. Some chemicals (e.g., bee venom) produce severe or life-threatening reactions in persons afflicted with allergies.

The term **chemical allergy** refers to "an immunologically mediated adverse reaction to a chemical resulting from previous sensitization to that chemical or to a structurally similar one."[6(p10)] Some of the alternative terms used to describe this response are hypersensitivity, allergic reaction, and sensitization reaction. After initial sensitization, small doses of a chemical may bring about an allergic reaction. Some genetically predisposed persons may be prone to *chemical idiosyncrasy*, meaning that they have either extreme sensitivity to low doses of a chemical or insensitivity to high doses.

Recall that the term *xenobiotics* refers to foreign chemicals that are introduced into the body. Although their effects may be beneficial, they also may produce adverse effects at the same time; these adverse effects may occur either directly or indirectly. A direct effect denotes an immediate impact upon the cells and tissues of the body or upon specific target organs. An indirect effect would be a change in the function of the body's biochemical processes. Examples of direct and indirect adverse effects are shown in **TABLE 3.3**.

Direct adverse effects of exposures to chemicals range from local effects to systemic effects to target organ effects. These terms are defined as follows:

Local effects—Damage at the site where a chemical first comes into contact with the body; examples are redness, burning, and irritation of the skin.

Systemic effects—Adverse effects associated with generalized distribution of the chemical throughout the body by the bloodstream to internal organs.

TABLE 3.3 Direct and Indirect Adverse Effects of Xenobiotics

Direct Adverse Effects	Indirect Adverse Effects
Cell replacement, such as fibrosis (buildup of scar tissue)	Modification of an essential biochemical function
Damage to an enzyme system	Interference with nutrition
Disruption of protein synthesis	Alteration of a physiological mechanism
Production of reactive chemicals in cells	
DNA damage	

Modified from National Library of Medicine, *Toxicology Tutor I. Basic Principles: Toxic Effects*. Available at: http://www.sis.nlm.nih.gov/enviro/toxtutor/Tox1/a31.htm. Accessed April 21, 2017.

Target organ effects—Some chemicals may confine their effects to specific organs; the most common organs affected by such chemicals are the liver, lungs, heart, kidneys, brain and nervous system, and the reproductive system.

FIGURE 3.8 differentiates between a systemic and organ toxicant.

Latency and Delayed Responses to Toxic Substances

The term *latency* refers to the time period between initial exposure and a measurable response. The latency period can range from a few seconds (in the case of acutely toxic agents) to several decades for agents that may be carcinogenic.

One example of a delayed effect of exposure to chemicals is carcinogenesis, the potential to induce cancerous growth of cells. The word **carcinogen** denotes a chemical (or substance) that causes or is suspected of causing cancer, a disease associated with unregulated proliferation of cells in the body. In fact, many forms of cancer are believed to have a latency period of 10 years up to 40 years (counting the time between first exposure to a carcinogen and the subsequent development of cancer). For example, mesothelioma (a rare form of thoracic or abdominal cancer) has a latency period as long as 40 years between first exposure to asbestos and subsequent development of the condition.

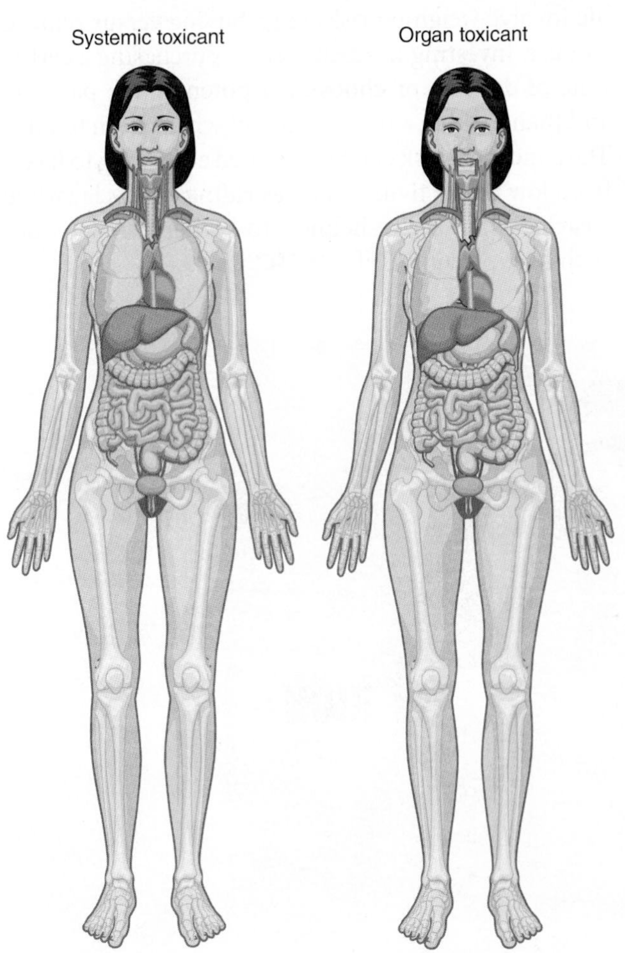

FIGURE 3.8 Systemic and organ toxicant.

The long latency for many of the health events studied in environmental research makes detection of the effects of exposures to toxic substances a methodologically difficult problem. For human exposures, toxicologists may be unable to differentiate among exposures to multiple chemicals or to rule out the impacts of confounding factors (discussed in the chapter on environmental epidemiology) that may be implicated in carcinogenesis.

Methods of Testing for Toxicity

The subjects used for testing the toxicity of chemicals include the following[21]:

- Volunteers who have had normal or accidental exposures
- Animals exposed purposively (in vivo experiments)
- Cells derived from human, animal, or plant sources (in vitro experiments)

Toxicity testing that uses these categories of subjects may take the form of epidemiologic investigations (described in the chapter on environmental epidemiology), formal clinical trials, or animal studies.

In the United States, randomized controlled clinical trials are regulated by the Food and Drug Administration (FDA). They are used to evaluate the safety of new drugs in humans, following tests in animals. The trials are conducted in three phases, beginning with small groups of patients and gradually expanding to larger patient populations after an earlier phase has demonstrated the safety and efficacy of the drug in question. Before the advent of rigorous clinical trials, new drugs could enter the marketplace without the benefit of evaluation. A notorious example is the drug thalidomide, which, when administered to pregnant women outside the United States (e.g., in Europe, Japan, Australia, and Canada) was found to be a potent teratogen (a drug or other substance that causes birth defects). Not approved for use in the United States, thalidomide was prescribed during the late 1950s to treat morning sickness in pregnant women (refer to the following textbox). As a result, these women gave birth to more than 10,000 children who had severe birth defects, such as missing limbs.[22]

Even with the use of clinical trials, new drugs entering the marketplace still may have potential to cause harm. An example that occurred in 2005

SOME QUICK FACTS ABOUT THALIDOMIDE: USED FOR MORNING SICKNESS

- Developed by Chemie Grünenthal GmbH in the former West Germany during 1950s.
- Demonstrated low toxicity in testing, suggesting it was safe.
- Not assessed for effects during pregnancy.
- Licensed in 1956 for over-the-counter sales in Germany and most European countries.
- Removed from marketplace in 1961 following widespread association with birth defects.

Data from Sciencemuseum.org. Science museum brought to life. Thalidomide. Available at: http://www.sciencemuseum.org.uk/broughttolife/themes/controversies/thalidomide. Accessed April 17, 2017.

involved Vioxx (used to control pain among arthritis patients), which was thought to increase the risk of heart problems.

When highly controlled and standardized conditions are desired, animal models are used in toxicologic testing. Another purpose for the use of animals is in preliminary testing of pharmaceuticals, pesticides, and substances that may present toxic hazards to human subjects. Standard protocols using animals evaluate the acute effects of chemicals, as well as toxicity to specific organs. Another concern of animal assays for toxicity is damage to the reproductive system or offspring. Animal protocols also are applied to measure the carcinogenic potential of chemicals. Research and toxicity testing may require the use of multiple animal species, including rats, mice, rabbits, dogs, and monkeys. Animal studies aid in preventing highly toxic chemicals from being introduced into human populations. Often alternatives such as in vitro studies have been employed as a means of reducing experimentation with live animals.

One focus of evaluation is the role of chemicals in causing toxicity to nucleic acids; some results might be mutations, chromosomal abnormalities, and DNA damage. In these studies, tests may be conducted with bacteria, fruit flies, and cell lines from animals or humans.

Links Between Toxicology and Risk Assessment

Once we have obtained toxicologic information regarding the nature and health effects of a chemical, we would like to be able to determine the risks to health that the chemical represents. Let us examine the concept of risk briefly. One can never be free from all risk to health and well-being. Even the most benign activities carry risk; for example, while riding on a busy street, a bicyclist may be struck by a car. Perhaps the person chose to ride a bicycle in order to avoid the financial risk of owning a car. Thus, many aspects of life involve weighing risks (e.g., buying versus renting a house, investing in stocks versus purchasing a certificate of deposit, or choosing a potential life partner) and then making a decision about what action to take. The concept of risk can be classified according to levels from low risk activities such as riding a bicycle (while wearing a protective helmet) to higher risk activities such as sky diving. Refer to **FIGURE 3.9**.

FIGURE 3.9 Risk levels.

In simple terms, a risk involves the likelihood of experiencing an adverse effect. The term **risk assessment** refers to "a process for identifying adverse consequences and their associated probability."[23(p611)] Risk assessment is "[t]he process of determining risks to health attributable to environmental or other hazards."[4] "Risk research addresses the identification and management of situations that might result in losses or harm, immediate or delayed, to individuals, groups, or even to whole communities or ecosystems, often as a result of the interaction of human activities with natural processes."[24(px)] Risk assessment provides a qualitative or quantitative estimation of the likelihood of adverse effects that may result from exposure to specified health hazards or from the absence of beneficial influences. "Risk assessment uses clinical, epidemiologic, toxicologic, environmental, and any other pertinent data."[4]

The meaning of the term *risk* varies greatly not only from one person to another, but also between lay persons and professionals; the latter characterize risk mainly in terms of mortality.[25] In a psychometric study, psychology professor Paul Slovic reported that lay persons classified risk according to two major factors. His methods enabled risks to be portrayed in a two-dimensional space so that their relative positions could be compared. The two factors that Slovic identified were the following:

> Factor 1, labeled "dread risk," is defined at its high (right-hand) end by perceived lack of control, dread, catastrophic potential, fatal consequences, and the inequitable distribution of risks and benefits. . . .
>
> Factor 2, labeled "unknown risk," is defined at its high end by hazards judged to be unobservable, unknown, new, and delayed in their manifestation of harm.[25(p283)]

Refer to **FIGURE 3.10**, which maps the spatial relationships among a large number of risks according to the two major factors shown in the figure. For example, nuclear reactor accidents fall in the space that defines uncontrollable dread factors that are of unknown risk. In other words, nuclear reactor accidents fall in the quadrant defined by both high levels of unknown risk and high levels of dread risk. In contrast, home swimming pools are perceived as falling in the quadrant in which risks are not dread and are known to those exposed.

Risk assessment generally takes place in four steps: (1) hazard identification, (2) dose–response assessment, (3) exposure assessment, and (4) risk characterization.[26,27] Refer to **FIGURE 3.11** for an illustration of the four step risk assessment process. The term **risk management** shown in the figure is discussed in the chapter on environmental policy and regulation.

Hazard Identification

Hazard identification (hazard assessment) "examines the evidence that associates exposure to an agent with its toxicity and produces a qualitative judgment about the strength of that evidence, whether it is derived from human epidemiology or extrapolated from laboratory animal data."[27(p286)] Evidence regarding hazards linked to toxic substances may be derived from the study of health effects among exposed humans and animals. These health effects may range from dramatic outcomes such as mortality or cancer to lower-level conditions such as developmental delays in children and reductions in immune status.[26]

A *hazard* is defined as the "[i]nherent capability of a natural or human-made agent or process to adversely affect human life, health, property, or activity, with the potential to cause a DISEASE, EPIDEMIC, ACCIDENT, or DISASTER."[4] Hazards may originate from chemicals, biological agents, physical and mechanical energy and force, and psychosocial influences. We have covered the toxic agents such as organic toxins and chemicals. Examples of other hazards will be covered in more detail elsewhere in the text. Some physical hazards arise from ionizing radiation from medical X-rays and naturally occurring background radiation.

Other hazards originate from nonionizing radiation—sunlight, infrared and ultraviolet light, and electromagnetic radiation from power lines and radio transmissions. In urban and work environments, mechanical energy is associated with high levels of noise that can be hazardous for hearing and for psychological well-being. Examples of psychosocial hazards are work-related stresses, combat fatigue, and post-traumatic stress disorder (topics not covered in this text).

Dose–Response Assessment

Dose–response assessment is the measurement of "the relationship between the amount of exposure and the occurrence of the unwanted health effects."[26(p38)] According to Russell and Gruber:

> Dose-response assessment examines the quantitative relation between the experimentally administered dose level of a toxicant and the incidence or severity or both of a response in test animals, and draws inferences for humans. The presumed human dosages and incidences in human populations may also be used in cases where epidemiological studies are available.[27(p286)]

Factor 2
Unknown risk

Laetrile
Microwave ovens
Water fluoridation
Saccharin
Nitrates
Water chlorination
Hexachlorophene
Polyvinyl Chloride
Coal tar hairdyes
Oral contraceptives
Diagnostic X-Rays
Valium
IUD
Darvon
Antibiotics
Rubber mfg.
Caffeine
Aspirin
Auto lead
Lead paint
Vaccines
DNA technology
SST
Electric fields
DES
Nitrogen fertilizers
Cadmium usage
Mirex
Trichloroethylene
2,4,5-T
Radioactive waste
Nuclear reactor accidents
Pesticides
Uranium mining
Asbestos Insulation
PCBs
Nuclear weapons fallout
Satellite crashes
Mercury
DDT
Fossil fuels
Coal burning (pollution)

Factor 1
Dread risk

Skateboards
Smoking (disease)
Power mowers
Snowmobiles
Trampolines
Tractors
Alcohol
Chainsaws
Elevators
Home swimming pools
Electric wir & appl (fires)
Downhill skiing
Recreational boating
Smoking
Electric wir & appl (shock)
Bicycles
Motorcycles
Bridges
Fireworks
Alcohol Accidents
Auto exhaust (CO)
D-CON
LNG storage & transport
Nerve gas accidents
Coal mining (disease)
Large dams
Skyscraper fires
Nuclear weapons (war)
Underwater construction
Sport parachutes
General aviation
Coal mining accidents
High construction
Railroad collisions
Commercial aviation
Auto racing
Auto accidents
Handguns
Dynamite

Factor 2

Not observable
Unknown to those exposed
Effect delayed
New risk
Risk unknown to science

Controllable
Not dread
Not global catastrophic
Consequences not fatal
Equitable
Individual
Low risk to future generations
Easily reduced
Risk decreasing
Voluntary

Uncontrollable
Dread
Global catastrophic
Consequences fatal
Not equitable
Catastrophic
High risk to future generations
Not easily reduced
Risk increasing
Involuntary

Factor 1

Observable
Known to those exposed
Effect immediate
Old risk
Risks known to science

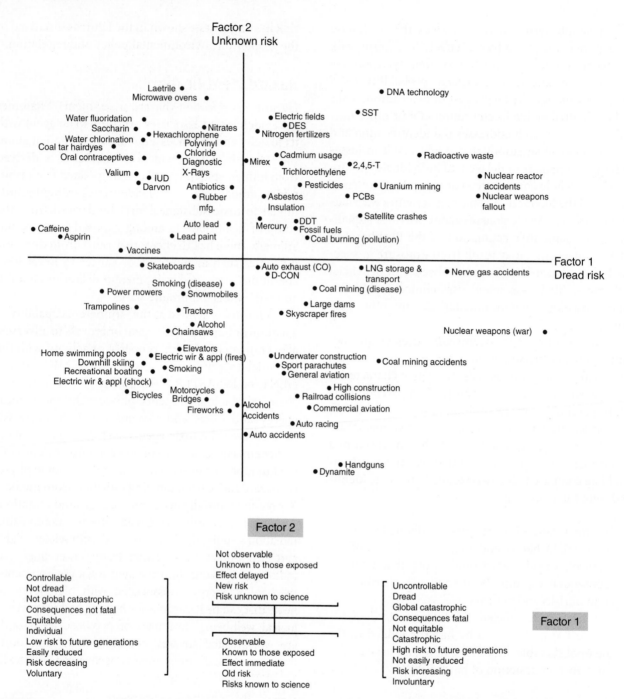

FIGURE 3.10 Location of 81 hazards on factors 1 and 2 derived from the relationships among 15 risk characteristics.
Reproduced with permission from Slovic P. Perception of risk. *Science*. 1987;236:282. Copyright 1987 AAAS. April 17 issue.

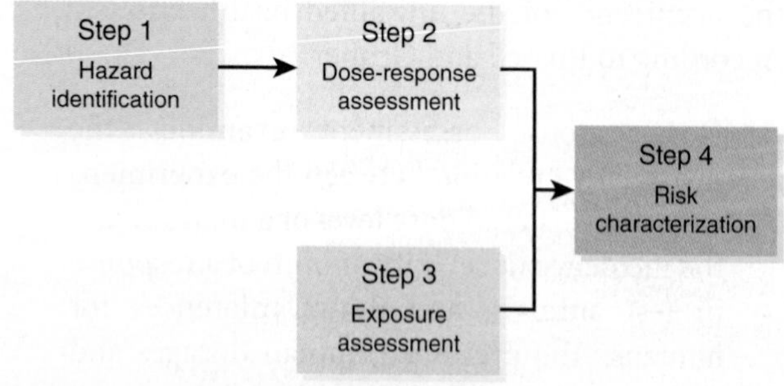

FIGURE 3.11 The four-step risk assessment process.
Modified from US EPA. Risk assessment. Human health risk assessment. Available at: https://www.epa.gov/risk/human-health-risk-assessment. Accessed February 8, 2017.

Exposure Assessment

Exposure assessment is defined as the procedure that "identifies populations exposed to the toxicant, describes their composition and size, and examines the roots, magnitudes, frequencies, and durations of such exposures."[27(p286)] **FIGURE 3.12** summarizes the three key steps in exposure assessment. The text box provides a more detailed overview of the steps involved in an exposure assessment.

The process of human exposure assessment is believed to be one of the weakest aspects of risk

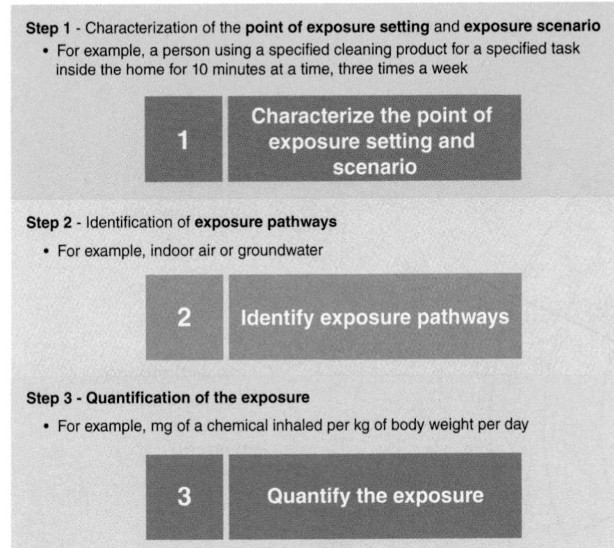

Step 1 - Characterization of the point of exposure setting and exposure scenario
• For example, a person using a specified cleaning product for a specified task inside the home for 10 minutes at a time, three times a week
1
Characterize the point of exposure setting and scenario
Step 2 - Identification of exposure pathways
• For example, indoor air or groundwater
2
Identify exposure pathways
Step 3 - Quantification of the exposure
• For example, mg of a chemical inhaled per kg of body weight per day
3
Quantify the exposure

FIGURE 3.12 The three-step process in exposure assessment.

Modified from ToxTutor. Available at: https://toxtutor.nlm.nih.gov/06-004.html. Accessed April 13, 2017.

assessment. Available methods are unable to provide adequate quantitative information regarding how much humans are exposed to toxic substances as well as the specific kinds and patterns of exposure.[28] The quality of exposure assessment data determines the accuracy of risk assessments and therefore is a limiting factor in the risk assessment process.[29]

When referring to a toxic substance, exposure assessment must take into account where the exposure occurs, how much exposure occurs, and how the substance is absorbed by the body. The process of human exposure assessment examines "the manner in which pollutants come into actual contact with the human body—the concentration levels at the points of contact and the sources of these pollutants making contact. The key word here is *contact*—the occurrence of two events at the same location and same time."[30(p449)] The methods by which human beings are exposed to toxic substances include encountering them in water,

WHAT IS AN EXPOSURE ASSESSMENT?

Suppose we would like to conduct an exposure assessment of an exposure to a toxic agent such as a chemical. For human populations, an exposure assessment with respect to toxic agents may be elucidated by the following components:

- **Magnitude:** How large is the exposure?
- **Frequency:** How often does an exposure take place?
- **Duration:** Over what time period has an exposure occurred?
- **Future exposures:** What is the estimate for future exposures?
- **Exposure pathways:** How does the agent move from its source to the individual?
- **Exposure routes:** How does the agent gain access to the body?

In addition, one needs to consider the characteristics of the population that is exposed. For example, how widespread is the exposure? And also, what are the exposed population's characteristics in terms of age, race, vulnerabilities, and related dimensions?

Exposure pathways include the processes for movement of substances from their sources to the people who are exposed. Exposure routes (concerned with contact with an agent) are modes of entry into the body. Examples of entry sites are through a body orifice (e.g., direct ingestion and inhalation) and via skin contact. Consider an employment-related example; workers might inadvertently ingest a toxic industrial chemical that has contaminated food or beverages brought to work. Community residents might inhale pesticides applied to agricultural fields. Another route is direct absorption of a toxic chemical through the skin when someone touches a toxic chemical.

An exposure assessment in the community could determine which people are being exposed as well as their varying levels of exposure. This procedure delineates the range of exposures. For example, let's assume that we would like to know a community's range of exposures to a toxic chemical at a factory. Hypothetically speaking, employees inside the factory might have the highest levels of exposure to the chemical. (It should be noted that most modern industrial concerns have developed protections for minimizing employee exposures.) The next highest levels of exposure would occur among persons who live or play downwind from the factory. Lower exposures would impinge upon residents in the geographic area surrounding the factory but not immediately downwind from the pollution source. (See **FIGURE 3.13**.)

Another crucial issue for exposure assessment is how to quantify exposures. One way this can be accomplished is by measuring exposures at their point of contact with the body; for example, by placing exposure measuring devices such as radiation monitors on the person of exposed individuals. Another method is to place monitoring devices strategically throughout the community. An example would be locating measuring devices for air pollution near sources of pollution, e.g., major highways. Also used are modeling techniques to predict exposures. Finally, at the individual level, one may also be able to quantify exposures by measuring biomarkers, excretions of chemicals from the body, and related methods that employ internal indicators of exposure. Internal indicators "reconstruct" exposures that have already happened.

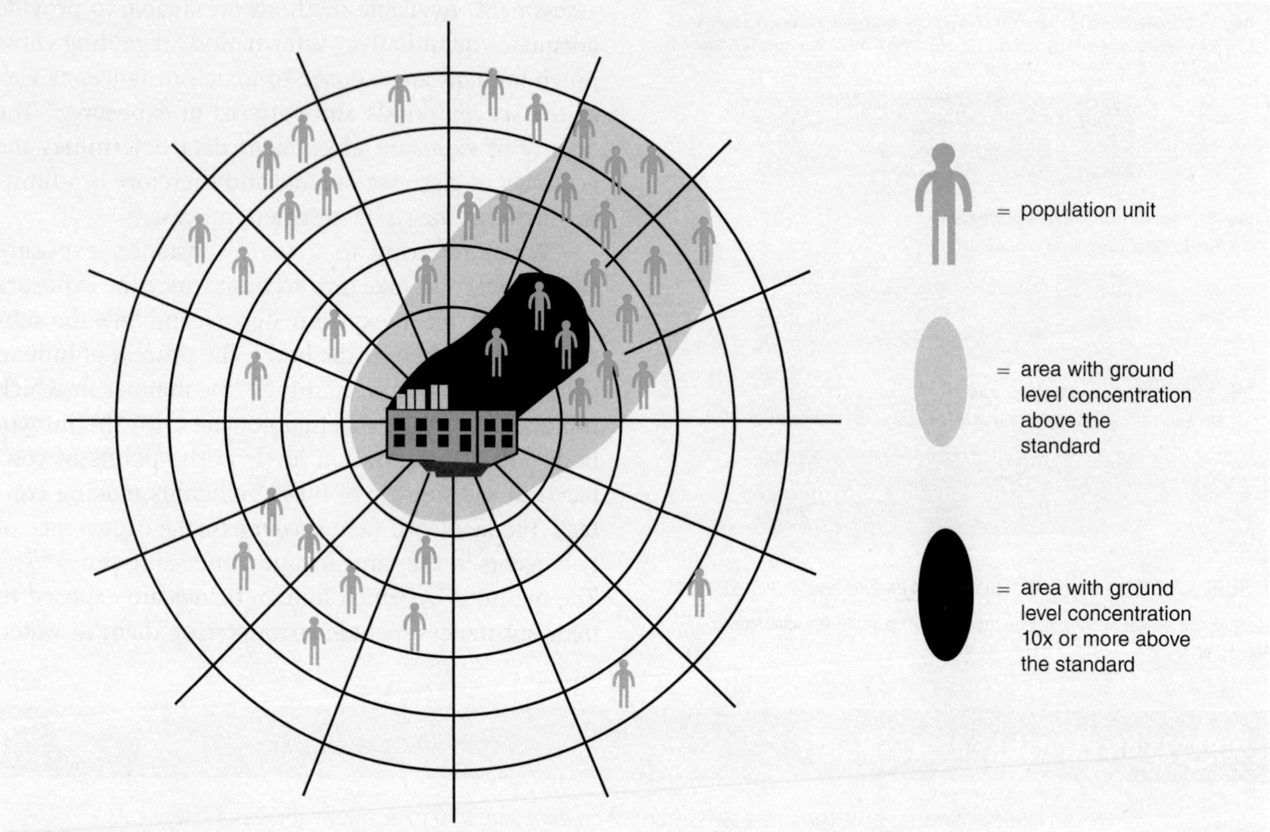

FIGURE 3.13 The range of exposure in a community.

Modified from US EPA. Risk assessment. Human health risk assessment. Available at: https://www.epa.gov/risk/human-health-risk-assessment. Accessed February 8, 2017.

Data from US Environmental Protection Agency. Conducting an exposure assessment. Step 3: Exposure Assessment. Available at: https://www.epa.gov/conducting-human-health-risk-assessment. Accessed March 14, 2017, and ToxTutor. Exposure Assessment. Available at: https://toxtutor.nlm.nih.gov/06-004.html. Accessed March 13, 2017.

air, food, soil, and various consumer products and medications.

As noted previously in the chapter on environmental epidemiology, high-quality data on exposure are necessary for making valid interpretations of a study's findings.[31] Several methods of exposure assessment (e.g., personal exposure monitoring and use of biological markers) are employed in toxicology, environmental epidemiology, and other environmental health disciplines. In some exposure assessments sampling devices may be placed in the community in order to determine the population's range of exposures. In tracking exposures to emissions from an industrial plant, researchers may find that the highest exposures occur downwind and closest to the plants. Figure 3.13 illustrates the range of exposure in a community. (Refer to the text box on exposure assessment for more information.)

During a review of records to assess exposure, the investigator may select a study population from personnel records maintained by a company. If the records of former and retired workers are retained by the company, a complete data set spanning long time periods may be available. Ideally, every previous and current worker exposed to the factor should be included. Selection bias may occur if some workers are excluded because their records have been purged from the company's database.[32] Data collected from employment records may include:

- Personal identifiers to permit record linkage to Social Security Administration files and retrieval of death certificates
- Demographic characteristics, length of employment, and work history with the company
- Information about potential confounding variables, such as the employee's medical history, smoking habits, lifestyle, and family history of disease

Some environmental studies use biomarkers that may be correlated with exposures to potential carcinogens and other chemicals. These biomarkers involve changes in genetic structure that are thought to be the consequence of an exposure.

One marker is called a **sister chromatid exchange (SCE)**. SCEs refer to reciprocal exchanges of deoxyribonucleic acid (DNA) between pairs of DNA molecules.[33] In a study of exposure to styrene gas (involved in boat building and plastics manufacture), researchers examined the utility of SCEs in comparison with environmental monitoring, and exhaled styrene levels as alternative measures of styrene exposure.[34] The use of biomarkers such as SCEs is growing in the environmental health field.

Risk Characterization

Risk characterization develops "estimates of the number of excess unwarranted health events expected at different time intervals at each level of exposure."[26(p38)] Risk characterization follows the three foregoing steps by integrating the information from hazard identification, dose–response assessment, and exposure assessment.[35] The process of risk characterization yields "[a] synthesis and summary of information about a hazard that addresses the needs and interests of decision makers and of interested and affected parties. Risk characterization is a prelude to decision making and depends on an iterative, analytic-deliberative process."[36(p216)] "Risk characterization presents the policy maker with a synopsis of all the information that contributes to a conclusion about the nature of the risk and evaluates the magnitudes of the uncertainties involved and the major assumptions that were used."[27(p286)]

Risk Management

Oriented toward specific actions, risk management "consists of actions taken to control exposures to toxic chemicals in the environment. Exposure standards, requirements for premarket testing, recalls of toxic products, and outright banning of very hazardous materials are among the actions that are used by governmental agencies to manage risk."[26(p37)] This topic is covered more fully in the chapter on environmental policy and regulation.

Conclusion

Toxicology, the science of poisons, is a powerful and important discipline for protecting human health and the environment. Often, toxicologists work collaboratively with specialists in other, overlapping fields, such as epidemiology. Toxicology has a lengthy and distinguished history, beginning with ancient civilizations. This chapter has covered some of the important terminology used to characterize the effects of xenobiotics on biological systems. Two of these concepts are the lethal dose and dose–response curves. Environmental chemicals have differing effects upon humans, depending upon a number of factors that include route of entry into the body, received dose, duration of exposure, interactions among chemicals, and individual sensitivity. Toxicology is an important component of risk assessment.

Despite the many helpful applications of toxicology, the field has some limitations as well. First, the laboratory models used (in vivo and in vitro) may not completely represent environmental conditions, which often are more complex than those conditions found in controlled settings. Also, toxicology, by looking only at adverse effects, ignores benefits important to human health and welfare that may also be present.

Study Questions and Exercises

1. Define environmental toxicology. Give an example of a study that would be relevant to this field.
2. What were the contributions of Paracelsus to toxicology?
3. Define the terms teratogen, xenobiotic, and carcinogen.
4. Explain the difference between toxins and toxicants. Give an example of each. How does a toxin differ from a poison?
5. State what is meant by the term LD_{50}.
6. What determines the toxicity of a chemical? To what extent do you agree with the assumption that "all substances are poisons"?
7. Describe methods for testing the toxicity of chemicals.
8. Explain the significance of duration of exposure from a toxicity point of view. What are the terms used to describe the different ranges of exposure?
9. What is latency? How does the issue of latency challenge environmental researchers who are trying to detect hazards associated with specific exposures?
10. Describe the typical shape of the population dose–response curve and explain why it is shaped this way.
11. Define the terms *dose* and *threshold*. Draw a figure to illustrate a threshold.
12. Explain the differences among synergism, antagonism, and potentiation. Give specific examples of each.
13. What is meant by the term *risk assessment*? Describe the steps that constitute risk assessment. Give an example of each step.
14. What is human exposure assessment? Describe some of the methods of exposure assessment.

For Further Reading

Newman C. 12 Toxic Tales, *National Geographic*, May 2005, 2-31.

References

1. US Department of Health and Human Services, Public Health Service, National Toxicology Program (NTP). About NTP. Available at: US Department of Health and Human Services, Public Health Service, National Toxicology Program. Accessed March 9, 2017.
2. Gallo MA. History and scope of toxicology. In: Klaasen CD, Watkins JB III, eds. *Casarett & Doull's Essentials of Toxicology*. 3rd ed. New York, NY: McGraw-Hill, 2015.
3. Henry CJ, Phillips R, Carpanini F, et al. Use of genomics in toxicology and epidemiology: findings and recommendations of a workshop. *Environ Health Perspect*. 2002;110:1047–1050.
4. Porta M, ed. *A Dictionary of Epidemiology*. 6th ed. New York, NY: Oxford University Press; 2014.
5. Weed DL. Environmental epidemiology basics and proof of cause-effect. *Toxicol*. 2002;181–182:399–403.
6. Eaton DL, Gilbert SG. Principles of toxicology. In: Klaasen CD, Watkins JB III, eds. *Casarett & Doull's Essentials of Toxicology*. 3rd ed. New York, NY: McGraw-Hill, 2015.
7. Borzelleca JF. Profiles in toxicology. Paracelsus: herald of modern toxicology. *Toxicol Sci*. 2000;53:2–4.
8. Landis WG, Sofield RM, Yu M-H. *Introduction to Environmental Toxicology*. 4th ed. Boca Raton, FL: CRC Press: 2010.
9. Briggs GB. Risk assessment policy for evaluating reproductive system toxicants and the impact of responses on sensitive populations. *Toxicol*. 1996;111:305–313.
10. Schwetz BA. Noncancer risk assessment: reproductive toxicology. *Drug Metab Rev*. 1996;28(1–2):77–84.
11. Online Medical Dictionary. Xenobiotics. Available at: http://www.online-medical-dictionary.org/definitions-x/xenobiotics.html. Accessed March 10, 2017.
12. National Library of Medicine. Welcome to ToxTutor. Available at: https://toxtutor.nlm.nih.gov/index.html. Accessed March 10, 2017.
13. Hyperdictionary. Meaning of toxicity. Available at: http://www.hyperdictionary.com/dictionary/toxicity. Accessed April 11, 2017.
14. Fox DA, Boyes WK. Toxic responses of the ocular and visual system. In: Klaasen CD, Watkins JB III, eds. *Casarett & Doull's Essentials of Toxicology*. 3rd ed. New York, NY: McGraw-Hill, 2015.
15. National Library of Medicine. *Toxicology Tutor I*. Basic principles: dose and dose response—dose. Available at: http://sis.nlm.nih.gov/enviro/toxtutor/Tox1/a21.htm. Accessed April 18, 2017.
16. National Institutes of Health, National Institute of Environmental Health Sciences. *Chemicals, the Environment, and You: Explorations in Science and Human Health*. Dose-response relationships. Available at: http://science.education.nih.gov/supplements/nih2/chemicals/guide/pdfs/lesson3.pdf. Accessed April 11, 2017.
17. Health Evaluation System and Information Service (HESIS), Occupational Health Branch, California Department of Public Health. Understanding toxic substances: an introduction to chemical hazards in the workplace. 2008 edition. Richmond, CA: HESIS; 2008.
18. Selikoff IJ, Hammond EC, Churg J. Asbestos exposure, smoking, and neoplasia. *JAMA*. 1968;204:104–112.
19. Australian Government, National Occupational Health and Safety Commission. Synergism and potentiation. Available at: http://www.safeworkaustralia.gov.au/sites/SWA/about/Publications/Documents/238/GuidanceNote_InterpretationOfExposureStandardsForAtmosphericContaminants_3rdEdition_NOHSC3008-1995_PDF.pdf. Accessed April 17, 2017.
20. Suk WA, Olden K. Multidisciplinary research: strategies for assessing chemical mixtures to reduce risk of exposure and disease. *Int J Occup Med Environ Health*. 2004;17:103–110.
21. National Library of Medicine. *Toxicology Tutor I*. Basic principles: toxicity testing methods. Available at: http://sis.nlm.nih.gov/enviro/toxtutor/Tox1/a51.htm. Accessed April 18, 2017.
22. Pannikar V. The return of thalidomide: new uses and renewed concerns. Available at: http://www.who.int/lep/research/Thalidomide.pdf. World Health Organization. Accessed April 17, 2017.
23. McKone TE. The rise of exposure assessment among the risk sciences: an evaluation through case studies. *Inhal Toxicol*. 1999;11:611–622.
24. Amendola A, Wilkinson DR. Risk assessment and environmental policy making. *J Hazard Mater*. 2000;78:ix–xiv.
25. Slovic P. Perception of risk. *Science*. 1987;236:280–285.
26. Landrigan PJ, Carlson JE. Environmental policy and children's health. *The Future of Children*. 1995;5(2):34–52.
27. Russell M, Gruber M. Risk assessment in environmental policy-making. *Science*. 1987;236:286–290.
28. US Department of Health and Human Services, National Toxicology Program. Human exposure assessment. Available at: http://webharvest.gov/peth04/20041020135705/ntp.niehs.nih.gov/index.cfm?objectid=06F6F41D-9B12-8FD0-63E4048B173CC36A. Accessed April 17, 2017.
29. Lippmann M, Thurston GD. Exposure assessment: input into risk assessment. *Arch Environ Health*. 1988;43:113–123.
30. Ott WR. Human exposure assessment: the birth of a new science. *J Expos Anal Environ Epidem*. 1995;5:449–472.
31. Gardner MJ. Epidemiological studies of environmental exposure and specific diseases. *Arch Environ Health*. 1988; 43:102–108.
32. Monson RR. *Occupational Epidemiology*. Boca Raton, FL: CRC Press; 1990.
33. Hyperdictionary. Meaning of sister chromatid exchange. Available at: http://www.hyperdictionary.com/dictionary/sister+chromatid+exchange. Accessed April 17, 2017.
34. Rappaport SM, Symanski E, Yager JW, et al. The relationship between environmental monitoring and biological markers in exposure assessment. *Environ Health Perspect*. 1995;103 (Suppl 3):49–54.
35. Duffus JH. Risk assessment terminology. *Chemistry International*. 2001;23(2):34–39.
36. Stern PC, Fineberg HV, eds. *Understanding Risk: Informing Decisions in a Democratic Society*. National Academy of Sciences' National Research Council, Committee on Risk Characterization. Washington, DC: National Academy Press; 1996.

© Jean-Luc Rivard/EyeEm/Getty Images

CHAPTER 4

Environmental Policy and Regulation

LEARNING OBJECTIVES

By the end of this chapter the reader will be able to:

- Contrast key environmental health regulatory agencies at three levels.
- State four principles that guide environmental policy development.
- Compare five major environmental laws.
- Describe environmental policies designed to protect vulnerable groups.
- Apply the steps in the policy-making process to a specific example.

Introduction

In this chapter you will learn how governments, advocacy groups and other organizations, and stakeholders create environmental policies for the mitigation of environmental hazards and protection of the health of the planet. (See **FIGURE 4.1** for the major topic of this chapter.) We will present terminology and concepts related to the environmental policy process. Also described are some of the major US governmental and international agencies charged with the development, adoption, and enforcement of environmental policies and regulations. The chapter also will cite some of the US laws and regulations that have been developed to protect air and water quality and natural resources and to safeguard the public against hazards that originate from toxic substances and wastes. Throughout this discussion you will need to keep in mind how policy development is an imperfect process that often reflects the tension that exists between political influences and scientific knowledge.

FIGURE 4.1 Environmental policy.

© Stuart Miles/Shutterstock

The Role of Policy and Environmental Challenges

During the current century, human illness and death linked to environmental exposures represent one of the most significant challenges to the world's inhabitants. To date, extensive resources have been expended in the implementation of public policies, regulations, and laws that are designed to protect the health of the public from environmentally caused diseases. Nevertheless, much more work needs to be completed. Referring specifically to the United States, Bailus Walker, the past president of the American Public Health Association, noted:

> Despite these investments—which amount to billions of dollars—we have not at all completed the task of preventing environmentally provoked disease and of providing more protection for the ecological system. Even an abbreviated list will show how many problems remain to be solved: an epidemic of childhood lead poisoning, the increasing incidence of genetic diseases exacerbated by environmental stressors, pesticides in food and water, too much ozone at ground level and too little in the stratosphere, global temperature warming, slow implementation of national toxic waste policies, and flaws in the institutions and processes the nation relies on to reduce environmental risks. Obviously, environmental health activities cannot be narrowly focused but must recognize the critical interrelationships between environmental media and between policies and programs.[1(p1395)]

Increasingly, protection from environmentally associated health hazards is regarded as a fundamental human right; policy making will need to take into account the reduction of disparities in health status that result from environmental sources. Momentum is gathering for the development of policies that protect the health of vulnerable population groups such as children, who may be even more susceptible to the actions of environmental toxins than adults. For example, data have shown an increasing occurrence of childhood asthma, which may be linked to air pollution.

Overview of the Environmental Policy Process

This overview will define and explain the following terms and concepts related to the policy process:

- Environmental policy
- Principles of environmental policy development
- The policy cycle
- The interplay between the evaluation process and policy development

Some argue that a systems approach—holistic thinking—is essential for policy development in the arena of the environment, which may be thought of as a set of interconnected elements or subsystems.[2] These include scientific, economic, cultural, and political dimensions that are relevant to the formulation and implementation of particular environmental policies. An example relates to the problem of agricultural pollution, which poses health hazards from animal wastes, fertilizers, and pesticides. Farming practices are designed to maximize output, using economic, climatic, and land use criteria. In addition to taking into account the health consequences of agricultural pollution, environmental policies also need to consider economic factors such as market costs that influence agricultural practices.[3] Another economic factor concerns stakeholders' willingness to pay for the costs necessitated by policy programs. Good policy decisions can be made if data on the economic costs and benefits of the policy package and its components are available.[4]

The environmental policy process takes place within the political context and reflects tension between political considerations and scientific knowledge. In some instances, the public may reject environmental policies that are justified scientifically. In illustration, research has suggested that logging of the nation's forests contributes to increases in greenhouse gases; as a result, some municipalities may decide to prohibit logging in order to protect the natural environment. However, in communities supported by lumber production, residents may oppose restrictions on logging because such prohibitions cause widespread unemployment. Another example is the reactivation of coal mining and fossil fuel extraction in order to create employment opportunities for miners.

In other instances, politics may be instrumental in the adoption of environmental policies that might not be scientifically justified. For example, scientific findings may demonstrate that the low levels of some pollutants in the water supply are unlikely to have adverse human health impacts. Nevertheless, consumers may demand that such pollutants be removed completely from the water supply even if their removal is very costly.

Public demonstrations in support of environmental causes is another example of the political process in action. Groups have demonstrated against attempts to exploit parks, government lands, and sensitive areas for resource extraction. Other demonstrations have been in support of environmental justice and in response to disasters such as oil spills. (See **FIGURE 4.2**.)

FIGURE 4.2 Demonstration against British Petroleum and in support of environmental justice.

Definition of Environmental Policy

An **environmental policy** is "[a] statement by an organization [either public, such as a government, or private] of its intentions and principles in relation to its overall environmental performance. Environmental policy provides a framework for action and for the setting of its environmental objectives and target."[5] The goal of environmental policy is "to reduce human risks or environmental damages resulting from pollution. Policy analysis facilitates social valuation of these risks and damages, by clarifying the costs of reducing environmental damages in terms of foregone economic returns."[3(pp244–245)] An example of an environmental policy in the United States is the National Environmental Policy Act of 1969, which sought to benefit the human population by preventing damage to the environment. The act is written, in part, as seen in the Congressional Declaration of National Environmental Policy box.

Principles of Environmental Policy Development

In the environmental health arena, a number of environmental principles or philosophies may guide the work of those who are charged with creation of policy (formal and informal policy actors, policy researchers, and policy analysts):

- The precautionary principle
- Environmental justice
- Environmental sustainability
- The polluter-pays principle

A significant ethical issue for policy developers involves setting an acceptable level of risk associated with a potential environmental hazard. There needs to be a "moral consensus regarding the level of risk that is sufficient to regard a substance in the environment as a potential threat to human and ecologic health."[7(p1786)]

Congressional Declaration of National Environmental Policy

Sec. 101 [42 USC § 4331].

(a) The Congress, recognizing the profound impact of man's activity on the interrelations of all components of the natural environment, particularly the profound influences of population growth, high-density urbanization, industrial expansion, resource exploitation, and new and expanding technological advances and recognizing further the critical importance of restoring and maintaining environmental quality to the overall welfare and development of man, declares that it is the continuing policy of the Federal Government, in cooperation with State and local governments, and other concerned public and private organizations, to use all practicable means and measures, including financial and technical assistance, in a manner calculated to foster and promote the general welfare, to create and maintain conditions under which man and nature can exist in productive harmony, and fulfill the social, economic, and other requirements of present and future generations of Americans.[6]

The **precautionary principle** states that "preventive, anticipatory measures . . . [should] be taken when an activity raises threats of harm to the environment, wildlife, or human health, even if some cause-and-effect relationships are not fully established."[8(p263)] The practice of risk assessment is a science of uncertainty, so environmental toxins often present risks that have not been ascertained completely. The precautionary principle suggests that policy makers should err on the side of "an ounce of prevention" and take protective measures even when full scientific certainty is lacking.

A specific case that could merit the application of the precautionary principle relates to endocrine disruptors. The term *endocrine disruptor* refers to "an exogenous substance or mixture that alters function(s) of the endocrine system and consequently causes adverse health effects in an intact organism, or its progeny, or (sub)populations."[9] For example, some chemicals are believed to have an estrogenic effect (acting as female hormones); two of the many varieties of these chemicals are DDT and the family of pesticides known as organochlorines. Scientists have speculated that endocrine-disrupting chemicals may be related to human sexual abnormalities such as low sperm counts in boys and animal abnormalities such as changes in the sexual functioning of aquatic animals.

The precautionary principle would advocate for the control of potential endocrine-disrupting chemicals, even though the focus of most research has been on whether or not they cause cancer (carcinogenesis). The absence of demonstrated carcinogenic effects does not rule out the possibility of other health hazards (e.g., endocrine disruption). Environmental research that focuses on risks of carcinogenic effects should not exclude from consideration other effects such as damage to the endocrine and immune systems; these health effects are overlooked frequently in environmental research. Further, exposures to such chemicals that occur early in life may pose greater health risks than exposures that occur later in life. Policies developed by the European Union adhere to the precautionary principle and advocate that preventive action be taken to reduce risks from potential environmental hazards.[10]

The principle is applied when the weight of scientific evidence suggests that a chemical is suspected of having adverse health consequences, even though such health effects have not been established definitively.

The concept of **environmental justice** denotes the equal treatment of all people in society irrespective of their racial background, country of origin, and socioeconomic status. (See **FIGURE 4.3**.) The presence of an environmental hazard may be the end product of disparities of power and privilege within a community. Consequently, unequal toxic exposures of adults and children from different racial, ethnic, or socioeconomic groups may occur. Children who reside in minority communities may receive especially high exposures to environmental toxins.

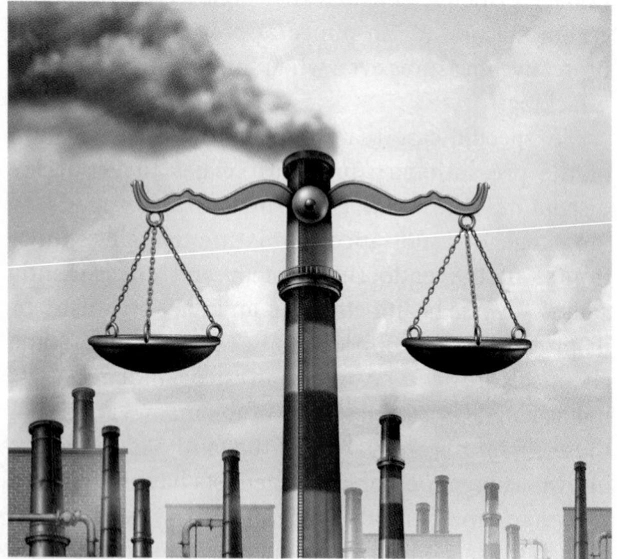

FIGURE 4.3 Environmental justice.

Here is a statement on environmental justice by the US Environmental Protection Agency (EPA):

> Environmental Justice is the fair treatment and meaningful involvement of all people regardless of race, color, national origin, or income with respect to the development, implementation, and enforcement of environmental laws, regulations, and policies. Fair treatment means that no group of people should bear a disproportionate share of the negative environmental consequences resulting from industrial, governmental and commercial operations or policies. Meaningful involvement means that: (1) people have an opportunity to participate in decisions about activities that may affect their environment and/or health; (2) the public's contribution can influence the regulatory agency's decision; (3) their concerns will be considered in the decision-making process; and (4) the decision makers seek out and facilitate the involvement of those potentially affected. EPA has this goal for all communities and persons across this Nation. It will be achieved when everyone enjoys the same degree of protection from environmental and health hazards and equal access to the decision-making process to have a healthy environment in which to live, learn, and work.[11]

As a goal of environmental policy, **environmental sustainability** adheres to the philosophical viewpoint "that a strong, just, and wealthy society can be consistent with a clean environment, healthy ecosystems, and a beautiful planet."[12(p5383)] The three components of sustainable development are materials and energy use, land use, and human development. Environmental sustainability means that resources should not be depleted faster than they can be regenerated; the concept also specifies that there should be no permanent change to the natural environment. Critics of sustainable development argue that the definition of the term is not entirely clear and is open to interpretation.[13] Individuals are able to contribute to environmental sustainability through sustainable living, which is adoption of a lifestyle that minimizes demands upon the environment. An example is to lower one's carbon footprint through energy conservation. (See **FIGURE 4.4**.)

A crucial aspect of any environmental policy concerns who should bear the costs of eliminating environmental hazards. As defined by the Organisation for Economic Co-operation and Development (OECD), the **polluter-pays principle** "means that the polluter should bear the expenses of carrying out the pollution

FIGURE 4.4 Sustainable living.

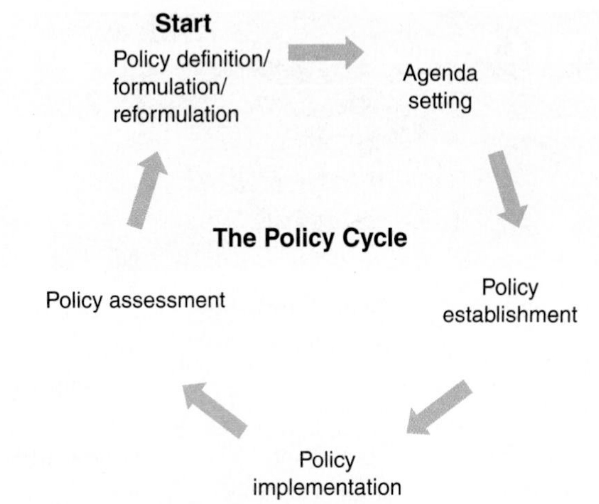

FIGURE 4.5 The policy cycle.

Modified from data presented in D@dalos, Policy Cycle: Teaching Politics. Available at: http://www.dadalos-d.org/politikdidaktik/politik/policy-zyklus.htm. Accessed February 14, 2017.

prevention and control measures introduced by public authorities in Member countries, to ensure that the environment is in an acceptable state."[14] The principle, implemented by the OECD in November 1974, applies to pollution that originates from a variety of sources including accidental releases during manufacturing processes and from installations that handle hazardous materials. The costs of controlling pollution generated during manufacturing and consumption are to be incorporated into the prices of associated goods and services.

The Policy Cycle

The **policy cycle** refers to the distinct phases involved in the policy-making process.[15] (Refer to **FIGURE 4.5**.) The policy cycle comprises several stages: (1) problem definition, formulation, and reformulation; (2) agenda setting; (3) policy establishment (i.e., adoption and legislation); (4) policy implementation; and (5) policy assessment.

The terms subsumed under the policy cycle are illustrated in **TABLE 4.1**.

Problem definition, *formulation*, and *reformulation* are terms that denote the processes of defining the problem for which the policy actors believe that policies are necessary. This early stage—problem definition and development of alternative solutions—often is regarded as the most crucial phase of the policy development process. The problems chosen should be significant for the health of the environment and have realistic and practical solutions. Poorly defined problems are unlikely to lead to successful policy implementation. Note that Figure 4.5 (the policy cycle) shows that following a process of assessment, problem definitions may need to be reformulated and the steps in the policy cycle repeated.

Agenda setting refers to setting priorities, deciding at what time to deal with the environmental problem, and determining who will deal with the problem. Policy makers need to establish priorities in order to reconcile budgetary constraints, resource restrictions, and the complexity of environmental problems against the need to develop those policies that are most feasible, realistic, and workable. Public support for protecting the environment is strong; opinion polls suggest that environmental policies are likely to gain acceptance in many sectors of society.

Barriers and difficulties in establishing priorities stem from the lack of information on risks[16] and lack of coordination among government agencies.[17] When the nature of the risks associated with an environmental hazard or toxin is uncertain, planners are left in a quandary about what aspects of the exposure require policy interventions. In addition, a criticism levied against the US Congress, which is a crucial policy-formulating body for the government of the United States, is its inability to set priorities due to fragmentation of authority among numerous committees and subcommittees that are involved with environmental policy.

A successful approach in developing priorities (and environmental policies in general) is to involve the community and stakeholders. This statement is especially true for the development of policies that impact arctic native territories and tribal lands. For such policies to be successful, planners should respect the cultural traditions of potentially affected indigenous peoples.[18] An illustration of an environmental action that could affect indigenous peoples occurred at the Standing Rock Sioux reservation in the United States. The Dakota Access Pipeline, a $3.8 billion project that would transport 500,000 barrels of oil per day from North Dakota to Illinois, would pass near the reservation. The tribe opposed construction of the pipeline because it could endanger water supplies and

TABLE 4.1 Components of the Environmental Policy Cycle

Component	Problem Definition, Formulation, and Reformulation	Agenda Setting	Policy Establishment	Policy Implementation	Assessment/ Evaluation
What happens?	Define problems and alternatives.	Set priorities. Involve stakeholders.	Formally adopt public policy. Legitimization.	Put the policy into practice.	Assess or evaluate effectiveness.
Who performs the function?	Formal and informal policy actors	Formal and informal policy actors	Formal decision makers	Government agencies	Arm of government responsible for assessment
What factors influence policy?	Research and science Interest groups Public opinion Social and economic factors	Research and science Interest groups Public opinion Social and economic factors	Research and science Interest groups Public opinion Social and economic factors	Research and science Interest groups Public opinion Social and economic factors	Research and science Interest groups Public opinion Social and economic factors
What problems are encountered?	Poorly defined problems	Lack of information on risk Lack of coordination	Inability to coordinate and assess research information	Lack of government support	Lack of sound scientific data

disrupt sacred sites on the reservation. The pipeline was first proposed in 2014.

Policy establishment involves the formal adoption of policies, programs, and procedures that are designed to protect the public from environmental hazards. A factor that impedes policy establishment is the unavailability of empirical information on the scope of risks associated with environmental hazards. According to Walker, "Limitations on our ability to coordinate, assess, and disseminate research information hampers efforts to translate policy into programs and services designed to reduce environmental risk."[16(p190)]

Policy implementation is the phase of the policy cycle that "focuses on achieving the objectives set forth in the policy decision."[16(p186)] Often this phase of the policy cycle is neglected in favor of the earlier phases of policy development. Barriers to policy implementation can arise from the government administration in power. In the case of the United States, whatever administration is in power may choose to weaken policy prescriptions due to political considerations. Organizations such as environmental groups, trade associations, and professional associations may stimulate public opinion and influence elected government officials with respect to environmental health policy.[19]

In order for a policy to be implemented successfully, policy developers may include economic incentives. For example, some policies for control of environmental pollutants use a market-based trading scheme.[20] This method assigns limits for pollution. Companies that fall below their assigned limits can accumulate credits for pollution control. These credits can be "banked" or resold to other companies that exceed their pollution limits.[21] The term *cap and trade* is used to describe a method for reducing greenhouse gas emissions from facilities such as electric generating plants.

In 2017, the state of California extended its unique cap-and-trade program, which aims to reduce greenhouse gases from facilities such as power plants and oil refineries. Each facility must obtain a permit for every metric ton of greenhouse gas that it emits. A certain number of these permits are available at no cost from the state. Other permits are available from state auctions. A facility that is successful in reducing its greenhouse gas emissions can sell its additional permits to other companies. The state proposes to use revenue generated by the cap-and-trade program to support projects (e.g., electric cars) for reducing greenhouse gases.

The political and social contexts may stimulate or impede the creation and implementation of environmental policy.[22] In many situations, public concern over a perceived hazard may provide the impetus to new environmental policies, regardless of whether scientific evidence supports their adoption. Research conducted in China has demonstrated the linkage among the social, environmental, and political elements as they affect policy goals and the implementation mechanisms for policy. (Refer to **FIGURE 4.6**.)

Assessment/evaluation, the final stage in the policy cycle, refers to assessment of the effectiveness of the policy. In order to facilitate assessment, environmental policies may incorporate **environmental objectives**, which "are statements of policy. . . intended to be assessed using information from a monitoring program. An environmental monitoring program has to be adequate in its quality and quantity of data so that environmental objectives can be assessed."[23(p144)] An example of an environmental objective is the statement that the amount of particulate matter in an urban area (e.g., Mexico City) will be reduced by 10% during the next 5 years.

Interplay between the Evaluation Process and Policy Development

Underlying the policy development cycle is environmental health research, which includes the identification of toxic substances and other hazards, assessment of mechanisms of environmental toxicity, and establishment of evaluations of interventions to mitigate hazards.[7] Some recent US governmental initiatives have emphasized the importance of research on pollutants as a specific component of environmental policies.[20]

Scientific data are linked closely to the policy cycle. "The creation of sound policy requires a foundation of sound data. It is through fundamental, mechanistically based research that environmental and public health officials can improve their understanding of risk and translate this knowledge into prevention strategies."[18(p118)] Epidemiologic studies provide the source of much valuable data for construction of environmental health policies. **FIGURE 4.7** demonstrates the interface between science and policy development.

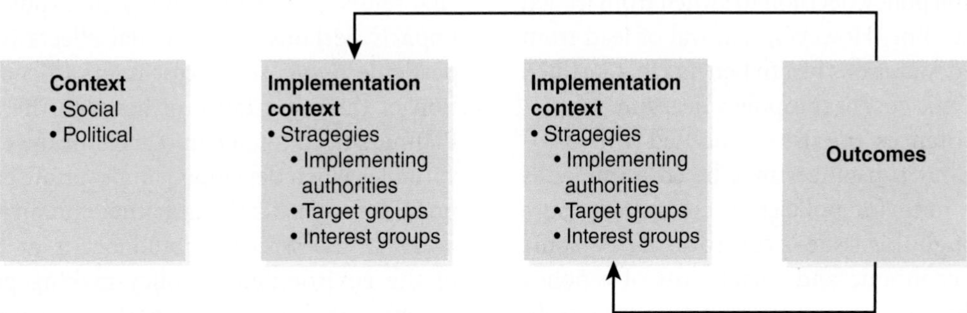

FIGURE 4.6 Relationships and interactions during policy implementation.

Data from Sinkule BJ, Ortolano L. *Implementing Environmental Policy in China*. Santa Barbara, CA: Praeger; 1995, p. 44.

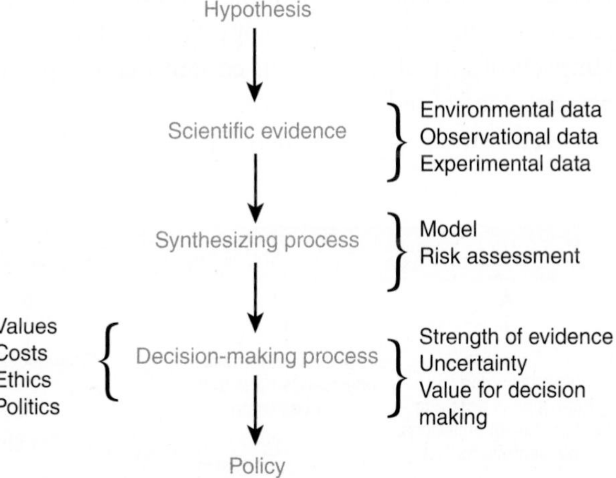

FIGURE 4.7 The interface of science and health policy.

Reproduced from Samet JM. Epidemiology and policy: the pump handle meets the new millennium. *Epidemiologic Reviews*. 2000;22(1):146, by permission of the Society for Epidemiologic Research.

The linkage begins with creation of hypotheses, moves to review of scientific evidence, and proceeds to development of models of exposures and health effects, and assessments of risks associated with exposures. The linkage concludes with the decision-making process in setting priorities and, ultimately, with formulation of specific policies.

Risk Assessment and Policy Development

In the chapter on environmental toxicology, risk assessment was defined as a process for determining health risks associated with environmental and other hazards. Risk assessment is closely aligned with the policy process through the balancing of economic and other costs with health and societal benefits that may accrue through specific policy alternatives. It should be obvious that the adoption of new environmental policies may involve the expenditure of considerable public and private funds, with potentially adverse economic consequences. For example, imagine the expense incurred when the United States and other countries implemented the policy decision to switch from leaded to unleaded gasoline. However, removal of lead from gasoline carried numerous health benefits that justified the high economic cost of this policy decision.

In some instances, the risks associated with exposure to hazardous substances may be so minimal as to obviate the need for policies to control exposure to them, although this issue is controversial. At some point the net economic and social costs of a policy may exceed the social benefits of pollution controls. An example is the presence of trace levels of pollutants in drinking water that has already undergone purification.

Policy adoption is difficult when an environmental issue is complex. An example comes from policies to control the potential environmental impacts of agricultural pesticide use.[24] Policy developers are confronted with several complex and often interrelated issues:

1. Pesticide use involves multiple contaminants that have differing characteristics.
2. Exposure of the population (both the general public and agricultural workers) to contaminants can arise from unspecified sources as well as clearly delineated sources; sometimes it is difficult to identify the sources of pesticide emissions.
3. The costs incurred to measure pesticide emissions are high.
4. There may be interactions among different types of pesticides.

The complicated nature of measuring pesticide exposures muddies the development of policies to limit the population's exposure to pesticides and to promote the appropriate uses of pesticides.

FIGURE 4.8 shows both a model of the four major factors connected with policy development and the interrelationships among them. The four components are hazard, risk, impacts, and social costs.[24] An example of their application is a scenario for the development of policies regarding the use of pesticides. The term *hazard* relates to chemical and physical properties (e.g., color, boiling point, volatility) of pesticides; *risk* refers to the probability of exposure; the term *impacts* pertains to the actual effects (e.g., harm) of pesticide usage; *social cost* refers to society's perception of the importance of harm to the environment. Planners need to take into account the entire panel of variables when developing appropriate environmental policies for pesticides and other chemicals.

Risk assessment should occur at the inception of the environmental policy-making process.[25] Risk assessment should be "a participatory procedure, in which the different stake-holders are involved early in the risk analysis process to 'characterize' risks, even before they are given a formal assessment."[10(px)] An example would be incorporating community-based participatory methods designed to assess biomarkers of contaminant exposures in children and adults. A biomarker is "[a] specific biochemical in the body which has a particular molecular feature that makes

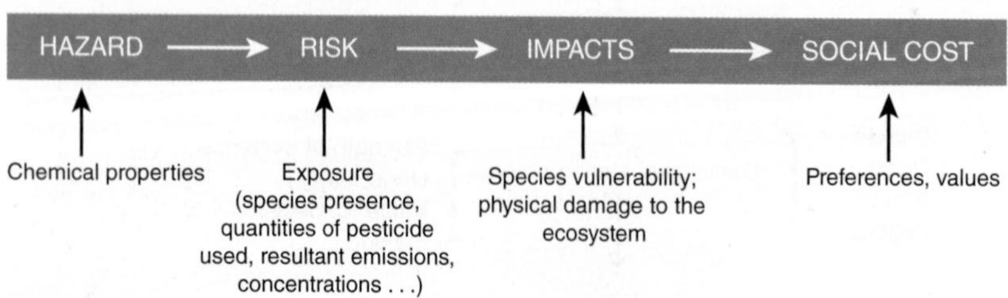

FIGURE 4.8 The links between hazard, risk, impacts, and social cost.

Reproduced from *Journal of Environmental Management*, vol. 65, Falconer K, Pesticide environmental indicators and environmental policy, p. 288, Copyright 2002, with permission from Elsevier.

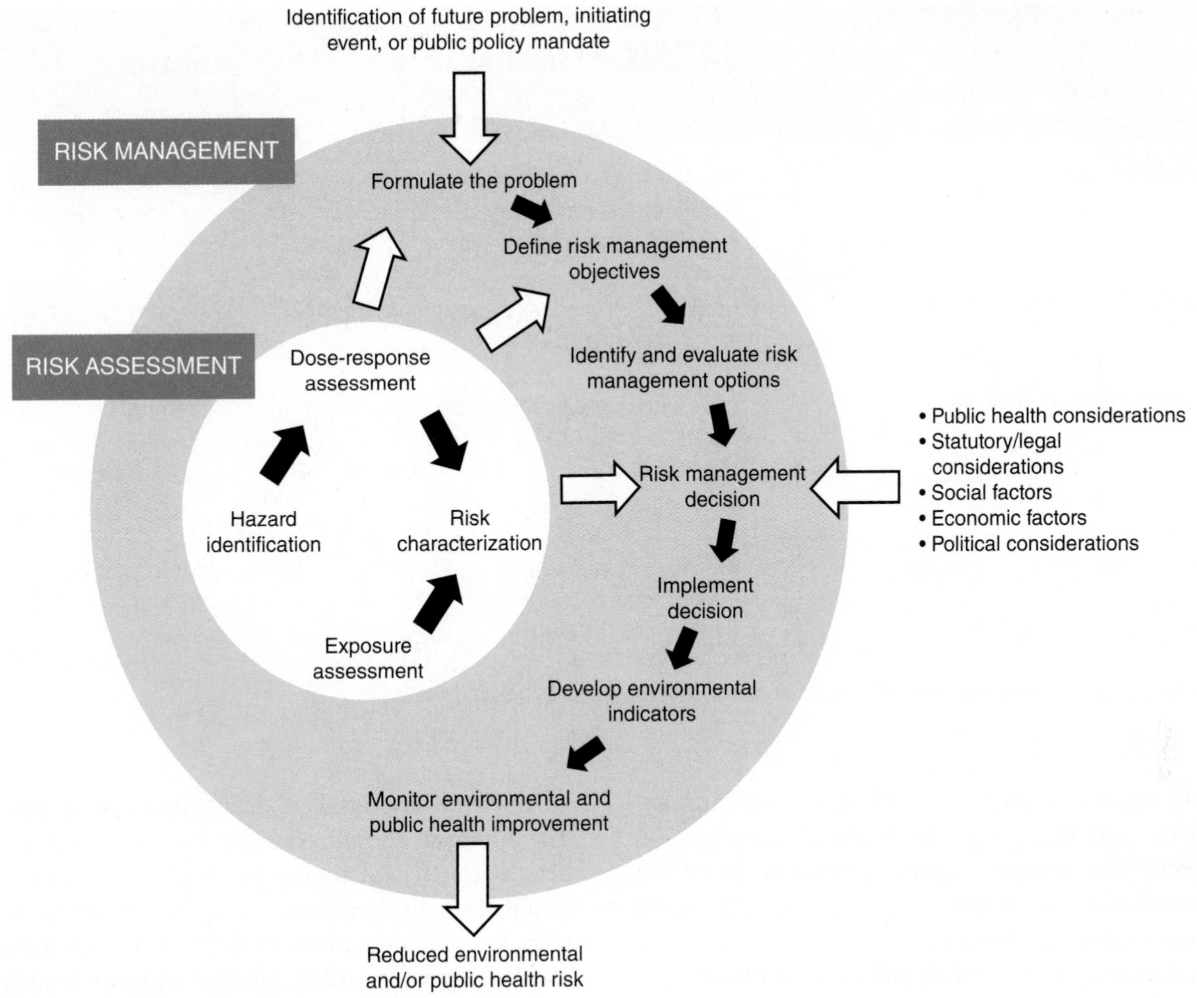

FIGURE 4.9 Risk assessment/risk management paradigm.

Reproduced with permission from National Research Council. *Risk Assessment in the Federal Government: Managing the Process, 1983* by the National Academy of Sciences. Courtesy of the National Academies Press, Washington, D.C.

it useful for measuring the progress of disease or the effects of treatment."[26] The participatory process aims "at eliciting the 'values' and the perspectives of the community involved so that the multiple dimensions of risk can be taken into account early on in the assessment."[10(pxi)]

Risk Management

The process of risk management involves the adoption of steps to eliminate identified risks or lower them to acceptable levels (often as determined by a government agency that has taken into account input from the public).[25] After a risk assessment has been completed, public policy mandates may be adopted for the management of the identifiable risk. Suppose, for example, that a risk assessment shows that exposure to lead is harmful to children. A government agency has adopted a policy that states that children shall be protected from environmental exposures to lead. The risk management plan involves the adoption and implementation of procedures that are designed to protect children from lead exposure. Refer to **FIGURE 4.9** for more information.

Other examples of risk management include the development of regulations that control the production, use, and disposal of toxic chemicals. In the United States, risk management of toxic substances is governed by the Toxic Substances Control Act (discussed later in this chapter). Methods of risk management include licensing laws, standard-setting laws, control-oriented measures, and monitoring.[25] (Refer to **TABLE 4.2**.)

Environmental Impact Assessment and Health Impact Assessment

An *environmental impact* is defined as "[a]ny change to the environment, whether adverse or beneficial, wholly or partially resulting from an organization's activities, products or services."[5] An **environmental impact assessment (EIA)** is a process that reviews

TABLE 4.2 Examples of Risk Management

Risk Management Procedure	Description
Licensing laws	Require licensing and registration for new and existing chemicals; include requirements for toxicity testing, including the Federal Insecticide, Fungicide, and Rodenticide Act (FIFRA).
Standard-setting laws	Establish standards of exposure for chemicals used in specific situations; an example is the Clean Air Act.
Control-oriented measures	Deal with explicitly identified chemicals, groups of chemicals, or chemical processes; an example is the design of packages so that they are childproof and prevent young children's access to harmful substances.
Monitoring	Measures of the level of an environmental toxin so that regulations can be enforced. Monitoring programs are in place for ozone and smog and for pesticide levels in foods, to name several examples.

Data from Landrigan PJ, Carlson JE. Environmental policy and children's health. *The Future of Children*. 1995;5(2):41–43.

the potential impact of anthropogenic (human-related) activities (e.g., a proposed construction project) with respect to their general environmental consequences. To give an example, an EIA might consider whether a proposed residential development would increase water and air pollution beyond acceptable limits or endanger threatened species of animals and plants. An EIA seeks abatement measures for pollution before it is produced; procedures for control of pollution should be feasible and cost effective.[22] An example of an abatement procedure is the construction of catch basins for containment of water runoff from new housing developments in order to prevent polluted water from reaching the ocean.

Although it is similar to an environmental impact assessment, a **health impact assessment (HIA)** refers to "a method for describing and estimating the effects that a proposed project or policy may have on the health of a population."[27(p198)] Refer to **TABLE 4.3** for a list of the types of projects that may require an HIA. Major construction projects can have a damaging effect upon the health of human populations and, accordingly, should require an HIA. An example would be the construction of a giant water reservoir in a densely populated, tropical country that has high levels of tropical diseases. The large body of standing water created by the reservoir could encourage the multiplication of disease-causing vectors such as mosquitoes.

Six stages have been identified for the completion of an HIA.[28] Two of these stages include review of research evidence and evaluation. An HIA makes an in-depth appraisal of data relevant to a project. The appraisal should consider how a project affects health-related and sociodemographic variables (e.g., population displacement). The HIA incorporates a plan for the meticulous evaluation and continuous monitoring of possible adverse outcomes such as increases in poverty, undue damage to the environment, detriments to the social fabric, and increases in morbidity from infectious and other diseases.[27]

Case Studies: Environmental Policies to Protect Human Health

The 2014–2018 EPA Strategic Plan

The mission of the US EPA is "To Protect Human Health and the Environment."[29] In order to support its mission, the EPA has developed a strategic plan that, according to the requirements of the Government Performance and Results Act of 1993, must be updated periodically. The FY 2014-2018 strategic goals and priorities include:

- "Addressing climate change and improving air quality;
- Reinvigorating water-quality-improvement efforts, including support for green infrastructure;
- Taking action on toxics and strengthening chemical safety;

TABLE 4.3 Applications of Health Impact Assessment

Activity	Examples
"Traditional" project EIA	Large dams, mines, power plants, airports
Regional or local project EIA	Development corridors, urban redevelopment
"Cumulative" environmental assessment	Air/watershed management
"Sectoral" environmental assessment	Energy/water sector planning
"Policy" environmental assessment	Infrastructure planning
Economic policy/ structural adjustment	Country-level impacts/ mitigation
International treaties	Climate/biodiversity
Special cases	Private sector/budget planning

Reproduced from Lerer LB. How to do (or not to do) . . . health impact assessment. *Health Policy and Planning*. 1999;14(2):199, by permission of Oxford University Press.

- Enhancing the livability and economic vitality of neighborhoods in and around brownfield sites;
- Aligning and incentivizing partnerships that spur technological innovations, reducing costs and pollution; and
- Advancing research efforts to provide relevant, robust and transparent scientific data to support the agency's policy and decision-making needs." [29(p1)]

Protection of the Arctic and Antarctic Environments

Significant contamination of the arctic environment with organic materials, toxic metals, and radioactivity has occurred as a result of pollution that originates in the industrialized areas of the world. The list of contaminants found in the traditional foods used by indigenous arctic peoples (and in their tissue samples) is extensive. (See **TABLE 4.4**; refer to Chapters 6 through 8 for definitions of the terms used in the table.)

In view of the future potential health impacts of toxic substances in the arctic environment, policies will need to be developed to protect this region. The use of a community advisory board working in conjunction with environmental scientists to protect the arctic environment has been recommended. The Arctic Monitoring and Assessment Program, with the input of community advisory boards, will aid in the development of appropriate policies. The Antarctic environment is also at risk and requires protection from pollution generated by military activities, tourism, and scientific exploration.[30]

Water Policy Reform in South Africa

South Africa has developed new policies—the National Water Act (1998) and the Water Services Act (1997)—to increase the equitability of water distribution and to protect aquatic ecosystems in the country.[31] Historically, before the introduction of these acts, water rights were held almost exclusively by large landowners. The purposes of the acts are to bring water rights under the control of the South African central government, to ensure that all citizens have sufficient water supplies to meet their basic needs, to make water affordable, to introduce controls over pollution, and to make sufficient water available for the maintenance of the aquatic environment.

Environmental Policies in Economies in Transition

In the Baltic and nearby regions of Central and Eastern Europe, former Soviet economies have changed to market economies as a result of the collapse of the Soviet Union. (Estonia, Poland, Hungary, and the Czech Republic are examples.) Previously, emphasis in these economies was placed on heavy industries that contributed greatly to the burden of pollution. Environmental protections were not enforced vigorously, resulting in heavy pollution of some regions. These economies inherited barriers to implementation of

TABLE 4.4 Environmental Contaminants Detected in Traditional Food Items and Human Tissue Samples

Industrial chemicals and by-products (e.g., dioxins, polychlorinated biphenyls, flame retardants)
Pesticides (e.g., DDT)
Polycyclic aromatic hydrocarbons (e.g., benzo(a)pyrene)
Heavy metals (e.g., mercury, lead)
Products of nuclear radiation

Modified and reproduced with permission from Suk WA, Avakian MD, Carpenter D, et al. Human exposure monitoring and evaluation in the Arctic: the importance of understanding exposures to the development of public health policy. *Environmental Health Perspectives*. 2004;112:115.

environmental policies that included economic and governmental inefficiencies and absence of environmental policy principles. New and effective environmental policies have been implemented that gradually have brought pollution under control. These policies include adoption of the polluter-pays principle and reduction in subsidies to inefficient heavy industries that cause pollution. Consequently, many of these countries have had substantial remediation of stationary sources of pollution such as those related to the coal mining industry.[32]

Control of Pollution across International Boundaries

Environmental impacts caused by the release of greenhouse gases such as carbon dioxide are global in scope. Control of such problems may not necessarily be confined to a single country but may involve actions at the international level. The New York Convention (1992) and Kyoto Protocol (1997) are two multilateral agreements that set forth international policies to reduce the emission of so-called greenhouse gases into the atmosphere. Other examples of environmental problems that have an international scope are climate change, ozone layer depletion, loss of **biodiversity**, and radioactive emergencies. Thus, international cooperation should be a feature of a single country's environmental policy. For example, three Baltic countries—Estonia, Latvia, and Lithuania—have designed multinational environmental agreements to control pollution in this region.[33]

Industrialization of Rural China

The shift from an agricultural economy to rapid industrialization, with consequential environmental quality degradation, is an increasingly common feature of rural China. Some of the causes of pollution are use of low-quality coal, energy inefficiencies, and lack of wastewater treatment facilities. Although strong environmental policies exist in China, numerous social, political, and economic barriers (e.g., insufficient funding) restrict their implementation or prevent their enforcement. Awareness is increasing among the general public regarding the adverse consequences of pollution. Nevertheless, environmental degradation remains a compelling issue in China.[22]

Protecting the Rights of Children and Special and Vulnerable Populations

Those who may be especially sensitive to environmental hazards include children, persons with genetic vulnerability, and minority groups such as blacks and American Indians. Although efforts are under way to intervene in childhood asthma, birth defects, and lead poisoning, increased attention needs to be given to other environmental issues such as exposure to carcinogens, neurotoxins, and endocrine disruptors.[34] Despite this need, according to some experts, the United States lacks a coherent policy to assure that the environment is free from children's exposure to environmental toxicants. Blacks in many cases reside in urban centers where they are exposed to high levels of air pollution and toxic metals such as lead. Many American Indians live on sovereign lands that have serious environmental problems. American Indian groups sometimes face difficulty in reaching consensus about the appropriate action steps needed to reduce environmental hazards on their lands.

Landrigan and Carlson state that "[u]nderstanding the differences in the effects of environmental contamination on children and adults is an important part of environmental policymaking; however, unless environmental health policies reflect the differences between adults and children, this knowledge will have little practical effect."[25(p34)] Children's vulnerability to environmental toxins is related to their reduced ability to metabolize and excrete toxicants, rapid growth and development that increase vulnerability to toxicants, and the fact that children have more time to develop chronic diseases than do adults.[35]

Nowadays children are exposed to thousands of new chemicals. As a result of medical advances, infectious diseases are taking a reduced toll in the developed world. Consequently, chronic diseases of noninfectious origin are replacing infectious diseases as major causes of death among children in the developed world. (This shift is known as the epidemiologic transition.) Long-term exposure to environmental toxicants may play a role in causing some of these chronic conditions. Note that asthma and lead poisoning are increasingly important examples of the new morbidity among children. Some researchers believe that asthma may be associated with factors such as outdoor and indoor air pollution and insect contamination; lead poisoning is associated with exposure to environmental lead from painted surfaces in older buildings and from leaded fuels (which presently are banned from use in the United States and most developed countries).

One method for controlling and intervening in children's exposures to environmental hazards is known as the lifecycle approach to identifying exposure pathways.[34] (See **FIGURE 4.10**.) Environmental policies can be directed toward the various lifestyle exposure pathways shown in the figure (e.g., prevention of children's oral, dermal, and inhalation exposures to toxicants in the home and outdoor environments).

Prenatal

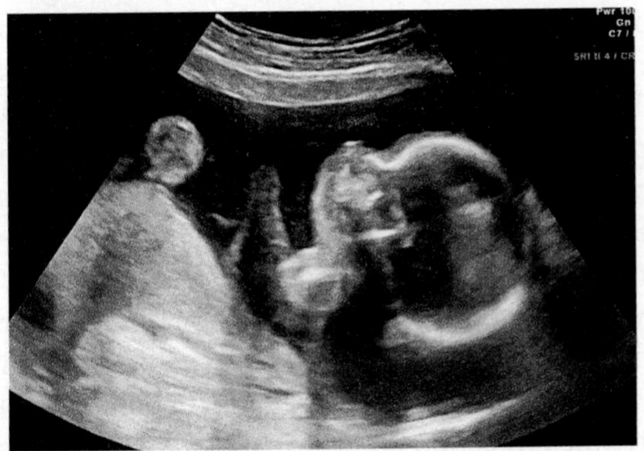

© PEDRE/E+/Getty

A stage when the developing fetus is very susceptible to the effects of pollutants; exposures can be transplacental and external.

Neonatal (birth to < 3 months)

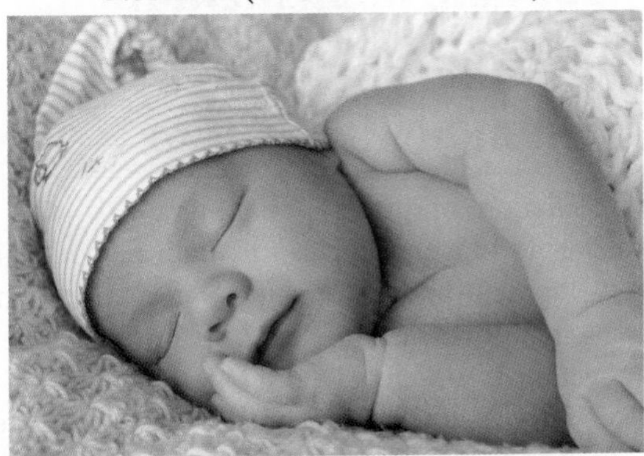

© Olga Max/Shutterstock

Oral and dermal exposures: breast and bottle feeding; hand-to-mouth activity.
Inhalation exposures: time spent sleeping; time spent in sedentary activities.

Infant/Crawler (3 to 12 months)

© Konstantin Chagin/Shutterstock

Oral and dermal exposures: consumption of solid food, increased floor mobility, hand-to-mouth activity.
Inhalation exposures: breathing close to floor; development of personal dust clouds.

Toddler (1 to < 2 years)

© NARONGRIT LOKOOLPRAKIT/Shutterstock

Oral and dermal exposures: consumption of range of solid foods; increased play activity and curiosity.
Inhalation exposure: time spent sleeping; time spent in sedentary activities.

Preschool (2 to < 6 years)

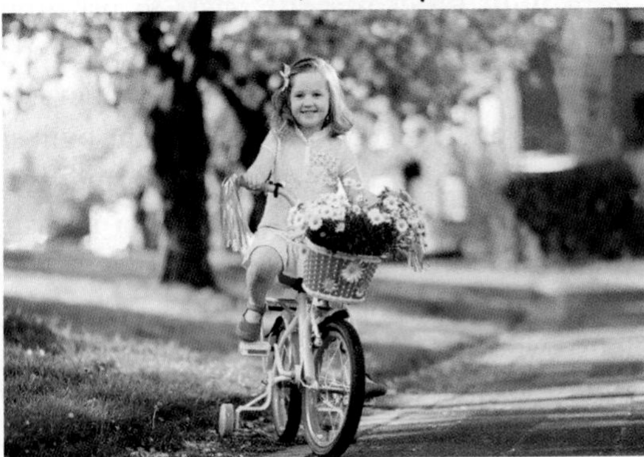

© FamVeld/Shutterstock

Oral and dermal exposures: wearing adult-style clothing; decreased hand-to-mouth activities.
Inhalation exposures: increased time spent outdoors.

School-Age (6 to < 11 years)

© Knot. P. Saengma/Shutterstock

Oral and dermal exposures: decreased oral contact with hands and objects; decreased contact with surfaces.
Inhalation exposures: time spent in school environments; participation in sports.

FIGURE 4.10 Pathways of children's exposure to environmental pollutants and hazards vary according to stage of development.

Data from United States Environmental Protection Agency. *A Decade of Children's Environmental Health Research: Highlights from EPA's Science to Achieve Results Program*. EPA/600/S-07/038. December 2007.

Turning from the developed world, consider the less-developed regions such as parts of Southeast Asia and the Pacific, where numerous environmental threats confront children. These regions comprise the most rapidly industrializing areas of the world and contain half the world's children.[35] The quantity and severity of threats are increasing due to rapid industrialization and the epidemiologic transition. Examples of environmental hazards are smoke from wood used in home cooking and heating, arsenic in ground water, pesticides, lead emissions, and methyl isocyanate.[35] (Methyl isocyanate is a highly poisonous chemical used in the manufacture of pesticides. On December 3, 1984, 40 tons (36.3 metric tons) of the gas was released in Bhopal, India, causing many fatalities.)

The Built Environment

The term **built environment** refers to urban areas and structures (e.g., roads, parks, and buildings) constructed by human beings, as opposed to undeveloped, rural areas. From the world perspective, the built environment is increasing dramatically as formerly rural and agricultural economies transition to industrialized economies. **FIGURE 4.11** shows a highly urbanized section of Caracas, Venezuela in 2006. Regarding the developed world, for example, in many sections of the United States, cities have taken over farmland and forests as metropolitan areas have expanded.

Policies for design of the built environment have great potential for influencing public health.[36] Examples of design features with implications for health consequences are land development, community design, and transportation patterns. Some land development methods inadvertently cause people to drive cars by increasing the distances that people must travel. Encouraging the use of cars by making them affordable and installing free parking lots may have the indirect consequences of increasing air pollution and a sedentary lifestyle. Adoption of innovative public policies may stimulate people to walk more and use public transportation. An example is the colocating of business facilities, shopping centers, and residences so that city inhabitants are able to walk to work, ride bicycles, or take public transportation. Other innovations include available bicycles and vehicles for sharing by the public. Finally, development policies should foster the protection of open space and the creation of public parks.

FIGURE 4.11 Built environment example: Caracas, Venezuela; 2006.

Agencies Involved in the Adoption, Implementation, and Enforcement of Environmental Policies

A constellation of international, national, state or territorial, and local agencies maintain responsibility for development and enforcement of environmental health regulations, and for investigation of incidents that affect environmental health.

FIGURE 4.12 shows a flow chart of some of the agencies involved with environmental health regulations.

Referring to Figure 4.12, you can see that the World Health Organization (WHO) is a major international agency that is responsible for environmental health at the global level. WHO provides leadership in minimizing adverse environmental health outcomes associated with pollution, industrial development, and related issues. Although WHO's primary mission is to control and prevent disease, its reach extends to environmental health, which is closely related to disease prevention. More information about WHO is presented later in this chapter.

At the national level, many countries have a federalist system of government. The concept of federalism implies a type of government "that is structured around a strong central (i.e., federal) government, with specified authorities retained by lower levels of government, such as states and local governments."[19(p34)] This situation characterizes the United States, where the legislative and judicial branches of government perform important functions in environmental policy formulation. The Congress of the US government (Senate and House of Representatives) retains significant responsibility for creation of environmental laws. Judicial bodies in the United States (i.e., the court system) support interpretation, strengthening, and

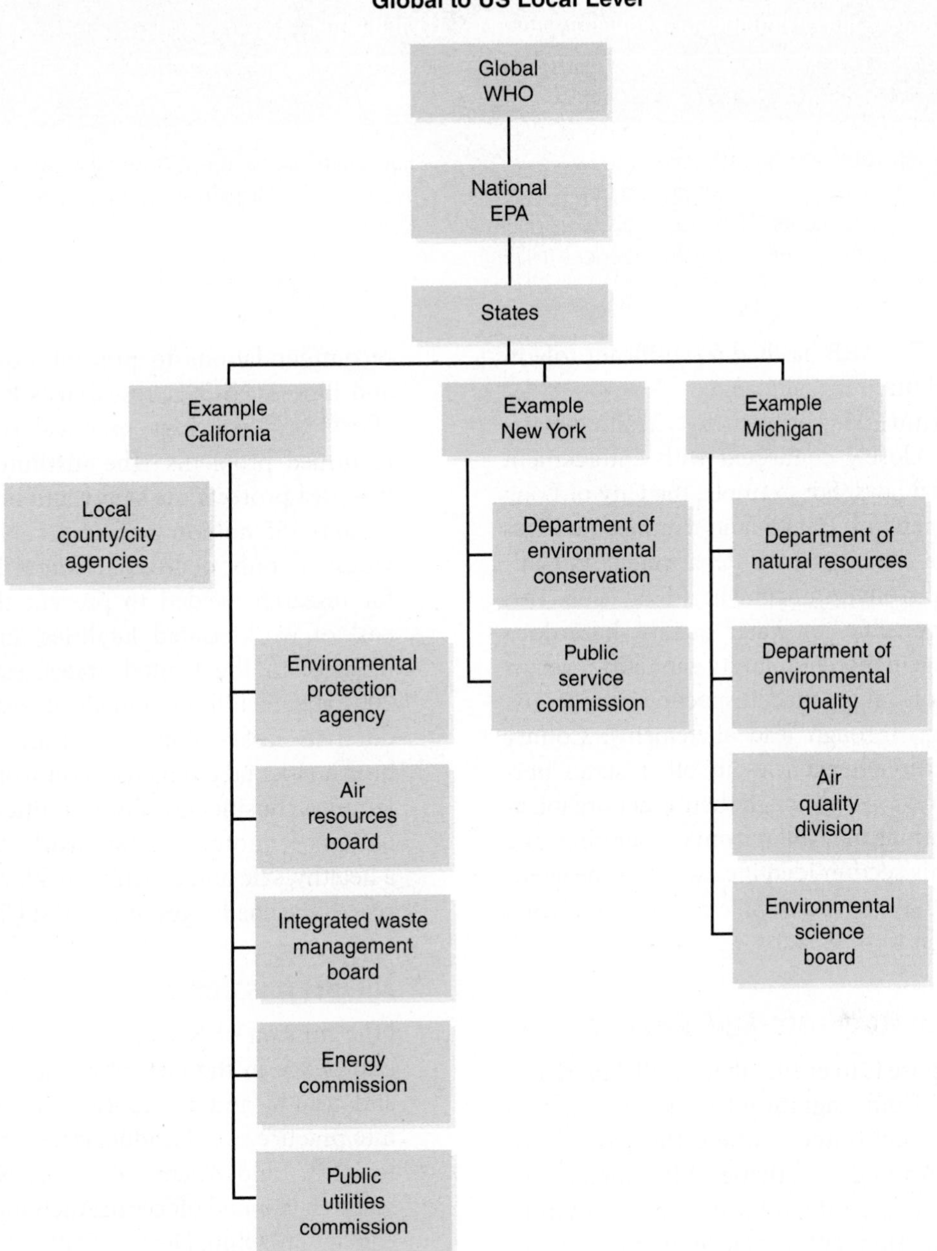

FIGURE 4.12 Overview of environmental health regulation agencies.

enforcement of environmental laws; an example is the interpretation of relevant environmental policies that are the focus of lawsuits.

At the national level, many government agencies regulate environmental health hazards in their respective countries. For example, in the United States, the EPA is responsible for protecting human health and safeguarding the natural environment, including air, land, and water. The EPA develops and enforces environmental regulations, which bring about cleaner air and purer water as well as protect the land. In addition, other national agencies (discussed later in this chapter) are charged with the responsibility of protecting the environment.

Within the United States, individual states and local governments also maintain responsibility for control of environmental health issues. Figure 4.12 shows some examples of agencies involved in environmental health regulation in three representative states—California, New York, and Michigan. As an example of one state agency's function, California's Air Resources Board (ARB) states as its mission: "To promote and protect public health, welfare and ecological resources through the effective and efficient reduction of air pollutants while recognizing and considering the effects on the economy of the state."[37] In large urban areas such as California's major cities, air pollution has continued to be a significant health and

TABLE 4.5 Some Major Environmental Regulatory Agencies: US and Other

United States	Other
US Environmental Protection Agency (EPA) National Institute for Occupational Safety and Health (NIOSH) Agency for Toxic Substances and Disease Registry (ATSDR) National Institute of Environmental Health Sciences (NIEHS)	European Environment Agency (EEA) World Health Organization (WHO)/Regional Office for Europe

aesthetic issue. The ARB has had a significant role in reducing air pollution in California.

Local government agencies such as health departments also are closely connected with enforcement of environmental laws. For example, the City of Long Beach Department of Health and Human Services (DHHS), one of the city health departments in California, operates an environmental health division. This division is responsible for water quality, hazardous materials management, community sanitation, vector control, lead poisoning, food inspection, and housing quality (e.g., through lead abatement). County and local government agencies in other states perform similar functions, although their exact organizational configurations may differ from one another. The remainder of this section identifies some of the agencies in the US that are responsible for environmental regulation. (Refer to **TABLE 4.5**.)

US Environmental Protection Agency (EPA)

"The EPA's purpose is to ensure that . . . all Americans are protected from significant risks to human health and the environment where they live, learn and work[.]"[38] Among its activities, EPA implements the best available scientific information for environmental risk reduction, enforces federal environmental laws fairly and effectively, supports dissemination of accurate information regarding human health and environmental risks, and maintains a leadership role with respect to protecting the global environment. Other functions of the EPA are to:

- "Give Grants . . .
- Study Environmental Issues . . .
- Sponsor Partnerships . . .
- Teach People About the Environment . . .
- Publish Information . . ."[38]

National Institute for Occupational Safety and Health (NIOSH)

> The National Institute for Occupational Safety and Health (NIOSH) is the U.S. federal agency that conducts research and makes recommendations to prevent worker injury and illness. NIOSH research is key to these efforts and provides practical solutions to identified problems. The Institute's work in this area protects the safety and health of the nation's 155 million workers. . . . NIOSH provides the only dedicated federal investment for research needed to prevent the societal cost of work-related fatalities, injuries and illnesses in the United States, estimated in 2007 at $250 billion in medical costs and productivity losses alone. . . . These safety and health risks take huge tolls on workers, their families, businesses, communities, and the nation's economy; NIOSH works to promote a healthy, safe and capable workforce that can rise to the challenges of the 21st Century. . . .

NIOSH Mission

[The mission of NIOSH is] [t]o develop new knowledge in the field of occupational safety and health, and to transfer that knowledge into practice. . . . . Headquartered in Washington, D.C. and Atlanta, Georgia, NIOSH has research labs and offices in Anchorage, Alaska; Cincinnati, Ohio; Denver, Colorado; Morgantown, West Virginia; Pittsburgh, Pennsylvania; and Spokane, Washington.

About NIOSH

The Occupational Safety and Health Act of 1970 established NIOSH. NIOSH is part of the U.S. Centers for Disease Control and Prevention, in the U.S. Department of Health and Human Services. It has the mandate to assure "every man and woman in the Nation safe and healthful working conditions and to preserve our human resources." NIOSH has more than 1,300 employees from a diverse set of fields including epidemiology, medicine, nursing, industrial hygiene, safety, psychology, chemistry, statistics, economics, and many branches of engineering. NIOSH works closely with the Occupational

Safety and Health Administration (OSHA) and the Mine Safety and Health Administration (MSHA) in the U.S. Department of Labor to protect American workers and miners.[39]

Agency for Toxic Substances and Disease Registry (ATSDR)

The Agency for Toxic Substances and Disease Registry (ATSDR) is the nation's public health agency for chemical safety. The agency's mission is to use the best science, take responsive action, and provide trustworthy health information to prevent and mitigate harmful exposures [to] toxic substances and related disease.

The discovery of contamination in New York State's Love Canal during the 1970s first brought the problem of hazardous wastes to national attention. Similarly, the health threat from sudden chemical releases came into focus in December 1984, when a cloud of Methyl isocyanate gas released from a Union Carbide facility in Bhopal, India, seriously injured or killed thousands of people.

Both events represent the kinds of issues at the core of ATSDR's congressional mandate. First organized in 1985, ATSDR was created by the Comprehensive Environmental Response, Compensation, and Liability Act (CERCLA) of 1980, more commonly known as the Superfund law. In 1986, Congress passed the Superfund Amendments and Reauthorization Act (SARA). Through these and other pieces of legislation, Congress responded to the public's demand for a more complete accounting of toxic chemicals and releases. In addition, Congress was—and remains—concerned by other pathways of potential exposure, including food, water, air, and consumer goods.

Since the creation of ATSDR, thousands of hazardous sites have been identified around the country. The Superfund program remains responsible for finding and cleaning up the most dangerous hazardous waste sites in the country. ATSDR has also been at the forefront in protecting people from acute toxic exposures that occur from hazardous leaks and spills, environment-related poisonings, and natural and terrorism-related disasters.

Under its CERCLA mandate, ATSDR's work falls into four functional areas:

- Protecting the public from toxic exposures;
- Increasing knowledge about toxic substances;
- Delivering health education about toxic chemicals; and
- Maintaining health registries.

Through our work in these areas, ATSDR continues to prevent and mitigate exposures and related health effects at hazardous waste sites across the nation.[40(p9)]

National Institute of Environmental Health Sciences (NIEHS)

The National Institute of Environmental Health Sciences (NIEHS) is one of 27 institutes and centers of the National Institutes of Health (NIH), which is a component of the U.S. Department of Health and Human Services (HHS). NIEHS is located in Research Triangle Park (RTP), North Carolina. NIEHS's broad focus on the environmental causes of disease makes the institute a unique part of the NIH. NIEHS is home to the National Toxicology Program (NTP), the nation's premier program for testing and evaluation of agents in our environment. . . .

NIEHS supports a wide variety of research programs directed toward preventing health problems caused by our environment.

Grants programs. The largest portion of the NIEHS budget goes to fund laboratory research, population-based studies, and training programs that are conducted at universities, hospitals, businesses, and organizations around the country and in other lands. This research is being supported through the NIH grants program — also known as the extramural program.

In-house laboratories. In-house, or intramural research, is done by scientists employed by the federal government who have laboratories at NIEHS. Research conducted at NIEHS includes epidemiology, biostatistics, molecular genetics, signal transduction, reproductive and developmental toxicology, respiratory biology, molecular carcinogenesis, and other environmental research areas. Our in-house scientists collaborate extensively with partners in other institutes, agencies, and academia. NIEHS also has a clinical research unit on site, to help translate basic research findings into human health gains.

National Toxicology Program. NTP is a federal, interagency program, headquartered at NIEHS, whose goal is to safeguard the public, by identifying substances in the environment that may affect human health. Current NTP initiatives are examining the effects of cell phone radiation, endocrine disruptors, and nanomaterials, as well as developing new approaches to advance high-throughput (high speed and high quantity) screening of chemicals, to reduce the number of animals used in research. NTP also conducts health hazard evaluations, such as the Report on Carcinogens, that are used by federal, state, and local health regulatory research agencies for decision-making.[41]

European Union (EU) and European Environment Agency (EEA)

The European Union (EU)'s environment program supports the development of environmental policies for protecting the health and quality of life for its residents of its vast geographic area. The European Environment Agency (EEA), which is located in Copenhagen, Denmark, is an arm of the European Union. The EEA is tasked with the mission of supplying information to 33 member countries regarding the environment. This information is used for environmental policy development, adoption, implementation, and evaluation. For more information about the EEA, consult https://www.eea.europa.eu/downloads/c10100ed9b3ec24746afc41b75a1c277/1502186235/who-we-are.pdf?direct=1.

Environmental Action Programs (EAPs) help to direct EU environmental policy and establish priority objectives. In 2013 the European Parliament and the Council of the European Union adopted the seventh Environmental Action Program, which extended until 2020. The nine priority objectives of the EAP are as follows:

1. "to protect, conserve and enhance the Union's natural capital.
2. to turn the Union into a resource-efficient, green, and competitive low-carbon economy.
3. to safeguard the Union's citizens from environment-related pressures and risks to health and wellbeing.
4. to maximise the benefits of the Union's environment legislation by improving implementation.
5. to increase knowledge about the environment and widen the evidence base for policy.
6. to secure investment for environment and climate policy and account for the environmental costs of any societal activities.
7. to better integrate environmental concerns into other policy areas and ensure coherence when creating new policy.
8. to make the Union's cities more sustainable.
9. to help the Union address international environmental and climate challenges more effectively."[42]

World Health Organization (WHO)/ Regional Office for Europe

WHO attributes up to 25% of the global burden of disease to environmental factors; according to WHO, a total of 20% of all deaths result from environmental factors (2007 data). Environmental determinants are manifested differentially across regions, demographic groups, and socioeconomic groups in a manner that demonstrates complex interactions with the environment.[43] The following passage describes WHO's Regional Office for Europe:

> The WHO Regional Office for Europe supports the efforts of its Member States and partners to understand and navigate this complexity and to identify policies and actions that can benefit the environment and human health, using the best available evidence to guide national and international decision-making in different sectors….In 1989, concerned about the growing evidence of the impact of hazardous environments on human health, the WHO Regional Office for Europe initiated the first-ever international environment and health process, developing a broad-based primary prevention public health approach for addressing environmental determinants of health.
>
> The European Environment and Health Process (EHP) is steered by ministerial conferences that are unique for bringing together different sectors and stakeholders to identify environment and health challenges, set priorities, agree on commitments and shape shared European policies and actions on environment and health.
>
> Representatives of WHO European Member States, the Commission of the European Union and their partners adopted the European Charter on Environment and Health at the first Ministerial Conference in Frankfurt in 1989, committing to basic principles, mechanisms and priorities for future action….The conference also called on WHO to establish the European Centre for Environment and Health, which remains the key institution of the EHP to this day.[43]

Beyond the Regional Office for Europe, WHO's global environmental health activities cover a broad range of issues in many countries. These public health and environmental topics include:

- Air pollution, including indoor air pollution
- Children's environmental health
- Environmental health impact assessment
- Climate change and human health
- Environmental health in emergencies
- Quantifying environmental health impacts
- Water, sanitation, and health

For more information, refer to the WHO website: http://www.who.int.

Environmental Advocacy Organizations

Although they are not components of government, environmental advocacy organizations constitute an important part of the fabric of environmental policy formulation and are a significant force in influencing public policy. Environmental advocacy organizations help to educate people and seek to mold public opinion regarding the organizations' positions on specific topics, including global warming, reduction of air pollution, and protection of consumers from potential carcinogens in food, water, and cosmetics. One of the recent concerns of environmental advocacy groups is the development of alternative and green energy sources. Refer to **FIGURE 4.13**, which shows the use of wind turbines for generation of electric power.

Many organizations lobby local, state, and federal governmental bodies in support of particular environmental causes such as protection of endangered species and purchase of wilderness areas for public access. In the United States and worldwide, numerous organizations function at various levels (from grassroots to national and international) to promote environmental causes. Some examples of the multitude of environmental advocacy groups are the following:

- Environmental Working Group
- Greenpeace
- National Resources Defense Council
- National Wildlife Federation
- The Nature Conservancy
- Sierra Club
- Union of Concerned Scientists

FIGURE 4.13 Wind turbine for generation of electric power.

Major US Environmental Health Laws

Several of the major environmental health laws adopted by the United States are presented in **TABLE 4.6**.

Clean Air Act (42 U.S.C. § 7401 et seq. [1970])

The Clean Air Act (CAA) is the comprehensive federal law that regulates air emissions from stationary and mobile sources. Among other things, this law authorizes EPA to establish National Ambient Air Quality Standards (NAAQS) to protect public health and public welfare and to regulate emissions of hazardous air pollutants.

One of the goals of the Act was to set and achieve NAAQS in every state by 1975 in order to address the public health and welfare risks posed by certain widespread air pollutants. The setting of these pollutant standards was coupled with directing the states to develop

TABLE 4.6 Major US Environmental Health Laws

Name of Law	Date Enacted
Clean Air Act	1970
Clean Water Act	1977, and later amendments
Safe Drinking Water Act	1974
National Environmental Policy Act (NEPA)	1969
Federal Insecticide, Fungicide, and Rodenticide Act (FIFRA)	1996
Toxic Substances Control Act (TSCA)	1976
Comprehensive Environmental Response, Compensation, and Liability Act (CERCLA-Superfund)	1980
Resource Conservation and Recovery Act (RCRA)	1976
Occupational Safety and Health Act	1970
Endangered Species Act	1973

state implementation plans (SIPs), applicable to appropriate industrial sources in the state, in order to achieve these standards. The Act was amended in 1977 and 1990 primarily to set new goals (dates) for achieving attainment of NAAQS since many areas of the country had failed to meet the deadlines. [See **FIGURE 4.14**.]

Section 112 of the Clean Air Act addresses emissions of hazardous air pollutants. Prior to 1990, CAA established a risk-based program under which only a few standards were developed. The 1990 Clean Air Act Amendments revised Section 112 to first require issuance of technology-based standards for major sources and certain area sources. "Major sources" are defined as a stationary source or group of stationary sources that emit or have the potential to emit 10 tons (9 metric tons) per year or more of a hazardous air pollutant or 25 tons (23.7 metric tons) per year or more of a combination of hazardous air pollutants. An "area source" is any stationary source that is not a major source.

For major sources, Section 112 requires that EPA establish emission standards that require the maximum degree of reduction in emissions of hazardous air pollutants. These emission standards are commonly referred to as "maximum achievable control technology" or "MACT" standards. Eight years after the technology-based MACT standards are issued for a source category, EPA is required to review those standards to determine whether any residual risk exists for that source category and, if necessary, revise the standards to address such risk.[44]

FIGURE 4.14 The 1990 Clean Air Act.

Reproduced from United States Environmental Protection Agency. *The Plain English Guide to the Clean Air Act*. EPA-456/K-07-001. Research Triangle Park, NC;2007. Cover page.

Clean Water Act (33 U.S.C. § 1251 et seq. [1972]) (Comprises Several Related Acts)

The Federal Water Pollution Control Act of 1948 was the first major US law to address water pollution. Growing public awareness and concern for controlling water pollution led to sweeping amendments in 1972. As amended in 1972, the law became commonly known as the Clean Water Act (CWA). The 1972 amendments:

Established the basic structure for regulating pollutant discharges into the waters of the United States.

Gave EPA the authority to implement pollution control programs such as setting wastewater standards for industry.

Maintained existing requirements to set water quality standards for all contaminants in surface waters.

Made it unlawful for any person to discharge any pollutant from a point source into navigable waters, unless a permit was obtained under its provisions.

Funded the construction of sewage treatment plants under the construction grants program.

Recognized the need for planning to address the critical problems posed by nonpoint source pollution.

Subsequent amendments modified some of the earlier CWA provisions. Revisions in 1981 streamlined the municipal construction grants process, improving the capabilities of treatment plants built under the program. Changes in 1987 phased out the construction grants program, replacing it with the State Water Pollution Control Revolving Fund, more commonly known as the Clean Water State Revolving Fund. This new funding strategy addressed water quality needs by building on EPA-state partnerships.

Over the years, many other laws have changed parts of the Clean Water Act. Title I of the Great Lakes Critical Programs Act of 1990, for example, put into place parts of the Great Lakes Water Quality Agreement of 1978, signed by the U.S. and Canada, where the two nations agreed to reduce certain toxic pollutants in the Great Lakes. That law required EPA to establish water quality criteria for the Great Lakes addressing 29 toxic pollutants with maximum levels that are safe for humans, wildlife, and aquatic life. It also required EPA to help the States implement the criteria on a specific schedule.[45]

Safe Drinking Water Act (42 U.S.C. § 300f et seq. [1974])

The Safe Drinking Water Act (SDWA) was established to protect the quality of drinking water in the U.S. This law focuses on all waters actually or potentially designed for drinking use, whether from above ground or underground sources.

The Act authorizes EPA to establish minimum standards to protect tap water and requires all owners or operators of public water systems to comply with these primary (health-related) standards. The 1996 amendments to SDWA require that EPA consider a detailed risk and cost assessment, and best available peer-reviewed science, when developing these standards. State governments, which can be approved to implement these rules for EPA, also encourage attainment of secondary standards (nuisance-related). Under the Act, EPA also establishes minimum standards for state programs to protect underground sources of drinking water from endangerment by underground injection of fluids.[46]

Refer to the text box for provisions of the Safe Drinking Water Act as amended in 1996.

1996 SDWA AMENDMENT HIGHLIGHTS

Consumer Confidence Reports All community water systems must prepare and distribute annual reports about the water they provide, including information on detected contaminants, possible health effects, and the water's source.

Cost-Benefit Analysis US EPA must conduct a thorough cost-benefit analysis for every new standard to determine whether the benefits of a drinking water standard justify the costs. . . .

Microbial Contaminants and Disinfection Byproducts US EPA is required to strengthen protection for microbial contaminants, including *Cryptosporidium*, while strengthening control over the byproducts of chemical disinfection. . . .

Operator Certification Water system operators must be certified to ensure that systems are operated safely. . . .

Public Information and Consultation SDWA emphasizes that consumers have a right to know what is in their drinking water, where it comes from, how it is treated, and how to help protect it. . . .

Small Water Systems Small water systems are given special consideration and resources under SDWA, to make sure they have the managerial, financial, and technical ability to comply with drinking water standards.

Source Water Assessment Programs Every state must conduct an assessment of its sources of drinking water (rivers, lakes, reservoirs, springs, and ground water wells) to identify significant potential sources of contamination and to determine how susceptible the sources are to these threats.

Reproduced from US Environmental Protection Agency, Office of Water (4606). Understanding the Safe Drinking Water Act. EPA 816-F-04-030. Washington, DC: US EPA; June 2004, p. 2. Available at: https://www.epa.gov/sites/production/files/2015-04/documents/epa816f04030.pdf. Accessed February 14, 2017.

National Environmental Policy Act (42 U.S.C. § 4321 et seq. [1969])

The National Environmental Policy Act (NEPA) was one of the first laws ever written that establishes the broad national framework for protecting our environment. NEPA's basic policy is to assure that all branches of government give proper consideration to the environment prior to undertaking any major federal action that significantly affects the environment.

NEPA requirements are invoked when airports, buildings, military complexes, highways, parkland purchases, and other federal activities are proposed. Environmental Assessments (EAs) and Environmental Impact Statements (EISs), which are assessments of the likelihood of impacts from alternative courses of action, are required from all Federal agencies and are the most visible NEPA requirements.[47]

Federal Insecticide, Fungicide, and Rodenticide Act (7 U.S.C. § 136 et seq. [1996])

The Federal Insecticide, Fungicide, and Rodenticide Act (FIFRA) provides for federal regulation of pesticide distribution, sale, and use. All pesticides distributed or sold in the United States must be registered (licensed) by EPA. Before EPA may register a pesticide under FIFRA, the applicant must show, among other things, that using the pesticide according to specifications "will not generally cause unreasonable adverse effects on the environment."

FIFRA defines the term "unreasonable adverse effects on the environment" to mean: "(1) any unreasonable risk to man or the environment, taking into account the economic, social, and environmental costs and benefits of the use of any pesticide, or (2) a human dietary risk from residues that result from a use of a pesticide in or on any food inconsistent with the standard under section 408 of the Federal Food, Drug, and Cosmetic Act."[48]

Toxic Substances Control Act (15 U.S.C. § 2601 et seq. [1976])

The Toxic Substances Control Act of 1976 provides EPA with authority to require reporting, record-keeping and testing requirements, and restrictions relating to chemical substances and/or mixtures. Certain substances are generally excluded from TSCA, including, among others, food, drugs, cosmetics and pesticides.

TSCA addresses the production, importation, use, and disposal of specific chemicals including polychlorinated biphenyls (PCBs), asbestos, radon and lead-based paint.

Various sections of TSCA provide authority to:

- Require, under Section 5, pre-manufacture notification for "new chemical substances" before manufacture.
- Require, under Section 4, testing of chemicals by manufacturers, importers, and processors where risks or exposures of concern are found.
- Issue Significant New Use Rules (SNURs), under Section 5, when it identifies a "significant new use" that could result in exposures to, or releases of, a substance of concern.
- Maintain the TSCA Inventory, under Section 8, which contains more than 83,000 chemicals. As new chemicals are commercially manufactured or imported, they are placed on the list.[49]

[Note: The foregoing list is abstracted from a list of nine items.]

Comprehensive Environmental Response, Compensation, and Liability Act (Superfund) (42 U.S.C. § 9601 et seq. [1980])

The Comprehensive Environmental Response, Compensation, and Liability Act—otherwise known as CERCLA [pronounced "SIR-cla"] or Superfund—provides a Federal "Superfund" to clean up uncontrolled or abandoned hazardous-waste sites as well as accidents, spills, and other emergency releases of pollutants and contaminants into the environment. Through CERCLA, EPA was given power to seek out those parties responsible for any release and assure their cooperation in the cleanup.

EPA cleans up orphan sites when potentially responsible parties cannot be identified or located, or when they fail to act. Through various enforcement tools, EPA obtains private party cleanup through orders, consent decrees, and other small party settlements. EPA also recovers costs from financially viable individuals and companies once a response action has been completed.

EPA is authorized to implement the Act in all 50 states and U.S. territories. Superfund site identification, monitoring, and response activities in states are coordinated through the state environmental protection or waste management agencies.[50]

Resource Conservation and Recovery Act (42 U.S.C. § 6901 et seq. [1976])

The Resource Conservation and Recovery Act (RCRA) [pronounced "rick-rah"] gives EPA the authority to control hazardous waste from the "cradle to grave." This includes the generation, transportation, treatment, storage, and disposal of hazardous waste. RCRA also set forth a framework for the management of nonhazardous solid wastes. The 1986 amendments to RCRA enabled EPA to address environmental problems that could result from underground tanks storing petroleum and other hazardous substances.

HSWA—the Federal Hazardous and Solid Waste Amendments—are the 1984 amendments to RCRA that focused on waste minimization and phasing out land disposal of hazardous waste as well as corrective action for releases. Some of the other mandates of this law include increased enforcement authority for EPA, more stringent hazardous waste management standards, and a comprehensive underground storage tank program.[51]

Occupational Safety and Health Act (29 U.S.C. § 651 et seq. [1970])

Congress passed the Occupational Safety and Health Act to ensure worker and workplace safety. Their goal was to make sure employers provide their workers a place of employment free from recognized hazards to safety and health, such as exposure to toxic chemicals, excessive noise levels, mechanical dangers, heat or cold stress, or unsanitary conditions.

In order to establish standards for workplace health and safety, the Act also created the National Institute for Occupational Safety and Health (NIOSH) as the research institution for the Occupational Safety and Health Administration (OSHA). OSHA is a division of the U.S. Department of Labor that oversees the administration of the Act and enforces standards in all 50 states.[52]

Endangered Species Act (16 U.S.C. § 1531 et seq. [1973])

The Endangered Species Act (ESA) provides a program for the conservation of threatened and endangered plants and animals and the habitats in which they are found. [Endangered species are those at risk of extinction; threatened species are those at risk of becoming endangered in the future. See **FIGURE 4.15**.] The lead federal agencies for implementing ESA are the U.S. Fish and Wildlife Service (FWS) and the U.S. National Oceanic and Atmospheric Administration (NOAA) Fisheries Service. The FWS maintains a worldwide list of endangered species. Species include birds, insects, fish, reptiles, mammals, crustaceans, flowers, grasses, and trees. [Refer to **FIGURE 4.16**, which is a cartoon about endangered species selection.]

The law requires federal agencies, in consultation with the U.S. Fish and Wildlife Service and/or the NOAA Fisheries Service, to ensure that actions they authorize, fund, or carry out are not likely to jeopardize the continued existence of any listed species or result in the destruction or adverse modification of designated critical habitat of such species. The law also prohibits any action that causes a "taking" of any listed species of endangered fish or wildlife. Likewise, import, export, interstate, and foreign commerce of listed species are all generally prohibited.[53]

FIGURE 4.15 Endangered species.

"We prefer to only protect the cutest animals."

FIGURE 4.16 Endangered species selection.

Conclusion

This chapter defined issues and principles that pertain to the development and implementation of public policies, regulations, and laws designed to protect the health of the public from environmentally caused diseases. The chapter presented terminology and concepts such as the policy cycle and risk assessment that are related to the environmental policy process. Also described were some of the major US and international agencies charged with the development, adoption, and enforcement of environmental policies and regulations.

Some of the specific US laws and regulations designed to reduce air pollution, maintain water quality, and protect natural resources also were reviewed. It was noted that a significant ethical issue for policy developers involves setting acceptable levels of risk associated with potential environmental hazards and balancing these risks against societal benefits that may accrue through specific policy alternatives. Despite the efforts made to develop and implement appropriate environmental policies, much more work is required in order to protect the population from environmental hazards. Particularly in need of greater protection from environmental hazards are vulnerable groups such as children, as well as residents of less-developed countries that do not have adequate environmental protections in place.

Although such efforts will be essential, promotion of human health in some cases may run counter to safeguarding the environment. Bioethicist David Resnik wrote that, "Policies that protect the environment, such as pollution control and pesticide regulation, also benefit human health. In recent years, however, it has become apparent that promoting human health sometimes undermines environmental protection. Some actions, policies, or technologies that reduce human morbidity, mortality and disease [e.g., increasing food production and draining swamps] can have detrimental effects on the environment. Since human health and environmental protection are sometimes at odds, political leaders, citizens, and government officials need a way to mediate and resolve conflicts between these values."[54(p261)]

Study Questions and Exercises

1. Define and give examples of the following terms:
   a. Risk assessment
   b. Risk management
   c. Environmental impact assessment
   d. Health impact assessment
   e. The built environment
2. What is meant by the term *environmental policy*? Give two examples of environmental policies that have been developed in the United States.
3. Discuss the precautionary principle. Describe the types of environmental situations to which this principle might apply.
4. Cite the EPA's statement on environmental justice. What does the term *fair treatment* mean?
5. List the components of meaningful involvement; indicate how they are correlated with equal access of communities to the decision-making process that leads to a healthful environment.
6. Define the term *environmental sustainability*. List the three components of sustainable development.
7. Describe the rationale for the polluter-pays principle.
8. List and define the stages of the policy cycle. Explain how environmental health research is a component of the policy cycle. How does environmental health research contribute to the implementation of sound environmental policies?
9. Explain how risk assessment aligns with policy development. Give an example.
10. Why is it important to incorporate community-based participatory methods into the process of environmental policy development?
11. Name some environmental problems that have an international scope. What policy initiatives have been developed to bring them under control?
12. Explain why children are more vulnerable to environmental toxins than adults. What types of policies are needed to protect vulnerable populations from environmental hazards?
13. Describe the role of each of the following agencies in policy formation and in protecting the environment:
    a. US Environmental Protection Agency (EPA)
    b. National Institute for Occupational Safety and Health (NIOSH)
    c. National Institute of Environmental Health Sciences (NIEHS)
    d. European Environment Agency (EEA)
    e. World Health Organization (WHO)
14. Define the following environmental acts and explain how they have helped to ensure a safe environment in the United States:
    a. Clean Water Act
    b. Safe Drinking Water Act
    c. National Environmental Policy Act
    d. Federal Insecticide, Fungicide, and Rodenticide Act
    e. Toxic Substances Control Act
    f. Comprehensive Environmental Response, Compensation, and Liability Act
    g. Resource Conservation and Recovery Act
    h. Occupational Safety and Health Act
    i. Endangered Species Act

For Further Reading

Johnson, BJ. The environmental health policy-making process. In Friis, RH, ed. *The Praeger Handbook of Environmental Health*, Volume 1. Santa Barbara, CA: ABC-CLIO, LLC, 2012, 305-332.

References

1. Walker B Jr. Environmental health and African Americans. *Am J Public Health*. 1991;81:1395–1398.
2. Edgens JG. Environmental policy through a systems approach. *J Med Assoc Georgia*. 1995;84:225–227.
3. House R, McDowell H, Peters M, et al. Agriculture sector resource and environmental policy analysis: an economic and biophysical approach. *Novartis Found Symp*. 1999;220:243–261.
4. Christie M. A comparison of alternative contingent valuation elicitation treatments for the evaluation of complex environmental policy. *J Environ Manage*. 2001;62:255–269.
5. Sturm A. The Global Development Research Center. Glossary of environmental terms. Available at: http://www.gdrc.org/uem/ait-terms.html. Accessed April 23, 2017.
6. Executive Office of the President of the United States. Council on Environmental Quality. *A Citizen's Guide to the NEPA: Having Your Voice Heard*. National Environmental Policy Act Sec. 101. Available at: https://energy.gov/sites/prod/files/nepapub/nepa_documents/RedDont/G-CEQ-CitizensGuide.pdf. Accessed April 23, 2017.
7. Sharp RR. Ethical issues in environmental health research. *Environ Health Perspect*. 2003;111:1786–1788.
8. Smith C. The precautionary principle and environmental policy: science, uncertainty, and sustainability. *Int J Occup Environ Health*. 2000;6:263–265.
9. EUROPA—European Commission. Endocrine disrupter research. What are endocrine disrupters? Available at: http://ec.europa.eu/environment/chemicals/endocrine/definitions/endodis_en.htm. Accessed April 23, 2017.
10. Amendola A, Wilkinson DR. Risk assessment and environmental policy making. *J Hazard Mater*. 2000;78:ix–xiv.
11. US Environmental Protection Agency. Environmental justice: basic information. Available at: https://www.epa.gov/environmentaljustice/learn-about-environmental-justice. Accessed April 24, 2017.
12. Thomas VM, Graedel TE. Research issues in sustainable consumption: toward an analytical framework for materials and the environment. *Environ Sci Technol*. 2003;37:5383–5388.
13. Maddox J. Positioning the goalposts: the best environmental policy depends on how you frame the question. *Nature*. 2000;403:139.
14. Organization for Economic Cooperation and Development. Recommendation of the council concerning the application of the polluter-pays principle to accidental pollution. July 7, 1989–C(89)88/Final. Available at: http://acts.oecd.org/Instruments/ShowInstrumentView.aspx?InstrumentID=38&InstrumentPID=305&Lang=en&Book=False. Accessed April 24, 2017.
15. Dadalos. Policy cycle: teaching politics. Available at: http://www.dadalos.org/politik_int/politik/policy-zyklus.htm. Accessed April 24, 2017.
16. Walker B Jr. Impediments to the implementation of environmental policy. *J Public Health Policy*. 1994;15:186–202.
17. Rabe BG. Legislative incapacity: the congressional role in environmental policy-making and the case of Superfund. *J Health Polit Policy Law*. 1990;15:571–589.
18. Suk WA, Avakian MD, Carpenter D, et al. Human exposure monitoring and evaluation in the Arctic: the importance of understanding exposures to the development of public health policy. *Environ Health Perspect*. 2004;112:113–120.
19. Johnson BJ. *Environmental Policy and Public Health*. Boca Raton, FL: Taylor & Francis Group, CRC Press; 2007.
20. Kaiser J. EPA gives science a bigger voice. *Science*. 2002;296:1005.
21. Schmidt CW. The market for pollution. *Environ Health Perspect*. 2001;109:A378–A381.
22. Swanson KE, Kuhn RG, Xu W. Environmental policy implementation in rural China: a case study of Yuhang, Zhejiang. *Environ Manage*. 2001;27:481–491.
23. Goudey R, Laslett G. Statistics and environmental policy: case studies from long-term environmental monitoring data. *Novartis Found Symp*. 1999;220:144–157.
24. Falconer K. Pesticide environmental indicators and environmental policy. *J Environ Manage*. 2002;65:285–300.
25. Landrigan PJ, Carlson JE. Environmental policy and children's health. *Future Child*. 1995;5(2):34–52.
26. Hyperdictionary. Meaning of biomarker. Available at: http://www.hyperdictionary.com/dictionary/biomarker. Accessed April 24, 2017.
27. Lerer LB. How to do (or not to do) . . . health impact assessment. *Health Policy Plan*. 1999;14(2):198–203.
28. Lock K, Gabrijelcic-Blenkus M, Martuzzi M, et al. Health impact assessment of agriculture and food policies: lessons learnt from the Republic of Slovenia. *Bull World Health Organ*. 2003;81:391–398.
29. US Environmental Protection Agency. *Fiscal Year 2014-2018 EPA Strategic Plan*. Washington, DC: U.S. Environmental Protection Agency; April 10, 2014.
30. Ensminger JT, McCold LN, Webb JW. Environmental impact assessment under the National Environmental Policy Act and the Protocol on Environmental Protection to the Antarctic Treaty. *Environ Manage*. 1999;24:13–23.
31. Tarmann A. South Africa's water policy champions rights of people and ecosystem. *Popul Today*. 2000;28(5):1–2.
32. Zylicz T. Environmental policy in economies in transition. *Scand J Work Environ Health*. 1999;25(suppl 3):72–80.
33. Kratovits A, Punning J-M. Driving forces for the formation of environmental policy in the Baltic countries. *Ambio*. 2001;30:443–449.
34. Goldman L, Falk H, Landrigan PJ, et al. Environmental pediatrics and its impact on government health policy. *Pediatrics*. 2004;113:1146–1157.
35. Suk WA, Ruchirawat KM, Balakrishnan K, et al. Environmental threats to children's health in southeast Asia and the western Pacific. *Environ Health Perspect*. 2003;111:1340–1347.
36. Pollard T. Policy prescriptions for healthier communities. *Am J Health Promot*. 2003;18(1):109–113.
37. California Air Resources Board. ARB mission and goals. Available at: http://www.arb.ca.gov/html/mission.htm. Accessed April 24, 2017.
38. US Environmental Protection Agency. About EPA. Available at: http://www.epa.gov/aboutepa/our-mission-and-what-we-do. Accessed April 24, 2017.
39. Centers for Disease Control and Prevention. National Institute for Occupational Safety and Health. Factsheet. DHHS(NIOSH) Publication No. 2013-140; October, 2015. Available at: https://www.cdc.gov/niosh/about/default.html. Accessed April 24, 2017.
40. US Department of Health and Human Services, Agency for Toxic Substances and Disease Registry. *FY 2008 ATSDR Annual Performance Report: Safer Healthier People*. Atlanta, GA: Agency for Toxic Substances and Disease Registry; 2008.
41. National Institutes of Health. National Institute of Environmental Health Sciences. NIEHS Priority Areas and

Programs. Available at: http://www.niehs.nih.gov/health/docs/niehs-overview.pdf. Accessed April 24, 2017.

42. European Commission. Living well, within the limits of our planet. 7th EAP—The new general Union Environment Action Program to 2020. Available at: http://ec.europa.eu/environment/pubs/pdf/factsheets/7eap/en.pdf. Accessed July 10, 2017.
43. WHO Regional Office for Europe. *Health and the Environment in the WHO European Region*. Copenhagen, Denmark: World Health Organization; 2013.
44. US Environmental Protection Agency. Summary of the Clean Air Act. Available at: https://www.epa.gov/laws-regulations/summary-clean-air-act. Accessed April 24, 2017.
45. US Environmental Protection Agency. History of the Clean Water Act. Available at: https://www.epa.gov/laws-regulations/history-clean-water-act. Accessed April 24, 2017.
46. US Environmental Protection Agency. Summary of the Safe Drinking Water Act. Available at: https://www.epa.gov/laws-regulations/summary-safe-drinking-water-act. Accessed April 24, 2017.
47. US Environmental Protection Agency. Summary of the National Environmental Policy Act. Available at: https://www.epa.gov/laws-regulations/summary-national-environmental-policy-act. Accessed April 24, 2017.
48. US Environmental Protection Agency. Summary of the Federal Insecticide, Fungicide, and Rodenticide Act. Available at: https://www.epa.gov/laws-regulations/summary-federal-insecticide-fungicide-and-rodenticide-act. Accessed April 24, 2017.
49. US Environmental Protection Agency. Summary of the Toxic Substances Control Act. Available at: https://www.epa.gov/laws-regulations/summary-toxic-substances-control-act. Accessed April 24, 2017.
50. US Environmental Protection Agency. Summary of the Comprehensive Environmental Response, Compensation, and Liability Act (Superfund). Available at: https://www.epa.gov/laws-regulations/summary-comprehensive-environmental-response-compensation-and-liability-act. Accessed April 24, 2017.
51. US Environmental Protection Agency. Summary of the Resource Conservation and Recovery Act. Available at: https://www.epa.gov/laws-regulations/summary-resource-conservation-and-recovery-act. Accessed April 24, 2017.
52. US Environmental Protection Agency. Summary of the Occupational Safety and Health Act. Available at: https://www.epa.gov/laws-regulations/summary-occupational-safety-and-health-act. Accessed April 24, 2017.
53. US Environmental Protection Agency. Summary of the Endangered Species Act. Available at: https://www.epa.gov/laws-regulations/summary-endangered-species-act. Accessed April 24, 2017.
54. Resnik DB. Human health and the environment: In harmony or in conflict? *Health Care Anal.* 2009;17:261–276.

PART II

Agents of Environmental Disease

© Jean-Luc Rivard/EyeEm/Getty Images

CHAPTER 5

Zoonotic and Vector-Borne Diseases

LEARNING OBJECTIVES

By the end of this chapter the reader will be able to:

- Provide a rationale for environmental change and infectious disease occurrence.
- Indicate how a disease may be transmitted from an animal reservoir to humans.
- Define the terms vector-borne and zoonotic diseases, giving examples.
- Compare three human diseases transmitted by arthropod vectors.
- Discuss methods used to control vector-borne and zoonotic diseases.

Introduction

Zoonotic and vector-borne diseases have substantial environmental components and contribute greatly to society's burdens of morbidity and mortality. Medicine and public health made great strides toward the control of infectious diseases during the late 19th century and the 20th century. In fact, by the mid-1960s, some health authorities proclaimed that society was moving toward the virtual elimination of infectious diseases as significant causes of morbidity and mortality.[1] Medical advances included immunizations, the use of antibiotics for the treatment of infectious diseases, declines in mortality from infectious and parasitic diseases, and the eradication of smallpox during the late 1970s. Public health achievements included improved environmental sanitation, disinfection of drinking water, and innovations in methods of food storage. Despite these notable accomplishments, infectious diseases once again have come to the forefront, especially with the occurrence of emerging and reemerging infections. In this chapter, you will learn about some of the notable examples of these conditions: malaria, dengue fever, and viral hemorrhagic fevers.

Terminology Used in the Context of Zoonotic and Vector-Borne Diseases

The pathogenic agents for diseases in this category may involve prions, viruses, bacteria, protozoa, and helminths. For example, a total of more than 200 zoonotic pathogens may cause disease in human beings. Refer to **TABLE 5.1** for a list of some of the diseases in this category.

TABLE 5.1 Examples of Zoonoses and Vector-Borne Diseases

Bacterial
Anthrax
Cat scratch disease
Escherchia coli O157:H7 infection (discussed in Chapter 11)
Lyme disease
Plague
Psittacosis
Salmonellosis (discussed in Chapter 11)
Tularemia
Viral
Dengue fever
Encephalitis
Eastern equine encephalitis (EEE)
Japanese encephalitis (JE)
St. Louis encephalitis (SLE)
Tick-borne viral encephalitis
Venezuelan equine encephalitis (VEE)
Western equine encephalitis (WEE)
Hand, foot, and mouth disease
Hantavirus
Human monkeypox
Influenza
Avian influenza (discussed in Chapter 1)
Swine flu
Rabies
Rift Valley fever
West Nile virus
Yellow fever
Parasitic
Cryptosporidiosis (discussed in Chapter 9)
Cysticercosis and taeniasis (discussed in Chapter 11)
Giardiasis (discussed in Chapter 9)
Leishmaniasis
Malaria
Trichinellosis (discussed in Chapter 11)
Rickettsial
Q fever
Rocky Mountain spotted fever (RMSF)
Nonconventional (e.g., prions)
Variant Creutzfeldt-Jakob disease (v-CJD); mad cow disease (discussed in Chapter 11)

Zoonosis

The term **zoonosis** refers to "an infection or infectious disease transmissible under natural conditions from vertebrate animals to humans."[2] The definition of *zoonosis* varies to include several different situations.[3] In some cases a zoonosis may be a disease-causing pathogen that maintains an infection cycle in a host that is independent from humans, who can become inadvertent hosts. Other definitions refer to organisms that can infect both humans and animals during their life cycles. Conceptions of zoonotic agents may also include pathogens that cause disease in a nonhuman host or situations in which an infected animal remains free from symptoms of the disease. Contact with the skin, the bite or scratch of an animal, direct inhalation or ingestion (e.g., eating contaminated foods such as infected meat), or the bite of an arthropod vector are some of the methods for transmission of zoonotic pathogens. Note that immunocompromised persons, infants, and children younger than 5 years old may be at increased risk of morbidity from zoonotic diseases (e.g., toxoplasmosis) transmitted by cats and dogs. **TABLE 5.2** provides examples of zoonotic diseases and animals that are associated with those diseases.

Vector

In the context of infectious diseases, a **vector** is defined as "an insect or any living carrier that transports an infectious agent from an infected individual or its wastes to a susceptible individual or its food or immediate surroundings."[2] Part of the chain in transmission of infectious disease agents, vectors include various species of rodents (rats and mice) and arthropods (mosquitoes, ticks, sand flies, and biting midges).

Vector-Borne Infection

The term **vector-borne infection** refers to "[s]everal classes of vector-borne infections . . . each with epidemiological features determined by the interaction between the infectious agent and the human host on the one hand and the vector on the other. Therefore, environmental factors, such as climatic and seasonal variations, influence the epidemiologic pattern by virtue of their effects on the vector and its habits."[2] Vector-borne infections spread by biological transmission, which refers to "[t]ransmission of the infectious agent to [a] susceptible host by bite of blood-feeding (arthropod) vector, as in malaria, or by other inoculation, as in *Schistosoma* infection."[2]

TABLE 5.2 Examples of Zoonotic Diseases (and Associated Animals)

Name/Type of Animal	Zoonotic Disease
Domestic animals	
Cats	Rabies, toxocariasis (a parasitic disease from *Toxocara* roundworms), cat scratch disease (cat scratch fever), toxoplasmosis
Dogs	Rabies, parasites (e.g., dog tapeworm, hookworm, roundworm), campylobacteriosis
Farm animals (in general)	Anthrax, brucellosis, *Escherichia coli* O157:H7, Q fever, cryptosporidiosis
Horses	Salmonellosis, ringworm; less common: anthrax, cryptosporidiosis, rabies
Poultry (e.g., chickens and ducks)	Avian influenza, salmonellosis
Sheep and goats	Anthrax, Q fever
Reptiles (e.g., pet turtles and snakes)	Salmonellosis
Wild animals	
Bats	Rabies
Bears and wild hogs	Trichinosis
Birds	Cryptococcosis, histoplasmosis
Mammals (e.g., beavers, bison, deer, foxes, raccoons, skunks)	Rabies (raccoons, skunks, foxes), giardiasis (beavers, deer), brucellosis (bison)
Rodents	Hantavirus, plague, tularemia

Data from Centers for Disease Control and Prevention. Healthy Pets Healthy People. Available at https://www.cdc.gov/healthypets/pets/index.html. Accessed July 18, 2017.

Examples of Vector-Borne Diseases

Names of diseases in this group are malaria, leishmaniasis, plague, Lyme disease, and Rocky Mountain spotted fever. Malaria is a disease of great significance for environmental health, even though it does not usually occur in developed regions such as North America, Europe, and Australia. Leishmaniasis affects residents of Middle Eastern countries, where it is endemic, and has been a particular concern of US military personnel stationed in endemic areas. The plague remains a serious threat to the world's population; during the Middle Ages, the "black death" (caused by plague) wiped out millions of people in Europe. Lyme disease, with an epidemic focus in the eastern and upper midwest United States and an expanding focus in the western United States, also occurs in other parts of the world.

Malaria

According to the Centers for Disease Control and Prevention (CDC), malaria is a disease that is found in more than 100 countries, with about 50% of the world's population at risk.[4] Typically, malaria is endemic to the warmer geographic areas of the globe. Endemic regions include Central and South America, Africa, South Asia, Southeast Asia, the Middle East, and Oceania. In 2015, the worldwide death toll for malaria was 438,000 persons, with an estimated 214 million clinical cases.[4]

The health impacts of malaria can be severe, as demonstrated by **FIGURE 5.1**, which shows a child who has developed edema (swelling) brought on by nephrosis associated with malaria. Nephrosis is a disease that impairs kidney function, resulting in a syndrome that produces edema, protein in the urine, and elevated levels of cholesterol in the blood.[5]

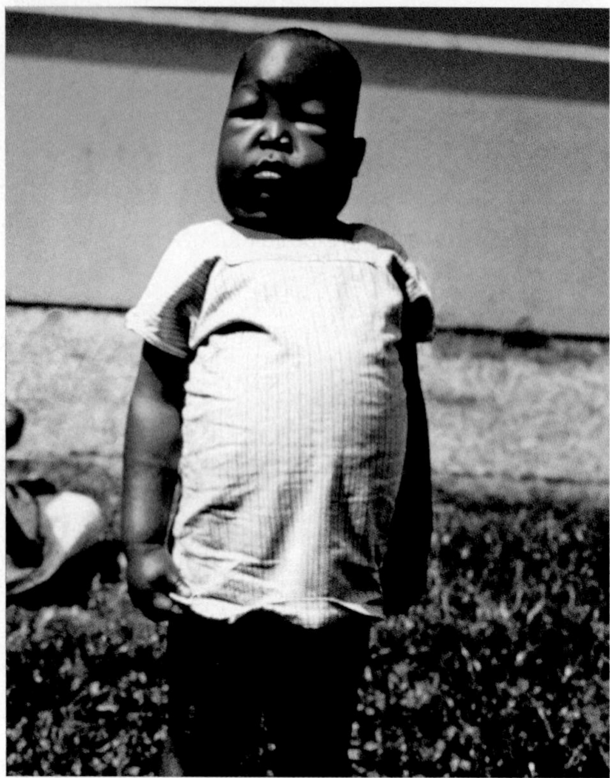

FIGURE 5.1 The edema exhibited by this African child was brought on by nephrosis associated with malaria (1975).
Courtesy of CDC/Dr. Myron G. Schultz

FIGURE 5.2 Ronald Ross, one of the discoverers of the malaria parasite.
Courtesy of CDC

Developing countries located in economically disadvantaged tropical and subtropical regions are especially affected by malaria. At greatest risk of dying are young children.[4] In addition to its human toll, malaria also exacts devastating economic impacts.[6] Sub-Saharan countries account for a disproportionate burden of morbidity and mortality from malaria.[7]

The average global, direct economic costs incurred by malaria are estimated to be $12 billion US annually.[6] The direct costs include those for treatment and prevention of the disease (e.g., medicine, hospitalization, and pesticide use). Other economic costs include lost productivity and earnings as well as negative impacts upon travel and tourism and agricultural labor.

The name *malaria* originates from the Italian *mala aria*, which referred to "bad air."[8] Originally, it was believed that malaria originated from the dank atmosphere around swamps. Now it is known that malaria is transmitted by mosquitoes that carry a unicellular parasite known as a plasmodium. Ronald Ross, an Indian Medical Service Officer (shown in **FIGURE 5.2**), is credited with discovering the malaria parasite in 1897 when he dissected mosquitoes and found the organism in their stomachs. He reported that he "saw a clear and almost perfectly circular outline before me of about 12 microns in diameter. The outline was much too sharp, the cell too small to be an ordinary stomach-cell of a mosquito. I looked a little further. Here was another, and another exactly similar cell."[9]

The four human forms of malaria are: *Plasmodium falciparum*, *P. vivax*, *P. ovale*, and *P. malariae*. Infection with plasmodia, the causative agent, can be life threatening; malaria symptoms include the occurrence of fever and flu-like symptoms such as headache, muscle aches, fatigue, and shaking chills. In some instances vomiting and diarrhea may accompany the disease. Another symptom of malaria is jaundice due to anemia, which is caused by the loss of red blood cells. *P. falciparum*, the most deadly type, may produce kidney failure, seizures, mental confusion, coma, and ultimately death.[4] *P. falciparum* is a very common variety of malaria in Africa south of the Sahara desert.

The *Anopheles* mosquito is the principal vector for malaria. The transmission of malaria involves the complex life cycle of mosquitoes (the vector) and human hosts (with human liver and human blood stages). The first stage in transmission involves the bite of an infected mosquito of the *Anopheles* type. (Refer to **FIGURE 5.3**.)

During the transmission cycle the malaria parasite (called the sporozoite form) is transferred to the human host when the mosquito takes a blood meal. In the human host the organism multiplies in the liver

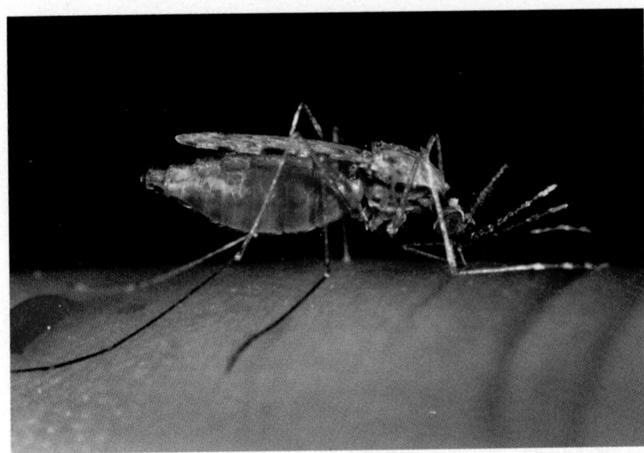

FIGURE 5.3 Female *Anopheles gambiae* mosquito feeding.
Courtesy of CDC/James Gathany

(exo-erythrocytic cycle) and red blood cells (erythrocytic cycle) and later develops into a form (gametocytic) that can be transmitted from the infected person to another host after a mosquito feeds on the infected host. In the sporogenic cycle in the mosquito, oocysts are produced that can be transmitted to a human host, continuing the life cycle of malaria. Symptoms of malaria occur approximately 9 to 14 days after the bite of an infected mosquito. **FIGURE 5.4** illustrates the cycle of transmission. (Malaria also may be transmitted by contaminated syringes or through blood transfusions.)

In the United States, malaria was endemic until the end of the 1940s. An antimalaria campaign was launched on July 1, 1947; one of the major activities consisted of spraying homes with DDT. By 1949, malaria was declared as having been brought under control as a major public health challenge in the United States.[10]

During the mid-20th century, efforts to control malaria by spraying with DDT and administering synthetic antimalaria drugs were found to be efficacious. Consequently, the World Health Organization (WHO) submitted a plan to the World Health Assembly in 1955 for the global eradication of malaria.[11] The plan

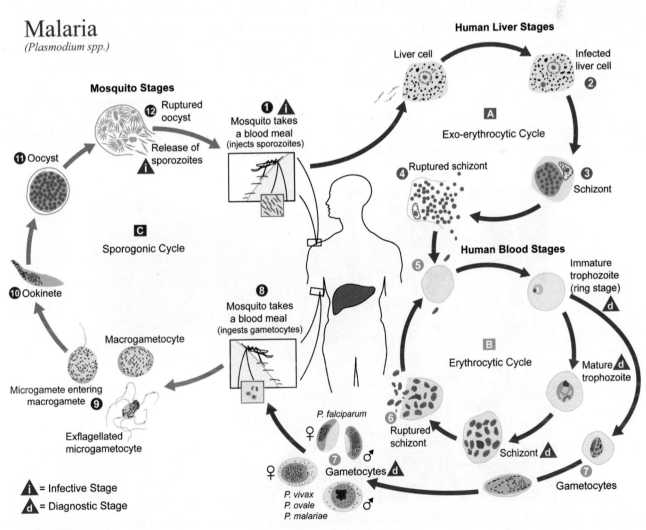

FIGURE 5.4 The life cycle of the malaria parasite (*Plasmodium* spp.).
Courtesy of CDC - DPDx/Alexander J. da Silva, PhD, Melanie Moser

FIGURE 5.5 A Stearman bi-plane spraying an insecticide during malaria control operations in Savannah, GA.

Courtesy of CDC

involved spraying, drug treatment, and surveillance. These efforts at eradication were successful in countries that had temperate climates and seasonal transmission of malaria but had limited success in other countries such as Indonesia, Afghanistan, Haiti, and Nicaragua.

DDT was highly effective in controlling the mosquito that can transmit malaria. For example, **FIGURE 5.5** and **FIGURE 5.6**, respectively, illustrate the use of bi-planes to spray for mosquitoes over wide areas and hand spraying methods to apply insecticide. At one time insecticides such as DDT were applied directly to military personnel (**FIGURE 5.7**). Over a period of several decades, environmental observers began to gather suggestive evidence that DDT was harmful to bird species and other species of wildlife; the use of DDT was eventually opposed by many developed nations, especially the United States. This opposition led to withdrawal of funding for malaria spraying programs.

Discontinuance of outdoor DDT spraying and mosquitoes' development of insecticide resistance are among the factors that are believed to have contributed to the resurgence of malaria-bearing mosquitoes. As an alternative to outdoor spraying, South Africa has used annual spraying of DDT inside of homes, a procedure that appears to reduce the number of malaria cases.[12] Another means to control malaria is the use of bed nets that are impregnated with insecticides.

In addition to the use of insecticides, other methods of malaria control include the regular use of suppressive drugs in endemic areas and the avoidance of blood from donors who reside in endemic areas. Sanitary improvements such as the filling and draining of swamps and removal of standing water help to reduce the breeding areas for malaria-carrying mosquitoes. More information on the topic of mosquito control is provided later in the chapter.

The equatorial and southern areas of the continent of Africa form one of the endemic regions for malaria, as shown in **FIGURE 5.8**. In Africa, which accounts for a total of 91% of the world's malaria deaths,[6] evidence

FIGURE 5.6 In 1958, the National Malaria Eradication Program used an entirely new approach, implementing DDT for spraying of mosquitoes.

Courtesy of CDC

FIGURE 5.7 US soldier is demonstrating DDT hand-spraying equipment while applying the insecticide.

Courtesy of CDC

points to the movement of malaria from the low-lying areas where it is usually found to higher elevations where it is uncommon. In the late 1980s and early 1990s, Kenya experienced an upward movement of malaria cases, possibly due to small changes in climate that provided suitable conditions for malaria to survive.[13]

Malaria cases are not confined to endemic areas, because they may be imported into nonendemic areas by refugees and immigrants.[14] For example, the US state of Minnesota saw an increase in the number of reported cases of malaria from 5 in 1988 to 76 in 1998. The sources of these cases were persons native to Minnesota who had traveled to foreign areas and returned home as well as immigrants whose travel originated in a foreign country. In nonendemic areas, importation of cases of malaria presents a significant challenge to healthcare providers, who may not be familiar with the symptoms of diseases that they may not ordinarily see in clinical practice or during their training.

Leishmaniasis

Three forms of leishmaniasis are the visceral, mucocutaneous, and cutaneous varieties; the latter is the focus of this section. Cutaneous leishmaniasis is transmitted by the bite of an infected sand fly.[15,16] After being bitten by the fly, the human host develops a characteristic sore on the skin that forms after an incubation period of several weeks or months. **FIGURE 5.9** illustrates a lesion due to leishmaniasis. Near the site of the bite, a localized reaction occurs that forms a papule, which is "a small, raised, solid pimple or swelling, often forming part of a rash on the skin and typically inflamed."[17] Subsequently, ulceration, healing, and a scar form at the site.

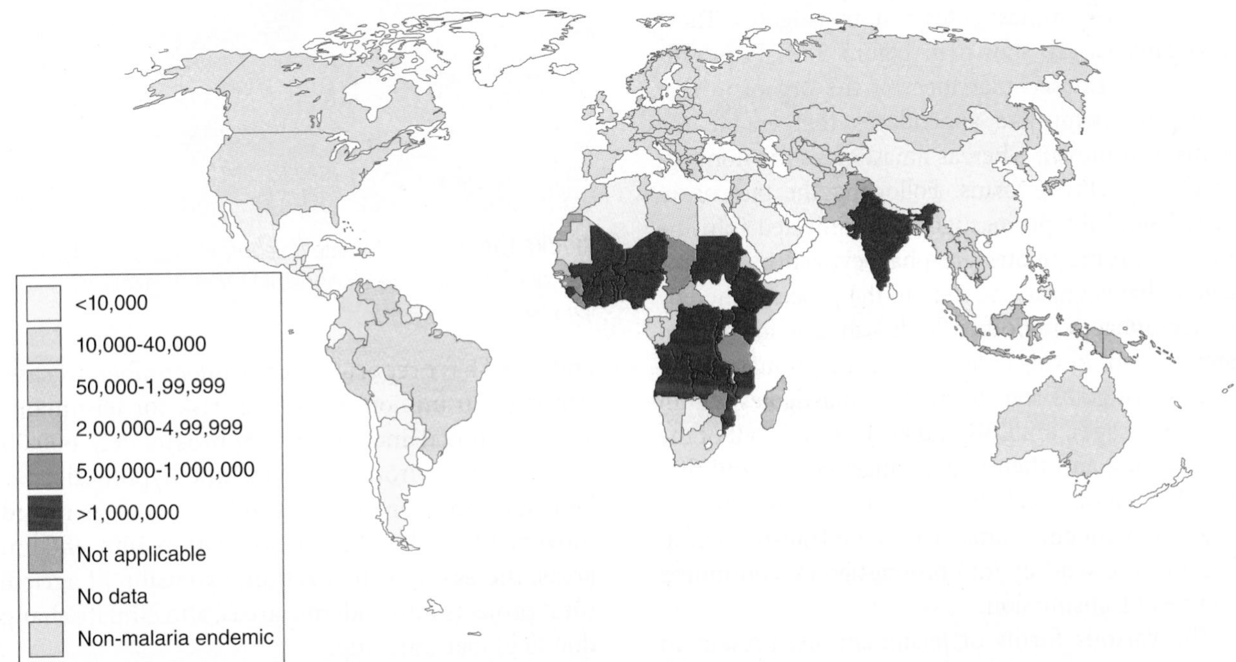

FIGURE 5.8 Global number of reported cases of malaria, 2014.

Modified and reproduced from World Health Organization. Number of malaria cases: Estimated cases 2000-2015. Available at: http://www.who.int/gho/malaria/epidemic/cases/en/. Accessed January 29, 2017. Copyright 2016.

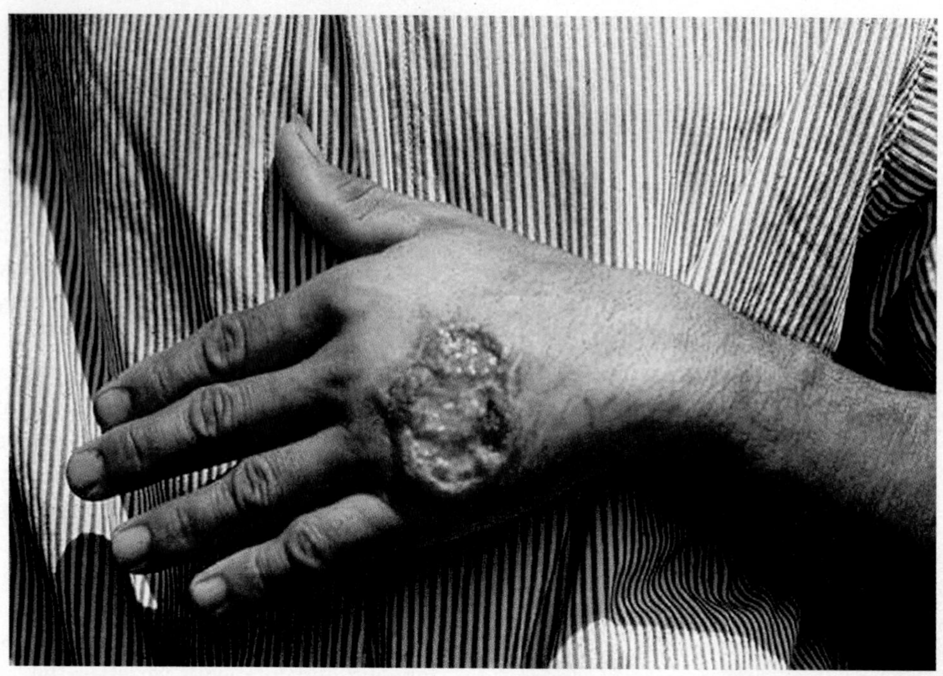

FIGURE 5.9 Skin ulcer due to leishmaniasis; hand of Central American adult.
Courtesy of CDC/Dr. D.S. Martin

The reservoir for the cutaneous form of leishmaniasis includes wild rodents, human beings, and carnivores (e.g., domestic dogs). The disease is transmitted from the reservoir to the human host by a sand fly known as the phlebotomus fly. (See **FIGURE 5.10**.)

FIGURE 5.11 shows the cycle of transmission from the sand fly to the human host and vice versa, during the sand fly and human stages. The causative agent for leishmaniasis is a protozoal organism, which exists in two forms—promastigotes and amastigotes. These two organisms are shown in Figure 5.11. Promastigotes are an extracellular form of the organism that has flagella (whip-like appendages that enable the organism to move), whereas amastigotes are nonflagellated intracellular forms. Following the bite of an infected sand fly, promastigotes are injected into the host. The promastigotes are phagocytized by macrophages; phagocytization refers to the process whereby the macrophages in bodily fluids scavenge for foreign materials and, in this case, capture the promastigotes. The promastigotes transform into amastigotes within the macrophages, multiply and later explode out of the macrophages, and then may be ingested by sand flies through a blood meal that contains infected macrophages. Subsequently, amastigotes are transformed in the gut of the sand fly into promastigotes, continuing the cycle of transmission.

The various forms of leishmaniasis, present in the countries surrounding the Mediterranean and endemic in another 82 countries, are showing an increasing incidence. For example, Syria, Tunisia, and Israel have reported increasing numbers of cases. Among 350 million persons at risk for leishmaniasis, the annual incidence is approximately 600,000 cases. The environmental factors hypothesized to be responsible for increases in leishmaniasis include movement of the human population into endemic areas, increasing urbanization, extension of agricultural projects into endemic areas, and climate change due to global warming.

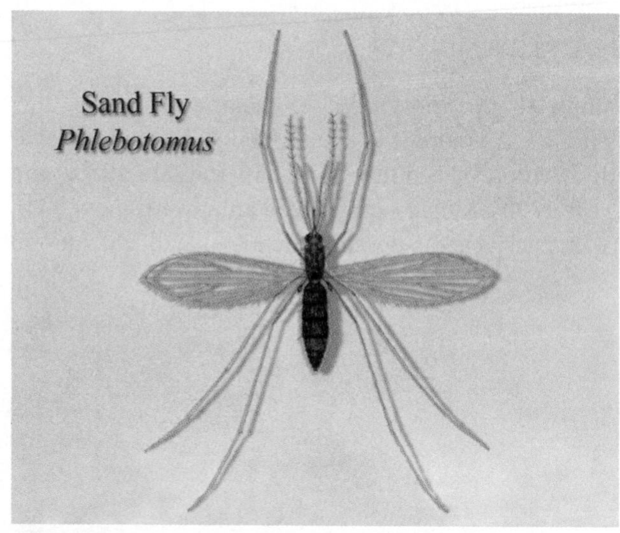

FIGURE 5.10 This is a female *Phlebotomus sp.* sand fly, a vector of the parasite responsible for leishmaniasis.
Courtesy of CDC

Cutaneous leishmaniasis is a concern of US military personnel stationed in endemic areas (e.g., Afghanistan, Iraq, and Kuwait). The condition has been subjected to

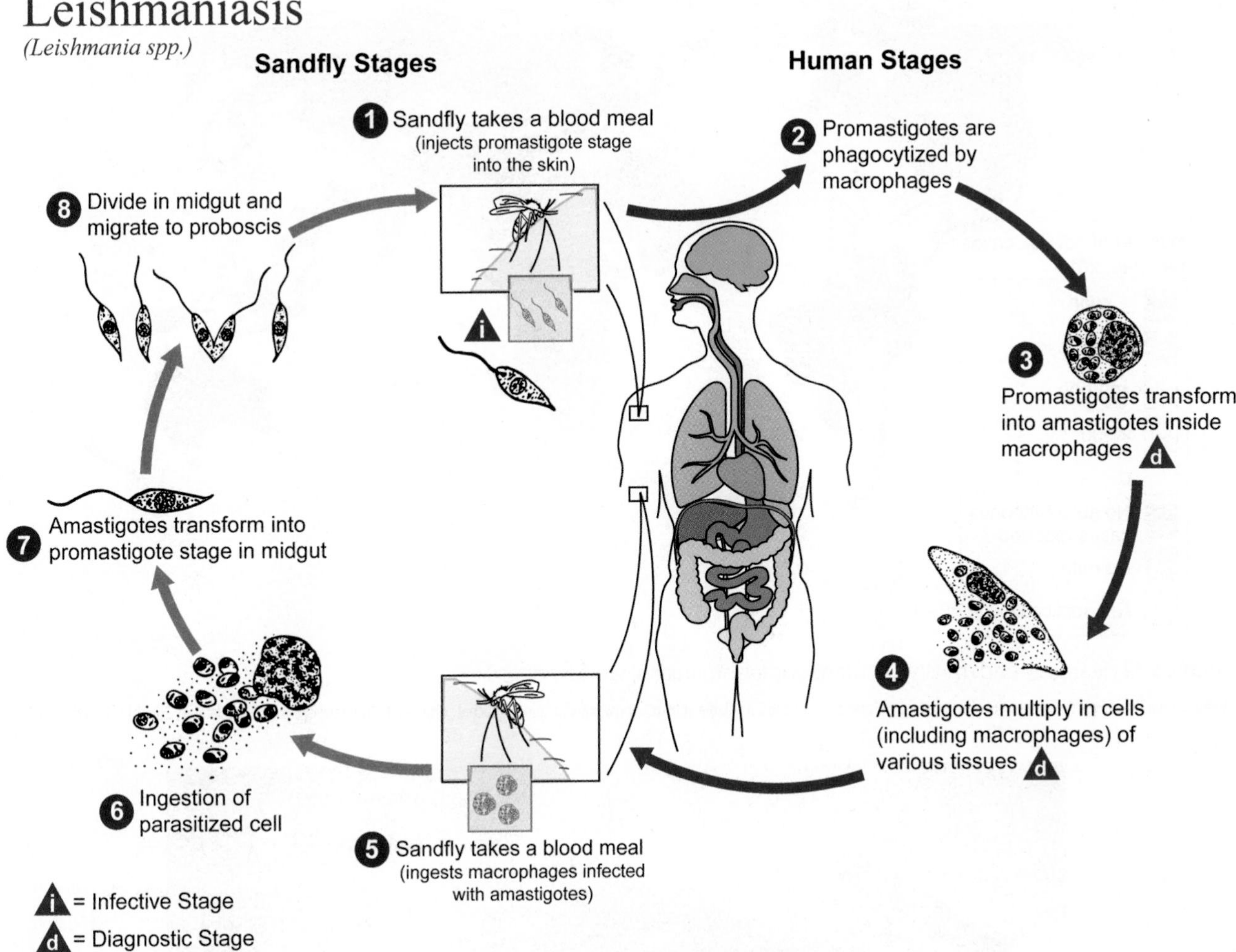
Leishmaniasis
(Leishmania spp.)
Sandfly Stages
Human Stages
1 Sandfly takes a blood meal (injects promastigote stage into the skin)
2 Promastigotes are phagocytized by macrophages
3 Promastigotes transform into amastigotes inside macrophages
4 Amastigotes multiply in cells (including macrophages) of various tissues
5 Sandfly takes a blood meal (ingests macrophages infected with amastigotes)
6 Ingestion of parasitized cell
7 Amastigotes transform into promastigote stage in midgut
8 Divide in midgut and migrate to proboscis
i = Infective Stage
d = Diagnostic Stage

FIGURE 5.11 This is an illustration of the life cycle of *Leishmania* spp., the causal agents of leishmaniasis.

Courtesy of CDC- DPDx/Alexander J. da Silva, PhD

extensive epidemiologic investigations in Israel, where leishmaniasis affects both military personnel and the general population. **FIGURE 5.12** demonstrates that in 2013 reported endemic cases were concentrated primarily in the Middle East and South America.

Methods of control of leishmaniasis include the following:

- Periodic application of long-acting insecticides to dwelling units
- Use of screens to prevent sand flies from entering housing
- Elimination of breeding areas (e.g., rubbish heaps) for the phlebotomus fly
- Destruction of rodent burrows and control of domestic dogs

Plague

The bacterium *Yersinia pestis* is the infectious agent for plague, a condition that infects both animals and humans. Plague may be transmitted by the bite of a flea harbored by rodents (see **FIGURE 5.13**, an oriental rat flea). Historians believe that the plague epidemic during the Middle Ages (the "black death") was caused by fleas from infested rats.

Although morbidity and case fatality rates among persons infected with the disease remain high, prompt treatment with antibiotics is efficacious.[18] Nevertheless, human infections with *Y. pestis* remain a matter of great concern for authorities. **FIGURE 5.14** illustrates how plague may be transmitted from the environment to humans. Plague is distributed widely across the world, including the United States, South America, Asia, and parts of Africa (see **FIGURE 5.15**). In some regions, plague is a zoonotic disease.

One of the natural reservoirs for plague is wild rodents, such as the ground squirrels that are at home in the western United States. Pets, such as house cats and dogs, may bring the wild rodents' fleas into homes or may even transmit plague on rare occasions from their bites or scratches. The condition known as bubonic plague begins with nonspecific

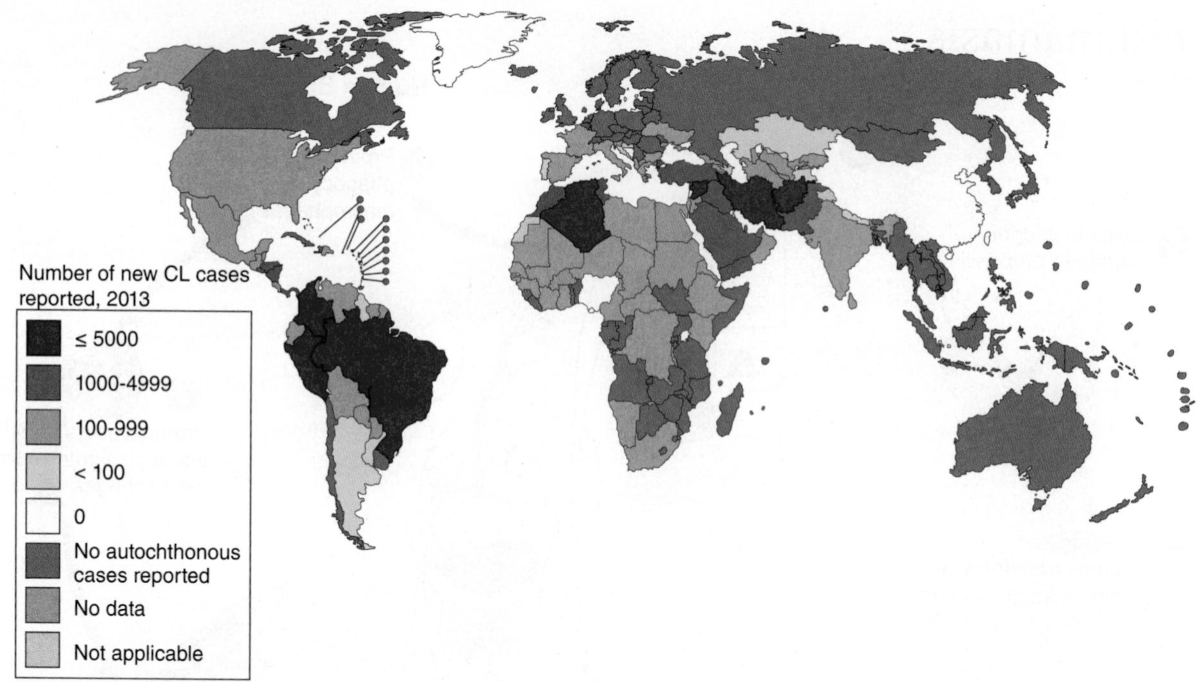

FIGURE 5.12 Status of endemicity of cutaneous leishmaniasis, worldwide, 2013.

Modified from World Health Organization. Leishmaniasis: Situation and trends. Available at http://www.who.int/gho/neglected_diseases/leishmaniasis/en/Accessed January 26, 2017. Copyright 2015.

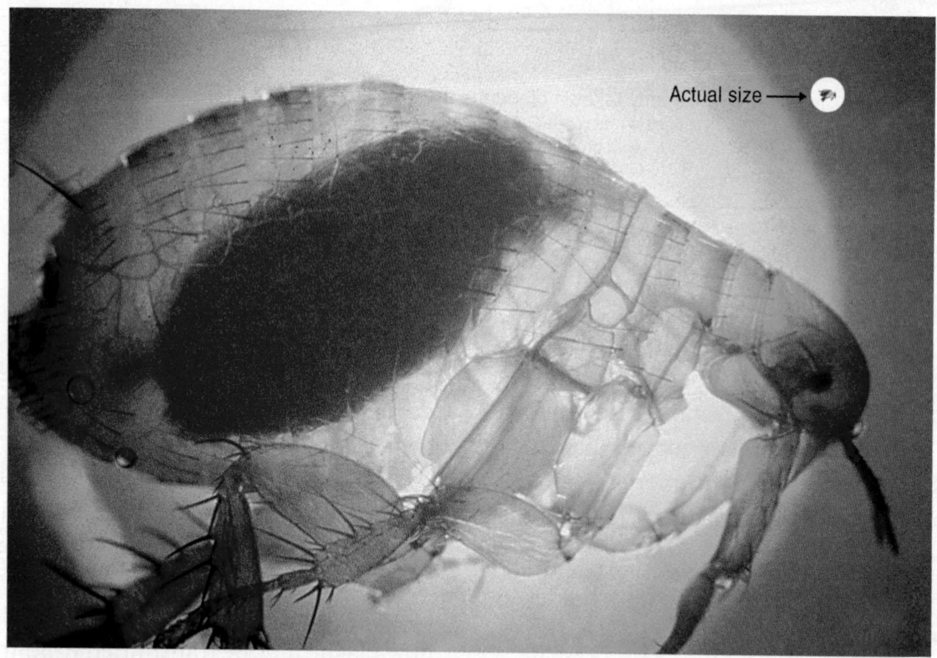

FIGURE 5.13 Male *Xenopsylla cheopis* (oriental rat flea) engorged with blood. This flea is the primary vector of plague in most large plague epidemics in Asia, Africa, and South America.

Courtesy of CDC

symptoms such as fever, chills, and headache and then progresses into lymphadenitis (infected lymph nodes) at the site of the initial flea bite. There may be secondary involvement of the lungs, known as pneumonic plague. The epidemiologic significance of this form of plague is that respiratory droplets from an infected person can transfer *Y. pestis* to other individuals. The case fatality rate for untreated bubonic plague is high, ranging from about 50% to 60%. Patients who are infected with the disease need to be placed in strict isolation and their clothing and other personal articles disinfected. Persons who have had contact with the patient should be placed under quarantine.

FIGURE 5.14 Transmission of plague from the environment to humans.

Reproduced from Centers for Disease Control and Prevention. Protect yourself from plague. C5235098-A. Available at: https://www.cdc.gov/plague/resources/235098_plaguefactsheet_508.pdf Accessed January 23, 2017.

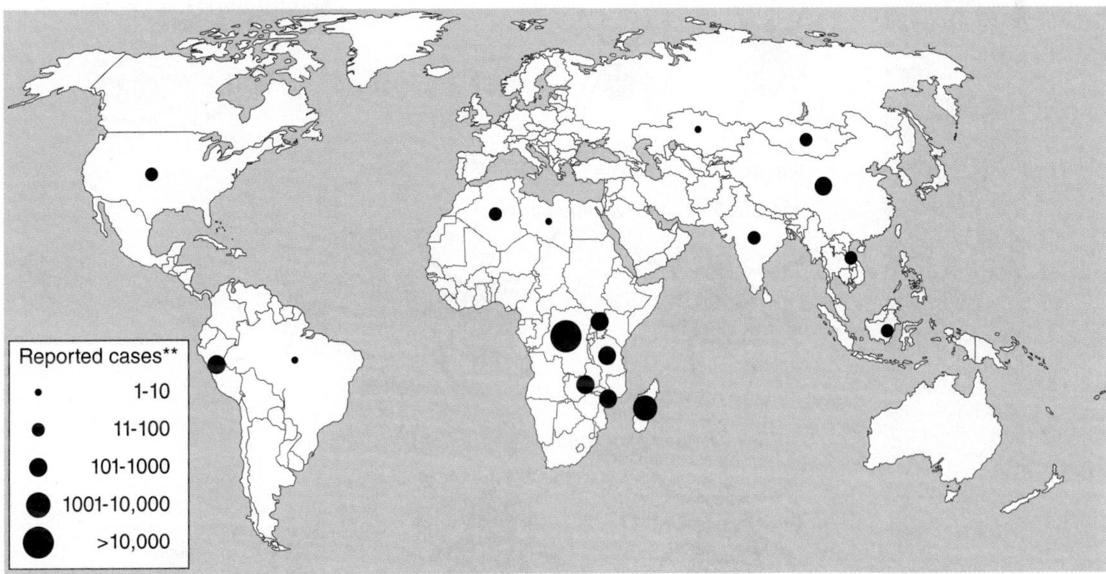

FIGURE 5.15 Worldwide distribution of reported* plague cases by country, 2000–2009.

Modified and reproduced from Centers for Disease Control and Prevention. Plague. Maps and statistics. Available at: https://www.cdc.gov/plague/maps/index.html. Accessed January 29, 2017.

* Data reported to World Health Organization (WHO).

** Dot placed in center of reporting country.

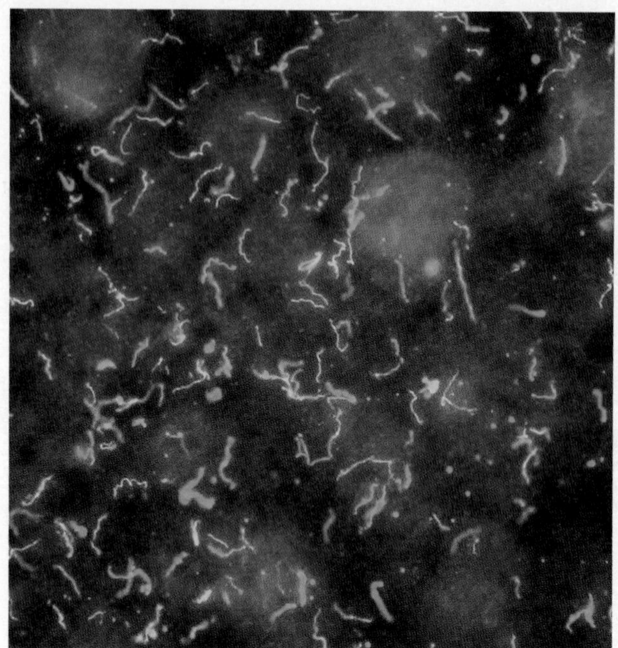

FIGURE 5.16 *Borrelia burgdorferi*, the spirochetal bacteria that cause Lyme disease.

Courtesy of National Institute of Allergy and Infectious Diseases (NIAID)

Environmental control of the disease may be accomplished by encouraging the public to avoid enzootic areas, especially rodent burrows, and direct contact with rodents. Unfortunately, some people ignore posted warning signs in nature parks and risk contracting plague when they feed squirrels and chipmunks. Also, the number of rodents needs to be kept in check. Public health officials should inform the public of the importance of preventing rats from entering buildings and should encourage the removal of food sources that could enable rats to multiply. Shipping areas and docks need to be patrolled because rats can be transferred to and from cargo containers and ships. Hunters and persons who handle wildlife should take care to wear gloves.[19]

Lyme Disease

Lyme disease is a condition that was identified in 1977 when a cluster of arthritis cases occurred among children around the area of Lyme, Connecticut. The causative agent for the disease is a bacterium (bacterial spirochete) known as *Borrelia burgdorferi*, shown in **FIGURE 5.16**.

Transmission of Lyme disease to humans is associated with infected black-legged ticks (*Ixodes scapularis*) that ingest blood by puncturing the skin of the host. (Refer to **FIGURE 5.17**.) In the Pacific coastal region of the United States, the western black-legged tick (*Ixodes pacificus*) has been identified as the vector. Also shown in the figure are the Lone Star tick and the

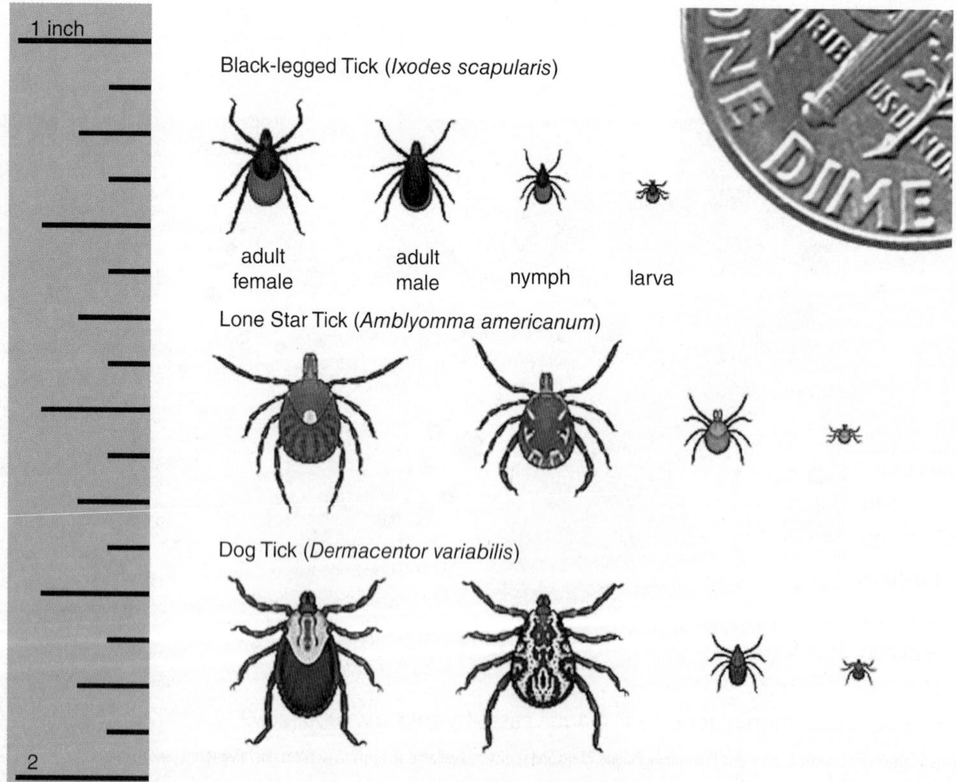

FIGURE 5.17 (Top) From left to right: The black-legged tick (*Ixodes scapularis*) adult female, adult male, nymph, and larva.

Modified and reproduced from Centers for Disease Control and Prevention. Lyme Disease. Available at: https://www.cdc.gov/ticks/life_cycle_and_hosts.html. Accessed January 25, 2017.

dog tick, which are not known to be transmitters of the Lyme disease bacterium to human beings.

A total of 28,453 confirmed cases were reported to the CDC in 2015.[20] From 2005 to 2014, the annual number of reported cases ranged from 19,931 (2006) to 29,959 (2009). Regarding the geographic distribution for Lyme disease, about 95% of cases in 2014 were concentrated in the northeastern, mid-Atlantic, and upper midwestern regions of the United States. (Refer to **FIGURE 5.18**.) The map in Figure 5.18 shows other geographic regions where Lyme disease occurred during that year.

Those at risk of developing Lyme disease include people who work in or come into contact with tick-infested areas. For example, persons who engage in recreational activities that take them outdoors into infested areas or who clear brush or landscape their own backyards may be at risk of infection with Lyme disease. Antibiotic therapy exists for successful treatment of most patients who are diagnosed in the early stages of the disease. Preventive measures are directed at reducing exposure to the ticks that carry the bacterium. For example, people who venture into endemic areas should be inspected for the presence of ticks. Wearing light-colored clothing helps to disclose the presence of ticks.

Rocky Mountain Spotted Fever

Rocky Mountain spotted fever (RMSF), among the most acute infectious diseases, is caused by *Rickettsia rickettsii*, a rickettsial agent. (Rickettsia are agents that are similar to viruses in that they reproduce within living cells and are similar to some bacteria in that they require oxygen and are susceptible to antibiotics.)

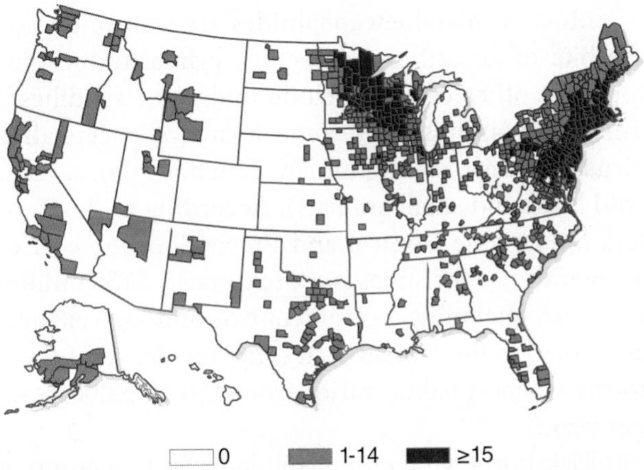

FIGURE 5.18 Incidence* of reported cases of Lyme disease by county—United States, 2014.

Modified and reproduced from Centers for Disease Control and Prevention. Summary of notifiable diseases—United States, 2014. *MMWR*. 2016;63(54):117.

* Per 100,000 population.

It is a febrile disease with sudden onset of moderate to high fevers that may last up to 3 weeks when untreated. Other symptoms include rash, headache, and chills. The case-fatality rate can range up to 25% among untreated patients, but drops dramatically when patients are treated with antibiotics early in the course of the disease. The bite of an infected tick (e.g., the American dog tick or the Rocky Mountain wood tick) is associated with transmission of RMSF. These species of ticks form the reservoir for *R. rickettsii*. Ticks also can transmit RMSF to wild animals and domestic animals such as dogs.

Viral Hemorrhagic Fevers

According to the Centers for Disease Control and Prevention, viral hemorrhagic fevers (VHFs) denote

> ...a group of illnesses that are caused by several distinct families of viruses. In general, the term "viral hemorrhagic fever [VHF]" is used to describe a severe multisystem syndrome.... Characteristically, the overall vascular system is damaged, and the body's ability to regulate itself is impaired. These symptoms are often accompanied by hemorrhage (bleeding); however, the bleeding is itself rarely life-threatening. While some types of hemorrhagic fever viruses can cause relatively mild illnesses, many of these viruses cause severe, life-threatening disease.[21]

Of particular interest to the field of environmental health are the characteristics that VHFs share. The viruses that cause most VHFs are zoonoses, referring to the fact that they require an animal or insect host as a natural reservoir; the viruses reside and multiply in their hosts and are dependent upon their hosts for survival. The principal reservoirs for VHFs are rodents and arthropods. Examples of rodents that form the animal host reservoir for VHFs include certain species of rats such as the cotton rat, mice including the deer mouse and house mouse, and field rodents. Mosquitoes and ticks serve as the main arthropod vectors, whereas other forms of VHFs have unknown hosts.

The causative viruses for VHFs can be found throughout much of the world. Nevertheless, many of the viruses responsible for VHFs tend to be localized to those geographic areas in which the host species reside. Humans may become infected when they encounter an infected host (e.g., through contact with excreta from rodents) or are bitten by an arthropod. Sometimes, person-to-person transmission of the viruses can occur; examples of viruses in this category are Ebola and Marburg viruses. In most instances, there are no known cures for viral hemorrhagic fevers with the exception of supportive therapy. Viruses that cause VHFs and are spread by arthropods are also classified as arboviruses (discussed in the next section).

Examples of VHFs discussed in this chapter are:

- Hantavirus pulmonary syndrome (see emerging zoonoses)
- Dengue hemorrhagic fever (covered in the section on emerging and reemerging infections)

Arthropod-Borne Viral Diseases (Arboviral Diseases)

The term **arboviral disease** (abbreviated from arthropod-borne viral disease) denotes viral diseases that can be acquired when blood-feeding arthropod vectors infect a human host. Typically, the complex cycle of transmission involves a nonhuman vertebrate (reservoir host) and an arthropod vector. The arthropod vectors that transmit arboviruses include ticks, sand flies, biting midges, and mosquitoes; however, most arboviruses are spread by mosquitoes. A total of more than 100 viruses are known at present as arboviruses that produce disease in humans. For some arboviral diseases (e.g., West Nile virus and Japanese encephalitis virus), humans are not usually an important component of the cycle of infection; other arboviruses (e.g., dengue fever virus and yellow fever virus) involve humans in virus amplification and infection of vectors.[19] The four main clinical symptoms or illnesses caused by arboviruses are:

1. Acute central nervous system (CNS) illness ranging in severity from mild aseptic meningitis to encephalitis or flaccid paralysis.
2. Acute self-limited fevers, with and without exanthum (rash), and often accompanied by headache; some may give rise to more serious illness with CNS involvement or hemorrhages.
3. Hemorrhagic fevers, often associated with capillary leakage, shock, and high case-fatality rates (these may be accompanied by liver damage with jaundice, particularly in cases of yellow fever).
4. Polyarthritis and rash, with or without fever and of variable duration, self-limited or with arthralgic sequelae lasting several weeks to years.[19(p34)]

Examples of diseases caused by arboviruses are arboviral encephalitides (e.g., mosquito-borne and tick-borne viral encephalitis) and West Nile virus.

Arboviral Encephalitides

This group of viral illnesses (encephalitis) is associated with acute inflammation of sections of the brain, spinal cord, and meninges. Generally of short duration, the majority of infections are asymptomatic. Other cases present as a mild illness with fever, headache, or aseptic meningitis. Severe illness also can occur, with headache; high fever; disorientation; coma; and, occasionally, convulsions, paralysis and death.[19]

Among the etiologic agents of viral encephalitis are the St. Louis encephalitis virus (SLEV), eastern equine encephalitis virus (EEEV), and LaCrosse encephalitis virus (LACV).[22] A form of viral encephalitis associated with West Nile virus is discussed in next section. St. Louis encephalitis occurs most commonly in the eastern and central states of the United States. Eastern equine encephalitis is found mostly in the Atlantic and Gulf Coast states. The geographic range for most cases of LaCrosse encephalitis spans the upper midwest, mid-Atlantic, and southeastern regions. Western equine encephalitis virus (WEEV) and Venezuelan equine encephalitis virus (VEEV) are two additional viral agents associated with encephalitis. An additional form of encephalitis known as Japanese encephalitis (JE) affects the western Pacific islands ranging from the Philippines to Japan, parts of eastern Asia including Korea and China, as well as India.[19]

Most arboviral encephalitides are transmitted by the bite of an arthropod vector—primarily mosquitoes, but other vectors include midges or sandflies.[19] The reservoir hosts for some forms of encephalitis viruses consist of nonhuman vertebrate hosts (e.g., wild birds and small animals). According to the Centers for Disease Control and Prevention, the cost of arboviral encephalitides is approximately $150 million per year, including vector control and surveillance activities. In the United States, the incidence of these forms of encephalitis varies from 150 to 3,000 cases per year.[22]

Tick-borne viral encephalitides refer to a group of viral encephalitides transmitted by ticks.[19] These forms of encephalitides, also caused by arboviruses, produce diseases that bear a similarity to mosquito-borne encephalitides; the geographic range for tick-borne encephalitides includes parts of the former Soviet Union, eastern and central Europe, Scandinavia, and the United Kingdom. They are found also in eastern Canada and the United States. The reservoir for this group of encephalitides consists of ticks, or ticks in combination with some mammalian species, rodents, and birds.

Other Mosquito-Borne Viral Fevers—West Nile Virus

The West Nile virus (WNV), an example of an arboviral disease, is classified as a type of mosquito-borne viral fever; the etiologic agent is a *Flavivirus*.[19] WNV has gained much notoriety in the United States

WEST NILE VIRUS (WNV) FACT SHEET

What Is West Nile Virus?

West Nile infection can cause serious disease. WNV is established as a seasonal epidemic in North America that flares up in the summer and continues into the fall.

What Can I Do to Prevent WNV?

The easiest and best way to avoid WNV is to prevent mosquito bites.

- When outdoors, use insect repellents containing DEET, picaridin, IR3535, some oil of lemon eucapyptus, or paramenthane-diol. Follow the directions on the package.
- Many mosquitoes are most active from dusk to dawn. Be sure to use insect repellent and wear long sleeves and pants at these times or consider staying indoors during these hours.
- Make sure you have good screens on your windows and doors to keep mosquitoes out.
- Get rid of mosquito breeding sites by emptying standing water from flower pots, buckets, and barrels. Change the water in pet dishes and replace the water in bird baths weekly. Drill holes in tire swings so water drains out. Keep children's wading pools empty and on their sides when they aren't being used.

What Are the Symptoms of WNV?

- **Serious symptoms in a few people.** About one in 150 people infected with WNV will develop severe illness [neuroinvasive disease]. The severe symptoms can include high fever, headache, neck stiffness, stupor, disorientation, coma, tremors, convulsions, muscle weakness, vision loss, numbness, and paralysis. These symptoms may last several weeks, and neurological effects may be permanent.
- **Milder symptoms in some people.** Up to 20% of the people who become infected will have symptoms that can include fever, headache, body aches, nausea, vomiting, and sometimes swollen lymph glands or a skin rash on the chest, stomach, and back. Symptoms can last for as short as a few days to as long as several weeks.
- **No symptoms in most people.** Approximately 80% of people who are infected with WNV will not show any symptoms at all.

How Does West Nile Virus Spread?

- **Infected mosquitoes.** WNV is spread by the bite of an infected mosquito. Mosquitoes become infected when they feed on infected birds. Infected mosquitoes can then spread WNV to humans and other animals when they bite.
- **Transfusions, transplants, and mother-to-child transmission.** In a very small number of cases, WNV also has been spread directly from an infected person through blood transfusions, organ transplants, breastfeeding, and during pregnancy from mother to baby.
- **Not through touching.** WNV is not spread through casual contact such as touching or kissing a person with the virus.

How Soon Do Infected People Get Sick?

People typically develop symptoms between 3 and 14 days after they are bitten by the infected mosquito.

How Is WNV Infection Treated?

There is no specific treatment for WNV infection. In cases with milder symptoms, people experience symptoms such as fever and aches that pass on their own, although illness may last weeks to months. In more severe cases, people usually need to go to the hospital where they can receive supportive treatment including intravenous fluids, help with breathing, and nursing care.

Reproduced from Centers for Disease Control and Prevention. West Nile virus (WNV) fact sheet. Available at: https://www.cdc.gov/westnile/resources/pdfs/wnvFactsheet_508.pdf. Accessed July 14, 2017.

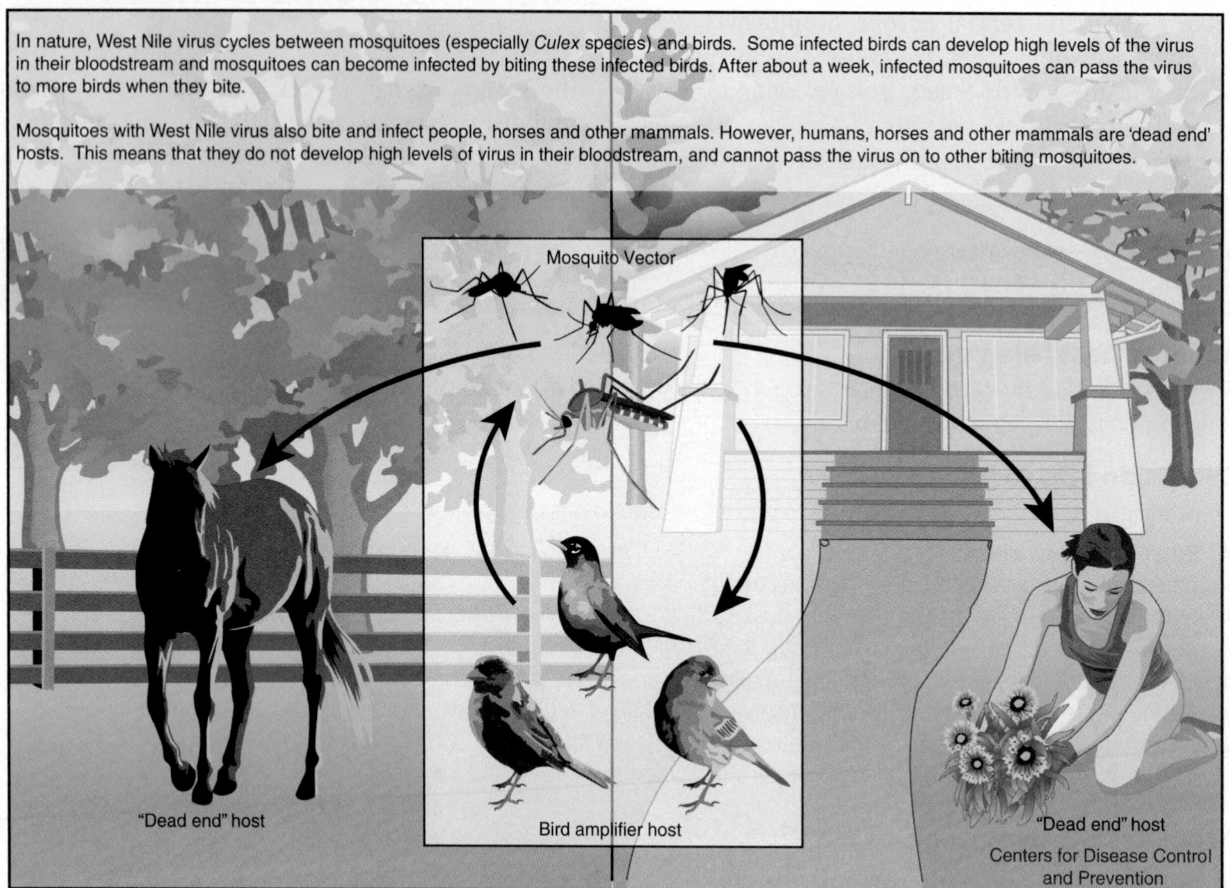
West Nile virus transmission cycle
In nature, West Nile virus cycles between mosquitoes (especially Culex species) and birds. Some infected birds can develop high levels of the virus in their bloodstream and mosquitoes can become infected by biting these infected birds. After about a week, infected mosquitoes can pass the virus to more birds when they bite.
Mosquitoes with West Nile virus also bite and infect people, horses and other mammals. However, humans, horses and other mammals are 'dead end' hosts. This means that they do not develop high levels of virus in their bloodstream, and cannot pass the virus on to other biting mosquitoes.
Mosquito Vector
"Dead end" host
Bird amplifier host
"Dead end" host
Centers for Disease Control and Prevention

FIGURE 5.19 West Nile virus transmission cycle.

Reproduced from Centers for Disease Control and Prevention. West Nile virus transmission cycle. Available at: https://www.cdc.gov/westnile/resources/pdfs/13_240124_west_nile_lifecycle_birds_plainlanguage_508.pdf. Accessed July 14, 2017.

because of its ability to spread rapidly across wide geographic regions. In addition, WNV caused the deaths of many wild birds and some humans. During summer 2004, WNV spread to the US state of California, having been associated previously with outbreaks in New York and other areas of the eastern United States. People most at risk of contracting WNV are those who spend a great deal of time outdoors, where they could be bitten by infected mosquitoes. Older people (age 50+) are more likely than younger people to develop serious symptoms from WNV infection. Refer to the text box for more information.

FIGURE 5.19 presents the West Nile virus transmission cycle. Neuroinvasive disease is the most severe form of WNV and occurs in somewhat less than 1% of cases. **FIGURE 5.20** shows the incidence of reported cases of neuroinvasive disease associated with West Nile virus in United States and US territories during 2014. The states with the highest reported incidence of neuroinvasive disease were Nebraska, North Dakota, California, South Dakota, Louisiana, and Arizona.

Emerging and Reemerging Infectious Diseases/Emerging Zoonoses

Emerging and reemerging infectious diseases include conditions that may have been unrecognized as well as those reappearing after a decline in incidence. The term **emerging zoonoses** refers to zoonotic diseases that are caused by either apparently new agents or by known agents that occur in locales or species that previously did not appear to be affected by these known agents.[23]

Several factors may have contributed to the resurgence of these conditions: migration of human populations within tropical and semitropical areas, increases in international travel, population growth with attendant crowding and urbanization, overuse of antibiotics and pesticides, lack of clean drinking water, and climate changes attributed to human activity.[1] In the northern hemisphere, rising temperatures

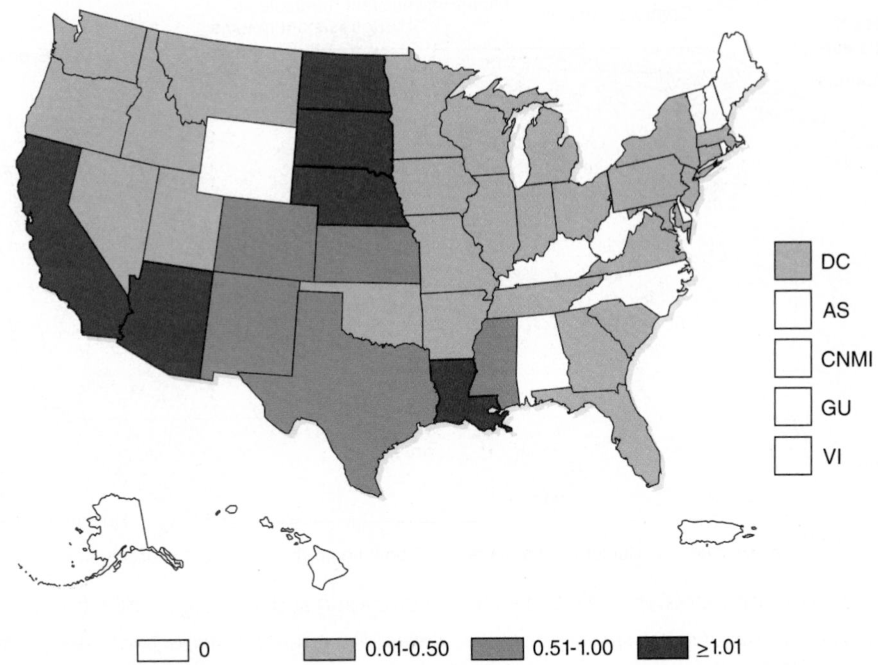

FIGURE 5.20 West Nile virus. Incidence* of reported cases of neuroinvasive disease—United States and US territories, 2014.

Modified and reproduced from Centers for Disease Control and Prevention. Summary of notifiable diseases—United States, 2014. *MMWR*. 2016;63(54):79.

* Per 100,000 population.

and increasing rainfall promote the growth and northward migration of mosquito populations, with corresponding increases in diseases transmitted by mosquitoes.

See **FIGURE 5.21** for the geographic locations where emerging and reemerging infections have been identified; note that the figure shows both zoonotic and nonzoonotic conditions. Among the most commonly identified factors associated with the rise of emerging zoonoses are ecological changes that result from agricultural practices (e.g., deforestation, conversion of grasslands, and irrigation).[23] Other factors include changes in the human population and human behavior (e.g., wars, migration, and urbanization). **TABLE 5.3** presents examples of emerging zoonoses and lists conditions that may be linked to their occurrence. In the following sections, some of these emerging zoonoses will be covered in more detail.

Hantaviruses and Hantavirus Pulmonary Syndrome

The hantavirus pulmonary syndrome (HPS) represents a very severe and sometimes fatal respiratory condition that is transmitted by rodent vectors.[24] The causative agent is the hantavirus, part of the bunyavirus family of viruses.[25] HPS may be transmitted when aerosolized (airborne) urine and droppings from infected rodents are inhaled. For example, vacationers who return to a mountain cabin after an extended absence may cause rodent droppings to be aerosolized when they sweep up in order to make the cabin habitable, or they may stir up virus particles when they walk across a floor contaminated with urine from rodents. The primary known vectors for transmission of the hantavirus are four species of rodents: the cotton rat, the rice rat, the white-footed mouse, and the deer mouse (shown in **FIGURE 5.22**). One of the most widely distributed vectors and the main host for the hantavirus is the deceptively cute deer mouse—*Peromyscus maniculatus*—which is found throughout North America.

Hantaviruses have existed for many years, yet they gained public attention in 1993 when a mysterious case (a so-called first outbreak) occurred. The CDC reports that "[i]n May 1993, an outbreak of an unexplained pulmonary illness occurred in the southwestern United States, in an area shared by Arizona, New Mexico, Colorado and Utah known as 'The Four Corners.' A young, physically fit Navajo man suffering from shortness of breath was rushed to a hospital in New Mexico and died very rapidly."[26] This case created an environmental health mystery in which an unknown causative agent also was believed to have caused the death of the man's fiancée and an

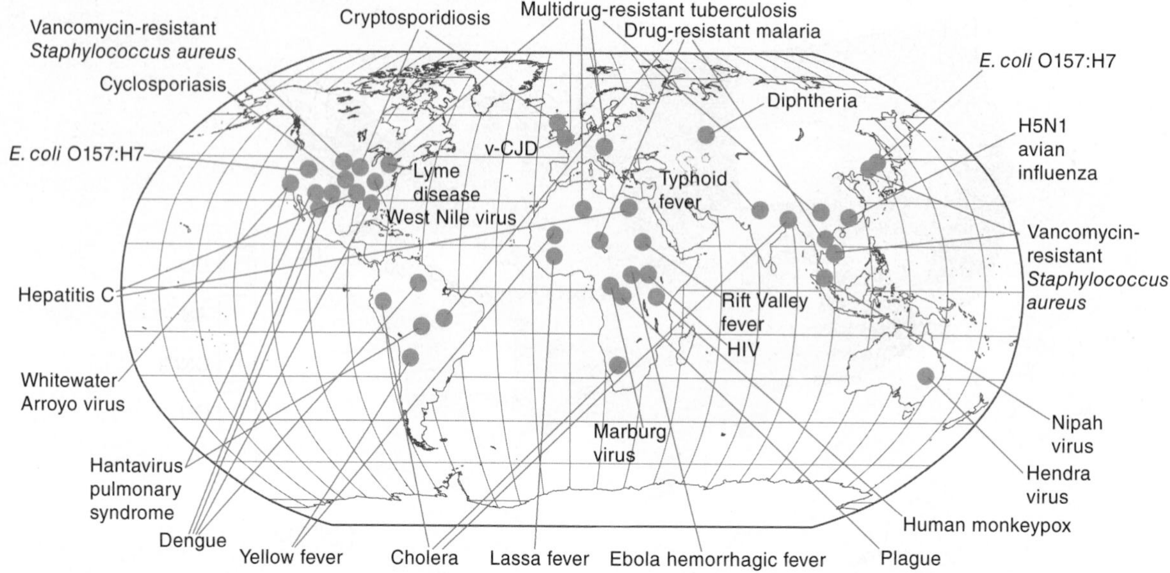

FIGURE 5.21 Range and recognized site(s) of origin of a variety of emerging and reemerging infections.

Modified and reproduced from AS Fauci. Infectious diseases: considerations for the 21st century. *Clinical Infectious Diseases*. 2001;32:677, with permission from the University of Chicago Press, © 2001 by the Infectious Diseases Society of America.

v-CJD = variant Creutzfeldt-Jakob disease; E. coli = *Escherichia coli.*

TABLE 5.3 Examples of Emerging Zoonoses and Factors Contributing to Emergence

Infection or Agent	Factor(s) Contributing to Emergence
Non-conventional agent	
Bovine spongiform encephalopathy (BSE—mad cow disease)[a]	Changes in rendering process
Viral	
Argentinian, Bolivian hemorrhagic fever	Changes in agriculture favoring rodent host
Dengue, dengue hemorrhagic fever	Transportation, travel and migration; urbanization
Ebola	Unknown, nosocomial
Hantaviruses	Ecological/environmental changes, increasing rodent contacts
Lassa fever	Urbanization favoring rodent host, increasing exposure
Rift Valley fever	Dam building, agriculture, irrigation; possible change in virulence or pathogenicity
Bacterial	
Heliobacter pylori (gastric ulcers)	Probably has been long widespread, now recognized
Escherichia coli O157:H7 (hemolytic-uremic syndrome)	Mass food processing technology allowing contamination of meat
Borrelia burgdorferi (Lyme disease)	Reforestation around homes, suburbanization, and other conditions favoring tick vector and deer population
Parasitic	
Cryptosporidium and other pathogens	Contaminated surface water, faulty water purification

[a] Discussed in Chapter 11.

Modified and reproduced from BB Chomel. New emerging zoonoses: a challenge and an opportunity for the veterinary profession. *Comparative Immunology, Microbiology & Infectious Diseases*. 1998;21:4. Copyright 1998; used with permission from Elsevier.

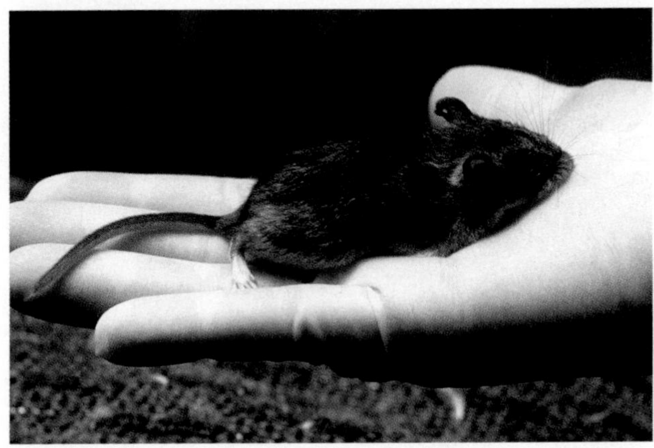

FIGURE 5.22 This is a deer mouse, *Peromyscus maniculatus*, a hantavirus carrier that becomes a threat when it enters human habitation in rural and suburban areas.

Courtesy of CDC/Brian W.J. Mahy, BSc, MA, PhD, ScD, DSc

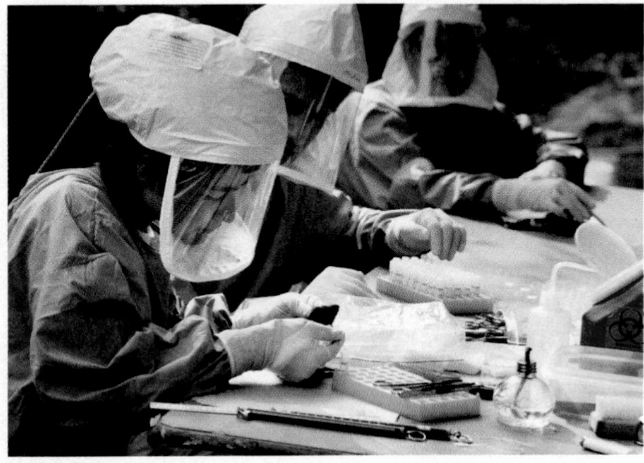

FIGURE 5.23 Three CDC health officials inspecting specimens suspected of being connected with a hantavirus outbreak.

Courtesy of CDC

additional five people. Epidemiologic investigations ruled out a new type of influenza or pesticide exposure. Eventually the deaths were linked to a heretofore unknown form of hantavirus, which was traced to the deer mouse via extensive rodent-trapping programs. To avoid a panic situation, the trappers did not wear protective devices. However, in the laboratory, researchers who dissected the trapped rodents wore special clothing, masks, and goggles as protection against biohazards (see **FIGURE 5.23**).

Further study suggested that rodents may have entered the households where the victims lived. During 1993, the deer mouse population increased to higher levels than normal. This population increase was attributed to heavy snows and rainfall in early 1993 that followed a lengthy drought in the Four Corners area. Because of the increased availability of food following the heavy precipitation, the mouse population increased by a factor of 10 between May 1992 and May 1993. Thus it was more likely that humans could contract hantavirus carried by the mice.

Although HPS is known to cause potentially deadly infections, fortunately the syndrome is rare. According to the CDC, "[t]hrough January 6, 2016, a total of 690 cases of hantavirus pulmonary syndrome have been reported in the United States. Of these, 659 cases occurred from 1993, onward, following identification of hantavirus pulmonary syndrome, whereas 31 cases were retrospectively identified."[27] The epidemiologic features of these cases in 35 states where the syndrome was reported were as follows: 63% male, 78% white, 18% American Indian, and mean age 38 years; the case-fatality rate for the disease was 36%.[27] (See **FIGURE 5.24**.) Among the states reporting the largest number of cases were New Mexico, Colorado, California, and Arizona.

Dengue Fever, Dengue Hemorrhagic Fever, and Dengue Shock Syndrome

Four related viruses (flavivirus serotypes) cause dengue infection, which produces a spectrum of illness that ranges from mild to severe. The mildest form of dengue causes a low-level illness with a nonspecific fever. Classic dengue fever (DF) is also a rather benign, self-limited disease; symptoms may include high fever, severe headache, eye pain, muscle pain, joint pain, rash, and mild bleeding. Dengue hemorrhagic fever (DHF) results in a life-threatening illness with fever (that lasts 2–7 days), abdominal pain, and bleeding phenomena. Dengue shock syndrome (DSS) includes the symptoms of DHF and is associated with shock (e.g., hypotension and rapid, weak pulse). DSS is a potentially fatal condition.[19,28]

As shown in **FIGURE 5.25**, the primary locations for dengue fever are the tropical and subtropical areas of the world, for example, southeast Asia, tropical Africa, and South America. The virus is thought to have existed among monkeys, and several centuries ago moved to humans in Africa and Asia. Dengue was uncommon until after the mid-20th century, with outbreaks occurring in the Philippines and Thailand in the 1950s and later in the Caribbean and Latin America.

The vector for transmission of the disease is a type of domestic day-biting mosquito known as the *Aedes aegypti*, a mosquito that prefers to feed on human hosts. (Refer to **FIGURE 5.26**.) The spread of the vector was linked to transport ships during World War II. At present, dengue is among the most significant mosquito-borne viral diseases that afflict humans. The CDC estimates that more than 100 million cases of dengue fever occur annually.[28]

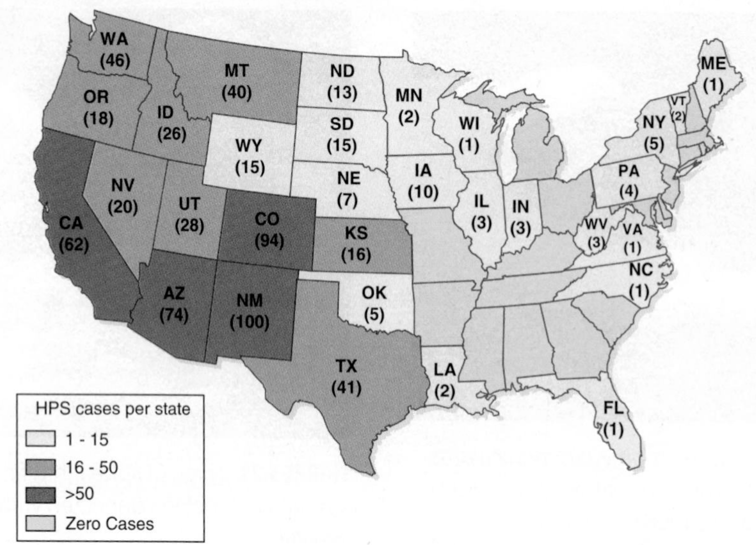

FIGURE 5.24 Hantavirus pulmonary syndrome (HPS) cases,* by state of residence: 1993—January 8, 2017.

Reproduced from Centers for Disease Control and Prevention. Hantavirus. Available at: https://www.cdc.gov/hantavirus/surveillance/state-of-exposure.html. Accessed January 26, 2017.

* N = 659 in 31 states.

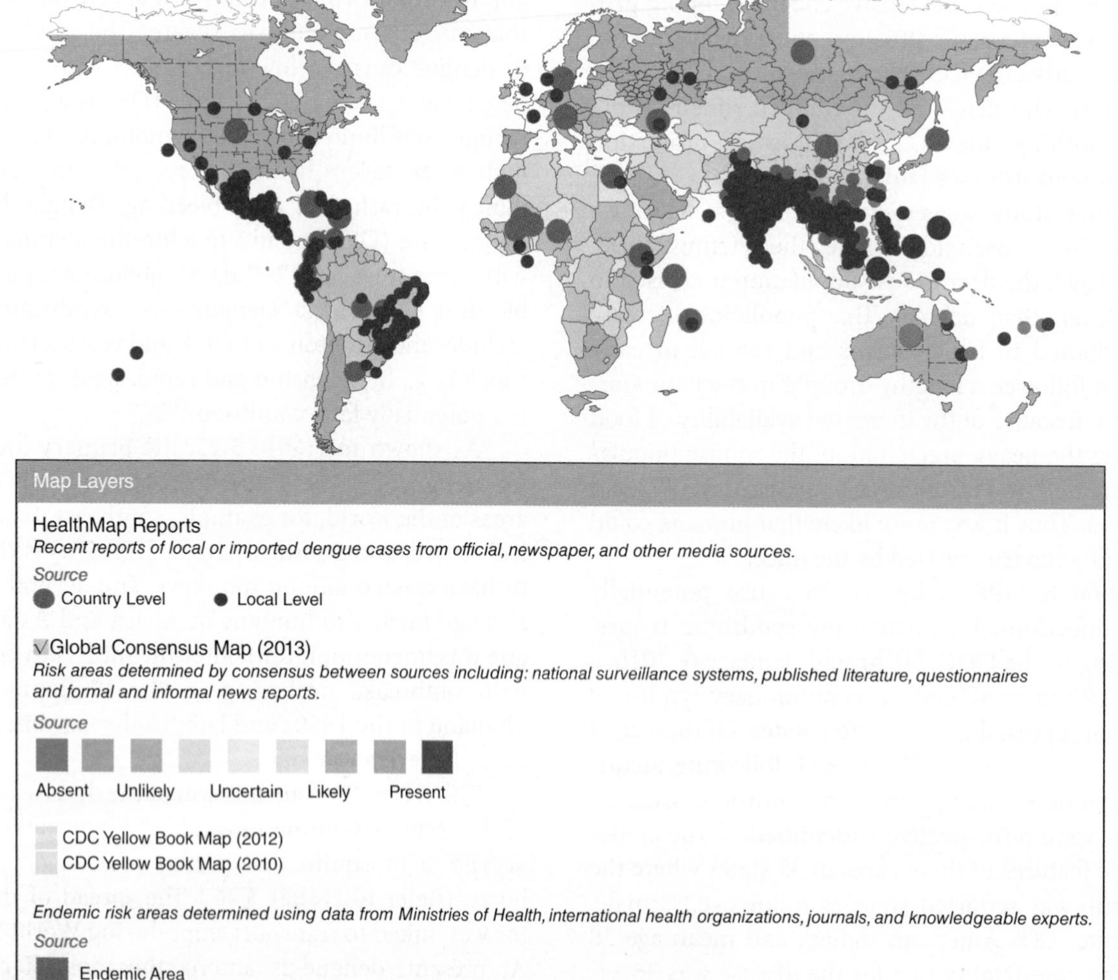

FIGURE 5.25 World distribution of dengue—2008.

Reproduced from Centers for Disease Control and Prevention. Dengue. Available at: http://www.healthmap.org/dengue/en/. Accessed October 19, 2017.

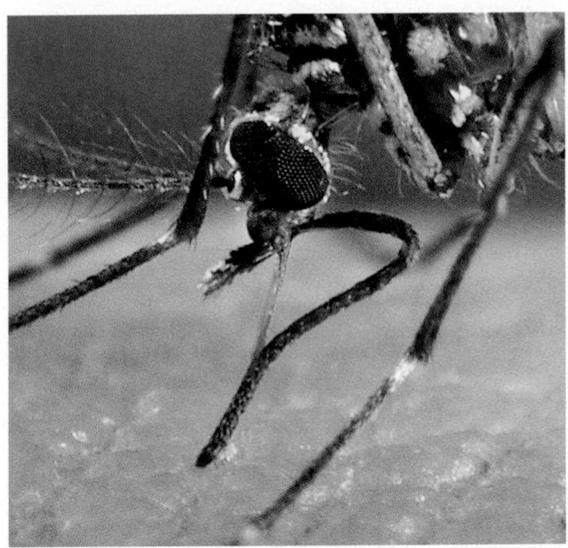

FIGURE 5.26 A female *Aedes aegypti* mosquito, which is the primary vector for the spread of dengue fever.
Courtesy of CDC/James Gathany

In the United States, most reported cases of dengue usually are imported by travelers who are returning from endemic areas or by immigrants. Although very uncommon, cases of dengue fever that originated in the continental United States have been documented. Southern Texas and the southeastern United States are at theoretical risk for transmission of dengue and for sporadic outbreaks.

Public health authorities have attributed the dramatic global emergence of dengue fever to lack of effective mosquito control in endemic areas, rapid population growth and urbanization, and poor or nonexistent systems for the treatment of water and human wastes. International travel is also thought to have contributed to increases in dengue fever by transmitting the viruses from endemic to nonendemic areas. Mosquito eradication programs that involved the use of DDT have been discontinued, causing the resurgence of *Aedes aegypti*. For example, between 1930 and 1970, the range of *A. aegypti* declined greatly in much of Central and South America; then, by 1998, the mosquito had reinvaded previous habitats.

Zika Virus

The mosquito-borne Zika virus, linked with development of a febrile rash, has been associated with adverse birth outcomes among pregnant women who become infected. The CDC asserts that congenital infection with the Zika virus is capable of producing microcephaly and abnormalities of the brain among infants born to infected pregnant women.[29] Since late 2014, health officials in Brazil observed increased Zika virus transmission that accompanied increased reports of microcephaly. As of January 2016, the total number of reports of suspected cases in Brazil reached 3,520.

Cases of Zika infection have also been reported in the US—279 reports as of early May 2016. Most of these were associated with travel to areas where local transmission of Zika had taken place. However, by midsummer of 2016, cases also were related to local occurrence in the continental United States (e.g., Florida and California) and US territories. **FIGURE 5.27** shows how the Zika virus is spread. In addition to being mosquito borne, the virus was found to be transmitted sexually and through blood transfusions. Epidemiologic techniques, for example, surveillance programs, have helped to track the Zika virus and reduce risks of its transmission.

Ebola Viral Hemorrhagic Fever

This condition refers to a dramatic, highly fatal, and acute disease associated with infection with Ebola virus (shown in **FIGURE 5.28**). Ebola hemorrhagic fever (HF) is one of a group of viral diseases known as Ebola-Marburg viral diseases.[19] Symptoms of infection with the virus can include sudden onset of fever, headache, diarrhea, and vomiting and may also involve external and internal bleeding. Ebola disease was first recognized in 1976 in the Sudan and in Zaire (now called the Democratic Republic of Congo—DRC). In 1995, Kikwit, DRC experienced an outbreak of 315 cases that caused 244 deaths. Outbreaks associated with a subtype of Ebola virus have occurred among monkeys housed in quarantine facilities in the United States.

The largest outbreak in history descended upon west Africa in 2014. By April 13, 2016, a total of 28,652 cases had been reported in Africa. Approximately two out of five persons with Ebola died. When the Ebola outbreak exploded in 2014, public health officials scrambled to meet the challenge. Epidemiologic methods contributed to bringing this massive outbreak under control.

Rift Valley Fever

The causative agent for Rift Valley fever (RVF) is a virus from the genus *Phlebovirus* in the family *Bunyaviridae*.[30] **FIGURE 5.29** presents an electron micrograph of the virus.

RVF can produce epizootic or widespread disease among domestic animals including cattle, buffalo, sheep, goats, and camels. Mosquitoes transmit the disease to domestic animals that can then spread the disease to one another. The disease is transmissible to humans and can produce epidemics among human populations. Infected mosquitoes are able to transmit RVF to humans, who also may contract the disease if

Most people get zika from a mosquito bite
A mosquito bites a person infected with zika virus
The mosquito becomes infected
A mosquito will often live in a single house during its lifetime
The infected mosquito bites a family member or neighbor and infects them
More mosquitoes get infected and spread the virus
More members in the community become infected
Other, less common ways, people get zika:
During pregnancy
A pregnant woman can pass zika virus to her fetus during pregnancy. Zika causes microcephaly, a severe birth defect that is a sign of incomplete brain development
Through sex
Zika virus can be passed through sex from a person who has zika to his or her sex partners
Through blood transfusion
There is a strong possibility that zika virus can be spread through blood transfusions

FIGURE 5.27 How Zika spreads.

Reproduced from Centers for Disease Control and Prevention. Protect your family and community: How Zika spreads. Available at: https://www.cdc.gov/zika/pdfs/Zika-Transmission-Infographic.pdf. Accessed July 14, 2017.

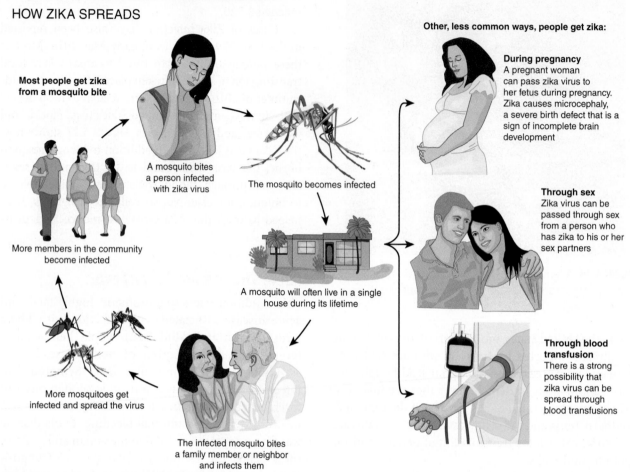

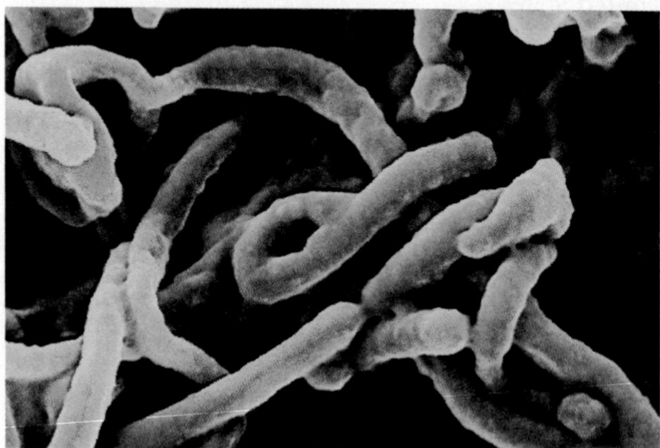

FIGURE 5.28 Scanning electron microscopic image of Ebola virus particles.

Courtesy of National Institute of Allergy and Infectious Diseases (NIAID)

they encounter the blood or body fluids of infected animals through slaughtering or handling infected meat. In humans, infection with RVF most typically produces either no obvious symptoms or mild illness characterized by fever and liver abnormality. Some patients may develop a more severe form in which they experience a hemorrhagic fever, encephalitis, or ocular diseases. An outbreak of Rift Valley fever in Kenya in 1950 through 1951 caused the death of approximately 100,000 sheep.

The endemic areas for RVF include southern Africa, most countries of sub-Saharan Africa, and Madagascar. (Refer to **FIGURE 5.30**.) Increases in Rift Valley fever follow periods of heavy rainfall when large numbers of mosquitoes develop. In west Africa in 1987, an outbreak of Rift Valley fever was linked to construction of the Senegal River project. This project modified the usual interactions between animals and humans when flooding occurred in the lower Senegal River area.

Other Zoonotic Diseases

Monkeypox

Monkeypox derives its name from 1958 outbreaks that occurred in laboratory monkeys.[31] A rare disease

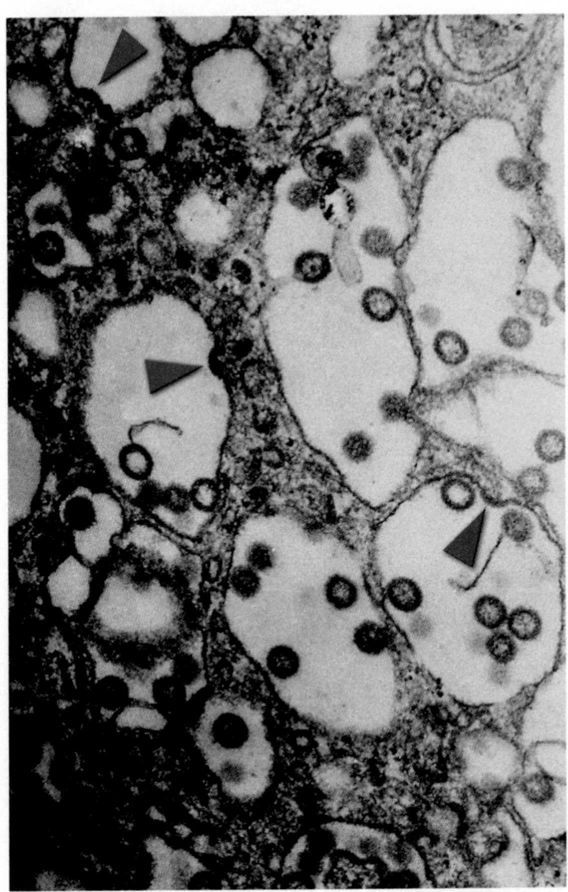

FIGURE 5.29 Electron micrograph of the Rift Valley Fever virus.
Courtesy of CDC/Dr. Fred Murphy

caused by the monkeypox virus, which is from the *Orthopoxvirus* genus in the family *Poxviridae*. This genus includes the smallpox virus (variola virus), the vaccinia virus (used for vaccinating against smallpox), and the cowpox virus. Endemic in the Democratic Republic of Congo, monkeypox has also occurred in other countries in central and west Africa. According to the CDC, "...symptoms of monkeypox are similar to but milder that the symptoms of smallpox. Monkeypox begins with fever, headache, muscle aches, and exhaustion. The main difference in symptoms of smallpox and monkeypox is that the latter causes lymph nodes to swell (lymphadenopathy) while smallpox does not."[31] Refer to the text box for a description of an outbreak that occurred in the midwestern United States during 2003—the only outbreak that has happened outside of Africa.

Tularemia

Also known as rabbit fever, tularemia is a bacterial disease associated with *Francisella tularensis*. The disease is broadly distributed in the United States, and occurs in all states except Hawaii. It is most often a disease of rural areas and is linked with wild animals such as rodents, rabbits, and hares. The bacterium may be transmitted in several ways, including the bite of an arthropod (e.g., tick or deer fly), coming into contact with infected animal carcasses, consuming food or water that has been contaminated with the bacterium, or even breathing in the bacterium.[32] The condition presents with a range of possible symptoms that are related to how the organism enters the body. Often, symptoms include skin ulcers, swollen lymph glands, painful eyes, and sore throat. Inhalation of the bacterium may cause sudden fever, chills, headaches, muscle aches, joint pain, dry cough, and progressive weakness; in some cases, pneumonia may occur.[19] Tularemia, which is treatable with antibiotics, may be a severe or fatal condition when untreated.

MULTISTATE OUTBREAK OF MONKEYPOX—ILLINOIS, INDIANA, AND WISCONSIN, 2003

The CDC has received reports of patients with a febrile rash illness who had close contact with pet prairie dogs and other animals. . . . A total of 51 patients reported direct or close contact with prairie dogs (*Cynomys* sp.), and one patient reported contact with a Gambian giant rat (*Cricetomys* sp.). One patient had contact with a rabbit (Family *Leporidae*) that became ill after exposure to an ill prairie dog at a veterinary clinic. Traceback investigations have been initiated to identify the source of monkeypox virus introduced into the United States and have identified a common distributor where prairie dogs and Gambian giant rats were housed together in Illinois. A search of imported animal records revealed that Gambian giant rats were shipped from Ghana in April to a wildlife importer in Texas and subsequently were sold to the Illinois distributor. The shipment contained approximately 800 small mammals of nine different species that might have been the actual source of introduction of monkeypox.

Modified from Centers for Disease Control and Prevention. Multistate outbreak of monkeypox—Illinois, Indiana, and Wisconsin, 2003. *MMWR*. 2003;52:537–540.

Rabies (Hydrophobia)

Rabies is an acute and highly fatal disease of the central nervous system caused by a virus transmitted most often through saliva from the bites of infected animals; globally, dog bites are the principal source of transmission of rabies to humans.[19] A disease that affects mammals, it causes encephalopathy and paralysis of the respiratory system. In the early stages of the disease, symptoms of rabies are nonspecific, consisting of apprehension, fever, headache, and

FIGURE 5.30 Rift Valley Fever distribution map.

Modified and reproduced from Centers for Disease Control and Prevention, National Center for Emerging and Zoonotic Infectious Diseases (NCEZID). Rift Valley Fever Distribution Map. Available at: https://www.cdc.gov/vhf/rvf/outbreaks/distribution-map.html. Accessed January 31, 2017.

malaise. The disease then progresses to paralysis, hallucinations, swallowing difficulties, and fear of water (called hydrophobia). Rabies is almost always fatal; only a few cases of survival from rabies have been documented worldwide. Once a person has developed clinical rabies, the current option for treatment consists of intensive medical care. However, vaccination (called postexposure prophylaxis [PEP]) can be administered to an individual who has been bitten by a suspected rabid animal in order to prevent clinical rabies. PEP should be given soon after the bite has occurred.

In the United States, human cases of rabies are rare. A case of abortive human rabies occurred in Texas beginning in early 2009. The case was termed "abortive" because that patient recovered without ever having received intensive care. The patient was a 17-year-old female who was seen in a hospital emergency room for symptoms that included severe headache, photophobia (sensitivity to light), neck pain, and fever. The patient was discharged after 3 days, when the symptoms resolved. Subsequently, after the headaches returned, she was rehospitalized and treated for suspected infectious encephalitis. During subsequent examinations, the patient revealed that while on a camping trip she had been bitten by flying bats. After serological tests were found to be positive for rabies, the patient was administered rabies

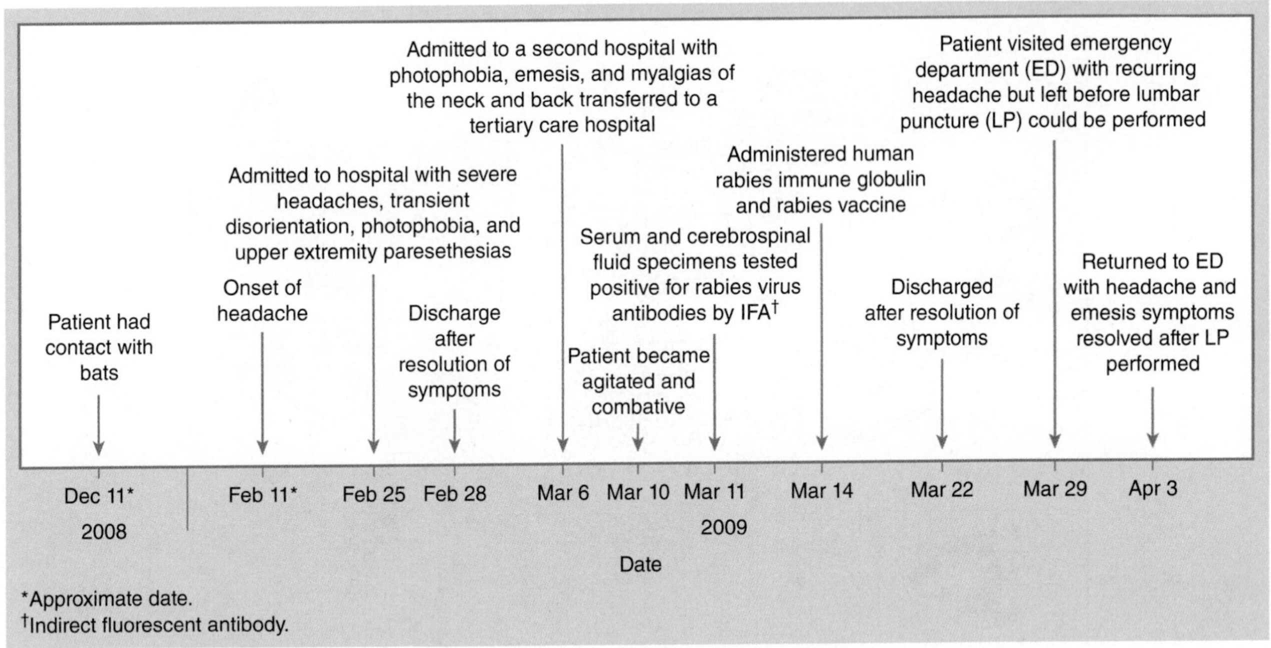

FIGURE 5.31 Timeline of course for a patient with presumptive abortive human rabies—Texas, 2009.

Reproduced from Centers for Disease Control and Prevention. Presumptive abortive human rabies—Texas, 2009. *MMWR* 2010;59:187.

immune globulin and rabies vaccine. Following these steps, the young woman was given supportive care until the symptoms resolved and she was discharged. Refer to **FIGURE 5.31** for a time line of the patient's illness.

The hosts for rabies are wild animals—carnivores and bats. Before 1960, the majority of US rabies cases occurred among domestic animals—dogs, cats, and cattle—now most cases occur in the wild. The CDC estimates that more than 90% of rabies cases occur in wild animals (e.g., skunks, raccoons, foxes, and coyotes) and the remainder in domestic animals. The locations of wild animal rabies reservoirs in the United States are shown in **FIGURE 5.32. FIGURE 5.33** presents the number of reported cases of rabies that occurred among wild and domestic animals during 1983 to 2014 (in the United States and Puerto Rico). The two most commonly affected species were raccoons and bats. In the United States, vaccination programs for domestic animals, measures to control animals, and public health laboratories for conducting rabies tests have changed the prevalence of rabies.

Environmental health programs have prevented human cases of rabies very successfully through post-exposure prophylaxis (PEP). In the United States, one reason that fatal human rabies cases have declined from about 100 annually in the early 1900s to about 1 to 2 cases annually at the end of the 20th century has been the introduction of PEP, which is nearly 100% successful; the remaining fatal cases usually can be attributed to failure to seek medical attention. PEP consists of a series of vaccinations that should be given as soon as possible after the occurrence of an animal bite from a suspected rabid animal. Outside the United States, rabid dogs—which cause 99% of human rabies deaths—constitute the most common source of rabies exposure. For this reason, an important component of rabies prevention is the control and vaccination of stray dogs. Within the United States, people are most likely to be exposed to rabies from bats.[19]

Anthrax

Anthrax gained notoriety in 2001 when it was distributed intentionally via the US mail system. Approximately one week after the September 11 terrorist attacks on the World Trade Center in New York City and the Pentagon in Washington, DC, letters containing anthrax spores were mailed to former NBC news anchor Tom Brokaw and to the editor of the *New York Post* in New York City, exposing employees to the spores. These exposures resulted in five confirmed cases of anthrax among media company employees or visitors. An isolated case was a 61-year-old hospital worker who subsequently died. Additional exposures also were reported in Washington, DC; Virginia, New Jersey, and Florida. Ultimately, there were a total of 22 confirmed cases of anthrax and 5 resulting deaths.[33–35]

The causative agent for anthrax, an acute infectious disease, is *Bacillus anthracis*, a spore-forming

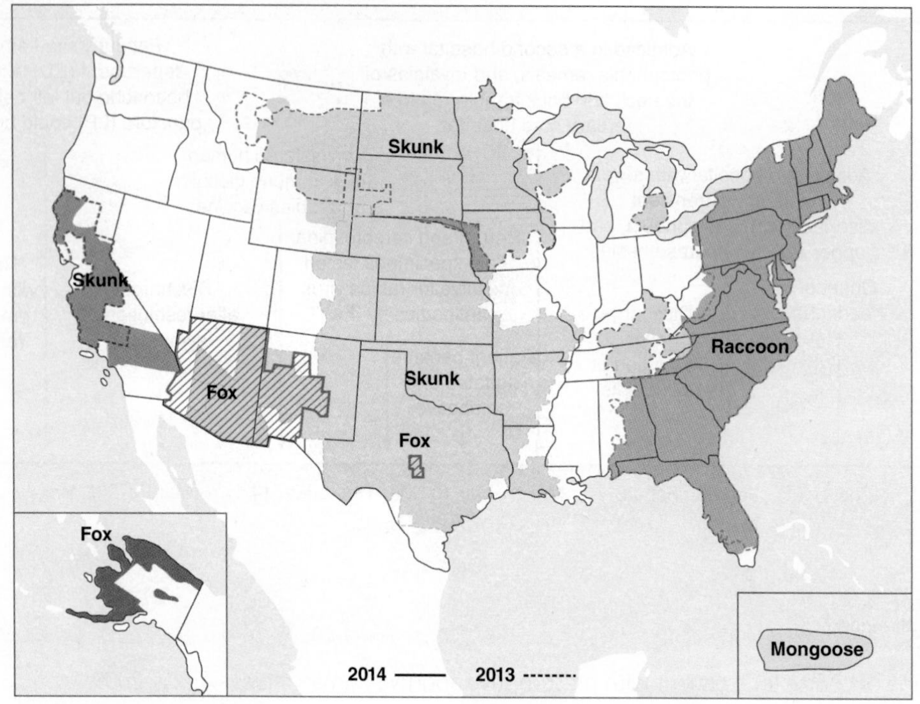

FIGURE 5.32 Distribution of major rabies virus variants among mesocarnivores* in the United States and Puerto Rico, 2008 to 2014.

Reproduced from Centers for Disease Control and Prevention. Rabies surveillance in the United States during 2014. March 2016. Available at: https://www.cdc.gov/rabies/pdf/2014-us-rabies-surveillance-508.pdf. Accessed January 25, 2017.

* An animal that consumes meat as 50% to 70% of its diet.
Black diagonal lines: fox rabies variants (Arizona gray fox and Texas gray fox).
Solid borders: 5-year rabies virus variant aggregates for 2009 through 2014.
Dashed borders: the previous 5-year aggregates for 2008 through 2013.

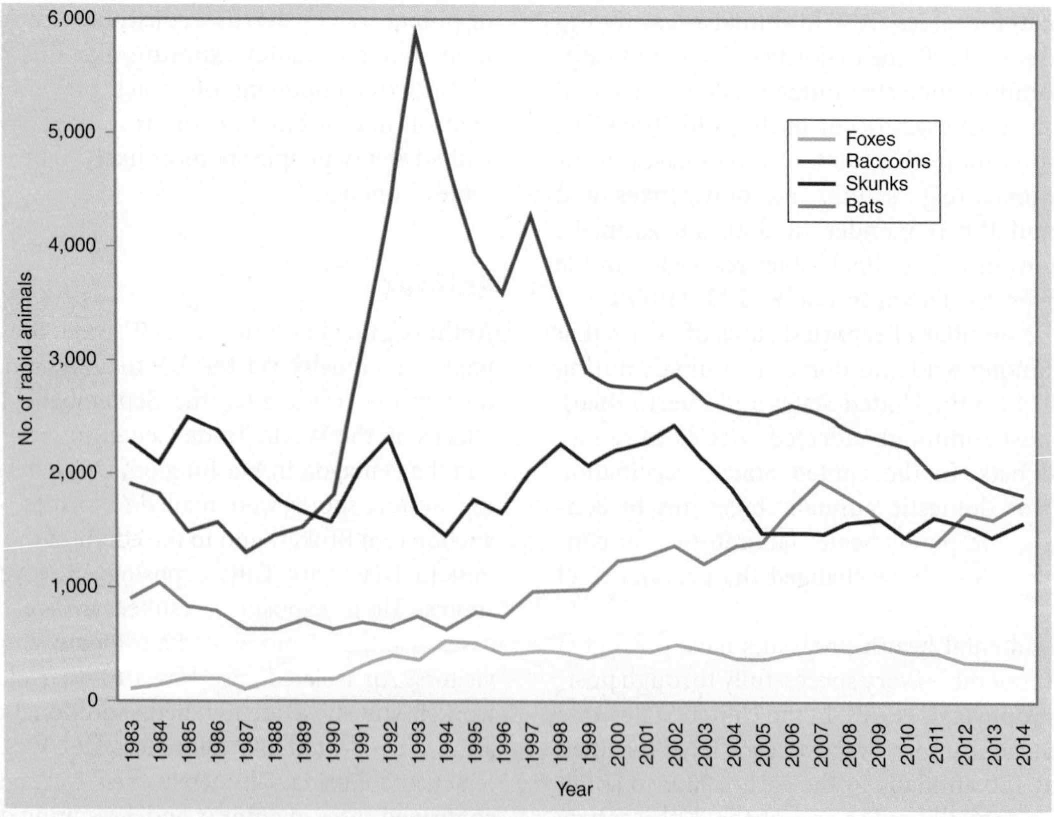

FIGURE 5.33 Cases of rabies among wildlife in the United States, by year and species, 1983 to 2014 (in thousands).

Reproduced from Centers for Disease Control and Prevention. Rabies surveillance in the United States during 2014. March 2016. Available at: https://www.cdc.gov/rabies/pdf/2014-us-rabies-surveillance-508.pdf. Accessed January 25, 2017.

bacterium. (Refer to **FIGURE 5.34**.) Anthrax is a disease that most commonly occurs among cattle, sheep, goats, and other herbivores.[36] Those normally at risk of anthrax are veterinarians, agricultural workers who come into contact with cattle, and those who are involved with the processing of animal products: hides, wool, and hair.[19] The general population is unlikely to become infected (with the exception of those who work on ranches or farms where livestock are present).

The three forms of anthrax are cutaneous, inhalational, and gastrointestinal. In humans, the cutaneous form is characterized by itching, skin lesions, development of a black eschar, and possible spread to lymph nodes. Although the case fatality rate for untreated cutaneous anthrax ranges from 5% to 20%, the disease can be treated effectively with antibiotics.[19]

The inhalational form of anthrax begins with nonspecific symptoms that mimic upper respiratory infections and then deepen into respiratory distress, fever, and shock. This form of anthrax is very severe and has a high case fatality rate.

The gastrointestinal form may be acquired from infected animals by eating contaminated meat that is undercooked. Gastrointestinal anthrax is the least common form of anthrax in the United States. Symptoms of gastrointestinal anthrax include nausea, anorexia, and abdominal pain; from 25% to 60% of cases are fatal. All three forms of anthrax are treatable with antibiotics that should be administered as soon as possible after suspected exposure. Anthrax infections can develop largely undetected until the patient becomes very ill and nearly dies. When the case is very advanced, treatment with antibiotics is not likely to be effective.

Before the use of anthrax as a terrorist weapon, environmental health experts were concerned mainly about outbreaks of the disease among cattle and other herbivores and among people who have contact with these animals and their products. Subsequently, public health departments were spurred to action by the potential threat from a terrorist-incited anthrax epidemic. The anthrax bacterium has the capacity to spread across a wide area and cause many cases of illness and numerous deaths. A challenge to environmental health professionals will be the development of effective methods to recognize a terrorist attack, respond to it in an appropriate and timely manner, and bring the threat under control.

Psittacosis

Psittacosis, a disease associated with the bacterial agent *Chlamydia psittaci*, is conveyed by dried bird droppings (refer to the text box for a description).

Influenza: Animal to Human Transmission of Influenza A Viruses

Influenza A viruses associated with human influenza epidemics and pandemics are linked to animal reservoirs (especially birds and swine). An example of an influenza A pandemic was the "Spanish flu" pandemic of 1918 to 1920 (responsible for the deaths of more

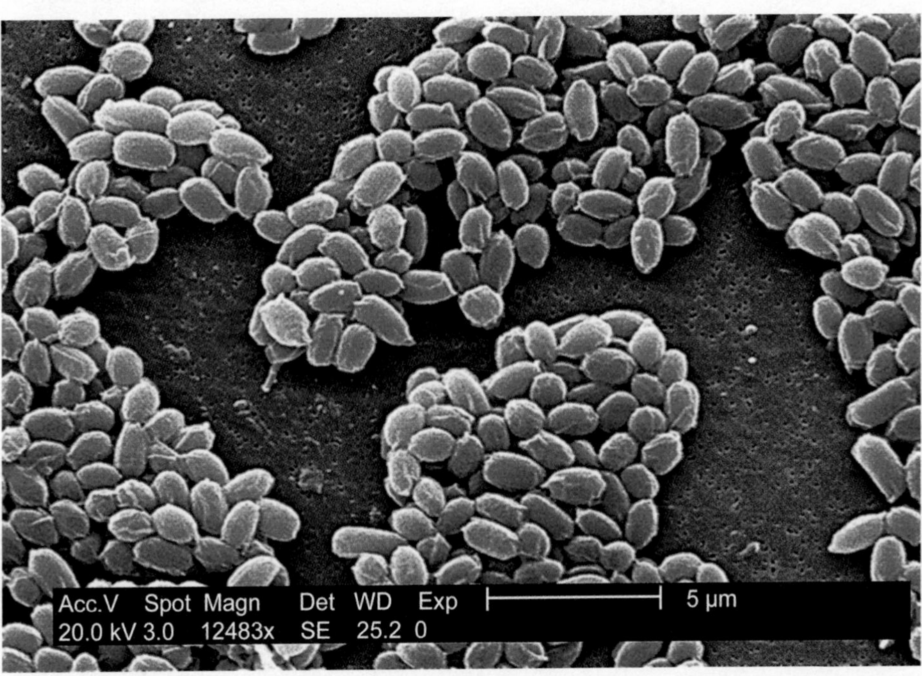

FIGURE 5.34 Spores from the Sterne strain of *Bacillus anthracis* bacteria.

Courtesy of CDC/Laura Rose

than 20 million people). Other influenza A pandemics were the outbreaks of the Asian flu in 1957 and the Hong Kong flu in 1968. Lesser outbreaks were the swine flu of 1976 and the chicken flu of 1997.

The three types of influenza viruses are A, B, and C, which have the following characteristics:

- Influenza A viruses—wild and aquatic birds are considered to be the main natural reservoir. In addition to humans, influenza A viruses can infect a variety of animals such as ducks, chickens, pigs, whales, horses, and seals. "However, certain subtypes of influenza A virus are specific to certain species, except for birds, which are hosts to all known subtypes of influenza A. . . . Influenza A viruses that typically infect and transmit among one animal species sometimes can cross over and cause illness in another species."[37] Examples are the transfer of influenza viruses from humans to pigs or from horses to dogs. Influenza A viruses can be transmitted indirectly from pigs (intermediate hosts) to humans or directly from birds to humans.

Direct infection of humans with avian influenza A is unusual. Human infections with avian influenza A viruses have been linked to direct contact with infected poultry. The spread of these influenza viruses may be promoted by the close proximity of live animals such as poultry and swine to residents of densely populated areas. In order to prevent epidemics, environmental health authorities need to maintain surveillance programs and conduct routine investigations of influenza outbreaks in animals and humans.

PSITTACOSIS

Clinical Features	In humans, fever, chills, headache, muscle aches, and a dry cough. Pneumonia is often evident on chest X-ray.
Etiologic Agent	*Chlamydia psittaci*, a bacterium.
Incidence	Since 2010, fewer than 10 confirmed cases are reported in the United States each year. More cases may occur that are not correctly diagnosed or reported.
Sequelae	Endocarditis, hepatitis, and neurologic complications may occasionally occur. Severe pneumonia requiring intensive-care support may also occur. Fatal cases have been reported but are rare.
Transmission	Birds are the natural reservoirs of *C. psittaci* and infection is usually acquired by inhaling dried secretions from infected birds. The incubation period is 5 to 19 days. Although all birds are susceptible, pet birds (parrots, parakeets, macaws, and cockatiels) and poultry (turkeys and ducks) are most frequently involved in transmission to humans. Personal protective equipment (PPE), such as gloves and appropriate masks, should be used when handling birds or cleaning their cages.
Risk Groups	Bird owners, aviary pet shop employees, poultry workers, and veterinarians. Outbreaks of psittacosis in poultry processing plants have been reported.
Surveillance and Treatment	Psittacosis is a reportable condition in most states. Tetracyclines are the treatment of choice.
Trends	Annual incidence varies considerably because of periodic outbreaks. A decline in reported cases since 1988 may be the result of improved diagnostic tests that distinguish *C. psittaci* from more common *C. pneumoniae* infections.
Challenges	Diagnosis of psittacosis can be difficult. Serologic tests are often used to confirm a diagnosis, but antibiotic treatment may prevent an antibody response, thus limiting diagnosis by serologic methods. Infected birds are often asymptomatic. Tracebacks of infected birds to distributors and breeders often is not possible because of limited regulation of the pet bird industry.

Modified and reproduced from Centers for Disease Control and Prevention. Pneumonia. Psittacosis. Available at: https://www.cdc.gov/pneumonia/atypical/psittacosis.html. Accessed January 22, 2017.

Influenza A viruses are categorized into subtypes according to the arrangement of two proteins on the surface of the virus. An example of a subtype is the 2009 H1N1 influenza virus (discussed in Chapter 1); another example is the H5N1 avian influenza A virus that occurred as early as 1997 in Hong Kong (Special Administrative Region).

- Influenza B viruses—confined to humans, who form the primary reservoir for these viruses.
- Influenza C viruses—a cause of mild illnesses and sporadic outbreaks among humans; they are not considered to be responsible for widespread epidemics.

Control and Prevention of Mosquito-Borne Diseases

One method to control mosquito-borne diseases is to monitor for the presence of viruses in sentinel chickens and birds. Some health departments position small flocks of chickens and other birds in strategic locations where mosquitoes may be active. These birds are tested periodically for the presence of viral antibodies. Examples of sentinel birds are shown in **FIGURE 5.35**.

Mosquitoes are dependent upon water to lead and complete their life cycle. Accordingly, one method for mosquito control consists of removing standing sources of water in which mosquitoes can multiply. Vector control experts advise homeowners to remove standing water from around their households, for example, water standing in buckets, old tires, or flowerpots. They also introduce mosquito-eating fish into small ponds and other bodies of standing water. Other preventive efforts include wearing long clothing to prevent being bitten by mosquitoes, using insect repellent, closing open windows, and repairing broken screens.

Conclusion

Zoonotic and vector-borne diseases represent a continuing and growing challenge. Among the recent conditions that have gained the attention of environmental health experts are the emerging and reemerging infections (e.g., dengue fever and hantavirus). In addition, scourges first observed and notorious during early historical times, such as plague, present hazards to modern society. Diseases brought under control during the mid-20th century (e.g., malaria) are showing resurgence. Several circumstances are responsible for the emergence of new diseases and the reemergence of formerly uncommon ones. Some of these factors are reduction in funding for environmental control programs, movement of human populations, increasing urbanization, and possible climatic changes. It is likely that zoonotic and vector-borne diseases will continue to challenge experts in the field of environmental health.

Study Questions and Exercises

1. Define the following terms:
   a. Emerging infectious disease
   b. Reemerging infectious disease
   c. Emerging zoonosis
   d. Zoonotic disease
   e. Vector
   f. Vector-borne infection

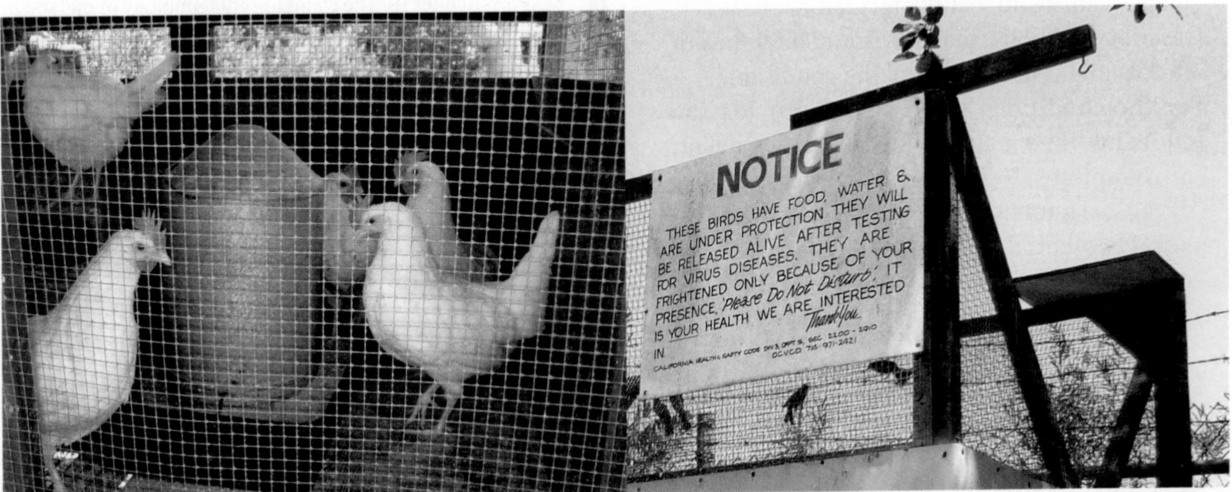

FIGURE 5.35 Sentinel birds used to monitor for mosquito-borne viruses.

2. State three factors that have contributed to the development of emerging infectious diseases. Give one example of an emerging disease and another example of a reemerging disease.
3. Give examples of three major zoonotic diseases and compare their modes of transmission. Using your own ideas, explain how transmission of these zoonotic diseases might be prevented.
4. Describe how vectors play a role in the transmission of disease.
5. List three characteristics of viral hemorrhagic fevers.
6. Discuss the impact of malaria on environmental health. Give an example of the economic impact this disease has on the developing world.
7. Summarize briefly the life cycle of malaria, describing the human and insect phases of the disease.
8. In what respects is leishmaniasis a vector-borne disease? Name some of the countries in the world where leishmaniasis is endemic and state why the disease is of particular concern to US military personnel.
9. Discuss the vector and agent factors involved in the transmission and causation of plague. Distinguish between bubonic and pneumonic plague. Describe procedures for control and prevention of plague.
10. What factors are involved in the chain of causation of Lyme disease? From the environmental health perspective, describe how the condition can be controlled and prevented.
11. What are arboviral encephalitides? Describe the reservoirs and modes of transmission of this group of diseases to humans.
12. What are the primary symptoms of West Nile virus (WNV)? List some of the ways that individuals can avoid being infected with WNV. Describe methods that environmental health officials might use to control the condition.
13. For the conditions listed below, provide the following information: (a) causative agent; (b) symptoms; (c) geographic distribution; (d) host factors; (e) responsible vectors, if any; (f) potential hazards to the human population; and (g) methods of control.

    Dengue fever and dengue hemorrhagic fever

    Rift Valley fever

    Rabies

    Anthrax

For Further Reading

The Great Influenza, John M Barry, 2005

References

1. Nicastri E, Girardi E, Ippolito G. Determinants of emerging and re-emerging infectious diseases. *J Biol Regul Homeost Agents*. 2001;15:212–217.
2. Porta M, ed. *A Dictionary of Epidemiology*. 6th ed. New York, NY: Oxford University Press; 2014.
3. Sellman J, Bender J. Zoonotic infections in travelers to the tropics. *Prim Care Clin Office Pract*. 2002;29:907–929.
4. Centers for Disease Control and Prevention. Malaria: frequently asked questions. The disease. What is malaria? Available at: https://www.cdc.gov/malaria/about/faqs.html. Accessed January 29, 2017.
5. Hyperdictionary. Meaning of nephrosis. Available at: http://www.hyperdictionary.com/search.aspx?define=nephrosis. Accessed January 29, 2017.
6. Centers for Disease Control and Prevention. Malaria. Impact of malaria. Available at: https://www.cdc.gov/malaria/malaria_worldwide/impact.html. Accessed January 29, 2017.
7. World Health Organization. Fact sheet: malaria. Available at: http://www.who.int/mediacentre/factsheets/fs094/en/. Accessed January 29, 2017.
8. *The American Heritage Dictionary of the English Language*. 4th ed. Boston, MA: Houghton Mifflin Company; 2000.
9. The Wellcome Trust. Ronald Ross and the treatment of malaria. Available at: https://wellcomecollection.org/articles/ronald-ross-and-treatment-malaria. Accessed April 24, 2017.
10. Centers for Disease Control and Prevention. Malaria: history. Elimination of malaria in the United States (1947–1951). Available at: https://www.cdc.gov/malaria/about/history/elimination_us.html. Accessed April 25, 2017.
11. Centers for Disease Control and Prevention. Malaria: history. Eradication efforts worldwide: success and failure (1955–1978). Available at: https://www.cdc.gov/malaria/about/history/. Accessed January 29, 2017.
12. Rosenberg T. What the world needs now is DDT. *The New York Times*; April 11, 2004;6:38.
13. Hay SI, Noor AM, Simba M, et al. Clinical epidemiology of malaria in the highlands of western Kenya. *Emerg Infect Dis*. 2002;8:543–548.
14. Seys SA, Bender JB. The changing epidemiology of malaria in Minnesota. *Emerg Infect Dis*. 2001;7:993–995.
15. Anis E, Leventhal A, Elkana Y, et al. Cutaneous leishmaniasis in Israel in the era of changing environment. *Public Health Rev*. 2001;29:37–47.
16. Wasserberg G, Abramsky Z, Anders G, et al. The ecology of cutaneous leishmaniasis in Nizzana, Israel: infection patterns in the reservoir host, and epidemiological implications. *Int J Parasitol*. 2002;32:133–143.
17. *The Oxford College Dictionary*. 2nd ed. New York, NY: Spark Publishing, by arrangement with Oxford University Press, Inc.; 2007.
18. Centers for Disease Control and Prevention. Plague: CDC plague home page. Available at: https://www.cdc.gov/plague/. Accessed January 30, 2017.
19. Heymann DL, ed. *Control of Communicable Diseases Manual*. 20th ed. Washington, DC: American Public Health Association; 2015.

20. Centers for Disease Control and Prevention. Lyme disease data tables. Reported cases of Lyme disease by state or locality, 2005–2015. Available at: https://cdc.goc/lyme/state/tables.html. Accessed April 25, 2017.
21. Centers for Disease Control and Prevention. Special Pathogens Branch. Viral hemorrhagic fevers. Available at: https://www.cdc.gov/vhf/index.html. Accessed January 30, 2017.
22. Centers for Disease Control and Prevention, Division of Vector-Borne Diseases. About Division of Vector-Borne Diseases. Available at: http://www.cdc.gov/ncezid/dvbd/about.html. Accessed July 15, 2017.
23. Chomel BB. Control and prevention of emerging zoonoses. *J Vet Med Educ*. 2003;30(2):145–147.
24. Centers for Disease Control and Prevention, National Center for Infectious Diseases, Special Pathogens Branch. Hantavirus. Hantavirus pulmonary syndrome (HPS). Available at: https://www.cdc.gov/hantavirus/hps/index.html. Accessed April 26, 2017.
25. Centers for Disease Control and Prevention, National Center for Infectious Diseases, Special Pathogens Branch. All about hantaviruses: hantaviruses. Available at: http://www.cdc.gov/ncidod/diseases/hanta/hps/noframes/hanta.htm. Accessed March 1, 2010.
26. Centers for Disease Control and Prevention, National Center for Infectious Diseases, Special Pathogens Branch. Hantavirus. Tracking a mystery disease: the detailed story of hantavirus pulmonary syndrome (HPS). Available at: https://www.cdc.gov/hantavirus/outbreaks/history.html. Accessed January 30, 2017.
27. Centers for Disease Control and Prevention. Hantaviruses. Reported cases of HPS. HPS in the United States. Available at: https://www.cdc.gov/hantavirus/surveillance/index.html. Accessed January 30, 2017.
28. Centers for Disease Control and Prevention. Dengue. Frequently asked questions. Available at: https://www.cdc.gov/dengue/faqfacts/index.html. Accessed April 26, 2017.
29. Centers for Disease Control and Prevention. Possible Zika virus infection among pregnant women—United States and territories, May 2016. *MMWR*. 2016;65(20):514–519.
30. Centers for Disease Control and Prevention. Rift Valley fever. https://www.cdc.gov/vhf/rvf/index.html. Accessed January 31, 2017.
31. Centers for Disease Control and Prevention. Monkeypox. Available at: http://www.cdc.gov/poxvirus/monkeypox/. Accessed January 31, 2017.
32. Centers for Disease Control and Prevention. Tularemia. Available at: http://www.cdc.gov/tularemia/. Accessed January 31, 2017.
33. Centers for Disease Control and Prevention. Update: investigation of bioterrorism-related anthrax, 2001. *MMWR*. 2001;50:1008–1010.
34. Jernigan DB, Raghunathan PL, Bell BP, et al. Investigation of bioterrorism-related anthrax, United States, 2001: epidemiologic findings. *Emerg Infect Dis*. 2002;8:1019–1028.
35. Hughes JM, Gerberding JL. Anthrax bioterrorism: lessons learned and future directions. *Emerg Infect Dis*. 2002;8:1013–1014.
36. Centers for Disease Control and Prevention. Anthrax. Basic information. Available at: https://www.cdc.gov/anthrax/basics/index.html. Accessed January 31, 2017.
37. Centers for Disease Control and Prevention. Influenza (flu). Information on avian influenza. Available at: https://www.cdc.gov/flu/avianflu/. Accessed January 31, 2017.

CHAPTER 6
Toxic Metals and Elements

LEARNING OBJECTIVES

By the end of this chapter the reader will be able to:

- Name five heavy metals that pose health hazards to humans.
- State mechanisms for exposure of humans to toxic metals.
- Distinguish between essential and toxic levels of trace metals.
- Describe occupational settings in which workers are exposed to toxic metals.
- Discuss methods for prevention of exposure to toxic metals.

Introduction

You may be surprised to learn about the widespread exposure of people to toxic metals and elements. Although some metals are essential for our well-being, others present grave hazards to the health of the human population and to some extent all living organisms. For example, you will learn about the significance of toxic heavy metals and how they can affect vulnerable population groups including women and children. A related issue concerns the pathways of metallic compounds into the human body. This chapter provides an overview of the major categories of metallic compounds found in the environment. Also supplied is a detailed review of some of the most important metals, their possible sources, and their effects. The chapter on pesticides and other organic chemicals will continue the discussion of chemical hazards, including pesticides, herbicides, and organophosphates.

Significance of Heavy Metals for the Human Population

Heavy metals (e.g., lead, mercury, and nickel) and other metallic compounds (e.g., aluminum, iron, and tin) are widely dispersed in the environment and arise from both natural and human-made sources. The entire human population is exposed to low levels of heavy metals from naturally occurring deposits of these elements in the earth's crust. Other sources of human exposure to metals are emissions from smelters and coal-fired power plants, contact with contaminants that are leaching from hazardous waste sites, ingestion of contaminated food and water, and contact with metals in occupations related to metal working.

Two examples of metals that pose hazards for human health are lead and mercury. The latter has elicited much public controversy as a cause of fish contamination that may be of special concern to women who are pregnant, nursing, or may become pregnant. Such women, according to guidelines from the US Environmental Protection Agency (EPA) and reports in the media, should avoid certain types of fish that contain high levels of mercury (examples are shark and swordfish) and limit intake of tuna, especially white tuna. It was also recommended that children should have a restricted intake of fish. These guidelines generated heated controversy: fish is a low-fat source of protein; tuna fish is the popular staple of many people's lunches.

Lead is a cause for alarm because of its potential to harm children's development. Not only do lead-based paints continue to pose a hazard to inner-city children, but lead also can present dangers to creatures in the rural environment. Lead bullets used by hunters remain in the game animals that they shoot and abandon. Scavengers such as endangered California condors may develop paralyzed digestive systems when they feed on the remains.

In another instance of lead exposure, the water supply of Flint, Michigan became contaminated with lead after the city changed its water supply in 2014 from Lake Huron to the Flint River as a cost-saving measure. The highly corrosive river water caused lead to leach from the water service lines in homes into household water.

The CERCLA Priority List of Hazardous Substances

Recall from the chapter on environmental policy and regulation, the functions of the Agency for Toxic Substances and Disease Registry (ATSDR), the US public health agency tasked with chemical safety. The Comprehensive Environmental Response, Compensation, and Liability Act (CERCLA) as a provision of the Superfund Amendments and Reauthorization Act (SARA) mandates that ATSDR and the EPA prioritize a list of hazardous substances. Note that the chapter on environmental policy and regulation also discusses CERCLA and SARA. The priority list ranks substances that "are most commonly found at facilities on the National Priorities List (NPL) and which are determined to pose the most significant potential threat to human health due to their known or suspected toxicity and potential for human exposure at these NPL sites."[1] The NPL, or Superfund List, is defined as "EPA's list of the most serious uncontrolled or abandoned hazardous waste sites [in the United States] identified for possible long-term remedial action under Superfund. The list is based primarily on the score a site receives from the Hazard Ranking System. EPA is required to update the NPL at least once a year. A site must be on the NPL to receive money

THE PRIORITY LIST OF HAZARDOUS SUBSTANCES THAT WILL BE THE CANDIDATES FOR TOXICOLOGICAL PROFILES

What is the Substance Priority List (SPL)?

The Comprehensive Environmental Response, Compensation, and Liability Act (CERCLA) section 104 (i), as amended by the Superfund Amendments and Reauthorization Act (SARA), requires ATSDR and the EPA to prepare a list, in order of priority, of substances that are most commonly found at facilities on the National Priorities List (NPL) and which are determined to pose the most significant potential threat to human health due to their known or suspected toxicity and potential for human exposure at these NPL sites. CERCLA also requires this list to be revised periodically to reflect additional information on hazardous substances.

This substance priority list is revised and published on a 2-year basis, with a yearly informal review and revision. (No list was published in 2009 while ATSDR transitioned to a new agency science database.) Each substance on the list is a candidate to become the subject of a toxicological profile prepared by ATSDR. The listing algorithm prioritizes substances based on frequency of occurrence at NPL sites, toxicity, and potential for human exposure to the substances found at NPL sites.

It should be noted that this priority list is not a list of "most toxic" substances, but rather a prioritization of substances based on a combination of their frequency, toxicity, and potential for human exposure at NPL sites.

Thus, it is possible for substances with low toxicity but high NPL frequency of occurrence and exposure to be on this priority list. The objective of this priority list is to rank substances across all NPL hazardous waste sites to provide guidance in selecting which substances will be the subject of toxicological profiles prepared by ATSDR.

The list below contains the top 20 hazardous substances from the CERCLA Priority List of Hazardous Substances for 2015. The complete list contains 275 substances ranked in order of their priority.

Top 20 Hazardous Substances From the ATSDR 2015 Substance Priority List

1. Arsenic
2. Lead
3. Mercury
4. Vinyl chloride
5. Polychorinated biphenyls
6. Benzene
7. Cadmium
8. Benzo[a]pyrene
9. Polycyclic aromatic hydrocarbons
10. Benzo[b]fluoranthene
11. Chloroform
12. Aroclor 1260
13. Dichlorodiphenyltrichloroethane (DDT), p, p'-
14. Aroclor 1254
15. Dibenzo[a, h]anthracene
16. Trichloroethylene
17. Chromium, hexavalent
18. Dieldrin
19. Phosphorus, white
20. Hexachlorobutadiene

Reproduced from Agency for Toxic Substance and Disease Registry (ATSDR): The ATSDR 2015 Substance Priority List. Available at: https://www.atsdr.cdc.gov/spl/. Accessed January 30, 2017.

from the Trust Fund for remedial action."[2] Refer to the text box for the 2015 list, which presents the top 20 hazardous substances. Note that list numbers 1, 2, 3, 7, and 17 are heavy metals or metallic compounds discussed in the present chapter. Many of the remaining compounds are covered in the chapter on pesticides and other organic chemicals.

Overview of Sources and Effects of Exposure to Metals

This chapter will use a three-category method for classifying metals, following Tokar et al.[3] This scheme is shown in **TABLE 6.1**, which classifies metals according to whether they are (1) major toxic metals; (2) **essential metals** with potential for toxicity; or (3) metals related to medical therapy.

The first group shown in Table 6.1, "Major Toxic Metals," comprises the toxic heavy metals, which are toxic to life-forms and are not necessary to sustain life. Examples of metals in this category are arsenic, lead, and mercury. A heavy metal is one that has a high atomic weight with a specific gravity that exceeds the specific gravity of water by five or more times. (Specific gravity is defined as "the **relative density** of a solid or liquid, usually when measured at a temperature of 20°C [68°F], compared with the maximum density of water (at 4°C [39.2°F]). For example, the specific gravity of carbon steel is 7.8, that of lead is 11.34, and that of pure gold is 19.32.")[4]

The second group consists of metals that are considered to be essential for life when present in trace amounts; examples are iron, copper, manganese, and zinc. However, concentrations greater than trace amounts have the potential for toxicity. The third group includes metals used in medical therapy such as aluminum and lithium (e.g., used in lithium therapy for depression).

Let's examine the first group of metals in Table 6.1—major toxic metals—in more detail. They not only are toxic at low concentrations (beyond trace amounts), but they also have the capacity to bioaccumulate (sometimes called **biomagnification**), which means that they become more concentrated and potentially more harmful as they move up the food chain. For example, lower organisms may ingest mercury-containing compounds in the water. These lower organisms are consumed by snails and then by small fish. Finally, the smaller fish provide a food source for larger fish that are harvested for human consumption.

Another example is the **bioaccumulation** of toxic substances such as heavy metals by zebra mussels. This species of mussels, which was introduced accidentally into the Great Lakes system of the United States, reproduces in great numbers. Concern has been raised about the ability of the mussels to filter out and concentrate pollutants in the water; some areas of the Great Lakes are known to be polluted with toxic contaminants. In the Great Lakes, biomagnification of toxic substances by zebra mussels could affect other life-forms such as freshwater fish and wildlife.

Sources of Exposure to Metals

There are several possible modes of exposure to metals, corresponding to the human portals and sites of entry that come into contact with the metals. These portals and sites include the lungs (through inhalation of dusts,

TABLE 6.1 Classification of the Toxic Effects of Metals

Major Toxic Metals with Multiple Effects	Essential Metals with Potential for Toxicity	Metals Related to Medical Therapy
Arsenic (As) [Beryllium (Be)[a] Cadmium (CD) [Chromium (Cr)][a] Lead (Pb) Mercury (Hg) Nickel (Ni)	Copper (Cu) Iron (Fe) Zinc (Zn)	Aluminum (Al) Lithium (Li) Platinum (Pt)

Data from Tokar EJ, Boyd WA, Freedman JH, Waalkes MP. Toxic effects of metals. In: Klaasen CD, Watkins JB III, eds. *Casarett & Doull's Essentials of Toxicology*. 3rd ed. New York, NY: McGraw-Hill, 2015:347.

[a] Goyer RA and Clarkson TW. Toxic effects of metals. In: Klaassen CD, ed. *Casarett and Doull's Toxicology: The Basic Science of Poisons*. 6th ed. New York, NY: McGraw Hill; 2001:811.

metal fumes, and vapors), skin (through contact with dusts), and the mouth (by ingestion). Contact with high concentrations of toxic metals is most likely to occur in an occupational setting (e.g., among persons who work with metals). Lower-level exposures may result from contact with the ambient environment. For example, children may ingest toxic metals that are present in paints in homes and playgrounds. Also, people may be exposed to levels of toxic metals in the foods that they eat.

Four major types of media have potential for exposure of the population to heavy metals. As shown in **TABLE 6.2**, these media are air, soil or dust, water, and biota (animal and plant life) or food. Some media are more relevant to the transport of a given type of heavy metal than are other media. For example, water is a medium for exposure of the population to arsenic and lead. Food is a medium for exposure of the population to mercury. Another possibility for exposure involves the cultural and religious practices of ethnic groups. These practices may involve the use of mercury for treatment of gastrointestinal symptoms and in candles and necklaces for spiritual cleansing.

Industrial facilities, metal smelters, and power plants that burn coal have the potential to release heavy metals into the environment. Refer to the text box about lead contamination from the Exide Technologies Plant (Los Angeles area). Examples of heavy metals emitted from these sources include cadmium and arsenic. In some communities, incinerators constructed to dispose of waste may disperse toxic heavy metals such as mercury and cadmium. Mines are another source of toxic heavy metal wastes, examples being arsenic leaching from gold mines and mercury from cinnabar mines. **Preservatives** used in wood may contain arsenic, which is an occupational hazard to carpenters. Also presenting toxic threats to the environment are chromium and mercury residues from some types of pesticides and cadmium from fertilizers. Finally, residents of older houses may be exposed to lead in paint and in plumbing systems. Many heavy metals are natural constituents of the earth's crust. One such heavy metal is arsenic, which is present in the groundwater in many areas of the world and poses a hazard to the residents who drink the groundwater. Other sources of exposure to heavy metals include air pollution, water that has been contaminated with wastes, contact with industrial wastes in the soil, and ingestion of contaminated food. Previously, the author noted that toxic metals can bioaccumulate in shellfish and pose a hazard when these mollusks are consumed.

LEAD CONTAMINATION FROM THE EXIDE TECHNOLOGIES PLANT, VERNON, CALIFORNIA

The Exide Technologies plant is located within a heavily populated residential area of 10,000 homes in Los Angeles County. Many of the area's residents are working-class Latinos. For decades, Exide recycled used lead batteries at the rate of approximately 25,000 batteries daily, 7 days a week. The plant emitted lead, arsenic, and other toxic materials into the surrounding community until 2015, when it closed permanently. Environmental assessments determined that soil in the neighborhood was contaminated with lead as far as 1.7 miles away. The state of California determined in 2012 that some of the children who lived near the plant had elevated blood lead levels. This environmental catastrophe will require years to remediate.

Data from Barboza T. How a battery recycler contaminated L.A.-area homes for decades. *Los Angeles Times*, December 21, 2015 and Barboza T, Poston B. What we know about California's largest toxic cleanup: Thousands of L.A. County homes tainted with lead. *Los Angeles Times*, August 6, 2017.

TABLE 6.2 Relevance of Different Media to Metal Exposures[a]

Metal	Air	Soil/Dust	Water	Biota/Food
Arsenic	+ Occupational Smelters	++ Mining Smelting	++++ Natural	++
Mercury	+ Occupational Cultural practices	++++	++++ Methylation in sediments	++++ Biomagnification MeHg in fish[b]
Iron	++ Occupational Mining	++ Mine tailings	++ Rust in pipes	+
Tin	+ Occupational smelters	0	+	+ Organotin pesticides[c], TBT in fish[d]
Lead	++ Occupational Smelters Leaded gasoline	+++ Toddlers ingestion	++ Lead pipes and solder	+++ Leaded glaze and glass Paprika Wheat
Chromium	+ Occupational	++	+	+

[a] ++++ Highly relevant; +++ medium relevant; ++ relevant; + less or infrequently relevant; 0 not relevant.
[b] MeHg = methylmercury.
[c] Organotin = a compound that contains both tin and carbon.
[d] TBT = tributyltin.
Modified and reproduced from Caussy D, Gochfeld M, Gurzau E, et al. Lessons from case studies of metals: investigating exposure, bioavailability, and risk. *Ecotoxicology and Environmental Safety*. 2003;56:46. Copyright 2003; used with permission from Elsevier.

Another source of exposure of the population to toxic heavy metals arises from used automobile tires, which are discarded every year by the millions worldwide. These tires are known to contain a number of toxic metals, including zinc, cadmium, and lead. When the tires are deposited in landfills, they degrade and may release toxic metals into the soil; in turn, the toxic metals may find their way into groundwater and also poison the air. Sometimes tires may ignite and cause a serious air pollution hazard. The toxic materials released from degrading automobile tires may persist for many years.

Common industrial processes may release toxic metals into the air, which impinge upon nearby residents and passersby who are unaware that they are breathing fumes that contain toxic metals. The jewelry districts found in many large US cities can produce toxic fumes during the smelting of metals and soldering operations that are connected with the manufacture of jewelry. These processes can release significant amounts of pollution into the surrounding areas.

Unsafe disposal of electronic equipment also has raised concerns about the release of toxic metals into the environment. Increasingly, electronic wastes are being tossed into the garbage without further thought about what happens to them when they are deposited in landfills. Examples of items that are being discarded with increasing frequency are fluorescent light bulbs, lithium ion batteries, cell phones, outmoded computers, computer monitors, and nonworking television sets. The toxic metals contained in these devices are lead, mercury, cadmium, chromium, lithium, arsenic, copper, and zinc. In addition, these devices contain toxic organic chemicals that may persist in the environment. (The topic of toxic organic chemicals will be discussed in the chapter on pesticides and other organic chemicals.) Two important reasons could explain why used electronic equipment is disposed of in an unsafe manner: There are few recycling programs for reuse or remanufacture of the equipment and little information is available to the consumer regarding the proper disposal of these devices. Unfortunately, less developed countries increasingly are being selected as the dumping grounds for electronic waste, which is scavenged by unwitting individuals who are not aware of the hazardous nature of this waste.

Effects of Exposure to Metals

The effects of acute poisoning from high levels of exposure to toxic metals can be differentiated from the effects of chronic low-level exposure. The symptoms of acute poisoning from exposure to metals generally have rapid onset—from a few minutes to approximately 1 hour. Depending upon the portal of entry, these symptoms may consist of gastrointestinal effects (vomiting and stomach pain) and neurological effects (headaches, suppression of normal breathing, and convulsions). In most cases, the symptoms are likely to be dramatic and severe when the person has had high levels of exposure.

Instances of acute poisoning from toxic metals have occurred when contaminated foodstuffs were consumed. For example, copper from copper cooking vessels may leach into foods that are left standing during meal preparation. Sometimes, errors in food manufacturing may accidentally introduce toxic metals into foods and beverages. An instance of such accidental poisoning occurred when Iraqi residents consumed homemade bread contaminated with mercury.[5] The source of the poisoning was organomercury compounds used as fungicides to treat grain intended for planting. In this incident 6,148 persons had to be hospitalized, and 452 patients died in hospitals. These hospitalization and mortality figures are thought to be underestimates of the actual scope of the incident.

The effects of long-term exposure to lower levels of toxic metals typically are more subtle than those of acute exposure. Often, the effects of exposure over extended time periods are difficult to differentiate from those of chronic medical conditions. In addition, the specific source of exposure may be impossible to determine. Symptoms of chronic exposure can include reduced cognitive functioning, such as learning impairment, as well as lethargy. Sometimes the symptoms show a pattern of symptom remission that causes the victim not to seek medical care.

Heavy Metal Exposure among Women and Children

The health effects of exposure to heavy metals (e.g., nickel, cadmium, lead, and mercury) are often different for women than they are for men. These differences in effects have been attributed to hormonal and metabolic processes related to menstruation, pregnancy, and menopause. Women usually have smaller body sizes than men, causing women to be more affected by a given amount of a metal than are men. As an example of their different responses to toxic metals, women are known to develop allergies to nickel and rashes from skin contact with nickel more frequently than men. In comparison to men, women are known to accumulate higher internal concentrations of cadmium and lead than men. During pregnancy, exposure to mercury is a particular hazard for the neurodevelopment of the fetus.[6]

The health effects of women's general environmental and occupational exposures to smelters have been studied.[7] In an investigation conducted in the north of Sweden, researchers examined the impact of emissions from the Rönnskär smelter, which was known to release heavy metals such as lead, mercury, and arsenic. Female smelter employees and a sample of female residents in the neighborhood surrounding the smelter were studied with respect to the occurrence of spontaneous abortions. No association between spontaneous abortions and employment in the smelter or residence near the smelter was reported.

For fetuses, infants, and children (especially young children), heavy metals are known to present serious hazards, which can include impairment of physical and mental development, damage to internal organs and the nervous system, some forms of cancer, and even mortality. For example, lead and mercury have the capacity to cross the placental barrier, with the potential to cause fetal brain damage. According to researchers, during growth of various body systems, children have increased sensitivity to heavy metals; this increased sensitivity may result in an increased frequency of adverse health effects.

Symptoms of exposure to heavy metals among children include nervous system damage as reflected in memory impairment and difficulty in learning, and a range of behavioral problems such as hyperactivity syndrome and overt aggressiveness. When children are exposed to high levels of toxic heavy metals, the effects may be irreversible. Because their body weight is smaller than that of adults, children consume more food in proportion to their body weight and consequently receive higher doses of heavy metals that may be present in their food.

One of the outcomes associated with exposure to toxic chemicals, including heavy metals, is the occurrence of developmental disabilities, which have trended upward in frequency during recent years.[8] "Such exposures are of particular concern for genetically susceptible individuals exposed during periods of developmental vulnerability."[8(pS14)] As an example of a particular disability associated with exposure to heavy metals, consider the effect that elevated blood lead levels have on intelligence. For example, doubling the blood lead level from 10 to 20 micrograms per deciliter would increase the demand for special programs for persons who have impaired cognitive ability by almost 60%. Thus, the reduction of blood lead levels to the lowest level possible is important for the population, especially children. Due to recognition of the adverse health effects of lead, especially on children, authorities in the United States have lowered the standard for blood lead levels periodically since the 1960s. (See the section titled *Lead* for more information.)

Toxic Heavy Metals

The metals in the group known as "major toxic metals with multiple effects" include arsenic, beryllium, cadmium, chromium, lead, mercury, and nickel. (Refer to Table 6.1.)

Arsenic

Arsenic in its pure form is a crystalline metalloid, an element with properties that are intermediate between those of a metal and a nonmetal. It is able to combine with other substances, metallic and nonmetallic, and also to form stable organic compounds.[9] An acutely poisonous material that is ubiquitous in nature—in soils and water—arsenic varies in toxicity depending upon its chemical form; it is also the by-product of refining gold and other metals. Although at one time arsenic was used for medicinal purposes, at present it is no longer widely employed for that application. Arsenic has been used for commercial applications as a pesticide and wood preservative (although no longer permitted for treatment of residential lumber), and in manufacturing processes.[10]

As shown in **TABLE 6.3**, arsenic exposure can come from many environmental sources, including industrial processes that release arsenic, products that contain arsenic, certain foods, and drinking water in

TABLE 6.3 Potential Sources of Arsenic Exposure[a]

Source	Examples
Natural environmental sources	Ground water, mineral ore, geothermal processes
Industrial sources	Smelting processes for cobalt, gold, lead, nickel, zinc
Arsenic-containing products	Insecticides such as ant killers and animal dips, algaecides[b]; wood preservatives used to pressure treat lumber for outdoor installations and playground equipment[c]
Arsenic in drugs	Some drugs used for cancer chemotherapy; some traditional medicines; Salvarsan (an arsenic compound) used to treat syphilis before antibiotics became available
Component uses of arsenicals	Some microwave devices, lasers, light-emitting diodes, photoelectric chemical cells, semiconductor devices
Other industrial processes that use arsenic	Generating electricity in coal-fired power plants, hardening metal alloys, purifying industrial gases
Foods that may contain arsenic	Seafood (e.g., clams, oysters, scallops, mussels, crabs, lobsters, certain bottom-feeding finfish, seaweed[d])
Drinking water in some regions	Artesian and tube wells supplied by geologically contaminated aquifers; contamination from mineral ores; contamination from mine and mill tailings

Notes:

[a] There has been no domestic production of arsenic since 1985, although the US was the world's largest consumer of arsenic in 2003.

[b] Regulatory restrictions for arsenic, especially for home products, have reduced its use.

[c] In 2003, production of wood preservatives accounted for more than 90% of domestic consumption of arsenic trioxide, one of the forms of arsenic. Since December 2003, residential lumber may no longer be treated with arsenic.

[d] The forms of arsenic found in seafood are generally considered to be nontoxic and are excreted in urine with 48 hours of ingestion.

Data from Agency for Toxic Substances and Disease Registry (ATSDR). Environmental Health and Occupational Medicine: case studies. Arsenic toxicity. Available at: https://www.atsdr.cdc.gov/csem/csem.asp?csem=1 and https://www.atsdr.cdc.gov/csem/arsenic/docs/arsenic.pdf. Accessed February 16, 2017.

some geographic areas. Ingestion of 100 mg of arsenic produces acute poisoning; however, many people are exposed to lower levels of arsenic in sources such as drinking water and food. Some areas of the world (e.g., Taiwan, Mexico, western South America [Chile and Argentina], Bangladesh, and Inner Mongolia) have high levels of naturally occurring arsenic in the drinking water. In some of these areas, the arsenic levels are as high as 1,000 μg per liter of water. Arsenic levels that range from 50 to 100 μg per liter are found in the drinking water in sections of the United States, for example, New Mexico, Arizona, Nevada, Utah, southern California, Idaho, and Nebraska.[11] **FIGURE 6.1** shows the distribution of arsenic in groundwater samples collected from wells in the United States.

The federal government has established standards for the amounts of arsenic that are permitted in the public drinking water supply. In 2001, the EPA announced the standard of 10 μg of arsenic per liter of drinking water, with the year 2006 set as the compliance date for this standard. Previously, a higher standard of 50 μg of arsenic per liter of water was permitted, but this standard was made more stringent when officials realized that arsenic at this level of concentration was a risk factor for cancer. **TABLE 6.4** provides a chronology of US standards for arsenic in drinking water.

Several massive arsenic poisoning incidents, due to the contamination of foodstuffs with arsenic, have been documented.[9] Arsenic-contaminated beer killed 6,000 people in England in 1900; several thousand people were poisoned in Japan in the mid-20th century due to arsenic-contaminated dry milk and soy sauce. Particularly at risk of arsenic exposure are some groups of mining and smelter workers as well as agricultural workers who come into contact with pesticides that contain arsenic.

Low-level chronic exposure to arsenic is associated with **melanosis**, a dermatologic condition that can cause the darkening of the skin of the entire body (diffuse melanosis). Other forms of melanosis are spotted melanosis, which causes spotted pigmentation of the skin, and buccal membrane melanosis, which appears on the tissues of the inside of the mouth, including the tongue. **FIGURE 6.2** shows a case of melanosis of the hand. Melanosis is a problem in some districts of Bangladesh and West Bengal, India, where the levels of arsenic in the groundwater exceed 50 μg per liter.[12] The Bangladeshi standard for the maximum permissible limit arsenic in drinking water is 50 μg per liter. The

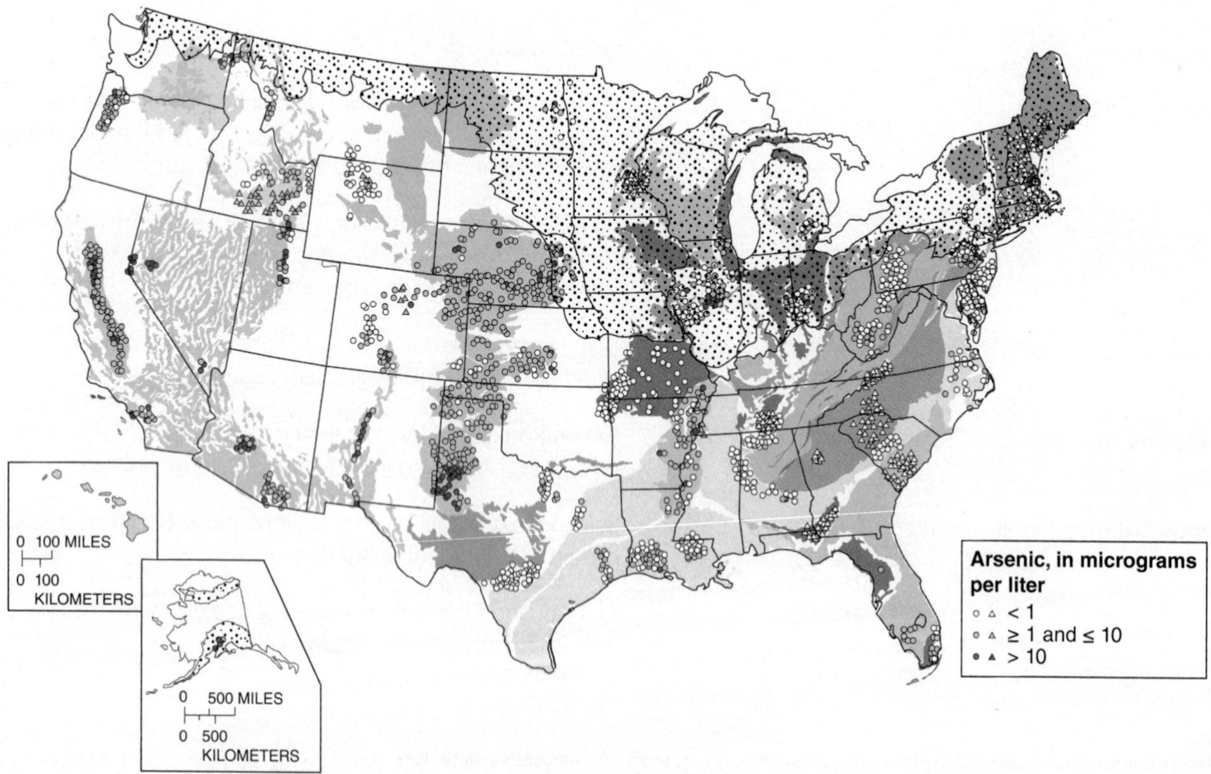

FIGURE 6.1 Geographic distribution of arsenic concentrations in groundwater collected from wells as part of the National Water-Quality Assessment Program, 1992–2003.

Modified and reproduced from Ayotte JD, Gronberg JM, Apodaca LE. Trace elements and radon in groundwater across the United States, 1992–2003. *U.S. Geological Survey Scientific Investigations Report 2011–5059*. 2011;5059:1=115. Accessed February 3, 2017.

TABLE 6.4 History of US Standards for Arsenic in Drinking Water

1942	US Public Health Service (USPHS) sets an interim drinking water standard of 50 μg [50 ppb] As/liter.[a]
1962	USPHS identifies 10 μg As/liter as the goal.[b]
1975	EPA adopts the interim standard of 50 μg As/liter set by the USPHS in 1942.[a]
1986	Congress directs EPA to revise the standard by 1989.[c]
1988	EPA estimates that the ingestion of 50 μg As/liter results in a skin cancer risk of 1 in 400.[d]
1992	Internal cancer risk estimated to be 1.3 per 100 persons at 50 μg As/liter.[e]
1993	World Health Organization (WHO) recommends lowering arsenic in drinking water to 10 μg As/liter.[f]
1996	Congress directs the EPA to propose a new drinking water standard by January 2000.[c]
1999	National Research Center (NRC) estimates cancer mortality risks to be about 1 in 100 at 50 μg As/liter.[g]
2000	EPA proposes a standard of 5 μg As/liter and requests comment on 3, 10, and 20 μg As/liter.[c]
2001	(January) EPA under Clinton lowers the standard to 10 μg As/liter.[c]
2001	(March) EPA under Bush delays lowering the standard.[c]
2001	(September) New NRC report concludes that EPA underestimated cancer risks.[h]
2001	(October) EPA announces it will adopt the standard of 10 μg As/liter.[c]
2002	(February) The effective date for new standard of 10 μg As/liter.[c]
2006	Compliance date for the new arsenic standard.[c]

Notes:

[a] See US Environmental Protection Agency (EPA). Office of Water. Drinking Water Standard for Arsenic. EPA 815-F-00-015.

[b] US Public Health Service (USPHS). Drinking water standards. *Federal Register*. 1962;27(42):2152.

[c] See www.house.gov/science/ets/oct04/ets_charter_100401.htm.

[d] US Environmental Protection Agency (EPA). *Special Report on Ingested Inorganic Arsenic: Skin Cancer; Nutritional Essentiality*. Washington, DC: EPA Risk Assessment Forum; 1988.

[e] Smith AH, Hopenhayn-Rich C, Bates MN, et al. Cancer risks from arsenic. *Environ Health Perspect*. 1992;97:259.

[f] World Health Organization (WHO). *Guidelines for Drinking-Water Quality, Vol. 1, Recommendations*. Geneva, Switzerland: World Health Organization; 1993.

[g] National Research Center (NRC). *Arsenic in Drinking Water*. Washington, DC: National Academies Press; 1999.

[h] NRC. *Arsenic in Drinking Water 2001 Update*. Washington, DC: National Academies Press; 2001.

World Health Organization has set a provisional guideline value for arsenic in drinking water of 10 μg per liter.

Accumulated evidence suggests that arsenic is a **carcinogen**. It is a cause of skin cancer when ingested and lung cancer when inhaled. Ingestion of arsenic has been linked with internal cancers such as bladder, kidney, and liver cancers.[13] Geographic regions that have high levels of naturally occurring arsenic in the drinking water (e.g., Taiwan and Argentina) also have elevated levels of bladder cancer mortality.[14] Some data suggest that there is a dose-response relationship between exposure to inorganic arsenic in drinking water and bladder cancer. Because many of the findings of an association between arsenic in drinking water and bladder cancer have used ecologic study designs, there is a need for additional analytical

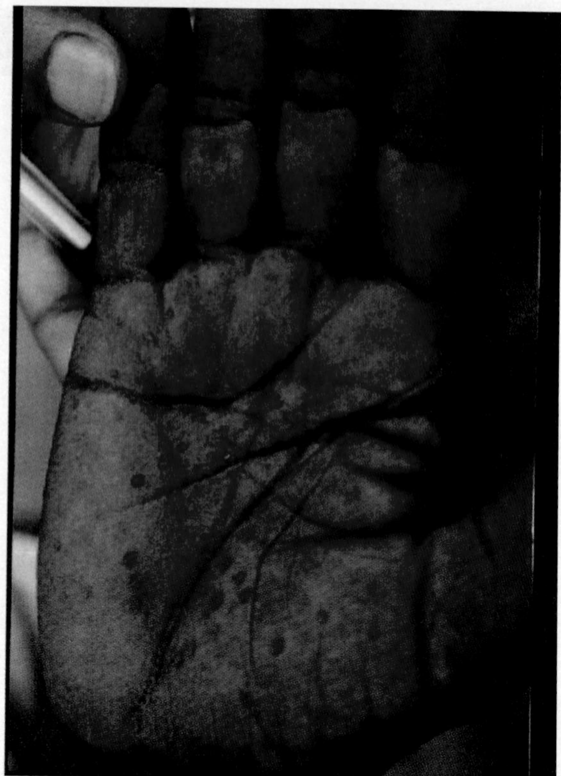

FIGURE 6.2 Melanosis of the hand from arsenic poisoning.

Reproduced with permission from Wilson R, Harvard University. Chronic arsenic poisoning: history, study and remediation. Available at: http://phys4.harvard.edu/~wilson/arsenic/pictures/5.%20Melanosis%20hand.jpg. Accessed May 27, 2017.

research on this topic. A factor that has impeded research on the carcinogenic aspects of arsenic is the absence of a clearly defined animal model of the effects of arsenic exposure; the use of laboratory animals would make possible studies that cannot be carried out with humans.[15]

In addition to the foregoing health outcomes, arsenic exposure has been connected with peripheral vascular disease, cerebrovascular disease, and cardiovascular disease (e.g., hypertensive heart disease). There also may be an association between long-term arsenic exposure and diabetes mellitus. Women exposed to arsenic in drinking water have demonstrated adverse pregnancy outcomes including spontaneous abortions, stillbirths, and preterm births.[16]

Beryllium

Beryllium is used widely in industry because of its special properties. It is not only lighter than aluminum, but also much stronger than steel. Beryllium also can be alloyed with other metals such as copper. Among those likely to be exposed to beryllium are employees in the metal processing industry. One of the most frequent methods of exposure to beryllium is inhalation. Exposure to beryllium can result in the disease berylliosis, also known as chronic beryllium disease (CBD). This condition can be extremely debilitating and in some cases fatal. Two factors that affect the development of CBD are beryllium exposure itself and genetic susceptibility, which has been shown to be associated with a specific gene.[17] Beryllium is classified as a class A carcinogen, the most hazardous level for a cancer-causing substance. Exposure of the human population to beryllium may result from the burning of coal, which contains beryllium compounds. Beryllium residues from emissions from coal-burning power plants may contaminate agricultural products.

Cadmium

Cadmium is found in all soils and rocks as well as coal and fertilizers derived from minerals. Typically cadmium occurs as a mineral in combination with other elements, such as oxygen (as in cadmium oxide).[18] As a result of absorption from environmental sources, fish, plants, and animals contain cadmium. Cadmium has a multitude of applications, e.g., batteries, coatings for metals, electronics, and (more recently) in the manufacture of nanoparticles destined for display screens.[19] It is released as a by-product of the mining industry, electroplating, iron and steel production, combustion of coal, and manufacture of fertilizers and pesticides. Other sources include smelting operations, as in the smelting of cadmium-containing zinc ores, and production of batteries.

Among the general population the primary sources of cadmium exposure are cigarette smoke and dietary cadmium. In addition, cadmium can reach our surface waters, where 90% of this metal originates from human sources.[19] Cadmium is known to bioaccumulate in shellfish and is found in some species of mushrooms.[20] Cadmium may be ingested from vegetables that are grown in cadmium-containing soil. For example, gardens near installations such as battery factories may contain soil that is contaminated with cadmium.

Another venue for cadmium exposure is one's job. Occupational exposure to cadmium comes from the production of nickel cadmium batteries, zinc smelting, paint manufacture, soldering, and employment in metal factories. In occupational settings, the modes for entry of cadmium into the body are primarily through inhalation and, secondarily, through the gastrointestinal tract.[21]

Cadmium is not regarded as essential for life processes. Moreover, various adverse health effects have been attributed to cadmium. Some of these are shown in **TABLE 6.5**.

The OSCAR study (Osteoporosis with Cadmium as a Risk Factor) involved the investigation of 1,000 persons who were exposed to cadmium from the environment or from an occupation. This study demonstrated that there was a relationship between environmental and occupational exposure to cadmium and kidney damage (renal tubular damage). Two other adverse health effects that have been attributed to cadmium exposure are osteoporosis in women and loss of height in men.[20]

Cadmium has been studied in relation to elevated blood pressure and the occurrence of cardiovascular diseases. "The cadmium hypothesis postulates that long-term environmental exposure to cadmium may lead to hypertension . . . and that cadmium via blood pressure elevation, or possibly via other mechanisms, may contribute to the pathogenesis of cardiovascular diseases in industrialized countries. . . ."[22(p257)] However, not all studies have been able to corroborate this hypothesis.

One of the conditions attributed to cadmium is known as **itai-itai disease**. Cases of this disease began occurring in the Jinzu River basin in Japan as early as 1912 and gradually increased until the number of cases peaked in the period of 1955 through 1959. A description of the disease is as follows: "The strange disease that appeared in the downstream basin of the Jinzu River around 1912 was called by locals 'itai-itai byo' ('itai' being what Japanese people say when inflicted with pain and 'byo' literally meaning disease). It came by this name because of the way victims cried out 'itai-itai' under the excruciating pain they endured. Women were mostly afflicted with pain across their entire body and more severe cases suffered broken bones when trying to move on their own."[23]

Various causes of the disease were studied, including malnutrition, infectious disease agents, and the effects of aging. Subsequent investigations revealed that the disease was caused by chronic cadmium poisoning due to the presence of the metal in the water of the Jinzu River. The cadmium contamination of the river was related to discharges from the Kamioka Mining Company, located upstream.[23,24]

TABLE 6.5 Health Effects Potentially Associated with Cadmium Exposure
Renal damage and renal dysfunction
Bone disease
Some forms of cancer
Hypertension
Cardiovascular diseases
Genotoxicity
Circulatory diseases
Prostate and lung cancers
Reproductive toxicity
Reduced life expectancy
Osteoporosis

Chromium

Found as a naturally occurring element in the earth's crust (e.g., in rocks, soils, and materials of volcanic origin), chromium occurs in various forms.[25] The most common forms are chromium(0), chromium(III), and chromium(VI). Found naturally in the environment, chromium(III) is as an essential nutrient related to the body's use of sugar, protein, and fat. The other two forms result from human activities such as industry. In view of its numerous industrial applications, chromium, particularly chromium(VI) and chromium(0), is a widespread environmental pollutant.

Human exposure to chromium in its various forms can occur as a result of occupational exposure (e.g., breathing chromium-contaminated air) and ingestion of food and drinking water. Early in the 20th century, occupational health specialists observed that employees in the chromium ore industry in Germany had higher rates of lung cancer than the general population. This finding has been replicated in numerous studies, which have suggested that chromium(VI) is a human carcinogen. Because of its potential to cause lung cancer, chromium(VI) has been classified by the World Health Organization and the EPA as a human carcinogen.

Regarding the environment, hexavalent chromium—chromium(VI)—seems to cause the greatest concern. Much attention has been given to chromium(VI), which can leach into drinking water. Consequences of ingestion of chromium(VI) are digestive problems and damage to organs such as the kidneys and liver. Chromium(VI) also may produce skin ulcers when it is applied to

HEXAVALENT CHROMIUM: ONE TOWN'S STORY

On December 7, 1987, officials from Pacific Gas and Electric Company (PG&E), the world's largest utility, advised California regulatory authorities that they'd detected hexavalent chromium (Cr[VI]) at levels of 580 micrograms per liter (µg/L)—over 10 times the state's 50-µg/L limit for total chromium—in a groundwater monitoring well. Cr(VI) was being used as an anticorrosive in the cooling towers of a PG&E gas compressor station in the Mojave Desert town of Hinkley.

People who lived in Hinkley had experienced a disturbing array of health problems: liver, heart, respiratory, and reproductive failure; cancer of the brain, kidney, breast, uterus, and gastrointestinal system; Hodgkin disease, frequent miscarriages, and more. Were these problems related to the compressor station's wastewater ponds? PG&E officials said no. But until 1972, PG&E had knowingly released 370 million gallons [1.4 billion liters] of Cr(VI)-contaminated wastewater into the unlined ponds, and the toxic compound had made its way into Hinkley's groundwater.

In 1993, 77 Hinkley plaintiffs filed a lawsuit against PG&E. The suit was a direct result of a massive communications effort mounted by Erin Brockovich (**FIGURE 6.3**), an employee in a local law firm. She had uncovered the utility's environmental misconduct and launched a personal investigation that ended in the largest settlement on record for a civil class-action lawsuit. PG&E filed a motion to strike all claims for preconception injuries (fear of cancer) as speculative. But the plaintiffs—648 by the end—ended up recovering for injury claims and settling with PG&E for $333 million. In addition, PG&E agreed to stop using Cr(VI) and clean up the contamination.

The case remains controversial among chromium experts because most of the Hinkley exposures involved drinking Cr(VI)-laced water. This route of exposure is widely believed to cause much less toxicity than inhalational exposures because ingested Cr(VI) is converted to inactive trivalent chromium in the stomach. Many experts also claim that the exposures were too low to cause health effects, and that there are few data linking Cr(VI) exposures to the Hinkley residents' symptoms.

But others counter that there are too many gaps in the data on chromium to dismiss the Hinkley residents' case. They believe the fact that this toxic form of chromium can enter all types of cells means that scientists may yet discover that it can damage many organ systems. Until more is known about how different doses and routes of exposure of Cr(VI) affect different populations, it is too soon to rule out high drinking water exposures as a health risk.

FIGURE 6.3 The advocacy efforts of Erin Brockovich, an employee of a California law firm (Masry & Vititoe), culminated in a successful lawsuit against a power company for groundwater pollution by chromium(VI). The plaintiffs were residents of Hinkley, California.

Modified and reproduced from Pellerin C, Booker SM. Reflections on hexavalent chromium: health hazards of an industrial heavyweight. *Environ Health Perspect*. 2000;108:A407.

the skin. The consequences of breathing chromium(VI) in high concentrations include respiratory problems, for example, nose bleeds, perforation of the nasal septum, and runny nose.

The film *Erin Brockovich* called attention to the potential hazards associated with groundwater pollution from chromium. The preceding text box titled, "Hexavalent Chromium: One Town's Story" discusses the successful advocacy efforts of Erin Brockovich, an employee of a California law firm, to win claims against a power company for groundwater pollution by chromium(VI)—shown in the text box as Cr(VI). The film is noteworthy because it brought the issue of environmental contamination with heavy metals such as chromium before the general public.

Mercury

Mercury is a naturally occurring metal that is highly toxic and exists in three major forms: metallic mercury (elemental mercury), inorganic mercury, and organic mercury. At room temperature metallic mercury is a silvery liquid metal. Inorganic mercury refers to compounds of mercury such as mercury oxide or mercury sulfide. The combination of carbon and mercury produces the so-called organic mercury compounds, the most common of which is methyl mercury.[26]

Mercury has been used medically to treat syphilis, as an agricultural fungicide, and in dental amalgams for filling cavities. It is released into the environment as a by-product of industrial processes. Even at low levels, when it becomes deposited in the beds of lakes, rivers, and other bodies of water, mercury represents a potential hazard to human health. Low levels of mercury can be magnified to higher levels as a result of methylation and bioaccumulation.

Through the process of methylation, microorganisms (bacteria and fungi) that ingest small amounts of mercury from sediments convert elemental mercury into methyl mercury; the process of bioaccumulation causes mercury levels to increase as the metal moves up the food chain and becomes more concentrated in aquatic invertebrates. Fish used for food may develop high levels of mercury in their tissues when they feed on these lower organisms. Bioaccumulation of mercury may result in mercury levels in fish that are unacceptably high for human consumption. For example, government agencies have warned that certain types of fish, such as shark, swordfish, tilefish, and king mackerel, may contain dangerously high levels of mercury. Canned albacore is suspected to contain unhealthful levels of mercury. As noted, mercury contamination is a particular hazard to the unborn children of pregnant women.

Mercury is extracted from cinnabar, a reddish ore composed of mercuric sulfide. (Refer to **FIGURE 6.4**.) Conversion of cinnabar into elemental mercury, a silvery liquid metal called quicksilver, involves crushing the ore and heating it at high temperatures in retorts in which the metal is distilled by condensation. During cinnabar processing, mercury fumes may be disseminated into the ambient environment and also may affect workers who breathe the fumes. Sediments from the mining operation can be transported into nearby bodies of water and persist as an environmental hazard. For more information, refer to the following text box about the New Almaden Mine in California.

High levels of mercury contamination from the effluents of factories have produced environmental disasters. In 1956, an environmental catastrophe occurred in Minamata Bay, Japan, where approximately 3,000 cases of neurologic disease resulted among people who ate fish contaminated with methyl mercury.[27] The neurologic condition, which became known as **Minamata disease**, was characterized by numbness of the extremities, deafness, poor vision, and drowsiness; the condition was unresponsive to medical intervention and frequently culminated in

FIGURE 6.4 Cinnabar ore.

Courtesy of Ron Horii, New Almaden Quicksilver Mining Museum, and Santa Clara County Parks and Recreation Department.

THE NEW ALMADEN MINE, CALIFORNIA

Named after the famous Almaden mines of Spain, the New Almaden Quicksilver Mine is located in the San Francisco Bay area of northern California. From the late 18th century until 1976 when mining operations terminated, large quantities of mercury were extracted from the mine. However, the legacy of the mining operation lives on: mercury contamination from this former mining operation poses a continuing hazard to fish and aquatic life in the affected geographic area; signs have been posted to warn persons not to consume fish caught in nearby streams and lakes due to potentially toxic levels of mercury. **FIGURE 6.5** shows different aspects of the mercury mining operations at the New Almaden mine.

FIGURE 6.5 The New Almaden Mine area. A: Sign advising against consumption of mercury-contaminated fish. B: Defunct equipment used for mercury mining and extraction. C: Cinnabar ware crafted by Chinese artisans.

(A, B) Courtesy of Ron Horii. Location: Almaden Quicksilver County Park.
(C) Courtesy of Ron Horii, New Almaden Quicksilver Mining Museum, and Santa Clara County Parks and Recreation Department.

death. (Refer to **FIGURE 6.6**.) The cause was attributed to discharges of mercury compounds into the bay by a plastics factory.

Lead

Sources of environmental lead include leaded gasoline, tap water from soldered pipes, and painted surfaces in older buildings. **FIGURE 6.7** illustrates these common sources of environmental lead. In the past, lead was added to gasoline in order to improve the performance of internal combustion engines and to paint in order to increase the coatings' durability and adhesive properties. Some of the paints used before the 1950s contained as much as 50% or more lead. Since 1978, the use of lead as an additive to paints has been outlawed in the United States. Also, lead has been banned as an additive in gasoline in the United States and other developed countries. The reason for this policy decision was the widespread environmental dissemination of lead by the combustion products of automobile engines. As a result of these measures, the number of children who have elevated blood lead levels has declined gradually.

FIGURE 6.6 Patient afflicted with Minamata disease.

FIGURE 6.8A illustrates how lead may peel from walls endangering the health of infants and children living in the house. Another common source of household lead exposure is imported pottery that is used in food service. Some countries still permit the use of lead as a component of paints and glazes for decorating pottery. Over time, this lead may leach into foods and be consumed; **FIGURE 6.8B** presents examples of pottery that may contain lead.

FIGURE 6.9 demonstrates the transfer of lead particles from the environment of workers to the home, where children and other family members may be exposed. Careful hand washing, personal hygiene, and laundering of clothing will prevent heavy metal–bearing dusts from leaving the workplace and endangering the health of family members.

Lead exposure is associated with serious central nervous system effects and other adverse health consequences, even when ingested at low levels. Lead poisoning is one of the most common environmental pediatric health problems in the United States. Especially damaging to the growing neurologic systems

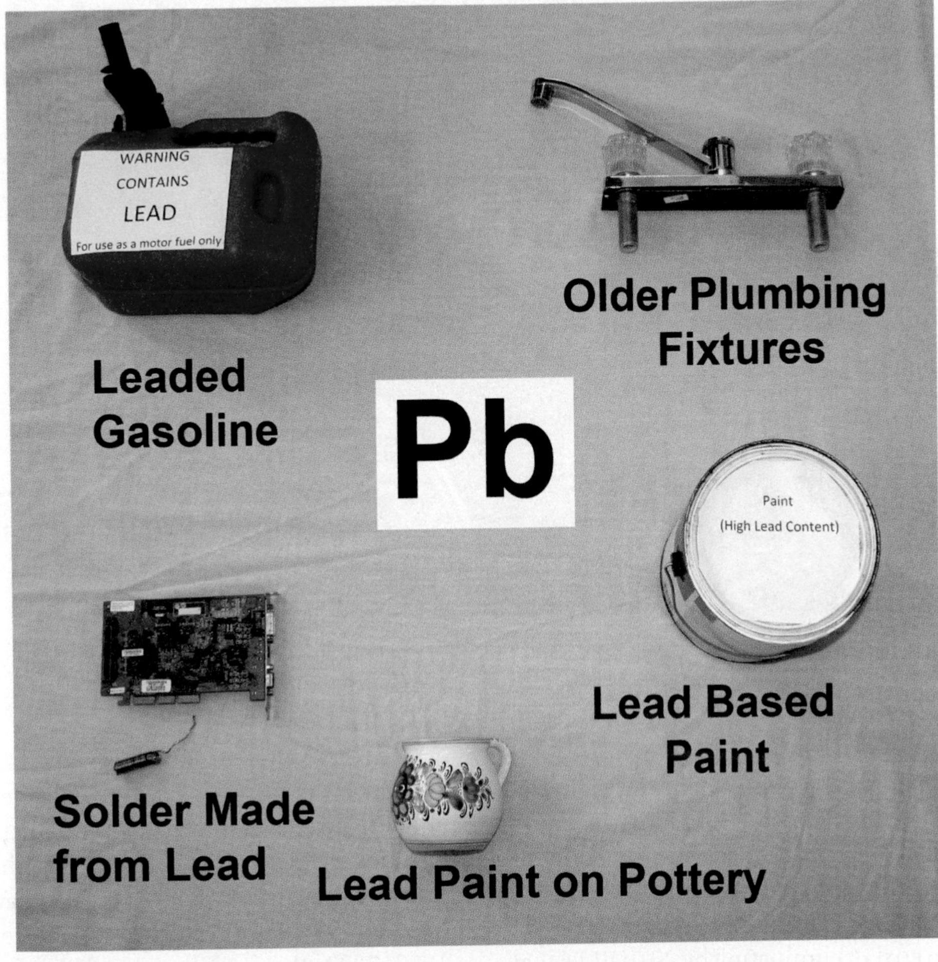

FIGURE 6.7 Examples of possible exposure to lead—a common environmental contaminant.

FIGURE 6.8A Lead from peeling paint may compromise the health of infants and children.

FIGURE 6.8B Lead-containing glazes pose a danger if used on dishes or containers intended for food use as lead may leach into the food itself.

of young children, lead is a neurotoxin.[28] Acutely elevated blood lead levels (BLLs)—defined as 70 µg/dL of blood or higher—are associated with severe neurologic effects, including coma and death. The US Department of Health, Education, & Welfare (DHEW) promulgated the goal of eliminating by 2010 BLLs that exceed 10 µg/dL among children younger than 6 years of age.

Since the period 1976 to 1980, the percentage of children who had BLLs that exceeded 10 µg/dL has declined markedly. The National Health and Nutrition Examination Survey (NHANES) estimated that during 1976 to 1980, the prevalence was more than 80%. By 1999–2000, the prevalence had dropped to 2.2% or 434,000 children in all. In 1997, a President's Task Force reported that approximately 775,000 children

FIGURE 6.9 Transfer of lead from the workplace to the home.
Reproduced from California Department of Health Services, Occupational Lead Poisoning Prevention Program, Childhood Lead Poisoning Branch. Don't Carry Lead into Your House. June 2004.

under age 6 were estimated to have blood lead levels above 10 μg/dL; the corresponding prevalence for blood lead levels above 15 μg/dL was about 190,000.[29]

Among children who have blood lead levels above 15 μg/dL, lead toxicity is associated with maladaptive behavior.[30] A study of inner-city pregnant women, nonpregnant women of childbearing age, and their children found a high prevalence of elevated blood lead levels compared with the general US population of whites and blacks.[31]

Evidence suggests that adverse effects of lead exposure among children occur at BLLs below 10 μg/dL; the threshold for harmful effects of lead exposure has not been determined.[28] Blood lead levels at or below 5 μg/dL are associated with lasting neurological damage and behavioral disorders.[32] Consequently, in May 2012 the CDC replaced its "level of concern" 10 μg/dL for blood levels among children younger than 6 years with a more stringent standard. The new standard is called a "reference value," which is derived from the upper level of the distribution of blood lead levels of US children younger than 6 years.[32] The reference value for blood lead levels of children aged 1 to 5 years is 5 μg/dL. Parents are notified if their children have this level or a higher blood lead level.

During the past three decades or so, children's blood levels have continued to decrease. Refer to **FIGURE 6.10** for trend data in these levels. The following text box presents a personal story of how elevated blood lead levels affected one family's child.

Cadmium and Lead

Because cadmium is a common contaminant in lead ores, environmental health researchers often study lead pollution paired with cadmium pollution. Cadmium pollution is often the by-product of

ONE FAMILY'S STORY

In April of 1996, my family and I managed to save enough to buy our own home. Within 4 months of moving in, our pride and joy evaporated when Samuel, then 10 months old, was diagnosed with a blood lead level of 32 micrograms per deciliter (μg/dL). I soon learned that my son's blood lead level was three times above the limit thought to cause future learning problems. A greater shock was that the lead paint, dust, and soil in and around our treasured home was the culprit.

Worse yet, a month later, Samuel's lead level had risen to 50 μg/dL. He was hospitalized that same afternoon and for three long, agonizing days he stayed in the hospital and began treatment. During Samuel's hospitalization, my husband and I spent many hours attempting to make our home lead safe, all the while keeping vigil over Sam. For nearly 4 years, Sam had his blood tested every 2 months. We continued to improve our home through repair loans to make it safe.

Today, our house has new windows, and lead abatement has been completed on the interior and exterior of our home. Samuel's blood lead level has dropped below 10 μg/dL. To see Samuel, now 4 years old, you would never know what this happy, beautiful little boy has had to endure.

Reproduced from President's Task Force on Environmental Health Risks and Safety Risks to Children. *Eliminating Childhood Lead Poisoning: A Federal Strategy Targeting Lead Paint Hazards*. Washington, DC: US Department of Housing and Urban Development; 2000:3. Available at: https://www.cdc.gov/nceh/lead/about/fedstrategy2000.pdf. Accessed February 15, 2017.

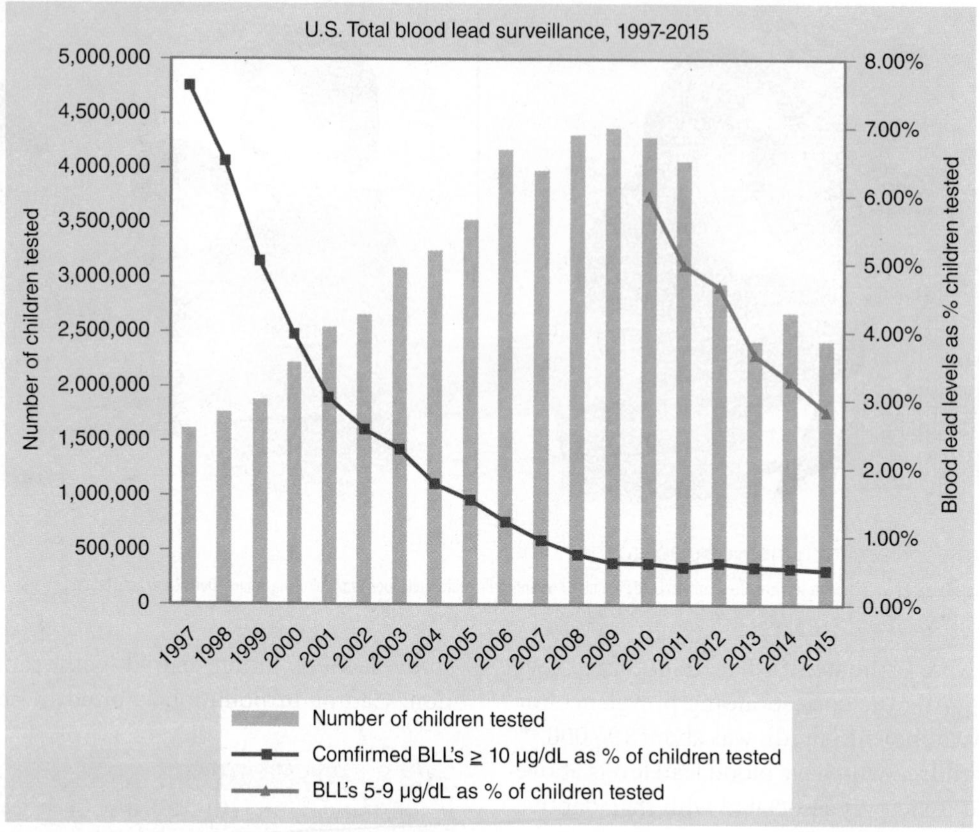

FIGURE 6.10 US totals blood lead surveillance, 1997–2015.

Reproduced from Centers for Disease Control and Prevention . Lead - CDC's National Surveillance Data (1997-2015). Available at: https://www.cdc.gov/nceh/lead/data/Chart_Website_StateConfirmedByYear_1997_2015.pdf. Accessed July 19, 2017.

smelting of lead-containing ores. A gigantic lead refinery located along the Ribeira River valley in the far south of the state of Sao Paulo, Brazil produced lead for a period of 50 years until it was closed in 1995. The refinery disseminated large amounts of lead and cadmium into the environment nearby as well as more distant areas. Children who lived close to the refinery were shown to have elevated BLLs.[33]

A German study demonstrated that cadmium and lead may be transported into people's houses through airborne dust.[34] Lead and cadmium concentrations in dust were followed in the German city of Hettstedt, a region in which mining and smelting activities were conducted. Normal household activities were associated with the movements of heavy metals into the home environment in the city, which had been contaminated with lead and cadmium.

Other research has examined the association between lead and cadmium exposure and human reproduction.[35,36] Some studies have reported the occurrence of increased rates of spontaneous abortions and preterm labor among women who live in areas contaminated with lead and cadmium. In sections of eastern Poland known as the Suwalki area, the soil contains high concentrations of lead and cadmium.[35] In addition, residents may be exposed to these heavy metals from air pollution. Nevertheless, the findings regarding the relationship between exposure to lead and cadmium and the frequency of spontaneous abortions have not been established definitively.

Another area where high levels of pollution from lead and cadmium have occurred is in the Asua Valley, which is on the outskirts of Bilbao, Spain.[37] This territory is in the Basque country located in northern Spain. There, during the 1960s through the 1980s, many industries such as foundries and metal-working plants produced air pollution laden with heavy metals. These heavy metals contaminated soil in food-growing areas. Consequently, residents, both children and adults, ingested very high levels of cadmium and lead, ranging from almost 200% to almost 700% of the tolerable daily intake of lead for children.

Children who live in areas heavily contaminated with lead and cadmium show concentrations of these metals in their hair and blood.[38] Children from Upper Silesia in southern Poland were studied for their blood and hair concentrations of lead and cadmium. This region, one of the most highly polluted in the country, contained Poland's largest zinc plant, which was responsible for heavy contamination of the locality with heavy metals. Very high concentrations of heavy metals were found in the bodily tissues of local children, especially among boys. Thought to spend more time outdoors than girls, boys were more likely to have high exposures to heavy metals. Because of high levels of lead and cadmium contamination of the soil in Poland, authorities have recommended that the consumption of foods produced locally, for example, vegetables and potatoes, be reduced.[39]

A Spanish epidemiologic study investigated whether job-related exposure to lead or cadmium was associated with elevated blood pressure levels among two groups of workers in Barcelona.[40] One group was employees who were exposed to lead (e.g., welders); the second group consisted of workers who were exposed to cadmium in pigment and resin factories. Researchers found that workers exposed to lead showed a significant increase in blood pressure, whereas those exposed to high levels of cadmium did not show such an increase. Apparently, in this research, high levels of exposure to these two heavy metals had differential effects upon blood pressure.

Nickel

Classified as a heavy metal, nickel is one of the constituents of the earth's crust. Because of its omnipresence, human exposure to low levels of nickel is probably universal and unavoidable. Nickel has a number of important qualities that make it a valuable metal for use in industry. One of nickel's useful qualities is that it can be machined and polished readily and resists corrosion from nonoxidizing acids and alkalis. It is employed in the production of many of the appliances and tools that are common in everyday life, e.g., nickel cadmium batteries. Environmental exposures to nickel wastes may occur when nickel cadmium batteries and other nickel-containing metals are discarded.

The occupational environment is a source for significant exposure to nickel. For example, some metal workers risk exposure to nickel and nickel compounds through inhalation or skin contact. One of the common reactions to skin contact with nickel is nickel allergy, manifested as contact dermatitis. Other effects of nickel exposure include cardiovascular-related and renal diseases as well as fibrosis of the lungs. One of the concerns about nickel is its potential carcinogenic action. For example, epidemiologic studies have suggested that nickel miners have a higher incidence of nasal cancer as well as lung cancer than the nonexposed population. Other occupational exposures to nickel occur in the production of nickel through refining, as well as processing the metal via electroplating and welding.

Nickel is present in many different chemical forms, some of which are compounds of nickel and other elements. These compounds are classified as either water soluble or water insoluble. The role of nickel compounds as a causal factor in carcinogenesis or as having a role in enhancing the exposure to other carcinogens has not been clearly established.[41] Overall, researchers believe that exposure to low doses of nickel is probably not harmful to human beings.[41,42]

Essential Metals with Potential for Toxicity

Copper, zinc, and iron, examples of metals that are essential for human nutrition, can be toxic if ingested in excessive amounts. An optimal range of these essential metals is necessary to maintain health. When insufficient amounts of these **trace metals** are consumed, dietary deficiencies can result. In fact, if these trace metals are absent from the diet, human beings and other organisms can fail to thrive. The middle range of consumption of essential nutrients is associated with normal health. At the other end of the continuum, when these levels are exceeded, toxic effects and lethality can occur.

Copper

A red-colored metal, copper is present naturally in the earth's rocks and soil. In trace amounts, copper is an essential nutrient. Sometimes copper combines with other elements to form compounds such as copper sulfate. Copper compounds often have a blue-green color. Elemental copper does not decompose in the environment.[43]

Because of its many applications, copper and its various forms have been distributed widely in the environment. Copper appears in electrical wires, pipes, in combination with other metals to form alloys, as a mildew inhibitor, and as a wood and leather preservative. The ATSDR estimates that in the year 2000 alone, approximately 1.4 billion pounds (635 million kilograms) of copper were released into the environment during industrial processing. In addition to the foregoing sources of environmental contamination by copper, others are metal smelting, mining, and waste disposal sites.

Avenues for human exposure to copper include inhalation, ingestion of copper-containing foods and water, and direct contact with the skin. Copper (along with lead) is a potential contaminant of tap water. When tap water stands in copper pipes overnight, small amounts of copper dissolve, causing copper levels to become more concentrated than when the water flows freely. By running the water for approximately half a minute in the morning before use, one can reduce the concentration of copper in the tap water. Workplace exposures to copper arise from breathing dust stirred up by mining and processing copper-rich ore. Other exposures result from inhalation of fumes and dust generated during the welding and grinding of copper metal.

Although copper is an essential nutritional element, at higher levels it is known to produce toxic effects. Exposure to concentrated amounts of copper (far above trace levels) can produce respiratory and gastrointestinal disturbances. The respiratory effects from copper dust include irritation of the respiratory tract (e.g., nose and mouth). The gastrointestinal effects associated with ingestion of copper include vomiting, diarrhea, nausea, and stomach cramping. Very high levels of copper are known to cause liver damage, renal damage, and death. Although it is associated with toxic effects, copper has not been classified as a human carcinogen.

Zinc

A frequently occurring element found in the earth's crust, zinc permeates air, soil, water, and, to some degree, all foods.[44] Zinc is used commercially as a coating for rust inhibition, as a component of batteries, and in combination with other metals to make brass, bronze, and other alloys. Refer to the following text box, which presents possible means of human exposure to zinc. In addition to its commercial uses, zinc is a nutritional element that is important for maintaining health. Among children, inadequate dietary intake of zinc is correlated with low socioeconomic status.[45] In comparison with children from well-to-do families, some poor children have a higher incidence of infectious diseases and other forms of morbidity, which may be related to inadequate intake of zinc.

The metal is believed to play a role in reducing the occurrence of common infections; it is theorized to be necessary for children's growth and for maintaining the health of pregnant women. Also, zinc is thought to influence fetal neurobehavioral development.[44,45] Inadequate zinc intake during pregnancy may affect the health of the developing fetus adversely, for example, by causing growth retardation. Among adults, insufficient intake of zinc is associated with anorexia, loss of the senses of taste and smell, dermatologic problems, and damage to the immune system.

Excessive levels of zinc exposure, which can have toxic effects, result from the ingestion of contaminated food and water and from inhalation of zinc dusts in occupational environments. Excessive amounts of zinc, defined as beginning at 10 to 15 times the levels required to maintain health, are thought to affect the normal functioning of blood-forming organs, biochemical processes, and the endocrine system. The consumption of large quantities of zinc is associated with gastrointestinal problems such as stomach cramps, nausea, and vomiting. Zinc also can cause anemia and damage to the pancreas. Children and pregnant women are at special risk for the toxic effects of zinc. These groups should use care in taking zinc as a nutritional supplement or for self-medication.[46] Breathing high concentrations of zinc in the workplace causes a disease known as metal fume fever. This condition appears to be an immune-mediated response that originates in the lungs.

HOW MIGHT I BE EXPOSED TO ZINC?

Zinc is an essential element needed by your body in small amounts. We are exposed to zinc compounds in food. The average daily zinc intake through the diet in the United States ranges from 5.2 to 16.2 milligrams (milligram = 0.001 gram). Food may contain levels of zinc ranging from approximately 2 parts of zinc per million (2 ppm) parts of some foods (e.g., leafy vegetables) to 29 ppm for other foods (meats, fish, poultry). Zinc is also present in most drinking water. Drinking water or other beverages may contain high levels of zinc if they are stored in metal containers or flow through pipes that have been coated with zinc to resist rust. If you take more than the recommended daily amount of supplements containing zinc, you may have higher levels of zinc exposure.

In general, levels of zinc in air are relatively low and fairly constant. Average levels of zinc in the air throughout the US are less than 1 microgram of zinc per cubic meter (μg/m^3) of air, but range from 0.1 to 1.7 μg/m^3 in areas near cities.

Air near industrial areas may have higher levels of zinc. The average zinc concentration for a 1-year period was 5 μg/m^3 in one area near an industrial source.

In addition to background exposure that all of us experience, about 150,000 people also have a source of occupational exposure to zinc that might elevate their total exposure significantly above the average background exposure. Jobs where people are exposed to zinc include zinc mining, smelting, and welding; manufacture of brass, bronze, or other zinc-containing alloys; manufacture of galvanized metals; and manufacture of machine parts, rubber, paint, linoleum, oilcloths, batteries, some kinds of glass and ceramics, and dyes. People at construction jobs, automobile mechanics, and painters are also exposed to zinc.

Reproduced from Agency for Toxic Substance and Disease Registry (ATSDR): Toxic Substances Portal – Zinc. How might I be exposed to zinc? Available at: https://www.atsdr.cdc.gov/ToxProfiles/tp60-c1-b.pdf. Accessed February 15, 2017.

Iron

Iron is one of the most ubiquitous metals in the earth's crust. Vital to human health, iron is important to the growth of cells and the transport of oxygen within the circulatory system. About 66% of the iron in the body is used in hemoglobin, the red blood cell protein responsible for transporting oxygen. Well-known symptoms of anemia (deficiency of hemoglobin) include fatigue and lowered immunity. **FIGURE 6.11** demonstrates the lower eyelid (conjunctiva) of a child suspected of having anemia, a condition that causes the conjunctiva to appear white.

Iron has the capacity to accumulate in the body, because little is excreted during metabolic processes. The disease known as hemachromatosis is associated with iron toxicity from excessive buildup of iron. Excessive amounts of iron can have toxic effects, such as vomiting, diarrhea, and damage to the intestines. Iron toxicity also may produce low blood pressure, lethargy, neurologic effects including seizures, and liver injury. Among those at greatest risk of iron toxicity are children. Acute iron intoxication (accidental iron poisoning) is among the most common childhood poisonings. Poisoning incidents sometimes occur when children accidentally swallow mineral supplements that contain concentrated amounts of iron. Other groups at risk from iron overload (iron toxicity) include adult men and postmenopausal women.

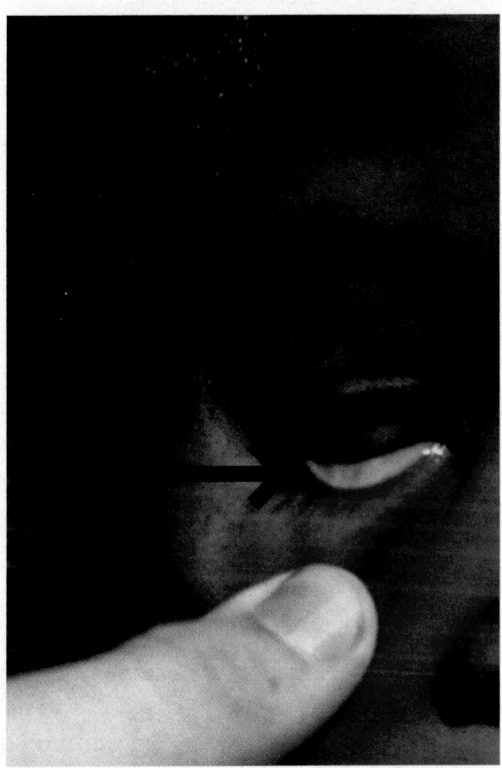

FIGURE 6.11 An Epidemic Intelligence Service (EIS) officer examining the palpebral conjunctiva of a Nigerian child suspected of having anemia.

Courtesy of CDC/ Dr. Lyle Conrad

Metals for Use in Medical Therapy

Metals in this group, which include aluminum, lithium, and platinum, are used for medical and other purposes. Aluminum is contained in medicines, for example, certain antacids. Because of its numerous useful properties and possible association with Alzheimer's disease, aluminum is discussed more fully in the following section.

Aluminum

The silver-white metal aluminum is used widely in food and beverage containers, in pots and pans, and in construction sites. Aluminum is an ingredient in various medicines and cosmetics, for example, to buffer aspirin and in antiperspirants. As a result of its widespread use, human exposure to aluminum is almost universal. One of the direct effects of exposure to aluminum dust is respiratory problems such as coughing or exacerbation of asthma. Low levels of aluminum exposure are not thought to be harmful.[47] The following text box provides information on sources of exposure to aluminum.

Aluminum has been studied as a factor that may be linked to Alzheimer's disease. Aluminum is known to be a powerful neurotoxicant.[48] The role of aluminum as a factor in the pathogenesis of Alzheimer's disease remains unconfirmed. One study reported an association between aluminum concentration in drinking water and Alzheimer's disease. The authors of this report suggested that the amount of residual aluminum in public drinking water supplies should be limited.[48,49] Another study reported an association between exposure to aluminum contained in intravenous feeding solutions and impaired neurologic development among preterm infants.[50] The ATSDR states, "Some studies show that people exposed to high levels of aluminum may develop Alzheimer's disease, but other studies have not found this to be true. We do not know for certain that aluminum causes Alzheimer's disease. Some people with kidney disease store a lot of aluminum in their bodies.... Sometimes these people developed bone or brain diseases that doctors think were caused by the excess aluminum."[47] There also have been some associations of large doses of aluminum with skeletal problems and the development of rashes among people who use deodorants containing aluminum compounds.

HOW MIGHT I BE EXPOSED TO ALUMINUM?

- Virtually all food, water, air, and soil contain some aluminum.
- The average adult in the US eats about 7–9 mg aluminum per day in his or her food.
- Exposure comes from breathing higher levels of aluminum dust in workplace air.
- It also occurs from living in areas where the air is dusty, where aluminum is mined or processed into aluminum metal, near certain hazardous waste sites, or where aluminum is naturally high.
- In addition, you might be exposed from eating substances containing high levels of aluminum (such as antacids) especially when eating or drinking citrus products at the same time.
- Children and adults may be exposed to small amounts of aluminum from vaccinations.
- Very little enters your body from aluminum cooking utensils.

Modified and reproduced from Agency for Toxic Substances and Disease Registry (ATSDR). Aluminum—ToxFAQs™. Available at: https://www.atsdr.cdc.gov/toxfaqs/tf.asp?id=190&tid=34. Accessed January 26, 2017.

Conclusion

Many heavy metals and other metallic compounds occur almost universally in the environment, either in the earth's crust or as a result of human activities. Consequently, the entire human population is exposed at one time or another to these metals. Although some metals in trace levels are essential for human nutrition, the same metals are toxic at higher levels. Other metals are toxic even at low levels of exposure. For example, arsenic, lead, and mercury are among the top three hazardous substances identified in ATSDR's CERCLA Priority List of Hazardous Substances. Arsenic, a potential carcinogen, is found in high concentrations in the drinking water of some areas of the world. Lead is of concern because of its adverse impact upon the health of children. Mercury, which is associated with neurologic disease, has been identified increasingly as a contaminant of fish and other foodstuffs. An important role for environmental policy makers is to reduce the level of toxic metals in the environment (e.g., air, water, and soil) and in foods, which may constitute an important avenue of exposure of the human population to these metals.

Study Questions and Exercises

1. Define and give examples of the following terms:
   a. Heavy metal
   b. "Itai-itai byo"
   c. Minamata disease
   d. Trace metals
2. Compare the three categories of metals shown in Table 6.1. Give an example of a metal from each category.
3. Why is lead contamination a concern for environmental health experts?
4. Name some of the portals of entry into the human body for toxic metals. Describe the methods by which these toxic metals may access specific portals of entry.
5. Give some examples of environmental sources of exposure of populations to toxic heavy metals.
6. Exposure to heavy metals presents a serious hazard to children.
   a. Give an example of a heavy metal to which children commonly are exposed in our environment.
   b. List the symptoms of exposure to this heavy metal among children.
   c. List the long-term effects of this heavy metal on children.
7. State two incidents of mass poisonings of humans that involved arsenic.
8. Provide the following information regarding cadmium:
   a. Possible sources of exposure
   b. Modes of entry into the body
   c. Two major health effects
   d. Name of an associated disease
9. Provide the following information regarding chromium:
   a. Where it occurs in the environment
   b. Its effects on human health
   c. Who is Erin Brockovich?
10. What are some of the medical uses of mercury?
11. Provide the following information regarding lead:
    a. Sources of exposure in the environment
    b. Adverse health effects of exposure to lead
12. Provide the following information regarding nickel:
    a. Two uses of nickel
    b. Two adverse effects of exposure to nickel
13. Name the metals that are essential for human nutrition.
14. How can you reduce your exposure to copper in tap water?
15. Describe the importance of iron to human health. Describe the side effects of ingesting excessive amounts of iron.

16. List at least three sources of exposure to aluminum. Describe the hypothesized association of aluminum and Alzheimer's disease.

For Further Reading

1. Power MC, Weisskopf MG. Lead: neurotoxic effects in adults. In Friis, RH, ed. *The Praeger Handbook of Environmental Health*, Volume 2. Santa Barbara, CA: ABC-CLIO, LLC, 2012, 201-231.

References

1. Agency for Toxic Substances and Disease Registry (ATSDR). CERCLA priority list of hazardous substances. Available at: https://www.atsdr.cdc.gov/spl/. Accessed January 30, 2017.
2. US Environmental Protection Agency. Vocabulary Catalog. Terms of Environment: Available at: https://iaspub.epa.gov/sor_internet/registry/termreg/searchandretrieve/glossariesandkeywordlists/search.do. Accessed February 4, 2017.
3. Tokar EJ, Boyd WA, Freedman JH, Waalkes MP. Toxic effects of metals. In: Klaasen CD, Watkins JB III, eds. *Casarett & Doull's Essentials of Toxicology*. 3rd ed. New York, NY: McGraw-Hill, 2015:347–359.
4. Dictionary.com. Specific gravity. Available at: http://www.dictionary.com/browse/specific-gravity. Accessed February 5, 2017.
5. Skerfving SB, Copplestone JF. Poisoning caused by the consumption of organomercury-dressed seeds in Iraq. *Bull World Health Organ*. 1976;54:101–112.
6. Vahter M, Berglund M, Akesson A, et al. Metals and women's health. *Environ Res*. 2002;88(3):145–155.
7. Wulff M, Högberg U, Stenlund H. Occupational and environmental risks of spontaneous abortions around a smelter. *Am J Ind Med*. 2002;41:131–138.
8. Stein J, Schettler T, Wallinga D, et al. In harm's way: toxic threats to child development. *J Dev Behav Pediatr*. 2002;23:S13–S22.
9. Peters GR, McCurdy RF, Hindmarsh JT. Environmental aspects of arsenic toxicity. *Crit Rev Clin Lab Sci*. 1996;33:457–493.
10. Jager JW, Ostrosky-Wegman P. Arsenic: a paradoxical human carcinogen. *Mutat Res*. 1997;386:181–184.
11. Brown KG, Ross GL. Arsenic, drinking water, and health: a position paper of the American Council on Science and Health. *Regul Toxicol Pharmacol*. 2002;36:162–174.
12. Rahman MM, Chowdhury UK, Mukherjee SC, et al. Chronic arsenic toxicity in Bangladesh and West Bengal, India—a review and commentary. *J Toxicol Clin Toxicol*. 2001;39:683–700.
13. Cantor KP. Arsenic in drinking water: how much is too much? [Editorial] *Epidemiol*. 1996;7:113–115.
14. Hopenhayn-Rich C, Biggs ML, Fuchs A, et al. Bladder cancer mortality associated with arsenic in drinking water in Argentina. *Epidemiol*. 1996;7:117–124.
15. Abernathy CO, Liu Y-P, Longfellow D, et al. Arsenic: health effects, mechanisms of actions, and research issues. *Environ Health Perspect*. 1999;107:593–597.
16. Ahmad SA, Sayed SU, Barua S, et al. Arsenic in drinking water and pregnancy outcomes. *Environ Health Perspect*. 2001;109:629–631.
17. Taylor TP, Ding M, Ehler DS, et al. Beryllium in the environment: a review. *J Environ Sci Health*. 2003;A38:439–469.
18. Agency for Toxic Substances and Disease Registry (ATSDR). ToxFAQs for cadmium. CAS #7440-43-9. October 2012. Available at: http://www.atsdr.cdc.gov/toxfaqs/index.asp. Accessed February 5, 2017.
19. United States Environmental Protection Agency. Aquatic life ambient water quality criteria. Update for cadmium—2016. Office of Water. EPA 822-F-003. March 2016.
20. Järup L. Cadmium overload and toxicity. *Nephrol Dial Transplant*. 2002;17(suppl 2):35–39.
21. Staessen JA, Lauweyrs RR, Ide G, et al. Renal function and historical environmental cadmium pollution from zinc smelters. *Lancet*. 1994;343:1523–1527.
22. Staessen J, Amery A, Bernard A, et al. Blood pressure, the prevalence of cardiovascular diseases, and exposure to cadmium: a population study. *Am J Epidemiol*. 1991;134:257–267.
23. International Center for Environmental Technology Transfer (ICETT). Preventative measures against water pollution Jinzu River, Toyama Prefecture. Discover [sic] of the itai-itai disease. Available at: https://www.icett.or.jp/english/abatement/toyama/disease.html. Accessed April 27, 2017.
24. Morikawa Y, Nakagawa H, Tabata M, et al. Study of an outbreak of itai-itai disease. [In Japanese] *Nippon Eiseigaku Zasshi*. 1992;46:1057–1062.
25. Agency for Toxic Substances and Disease Registry (ATSDR). ToxFAQs for chromium. CAS # 7440-47-3. October 2012. Available at: https://www.atsdr.cdc.gov/index.html. Accessed April 28, 2017.
26. Agency for Toxic Substances and Disease Registry (ATSDR). ToxFAQs for mercury. CAS # 7439-97-6. April 1999. Available at: https://www.atsdr.cdc.gov/index.html. Accessed April 28, 2017.
27. Powell PP. Minamata disease: a story of mercury's malevolence. *South Med J*. 1991;84:1352–1358.
28. Meyer PA, Pivetz T, Dignam TA, et al. Surveillance for elevated blood lead levels among children—United States, 1997-2001. In: *Surveillance Summaries*, September 12, 2003. *MMWR*. 2003;52(No. SS-10):1–9.
29. President's Task Force on Environmental Health Risks and Safety Risks to Children. *Eliminating Childhood Lead Poisoning: A Federal Strategy Targeting Lead Paint Hazards*. Washington, DC: US Department of Housing and Urban Development; 2000.
30. Sciarillo WG, Alexander G, Farrell KP. Lead exposure and child behavior. *Am J Public Health*. 1992;82:1356–1360.
31. Flanigan GD Jr, Mayfield R, Blumenthal HT. Studies on lead exposure in patients of a neighborhood health center: Part II. A comparison of women of childbearing age and children. *J Natl Med Assoc*. 1992;84:23–27.
32. Raymond J, Brown MJ. Childhood blood lead levels—United States, 2007–2012. *MMWR*. 2015;62(54):76–80.
33. Paoliello MM, De Capitani EM, da Cunha FG, et al. Exposure of children to lead and cadmium from a mining area of Brazil. *Environ Res*. 2002;88(2):120–128.
34. Meyer I, Heinrich J, Lippold U. Factors affecting lead and cadmium levels in house dust in industrial areas of eastern Germany. *Sci Total Environ*. 1999;234:25–36.
35. Laudanski T, Sipowicz M, Modzelewski P, et al. Influence of high lead and cadmium soil content on human reproductive outcome. *Int J Gynecol Obstet*. 1991;36:309–315.

36. Fagher U, Laudanski T, Schütz A, et al. The relationship between cadmium and lead burdens and preterm labor. *Int J Gynecol Obstet*. 1993;40:109–114.
37. Alonso E, Cambra K, Martinez T. Lead and cadmium exposure from contaminated soil among residents of a farm area near an industrial site. *Arch Environ Health*. 2001;56:278–282.
38. Chlopicka J, Zachwieja Z, Zagrodzki P, et al. Lead and cadmium in the hair and blood of children from a highly industrial area in Poland. *Biol Trace Elem Res*. 1998;62:229–234.
39. Gzyl J. Assessment of Polish population exposure to lead and cadmium with special emphasis to the Katowice province on the basis of metal concentrations in environmental compartments. *Cent Eur J Public Health*. 1997;5(2):93–96.
40. Schuhmacher M, Bosque MA, Domingo JL, et al. Effects of chronic lead and cadmium exposure on blood pressure in occupationally exposed workers. *Biol Trace Elem Res*. 1994;41:269–278.
41. Oller AR. Respiratory carcinogenicity assessment of soluble nickel compounds. *Environ Health Perspect*. 2002;110(suppl 5):841–844.
42. Denkhaus E, Salnikow K. Nickel essentiality, toxicity, and carcinogenicity. *Crit Rev Oncol Hematol*. 2002;42:35–56.
43. Agency for Toxic Substances and Disease Registry (ATSDR). Public health statement for copper. September 2004. Available at: https://www.atsdr.cdc.gov/phs/phs.asp?id=204&tid=37. Accessed July 19, 2017.
44. Agency for Toxic Substances and Disease Registry (ATSDR). ToxFAQs for zinc. August 2005. Available at: https://www.atsdr.cdc.gov/phs/phs.asp?id=300&tid=54. Accessed January 30, 2017.
45. Hotz C, Lowe NM, Araya M, et al. Assessment of the trace element status of individuals and populations: the example of zinc and copper. *J Nutr*. 2003;133(5 suppl 1):1563S–1568S.
46. Piao F, Yokoyama K, Ma N, et al. Subacute toxic effects of zinc on various tissues and organs of rats. *Toxicol Lett*. 2003;145(1):28–35.
47. Agency for Toxic Substances and Disease Registry (ATSDR). ToxFAQs for aluminum. September 2008 (updated 02/18/2010). Available at: https://www.atsdr.cdc.gov/phs/phs.asp?id=1076&tid=34. Accessed July 19, 2017.
48. Flaten TP. Aluminum as a risk factor in Alzheimer's disease, with emphasis on drinking water. *Brain Res Bull*. 2001;55:187–196.
49. McLachlan DRC, Bergeron C, Smith JE, et al. Risk for neuropathologically confirmed Alzheimer's disease and residual aluminum in municipal drinking water employing weighted residential histories. *Neurology*. 1996;46:401–405.
50. Bishop NJ, Morley R, Day JP, et al. Aluminum neurotoxicity in preterm infants receiving intravenous-feeding solutions. *New Engl J Med*. 1997;336:1557–1561.

© Jean-Luc Rivard/EyeEm/Getty Images

CHAPTER 7
Pesticides and Other Organic Chemicals

LEARNING OBJECTIVES

By the end of this chapter the reader will be able to:

- Compare classes of pesticides.
- Name three commonly used insecticides and one commonly used herbicide.
- Discuss the health effects of exposure to pesticides.
- Name one chemical used in the manufacture of plastics.
- Describe the potential health effects of exposure to household cleaning products.

Introduction

In this chapter, you will learn about the benefits and hazards of chemicals such as **pesticides** and other **organic chemicals** and how they may enter the environment. For example, regarding benefits, some chemicals are essential to modern society. Others may present hazards such as human reproductive difficulties.[1] We will review situations in which first responders to chemical accidents are confronted with the practical issue of how to protect the environment, the public, and themselves from potentially toxic substances. Additional topics are the use of pesticides, the impacts of toxic chemicals such as dioxins, and dangerous chemicals that persist in the environment. Remaining issues concern organic solvents, chemicals used for the manufacture of plastics, and home cleaning products. The chapter concludes with coverage of how organic chemicals that are released into the environment may have estrogenic activity.

How Likely are We to be Exposed to Hazardous Chemicals?

Hazardous chemicals, omnipresent in the environment, are vital to society yet simultaneously raise the specter of potential harm to all living organisms on earth. One expert has stated, "The ubiquity of toxic substances in the environment continues to be a significant public health concern. Body burdens of chemicals such as dioxins, methyl mercury, PCBs, and a range of the 'usual suspects' are the increasingly represented norm rather than the exception in the US population."[2(pA374)] An example of the omnipresence of chemicals close to home can be found many garages, which are used for storage of numerous potentially dangerous chemicals.

The National Center for Environmental Health, a branch of the Centers for Disease Control and Prevention (CDC), developed an inventory of biomonitoring exposure data for 212 environmental chemicals to which the noninstitutionalized, civilian US population was exposed over a 6-year period ranging from 1999 through 2004.[3] Although it would seem that the population has had declining levels of exposure to environmental chemicals, "serious concerns remain regarding the exposure of the population to a wide spectrum of chemicals in the environment."[2(pA374)] The Fourth National Report on Human Exposure to Environmental Chemicals reported widespread exposure to chemicals found in fire retardants, bisphenol A (BPA), plastics, and nonstick coatings used in cookware.[3]

Advantages and Disadvantages of Hazardous Chemicals

One should take a balanced view of the advantages and disadvantages of hazardous chemicals. First consider the advantages: Although chemicals have the potential to cause harm, they are essential to the functioning of modern society. In fact, humanity could not live without them. According to the National Institute of Environmental Health Sciences (NIEHS):

> Chemicals are the basis of our way of life—and health—today. There are about 15,000 chemicals made and used in high volume in the United States. (. . . when we roast coffee for our morning jump-start, we end up with more than 1,000 chemicals right there in our cup!) And there are many natural products—herbal products, for example—that have become widely marketed and used without testing.
>
> Synthetic chemicals, used on our farms, not only help feed us cheaply and well, they help feed much of the rest of the world as well. Chemical fibers clothe us. Chemicals cure us. They form key parts of our cars and our phones and computers, many building materials, rugs and other furnishings—you name it. Little wonder that US production of these synthetic chemicals has climbed, almost without pause, from 10 million pounds (4.5 million kilograms) in 1918 to well over 300 billion pounds (136 billion kilograms) in recent years.[4]

Regarding the disadvantages of hazardous chemicals, some [e.g., those called polychlorinated biphenyls (PCBs) and dioxins] have been implicated as human health hazards, for example, in the etiology of cancer and adverse birth outcomes. As another, more extreme, example, consider the public's anxiety surrounding the alleged use of chemical weapons during the 1991 Persian Gulf War. Some of the returning soldiers developed symptoms of what came to be known as the "Gulf War Syndrome." Were the symptoms (e.g., multiple chemical sensitivity, chronic fatigue syndrome, and fibromyalgia) reported by afflicted veterans attributable to exposure to chemical weapons, or could other explanations account for those symptoms?

Silent Spring, the classic work by Rachel Carson, is credited with sensitizing the public to the potential hazards of chemicals that were being disseminated into the environment.[5] This issue continues to be a cause for alarm. On a typical day, people encounter numerous chemicals, many of which are known to be injurious to human health and the environment. A large percentage of these chemicals have not been tested at all. Possible adverse health consequences that have been linked to those chemicals that have been tested include impairment of fertility and reproductive outcomes, psychological and behavioral deficits such as autism, and carcinogenesis.

The US Environmental Protection Agency (EPA) expanded its effort to measure the blood and urine concentrations of chemicals by increasing the number from 116 to 148 chemicals; these include several of the major pesticides and organic chemicals. According to a 2005 report from EPA, the levels of three organochlorine pesticides (aldrin, endrin, and dieldrin) are low or undetectable in the United States population.[6] The fact that the levels of these three pesticides are low (due to their discontinued use) reflects progress in the control of toxic chemicals.

Another issue of concern for our environment is the impact of accidental spills and releases of toxic chemicals. One illustration is chemical incidents that cause the dispersal of hazardous substances from industrial facilities and other sources. (Refer to the text box "Acute Chemical Incidents Surveillance.") A second example is dissemination of chemicals by illegal methamphetamine labs and how first responders must be prepared to cope with these events (refer to the text box "Injuries from Methamphetamine-Related Chemical Incidents—Five States, 2001–2012").

ACUTE CHEMICAL INCIDENTS SURVEILLANCE—HAZARDOUS SUBSTANCES EMERGENCY EVENTS SURVEILLANCE, NINE STATES, 1999–2008

Although they are infrequent, acute chemical incidents (i.e., uncontrolled or illegal release or threatened release of hazardous substances lasting < 72 hours) with mass casualties or extraordinary levels of damage or disruption severely affecting the population, infrastructure, environment, and economy occur, and thousands of less damaging chemical incidents occur annually. Surveillance data enable public health and safety professionals to better understand the patterns and causes of these incidents, which can improve prevention efforts and preparation for future incidents. . . .

The Hazardous Substances Emergency Events Surveillance (HSEES) system was operated by the Agency for Toxic Substances and Disease Registry (ATSDR) during January 1991–September 2009 to describe the public health consequences of chemical releases and to develop activities aimed at reducing the harm. This report. . . . summarizes incidents from the nine states (Colorado, Iowa, Minnesota, New York, North Carolina, Oregon, Texas, Washington, and Wisconsin) that participated in HSEES during its last 10 full years of data collection (1999–2008). . . .

During 1999–2008, a total of 57,975 chemical incidents occurred: 41,993 (72%) occurred at fixed facilities, and 15,981 (28%) were transportation related. Chemical manufacturing. . . (23%) was the industry with the most incidents; however, the number of chemical incidents in chemical manufacturing decreased substantially over time. . ., whereas the educational services category. . . and crop production category. . . had a consistently increasing trend. The most common contributing factors for an incident were equipment failure (n = 22,535, 48% of incidents) and human error (n = 16,534, 36%). The most frequently released chemical was ammonia 3,366 (6%). Almost 60% of all incidents occurred in two states, Texas and New York. [Refer to **TABLE 7.1** for the distribution of injuries associated with the top five chemicals released. As shown in the table, carbon monoxide exposure was the source of the greatest numbers of injuries and deaths during the period of observation.]

TABLE 7.1 Number and Percentage of Persons Injured by Top Five Chemicals Released, by Disposition of Injured Person—Hazardous Substances Emergency Events Surveillance System, Nine States,* 1999–2008

	Carbon Monoxide		Ammonia		Chlorine		Hydrochloric Acid		Sulfuric Acid	
Disposition	**No.**	**(%)**	**No.**	**(%)**	**No.**	**(%)**	**No.**	**(%)**	**No.**	**(%)**
Treated at hospital (not admitted)	1,608	(68)	715	(62)	450	(59)	204	(63)	184	(58)
Treated on scene with first aid	391	(17)	154	(13)	158	(21)	68	(21)	41	(13)
Treated at hospital (admitted)	199	(8)	96	(8)	41	(5)	14	(4)	29	(9)
Seen by private physician	27	(1)	84	(7)	34	(5)	19	(6)	43	(14)
Injury reported by an official	4	(0)	50	(4)	36	(5)	6	(2)	14	(4)

(continues)

TABLE 7.1 Number and Percentage of Persons Injured by Top Five Chemicals Released, by Disposition of Injured Person—Hazardous Substances Emergency Events Surveillance System, Nine States,* 1999–2008 *(continued)*

	Carbon Monoxide		Ammonia		Chlorine		Hydrochloric Acid		Sulfuric Acid	
Disposition	**No.**	**(%)**	**No.**	**(%)**	**No.**	**(%)**	**No.**	**(%)**	**No.**	**(%)**
Observed at hospital (no treatment)	11	(1)	35	(3)	39	(5)	6	(2)	5	(2)
Death	71	(3)	7	(1)	3	(0)	4	(1)	2	(1)
Unknown	53	(2)	12	(1)	2	(0)	4	(1)	0	(0)
Total†	**2,364**	**(100)**	**1,153**	**(99)**	**763**	**(100)**	**326**	**(100)**	**318**	**(101)**

Reproduced from Centers for Disease Control and Prevention. CDC Surveillance Summaries, Hazardous substances emergency events surveillance, nine states, 1999–2008. *MMWR.* 2015;64(SS-2):42.

Reproduced from Centers for Disease Control and Prevention. CDC Surveillance Summaries, Hazardous substances emergency events surveillance, nine states, 1999–2008. *MMWR.* 2015;64(SS-2):1.

INJURIES FROM METHAMPHETAMINE-RELATED CHEMICAL INCIDENTS—FIVE STATES, 2001–2012

Illegal methamphetamine production results in fires, explosions, spills, or air releases of hazardous chemicals (meth-chemical incidents), placing the meth producer and others nearby, including children, workers, and responders, at risk for injury or death and causing environmental contamination.

During 2001–2012, a total of 1,325 meth-related chemical incidents were reported in the five states.... [The] percentage of incidents increased each year from 2001 through 2004, then decreased each year through 2007, and increased again through 2012....

In 87 (7%) of the meth-related chemical incidents, 162 persons were injured, including at least 26 (16%) children.... Among those injured, 136 (84%) were treated at a hospital, including 19 (73%) children; 36 (22%) injured persons, including 19 (73%) children, required hospital admission. The percentage of injured persons who went to a hospital increased over time.... Two adults died: one, who might have been a meth cook, was found dead in a meth laboratory; the second was a law enforcement official....

The most commonly reported injuries were respiratory irritation (44%), chemical and thermal burns (27%), and eye irritation (22%).... Chemical and thermal burns significantly increased, from 7% during 2005–2007 to 44% during 2008–2012..., temporally associated with new, hazardous production methods. During the same time, skin irritation injuries decreased from 20% to 2%..., eye symptoms decreased from 18% to 11%, and respiratory symptoms decreased from 57% to 31%....

Most injuries were to members of the general public (97) and law enforcement officials (42), followed by employees working in areas where meth contamination occurs, including hotels and motels, abandoned buildings, and treatment centers (14); and firefighters (7).... The most commonly reported injuries among the general public were burns (43%) and respiratory irritation (37%); among injured law enforcement officials, respiratory irritation (64%), and eye irritation (38%) were most frequently reported.... Among the 14 injured employees, nine reported headache, seven respiratory irritation, and seven eye irritation....

Data from five states suggest that, beginning in 2005, when state and federal legislative efforts to restrict meth precursors were enacted, meth-related chemical incidents temporarily declined in those states. However, in 2008, as meth producers learned to circumvent laws and obtain restricted precursor drugs, and introduced the hazardous "shake-and-bake" meth-making method, such incidents began to rise, as did the percentage of events with injuries, particularly burns.

Modified and reproduced from Centers for Disease Control and Prevention. Injuries from methamphetamine-related chemical incidents—five states, 2001–2012. *MMWR.* 2015;64:909-912.

TABLE 7.2 30-Second Organic Chemistry Lesson

Term	Definition
Aromatic compound	An organic molecule that contains a benzene ring, for example, benzene and toluene.[a]
Hydrocarbon	An organic compound (as acetylene, benzene, or butane) containing only carbon and hydrogen and often occurring in petroleum, natural gas, coal, and bitumens.[b]
Organic chemicals/ compounds	Naturally occurring (animal- or plant-produced or synthetic) substances containing mainly carbon, hydrogen, nitrogen, and oxygen.[c] Example: table sugar.
Persistent organic pollutants (POPs)	Toxic chemicals that adversely affect human health and the environment around the world. Because they can be transported by wind and water, most POPs generated in one country can and do affect people and wildlife far from where they are used and released. They persist for long periods of time in the environment and can accumulate and pass from one species to the next through the food chain.[d] Example: the pesticide DDT.
Polycyclic aromatic hydrocarbons (PAHs)	A group of over 100 different chemicals that are formed during the incomplete burning of coal, oil and gas, garbage, or other organic substances like tobacco or charbroiled meat.[e]
Volatile organic compounds (VOCs)	Organic compounds that evaporate readily into the air. VOCs include substances such as benzene, toluene, methylene chloride, and methyl chloroform.[f]

[a]About.com. Chemistry. Aromatic compound definition. Available at: http://chemistry.about.com/od/chemistryglossary/a/aromaticcmpndef.htm. Accessed February 19, 2017;
[b]Medline Plus ®/Merriam-Webster. Medical Dictionary. Available at: http://c.merriam-webster.com/medlineplus/hydrocarbon. Accessed February 19, 2017;
[c]US Environmental Protection Agency. Vocabulary catalog. Terms of environment. Available at: https://ofmpub.epa.gov/sor_internet/registry/termreg/searchandretrieve/glossariesandkeywordlists/search.do?details=&glossaryName=Terms%20of%20Env%20(2009)#formTop. Accessed February 19, 2017;
[d]US Environmental Protection Agency. Persistent organic pollutants: A global issue, a global response. Available at: https://www.epa.gov/international-cooperation/persistent-organic-pollutants-global-issue-global-response. Accessed February 19, 2017;
[e]Agency for Toxic Substances and Disease Registry (ATSDR). ToxFAQs: Polycyclic Aromatic Hydrocarbons (PAHs). Available at https://www.atsdr.cdc.gov/toxfaqs/tfacts69.pdf. Accessed February 19, 2017;
[f]Agency for Toxic Substances and Disease Registry (ATSDR). Glossary of terms. Available at: https://www.atsdr.cdc.gov/glossary.html. Accessed February 19, 2017.

Terms Used to Describe Organic Chemicals

TABLE 7.2 lists some definitions that will be helpful in reading the chapter.

Chemicals Among the Top 20 in the ATSDR List of 275 Hazardous Substances

Referring back to the list of hazardous substances presented in the chapter on toxic metals and elements, among the most important chemical hazards identified by the ATSDR are:

- Polychlorinated biphenyls (PCBs)—Used in electrical cable insulation and transformers.
- Benzene—Used as a solvent in the manufacture of plastics and other materials and as a constituent of fuels.
- Polycyclic aromatic hydrocarbons (PAHs)—Individual members of the ATSDR list; they include benzo(a)pyrene (BaP), benzo(b)fluoranthene, and dibenzo(a,h)anthracene. PAHs are not used commercially. As noted previously, they result from incomplete combustion of petroleum-based chemicals and other organic substances.
- Chloroform—Formerly used as an anesthetic; now used to manufacture other chemicals.
- DDT; DDT, p, p' (DDT, p, p' is one of the synonyms for DDT)—Used as an insecticide. (DDT no longer is used as an insecticide in the United States.)
- Aroclor 1254 and Aroclor 1260—Both are a type of polychlorinated biphenyl (PCB); used in electrical transformers and other applications such as hydraulic fluids, plastics manufacture, and pesticide extenders.
- Trichloroethylene (TCE)—A solvent used by industries and the military.
- Dieldrin—Used as a pesticide.
- Chlordane—Used as a pesticide.

Pesticides

This section covers historical methods for controlling insects, development of organic pesticides, the major classes of pesticides, and examples of their health effects. Spraying pesticides around people's residences is a method for controlling invading pests. However, spraying also exposes the people who live in pesticide-treated homes to possible health hazards and can have adverse effects upon the environment.

The repertoire of available pesticides has changed remarkably over the past four centuries. During the middle of the 18th century, control of insects was accomplished by manually removing them or by using inorganic poisons that frequently were not very effective. Inorganic chemicals used in the past included compounds of sulfur, zinc, copper, mercury, lead, and arsenic (i.e., arsenicals) in addition to boric acid. The use of various natural compounds such as nicotine and petroleum happened during the 19th century. These inorganic chemicals and natural substances formed the arsenal of **pest** control weapons.

The application of pesticides that are derived from organic chemicals flourished during the 20th century. An example was the category of organophosphate pesticides. The "Golden Age of Discovery" of insecticides occurred in the mid-1900s, with the invention of neuroactive chemicals, such as chlorinated hydrocarbons, organophosphates, methylcarbamates, and pyrethroids.[7] (These terms are defined later in the chapter.) About the middle of the 20th century, use of the pesticide DDT was purported to have led to a complete victory over harmful insects. However, by the 1970s, use of DDT was restricted in the United States because of concerns about this chemical's effects on health and the environment. A significant question for environmental policy is "How should society regulate pesticide use to ensure a bountiful harvest of diverse crops that is needed for a healthy diet while, at the same time, making sure that pesticide uses do not have adverse effects on health?"[8(p725)]

On a global scale, estimates of annual use of pesticides at the beginning of the 21st century were about 3 million metric tons (3.3 million tons); the United States alone accounted for about 1.2 billion pounds (544 million kilograms) of this total.[8] According to the US Department of Agriculture:

> Pesticides are a vital input in today's agriculture, protecting food and fiber from damage by insects, weeds, diseases, nematodes, and rodents. US agriculture spends about 8 billion dollars annually on pesticides—about 70 percent of all pesticides sold in the country. It is estimated that each dollar invested in pesticide control returns approximately 4 dollars in crops saved. Nevertheless, pests still destroy nearly 13 percent of all potential food and fiber crops in the US[.] Farmers' expenditures on pesticides are about 4–5 percent of total farm production costs.[9]

Sometimes the term *pesticide* is used interchangeably by the public with **insecticide**.[10] Nevertheless, these terms relate to distinct purposes. A pesticide is "any substance or mixture of substances intended for preventing, destroying, repelling, or mitigating pests. Pests can be insects, rodents, weeds, and a host of other unwanted organisms. . . . Pesticides are usually divided on the basis of their target. The major classes are insecticides, herbicides, rodenticides, and fungicides."[11(p251)] In contrast with pesticides, insecticides are pesticide compounds that are used specifically to control insects. **TABLE 7.3** presents definitions of terms that are related to the category of agents (i.e., pesticides) presented in this section.

Pesticides are applied by a number of methods including spraying from tractors and widespread broadcasting by aerial spraying. **FIGURE 7.1** illustrates an airplane used for spraying a field. Although highly efficient, these methods of spraying can cause pesticide drift and lead to contamination of waterways.

Classes of Pesticides and Insecticides

Several categories of pesticides are in use at present; these include some types of inorganic chemicals, organic chemicals, and a variety of other substances including insecticidal soaps, contraceptives, insect growth regulators and hormones, as well as warfarin (an anticoagulant for control of rodents).

Four major classes of pesticides and insecticides are derived from organic chemicals: **organophosphates (OPs)**, organocarbamates (also called carbamates), organochlorides (also known as **organochlorines**), and pyrethroids (from the class of pyrethrins).[12] Other chemicals not included in these classes are used as pesticides, such as chloropicrin (covered later in this chapter). Chloropicrin is an unclassified pesticide used as a soil fumigant.

Because they do not appear to persist in the environment, pesticides from the OP group are among those chosen most widely for agricultural purposes; yet, at the same time, OPs are among the most common sources of pesticide poisoning. Similarly, another group of so-called nonpersistent pesticides are the carbamates group, an example being aldicarb. Although pesticides of the OP class are implicated frequently

TABLE 7.3 Definition of Pesticide and Related Terms

Term	Definition
Pesticide	Substances intended to repel, kill, or control any species designated a "pest" including weeds, insects, rodents, fungi, bacteria, or other organisms. The family of pesticides includes herbicides, insecticides, rodenticides, fungicides, and bactericides.[a]
Insecticide	A pesticide compound specifically used to kill or prevent the growth of insects.[a]
Herbicide	A pesticide designed to control or kill plants, weeds, or grasses. Almost 70% of all pesticides used by farmers and ranchers are herbicides. These chemicals have wide-ranging effects on nontarget species.[a]
Fungicide	A pesticide used to control fungi.[a]
Nematocide	A pesticide used to kill nematodes (microscopic, worm-like organisms that feed on plant roots).[a]
Rodenticide	A pesticide or other agent used to kill rats and other rodents or to prevent them from damaging food, crops, or forage.[a]
Fumigant	A pesticide that is vaporized to kill pests. They often are used in buildings and greenhouses.[b]

[a]Modified and reproduced from US Environmental Protection Agency. Vocabulary Catalog. Pesticides glossary. Available at: https://iaspub.epa.gov/sor_internet/registry/termreg/searchandretrieve/glossariesandkeywordlists/search.do?details=&vocabName=Pesticides%20Glossary#formTop. Accessed February 17, 2017.

[b]Waste and cleanup risk assessment glossary. Available at: https://ofmpub.epa.gov/sor_internet/registry/termreg/searchandretrieve/glossariesandkeywordlists/search.do?details=&glossaryName=Waste%20and%20Cleanup%20Risk%20Assess#formTop. Accessed February 17, 2017.

FIGURE 7.1 An airplane used for spraying pesticides on crops.

Reproduced courtesy of CA Dept. of Pesticide Regulation.

in fatal poisoning incidents, carbamates are rarely a source of mortality from poisoning. Most of the pesticides from the third group, organochlorines, have been banned in the United States because of their propensity to remain in the environment.

The fourth group, pyrethrins, refers to an organic insecticide obtained from naturally occurring sources; often, pesticides derived from such natural sources are called botanicals. The insecticides from the pyrethrin group are useful for controlling insects and also have low toxicity for mammals.

Organophosphate Pesticides (Anticholinesterases)

The EPA reported that each year in the US the agricultural usage of organophosphates declined from 70 million pounds (32 million kilograms) in 2000 to 20 million pounds (9 million kilograms) in 2012.[13] During this 12-year period, organophosphate use tumbled from 71% to 33% of all insecticides. The use of OPs declined during the 21st century in many states, as lower-risk alternatives were implemented. The agricultural uses of OPs are for control of insect pests on crops in fields, orchards, and vegetable plots. Nonagricultural uses include control of insects in homes, offices, and outdoors; for example, OP pesticides are used for mosquito and termite abatement. The popularity of OP pesticides arises from the facts that they are (1) inexpensive in comparison to alternatives; (2) can be used to control a wide range of insects, thus eliminating the need for multiple applications of different pesticides; and (3) have not been weakened by the resistance of insects. In addition, OPs tend not to persist in the environment.

Organophosphate pesticides are related to "nerve gases" such as sarin, soman, and tabun. In March 1995, the Aum Shinrikyo doomsday cult released sarin gas into the Tokyo subway system during the morning rush hour. As a result, 13 people died and more than 6,000 were injured.[14] In common with nerve gases, OP pesticides act on the nervous systems of insects and higher organisms by disrupting the transmission of nerve impulses. This effect occurs when OPs reduce the ability of the enzyme cholinesterase to regulate the neurotransmitter acetylcholine, which aids in cell-to-cell transfer of nerve impulses.[11]

Organophosphate poisoning can produce both acute and long-term effects. In the former case, the anticholinesterase activity happens soon after exposure and causes impairment of the neural impulse transfer mechanism. A group of immediate symptoms may occur: Nerves and muscles may become uncoordinated, producing respiratory paralysis and weakness; neurological symptoms may include seizures and loss of consciousness; cramping of the abdominal area may occur; and in extreme cases, death may ensue. The effects of long-term exposure may include a condition known as organophosphate-induced delayed polyneuropathy, which is manifested by numbness, loss of sensory abilities, and weakness. Clinicians believe that in most cases this delayed form of polyneuropathy is irreversible.

Examples of OP pesticides are diazinon, malathion, methyl parathion, and parathion. Pesticides from the organophosphate group appear in a wide variety of products: sprays, baits, indoor foggers and bombs, flea collars, pet shampoos, powders, animal dips, and granules. Some OP pesticides (e.g., malathion) are approved for direct application to food crops such as fruits (e.g., apples and grapes), row crops, and vegetables (e.g., tomatoes). A more extensive list of OP pesticides is provided in **TABLE 7.4**, with the trade names of some of the pesticides shown in brackets.

Despite the benefits that OP pesticides provide in the control of damage by insects to food crops and in the abatement of insect vectors, they also present

TABLE 7.4 Partial List of Organophosphate Pesticides (Insecticides)

Highly Toxic Commercial Products	Moderately Toxic Commercial Products
Azinphos-methyl	Acephate
Carbophenothion	Bensulide
Coumaphos	Chlorpyrifos [Dursban]
Dialifor	Diazinon
Dicrotophos	Dichlorvos [DDVP]
Dioxathion	Ethion
Ethyl parathion	Ethoprop
Fenamiphos	Fenitrothion
Fonofos	Fenthion
Fosthietan	Isoxathion
Isofenphos	Malathion
Methamidophos	Naled [Dibrom]
Methidathion	Oxydemeton-methyl
Methyl parathion	Phosalone
Mevinphos	Pirimiphos-methyl
Monocrotophos	Propetamphos
Phorate	Sulprofos
Phosphamidon	Temephos
Sulfotep	Tetrachlorvinphos
Terbufos	Trichlorfon

Modified and reproduced from Roberts J, Reigart R, eds. *Recognition and Management of Pesticide Poisonings*. 6th ed. Washington, DC: US Environmental Protection Agency; 2013, pp 44-45, 46-47.

several noteworthy hazards. Two of these hazards are the potential for harm to the persons who apply the pesticide and, more generally, the potential for harm to the entire population, particularly children. The extensive use of organophosphate pesticides worldwide increases the likelihood of pesticide poisoning, especially in the developing world, and thus is a major concern for environmental health officials who work

METAM SODIUM SPILL ON THE UPPER SACRAMENTO RIVER

The upper Sacramento River is located in northern California. This picturesque region, known for its natural beauty and abundant wildlife, attracts fishermen and vacationers from all over the world. On July 14, 1991, a massive spill of approximately 19,000 gallons [971,923 liters] of the soil fumigant and herbicide metam sodium occurred when a freight train derailed on a section of the Southern Pacific Railroad tracks known as the Cantara Loop. (See **FIGURE 7.2**.) Seven derailed railroad tank cars cascaded directly into the river from an area on a trestle that was immediately above the river. One of them landed in the river at a point where the water was about one meter [3.28 feet] deep.[a, b]

The spill necessitated the evacuation of the nearby town of Dunsmuir. The effects of the spill were devastating to aquatic and terrestrial wildlife, resulting in a lengthy swath of destruction along the upper Sacramento River, which is a vital watershed for the state of California. A pea-green foam was reported to be coursing down the river, leaving in its poisonous wake numerous dead trout. Virtually all aquatic life was exterminated downstream from the spill.[c]

In addition to eradicating aquatic life, the spilled pesticide seeped into shallow aquifers and released toxic gases into the air.[a–c] Soon after the spill, eyewitnesses who were near Dunsmuir reported an obnoxious odor that made breathing difficult. Numerous exposed persons sought medical assistance for health complaints that ranged from headache and eye irritations to nasal irritation and chest tightness.

FIGURE 7.2 Metam sodium spill into the upper Sacramento River.

Courtesy of Cantara Trustee Council. Final report on the recovery of the upper Sacramento River—subsequent to the 1991 Cantara spill, p. 2. Available at: https://nrm.dfg.ca.gov/FileHandler.ashx?DocumentID=17248&inline=true. P. 2. Accessed May 1, 2017.

From July 21 to July 22, 1991, inmates from the Shasta County jail and crew leaders were dispatched to pick up dead fish along the site of the spill. Among this group of 42 persons, 27 (64%) developed dermatitis on their lower extremities and some other sections of the body that came in contact with the river water. Nevertheless, according to the CDC, the pesticide was not identified definitely as the cause of the dermatitis cases, as the pesticide had degraded to very low concentrations by the time the workers contacted the river water.[d]

In 1995, following a legal settlement with Southern Pacific Railroad and other parties alleged to be responsible for the spill, the Cantara Trustee Council (CTC) was formed and extensive restoration of the affected areas commenced.[c] The CTC continued between 1995 and 2007.

[a] Gherman E. A toxic nightmare: the Dunsmuir metam sodium spill revisited. *Sonoma County Free Press.* July 1997. Available at: http://sonomacountyfreepress.blogspot.com/search?q=A+toxic+nightmare. Accessed May 1, 2017.

[b] US Fish and Wildlife Service. Restoring our resources: California's upper Sacramento River Cantara loop. Available at: http://www.pskf.ca/publications/cheakums05/Sacramento%20River%20Spill.pdf. Accessed July 23, 2017.

[c] Cantara Trustee Council. Final Report on the Recovery of the Upper Sacramento River—subsequent to the 1991 Cantara spill. Available at: https://nrm.dfg.ca.gov/FileHandler.ashx?DocumentID=17248&inline=true. Accessed May 1, 2017.

[d] Centers for Disease Control and Prevention. Dermatitis among workers cleaning the Sacramento River after a chemical spill—California, 1991. *MMWR.* 1991;40:825–827, 833.

for public health departments and other environmental agencies.[15] Presently, diazinon is banned in the United States for use as a consumer product because of its toxicity to wildlife.

Methyl parathion is an example of an organophosphate pesticide that is used to control insects on crops. It is classified by the US EPA as a category I pesticide, which means that it is among the most toxic. A major incident of poisoning with methyl parathion occurred in Lorain County, Ohio, as a consequence of the pesticide being sprayed illegally for pest control in private residences over a period of years during the late 1980s and early 1990s until the practice was reported to authorities. Beginning in 1994, the EPA conducted an investigation that evaluated 254 area homes and 747 individuals. Results showed that residents in the exposed households reported that they were afflicted by a wide range of symptoms and that their pets died. Soon after their homes were sprayed, 49 residents were hospitalized and at least one died. Children were more likely to be affected than adults. EPA officials noted:

> Health care providers apparently did not consider pesticide poisoning when confronted with acute generic presentations, and although veterinarians may have diagnosed pesticide poisoning in pets, this information did not reach the public health system. . . . [A]cute and short-term organophosphate poisoning occurred but was not recognized in Lorain County during the years that methyl parathion was sprayed in private residences.[16(p1050)]

Carbamates

The pesticides in the class of carbamates are close relatives of the organophosphate pesticides. As is true of organophosphates, carbamates dissipate quickly from the environment as a result of breaking down into other substances. Farmers use carbamate-based pesticides as insecticides, fungicides, and herbicides. Some varieties are also in use around the home, an example being the carbamate insecticide carbyl (trade name Sevin), which is used to control garden pests (e.g., wasps, hornets, and snails) and is an ingredient in some products applied to furry pets to control ticks and fleas. Common types of the approximately 30 EPA-registered carbamates are aldicarb, fenoxycarb, and propoxur. The form of carbamates known as thiocarbamate is used mainly as an herbicide; the form known as dithiocarbamate is primarily for use as a fungicide. Metam sodium is a carbamate pesticide from the dithiocarbamate group. One of the pesticides used most frequently in the United States (with approximately 60 million pounds [27 million kilograms] applied annually), metam sodium is a common soil fumigant. (Refer to the text box for information on the 1991 chemical spill into the Sacramento River in the United States.)

FIGURE 7.3 Bhopal, India: a view of the area of the gas plant that leaked methyl isocyanate (MIC) gas.

Methyl isocyanate (MIC) is an intermediate chemical used for the manufacture of carbamate pesticides. Also, MIC is created as metam sodium breaks down. When acute exposure occurs, MIC is extremely toxic to life-forms (e.g., human beings, aquatic organisms, and plants). A notorious incident that involved the pesticide was the accidental release of MIC during a 1984 industrial accident in Bhopal, India that killed more than 3,800 people. **FIGURE 7.3** shows the section of a plant that released MIC. Refer to the text box for a detailed account of the Bhopal incident.

Organochlorines

The third group of pesticides is derived from chlorinated hydrocarbons. *Chlorinated hydrocarbon* is

> ... a generic term given to compounds containing chlorine, carbon and hydrogen. The term can be used to describe organochlorine pesticides such as lindane and DDT, industrial chemicals such as polychlorinated biphenyls (PCB[s]), and chlorine waste products such as dioxins and furans. These compounds are persistent in the environment and most bioaccumulates [sic] [defined in the chapter on toxic metals and elements] in the food chain. The human and environmental health risks of chlorinated hydrocarbons depend on the compound in question. As a general statement, exposure to chlorinated hydrocarbon[s] has been associated with suppression of the immune system and cancer.[17]

THE BHOPAL, INDIA, INCIDENT

The Bhopal, India incident refers to history's most severe industrial catastrophe. In this disaster, a Union Carbide pesticide factory in Bhopal released methyl isocyanate (MIC) gases on the night of December 2, 1984.[a] Water that intruded accidentally into a storage tank for MIC produced a chemical reaction that unleashed a suffocating toxic plume into the atmosphere. The Union Carbide factory was upwind of the city of Bhopal, which had a population of 900,000 at the time. Unsuspecting residents were exposed to a deadly cloud from the discharge of around 30 to 40 tons (27 to 36 metric tons) of poisonous MIC. In total at least 600,000 persons came into contact with MIC. This catastrophe resulted in an immediate death toll estimated to exceed 3,800 residents; in addition, about 170,000 persons exposed to MIC were left with adverse health effects.[b] Thirty years subsequent to the disaster, a total of approximately 15,000 additional deaths resulted from MIC exposure over time, although estimates vary. Not confined to the acute effects of MIC release, the impact of the Bhopal incident remains today, with continuing adverse effects upon the health of those directly exposed as well as their children born after the disaster.

Unfortunately, installed safety measures that were designed to prevent a gas leak and a siren to alert the community of dangers from the plant were turned off or malfunctioning. According to eyewitness reports, "Many died in their beds, others staggered from their homes, blinded and choking, to die in the street. Many more died later after reaching hospitals and emergency aid centres. The early acute effects were vomiting and burning sensations in the eyes, nose and throat, and most deaths have been attributed to respiratory failure. For some, the toxic gas caused such massive internal secretions that their lungs became clogged with fluids, while for others, spasmodic constriction of the bronchial tubes led to suffocation. Many of those who survived the first day were found to have impaired lung function. Further studies on survivors have also reported neurological symptoms including headaches, disturbed balance, depression, fatigue and irritability. [There also have been continuing reports of] [a]bnormalities and damage to the gastrointestinal, musculoskeletal, reproductive and immunological systems."[c]

[a] Taylor A. Bhopal: the world's worst industrial disaster, 30 years later. *The Atlantic.* Dec 2, 2014.

[b] US Environmental Protection Agency. Methyl Isocyanate. Available at: https://www.epa.gov/sites/production/files/2016-09/documents/methyl-isocyanate.pdf. Accessed July 21, 2017.

[c] Greenpeace. Bhopal: the ongoing disaster 1984–2001. Available at: http://www.greenpeace.org/belgium/Global/belgium/report/2001/11/bhopal-the-ongoing-disaster.pdf. Accessed July 22, 2017.

Organochlorines include

> … a wide range of chemicals that contain carbon, chlorine and, sometimes, several other elements. A range of organochlorine compounds have been produced including many herbicides, insecticides, fungicides as well as industrial chemicals such as polychlorinated biphenyls (PCBs). The compounds are characteristically stable, fat-soluble and bioaccumulate. Organochlorines pose a range of adverse human health risks and some are carcinogens.[18]

Examples of organochlorine pesticides are:

- DDT
- Lindane
- Chlordane
- Mirex
- Hexachlorobenzene
- Methoxychlor

The organochlorine pesticide DDT (dichlorodiphenyltrichloroethane) is similar in chemical structure to DDE (dichlorodiphenyldichloroethylene) and DDD (dichlorodiphenyldichloroethane); frequently, both DDE and DDD are present as contaminants in DDT pesticides. DDE does not have commercial value; DDD was used in the past as a pesticide.[19] A metabolite of DDT, DDE is formed when DDT is ingested. For example, fish convert DDT to DDE when they consume DDT that is present in lower aquatic organisms.

In the United States, widespread use of organochlorine pesticides such as DDT began during the early 1940s and reached a maximum during the 1960s. Due to concerns about the possible adverse effects of DDT upon the health of humans and wildlife, application of DDT was prohibited in 1972 in the United States. (Refer to the following text box.) Although most developed nations at present ban the use of DDT, some countries continue to use this pesticide.

DDT is not regarded as a highly toxic pesticide—a consideration that in the mid-20th century supported its use. For many years, DDT was employed worldwide to control insects and harmful mosquitoes that carry malaria. It was credited at one time for freeing a substantial segment of the world's population from the scourge of malaria and, consequently, with saving millions of persons from death due to malaria.

The downside of DDT is the fact that, in common with the organochlorines, it is a highly stable chemical

DDT BAN TAKES EFFECT (1972)

The general use of the pesticide DDT will no longer be legal in the United States after today, ending nearly three decades of application during which time the once-popular chemical was used to control insect pests on crop and forest lands, around homes and gardens, and for industrial and commercial purposes.

An end to the continued domestic usage of the pesticide was decreed on June 14, 1972, when William D. Ruckelshaus, Administrator of the Environmental Protection Agency, issued an order finally canceling nearly all remaining federal registrations of DDT products. Public health, quarantine, and a few minor crop uses were excepted, as well as export of the material. . . .

The cancelation decision culminated 3 years of intensive governmental inquiries into the uses of DDT. As a result of this examination, Ruckelshaus said he was convinced that the continued massive use of DDT posed unacceptable risks to the environment and potential harm to human health. . . .

DDT was developed as the first of the modern insecticides early in World War II. It was initially used with great effect to combat malaria, typhus, and the other insect-borne human diseases among both military and civilian populations. . . .

A persistent, broad-spectrum compound often termed the "miracle pesticide," DDT came into wide agricultural and commercial usage in this country in the late 1940s. During the past 30 years [i.e., before 1972], approximately 675,000 tons (612,000 metric tons) have been applied domestically. The peak year for use in the United States was 1959 when nearly 80 million pounds (36 million kilograms) were applied. From that high point, usage declined steadily to about 13 million pounds (5.9 million kilograms) in 1971, most of it applied to cotton.

The decline was attributed to a number of factors including increased insect resistance, development of more effective alternative pesticides, growing public and user concern over adverse environmental side effects—and governmental restriction on DDT use since 1969.[a]

New DDT Report Confirms Data Supporting 1972 Ban, Finds Situation Improving

Residues of the pesticide DDT in the food supply, human tissues and in the environment have declined in recent years especially since the chemical was banned for major uses by EPA in 1972. . . . [T]he ban has contributed to the decline of DDT levels in fish. For example, one Federal study of Lake Michigan lake trout showed DDT levels decreased from 19.19 ppm in 1970 to 9.96 ppm in 1973. DDT levels in coho salmon declined from 11.82 ppm in 1969 to 4.48 . . . [ppm] in 1973. . . . DDT should still be considered a potential human cancer agent based on the results of animal studies. . . . DDT is stored in human fatty tissue, wildlife and fish; DDT is toxic to fish and birds and interferes with the reproduction of some species; DDT persists in soil and water for years.[b]

[a]US Environmental Protection Agency. EPA's Web Archive. DDT Ban Takes Effect. Available at: https://archive.epa.gov/epa/aboutepa/ddt-ban-takes-effect.html. Accessed February 17, 2017; and
[b]US Environmental Protection Agency. EPA's Web Archive. New DDT Report Confirms Data Supporting 1972 Ban, Finds Situation Improving. Available at: https://archive.epa.gov/epa/aboutepa/new-ddt-report-confirms-data-supporting-1972-ban-finds-situation-improving.html. Accessed February 17, 2017.

that persists in the environment. This characteristic is associated with the phenomenon of bioaccumulation, the concentration of a chemical as it moves up the food chain. Amplification of the chemical's concentration may increase the likelihood of adverse health effects among animals at the upper end of the food chain. Specifically, DDT, which has the ability to become concentrated in the adipose (fatty) tissues of the body, has an estimated half-life of approximately 10 years; the term *half-life* indicates the time required for one half of the pesticide to be cleared from the body. So persistent and widespread is contamination of the environment with DDT, all living organisms on earth contain some levels of this pesticide. Nevertheless, it is noteworthy that average human body burdens of DDT have tended to decline with its decreasing use. US residents were reported to have the following average body fat concentrations of the pesticide for the years 1950, 1956, and 1980, respectively: 5 parts per million (ppm), 15.6 ppm, and 3 ppm.[20] However, levels of DDE, the DDT metabolite, have tended to remain constant, as people continue to ingest the DDE that is present in some species of fish and other DDE-containing foodstuffs.

In the developing world, people continue to consume foods that contain much higher levels of DDT and organocholorine pesticides than are permitted in the developed world. Exposure to high levels of DDT in developing countries is of particular concern for the health of children.

The various human health effects that have been linked to DDT include cancer, reproductive effects, impaired lactation, falling sperm counts, and impaired neurologic function. Among the forms of cancer associated with DDT in some studies are pancreatic cancer, non-Hodgkin's lymphoma, and breast cancer. DDT also causes nervous system abnormalities among pesticide workers; these symptoms include irritability, dizziness, and numbness. An extensive review of the literature suggested that "Other than the long-recognized neurologic toxicity associated with DDT poisoning, and laboratory abnormalities [e.g., abnormal results from blood tests] in DDT-exposed workers, human health effects of DDT exposure are not established."[21(p222)]

Pyrethrins

A pyrethrin is an insecticide derived from natural sources, namely certain varieties of chrysanthemum flowers.[22] Pesticides from this group have great ability to paralyze and kill flying insects, although some insects may recover because of their own defensive detoxifying enzymes. The paralyzing action of pyrethrins occurs through interference with the transmission of neural impulses via action on sodium channels.

Manufactured chemicals that have a composition similar to pyrethrins are called pyrethroids.[22] Although more than 1,000 types of pyrethroids have been synthesized, only a few are in use as insecticides in the United States. The most common pyrethroid is permethrin, with over 1 million pounds (453 thousand kilograms) applied in the United States in 1997; other common pyrethroids are bifenthrin, cyfluthrin, cypermethrin, and telfluthrin.

Pyrethrin-based insecticides, having low concentrations of the active ingredient, are used inside the home in aerosol cans, insecticide bombs, insecticidal pet shampoos, treatments for lice applied directly to humans, and mosquito repellents. They degrade rapidly by the action of sunlight and are not very toxic when applied to animals, but are highly toxic to some beneficial insects (e.g., honeybees) as well as fish and other forms of aquatic life. People may inhale these insecticides as a result of spraying and may ingest them in foods. High levels of exposure to pyrethrin-based insecticides produce characteristic acute effects such as nausea, headache, dizziness, and neurologic symptoms. The text box documents episodes of human illness due to exposure to pyrethrins and pyrethroids.

ILLNESSES ASSOCIATED WITH USE OF AUTOMATIC INSECTICIDE DISPENSER UNITS—SELECTED STATES AND UNITED STATES, 1986–1999

To control indoor flying insects, restaurants and other businesses commonly use pyrethrin and pyrethroid insecticides sprayed from automatic dispensing units. Usually placed near entrances, these units are designed to kill flying insects in food service or work areas. On May 18, 1999, the Florida Department of Health (FDH) was notified by the Florida Department of Business and Professional Regulation (DBPR) that during May 12–17, three persons developed pesticide-related illnesses associated with improperly placed automatic insecticide dispensers. . . .

Cases 1–3

A 42-year-old cook working at a Florida restaurant developed a sore throat, dyspnea, headache, and dizziness on May 12, 1999, after a several-hour exposure to mist released from insecticide dispensers in the food preparation area. The insecticide dispensers had been installed on May 10, but it is unknown on what day the cook was first exposed. The cook removed the dispensers on May 12 and noted relief of his symptoms. However, the restaurant management reinstalled the dispensers on May 14, and on May 15, a 40-year-old male customer developed headache and shortness of breath within 1 hour of entering the restaurant. These symptoms lasted approximately 4 hours. On May 17, approximately 45 minutes after leaving this restaurant, a 47-year-old male customer experienced a sharp burning sensation in his left eye and noted swelling, redness, and irritation of the eyelid that persisted approximately 24 hours. The implicated pesticide dispenser was within 6 feet of the booth where this customer had been sitting, and it faced his left eye. This person reported his symptoms to DBPR on May 18. None of the three persons sought medical attention for their symptoms. The active ingredients released by these dispensers were pyrethrin and piperonyl butoxide.

Case 4

On August 20, 1995, a 17-year-old male restaurant employee in California was changing the cartridge of an automatic insecticide dispenser. When he closed the dispenser panel, the firing mechanism was activated and discharged a pyrethrin-containing mist into his right eye. The employee immediately experienced burning in the eye and promptly sought medical attention at the emergency department of a local hospital. He was diagnosed with chemical conjunctivitis and treated symptomatically.

Modified and reproduced from Centers for Disease Control and Prevention. Illnesses associated with use of automatic insecticide dispenser units—selected states and United States, 1986–1999. *MMWR*. 2000;49:492–493.

Herbicides/Defoliants

Examples of chemicals in the category of herbicides and defoliants are atrazine, paraquat, and Agent Orange (2,4-D [2,4-dichlorophenoxyacetic acid] and 2,4,5-T [2,4,5 trichlorophenoxyacetic acid]). Consider the first herbicide in the foregoing list, atrazine, which is one of the most widely used weed killers in the United States. When atrazine is applied as a weed killer on field crops such as corn, runoff from irrigation and rainfall may carry this herbicide into waterways. Some experts believe that atrazine has an estrogenic effect on frogs, turning males into females.[23] Estrogenic activity refers to the observation that a chemical has an effect that is similar to that of the female sex hormone estrogen. One study suggested that there was an association between atrazine and prostate cancer among workers exposed at an atrazine manufacturing plant.

The second herbicide, paraquat, is a standard herbicide used on many crops.[24] Some evidence indicates that paraquat, a highly toxic chemical, was applied in the past to marijuana plants; some marijuana in the United States showed paraquat contamination. The herbicide paraquat could cause lung damage when smoke from contaminated marijuana is inhaled.[25]

The third herbicide, Agent Orange, was applied by the US military as a defoliant in Vietnam during the Vietnam War of the 1960s and early 1970s to prevent the thick jungle foliage from concealing the enemy. During Operation Ranch Hand, which spanned the years from 1962 to 1971, approximately 19 million gallons (72 million liters) of defoliants were sprayed on 3.6 million acres (1.45 million hectares) located in Vietnam and Laos.[26] Although the defoliants were mixtures of various herbicides, most contained the phenoxyacetic acid–based herbicides 2,4-D and 2,4,5-T. One of the most commonly used mixtures consisted of equal amounts of 2,4-D and 2,4,5-T. This herbicide mixture was shipped in drums marked with orange stripes; hence, it was given the name Agent Orange. Of particular concern was the fact that Agent Orange contained very small amounts of dioxins that originated from the manufacture of the herbicide.[26] (The health effects of dioxins will be covered later in this chapter.)

Around 1.5 million military personnel served during the time interval that the spraying with Agent Orange occurred. However, the exact number of persons exposed and the degree of their exposure to the herbicides can only be estimated. Returning veterans disclosed a number of unusual health outcomes that affected themselves (e.g., cancer and skin rashes) and their children (e.g., birth defects of various types). Some veterans attributed these occurrences to exposure to Agent Orange during the war.

Subsequently, a large number of epidemiologic studies have explored the health effects of veterans' exposures to Agent Orange. A report issued by the Institute of Medicine, an arm of the US National Academies, concluded that there was sufficient evidence that Agent Orange was associated with several forms of cancer (soft tissue sarcoma, non-Hodgkin's lymphoma, Hodgkin's disease, and chronic lymphocytic leukemia).[26] The linkage between Agent Orange and other health effects was reported to be less certain, although there have been some putative findings regarding this association. For example, another report from the Institute of Medicine stated that "there is limited/suggestive evidence of an association between exposure to the herbicides used in Vietnam or the contaminant dioxin and type 2 diabetes."[27(p2)]

A list of herbicides is shown in **TABLE 7.5**. This list is provided for information only; the additional herbicides are not discussed in the text.

TABLE 7.5 Ten Herbicides Used Commonly in US Agriculture in 2012

Name of Herbicide[a]	Chemical Family[b]
2,4-D	Phenoxy-carboxylic acid; [phenoxyacetic acid-based herbicide; 2,4,5-T is from the same family.]
Acetochlor	Chloroacetamide
Atrazine	Triazine
Dicamba	Benzoic acid
Glyphosphate	Glycine [most commonly used herbicide in the U.S.; trade names include Roundup®.]
Paraquat	Bipyridylium [highly toxic to humans in small doses; diquat is from the same chemical family].
S-Metolachlor	Chloroacetamide
Pendimethalin	Dinitroaniline
Propanil	Amide
Trifluralin	Dinitroaniline

[a]Data from Atwood D, Paisley-Jones C. *Pesticide Industry Sales and Usage: 2009-2012 Market Estimates*. Washington, DC: US Environmental Protection Agency, 2017, p 14; [b]Data from Weed Science Society of America. Herbicide site of action (SOA) classification list, August 16, 2017. Available at: http://wssa.net/wp-content/uploads/WSSA-Herbicide-SOA-20170816.xlsx. Accessed January 8, 2017.

Other Health Effects of Pesticide Exposure

Pesticide exposures affect many classes of living beings, including agricultural workers, animals, and the general population. An English toxicologist concluded "that although we can be sure that pesticides pose no gross threat to health in the general population, subtle effects on more highly exposed subpopulations are, as yet, more difficult to rule out."[28(p219)]

Those most likely to experience the harmful effects of pesticides include agricultural workers and other people who are unwittingly exposed to or misuse them. In addition, pesticides sometimes are used as a means of attempting suicide in agricultural areas.[28] Employees who have been exposed to pesticides in the workplace unknowingly may endanger their family members as well as themselves; workers' clothing may accumulate pesticides, which later may be carried into the workers' homes and contaminate unsuspecting family members.[29] **FIGURE 7.4** demonstrates agricultural workers hoeing weeds around plants. During the growing and processing stages of agriculture, workers may come into contact with pesticides that have been applied to plants.

During the 1970s, in order to protect themselves from pesticides, some workers involved in spray applications donned the protective gear shown in **FIGURE 7.5**.

FIGURE 7.5 This individual was demonstrating the types of protective clothing worn by pesticide workers during the 1970s.

Courtesy of CDC

As noted previously, many products designed to control insect infestations on pets contain pesticides. The National Resources Defense Council points out that it is likely that thousands of pets have been killed by pesticides intended to control fleas and ticks.[30] Even though these products have been labeled as safe for adults, they may pose hazards to children, whose developing bodies are more sensitive to toxic chemicals than are adults' bodies. Some children are exposed to pesticides applied to pets when the children play on the floor, put their pets' toys in their own mouths, and put their hands in their mouths after petting the animals.

FIGURE 7.4 Agricultural workers hoeing weeds around plants.

The pesticides used in agriculture may have undesirable effects upon cattle and other livestock.[31] Exposure of livestock to pesticides may arise from contaminated feed, pesticides applied to fields, and agents used to control insect pests that plague food animals. The effects that animals may experience include acute and chronic toxicity. Pesticide residues also may build up in meat, poultry, and dairy products that subsequently are consumed by humans.

The widespread application of pesticides to food crops could have unintentional and adverse effects upon helpful insects. Certain species of insects called pollinators perform vital functions, including pollination of essential crops, within natural and agricultural ecosystems in North America. [32] Honey bees are among the most crucial pollinating insects for the world's economy. The Food and Agricultural Organization of the United Nations notes that honey bees pollinate more than 70% of the 100 crop species that contribute 90% of global foods.[33]

Colony collapse disorder (CCD) occurs when worker bees abandon their hive leaving behind the queen bee and young bees, resulting in the eventual death of the bee hive. For example, scientists in the United States observed the deaths of 30 to 90% of bee hives during the 2006 to 2007 winter season. [34] Although the percentage of CCD cases dropped substantially as of 2015, scientists continue to be concerned about CCD.

An invasive mite (a pest that attacks honey bees) has been studied as one of the possible causes of CCD. A range of other factors is being investigated, including the possible action of pesticides. Overexposure to pesticides can cause beekills. The EPA notes that an acute beekill from pesticide poisoning is not the same phenomenon as CCD, which appears to have a variety of environmental causes. In order to prevent beekills, the EPA requires warning labels on pesticides that are harmful to bees.

Another concern regarding pesticides is what may happen when they drift into schools and homes located near agricultural fields.[35] (Refer to the chloropicrin text box.) Pesticide drift may pose a greater hazard to children and sensitive subpopulations (e.g., pregnant women) than to the general population.

ILLNESS ASSOCIATED WITH DRIFT OF CHLOROPICRIN SOIL FUMIGANT INTO A RESIDENTIAL AREA—KERN COUNTY, CALIFORNIA, 2003

Chloropicrin is the fourth most commonly used soil fumigant in California. Exposure to chloropicrin causes eye and respiratory tract irritation, vomiting, and diarrhea. . . . This report describes an investigation by the California Department of Pesticide Regulation (CDPR) and the Kern County Agriculture Commissioner (KCAC) into illnesses associated with the offsite drift of chloropicrin in Kern County. A total of 165 persons experienced symptoms consistent with chloropicrin exposure. The findings underscore health risks associated with fumigants and the usefulness of procedures adopted in California to ensure both prompt identification of exposure events and timely notification of the affected public.

On October 3, 2003, an agricultural pest control service began applying 100% chloropicrin at a concentration of 80 pounds [36.3 kilograms]/acre [0.4 hectare] to 34 [13.7 hectares] acres of fallow land in Kern County. Chloropicrin was injected 17–18 inches [43–46 centimeters] into the soil; a weighted board was used to compact the soil, treating 18 acres [7.3 hectares]. That evening, residents living one-quarter mile [0.4 kilometers] west of the application site experienced irritant symptoms. The Kern County Fire Department [KCFD] was contacted to investigate; however, darkness, distance from the treated field, and absence of chloropicrin odor prevented firefighters from identifying the source of the irritation. Records from a weather station approximately 7 miles [11.3 kilometers] southeast of the application site indicated low wind speeds and stable atmospheric conditions but also that the wind direction had changed that evening, blowing from the field toward the residential dwellings.

The next day, chloropicrin was applied to the remaining 16 acres [6.5 hectares]. A 60-foot [18.3-meter], chloropicrin-free buffer was maintained around the perimeter of the field because workers noted a persistent odor when they arrived. Residents one-quarter mile [0.4 kilometers] west and south of the field complained about irritant symptoms that evening. Residents notified KCFD; several responding firefighters experienced eye irritation. The wind had changed again that evening and begun blowing from the field toward the residential dwellings. Suspecting a pesticide release, KCFD notified KCAC. The field was recompacted, and the odor ceased.

On October 6, KCAC notified CDPR about the incident. KCAC and CDPR conducted in-person interviews at 35 households located approximately one-quarter mile [0.4 kilometers] west and south of the field and at a day care center; additional interviews were conducted on October 15. The 35 households and day care center had a total of 172 persons present during the exposure period. Representatives from each household and the day care center were interviewed by using a standardized questionnaire. . . . In addition, five workers involved with the fumigation were questioned informally, and KCFD records were reviewed to identify affected firefighters.

The investigation determined that 165 persons reported symptoms compatible with illness caused by chloropicrin; median age of the persons was 16 years (range: 3 months–63 years). Nearly all (99%) had irritant symptoms (e.g., eye or upper respiratory) . . . ; nine (5%) received medical evaluations. Seven had persistent respiratory symptoms when interviewed 11 days after the event. Follow-up medical care was limited because most of the affected persons lacked health insurance.

Modified and reproduced from Centers for Disease Control and Prevention. Brief report: illness associated with drift of chloropicrin soil fumigant into a residential area—Kern County, California, 2003. *MMWR.* 2004;53:740.

DESCRIPTION OF BIOPESTICIDES

Biopesticides are certain types of pesticides derived from such natural materials as animals, plants, bacteria, and certain minerals. For example, canola oil and baking soda have pesticidal applications and are considered biopesticides. . . .

- Microbial pesticides consist of a microorganism (e.g., a bacterium, fungus, virus, or protozoan) as the active ingredient. [An example is *Bacillus thuringiensis* (Bt), which can kill insect larvae.]
- Plant-incorporated protectants (PIPs) are pesticidal substances that plants produce from genetic material that has been added to the plant. For example, scientists can take the gene for the Bt pesticidal protein and introduce the gene into the plant's own genetic material.
- Biochemical pesticides are naturally occurring substances that control pests by nontoxic mechanisms. . . . [Examples] include substances, such as insect sex pheromones, that interfere with mating, as well as various scented plant extracts that attract insect pests to traps.

Modified and reproduced from US Environmental Protection Agency. What are biopesticides? Available at: https://www.epa.gov/ingredients-used-pesticide-products/what-are-biopesticides. Accessed February 17, 2017.

However, one California study reported no strong association between residential proximity to locations where pesticides were applied and fetal deaths.[36] Children appear to be more prone to pesticide poisoning than adults because children have large skin-surface to body-mass ratios.[37] Children also demonstrate personal hygiene deficiencies such as hand–mouth behaviors that could cause the ingestion of pesticides.

Biopesticides

In addition to chemical pesticides, other types include biopesticides, antimicrobials, and pest control devices. The text box provides additional information. (Instead of using pesticides, some farmers have been able to control pests by growing several types of crops in the same area at the same time, rotating crops, and applying compost; further discussion of this topic is beyond the scope of this text.)

Dioxins

The term **dioxin** refers to "a family of chemical compounds that are unintentional byproducts of certain industrial, non-industrial and natural processes, usually involving combustion."[38] The word *dioxin* usually means the chemical TCDD (2,3,7,8-tetrachlorodibenzo-*p*-dioxin),[39] which is the most toxic[40,41] and one of the most widely researched forms of dioxin. Researchers have observed that dioxins are among the most toxic chemicals ever used in experiments with laboratory animals.[39]

Other chemicals that are called "dioxin-like compounds" or simply "dioxins" come from three related chemical families: (1) chlorinated dibenzo-*p*-dioxins (CDDs), (2) chlorinated dibenzofurans (CDFs), and (3) certain polychlorinated biphenyls (PCBs).[42] A total of 419 compounds in the dioxin family are known to exist; however, scientists regard only about 30 of these as being the most poisonous.[40]

Both natural and human activities produce dioxins. Natural events such as forest fires and volcanic eruptions cause the emission of dioxins into the environment.[41] Dioxins that originate from these sources contribute to natural background levels of dioxins. With respect to human activities, background burning of trash creates dioxins,[42] as do incineration of industrial and municipal wastes and the burning of some fuels. Dioxins are produced when wood pulp is bleached in the paper manufacturing process and during the manufacture and application of some herbicides (e.g., dioxins were a contaminant in trace amounts in the herbicide Agent Orange). Burning of tobacco gives rise to minute amounts of dioxin in cigarette smoke.[43]

Dioxins, almost universal contaminants, are very stable chemicals that take many years to decompose. The World Health Organization refers to dioxins as members of the "dirty dozen club."[40] The club consists of persistent organic chemicals (POPs), so named because of their tendency to remain in the environment. (Refer to Table 7.2 for more information.) When they are released into the atmosphere from combustion, dioxins may drift over vast areas before settling on the earth's surface. They also may reach bodies of water as contaminants that unite with the sediments in the water. Aquatic organisms that feed on lake sediments may ingest these dioxins, which become ever more concentrated as they pass up the food chain to fish. The latter, in turn, may be consumed by organisms such as animals or humans at the top of the food chain. Because dioxins are fat soluble, the process of bioconcentration may result in the accumulation of significant levels in animals' fatty tissues, the locus where dioxins tend to remain.

The health effects of exposure to dioxins depend on several factors: duration of exposure, frequency of exposure, when the exposure occurred, concentration of the agent, and route of entry into the body.

2,3,7,8-TETRACHLORODIBENZO-P-DIOXIN EXPOSURE TO HUMANS—SEVESO, ITALY

At approximately noon on Saturday, July 10, 1976, an explosion occurred during the production of 2,4,5-trichlorophenol in a factory in Meda, about 25 km [15.5 miles] north of Milan, in the Lombardia region of Italy. A cloud of toxic material was released and included 2,3,7,8-tetrachlorodibenzo-p-dioxin (TCDD). Debris from this cloud fell south-southeast of the plant on an area of about 2.8 km^2 [1 square mile], including parts of the towns of Seveso, Meda, Cesano Maderno, and Desio.

The size of the contaminated area was estimated primarily by measuring TCDD in the soil; additional criteria included the presence of dead animals (e.g., birds, rabbits, chickens) and detection of dermal lesions among persons in the area. The contaminated area was divided into three zones (A, B, and R) depending on the concentration of TCDD in the soil. . . . An additional zone, Zone S, outside the contaminated area was examined as a control zone. . . . Zone A, the most heavily contaminated section, was further divided into seven subzones, A1–A7, based on increasing distance from the factory.

The total amount of TCDD deposited in the contaminated area was initially estimated at about 165 g [5.8 ounces]. . . ; subsequently, it has been estimated to be at least 1.3 kg [2.7 pounds]. . . . Within 20 days of the explosion, the Italian authorities had evacuated the 211 families (735 persons) from the area later defined as Zone A and had taken immediate measures to minimize the risk of exposure to residents in nearby areas (primarily those in Zone B).

The Italian authorities were assisted by several national and international technical commissions in assessing adverse health effects. Residents of zones A, B, and R underwent extensive medical examinations from 1976 to 1985; chloracne, detected in a small segment of the population, was the only abnormal finding. . . . Only one potentially exposed person was measured for TCDD; she was a 55-year-old woman residing in a portion of Zone A . . ., who died from pancreatic adenocarcinoma 7 months after the explosion. Her TCDD whole-weight levels varied from 6 parts per trillion (ppt) in blood to 1840 ppt in adipose tissue. . . .

Modified and reproduced from Centers for Disease Control and Prevention. International notes preliminary report: 2,3,7,8-tetrachlorodibenzo-p-dioxin exposure to humans—Seveso, Italy. *MMWR.* 1988;37:733–734.

Depending on these factors, the health effects of exposure to dioxins include:

- Chloracne
- Skin rashes
- Skin discoloration
- Growth of excessive body hair
- Liver damage
- Possible cancer risks
- Endocrine effects
- Reproductive and developmental effects

One of the most noteworthy health consequences of dioxin exposure at high concentrations is chloracne, a skin condition typified by a very severe and disfiguring type of acne.[42] Other dermal effects of high levels of exposure include patchy discoloration and increased growth of hair. Long-term exposure to dioxins has been associated with disruption of the immune, endocrine, reproductive, and nervous systems. Dioxins have been reported to cause cancer in laboratory animals.

There have been several documented examples of exposure of the population to potentially hazardous levels of dioxin. One is known as the Belgian PCB/dioxin incident, in which dioxin-contaminated chicken was distributed inadvertently to consumers in Belgium in 1999.[44] As a result of a processing accident, PCBs (that were secondarily contaminated with dioxins) became infused into a batch of recycled fats destined for use in animal fodder. The feeds, sold to more than 2,500 farms, apparently caused illnesses in the chickens but could not be documented as the cause of any illnesses in humans. Due to uncertainty about who consumed the tainted chickens and a long delay in discovering the contamination (until after the feeds had been distributed), the ultimate human health effects of this incident may be difficult to ascertain.[44] Two other dioxin exposure events occurred in Times Beach, Missouri and Seveso, Italy. (Refer to the text box for more information.)

Polychlorinated Biphenyls

Often, the group referred to as polychlorinated biphenyls (PCBs) is presented as a single chemical. However, PCBs comprise an extensive class of organochlorines with 209 congeners (members of the same class), some of which are found more commonly in industrial chemicals than others.[45] Also, PCBs are classified as dioxin-like chemicals. Trade names for PCBs are Arochlor (formulated as nine common mixtures), Askarel, and Therminol.

Estimates indicate that from the 1920s to the late 1970s, about 1.5 billion pounds (ca. 680,400,000 kgs) of PCBs were manufactured in the United States.[46]

PCBs won favor because of their chemical stability, low flammability, inexpensive cost, and insulating properties. Two of the common uses of PCBs were as an insulating fluid in transformers and capacitors and as a lubricant, although the use of PCBs for these purposes has been discontinued. Leakage of PCBs from aging transformers can contaminate the soil and environment. PCBs are known to be present at as many as 500 sites denoted on the 1,598 sites that the EPA has provided on the National Priorities List.[45] (*Note:* The National Priorities List was defined in the chapter on toxic metals and elements.) Because of concerns about their environmental and health effects, the manufacture of PCBs was terminated in 1977.[45]

One of the concerns voiced about PCBs is that they remain for long periods in the environment, for example, by binding to sediments. PCBs present in the environment tend to bioaccumulate in fish and other animals used for food and, in turn, impact human health. The ATSDR reports that human beings, their communities, and the environment will continue to face the effects of exposure to PCBs for many decades into the future.[47] This exposure will take place despite the slow, natural process of chemical degradation of PCBs and environmental cleanup efforts. Heavily contaminated US regions are the Hudson River in New York and the Lower Fox River in the Green Bay area of Wisconsin. For example, in New York, the Hudson Falls and Fort Edward General Electric Company plants discharged approximately 1.3 million pounds (600,000 kilograms) of PCBs into the Hudson River over a 30-year period that ended in 1977.[48] A 200-mile (approximately 340-km) stretch has been designated at a Superfund site. See **FIGURE 7.6** for a picture of the Hudson Falls plant.

PCBs have been shown conclusively to cause cancer in animals and are designated as probable human carcinogens.[49] Several mortality studies conducted among employees who had been exposed to PCBs suggested that malignant melanoma and liver, biliary tract, and intestinal cancers were associated with these substances. Effects of exposure to PCBs have been investigated in relation to the risk of breast cancer, although epidemiologic studies did not support this proposed association.[50]

PCBs also have been investigated with respect to their impact on the immune system, reproductive system, and children's intellectual development. PCB exposure may limit the development of immune responses to the Epstein-Barr virus and other viral and bacterial infections.[49] Exposure to PCBs is a possible risk factor for non-Hodgkin's lymphoma.[49] Findings suggest that PCBs cause menstrual disturbances and influence male fertility.[51] Prenatal PCB exposure has been linked to abnormal sperm morphology.[52] A number of studies indicate that prenatal exposure to PCBs is associated with children's lower performance on IQ tests.[53] One of the most strongly affected dimensions was performance tests of memory and attention.[54]

FIGURE 7.6 The General Electric Hudson Falls plant.

Reproduced from US Environmental Protection Agency. Available at: https://www3.epa.gov/hudson/images/gehf1.jpg. Accessed July 24, 2017.

Organic Solvents

The term **solvent** refers to "a liquid substance capable of dissolving other substances; 'the solvent does not change in forming a solution.'"[55] How do people become exposed to solvents? There are several modes of exposure, including breathing their vapors directly, ingesting them in foods and water that have low levels of contamination, using foods and cosmetics packed in certain types of plastics, and sometimes by smoking cigarettes. Workers may be exposed chronically to solvents that are used by factories. When pregnant female workers are exposed to solvents, these substances are likely to cross the placental barrier and affect the developing fetus. Exposure to solvents among pregnant women has been linked to spontaneous abortions.[56]

Industrial facilities may release vapors from solvents into the surrounding neighborhoods where residents could inhale them. In addition to being emitted into the air from factories, solvents may enter the soil and groundwater through improper disposal or when they leach out of waste sites. Generally, most solvents do not persist in the environment (although there are some exceptions) because of their volatility. Also, research has not resolved fully the issue of their long-term environmental impacts. Only a few members of the solvent group are suspected carcinogens.[57] For example, the solvent benzene is a known carcinogen.

Following is an abbreviated list of solvents and a brief discussion of each one. The selection of compounds presented in this section is somewhat arbitrary, as there are thousands of such chemicals, many of which have both positive and negative impacts on the environment. Thus, this review cannot in any sense be considered comprehensive, but rather is generally representative of some of the issues that occur in the context of these chemicals. Examples of important solvents are the following:

- Tetrachloroethylene
- Trichloroethane
- Trichloroethylene (TCE)
- Toluene
- Acetone
- Benzene

Tetrachloroethylene (synonyms: perchloroethylene, PCE, perchlor, tetrachloroethene) is a synthetic, nonflammable solvent used in the dry cleaning industry and as a metal degreaser. This compound evaporates easily and has a distinctive odor.[58] Exposure to high levels causes neurologic effects and is capable of producing unconsciousness and death. The health effects of exposure to low levels have not been determined. Tetrachloroethylene exposure at very high levels caused liver and kidney cancer in studies of laboratory animals.[58] The International Agency for Research on Cancer (IARC) reported that there is

> sufficient evidence to designate PCE as carcinogenic in animals, with limited evidence in humans. With regard to occupational exposure through dry cleaning, PCE is considered to be possibly carcinogenic to humans. . . . The current epidemiologic evidence does not support a conclusion that occupational exposure to PCE is a risk factor for cancer of any specific site.[59(p473)]

Trichloroethane exists in two common forms: 1,1,1-trichloroethane and 1,1,2-trichloroethane. The form known as 1,1,1-trichloroethane (synonym: methylchloroform) is a solvent found in degreasing compounds and household cleaners that has had many other uses.[60] Depending upon the level of exposure, symptoms range from dizziness to loss of consciousness. Although not presently classified as a carcinogen, its manufacture has been discontinued since January 1, 2002, because of potential to harm the ozone layer.

The variant called 1,1,2-tricholorethane is a colorless, sweet-smelling solvent that has low flammability and is used as an intermediate chemical for the production of other chemicals. [61] Direct contact with the skin produces localized irritation. The compound breaks down very slowly after being released into the environment and is believed to persist indefinitely if it seeps into subsurface groundwater. To date, this compound has not been classified as a human carcinogen.

Trichloroethylene (TCE) is a grease-dissolving solvent that evaporates readily.[62] In addition to its use as a solvent, TCE is added to some cleaning agents, paint removers, and adhesives. In the environment, TCE may persist in underground water supplies for long periods. Low exposure levels produce neurologic effects such as dizziness and headaches. Inhalation or ingestion of large amounts of TCE can cause unconsciousness and death. The consumption of minute quantities of TCE present in public water supplies may lead to adverse health effects, although the consequences of long-term, low-level exposures have not been verified definitively. Research has suggested that TCE is a probable human carcinogen. In the chapter on water quality, information on contamination of public drinking water supplies with TCE will be presented.

Toluene, used in the manufacture of nail polish, paints, and adhesives, is a clear, colorless liquid that has a characteristic odor.[63] Low to moderate degrees of exposure to toluene are known to cause reversible neurologic effects including anorexia, nausea, and confusion. Heavy exposures are capable of producing loss of consciousness and possibly death. Toluene is suspected of having adverse effects upon the renal system.

Acetone, a colorless liquid with a distinctive sweet odor, is used as an ingredient in nail polish, cleaners, and paints.[64] This highly volatile solvent is flammable and dissolves in water. Moderate levels of exposure to acetone can cause irritation of the eye (e.g., tearing) and the respiratory tract (e.g., nose, throat, and lungs). High levels are capable of causing headaches, vomiting, and loss of consciousness. Most volatile compounds have a threshold for irritating effects that is much higher than the human odor detection threshold. Habituation (desensitization) sometimes occurs among workers who are exposed repetitively; this phenomenon makes it difficult to establish appropriate exposure guidelines for volatile chemicals.[65] The EPA does not classify acetone as a carcinogen.

Benzene is a sweet-smelling, flammable, colorless liquid that evaporates rapidly.[66] Benzene is used as an intermediary chemical for the manufacture of plastics, resins, and synthetic fibers. It also is used as a solvent for waxes, paints, resins, oils, plastics, and rubber; it has been added as a component of fuels for motor vehicles.[67] Benzene occurs naturally in gasoline and crude oil, and as a component of cigarette smoke.

Benzene is an extremely toxic chemical: inhalation of benzene vapors may cause headache and nausea as well as neurologic symptoms such as dizziness; heavy exposures can lead to unconsciousness.[68] In addition to damaging the blood and blood-forming organs, benzene is a known carcinogen that is causally related to the etiology of a form of leukemia called acute myelogenous leukemia.[69]

Muzaffer Aksoy, a Turkish hematologist, observed that many patients afflicted with leukemia were shoemakers.[70] At the time that he made this initial observation, benzene was used extensively in Turkey as a solvent in the manufacture of leather goods. Following this clinical observation, Aksoy conducted an epidemiologic study (from 1967 to 1974) of 28,500 employees involved with the manufacture of shoes, slippers, and handbags. In 1974, Aksoy reported a positive association between benzene exposure and leukemia on the basis of this epidemiologic research. His work led to the restriction in many countries of occupational benzene exposures, which now must be kept at negligible levels in the United States.

Chemicals Used in the Manufacture of Plastics

Included in the category of chemicals used in the manufacture of plastics are styrene and vinyl chloride. Chemicals called phthalates are used as plasticizers (a term that will be defined subsequently).

Styrene (styrene monomer) is a clear, colorless liquid that has an aromatic odor.[71] It is commercially important for the manufacture of polystyrene resins (used for many types of plastics, such as packaging, disposable drinking glasses, and building insulation). Short-term inhalation of styrene can produce central nervous system effects such as muscle weakness, and difficulty in concentrating on tasks; irritation of the respiratory tract also can result. According to the IARC, styrene has been listed as possibly carcinogenic, although the evidence supporting this effect among humans and in animal studies is limited.

At room temperature, *vinyl chloride* (synonyms: chloroethene, chloroethylene, ethylene monochloride, monochloroethylene) is a colorless, flammable gas that has a distinctive sweet odor and is highly volatile.[72] When kept under pressure or at low temperatures, vinyl chloride becomes liquefied. In the United States, vinyl chloride is used mainly for the manufacture of polyvinyl chloride (PVC), which is an ingredient of a multitude of plastic products such as pipes, vinyl siding for houses, plastic coatings, and upholstery.

Depending upon the concentration, exposure to vinyl chloride can cause central nervous system effects, loss of consciousness, or death.[72] Workers who are exposed chronically to high levels of vinyl chloride may develop nerve and immune system damage, poor circulation in the hands and forearms, and damage to the bones at the tips of the fingers. The US Department of Health and Human Services classifies vinyl chloride as a human carcinogen. Numerous epidemiologic studies as well as case reports have verified that long-term chronic occupational exposure to vinyl chloride is associated with liver cancer (angiosarcoma of the liver).[73,74]

The term **plasticizer** denotes a substance that is added during the manufacture of a plastic to increase its durability as well as make it softer and more pliable. This section will focus on **phthalates** (e.g., diethyl phthalate [DEH]; di(2-ethylhexyl) phthalate [DEHP]; di-*n*-butyl phthalate [DBP]). Phthalates are a class of plasticizer that is used to increase the flexibility of plastics. They are used in plastic components for automobiles, packaging for foodstuffs, toothbrushes, and toys.[75] In addition, phthalates are used as additives in cosmetics, aspirin, and insecticides. DBP is used in nail polish,[76] which is especially popular among women between the ages of 20 and 40 years. Many women of childbearing age have significant concentrations of a DBP metabolite in their bodies. The importance of phthalates in causing adverse human health effects is unclear; the impact of DBP, in particular, on the health of unborn children is speculative.[77] DBP is thought to have an antiandrogenic effect (suppresses hormones related to the development of male sexual characteristics) in rats.

Phthalates are released rather easily from plastic products and can contaminate food wrapped in plastic packaging. When young children chew on toys made from polyvinyl chloride, some of the phthalate additives may be ingested and, conceivably, pose a health risk. In 2006, the European Union prohibited the use of several types of phthalates as additives to plastics in toys that can be placed in the mouth by toddlers younger than 3 years of age. The US in 2009 banned the use of phthalates for children's toys and child care products.[78]

The phthalate plasticizer DEHP is used in medical products formulated from polyvinyl chloride. Examples of such products are plastic bags used to contain intravenous solutions, blood products in plastic packaging, plastic tubing, and protective gloves. Concern has been raised over the possible toxic hazards associated with the use of polyvinyl products in the medical environment. Current information supports the view that the use of products that contain DEHP is generally safe, although some patients (e.g., children) may face the possibility of risks connected with DEHP.[79] The ATSDR asserts that "At the levels found in the environment, DEHP is not expected to cause harmful health effects in humans. Most of what we know about the health effects of DEHP comes from studies of rats and mice given high amounts of DEHP."[80]

Another chemical used in the manufacture of plastics is bisphenol A (BPA). The public has been concerned about BPA because its use is widespread and there have been some reports of animal studies that showed reproductive and developmental effects. Refer to the text box for more information.

BISPHENOL A (BPA)

Bisphenol A, more commonly known as BPA, is a chemical widely used to make polycarbonate plastics and epoxy resins.... Polycarbonate plastics have many applications including use in some food and drink packaging such as water and baby bottles, compact discs, impact-resistant safety equipment, and medical devices including those used in hospital settings....

- BPA can leach into food from the epoxy resin lining of cans and from consumer products such as polycarbonate tableware, food storage containers, water bottles, and baby bottles. [Another possible source of exposure is] by handling cash register receipts.
- [People are] concerned about BPA because human exposure to BPA is widespread.... [Also,] some laboratory animal studies report subtle developmental effects in fetuses and newborns exposed to low doses of BPA.
- The NTP [National Toxicology Program] has *"some concern"* for BPA's effects on the brain, behavior, and prostate gland in fetuses, infants, and children at current exposure levels. (Refer to **FIGURE 7.7** for other conclusions.)
- BPA-based polycarbonate resins may no longer be used in baby bottles and sippy cups.

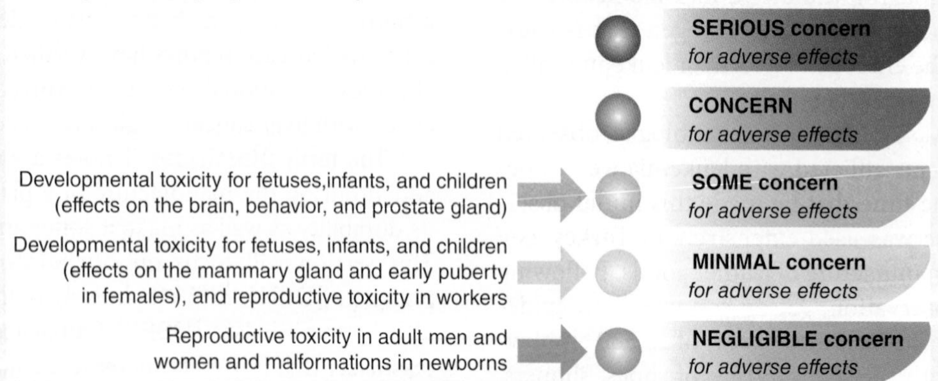

FIGURE 7.7 National Toxicology Program conclusions regarding bisphenol A (BPA).

Modified and reproduced from National Institute of Environmental Health Sciences. Bisphenol A (BPA). National Toxicology Program. Available at: https://www.niehs.nih.gov/health/assets/docs_a_e/bisphenol_a_bpa_508.pdf. Accessed January 26, 2017.

Modified and reproduced from National Institute of Environmental Health Sciences. National Toxicology Program. Bisphenol A (BPA) Available at: https://www.niehs.nih.gov/health/assets/docs_a_e/bisphenol_a_bpa_508.pdf. Accessed January 26, 2017; and U.S. Food and Drug Administration. Bisphenol A (BPA): use in food contact application. Available at: https://www.fda.gov/newsevents/publichealthfocus/ucm064437.htm. Accessed February 17, 2017.

Cleaning and Household Products

The list of chemicals used commonly in household products is extensive. **TABLE 7.6** provides some examples. Exposure to these products may be hazardous if they are used improperly or disposed of in an unsafe way. For example, care should be taken not to mix chlorine bleach and ammonia, because toxic chloramine gases will be released.

Environmental Estrogens

Some organic chemicals (e.g., chlorinated hydrocarbon pesticides) are alleged to have estrogenic activity (defined previously). Sometimes DDT (and metabolites such as DDE) is referred to as an endocrine disruptor with respect to its ability to act as an antagonist to androgen, the male sex hormone. Another example comes from animal studies of the estrogenic pesticide methoxychlor, which has been shown to have negative consequences for fertility, pregnancy, and development.[81] It is possible that when estrogenic-like chemicals are present in the environment, they may have abnormal influences on the reproductive systems of exposed humans and animals. In addition, these chemicals may act as cancer promoters by having an influence on the onset of female cancers that are thought to be caused by estrogenic activity. The following text box, titled Women's Health and the Environment, presents speculations regarding the possible role of estrogenic and other environmental chemicals in women's health.

Conclusion

Pesticides and other organic chemicals are indispensable to modern society; nevertheless, they have the potential to cause great harm. The average person is exposed to a multitude of chemicals—even a cup of coffee contains more than 1,000 chemicals. The ATSDR has compiled a list of 275 hazardous substances. Among the top 20 hazardous substances, 9 are from chemical groupings such as PCBs, pesticides, and solvents. Throughout the world millions of tons of pesticides are used to control insects, weeds, and rodents. Organophosphate pesticides are used commonly for a number of reasons, including the fact that they tend not to persist in the environment; however,

TABLE 7.6 Toxic Household Products

Product Name	Use(s)
Sodium hypochlorite	Chlorine bleach
Petroleum distillates	Metal polishes
Ammonia	Glass cleaner
Phenol and cresol	Disinfectants
Nitrobenzene	Furniture and floor polishes
Perchloroethylene or 1-1-1 trichloroethane	Solvents for use in dry cleaning fluids, spot removers, and carpet cleaners
Naphthalene or paradichlorobenzene	Mothballs and toilet bowl cleaners
Hydrochloric acid or sodium acid sulfate	Toilet bowl cleaners
Formaldehyde	Preservative in many household products; glue in particleboard and plywood furniture
Formaldehyde, phenol, and pentachlorophenol	Spray starch

Data from Environmental Law Centre. Environment. Available at: http://www.elc.org.uk/pages/envirohome.htm. Accessed February 18, 2017.

WOMEN'S HEALTH AND THE ENVIRONMENT

Women have a particular stake in environmental health research. Not only do they share many of the same diseases as men and children—in which the environment, along with genetic susceptibility, has an important role—but women also have particular environmental diseases related to their gender. . . . Women tend to carry more fat, in which substances introduced lower in the food chain may accumulate. . . .[a]

Breast Cancer Risk and Environmental Factors

For millions of women whose lives have been affected by breast cancer, the 1994 discovery of the first breast cancer gene by researchers from the National Institute of Environmental Health Sciences (NIEHS) and their collaborators was a welcome sign of progress in the fight against this disease. (However, the breast cancer gene, BRCA1, is thought to account for less than one-third of breast cancer risk.) While this discovery and others like it are certainly encouraging, statistics tell us that breast cancer is still a major health concern for women everywhere. According to the American Cancer Society, breast cancer is the most common cancer among women in the United States, other than skin cancer. . . .

Although scientists have identified many risk factors that increase a woman's chance of developing breast cancer, they do not yet know how these risk factors work together to cause normal cells to become cancerous. Most experts agree that breast cancer is caused by a combination of genetic, hormonal, and environmental factors. . . .

During early childhood and adolescence, the breast tissue is developing and maturing, and recent studies suggest that environmental exposures, such as certain chemicals, diet, and social factors, during these critical stages of development, may affect breast cancer risk later in life.[b]

Modified and reproduced from [a]Environment, Health and Safety Online. Women's health and the environment. Available at: http://www.ehso.com/ehshome/womenshealth2.htm. Accessed February 17, 2017; and [b]National Institute of Health. National Institute of Environmental Health Sciences. Breast cancer risk and environmental factors. Available at: https://www.niehs.nih.gov/health/assets/docs_a_e/environmental_factors_and_breast_cancer_risk_508.pdf. Accessed February 17, 2017.

they can be hazardous when used inappropriately. Use of organochlorine pesticides such as DDT has been discontinued in many nations because of concerns about possible adverse health effects associated with the persistence of DDT in the environment. Although chemicals classified as solvents produce acute health effects, only a few of the chemicals in this group are documented carcinogens. In many cases, the long-term health effects of solvents are unknown. Continuing research is needed to ascertain the long-term health effects of pesticides and other organic chemicals.

Study Questions and Exercises

1. Define the following terms:
   a. Pesticide
   b. Insecticide
   c. Herbicide
   d. Pyrethroid
2. Describe the four major classes of pesticides. What are the advantages and disadvantages of each type?
3. What measures would you take to reduce the number of emergencies that involve hazardous substances? What are the three most common substances that produce personal injuries?
4. What was history's most severe incident that involved toxic chemicals? How could this accident have been prevented? Was the accident a one-time event, or are there continuing health problems?
5. What were some of the concerns that led to the banning of DDT in the United States and other developed countries? Do you believe that the hazards associated with DDT outweigh the pesticide's benefits?
6. Support or refute the following statement: Pyrethrin-based insecticides are unlikely to produce severe human health effects.
7. Describe Operation Ranch Hand, and state how Agent Orange received its name. Why is Agent Orange alleged to be a potential health hazard?
8. What type of agent is chloropicrin? Describe the hazards to the community associated with chloropicrin.
9. Name one of the most toxic groups of chemicals ever used in animal experiments. Describe the health effects of this family of chemicals.
10. Name three chemicals that the chapter identified as human carcinogens. Give some examples of accidents/incidents that involved these chemicals. What is being done to limit the human population's exposure to these chemicals?
11. Define the term botanical insecticide and give an example. What is an advantage of using botanicals?
12. Describe the phenomenon of bioaccumulation of toxic chemicals. Give examples of at least two chemicals that tend to bioaccumulate.

13. Give some examples of common household chemicals that are potentially hazardous. Using you own ideas, suggest how these chemicals can be used and stored safely.
14. What is the significance of antiandrogenic chemicals for the environment? Name at least one such chemical.

For Further Reading

Silent Spring, Rachel Carlson, 1962

References

1. Gardella JR, Hill JA. Environmental toxins associated with recurrent pregnancy loss. *Semin Reprod Med.* 2000;18:407–424.
2. De Rosa CT. Restoring the foundation: tracking chemical exposures and human health. [Editorial] *Environ Health Perspect.* 2003;111:A374–A377.
3. National Center for Environmental Health. *Fourth National Report on Human Exposure to Environmental Chemicals.* Atlanta, GA: Centers for Disease Control and Prevention; 2009.
4. National Institute of Environmental Health Sciences. *The New Environmental Health: Our Chemical World, and Our Dilemma.* NIH Publication #02-5081. Available at: http://www.dphu.org/uploads/attachments/books/books_1464_0.pdf. Accessed July 21, 2017.
5. Carson R. *Silent Spring.* New York, NY: Houghton Mifflin Company; 1962.
6. National Center for Environmental Health. *Third National Report on Human Exposure to Environmental Chemicals.* NCEH Publication No.: 05-0725. Atlanta, GA: Centers for Disease Control and Prevention; 2005.
7. Casida JE, Quistad GB. Golden age of insecticide research: past, present, or future? *Ann Rev Entomol.* 1998;43:1–16.
8. Lockwood AH. Organophosphate pesticides and public policy. [Editorial] *Curr Opin Neurol.* 2002;15:725–729.
9. Kellogg RL, Nehring R, Grube A, et al. Environmental indicators of pesticide leaching and runoff from farm fields. Paper presented at: Conference on Agricultural Productivity: Data, Methods, and Measures; March 9–10, 2000; Washington, DC: United States Department of Agriculture, Natural Resources Conservation Service. Available at: https://www.nrcs.usda.gov/wps/portal/nrcs/detail/national/technical/?cid=nrcs143_014053. Accessed April 30, 2017.
10. Reigart JR, Roberts JR. Pesticides in children. *Pediatr Clin North Am.* 2001;48:1185–1198.
11. Costa LG. Basic toxicology of pesticides. *Occup Med.* 1997;12(2):251–268.
12. Peter JV, Cherian AM. Organic insecticides. *Anaesth Intensive Care.* 2000;28:11–21.
13. Atwood D, Paisley-Jones C. Pesticide industry sales and usage. Washington, DC: US Environmental Protection Agency, 2017. Available at: https://www.epa.gov/sites/production/files/2017-01/documents/pesticides-industry-sales-usage-2016_0.pdf. Accessed April 30, 2017.
14. *The Japan Times* Online. Deadly sarin attack on Tokyo subway system recalled 20 years on. March 20, 2015. Available at: http://www.japantimes.co.jp/news/2015/03/20/national/tokyo-marks-20th-anniversary-of-aums-deadly-sarin-attack-on-subway-system/. May 1, 2017.
15. Kwong TC. Organophosphate pesticides: biochemistry and clinical toxicology. *Ther Drug Monit.* 2002:24:144–149.
16. Rubin C, Esteban E, Kieszak S, et al. Assessment of human exposure and human health effects after indoor application of methyl parathion in Lorain County, Ohio, 1995–1996. *Environ Health Perspect.* 2002;110(suppl 6):1047–1051.
17. European Environment Agency. Chlorinated hydrocarbon. Available at: http://glossary.eea.europa.eu/EEAGlossary/C/chlorinated_hydrocarbon. Accessed May 1, 2017.
18. European Environment Agency. Organochlorines. Available at: http://glossary.eea.europa.eu/EEAGlossary/O/organochlorines. Accessed May 1, 2017.
19. Agency for Toxic Substances and Disease Registry (ATSDR). ToxFAQs for DDT, DDE, and DDD. http://www.atsdr.cdc.gov/toxfaqs/index.asp. Accessed May 1, 2017.
20. Turusov V, Rakitsky V, Tomatis L. Dichlorodiphenyltrichloroethane (DDT): ubiquity, persistence, and risks. *Environ Health Perspect.* 2002;110:125–128.
21. Longnecker MP, Rogan WJ, Lucier G. The human health effects of DDT (dichlorodiphenyltrichloroethane) and PCBs (polychlorinated biphenyls) and an overview of organochlorines in public health. *Annu Rev Public Health.* 1997;18:211–244.
22. Agency for Toxic Substances and Disease Registry (ATSDR). ToxFAQs for pyrethrins and pyrethroids. Available at: https://www.atsdr.cdc.gov/toxfaqs/tfacts155.pdf. Accessed May 1, 2017.
23. Biello D. A common herbicide turns some male frogs into females. *Scientific American.* March 2, 2010. Available at: https://www.scientificamerican.com/article/common-herbicide-turns-male-frogs-into-females/. Accessed May 3, 2017.
24. US Environmental Protection Agency. Terms & acronyms. Available at: https://ofmpub.epa.gov/sor_internet/registry/termreg/searchandretrieve/termsandacronyms/search.do. Accessed May 1, 2017.
25. Centers for Disease Control and Prevention. Facts about paraquat. Available at: https://emergency.cdc.gov/agent/paraquat/basics/facts.asp. Accessed May 2, 2017.
26. Frumkin H. Agent Orange and cancer: an overview for clinicians. *CA Cancer J Clin.* 2003;53:245–255.
27. National Academies, Institute of Medicine. *Veterans and Agent Orange: Herbicide/Dioxin Exposure and Type 2 Diabetes.* Washington, DC: National Academies Press; 2000.
28. Ray DE. Pesticide neurotoxicity in Europe: real risks and perceived risks. *Neurotoxicol.* 2000;21(1–2):219–221.
29. Hood E. Bringing home more than a paycheck: workers and pesticides. *Environ Health Perspect.* 2002;110:A765.
30. Wallinga D, Greer L. *Poisons on Pets: Health Hazards from Flea and Tick Products.* Natural Resources Defense Council; 2000. Available at: https://www.nrdc.org/sites/default/files/pets.pdf. Accessed May 2, 2017.
31. Shull LR, Cheeke PR. Effects of synthetic and natural toxicants on livestock. *J Anim Sci.* 1983;57(suppl 2):330–354.
32. National Academy of Sciences. The National Academies of Sciences, Engineering, and Medicine. *Status of Pollinators in North America (2007).* Washington, DC: National Academies Press; Available at: http://dels.nas.edu/Report/Status-Pollinators-North-America/11761. Accessed July 26, 2017.
33. Organization for Economic Co-operation and Development (OECD). Managing pesticide risk to insect pollinators. Available at: http://www.oecd.org/chemicalsafety

/risk-mitigation-pollinators/insect-pollinators-risks.htm. Accessed July 26, 2017.

34. U.S. Environmental Protection Agency. Colony collapse disorder. Available at: https://www.epa.gov/pollinator-protection/colony-collapse-disorder. Accessed July 26, 2017.
35. Ames RG. Pesticide impacts on communities and schools. *Int J Toxicol.* 2002;21:397–402.
36. Bell EM, Hertz-Picciotto I, Beaumont JJ. Case-cohort analysis of agricultural pesticide applications near maternal residence and selected causes of fetal death. *Am J Epidemiol.* 2001;154:702–710.
37. Sanborn MD, Cole D, Abelsohn A, et al. Identifying and managing adverse environmental health effects: 4. Pesticides. *CMAJ.* 2002;166:1431–1436.
38. DioxinFacts.org. Questions and answers: dioxins, furans, TCDD, PCBs. Available at: http://www.dioxinfacts.org/questions_answers/. Accessed May 3, 2017.
39. Vanden Heuvel JP, Lucier G. Environmental toxicology of polychlorinated dibenzo-*p*-dioxins and polychlorinated dibenzofurans. *Environ Health Perspect.* 1993;100:189–200.
40. World Health Organization. Fact sheet No. 225: dioxins and their effects on human health. Available at: http://www.who.int/mediacentre/factsheets/fs225/en/. Accessed May 3, 2017.
41. National Institutes of Health. National Institute of Environmental Health Sciences. Dioxins. Available at: https://www.niehs.nih.gov/health/topics/agents/dioxins/. Accessed May 3, 2017.
42. US Environmental Protection Agency. Learn about dioxin. Available at: https://www.epa.gov/dioxin/learn-about-dioxin. Accessed May 3, 2017.
43. Agency for Toxic Substances and Disease Registry (ATSDR). Public health statement for chlorinated dibenzo-*p*-dioxins (CDDs). Available at: https://www.atsdr.cdc.gov/toxfaqs/tfacts104.pdf. Accessed May 3, 2017.
44. Bernard A, Fierens S. The Belgian PCB/dioxin incident: a critical review of health risks evaluations. *Int J Toxicol.* 2002;21:333–340.
45. Agency for Toxic Substances and Disease Registry (ATSDR). ToxFAQs for polychlorinated biphenyls (PCBs). Available at: https://www.atsdr.cdc.gov/toxfaqs/tfacts17.pdf. Accessed May 2, 2017.
46. US Department of Commerce. National Oceanic and Atmospheric Administration. National Ocean Service. What are PCBs? Available at: http://oceanservice.noaa.gov/facts/pcbs.html. Accessed May 3, 2017.
47. Faroon OM, Keith S, Jones D, et al. Carcinogenic effects of polychlorinated biphenyls. *Toxicol Ind Health.* 2001;17:41–62.
48. US Environmental Protection Agency. Hudson River PCBs superfund site. Available at: https://www3.epa.gov/hudson/cleanup.html. Accessed May 2, 2017.
49. US Environmental Protection Agency. Polychlorinated biphenyls (PCBs). Health effects of PCBs. Available at: https://www.epa.gov/pcbs/learn-about-polychlorinated-biphenyls-pcbs#healtheffects. Accessed May 2, 2017.
50. Negri E, Bosetti C, Fattore E, et al. Environmental exposure to polychlorinated biphenyls (PCBs) and breast cancer: a systematic review of the epidemiological evidence. *Eur J Cancer Prev.* 2003;12:509–516.
51. Faroon OM, Keith S, Jones D, et al. Effects of polychlorinated biphenyls on development and reproduction. *Toxicol Ind Health.* 2001;17:63–93.
52. Hsu PC, Huang W, Yao WJ, et al. Sperm changes in men exposed to polychlorinated biphenyls and dibenzofurans. *JAMA.* 2003;289:2943–2944.
53. Schantz SL, Widholm JJ, Rice DC. Effects of PCB exposure on neuropsychological function in children. *Environ Health Perspect.* 2003;111: 357–376.
54. Jacobson JL, Jacobson SW. Intellectual impairment in children exposed to polychlorinated biphenyls in utero. *New Engl J Med.* 1996;335:783–789.
55. Hyperdictionary. Meaning of solvent. Available at: http://www.hyperdictionary.com/dictionary/solvent. Accessed May 2, 2017.
56. Lindbohm ML, Taskinen H, Sallmen M, et al. Spontaneous abortions among women exposed to organic solvents. *Am J Ind Med.* 1990;17:449–463.
57. Wernke MJ, Schell JD. Solvents and malignancy. *Clin Occup Environ Med.* 2004;4:513–527,vii.
58. Agency for Toxic Substances and Disease Registry (ATSDR). ToxFAQs for tetrachloroethylene. Available at: https://www.atsdr.cdc.gov/toxfaqs/tfacts18.pdf. Accessed May 2, 2017.
59. Mundt KA, Birk T, Burch MT. Critical review of the epidemiological literature on occupational exposure to perchloroethylene and cancer. *Int Arch Occup Environ Health.* 2003;76:473–491.
60. Agency for Toxic Substances and Disease Registry (ATSDR). ToxFAQs for 1,1,1-trichloroethane. https://www.atsdr.cdc.gov/toxfaqs/tfacts70.pdf. Accessed May 2, 2017.
61. Agency for Toxic Substances and Disease Registry (ATSDR). ToxFAQs for 1,1,2-trichloroethane. Available at: https://www.atsdr.cdc.gov/toxfaqs/tfacts148.pdf. Accessed May 2, 2017.
62. Agency for Toxic Substances and Disease Registry (ATSDR). ToxFAQs for trichloroethylene (TCE). Available at: https://www.atsdr.cdc.gov/toxfaqs/tfacts19.pdf. Accessed May 2, 2017.
63. Agency for Toxic Substances and Disease Registry (ATSDR). ToxFAQs for toluene. Available at: https://www.atsdr.cdc.gov/toxfaqs/tfacts56.pdf. Accessed May 2, 2017.
64. Agency for Toxic Substances and Disease Registry (ATSDR). ToxFAQs for acetone. Available at https://www.atsdr.cdc.gov/toxfaqs/tfacts21.pdf. Accessed May 2, 2017.
65. Dalton P, Wysocki CJ, Brody MJ, et al. Perceived odor, irritation, and health symptoms following short-term exposure to acetone. *Am J Ind Med.* 1997;31:558–569.
66. Agency for Toxic Substances and Disease Registry (ATSDR). ToxFAQs for benzene. Available at: https://www.atsdr.cdc.gov/toxfaqs/tfacts3.pdf. Accessed May 2, 2017.
67. US Environmental Protection Agency. Health effects notebook for hazardous air pollutants. Available at: https://www.epa.gov/haps/health-effects-notebook-hazardous-air-pollutants. Accessed May 2, 2017. Benzene. Available at: https://www.epa.gov/sites/production/files/2016-09/documents/benzene.pdf. Accessed May 2, 2017.
68. Canadian Centre for Occupational Health and Safety (CCOHS). Benzene: OHS answers. Available at: http://www.ccohs.ca/oshanswers/chemicals/chem_profiles/benzene.html. Accessed May 2, 2017.
69. Pyatt D. Benzene and hematopoietic malignancies. *Clin Occup Environ Med.* 2004;4:529–555, vii.
70. Yaris F, Dikici M, Akbulut T, et al. Story of benzene and leukemia: epidemiologic approach of Muzaffer Aksoy. *J Occup Health.* 2004;46:244–247.
71. Agency for Toxic Substances and Disease Registry (ATSDR). ToxFAQs for styrene. https://www.atsdr.cdc.gov/toxfaqs/tfacts53.pdf. Accessed May 2, 2017.

72. Agency for Toxic Substances and Disease Registry (ATSDR). Public health statement for vinyl chloride. July 2006. Available at: https://www.atsdr.cdc.gov/toxfaqs/tfacts20.pdf. Accessed May 2, 2017.
73. Bosetti C, La Vecchia C, Lipworth L, et al. Occupational exposure to vinyl chloride and cancer risk: a review of the epidemiologic literature. *Eur J Cancer Prev*. 2003;12:427–430.
74. National Institutes of Health. National Institute of Environmental Health Sciences, National Toxicology Program. 14th report on carcinogens: substance profiles: vinyl halides (selected) vinyl chloride. Available at: https://ntp.niehs.nih.gov/ntp/roc/content/profiles/vinylhalides.pdf. Accessed May 2, 2017.
75. Agency for Toxic Substances and Disease Registry (ATSDR). ToxFAQs for diethyl phthalate. Available at: https://www.atsdr.cdc.gov/toxfaqs/tfacts73.pdf. Accessed May 2, 2017.
76. Agency for Toxic Substances and Disease Registry (ATSDR). ToxFAQs for di-*n*-butyl phthalate. Available at: https://www.atsdr.cdc.gov/toxfaqs/tfacts135.pdf. Accessed May 2, 2017.
77. Washam C. Baby ills from beauty aids? *Environ Health Perspect*. 2001;109:A202.
78. Brown P, Krenn-Hrubec K. Issue Brief: Phthalates and children's products. Washington, DC: National Research Center for Women & Families; 2009.
79. Fanelli R, Zuccato E. Risks and benefits of PVC in medical applications. *Boll Chim Farm*. 2002;141:282–289.
80. Agency for Toxic Substances and Disease Registry (ATSDR). ToxFAQs for di(2-ethylhexyl) phthalate (DEHP). https://www.atsdr.cdc.gov/toxfaqs/tfacts9.pdf. Accessed May 2, 2017.
81. Cummings AM. Methoxychlor as a model for environmental estrogens. *Crit Rev Toxicol*. 1997;27:367–379.

© Jean-Luc Rivard/EyeEm/Getty Images

CHAPTER 8

Ionizing and Nonionizing Radiation

LEARNING OBJECTIVES

By the end of this chapter the reader will be able to:

- Define the terms ionizing radiation and nonionizing radiation.
- State the differences between ionizing and nonionizing radiation.
- Describe sources and types of ionizing and nonionizing radiation.
- Discuss the health effects of exposure to ionizing and nonionizing radiation.
- Describe major incidents in which the population was exposed unexpectedly to ionizing radiation.

Introduction

For many people the very term radiation evokes a feeling of trepidation. The proliferation of nuclear weapons and mishaps at nuclear power stations tend to amplify this sense of unease. Most of the time people think of ionizing radiation when referring to radiation. Nevertheless, radiation includes both ionizing and nonionizing radiation. Both forms of radiation share certain similarities but also demonstrate major differences regarding their sources and effects. In the first half of the chapter, we will focus on ionizing radiation, specifically the types of ionizing radiation, nomenclature used to describe its principal forms, and terms and methods related to its measurement. The chapter will review the major environmental sources of ionizing radiation and their possible health impacts. The second half of the chapter will cover the terminology used to describe the major categories of nonionizing radiation. The concluding topics will be the environmental sources of nonionizing radiation and their hypothesized health effects.

Some Preliminary Definitions

Let's begin with a few preliminary definitions. Generally speaking, radiation refers to energy that travels through space. The term **ionizing radiation** denotes radiation that "has so much energy it can knock electrons out of atoms, a process known as ionization. Ionizing radiation can affect the atoms of living things, so it poses a health risk by damaging tissue and DNA in genes. Ionizing radiation comes from radioactive elements, cosmic particles from outer space and x-ray machines."[1] Sources of ionizing radiation include those decaying radioactive materials that emit alpha and beta particles and gamma rays, medical procedures that employ radiopharmaceuticals and devices

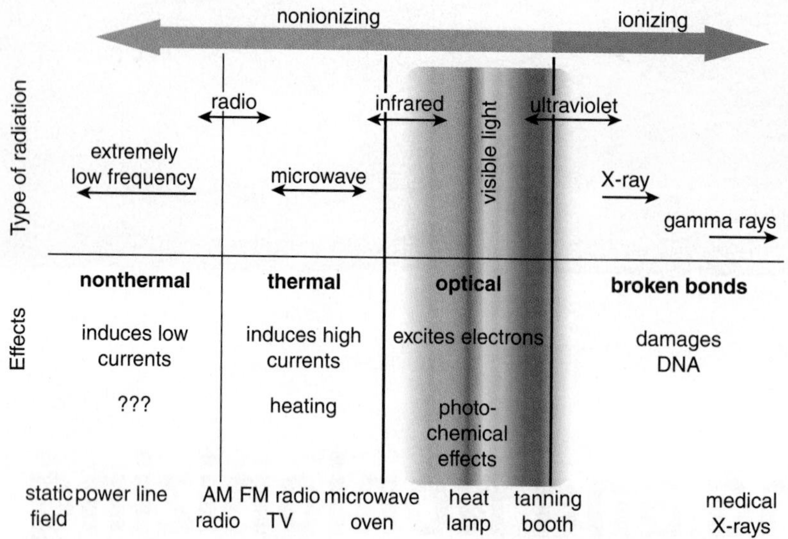

FIGURE 8.1 Types of radiation in the electromagnetic spectrum: ionizing and nonionizing radiation.

such as X-ray machines, and bodies in outer space that contribute to the earth's background radiation. (These types of radiation will be defined in more depth later in the chapter.)

In contrast to ionizing radiation, the term **nonionizing radiation** refers to "radiation [that] has enough energy to move atoms in a molecule around or cause them to vibrate, but not enough to remove electrons. . . . Examples of this kind of radiation are sound waves, visible light, and microwaves."[1] Other examples include radio frequency energy (e.g., radiation from radio antennas), optical and ultraviolet radiation (e.g., from the sun), infrared (heat) radiation, and low-frequency radiation (e.g., radiation from electric power lines). Potential health effects of exposure to nonionizing radiation range from annoyance to acute symptoms such as pain and swelling to serious eye injury and long-term consequences such as cancer.

As shown in **FIGURE 8.1**, ionizing radiation and nonionizing radiation together make up the electromagnetic spectrum, which forms an energy continuum from the low-level energy fields associated with electric power lines, to mid-level energy from radio stations, to high levels of energy from medical X-rays. The highest levels of ultraviolet radiation demarcate the dichotomy between ionizing radiation and nonionizing radiation. The electromagnetic spectrum covers four major energy levels described in terms of their effects: nonthermal, thermal, optical, and broken chemical bonds. We will begin with electromagnetic radiation on the right-hand side of the spectrum—ionizing radiation, which has sufficient energy to break chemical bonds.

Overview of Ionizing Radiation

Since Roentgen's discovery of X-rays in 1895, applications of radiation have proliferated in industry, biology, chemistry, medicine, and other fields such as nuclear weapons design. Exposure of human populations to ionizing radiation arises from two main sources: anthropogenic (human-made) and natural. Although many opportunities for the population's exposure to human-made ionizing radiation exist, natural ionizing radiation is responsible for most of the annual exposure and anthropogenic radiation for a minority of such exposure. Sources of natural ionizing radiation are cosmic rays, other forms of radiation that impinge upon earth from outer space, and radiation from geologic formations that contain radioactive elements (radioelements) such as uranium, from which radon gas is formed as a decay product. Also, radioelements are present naturally in the human body (an example is potassium-40). As noted previously, synthetic sources of ionizing radiation are medical X-rays and other procedures used in medicine (medical tests and therapies), consumer products, radioactive substances used in industry, nuclear power generators, and radioactivity (e.g., radioactive fallout) from the production and detonation (primarily for testing purposes) of nuclear weapons. Consumer products that sometimes contain radioactive substances include:

- Radioluminescent paints and signs
- Mantles of camping lanterns
- Colorful ceramic glazes

- Potassium-based salt substitutes
- Dentures
- Tobacco products
- Ionization-type smoke detectors

With the increasingly numerous applications of radiation, a good understanding of the health effects of radiation exposure has developed. Quite early in the use of X-ray machines, health authorities and others observed the occurrence of radiation burns and radiation sickness among technicians. Other early observations included the recognition of health problems such as skin cancer and other cancers associated with ionizing radiation exposure. During the early 1900s, the radioactive element radium was used as a paint to create "glow-in-the-dark" watch and other instrument dials. The dials were handpainted by workers who, before applying the paint, used "mouth tipping" to create a point on the brushes that were coated with radium. After these workers began to develop a form of bone cancer known as "radium jaw," the practice of mouth tipping was discontinued during the mid-1920s.[2] From these beginnings, the body of research information (from experimental and epidemiologic studies) regarding health effects of radiation exposure has become quite extensive, spanning a time period of more than a century.[3]

Types of Ionizing Radiation

Radiation and radioactivity are two important terms used in the context of ionizing radiation. The term **radiation** refers to "[t]he propagation of energy through space, or some other medium, in the form of electromagnetic waves or particles."[4] Radiation also includes neutrons, which are part of the nucleus of an atom and do not have an electrical charge. The phenomenon of **radioactivity** occurs during **radioactive decay**, which is "[r]eduction in activity of a quantity of radioactive material by disintegration of its atoms. Elements that undergo radioactive decay are said to be radioactive."[4]

Related to the concept of radioactivity are the terms *isotope, nuclide, radioisotope*, and *radionuclide*. An **isotope** is "each of two or more forms of the same element that contain equal numbers of protons but different numbers of neutrons in their nuclei, and hence differ in atomic mass but not in chemical properties; in particular, a radioactive form of an element."[5] For example, carbon is an atom (atomic number 6), and two isotopes of carbon are carbon-12 and carbon-14. There are more than one dozen radioisotopes of uranium (atomic number 92) plus several nonradioactive isotopes.

The term **nuclide** refers to "[a]ny species of atom that exists for a measurable length of time. A nuclide can be distinguished by its atomic mass, atomic number, and energy state."[6] Nuclides include radioactive and nonradioactive atoms. Carbon-14 is an example of a nuclide as well as an isotope.

A **radioisotope** is "[a] radioactive isotope. A common term for a radionuclide."[6] An example is the radioisotope of iodine, iodine-131 (^{131}I). A **radionuclide** is "[a] radioactive nuclide. An unstable isotope of an element that decays or disintegrates spontaneously, emitting radiation." In illustration, ^{131}I also is a radionuclide as well as a radioisotope. Often radionuclides are described in terms of their half-life, meaning "[t]he time in which half the (a large number of) atoms of a particular radioactive nuclide disintegrate."[6] The half-life of a radionuclide can range from a very brief time to millions of years. For example, ^{131}I has an eight-day half-life; the half-life of ^{227}U is 1.3 minutes and of ^{238}U, 4.47 billion years.

The definitions of the terms **alpha particle, beta particle, gamma rays**, and **X-rays** are shown in the following text box.

DEFINITIONS: ALPHA PARTICLE, BETA PARTICLE, GAMMA RAYS, AND X-RAYS

- *Alpha particle:* Alpha particles (α) are positively charged and made up of two protons and two neutrons from the atom's nucleus. Alpha particles come from the decay of the heaviest radioactive elements, such as uranium, radium, and polonium.... Alpha particles lack the energy to penetrate even the outer layer of skin, so exposure to the outside of the body is not a major concern. Inside the body, however, they can be very harmful. If alpha-emitters are inhaled, swallowed, or get into the body through a cut, the alpha particles can damage sensitive living tissue.
- *Beta particle:* Beta particles (β) are small, fast-moving particles with a negative electrical charge that are emitted from an atom's nucleus during radioactive decay..... Beta particles are more penetrating than alpha particles, but are less damaging to living tissue and DNA because the ionizations they produce are more widely spaced. They travel farther in air than alpha particles, but can be stopped by a layer of clothing or by a thin layer of a substance such as aluminum.... However, as with alpha-emitters, beta-emitters are most hazardous when they are inhaled or swallowed.

- *Gamma rays, or gamma radiation:* Gamma rays (γ) are weightless packets of energy called photons. Unlike alpha and beta particles, which have both energy and mass, gamma rays are pure energy. Gamma rays are similar to visible light, but have much higher energy.... [They form part of the electromagnetic spectrum.] Gamma rays are a radiation hazard for the entire body. They can easily penetrate barriers that can stop alpha and beta particles, such as skin and clothing. Gamma rays have so much penetrating power that several inches of a dense material like lead, or even a few feet of concrete, may be required to stop them. Gamma rays can pass completely through the human body; as they pass through, they can cause ionizations that damage tissue and DNA.
- *X-rays:* X-rays are similar to gamma rays in that they are photons of pure energy. X-rays and gamma rays have the same basic properties but come from different parts of the atom. X-rays are emitted from processes outside the nucleus, but gamma rays originate inside the nucleus. They also are generally lower in energy and, therefore, less penetrating than gamma rays. X-rays can be produced naturally or by machines using electricity.

Modified from US Environmental Protection Agency. Radiation protection. Radiation basics. Ionizing and non-ionizing radiation. Available at: https://www.epa.gov/radiation/radiation-basics. Accessed September 22, 2017.

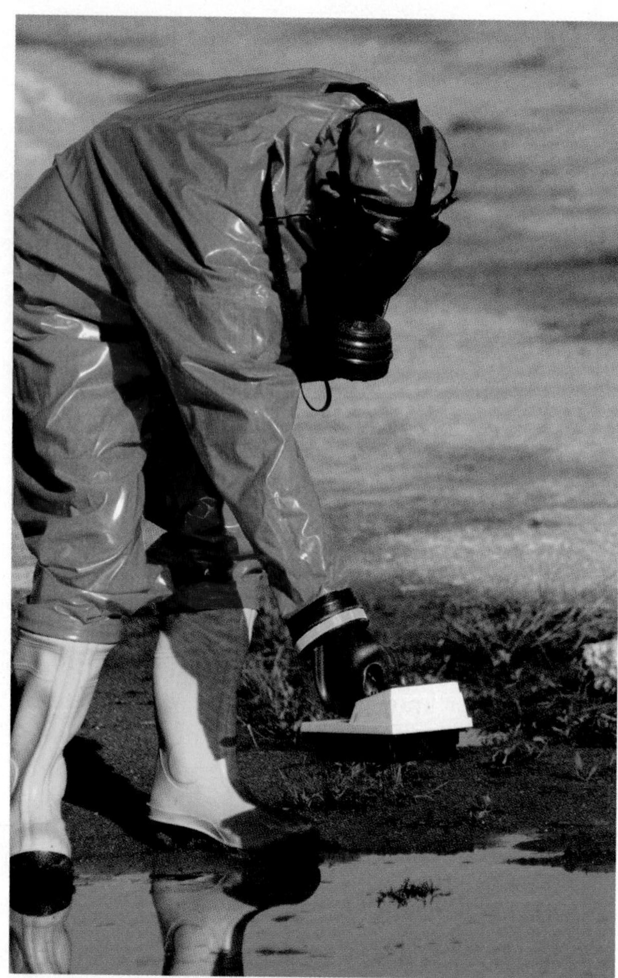

FIGURE 8.2 Measuring radiation during a NATO exercise.

Reproduced from North Atlantic Treaty Organization (NATO). NATO Media Library. Available at: http://www.nato.int/multi/photos/2003/m031007a.htm. Accessed February 5, 2017.

Measurement of Ionizing Radiation Dose Units

One widely used device for measuring radiation is the Geiger counter, a hand-held device shown in **FIGURE 8.2**. The frequency of audible clicks that the device emits when gas in a detector tube is ionized is correlated with the intensity of radiation exposure. Also, a visual readout meter indicates the rate of exposure to radiation. Other instruments may be used for monitoring forms of ionizing radiation such as gamma rays. A device that is used most commonly in the occupational environment is a film badge, which is worn by workers who are potentially exposed to gamma radiation. The badge contains dental X-ray film that darkens in proportion to the degree of exposure to radiation.[7]

The unit of energy associated with emissions from a radioactive material (radionuclide) is the electron volt (eV). The units used for measurement of the amount of radioactivity, dosage, and exposure are shown in **TABLE 8.1**. The term **common units**, which refers to an earlier system of radiation measurement, is used often in the United States. The abbreviation **SI (System International) units**, which denotes the International System of Units, has been implemented more recently and has replaced the older system of common units. The terms shown in the table are discussed in the text. Going across the top row of Table 8.1, we see the following terms: *radioactivity, absorbed dose, effective dose (dose equivalent)*, and *exposure*. These terms are defined in **TABLE 8.2**.

Returning to the first row in Table 8.1, we see the term *common units*. These units are the **curie**, **rad**, **rem**, and **roentgen**. The common units are defined in **TABLE 8.3**.

Table 8.1 contains the term *SI units*, which stands for the International System of Units. The SI units employ the terms *becquerel, gray*, and *sievert*. The definitions of the SI units and their equivalency in common units are shown in **TABLE 8.4**. Not shown in the table is the SI unit for exposure, coulomb/kilogram (C/kg).

TABLE 8.1 Ionizing Radiation Measurements

	Radioactivity	Absorbed Dose	Effective Dose (Dose Equivalent)	Exposure
Common units (US unit)	Curie (Ci)	Rad	Rem	Roentgen (R)
SI units	Becquerel (Bq)	Gray (Gy)	Sievert (Sv)	Coulomb/kilogram (C/kg)

Data from US Environmental Protection Agency. Radiation Protection: Radiation Terms and Units. Available at: https://www.epa.gov/radiation/radiation-terms-and-units. September 23, 2017.

TABLE 8.2 Definitions of Dose, Exposure, and Radioactivity

Term	Definition
Absorbed dose	Describes the amount of radiation absorbed by an object or person.
Effective dose (dose equivalent)†	Describes the amount of radiation absorbed by person, adjusted to account for the type of radiation received and the effect on particular organs.
Exposure	Describes the amount of radiation traveling through the air. Many types of radiation monitors [e.g., Geiger counters] measure exposure.
Radioactivity	Refers to the amount of ionizing radiation released by a material. Whether it emits alpha or beta particles, gamma rays, X-rays, or neutrons, a quantity of radioactive material is expressed in terms of its radioactivity (or simply its activity). This represents how many atoms in the material decay in a given time period.

† United States Nuclear Regulatory Commission. Measuring radiation. Available at: https://www.nrc.gov/about-nrc/radiation/health-effects/measuring-radiation.html. Accessed September 27, 2017.
Modified from US Environmental Protection Agency. Radiation Protection: Radiation Terms and Units. Available at: https://www.epa.gov/radiation/radiation-terms-and-units. September 23, 2017.

TABLE 8.3 Definitions of Common Units of Radiation

Common Unit	Definition
Curie (Ci)	The original unit used to describe the intensity of radioactivity in a sample of material. [1 Ci equals 3.7×10^{10} disintegrations per second.]
Rad (radiation absorbed dose)	A former unit of absorbed dose of ionizing radiation.
Rem (roentgen equivalent in man)	A measure of dose deposited in body tissue, averaged over the body. One rem is approximately the dose from any radiation corresponding to one roentgen of gamma radiation. [The term Mrem indicates one-thousandth of a rem. It is often used to measure radiation received from medical X-rays and background sources.]
Roentgen (R)	Unit of exposure measuring the ionizing ability of gamma radiation. [The term *ionizing* refers to the capability of producing ions, which are electrically charged atomic particles.]

Data from Lawrence Berkeley National Laboratory. Nuclear Science Glossary, Appendix A: Glossary of Nuclear Terms. Available at: http://www.lbl.gov/abc/wallchart/glossary/glossary.html. Accessed February 11, 2017.

TABLE 8.4 Definitions of the International System of Units (SI Units) and Their Equivalency in Common Units

SI Unit	Definition	Equivalency in Common Units
Becquerel (Bq)	(Corresponds to radioactivity) The SI unit of activity, which is defined as one disintegration per second	37 billion Bq (37 gigabecquerels) = 1 curie
Gray (Gy)	(Corresponds to absorbed dose) The SI unit of absorbed dose	1 gray = 100 rad; 1 rad = 1 centigray (cGy) = 0.01 Gy
Sievert (Sv)	(Corresponds to dose equivalent) The SI unit of dose equivalent	1 Sv = 100 rem [for X-ray or gamma radiation]

Data from Oak Ridge Institute for Science and Education, Radiation Emergency Assistance Center/Training Site (REAC/TS). *The Medical Aspects of Radiation Incidents*. 4th Edition. Oak Ridge, Tennessee: July, 2017.

The size or weight of a container or shipment does not indicate how much radioactivity is in it.

The amount of radioactivity in a quantity of material can be determined by noting how many curies of the material are present. This information should be found on labels and/or shipping papers.

More curies = a greater amount of radioactivity

A large amount of material can have a very small amount of radioactivity; a very small amount of material can have a lot of radioactivity.

For example, uranium-238 has 0.00015 curies of radioactivity per pound (0.15 millicuries), while cobalt-60 has nearly 518,000 curies per pound.

FIGURE 8.3 Measurement: how much radiation is present?

Reproduced from Oak Ridge Institute for Science and Education. Guidance for Radiation Accident Management, Radiation Emergency Assistance Center/Training Site (REAC/TS). Measurement: Activity: How much is present? Available at: http://orise.orau.gov/reacts/guide/measure.htm. Accessed March 24, 2010. These works were produced under contract number DE-AC05-06OR23100 between the U.S. Department of Energy and ORAU and is subject to the following license: A paid-up nonexclusive, irrevocable, worldwide license in such work to reproduce, prepare derivative works therefrom, distribute copies to the public and perform or publicly display by or for the government. All other rights relating to the works are retained by ORAU and/or the U.S. Department of Energy.

It is possible to express fractional units of radiation doses by using the following prefixes: milli (m) and micro (μ). In addition, the prefix kilo (k) is used. For example, when these prefixes are applied to a rad, an mrad is 1/1,000 of a rad; a μrad is 1/1,000,000 of a rad; a krad is 1,000 times one rad, or 1,000 rads. These prefixes may be applied to other measures of radiation (e.g., sievert and rem). Note that the amount of radioactivity (defined previously as curies or becquerels) present in a material does not depend on its weight or size but on the particular radioactive element or isotope. (See **FIGURE 8.3**.)

Health Effects of Exposure to Ionizing Radiation

The dose of radiation is associated with the amount of bodily tissue damage that may occur. Factors that govern the amount of exposure to radiation that a person receives (dose delivered) include:

- The total amount of time exposed to the radioactive source
- Distance from the radioactive source
- Degree of radioactivity (rate of energy emission) of a radioactive material

The amount of radiation dose that is absorbed—expressed as either grays or rads—may or may not be associated with acute health effects, depending upon individual susceptibility and amount of exposure as determined by the foregoing factors. The health effects of radiation exposure are described as nonstochastic (acute) and stochastic effects.

Nonstochastic Effects

TABLE 8.5 shows the nonstochastic (acute) effects of varying levels of radiation exposure. Among the acute effects levels of radiation exposure are tissue burns and radiation sickness (e.g., nausea, weakness, and loss of hair). At low levels, for example an exposure of 5 rems, radiation usually does not produce immediately detectable harm. The term **hormesis** refers to the belief that "[s]ome nonnutritional toxic substances also may impart beneficial or stimulatory effects at low doses, but at higher doses they produce adverse effects."[8(p13)]

TABLE 8.5 Nonstochastic Health Effects of Whole-Body Exposure for an Average Person and Time to Onset Following Acute Exposure

Exposure (rem)	Health Effect	Time to Onset
5–10	Changes in blood chemistry	
50	Nausea	Hours
55	Fatigue	
70	Vomiting	
75	Hair loss	2–3 weeks
90	Diarrhea	
100	Hemorrhage	
400	Possible death	Within 2 months
1,000	Destruction of intestinal lining	
	Internal bleeding	
	Death	1–2 weeks
2,000	Damage to central nervous system	
	Loss of consciousness	Minutes
	Death	Hours to days

Reproduced from US Environmental Protection Agency.

Consistent with the notion of hormesis, some scientists assert that low levels of radiation are beneficial to one's health. At high levels (e.g., an exposure of 350 to 400 rems or higher), ionizing radiation is capable of producing fatal injuries. The acute effects shown in Table 8.5 are cumulative, meaning that when a person experiences the health effect associated with an exposure at a higher level, the effects that occur at a lower level also will have been experienced. For example, subjects who experience hemorrhage at an exposure of 100 rems of radiation also will have had changes in blood chemistry.

Stochastic Effects

Stochastic effects are those associated with low levels of exposure to radiation over long time periods. The term *stochastic* refers to an increased probability of the occurrence of an adverse health event. Among the effects linked potentially to low levels of exposure to radiation are carcinogenesis and genetic damage such as changes in DNA. These changes may be inherited by the progeny of exposed individuals who have experienced DNA damage. In some cases the damage to DNA is so severe that it is not transmitted (i.e., the germinal cells are incapable of viable cell division).

Sources of Environmental Exposure to Ionizing Radiation

The human population is exposed to ionizing radiation from many sources. Ionizing radiation, produced from natural origins such as cosmic radiation (radiation from space) and uranium deposits, causes increases in background radiation levels in some geographic areas. A decay product of uranium, radioactive radon gas may seep into homes, thus exposing the residents. Although they are generally regarded as safe, nuclear power plants have been known to leak ionizing radiation into the environment under unusual conditions. Radioactive fallout from aboveground nuclear testing has exposed large populations to radiation. Still other exposure sources are decommissioned and abandoned nuclear weapons facilities and storage of nuclear wastes. Finally, medical X-rays and radiation diagnostic and therapy procedures produce exposures to ionizing radiation. **FIGURE 8.4** shows the relative proportions of different radiation sources to which the US population is exposed.

Naturally Occurring Radiation

The estimated average annual exposure of the world's population to naturally occurring radiation is 2.4 mSv[9] (0.0024 Sv = 0.24 rem, or 240 mrem). The sources of natural radiation include cosmic rays, terrestrial gamma rays, radionuclides that are present in the body, and radon gas. The relative distribution of the exposures from these sources is shown in **FIGURE 8.5**.

Uranium

A very common element in the earth's crust, uranium is present in soil and geologic formations almost universally. The uranium ores uraninite and pitchblende are found in large amounts in North America, Africa, and Australia and in lesser amounts on other continents. Nearly all plants, animals, and aquifers contain tiny amounts of uranium. Coal used to fuel electric-generating plants has numerous impurities including radionuclides of uranium (plus thorium and radioactive decay products), which are not destroyed during burning and become concentrated in leftover ash. Fly ash that escapes from power plants into the environment can expose nearby residents to

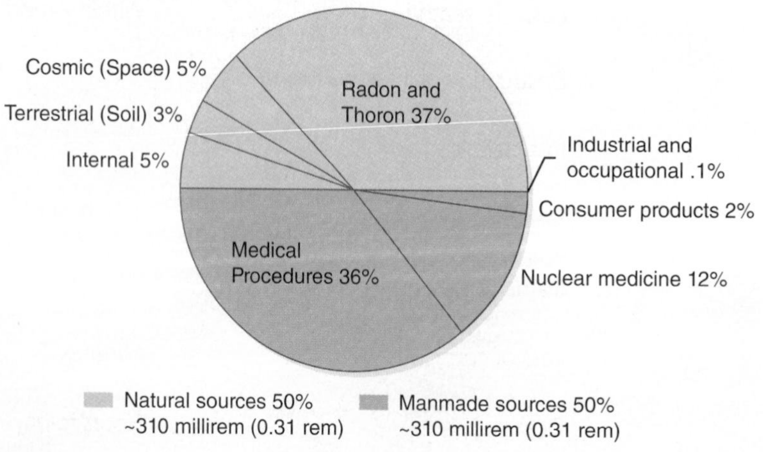

FIGURE 8.4 Sources of radiation exposure in the United States.

Reproduced from United States Nuclear Regulatory Commission. Available at: http://www.nrc.gov/images/reading-rm/basic-ref/glossary/ionizing-radiation.jpg. Accessed February 5, 2017.

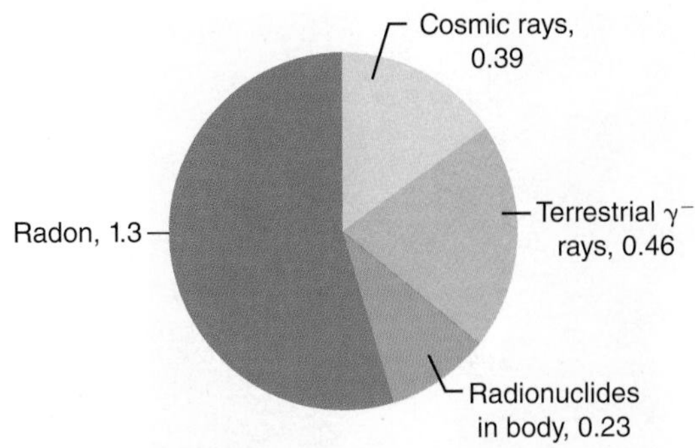

FIGURE 8.5 Annual effective doses to adults from natural sources of radiation (mSv).

Data from World Health Organization, International Agency for Research on Cancer. Ionizing Radiation, Part 1: X- and Gamma (γ)-Radiation, and Neutrons. *IARC Monographs*. 2000;75:78.

radioactivity from radionuclides; pollution control technology aids in removal of fly ash and radioactive particles from power plant emissions.

Estimates suggest that the bodies of typical non-occupationally exposed persons who live in Europe and the United States contain about 22 μg of uranium. The three isotopes of uranium found in the natural environment are ^{234}U, ^{235}U, and ^{238}U, the last being the most common form. When ^{238}U that is present in the foundations of building sites and in building materials decays, it produces radon gas.

Radon

An inert, colorless, and extremely toxic gas, radon is produced by the decay of radium and uranium. One of the radioactive decay products of uranium is radium-226 (^{226}Ra), which decays into radon-222 (^{222}Rn). Suspicions regarding the carcinogenic effect of radon were aroused by studies of underground miners who were exposed to high levels of radon gas and had elevated risks of lung cancer.[10] Classified as a Class A carcinogen (known human carcinogen) by the US Environmental Protection Agency (EPA) and other agencies, radon gas causes no immediate symptoms but has been associated with long-term adverse health effects. Environmental radon produces one of the largest sources of human exposure to ionizing radiation and is thought to be one of the leading causes of lung cancer in the United States. Up to 20% of US lung cancer cases may be caused by radon despite evidence that cigarette smoking, asbestos exposure, and urban air pollution are the leading causes of lung cancer. According to the World Health Organization (WHO), lifetime risk of lung cancer shows a dose-response increase with increasing radon exposure and no known threshold at the low end of exposure.[11]

In addition to the association demonstrated between lung cancer and radon exposure among miners as well as among experimentally exposed laboratory animals, radon exposure that occurs in people's residences has been linked to lung cancer. An Iowa case-control study of lung cancer reported a positive association with cumulative radon exposure.[12] The area was selected for study because some Iowa homes—especially in the basement and on the first floor levels—have the highest indoor radon levels in the United States.

An estimated 15,000 to 22,000 lung cancer deaths are believed to be caused by radon in indoor air. According to EPA documents, approximately 50% of homes in some areas have high levels of radon gas. During the winter when windows are kept closed, radon levels increase. The EPA recommends that homeowners have their residences tested to verify that unsafe levels of radon gas are not present.

Another possible source of exposure to radon is drinking water. Waterborne radon may be released into the air, increasing the risk of lung cancer if inhaled; ingestion of water that contains radon may increase the risk of cancers in internal organs such as the stomach. The primary origins of waterborne radon are underground aquifers in some locales; not all water from underground sources contains radon, and surface water is generally free from radon contamination. Only a small percentage of the radon found in air originates from drinking water.

Extraterrestrial Radiation

Primary cosmic rays originate from outer space (e.g., the earth's galaxy and the sun) and interact with the earth's atmosphere to produce secondary cosmic rays.[9] Cosmic rays have the ability to penetrate and cross the human body easily. Radiation from the sun, called solar cosmic radiation, coincides with solar flares and sunspots on the surface of the sun. These solar events have the ability to increase ground-level radiation on the earth's surface. On February 23, 1956, the largest solar event observed to date caused the levels of neutron radiation to increase by a factor of 3,600% over the normal background radiation level.[9] Passengers and crew members of high-altitude aircraft receive higher exposures to the radiation from solar particle events than do persons who remain on the ground. However, a European study of airline pilots failed to detect adverse health effects associated with exposures to ionizing radiation.[13]

Nuclear Facilities—Nuclear Accidents

Other potential sources for exposure of the population to ionizing radiation are nuclear facilities, which include weapons production plants, test sites, and nuclear power plants. **FIGURE 8.6** illustrates the nuclear power plant on the Susquehanna River at Three Mile Island, Pennsylvania, which was the location of an

FIGURE 8.6 Three Mile Island nuclear power plant.

unintentional release of radiation into the environment on March 28, 1979.

The Three Mile Island accident involved a series of mechanical failures and mistakes that led to the accidental release of radiation into the environment when the core of the reactor experienced a partial meltdown. Following the incident, research studies suggested that the amount of human exposure to ionizing radiation was very low (less than 100 mrems).[14,15]

The association between exposure to ionizing radiation that occurred among residents who lived downwind of the nuclear reactor and health effects such as cancer remains a controversial issue. One study reported a modest association between postaccident cancer rates and proximity to the power plant. Although postaccident cancer rates increased during 1982 and 1983, and subsequently declined, plant emissions did not appear to account for the observed increase.[16] The major consequence of the Three Mile Island accident was that no new nuclear power plants have been built in the United States since 1979.[17] Despite the fact that the incident at Three Mile Island has become a symbol for the antinuclear movement, the nuclear power industry argues that the containment of the meltdown reinforced the notion that the installed safety systems were effective.

In contrast to the Three Mile Island release, the nuclear power plant accident at Chernobyl, Ukraine on April 26, 1986 was a major public health disaster that exposed a large portion of the European population to ionizing radiation. One consequence of the disaster was to call into question the safety of older nuclear reactors that use a similar design and are operating in eastern Europe. Two explosions at the power plant and an associated fire required 9 days to extinguish. One report stated that "a cloud of radioactivity rained down on western Europe as the fallout from the Chernobyl disaster spread with the wind. Plumes of radioactive waste covered the countryside, homes, people, land and food crops."[18(p16)]

Radioactive materials, primarily iodine and cesium, were deposited over wide swaths of Europe, including sections of the former Soviet Union, Sweden, Austria, Switzerland, and parts of Germany and northern Italy.[18] According to one account, "Half the reactor's caesium-137 and iodine-131 escaped and 3–5% of the remaining radioactive material escaped. . . . The accident killed 31 people; a further 209 suffered acute radiation sickness. About 185,000 people received radiation doses of over four times the internationally accepted limit; a much larger number received some increased radiation dose."[18(p16)] An exclusion zone with a 30-km (18.6-mile) radius around the reactor was created eventually. In some parts of the exclusion zone, ground radiation levels from the deposited ^{137}cesium were extremely high. In order to limit the spread of radioactivity from the damaged reactor, officials constructed a steel and concrete sarcophagus that encases the reactor. Noteworthy is the fact that the Chernobyl facility lacked an effective containment structure, which is not the case for US nuclear power reactors.

Carcinogenesis is one of the main health phenomena investigated as a consequence of the release of radioactivity from the Chernobyl power plant. Researchers have examined the occurrence of excess cases of cancer in neighboring sections of Europe that received radioactive fallout as well as those in close proximity to the reactor. A number of methodological issues cloud possible associations between cancer and exposure to radioactivity from Chernobyl. No method is available to distinguish between cancer (e.g., thyroid cancer and leukemia) that has resulted from the Chernobyl disaster and cancer that originates from other exposures. Exposures to natural background radiation and medical radiation among some European populations may exceed the radiation doses from the Chernobyl release. In addition, the residents of Europe generally were exposed to low doses of radiation. In summary, researchers have been unable to demonstrate any increase in rates of morbidity and mortality from cancer that can be assigned definitively to radioactivity from Chernobyl.[19] Another health effect of the Chernobyl accident could be the production of birth defects such as congenital anomalies. However, data from the EUROCAT epidemiologic surveillance of congenital anomalies have not suggested an increase in central nervous system anomalies or Down syndrome in western Europe.[20]

Closer to the Chernobyl nuclear facility, however, there were marked increases in thyroid cancer as soon as 4 to 5 years after the accident. The Gomel region of Belarus in the former Soviet Union lies immediately to the north of Chernobyl. There was a sharp increase in the number of thyroid cancer cases among children in Gomel from 1 to 2 per year during 1986–1989 to 38 in 1991.[21] Subsequent research also demonstrated increases in the number of thyroid cancer cases in three radiation-affected countries (Belarus, Ukraine, and western parts of Russia). The number of such cases had totaled 5,000 by the year 2010, with 15,000 additional cases expected in the next 50 years among those exposed as children. Persons exposed as adults do not appear to demonstrate an increase in thyroid cancer rates.[22] Another outcome has been increases in the number of congenital abnormalities among children.[18]

The 30th anniversary of the Chernobyl reactor meltdown was observed during 2016. Children and adolescents who resided in areas most contaminated with radioactivity have been the subject of continuing follow-up since the meltdown. For several months following the catastrophe, these youngsters consumed milk laden with radioactive iodine. Among this exposure group, approximately 11,000 thyroid cancer cases developed as of 2016. However, scientists are uncertain regarding the proportion of these cases that can be linked to consumption of radioactive iodine; this is because thyroid cancer can have other causes.

Noncancer morbidity among radiation clean-up workers at the reactor site included radiation-induced cataracts among those who were exposed to radiation. In addition, information from 2006 suggested elevated cardiovascular mortality among employees tasked with clean-up of radioactive materials. More recent analyses suggested that cardiovascular mortality increased among persons who had low doses of exposure to ionizing radiation.[23]

With respect to proximity to nuclear installations other than Chernobyl, two studies did not show a relationship between cancer rates and location near a nuclear facility. For example, no excess of cancer deaths was found among populations near the Rocky Flats nuclear weapons plant in Colorado[24] or California's San Onofre nuclear power plant (which has a formidable containment structure).[25] A systematic literature review concluded that although many studies of the health effects of community exposure to radiation were statistically sound, they did not provide adequate quantitative estimates of radiation dose that could be used to assess dose-response relationships.[26]

A major release of widespread radioactivity occurred in the Urals region of Russia.[27] In 1948, the former Soviet Union began production of its first atomic bomb at a plutonium (defined in the textbox) separation plant near the city of Chelyabinsk. From 1949 to 1956, the plant discharged substantial amounts of nuclear waste into the nearby Techa River. An open depot for the storage of radioactive wastes contaminated Lake Karachay, which later evaporated and left residual radioactive dusts that affected the population downwind from the site. In 1957, a tank that stored nuclear wastes suffered a thermochemical explosion that spread radioactivity into the surrounding area. Exposed residents developed chronic radiation sickness, changes in blood parameters, and increased morbidity and mortality rates from leukemia and cancer.

On March 11, 2011, a strong earthquake followed by a tsunami caused a major disaster at the Fukushima

WHAT IS PLUTONIUM?

"Plutonium is a radioactive, metallic element with the atomic number 94.... Plutonium is created in a nuclear reactor when uranium atoms, specifically uranium-238 absorb neutrons. Nearly all plutonium is [human]-made." An isotope of plutonium—Plutonium-239—is a fissile material used as one of the primary fuels in nuclear weapons.

Data from United States Nuclear Regulatory Commission. The United States Nuclear Regulatory Commission's Science 101: what is plutonium? Available at: https://www.nrc.gov/reading-rm/basic-ref/students/science-101/what-is-plutonium.pdf. Accessed May 9, 2017.

CATALOG OF NUCLEAR ACCIDENTS

Since the 1940s, there have been numerous mishaps that involved nuclear materials. In some instances, these resulted in fatalities or exposed many people to ionizing radiation. Some involved the manufacture of nuclear weapons, weapons testing, transport of nuclear devices, and accidents at electric power generating stations. Estimates indicate that more than 50 nuclear warheads and 10 nuclear reactors lie on ocean floors as a consequence of ships that have sunk and aircraft that have crashed or unintentionally dropped nuclear bombs. A few examples follow:

- In the 1950s, several bomber aircraft accidentally dropped nuclear weapons or crashed with such weapons on board. In one incident on March 11, 1958, a B-47E en route to England dropped a nuclear weapon on Mars Bluff, South Carolina. The unarmed device landed on the property of Walter Gregg, destroying his house. Several of Gregg's family members were injured when the bomb's TNT-based trigger exploded on impact with the ground, causing a giant crater. Several nearby structures also were damaged.[a]
- On April 10, 1963, the nuclear submarine USS *Thresher* with its reactor core sank to the ocean floor in the vicinity of Boston, Massachusetts; all crew members perished.[b]
- During the 1970s, incidents included the sinking of a Soviet nuclear submarine, releases of radiation from nuclear power plants (e.g., Three Mile Island accident, discussed in text), and leakage of radioactive water from a nuclear fuel reprocessing plant.
- On April 26, 1986, the Chernobyl accident occurred (refer to text).
- On September 30, 1999, Japan's worst nuclear accident up to that time occurred when a tank located at a uranium reprocessing plant in Tokaimura exploded, killing two workers. The incident exposed reactor workers, rescue personnel, and community residents to excess levels of radiation.[c]
- On August 12, 2000, the Russian nuclear submarine *Kursk* sank in the Barents Sea.[d]
- On March 11, 2011, a strong earthquake followed by a tsunami caused a major disaster at the Fukushima Daiichi Nuclear Power Station in northern Japan. As a result of damage to the cores of three of the six reactors at the site, radiation and hydrogen were released. The hydrogen caused three of the buildings that housed the damaged reactors to explode. The nearby populations were evacuated and all nuclear power plants in the country were ultimately shut down.[e]

Data from [a]*Columbia Star*. Atomic Bomb dropped on Florence, S.C., March 11, 1958. Available at: http://www.thecolumbiastar.com/news/2008-03-21/news/036.html. Accessed February 10, 2017;
[b]Arlington National Cemetery. USS Thresher (SSN-593), 1961–1963. Available at: http://www.arlingtoncemetery.net/uss-thresher.htm. Accessed February 10, 2017;
[c]R Toohey. The criticality accident at Tokaimura, Japan. *Health Phys*. 2002;82(suppl):S163;
[d]Guardian. The wreck of the Kursk begins its journey home. Available at: http://www.guardian.co.uk/world/2001/oct/09/kursk.russia. Accessed February 10, 2017; and
[e]National Research Council. *Lessons Learned from the Fukushima Nuclear Accident for Improving Safety of U.S. Nuclear Plants*. Washington, DC: The National Academies Press; 2014.

Daiichi Nuclear Power Station in northern Japan. Refer to the text box for more information on this event and selected other nuclear incidents. Although the reactor failure at Fukushima was a major event, it produced much lower emissions of radioactive materials than at Chernobyl. Also, most of the radionuclides fell onto the Pacific Ocean instead of on the land. No deaths resulted from the acute effects of radiation in Fukushima; this was not the case in Chernobyl.[28]

In summary, the preceding box titled "Catalog of Nuclear Accidents" provides an overview of mishaps at nuclear power plants and several other incidents that involved nuclear weapons and nuclear submarines.

Nuclear Bomb Explosions

There are several types of nuclear bombs, including atomic bombs and hydrogen bombs. Explosion of an atomic bomb is a fission reaction (splitting apart of uranium or plutonium atoms). In contrast, hydrogen bombs exploit the fusion reaction of hydrogen. The following text box describes the effects of a nuclear explosion.

Information presented in the text box indicates that persons near the site of a nuclear explosion would experience thermal burns and radiation-induced skin injuries. **FIGURE 8.7** illustrates injuries to the skin caused by thermal burns and radiation burns (e.g., beta burns from beta radiation emitters). For a beta injury to occur, beta particle emitters (e.g., from the products of nuclear fission) must come into direct contact with skin and remain there for an extended time period. The injuries shown in Figure 8.7 resulted from the nuclear explosions that occurred in Japan during World War II.

Also as noted in the following text box, one of the major by-products of the detonation of a nuclear weapon is radioactive fallout. When a nuclear weapon explodes above ground, it produces large amounts of radioactive dusts that later become dispersed over the earth downwind from the burst. Some of the materials may even be injected into the upper atmosphere and, as they settle out, persist as

EFFECTS OF A NUCLEAR EXPLOSION

The effects of a nuclear weapon explosion are the blast (damage to or destruction of buildings and those in them), heat (destruction or injury by high temperatures or fire), intense light (damage to eyesight), and ionizing radiation (causing acute radiation syndromes of different degrees of severity). A detonation similar in size to the bomb that was dropped on Hiroshima (15 kilotons) could result in a high damage zone of a few kilometers radius where there would be few or no survivors. There would also be a significant number of injured people over a considerably larger area. However, the effects of nuclear explosion generated by the blast could be reduced somewhat by preparation in advance and prompt action during the event.

Detonating a nuclear weapon generates an intense and immediate pulse of radiation, and gives rise to longer-lasting radioactive contamination. Objects close to the explosion can be made radioactive by neutrons from the explosion, but farther away objects become radioactive from fallout of radioactive debris generated by the weapon. The initial "prompt" radiation would directly affect everybody up to a few kilometers, whereas fallout can irradiate people over a much larger area in a number of ways. A radioactive cloud is spread by the wind, possibly over large distances. Rainfall can wash out some of the radioactive materials from the air and enhance fallout deposits onto surfaces where it has rained. Once this radioactive cloud deposits as particles on surfaces, it can be picked up on clothing and other objects or inhaled or spread further. The radioactive fallout can also contaminate food and water supplies if it deposits on crops, animal feed, or in drinking water sources. The hazard from this fallout reduces with time but could last for many months or more.

Radioactive materials may be inhaled from the air or ingested in food causing internal radioactive contamination and damaging internal organs of the body. External exposure can be due to radioactive materials deposited on the ground, buildings, or even on our clothes or skin. If radioactive materials come into contact with the skin, they may cause radiation burns.

Persons near the site of detonation surviving the initial blast and thermal effects would be exposed to high levels of radiation and could develop symptoms of radiation sickness. Early symptoms of high radiation exposure include nausea, vomiting, diarrhea, fatigue, and headache. While thermal burns may appear within minutes, radiation induced skin injury and other early symptoms develop over days and weeks. Depending on the severity of exposure, victims may experience different symptoms and medical care will be needed. Following years or even some decades after exposures, an increased risk of cancer among the wider exposed population could be expected.

There will be considerable damage to the infrastructure of society. Communication systems may be severely disrupted and normal modes of transport may not be available. Consequently, there would be great difficulty in providing effective medical treatment to the large numbers of casualties.

Modified with permission from World Health Organization Radiation and Environmental Health Unit. Health Protection Guidance in the Event of a Nuclear Weapons Explosion, WHO/RAD Information Sheet, February 2003. Available at: http://helid.digicollection.org/pdf/s13464e/s13464e.pdf. Accessed February 10, 2017.

residual hazards. A fission reaction may produce more than 300 radioactive products that have varying half-lives, ranging from a fraction of a second to many months or years. Integral to a nuclear explosion is the emission of neutrons that induce radioactivity in elements in the soil near the site of the explosion (ground zero). This radioactivity, which generally is confined to a limited area, decays gradually over time as beta and gamma radiation are released from the activated elements. Very fine particles (size range 10 nanometers to 20 micrometers) may be pushed up high into the atmosphere and can be dispersed worldwide. The heavier particles, such as vaporized water and soils, are incorporated into a radioactive cloud that settles over a wide geographic area of several hundred square miles, depending on wind and weather conditions. The initial level of radioactivity downwind and close to the explosion may be approximately 300 Gy per hour, the LD_{50} being a cumulative dose of 4.5 Gy. After the radioactivity from a nuclear weapon dissipates over several weeks, the nearby fallout zone becomes safe to enter, but crops and food animals taken from the area may be unsuitable to consume.

Among the components of fallout from weapons testing is radioactive iodine, which may become concentrated in the thyroid gland, potentially increasing the rate of thyroid cancer. There have been many studies of the health effects of fallout from aboveground atmospheric testing of nuclear weapons at the Nevada test site in the United States during 1951 to 1958. One cohort study followed young people who lived in proximity to the test site during infancy and childhood.[29] Three cohorts, similar in demographic and lifestyle characteristics, were selected. Two cohorts (exposure cohorts) were from Washington County, Utah, and Lincoln County, Nevada, both on the west side of the test site, close to the site, and in the pathway of the

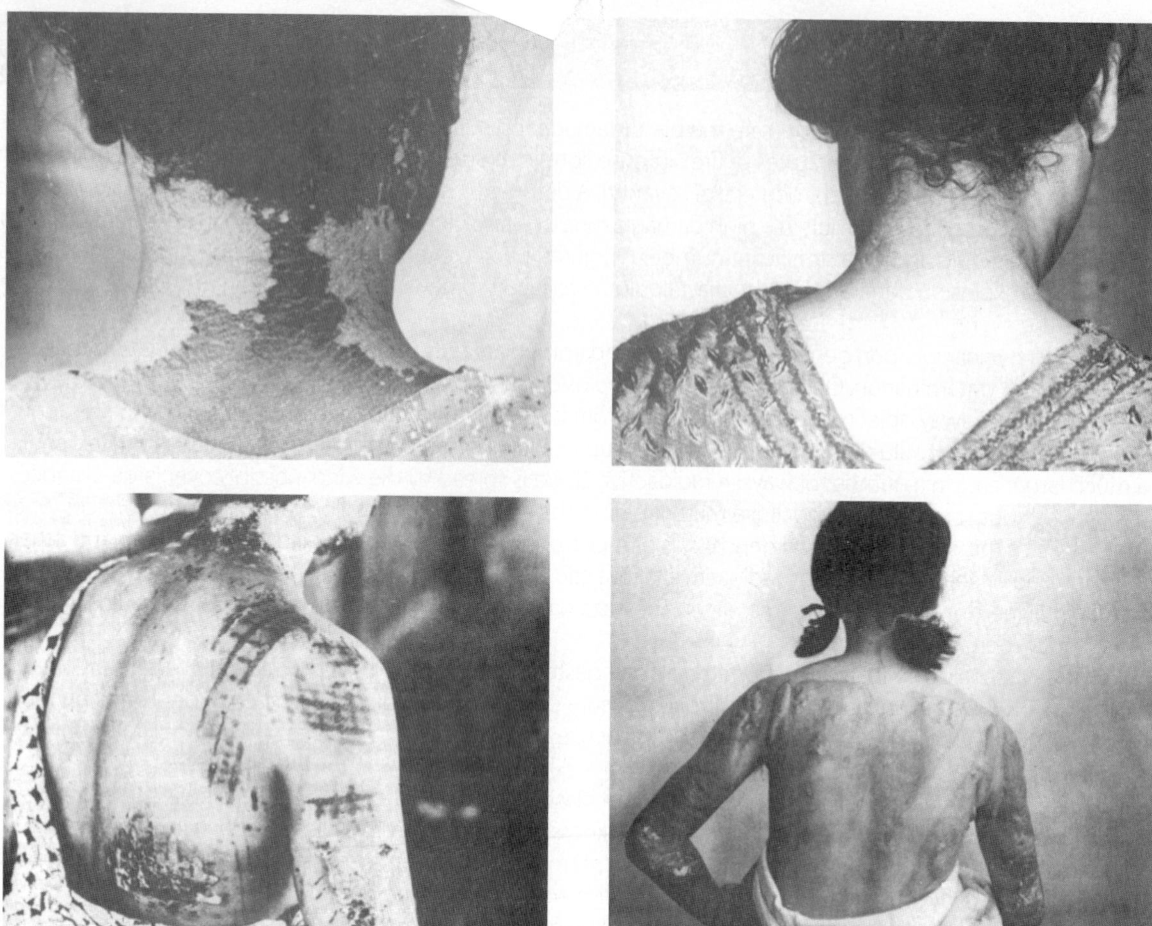

FIGURE 8.7 Upper left: Beta burn on the neck 1 month after exposure. Upper right: Beta burn 1 year after exposure. Lower left: Thermal (flash) burn—the patient's skin is burned in a pattern corresponding to the dark portions of a kimono worn at the time of the explosion. Lower right: Thermal (flash) burn—the skin under the areas of contact with clothing is burned and the protective effect of thicker layers of clothing can be seen on the shoulders and across the back.

Modified and reproduced from Glasstone S, Dolan PJ, eds. *The Effects of Nuclear Weapons*. 3rd ed. Washington, DC: US Department of Defense and the Energy Research and Development Administration; 1977:568 (Figure 12.71), 569 (Figure 12.72), 595 (Figure 12.158a), 597 (Figure 12.161a)

heaviest fallout. A third unexposed cohort (control cohort) was selected from Graham County, Arizona, far to the south of the test site. At 12 to 15 years and 30 years after the heaviest fallout, there was a slight but nonsignificant increase in rates of thyroid cancer among the two exposure cohorts in comparison with the control cohort. Thus, it was concluded that living near the Nevada test site did not produce a statistically significant increase in thyroid neoplasms.

Studies of the long-term consequences of exposure to radiation from the atomic bombs dropped in Japan confirmed increases in the risk of breast cancer, especially if exposure occurred between the ages of 10 and 19 years.[30] Yet, some individuals seem to be resistant to the effects of radiation. Ms. Asa Takii, who survived the Hiroshima atomic bomb blast, was Japan's oldest person when she died in 1998 at the age of 114 years.[31] The detonation (August 6, 1945) killed her husband and family members and buried her for several days before she was rescued. Hiroshima commemorated the 71st anniversary of the detonation on August 6, 2016. On that date, the 174,080 survivors of the detonation were on average about 80 years old.[32]

Dirty Bombs

After the September 11, 2001 terrorist attacks in New York and Washington, DC, concern was raised over the possibility of additional incidents, particularly from a device known as a dirty bomb. A dirty bomb differs from an atomic bomb or other nuclear bomb in that the explosion, which disperses radioactive materials, is caused by conventional explosives and not by nuclear fission or fusion. In order to construct a dirty bomb, a terrorist would combine explosives with radioactive material from various sources and explode the bomb in a densely populated urban zone. Most likely, the effects of such an explosion would be confined to a small area. As a result, the most significant injuries likely to be caused by a dirty bomb would be from the explosives and explosion itself, rather than from radiation exposure. The main impact of the dirty bomb would probably be psychological by causing the population to panic. The nuclear material, which would be dispersed, possibly could contaminate buildings and land for a period of time. However, the amount of radiation produced probably would not be sufficient to cause severe acute illnesses. Nevertheless, the long-term effects of radiation exposure might include increased incidence of cancers among persons exposed to the radiation. The cost of cleaning up after the explosion could also be extremely expensive.[33–35] The text box describes a dirty bomb, or **radiological dispersal device (RDD)**, which has been the subject of much attention from the media.

FACT SHEET: DIRTY BOMBS

A "dirty bomb" is one type of a radiological dispersal device (RDD) that combines conventional explosives, such as dynamite, with radioactive material. The terms *dirty bomb* and *radiological dispersal device* (RDD) are often used interchangeably in the media. Most RDDs would not release enough radiation to kill people or cause severe illness—the conventional explosive itself would be more harmful to individuals than the radioactive material. However, depending on the situation, an RDD explosion could create fear and panic, contaminate property, and require potentially costly cleanup. Making prompt, accurate information available to the public may prevent the panic sought by terrorists.

A dirty bomb is in no way similar to a nuclear weapon or nuclear bomb. A nuclear bomb creates an explosion that is millions of times more powerful than that of a dirty bomb. The cloud of radiation from a nuclear bomb could spread tens to hundreds of square miles, whereas a dirty bomb's radiation could be dispersed within a few blocks or miles of the explosion. A dirty bomb is not a "weapon of mass destruction" but a "weapon of mass *disruption*," where contamination and anxiety are the terrorists' major objectives....

The extent of local contamination would depend on a number of factors, including the size of the explosive, the amount and type of radioactive material used, the means of dispersal, and weather conditions. Those closest to the RDD would be the most likely to sustain injuries due to the explosion. As radioactive material spreads, it becomes less concentrated and less harmful. Prompt detection of the type of radioactive material used will greatly assist local authorities in advising the community on protective measures, such as sheltering in place, or quickly leaving the immediate area. Radiation can be readily detected with equipment already carried by many emergency responders. Subsequent decontamination of the affected area may involve considerable time and expense.

Immediate health effects from exposure to the low radiation levels expected from an RDD would likely be minimal.

Reproduced from United States Nuclear Regulatory Commission. Fact Sheet on Dirty Bombs. Available at: https://www.nrc.gov/reading-rm/doc-collections/fact-sheets/fs-dirty-bombs.pdf. Accessed February 10, 2017.

Medical Uses of Ionizing Radiation

After accounting for the amount of radiation received from natural background sources, experts note that exposure of the population to ionizing radiation from medical procedures is the second largest source of exposure.[9] Approximately half of radiation exposure comes from natural background sources; about 40% of exposure has been attributed to medical exposure.[36] The risks of exposure to medically related ionizing radiation need to be balanced against the benefits that can be accrued from its helpful applications.

Medical devices that use radioactivity are essential for many common diagnostic and therapeutic regimens. These medical procedures include the use of X-ray machines, nuclear medicine, and radiation therapy. Not surprisingly, the number of procedures that employ radiation is substantially greater in developed countries than in less developed countries. In addition, people over the age of 40 in comparison with younger persons tend to be the recipients of a greater number of procedures that involve radiation. Radiation exposure during medical procedures does not affect only the patient; the medical personnel who administer the procedures also may be exposed. Some members of the public may be exposed to radiation that emanates from patients who have undergone treatment. Nevertheless, the life-serving effects of nuclear medicine procedures appear to greatly exceed their risks to the public.

Medical X-rays and Fluoroscopy

The X-ray machine has been employed since the early 1900s for diagnosis and therapy in medicine and dentistry. A dental X-ray produces a typical effective dose of radioactivity of 2 to 3 mrems; a chest X-ray delivers

5 to 10 mrems. More recently, newer diagnostic procedures (e.g., computed tomography [CT]) have been developed that require exposure to relatively high doses of radiation. A CT scan of the abdomen exposes the subject to 1,000 mrems. In developed countries, technical developments such as the use of shielding have helped to minimize the amount of radiation exposure from X-ray machines.

Another diagnostic procedure known as fluoroscopy uses X-rays to generate real-time images of the interior of the body. Fluoroscopy has a number of important applications, such as angiography, biopsy, placement of drainage tubes, and use during orthopedic procedures. An example of a procedure that uses fluoroscopy involves the administration of barium for visualization of the colon. During the procedure, the patient receives a barium enema, which involves the placement of a solution that contains barium into the colon. The barium solution is a strong absorber of X-rays and helps to produce images that can be captured on film.[9]

Diagnostic Nuclear Medicine

In this application, healthcare personnel inject liquid radionuclides directly into the patient's body. Also, the patient may inhale gaseous radionuclides. Particular radionuclides have an affinity for specific target organs. After the nuclides are injected or inhaled, one may determine whether an organ is functioning properly or whether it is infused properly with blood. Among the uses of such techniques are diagnosis of cancer and tests of thyroid function. The most commonly used radionuclides include ^{123}iodine and ^{131}iodine (liquids) and ^{133}xenon (gas).

Radiation Therapy

The objective of radiation therapy is to kill cancerous cells in specific bodily sites, called the target volume. The three types of radiation therapy are teletherapy, brachytherapy, and nuclear therapy. In teletherapy, a beam of radiation originating from outside the patient is aimed at a lesion. Brachytherapy involves the placement of encapsulated radioactive material into a tumor. Nuclear therapy requires the oral ingestion or intravenous injection of radionuclides that travel to the site of a lesion in a target organ. A possible adverse side effect of radiation therapy is the increased risk of cancer developing outside the immediate area targeted for treatment. For example, radiation treatments for acne in the 1950s were linked to bone cancers.

THE GOIÂNIA INCIDENT

In the Goiânia incident, authorities believe that scavengers dismantled a metal canister from a radiotherapy machine at an abandoned cancer clinic and left it in a junkyard. During the dismantling procedure the metal capsule that contained the caesium-137 source was ruptured. Over the next week, several hundred people in Goiânia were exposed to the caesium-137, but did not know it. Some children and adults, thinking the caesium powder was "pretty," even rubbed it over their bodies. Others inadvertently ate food that had been contaminated with the radioactive powder. After 1 week, a public health worker correctly diagnosed radiation syndrome when a sufferer visited a clinic. The Brazilian Nuclear Energy Commission sent in a team and they discovered that over 240 persons were contaminated with caesium-137, four of whom later died. The accident also contaminated homes and businesses and this required a major clean-up operation.

Reproduced from International Atomic Energy Agency. IAEA Press Release 2002/09, Inadequate Control of World's Radioactive Sources. Available at: https://www.iaea.org/newscenter/pressreleases/inadequate-control-worlds-radioactive-sources. Accessed February 10, 2017.

Nuclear Waste Disposal

Nuclear wastes can present a significant public health hazard if not properly secured. In Goiânia, Brazil, during September 1987 local scavengers removed radioactive ^{137}cesium from a radiotherapy machine. An account of the incident is presented in the text box titled "The Goiânia Incident."

Nonionizing Radiation

Nonionizing radiation is classified as the form of radiation on the left half of the electromagnetic spectrum shown in Figure 8.1, which portrays the following types of nonionizing radiation as we move up the energy spectrum: extremely low frequency (power line), radio (AM, FM, TV), microwave, infrared, visible light, and ultraviolet. The upper limit of ultraviolet radiation merges with the ionizing radiation portion of the electromagnetic spectrum.

Radiation from the electromagnetic spectrum is measured in wavelengths and cycles per second (called hertz). Variations in the units assigned to hertz are as follows:

- Hertz (Hz)—cycles per second (e.g., 60 Hz)
- Kilohertz (KHz)—Hz times 1,000 (e.g., 100 KHz = 100,000 Hz)

- Megahertz (MHz)—Hz times 1 million (e.g., 50 MHz = 50 million Hz)
- Gigahertz (GHz)—Hz times 1 billion (e.g., 75 GHz = 75 billion Hz)

Energy at the low end of the electromagnetic spectrum has long wavelengths and low hertz; conversely, short wavelengths and high frequency characterize the energy at the upper end of the spectrum.[1] Here are some additional examples of nonionizing radiation:

- Extremely low frequency (ELF) radiation—ELF radiation has frequencies of 100 Hz or less and wavelengths of 100 million meters or longer. One source of ELF radiation is high-tension power lines that run overhead; ELF radiation from power lines is in the range of 50 to 60 hertz. The frequency of alternating current electric power transmission in Europe and many countries is 50 Hz; in the United States, it is 60 Hz. When alternating currents pass through electric power transmission lines, sometimes they produce weak electromagnetic fields (EMFs).
- AM, FM, VHF-TV radio waves—These have frequencies between 1 million and 100 million hertz (1 MHz to 100 MHz) and wavelengths that range from 1 to 100 meters.
- Microwave radiation—Occurs in telecommunications equipment, radar, and ovens for heating food; frequencies are approximately 2.5 billion hertz (2.5 GHz) and wavelengths are approximately 1 hundredth of a meter long.
- Infrared (IR) radiation (IRA, IRB, IRC)—The source of heat in heat lamps used for keeping food warm.
- Visible light radiation.
- Ultraviolet (UV) radiation (UVA, UVB, UVC).

The human population is exposed ever more frequently to anthropogenic (human-made) sources of nonionizing radiation, such as radios and radio communications transmitters, cellular phones, medical equipment, microwave ovens, heat sealers, and radar transmitters. Even without the anthropogenic sources, nonionizing radiation impinges upon the human population from outer space and from the earth itself. According to one expert, "Environmental exposure to anthropogenic magnetic fields and radiation (EMFs) has steadily increased throughout this century. Today everyone is exposed to a complex mix of EMFs, both at home and at work, from the generation and transmission of electricity, from domestic appliances and industrial plant[s], and from telecommunications and broadcasting."[37(p185)]

Overview of Health Effects Associated with Nonionizing Radiation

The electromagnetic spectrum (nonionizing radiation) is subdivided into the nonthermal, thermal, and optical segments. The health effects of exposure to nonionizing radiation vary somewhat depending upon the segment of the spectrum considered and may be differentiated by their established and hypothesized short-term and long-term effects. Consider the findings that have accumulated regarding specific types of nonionizing radiation (e.g., extremely low frequency radiation, radiofrequency radiation, and microwave radiation).

Extremely low frequency (ELF) radiation does not appear to produce many discernible short-term health effects. However, when we consider forms of nonionizing radiation that have higher frequency levels in comparison with ELF radiation (i.e., radiofrequency and microwave radiation), we find that the primary immediate health effect produced is heating of the body. Radiofrequency radiation produces effects on or near the skin's surface; in comparison, microwave radiation penetrates more deeply into the body, causing internal layers to heat up. Microwave radiation also can cause interference with implanted pacemakers. The fact that nonionizing radiation in the radiofrequency to microwave range produces heat makes this type of radiation useful for many applications such as heating food and in manufacturing processes, for example, those that use microwaves to power heat-sealing tools.

Sources of Exposure to Nonionizing Radiation

Extremely Low Frequency Radiation

Extremely low frequency (ELF) radiation is of the type in the range of 50 to 60 hertz that originates from electric power poles, wiring in the walls of buildings, and some electrical appliances. When associated with electric wiring and high-tension electrical lines, the source of ELF radiation is alternating electrical currents that flow through the circuits. Electricity that flows through transmission lines may produce electromagnetic fields (EMFs). For example, **FIGURE 8.8** illustrates electric power lines that can produce extremely low frequency–associated EMFs. Because electrical power lines are designed to minimize power loss by keeping the energy inside the line, only very minute amounts of energy are released from the lines. Nevertheless, most communities prohibit

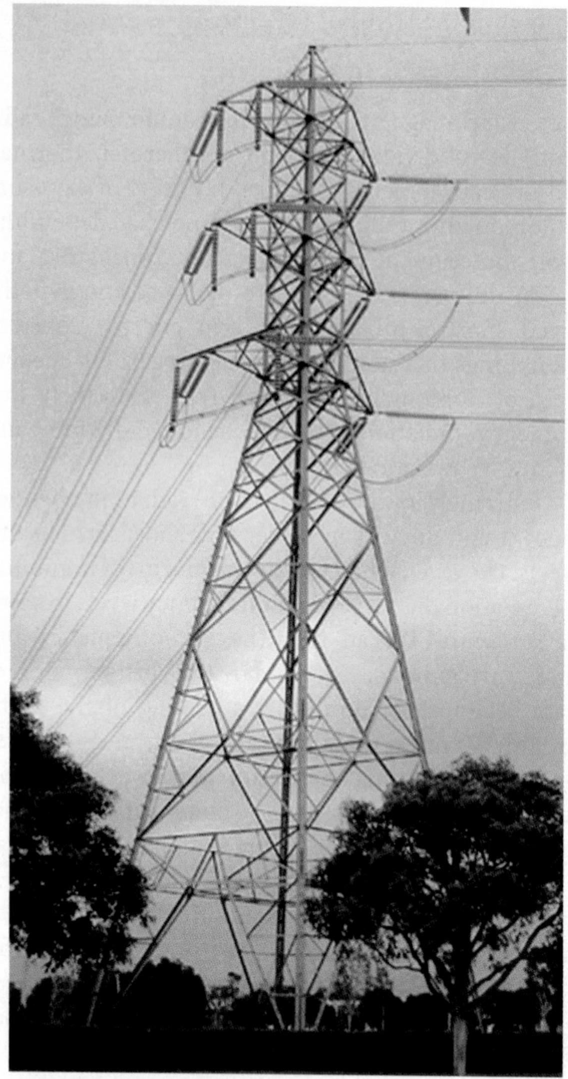

FIGURE 8.8 High-tension power lines. Environmental health studies have examined their potential health effects.

construction of homes directly under high-tension lines because of concerns about health effects found in some epidemiologic studies. These findings have suggested that there are associations between EMFs from power lines and childhood cancers. The amount of ELF radiation emitted from the wiring and appliances inside homes is probably negligible.

Epidemiologic studies that have examined the health effects of exposure to ELF radiation have produced conflicting findings. Although many of the studies have reported negative findings, some research has reported low-level associations between ELF radiation exposure among persons who live near power lines and cancer (e.g., childhood leukemia and brain cancers in children and adults). The International Agency for Research on Cancer considers EMFs emitted from ELF sources to be potentially carcinogenic.[38] The US EPA noted "that it can no longer categorically state that there is no effect from this radiation. Growing public concern about reported leukemia risk (especially in children), a possible connection with brain cancer, and potential interference with pacemakers has raised serious questions about the carcinogenicity of electromagnetic radiation."[39(p1447)]

Investigators have pursued the following three lines of research on possible linkages between cancer and EMFs from high-voltage power lines:[40]

- Associations between childhood cancer and residence near power lines
- Correlations between adult cancers and exposure to such radiation in occupations—an example being electrical contracting
- Associations between cancers among adults and exposure to power lines

The EMF and Breast Cancer on Long Island Study (EBCLIS) explored the possible relationship between breast cancer and women's exposure to EMFs.[41] During the course of this case-control study, researchers visited the Long Island, New York homes of the study participants (576 breast cancer patients and 585 controls). EMF measurements in several areas of the houses (e.g., bedroom, front door, and the most lived-in room) were taken. These measurements were used to estimate the strengths of EMFs in the 60-Hz range. The EBCLIS study reported no association between the strength of EMF levels and risk of breast cancer.

Norwegian investigators performed a case-control study of women to determine whether exposure to electromagnetic fields such as those found at home and at work was associated with female breast cancer.[38] The study examined women who had residential exposure to EMFs from electric power lines located close to their homes and those who had occupational exposure to magnetic fields generated by 50-Hz currents. Rather than measuring residential exposure directly, researchers assessed exposure to EMFs on the basis of residence in a defined geographic corridor in proximity to a high-voltage power line. A total of 1,830 breast cancer cases and 3,658 controls participated in the study. The investigators reported a 60% increase in risk of breast cancer among women who had residential exposure to EMFs. Occupational exposure to EMFs did not appear to be related to breast cancer.

Another study, also conducted in Norway, assessed whether exposure to EMFs from high-voltage power lines was associated with increased incidence of hematological cancers in adults.[42] This research was a case-control study in which persons (n = 1,063) who resided in a corridor that had exposure to high-voltage power lines were matched with controls according to year of birth, sex, and municipality of residence. As there were only a few nonsignificant associations with some forms of hematological cancers (e.g., chronic lymphocytic leukemia), the researchers indicated that no firm conclusions could be drawn.

A nationwide cohort study of Finnish adults searched for an association between residence near high-voltage power lines and risk of cancer.[40] The study examined a cohort of 383,700 adult subjects who had lived for a period of 20 years within 500 meters (547 yards) of overhead power lines. The outcome variable was cancer of all sites as well as specific cancer diagnoses (e.g., stomach, colon, and nervous system). The researchers concluded that there was no association between ELF magnetic fields and adult cancers. The study's authors state, "The results of the present study suggest strongly that typical residential magnetic fields generated by high voltage power lines are not related to cancer in adults."[40(p1050)]

New Zealand researchers studied the relationship between EMF exposures and risk of childhood cancer.[43] Their case-control study examined 303 cases among children who ranged from newborn to 14 years of age. The cases were matched with an equivalent number of controls. Measurements of EMFs were taken in the children's bedrooms and the rooms in which they spent most of their time. Other EMF exposure sources that were considered were mothers' use of electric blankets and other electrical appliances during pregnancy. Similarly, children's use of such appliances also was evaluated. Risk of leukemia was one of the few outcomes associated significantly with EMF exposure, and the association was significant only for children who had the highest exposure levels. The investigators indicated that the findings should be interpreted with caution because of the small sample sizes employed in the investigation.

In summary, the research literature suggests that findings are inconsistent, not definitive, with respect to EMF exposure and cancer. "Overall, the few biological effects that may result from exposure to EMFs at levels normally encountered by members of the public have a tendency to be small in magnitude, short in duration and reversible. Effects may also be seen under very specific exposure and test conditions. This suggests that any risk to health would be small and possibly negligible."[37(p191)]

The long-term health effects ascribed to EMF exposure differ from the immediate effects and are open to debate, or, at the very least, not well understood. Jauchem and Merritt reported, "In general, the biological effect of EMF has been a controversial subject. Many highly speculative notions have become entrenched in the popular press; some unsubstantiated claims in the scientific literature are widely referenced and often misquoted in the lay literature."[44(p895)]

Radio Frequency Radiation

When an electric charge alternates (oscillates back and forth) very rapidly in a radio antenna, it produces radio waves, which move at the speed of light. The effect is similar to dropping a pebble into a smooth lake and causing ripples on the surface. The analogy breaks down in the sense that, in comparison with ripples on bodies of water, the electromagnetic waves oscillate at much higher frequencies; these waves also are referred to as electromagnetic radiation. **Radiofrequency (RF) radiation** covers the range of approximately 0.5 MHz to 300,000 MHz. (Refer to **TABLE 8.6**.) In comparison with ELF radiation, RF radiation produces higher amounts of energy. As a result, one who ventures too closely to a strong source of RF radiation may experience thermal effects such as heating.[45] Under certain circumstances, irreversible effects such as eye damage (cataracts) can occur.

Cancer is one health outcome that has been investigated in relationship to RF radiation. Several of these investigations have suggested that there may be an increased risk of some forms of cancer among persons who have had RF exposure; however, these findings are inconsistent.[46] "The evidence is weak in regard to

TABLE 8.6 Examples of Frequencies in the Radio Range

Radiofrequency Radiation	Frequencies
AM radio	0.5 to 1.5 MHz
Amateur radio	3 to 30 MHz
FM, VHF-TV	50 to 150 MHz
UHF-TV	300 to 3,000 MHz
Microwave ovens	2,450 MHz
Radar, microwave communication	3,000 to 300,000 MHz or 3 to 300 GHz

Data from Valberg PA. Radio frequency radiation (RFR): the nature of exposure and carcinogenic potential. *Cancer Causes Control*. 1997;8:324.

its inconsistency, the weak design of the studies, the lack of detail on actual exposures, the limitations of the studies in their ability to deal with other likely factors, and in some studies there may be biases in the data used. Whereas the current epidemiologic evidence justifies further research to clarify the situation, there is no consistent evidence of any substantial effect on human cancer causation."[46(p166)]

Cell Phones

The cell phone is a very low-power apparatus that transmits in the radiofrequency range of 900 to 1,800 MHz. As of 2000, there were an estimated 92 million and 500 million cellular telephones in use in the United States and worldwide, respectively.[47] Since 2000. these numbers have risen dramatically: As of 2013, surveys revealed that 91% of the population in the United States used cell phones.[48] Approximately 6.9 billion wireless telephone subscriptions were tallied worldwide in 2014.[49] The widespread popularity of cell phones means that even small adverse health effects could have substantial implications for population health.[50] Cell phone antenna banks are themselves a source of RF radiation. (See **FIGURE 8.9**.)

FIGURE 8.9 Cell phone antenna.

When communicating via a cell phone, the individual places the device close to his or her head. It has been suggested that RF radiation from cell phones can cause "hot spots" inside the head.[45] The use of cell phones has raised the public's concern regarding several adverse effects, such as headaches, memory loss, sleep disturbance, nausea, dizziness, and forms of cancer such as brain tumors.[51] One legal case that reached national attention in the United States was a 1993 lawsuit filed in Florida by a man on behalf of his wife. The husband alleged that cell phone radiation caused his wife to die from a brain tumor. The case was subsequently dismissed when the court determined that the evidence was insufficient to show a causal link between cell phone use and brain tumors.[50] At present the hard evidence on the health effects of cell phone use is scant.[46]

In general, epidemiologic research findings do not support a clear relationship between use of cellular phones and morbidity and mortality. One study examined mortality differences between users of cellular phones, which usually require that the transmission antenna be held close to the head, and users of mobile telephones, which have an antenna that is located at a distance from the head. The age-specific mortality rates of the two types of telephone users were similar.[52] A case-control study of intracranial tumors of the nervous system did not substantiate a causal relationship between these forms of cancer and use of handheld cellular telephones.[47] Risk of intracranial tumors was not shown to increase with increasing use of cellular phones (e.g., more than 60 minutes per day for 5 years or more). Nevertheless, these and many other epidemiologic studies have not had an opportunity to examine health effects over a very long time period. A group of Swedish researchers conducted case-control studies on the relationship between certain types of brain tumors and use of mobile phones and cordless phones. The researchers found an association between the use of such phones and certain types of brain tumors (i.e., astrocytoma and acoustic neuroma). The highest risk level was found among subjects who had latency periods greater than 10 years.[53]

The National Toxicology Program asserted that the continuing study of radiofrequency radiation is essential for the following reasons:

- "Widespread human exposure.
- Current exposure guidelines are based largely on protection from acute injury from thermal effects.
- Little is known about potential health effects of long-term exposure to radiofrequency radiation.
- Data from human studies are inconsistent. Additional studies are being conducted.

Many people are concerned that cell phone radiation will cause cancer or other serious health effects. While current scientific evidence has not conclusively linked cell phones with any health problems, NTP and other scientific organizations recognize that additional data are needed."[48]

In summary, "[t]oday, the only health problem known to be related to the use of cell phones is an increase in traffic accidents among voluble drivers. Nevertheless, the popular press tends to put a sensational spin on the idea that cell phones contribute to brain cancer, and anecdotal reports of headaches, skin numbing, and memory loss due to cell phone use have helped to heighten public fears."[50(pA475)]

FIGURE 8.10 California sun worshiper.

Microwave Radiation

Microwave radiation, which is produced by radar, gained popularity during World War II because of its applications in devices used to track aircraft. This form of radiation also is used to heat foods. When the use of radar became more frequent, concerns about the health effects of radar exposure in occupational settings spurred an increase in the number of health research efforts. Over the years, the power emitted by radar devices has continued to increase, raising alarm about the possible health effects that might occur among exposed personnel. Standards now limit whole-body exposure to radar in occupational settings to 100 W/m^2.[54] (W/m^2 refers to watts per square meter—the amount of energy impacting the surface of the skin.) The exposure limit is believed to protect workers from ocular effects (lens opacities such as cataracts and clouding of the lenses of the eyes), one of the principal adverse health outcomes associated with microwave radiation exposure.

Ultraviolet Radiation

Ultraviolet radiation (UVR) is a form of nonionizing radiation in the optical range. Sources of UVR include welders' arcs, lamps used for tanning beds, some flood lamps used in photography, halogen desk lamps, lightning, and electrical sparks. However, one of the most common sources of exposure to ultraviolet radiation is sunbathing, which is regarded as a popular and healthful pastime by many residents of sunny climes, for example, California and Florida. (Refer to **FIGURE 8.10**.)

Evidence from epidemiologic research suggests strongly that sun exposure is associated with skin cancers of various forms.[55] (More information on skin cancer will be provided later in this section.) In the United States, areas located in lower geographic latitudes (e.g., New Orleans, Atlanta) that have higher levels of ambient UVR tend to have higher incidence rates for skin cancer than areas that fall in higher latitudes and have lower levels of UVR (e.g., Seattle, Minneapolis-St. Paul).[55] The thinning of the earth's ozone layer also may be responsible for increasing ambient levels of UVR.

UVR coming from the sun is subdivided into UVA, UVB, and UVC, depending upon the wavelength of the light. The wavelengths of these three forms are as follows: UVA—315 to 400 nanometers (nm); UVB—280 to 315 nm; and UVC—100 to 280 nm.[56] (Note that 1 nanometer equals 1 billionth of a meter.)

Many people enjoy some amount of sun exposure. However, as noted, excessive exposure can produce harmful consequences for one's health. To help protect exposed persons from excessive amounts of sunlight, the ultraviolet (UV) index provides a measure of the amount of UV exposure that one is likely to have on a given day. (Refer to the following text box.)

The health effects suspected of being associated with UVR include both dermal and ocular conditions. The former encompass temporary effects such as burns and longer-term consequences such as photoaging of the skin, nonmelanoma skin cancer (NMSC), and malignant melanoma (MM). The former encompass temporary effects such as burns and longer-term consequences such as photoaging of the skin. The latter (i.e., ocular effects) include acute temporary blinding, retinal damage, lens opacities and MM of the pigmented portion of the eye. Of the three forms of UVR, UVB is considered to be the most harmful to human health because it is the most effective in producing short-term and long-term adverse health effects.

THE UV INDEX

The UV Index provides a forecast of the expected risk of overexposure to UV radiation from the sun. [See **TABLE 8.7**.] The National Weather Service calculates the UV Index forecast for most zip codes across the US, and EPA publishes this information. The UV Index is accompanied by recommendations for sun protection and is a useful tool for planning sun-safe outdoor activities.

Ozone depletion, as well as seasonal and weather variations, cause different amounts of UV radiation to reach the earth at any given time. Taking these factors into account, the UV Index predicts the level of solar UV radiation and indicates the risk of overexposure on a scale from 0 (low) to 11 or more (extremely high). A special UV Alert may be issued for a particular area, if the UV Index is forecasted to be higher than normal.

TABLE 8.7 UV Index Scale

UV Index Number	Exposure Level
0 to 2	Low
3 to 5	Moderate
6 to 7	High
8 to 10	Very high
11 or more	Extreme

Data from US Environmental Protection Agency. UV Index Scale. Available at: https://www.epa.gov/sunsafety/uv-index-scale-0. Accessed February 10, 2017.

Data from US Environmental Protection Agency. UV Index Scale. Available at: https://www.epa.gov/sunsafety/uv-index-scale-0. Accessed February 10, 2017.

Dermal Effects

The acute effects of UVR exposure upon the skin include erythema (skin reddening), inflammation, pain, swelling, tanning, photosensitization, phototoxicity, and photoallergic reactions. These reactions develop several hours after exposure. Among sensitive persons, intense sunlight exposure for short time periods may produce reddening of the skin (similar to a first-degree burn). Longer exposures to intense sunlight may result in more severe reactions such as swelling, blistering, and peeling of the skin. Thickening of the skin and tanning caused by changes in melanin are defensive responses that protect the skin from further injury.

Sunlight-associated photosensitivity refers to heightened sensitivity to sunlight; one of the causes of photosensitivity of the skin is previous exposure to sunlight, particularly the ultraviolet component. A sunlight-associated **photoallergy** is a type of dermatitis (intense itching rash) that is activated by sunlight, which sensitizes the skin to allergens. Sunlight-associated **phototoxicity** denotes a severe reaction to sunlight that occurs sometimes in conjunction with drugs (e.g., antibiotics and diuretics) that a person may be taking. A sunburned area of skin that blisters, indicating that a second-degree burn has occurred, is one symptom of phototoxicity.

UVR exposure is linked to a number of long-term effects, which include photoaging, premalignant lesions, and malignant lesions such as skin cancer.[57] These effects may not appear for many years and tend to be cumulative. Persons who have spent a great deal of time outdoors over a period of years and who have not had adequate sun protection will be more likely to experience adverse effects from sunlight exposure than those who either tend to stay indoors or protect themselves with sun-blocking lotions and appropriate clothing. Photoaging of the skin is characterized by a leathery appearance and several important changes such as the occurrence of brown spots and loss of elasticity. The effects of photoaging are different from those of normal aging and can be observed when the areas of skin exposed to sunlight are compared with nonexposed areas in the same person. The exposed skin areas such as the faces of some elderly persons may have developed brown spots (hyperpigmentation), whereas nonexposed areas such as the legs appear smooth and without signs of hyperpigmentation.

Another adverse long-term health effect attributed to UVR exposure is skin cancer. The three types of skin cancer are squamous cell carcinoma (SCC), basal cell carcinoma (BCC), and malignant melanoma (MM). The two forms of nonmelanomatous skin cancer—SCC and BCC—account for about 40% of all cancers in the United States.[58] The occurrence of all three forms of skin cancer has shown an increasing trend.

Malignant melanoma is the most serious and potentially fatal form of skin cancer. In 2017, the incidence of malignant melanoma had risen to more than 87,000 cases annually and caused the deaths of more than 9,700 patients. **FIGURE 8.11** presents data on the incidence and mortality from MM in the United States between the years 1975 and 2013. The highest incidence and mortality rates occurred among white men and women.

This most dangerous variety of skin cancer also is one of the most common types of cancer among young adults. Melanomas demonstrate well-known

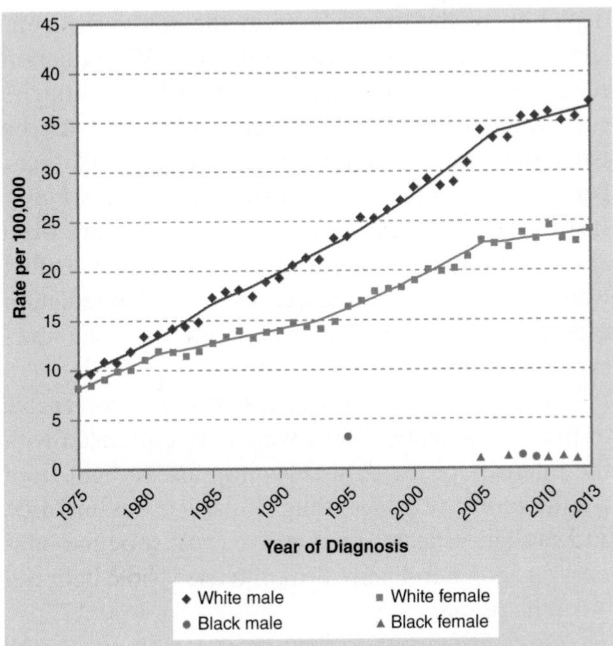

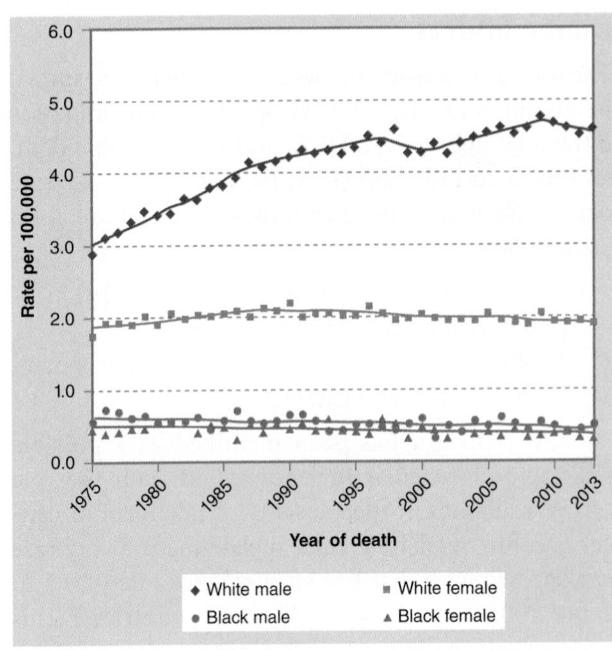

FIGURE 8.11 Incidence and mortality of melanoma of the skin, 1975–2013.

Reproduced from National Cancer Institute. Surveillance, Epidemiology, and End Results Program. Available at: https://seer.cancer.gov/faststats/. Accessed February 13, 2017.

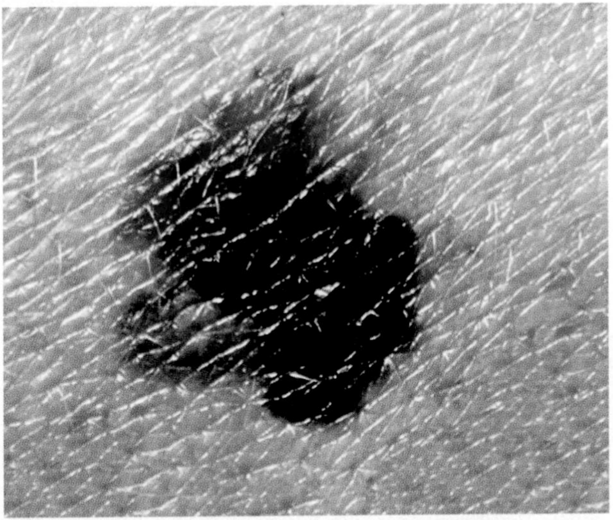

FIGURE 8.12 Malignant melanoma.

Courtesy of National Cancer Institute.

characteristics including lesions with irregular borders that may present with multiple colorations. (Refer to **FIGURE 8.12**.) The lesions, which may progress rapidly, often are vulnerable to injury and bleed easily. In many instances, the originating site of a malignant melanoma is a mole that demonstrates some of the foregoing characteristics. Men who develop MM are more likely than women to develop lesions on the head, neck, and trunk; women are more likely to develop MM on the lower limbs.[59]

The causes of malignant melanoma are not well understood. However, several risk factors have been established, including skin tone and latitude of residence. White race is a risk factor for MM. People who have fair complexions with blue eyes and light hair (e.g., blond or red) are at higher risk for MM than those who have darker skin. MM is uncommon among Asians and blacks. Apparently, the amount of melanin (which causes brown pigmentation) in the skin is correlated with protection from UVR. As noted previously, another risk factor for MM is latitude of residence. Among fair-skinned persons, the incidence of MM increases with decreasing latitude (living closer to the equator). The world's highest incidence of MM is found among Australia's white population. Certain occupations carry risk of skin cancer. For example, roofers who spend a great deal of time outdoors may incur increased risk of skin cancer. The reflective materials used in roofing surfaces tend to amplify the amount of exposure to UVR.

Spending too much time in the sun without adequate protection carries many health risks for the skin and eyes. Nevertheless, tanning booths remain a popular way for people to acquire a "healthy" bronzed appearance. In the United States, the indoor tanning industry generates over $1 billion in annual sales. Most tanning booth operators now use tanning machines that emit UVA primarily and have discontinued the once widely used UVB. Despite these modifications, the use of tanning booths is discouraged because of the risks of photoaging of the skin, damage to the eyes, and some of the forms of skin cancer. Phototoxic and photoallergic reactions can be activated by UVA radiation from tanning booths.

Ocular Effects

The optical spectrum and its adverse effects are shown in **FIGURE 8.13**. The optical spectrum encompasses ultraviolet light (UVA, UVB, and UVC), visible light, and infrared radiation (IRA, IRB, and IRC). Adverse ocular effects associated with these forms of radiation are as follows:

- UV radiation (particularly UVB radiation)—photokeratitis, a burn of the cornea
- Visible and infrared radiation—retinal burns, corneal burns, and cataracts

UV radiation has been identified as a possible risk factor for ocular melanoma, although the role of UV radiation in the causality of this form of cancer is controversial.[60] Ocular melanoma is a very rare cancer diagnosis that has an incidence rate of 0.5 to 1 per 100,000 per year in Western countries.[60] This very serious malignancy is the most common cancer of the eye among adults. As is true for cutaneous melanoma, ocular melanoma affects fair-skinned persons who have light eye color more frequently than dark-skinned persons.[61] Ocular melanoma must be treated aggressively by removing the eye or by radiation therapy. Further suggesting its linkage to ultraviolet radiation is the observation of an elevated risk for ocular melanoma among welders who may be exposed to ultraviolet light from arc welding.[60]

Usually, the ozone layer protects the eyes from harmful levels of UV radiation. It is believed that depletion of the ozone layer in the southern hemisphere has increased the levels of UVB radiation reaching the earth's surface, particularly in countries such as Chile.[62] These increased UVB levels may be harmful to the eyes of human and animal populations, particularly ruminants that graze outdoors in pastures. In the area of Punta Arenas, Chile, there have been anecdotal reports of domestic and wild animals that were blinded by the increasing levels of UVB radiation caused by the ozone hole over southern Chile; however, these reports have not been substantiated.

Sunlight and other sources of light are capable of producing acute effects, which are correlated with the intensity of the light. Viewing intense light from a light source (e.g., searchlights, lasers, welding arcs, and sunlight reflected by snow) may cause ocular injuries. As shown in Figure 8.13, intense visible light can burn the retina.

One of the most common causes of injury to the eyes from radiation in the ocular spectrum is UVR. Intense UVR is hazardous to the cornea, lens, and sometimes the retina.[63] Acute UVB exposure can cause actinic keratitis, a condition also known as snow blindness.[62] Research studies have suggested that sunlight exposure may be associated with longer-term consequences such as cataracts (lens opacities), which are linked to age and cumulative amount of exposure to sunlight.[64] Ordinary, inexpensive sunglasses are effective in reducing the amount of ultraviolet radiation that reaches the eye.[65]

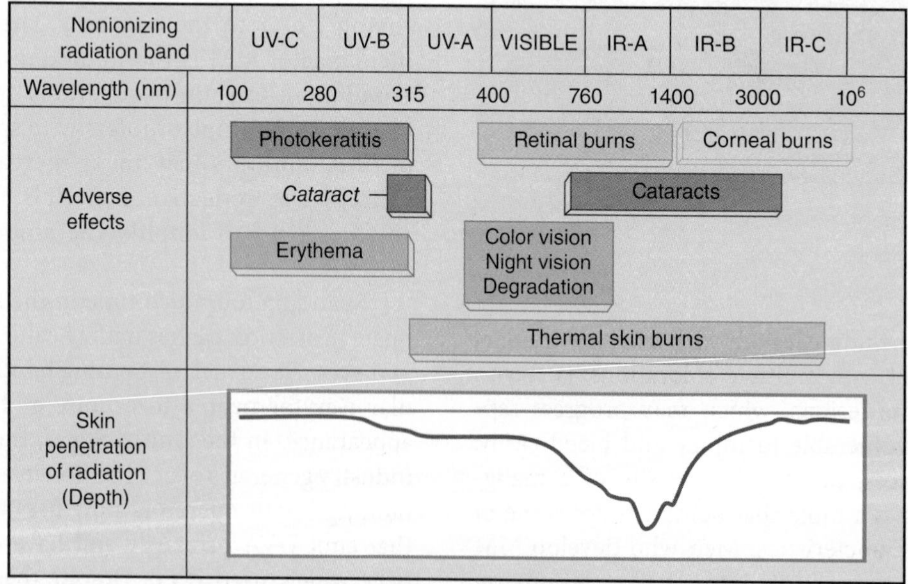

FIGURE 8.13 The optical spectrum and ocular effects. Photochemical mechanisms dominate in the UVR bands and in the short wavelength region of the visible spectrum. Thermal injury only occurs from brief, intense acute exposures and is more characteristic at longer wavelengths.

Reproduced from Sliney DH. Photoprotection of the eye—UV radiation and sunglasses. *J Photochem Photobiol B: Biol*. 2001;64(2–3):167. Copyright 2001, with permission from Elsevier.

In summary, the health effects of UVR exposure are not all adverse. UVR in sunlight promotes vitamin D synthesis and reduction of seasonal affective disorder. Certain skin conditions (e.g., acne, psoriasis, and eczema) appear to be alleviated by UVR. Studies also suggest that UVR provides a protective effect against some internal cancers such as prostate cancer.[66]

Conclusion

Sharing the characteristic that they are components of the electromagnetic spectrum, ionizing and nonionizing radiation differ in their levels of energy. Ionizing radiation is the kind of radiation that most people think of when they refer to radiation. This form of radiation has the capacity to remove electrons from atoms, whereas nonionizing radiation does not. Both ionizing and nonionizing radiation are integral for the functioning of modern society; for example, ionizing radiation is used in medical devices, electric power plants, and many other helpful applications, and nonionizing radiation is emitted by communication devices and many other sources. Although both forms of radiation present some documented hazards to the world's population, they also carry many benefits; it is necessary to balance their benefits against their potential harm. Protection of the population from radiation hazards can be accomplished through research that assesses the acute and long-term risks and hazards. Through well-conducted research studies, it may be possible to protect the health of the population, enjoy the benefits of radiation, and eliminate unnecessary alarm associated with any unsubstantiated claims about the adverse health effects of ionizing and nonionizing radiation.

Study Questions and Exercises

1. Define the following terms in your own words:
   a. Radiation
   b. Radioactivity
   c. Radioisotope
   d. Hertz
   e. Extremely low frequency radiation
   f. UV index
2. Describe the similarities and differences between ionizing and nonionizing radiation.
3. Define the term *stochastic*. How do the stochastic effects of ionizing radiation differ from the nonstochastic effects?
4. Match the term and its definition.

a. Becquerel	SI unit measure of amount of radioactivity present in a material
b. Coulomb/kg	SI unit measure of amount of exposure to radioactivity
c. Curie	SI unit measure of absorbed dose of radiation
d. Gray	SI unit measure of radiation dose related to biological effect
e. Rad	Common unit measure of amount of radioactivity present in a material
f. Rem	Common unit measure of amount of exposure to radioactivity
g. Roentgen	Common unit measure of absorbed dose of radiation
h. Sievert	Common unit measure of radiation dose related to biological effect

5. Distinguish among α, β, and γ radiation. Explain how α radiation can be a health hazard even though it is unable to penetrate the skin.
6. Support or refute the statement that leakage from nuclear power plants is the greatest source of exposure of the human population to ionizing radiation.
7. How do the immediate effects of a nuclear explosion differ from the long-term effects?
8. Why might radar installations present a hazard to human health? Give two examples.
9. Define the three types of ultraviolet radiation and discuss their health effects.
10. Discuss the risks and benefits to society associated with the use of devices that produce radiation (ionizing and nonionizing). What are the risks and benefits associated with radioisotopes?
11. Describe the ocular effects of nonionizing radiation in the optical spectrum.
12. What is meant by the term *dirty bomb*? What would be the main consequences of detonation of a dirty bomb in a large city?
13. Review the section on nuclear facilities. List at least two adverse consequences regarding the use of nuclear power plants for generating electricity.
14. What form of cancer was an issue of concern for children and adolescents who consumed milk

contaminated with radioactive iodine following the Chernobyl accident?

15. Describe some of the types of extraterrestrial radiation. Give two examples of groups of people who might be affected by such radiation.

For Further Reading

The Radium Girls: The Dark Story of America's Shining Women, Kate Moore, 2017

References

1. US Environmental Protection Agency. Radiation protection. Radiation basics. Available at: https://www.epa.gov/radiation/radiation-basics. Accessed February 5, 2017.
2. Bistline RW. What have epidemiologic studies told us about radiation health effects? *Colo Med*. 1980;77(10):351–353.
3. Samet JM. Epidemiologic studies of ionizing radiation and cancer: past successes and future challenges. *Environ Health Perspect*. 1997:105(suppl 4):883–889.
4. Oak Ridge Institute for Science and Education (ORISE). Radiation Emergency Assistance Center/Training Site (REAC/TS). Quick reference information—radiation. Available at: https://orise.orau.gov/reacts/documents/radiological-terms-quick-reference.pdf Accessed May 5, 2017.
5. *The Oxford College Dictionary*. 2nd ed. New York, NY: Spark Publishing, by arrangement with Oxford University Press, Inc.; 2007.
6. Lawrence Berkeley National Laboratory. Nuclear Science Glossary, Appendix A: Glossary of nuclear terms. Available at: http://www.lbl.gov/abc/wallchart/glossary/glossary.html. Accessed February 11, 2017.
7. Health Physics Society. Radiation basics. Available at: http://www.hps.org/publicinformation/ate/faqs/radiation.html. Accessed May 5, 2017.
8. Eaton DL, Klaassen CD. Principles of toxicology. In: Klaasen CD, Watkins JB III, eds. *Casarett & Doull's Essentials of Toxicology*. 3rd ed. New York, NY: McGraw-Hill, 2015.
9. World Health Organization, International Agency for Research on Cancer. Ionizing radiation, Part 1: X- and gamma (γ)-radiation, and neutrons. *IARC Monographs*. 2000;75:35–115.
10. Centers for Disease Control and Prevention. Agency for Toxic Substances & Disease Registry. Radon toxicity. Who is at risk of radon exposure. Available at: https://www.atsdr.cdc.gov/csem/csem.asp?csem=8&po=7. Accessed September 28, 2017.
11. World Health Organization. Radon and health. Available at: http://www.who.int/mediacentre/factsheets/fs291/en/. Accessed September 29, 2017.
12. Field RW, Steck DJ, Smith BJ, et al. The Iowa radon lung cancer study—phase I: residential radon gas exposure and lung cancer. *Sci Total Environ*. 2001;272:67–72.
13. Langner I, Blettner M, Gundestrup M, et al. Cosmic radiation and cancer mortality among airline pilots: results from a European cohort study (ESCAPE). *Radiat Environ Biophys*. 2004;42:247–256.
14. PBS Online. Frontline. Three Mile Island: the judge's ruling. June 7, 1996. Available at: http://www.pbs.org/wgbh/pages/frontline/shows/reaction/readings/tmi.html. Accessed May 6, 2017.
15. Warrick J. Study links Three Mile Island releases to higher cancer rates. Washington Post.com. February 24, 1997. Available at: http://www.washingtonpost.com/wp-srv/national/longterm/tmi/stories/study022497.htm. Accessed May 6, 2017.
16. Hatch M, Wallenstein S, Beyea J, et al. Cancer rates after the Three Mile Island nuclear accident and proximity of residence to the plant. *Am J Public Health*. 1991;81:719–724.
17. Stencel M. 20 years later: a nuclear nightmare in Pennsylvania. Washington Post.com. March 27, 1999. Available at: http://www.washingtonpost.com/wp-srv/national/longterm/tmi/tmi.htm. Accessed May 6, 2017.
18. Carlisle D. Bitter rain. *Nurs Times*. 1996;92(16):16–17.
19. Cardis E, Krewski D, Boniol M, et al. Estimates of the cancer burden in Europe from radioactive fallout from the Chernobyl accident. *Int J Cancer*. 2006;119:1224–1235.
20. Dolk H, Lechat MF. Health surveillance in Europe: lessons from EUROCAT and Chernobyl. *Int J Epidemiol*. 1993;22:363–368.
21. Kazakov VS, Demidchik EP, Astakhova LN. Thyroid cancer after Chernobyl [letter]. *Nature*. 1992;359:21.
22. Reiners C. Radioactivity and thyroid cancer. *Hormones*. 2009;8(3):185–191.
23. World Health Organization. 1986–2016: Chernobyl at 30. An update. April 25, 2016.
24. Johnson CJ. Cancer incidence in an area contaminated with radionuclides near a nuclear installation. *Ambio*. 1981;10:176–182.
25. Enstrom JE. Cancer mortality patterns around the San Onofre nuclear power plant, 1960–1978. *Am J Public Health*. 1983;73:83–91.
26. Shleien B, Ruttenber AJ, Sage M. Epidemiologic studies of cancer in populations near nuclear facilities. *Health Phys*. 1991;61:699–713.
27. Akleyev AV, Kossenko MM, Silkina LA, et al. Health effects of radiation incidents in the southern Urals. *Stem Cells*. 1995;13(suppl 1):58–68.
28. Steinhauser G, Brandi A, Johnson TE. Comparison of the Chernobyl and Fukushima nuclear accidents: a review of environmental impacts. *Sci Total Environ*. 2014;470–471.
29. Rallison ML, Lotz TM, Bishop M, et al. Cohort study of thyroid disease near the Nevada test site: a preliminary report. *Health Phys*. 1990;59:739–746.
30. Tokunaga M, Norman JE, Asano M, et al. Malignant breast tumors among atomic bomb survivors, Hiroshima and Nagasaki, 1950–1974. *J Natl Cancer Inst*. 1979;62:1347–1359.
31. Associated Press. Japan's oldest person dies at 114. July 31, 1998. Available at: http://www.apnewsarchive.com/1998/Japan-s-Oldest-Person-Dies-at-114/id-77c47b2f3a3f7d299dace61ebf7005af. Accessed May 9, 2017.
32. Japan Times. On 71st anniversary of atomic bombing, Hiroshima mayor urges world leaders to follow Obama in visiting city. August 6, 2016. Available at: http://www.japantimes.co.jp/news/2016/08/06/national/hiroshima-marks-71st-anniversary-of-atomic-bombing/#.WRID4Rqo6Rs. Accessed May 9, 2017.
33. BBC News. Analysis: effects of a dirty bomb. June 11, 2002. Available at: http://news.bbc.co.uk/2/hi/health/2037769.stm. Accessed May 6, 2017.
34. Centers for Disease Control and Prevention. Radiation emergencies: frequently asked questions (FAQs) about dirty

bombs. Available at: https://emergency.cdc.gov/radiation/dirtybombs.asp. Accessed May 7, 2017.

35. United States Nuclear Regulatory Commission. Fact sheet on dirty bombs. Available at: https://www.nrc.gov/reading-rm/doc-collections/fact-sheets/fs-dirty-bombs.pdf. Accessed May 7, 2017.
36. Jablon S, Bailar JC III. The contribution of ionizing radiation to cancer mortality in the United States. *Prev Med.* 1980;9:219–226.
37. Sienkiewicz Z. Biological effects of electromagnetic fields and radiation. *J Radiol Prot.* 1998;18:185–193.
38. Kliukiene J, Tynes T, Andersen A. Residential and occupational exposures to 50-Hz magnetic fields and breast cancer in women: a population-based study. *Am J Epidemiol.* 2004;159:852–861.
39. The hazards of extremely low frequency radiation. *Am J Public Health.* 1991;81:1447.
40. Verkasalo PK, Pukkala E, Kaprio J, et al. Magnetic fields of high voltage power lines and risk of cancer in Finnish adults: nationwide cohort study. *BMJ.* 1996;313:1047–1051.
41. Schoenfeld ER, O'Leary ES, Henderson K, et al. Electromagnetic fields and breast cancer on Long Island: a case-control study. *Am J Epidemiol.* 2003;158:47–58.
42. Tynes T, Haldorsen T. Residential and occupational exposure to 50 Hz magnetic fields and hematological cancers in Norway. *Cancer Causes Control.* 2003;14:715–720.
43. Dockerty JD, Elwood JM, Skegg DC, et al. Electromagnetic field exposures and childhood cancers in New Zealand. *Cancer Causes Control.* 1998;9:299–309.
44. Jauchem JR, Merritt JH. The epidemiology of exposure to electromagnetic fields: an overview of the recent literature. *J Clin Epidemiol.* 1991;44:895–906.
45. Verschaeve L, Maes A. Genetic, carcinogenic and teratogenic effects of radiofrequency fields. *Mutat Res.* 1998;410:141–165.
46. Elwood JM. A critical review of epidemiologic studies of radiofrequency exposure and human cancers. *Environ Health Perspect.* 1999;107(suppl 1):155–168.
47. Inskip PD, Tarone RE, Hatch EE, et al. Cellular-telephone use and brain tumors. *New Engl J Med.* 2001;344:79–86.
48. National Institute of Environmental Health Sciences, National Toxicology Program. Cell phone radiofrequency radiation studies. https://www.niehs.nih.gov/health/assets/docs_a_e/cell_phone_radiofrequency_radiation_studies_508.pdf. Accessed May 7, 2017,
49. World Health Organization. Electromagnetic fields and public health: mobile phones. Fact Sheet No 193. Available at: http://www.who.int/mediacentre/factsheets/fs193/en/. Accessed October 3, 2017.
50. Forum: of mobile phones and morbidity. *Environ Health Perspect.* 1998;106:A474–A475.
51. Maier M, Blakemore C, Koivisto M. The health hazards of mobile phones. *BMJ.* 2000;320:1288–1289.
52. Rothman KJ, Loughlin JE, Funch DP, et al. Overall mortality of cellular telephone customers. *Epidemiol.* 1996;7:303–305.
53. Hardell L, Carlberg M. Mobile phones, cordless phones and the risk for brain tumours. *Int J Oncol.* 2009;35:5–17.
54. Osepchuk JM, Petersen RC. Historical review of RF exposure standards and the International Committee on Electromagnetic Safety (ICES). *Bioelectromagnetics.* 2003;suppl 6:S7–S16.
55. Armstrong BK, Kricker A. The epidemiology of UV induced skin cancer. *J Photochem Photobiol B.* 2001;63:8–18.
56. Tenkate TD. Ultraviolet radiation: human exposure and health risks. *J Environ Health.* 1998;61:9–15.
57. Urbach F. Ultraviolet radiation and skin cancer of humans. *J Photochem Photobiol B.* 1997;40:3–7.
58. Gilchrest BA, Eller MS, Geller AC, et al. The pathogenesis of melanoma induced by ultraviolet radiation. *New Engl J Med.* 1999;340:1341–1348.
59. Miller BA, Kolonel LN, Bernstein L, et al, eds. *Racial/Ethnic Patterns of Cancer in the United States, 1988–1992.* Bethesda, MD: National Cancer Institute; 1996. NIH Pub. No. 96–4104.
60. Guénel P, Laforest L, Cyr D, et al. Occupational risk factors, ultraviolet radiation, and ocular melanoma: a case-control study in France. *Cancer Causes Control.* 2001;12:451–459.
61. Neugut AI, Kizelnik-Freilich S, Ackerman C. Black-white differences in risk for cutaneous, ocular, and visceral melanomas. *Am J Public Health.* 1994;84:1828–1829.
62. Schein OD, Vicencio C, Muñoz B, et al. Ocular and dermatologic health effects of ultraviolet radiation exposure from the ozone hole in southern Chile. *Am J Public Health.* 1995;85:546–550.
63. Sliney DH. Photoprotection of the eye—UV radiation and sunglasses. *J Photochem Photobiol B: Biol.* 2001;64:166–175.
64. West SK, Duncan DD, Muñoz B, et al. Sunlight exposure and risk of lens opacities in a population-based study: the Salisbury Eye Evaluation Project. *JAMA.* 1998;280:714–718.
65. Rosenthal FS, Bakalian AE, Lou C, et al. The effect of sunglasses on ocular exposure to ultraviolet radiation. *Am J Public Health.* 1988;78:72–74.
66. Bodiwala D, Luscombe CJ, Liu S, et al. Prostate cancer risk and exposure to ultraviolet radiation: further support for the protective effect of sunlight. *Cancer Lett.* 2003;192:145–149.

PART III

Applications of Environmental Health

CHAPTER 9
Water Quality

LEARNING OBJECTIVES

By the end of this chapter the reader will be able to:

- Describe sources of potable water.
- Define what is meant by the hydrological cycle.
- List hazardous substances that may be found in drinking water.
- Describe how water is made safe for human consumption.
- Discuss hazards to the aquatic environment (oceans, lakes, and rivers) associated with environmental pollution.

Introduction

This chapter covers the topics of water quality (e.g., freedom from waterborne diseases and hazards) and the water supply (e.g., sources and availability of water). Water quality is a crucial issue for environmental health, given that water is essential for life on earth. Residents of the United States and other developed countries assume that they will be able to turn on a faucet and draw a refreshing glass of water that is free from dangerous contaminants and microbial agents. (Refer to **FIGURE 9.1**.) In contrast, a safe water supply is not always available in the less developed regions of the world, where waterborne diseases represent a significant public health threat. In this chapter, you will learn about how water is treated to make it safe for residential consumption, microbial waterborne pathogens, chemicals in the water supply, and beach and costal pollution (for example, from oil spills).

Water Quality and Public Health

Safe, high-quality drinking water is an essential aspect of public health. One of the most significant measures to protect the health of the public was the introduction of chlorination of drinking water. Since the early 20th century, drinking water chlorination has resulted in drastic reductions in waterborne infections (e.g., cholera and typhoid) in the United States.

Vulnerable groups such as children, the elderly, and immunocompromised patients (e.g., those who are undergoing chemotherapy, taking steroids, or afflicted with HIV/AIDS) are at special risk of diseases caused by water contamination.[1,2] In the United States, water quality regulations are designed to protect the public from contaminated drinking water and from other forms of water pollution. Two major water quality regulations are the Safe Drinking Water

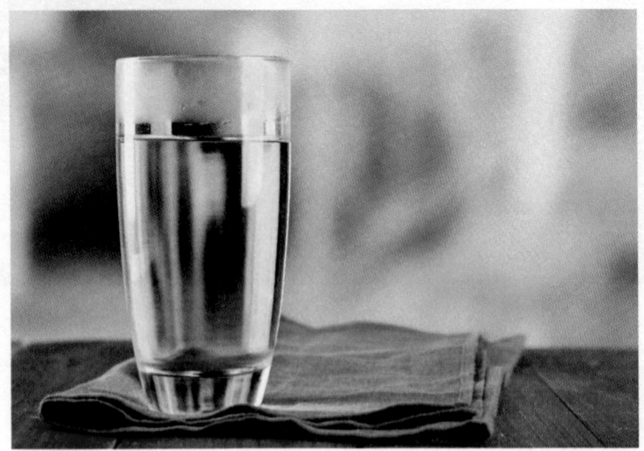

FIGURE 9.1 A glass of pure water.

Act and the Clean Water Act, both discussed elsewhere in the chapter on environmental policy and regulation.

The US Environmental Protection Agency (EPA) notes some interesting facts about water: Although it is possible for a human being to live up to a month without food, that same individual can survive for only about a week without water.[3] The average requirement for human consumption of water per day is approximately 2.5 liters (about 2.5 quarts). This level is necessary to maintain health and includes water from all sources, including food sources. It is also noteworthy that approximately two-thirds of the human body is made up of water; this amount increases to about three-quarters for human brain tissue.

In the United States the average person uses about 100 gallons (about 400 liters) of water per day, and the average residence uses over 100,000 gallons (about 400,000 liters) during a typical year. From 50% to 70% of this household water is used for outdoor purposes such as watering lawns and washing cars. **FIGURE 9.2** illustrates the relative percentage of different types of residential uses for water in a southern California city in the United States. Because of its semiarid climate, this community uses more than half of its water for landscape maintenance; 5% is lost to system leaks.

In addition to its life-supporting role, water is required in immense quantities by agriculture and industry, such as new car manufacture, steel production, and production of canned foods. According to the EPA,[3] examples of the amounts of water required for industrial purposes are as follows:

- Manufacture of a new automobile and its four tires: 39,090 gallons (about 160,000 liters)
- Production of one barrel of beer: 1,500 gallons (about 6,000 liters)
- Production of one ton of steel: 62,600 gallons (about 250,000 liters)
- Processing one can of fruit or vegetables: 9.3 gallons (about 38 liters)

Despite the fact that water is a necessity of human life, about 20% of the world's population lacks safe drinking water. With the increasing world population, the problem of lack of access to safe drinking water is likely to worsen.[2] Although most developed nations treat industrial discharges and **sewage**, contamination of water supplies with chemicals such as arsenic and some microorganisms place public water supplies at risk. In the wealthier nations (e.g., Japan, western European countries, and the United States), the public water supplies, although generally safe, may carry organic chemicals, lead, and arsenic, as well as bacteria, viruses, and parasites from fecal contamination.

The problem of water quality is especially acute for developing nations. In these countries up to 90% of the cities discharge their untreated sewage into rivers and streams. (Refer to **FIGURE 9.3**.) These surface

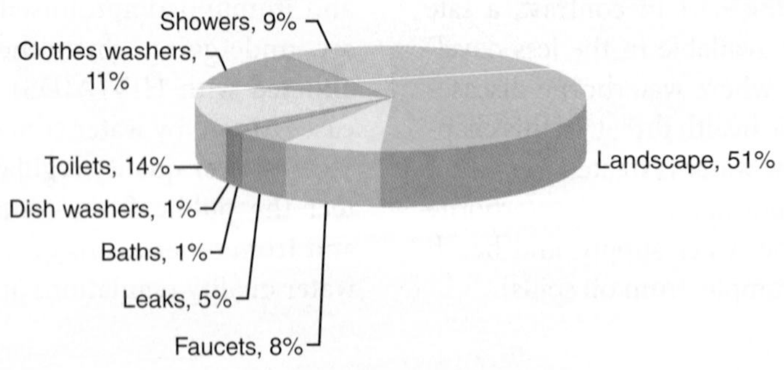

FIGURE 9.2 Residential uses for water in a southern California community.

Data from Irvine Ranch Water District, Irvine, California.

FIGURE 9.3 A slum in Ecuador that was heavily affected by cholera due to its proximity to unsafe water sources.
Courtesy of CDC

FIGURE 9.4 Examples of surface waters.

waters in turn may be used for drinking and personal sanitation purposes. In the developing world, there are several large urban areas known as megacities, defined as cities with 10 million or more people. The unsanitary water supplies found in many of these densely populated areas foster the spread of gastrointestinal illnesses and serious infectious diseases that contribute greatly to the burden of morbidity and mortality in these regions.

The lack of a safe water supply is compounded in some parts of the developing world by chronic water shortages. Regions that face a water shortfall include north Africa, the Middle East, and parts of sub-Saharan Africa. The residents of these areas are faced periodically with a chronic shortage of freshwater.

The Water Supply

The two main sources of drinking water used by the human population are surface water and **groundwater**.[4] (Refer to the text box for definitions.) Surface waters are from rivers and lakes, as shown in **FIGURE 9.4**.

The second type of water available for human consumption is groundwater, which is stored naturally in underground aquifers. (See **EXHIBIT 9.1** for a definition of the term **aquifer** and other key hydrological terms.) Wastewater from human activities can find its way back into aquifers. In some regions of the world, communities treat wastewater and store it in percolation ponds so that it may recharge aquifers. Refer to **FIGURE 9.5**, which demonstrates that communities near the sea run the risk of saltwater intrusion into the aquifer when excessive amounts of water are withdrawn. Saltwater intrusion eventually may render the water unsuitable for human consumption.

The field of water science is known as hydrology. Some of the terms that relate to hydrology are *aquifer*, the **hydrological cycle**, **water scarcity**, and **water stress**. Some of these terms are discussed later in the chapter.

The Hydrological Cycle

The hydrological cycle is akin to a pumping system, driven by the sun, which moves freshwater from the oceans to landmasses and then returns it to the ocean.[5] This cycle describes the process by which the

WATER DEFINITIONS USED BY FEDERAL AND STATE WATER AUTHORITIES

Finished water: the water (e.g., drinking water) delivered to the distribution system after treatment, if any.
Groundwater: water that is contained in the interconnected pores in an aquifer.
Groundwater system: a system that uses water extracted from an aquifer (i.e., a well or spring) as its source.
Groundwater under the direct influence of surface water: as defined by the US Environmental Protection Agency (EPA), any water beneath the surface of the ground with substantial occurrence of insects or other macroorganisms, algae, or large-diameter pathogens (e.g, *Giardia intestinalis* or *Cryptosporidium*), or substantial and relatively rapid shifts in water characteristics (e.g., turbidity, temperature, conductivity, or PH) that closely correlate with climatologic or surface water conditions. Direct influence must be determined for individual sources in accordance with criteria established by the state (where the source is located).
Source water: untreated water (i.e., raw water) used to produce drinking water.
Surface water: all water on the surface (e.g., lakes, rivers, reservoirs, ponds, and oceans) as distinguished from subsurface or groundwater.

Note: On December 16, 1998, the EPA's Interim Enhanced Surface Water Treatment Rule (IESWTR) appeared in the *Federal Register*. The purpose of the rule was to "[i]mprove control of microbial pathogens including specifically the protozoan *Cryptosporidium,* in drinking water"[a(p.69478)] as well as minimize risks associated with agents used to disinfect water. The control of *Cryptosporidium* applied both to surface waters and groundwater under the influence of surface water. The IESWTR mandated that "[s]tates are required to conduct sanitary surveys for all public water systems using surface water or ground water under the direct influence of surface water. . . Sanitary surveys are required no less frequently than every three years for community systems. . . ."[a(p69484)] EPA standards for evaluation and processing of groundwater under the influence of surface water are stricter than those applied to groundwater not under such influence.[b]

[a] Environmental Protection Agency. National Primary Drinking Water Regulations: Interim Enhanced Surface Water Treatment. *Federal Register*. December 16, 1998;63(241):69478–69521.
[b] Gostin LO, Lazzarini Z, Neslund, VS Osterholm MT. Water quality laws and waterborne diseases: *Cryptosporidium* and other emerging pathogens. *Am J Public Health*. 2000;90:847–853.

Water definitions modified and reproduced from Centers for Disease Control and Prevention, Surveillance Summaries. Surveillance for waterborne disease and outbreaks associated with drinking water and water not intended for drinking—United States, 2005–2006. *MMWR*. 2008;57(SS-9):64–65.

EXHIBIT 9.1

Glossary of Key Hydrological Terms

Aquifer: a layer or section of earth or rock that contains freshwater, known as groundwater (any water that is stored naturally underground or that flows through rock or soil, supplying springs and wells).[a]
Freshwater lakes: most freshwater lakes are located at high latitudes, with nearly 50% of the world's lakes in Canada alone. Many lakes, especially those in arid regions, become salty through evaporation, which concentrates the inflowing salts. The Caspian Sea, the Dead Sea, and the Great Salt Lake are among the world's major salt lakes.[b]
Glaciers and icecaps: glaciers and icecaps cover about 10% of the world's landmass. These are concentrated in Greenland and Antarctica and contain ~70% of the world's freshwater. Unfortunately, most of these resources are located far from human habitation and are not readily accessible for human use.

According to the United States Geological Survey (USGS), 96% of the world's frozen freshwater is at the South and North Poles, with the remaining 4% spread over 550,000 km^2 [212,000 mi^2] of glaciers and mountainous icecaps measuring about 180,000 km [43,000 mi^3].[b]
Hydrological (water) cycle: the natural cycle by which water evaporates from oceans and other water bodies, accumulates as water vapor in clouds, and returns to oceans and other water bodies as precipitation.[a]
Nonrenewable water: water in aquifers and other natural reservoirs that . . . [is] not recharged by the hydrological cycle or . . . [is] recharged so slowly that significant withdrawal for human use causes depletion. Fossil aquifers are in this category: they recharge so slowly over centuries that they are, in effect, a nonrenewable resource.[a]
Renewable water: freshwater that is continuously replenished by the hydrological cycle for withdrawal within reasonable time limits, such as water in rivers, lakes, or reservoirs that fill from precipitation or from runoff. The renewability of a water source depends both on its natural rate of replenishment and the rate at which the water is withdrawn for human use.[a]
Reservoirs: artificial lakes, produced by constructing physical barriers across flowing rivers, which allow the water to pool and be used for various purposes. The volume of water stored in reservoirs worldwide is estimated at 4,286 km^3 [1,028 mi^3].[b]
Runoff: water originating as precipitation on land that then runs off the land into rivers, streams, and lakes, eventually reaching the oceans, inland seas, or aquifers, unless it evaporates first. That portion of runoff that can be relied on year after year and easily used by human beings is known as stable runoff.[a]

Water withdrawal: removal of freshwater for human use from any natural source or reservoir, such as a lake, river, or aquifer. If not consumed, the water may return to the environment and can be used again.[a]
Water scarcity: according to a growing consensus among hydrologists, a country faces water scarcity when its annual supply of renewable freshwater is less than 1,000 cubic meters [35,000 cubic feet] per person. Such countries can expect to experience chronic and widespread shortages of water that hinder their development.[a]
Water stress: a country faces water stress when its annual supply of renewable freshwater is between 1,000 and 1,700 cubic meters [35,000 and 60,000 cubic feet] per person. Such countries can expect to experience temporary or limited water shortages.[a]
Wetlands: wetlands include swamps, bogs, marshes, mires, lagoons, and floodplains. (In some cases, wetlands are associated with rivers.) According to the United Nations Environment Programme, the earth's 10 largest wetlands are: West Siberian Lowlands, Amazon River, Hudson Bay Lowlands (in Canada), Pantanal (in mid-South America), Upper Nile River, Chari-Logone River (in Africa), Hudson Bay Lowlands in the South Pacific, Congo River, Upper Mackenzie River (in northwestern Canada), and North America prairie potholes (wetlands made up of shallow depressions in the northern Great Plains).

The total global area of wetlands is estimated at ~2,900,000 km² [1,100.000 mi²]. . . . Most wetlands range in depth from 0–2 meters [0-6.6 feet]. Estimating the average depth of permanent wetlands at about one meter, the global volume of wetlands could range between 2,300 km³ [552 mi³] and 2,900 km³ [696 mi³].[b]

[a] Reproduced from Hinrichsen D, Robey B, Upadhyay UD. Solutions for a Water-Short World. *Population Reports*. Series M, Number 14. Baltimore, Johns Hopkins University School of Public Health, Population Information Program, September 1998.
[b] Modified and reproduced from Vital Water Graphics: Freshwater Resources, United Nations Environment Programme, © 2002 United Nations. Reproduced with the permission of the publisher. Available at: http://staging.unep.org/dewa/assessments/ecosystems/water/vitalwater/02.htm. Accessed June 5, 2017.

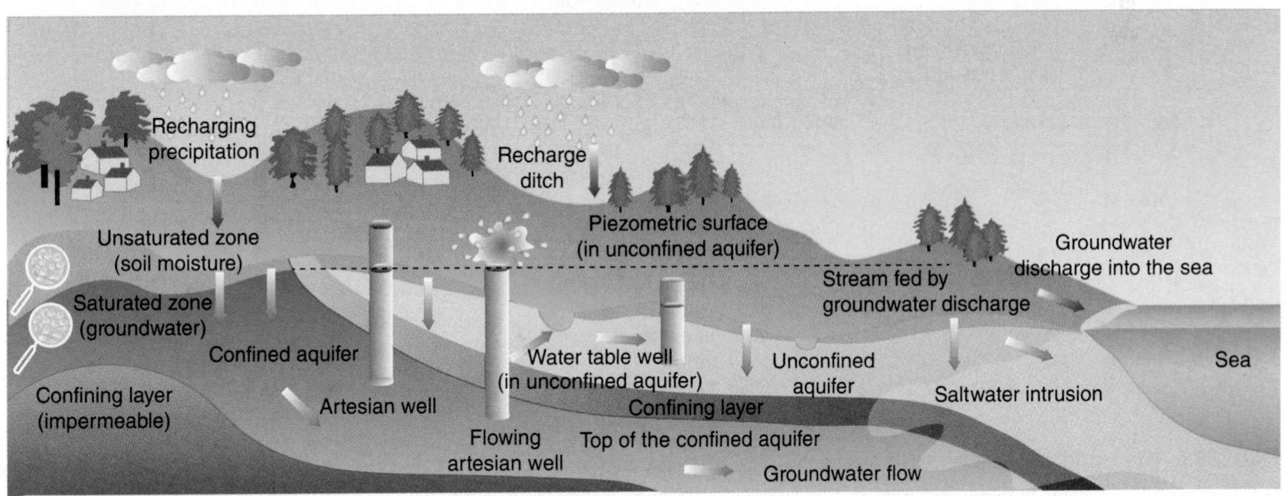

FIGURE 9.5 Types of aquifers, wells, and groundwater flow.

Reproduced from Vital Water Graphics: Freshwater Resources, United Nations Environment Programme, © 2002 United Nations. Reproduced with the permission of the publisher. Available at: http://staging.unep.org/dewa/assessments/ecosystems/water/vitalwater/a1.htm. Accessed June 5, 2017.

freshwater supply is continuously replenished. Refer to **FIGURE 9.6**.

The United Nations Environment Program provides the following description of the hydrological cycle:

> Water is transported in different forms within the hydrological cycle or "water cycle." . . . [E]ach year about 502,800 km³ [120,600 mi³] of water evaporates over the oceans and seas, 90% of which (458,000 km³) [110,000 mi³] returns directly to the oceans through precipitation, while the remainder (44,800 km³) [10,700 mi³] falls over land. With evapo-transpiration totaling about 74,200 km³ [17,800 mi³], the total volume in the terrestrial hydrological cycle is about 119,000 km³ [28,500 mi³]. About 35% of this, or 44,800 km³ [10,700 mi³], is returned to the oceans as run-off from rivers, groundwater and glaciers. A considerable portion of river flow and groundwater percolation never reaches the ocean, having evaporated in internal runoff areas or inland basins lacking an outlet to the ocean. However, some groundwater that bypasses the river systems reaches the oceans. Annually the hydrological cycle circulates nearly 577,000 km³ [138,000 mi³] of water. . . .[6]

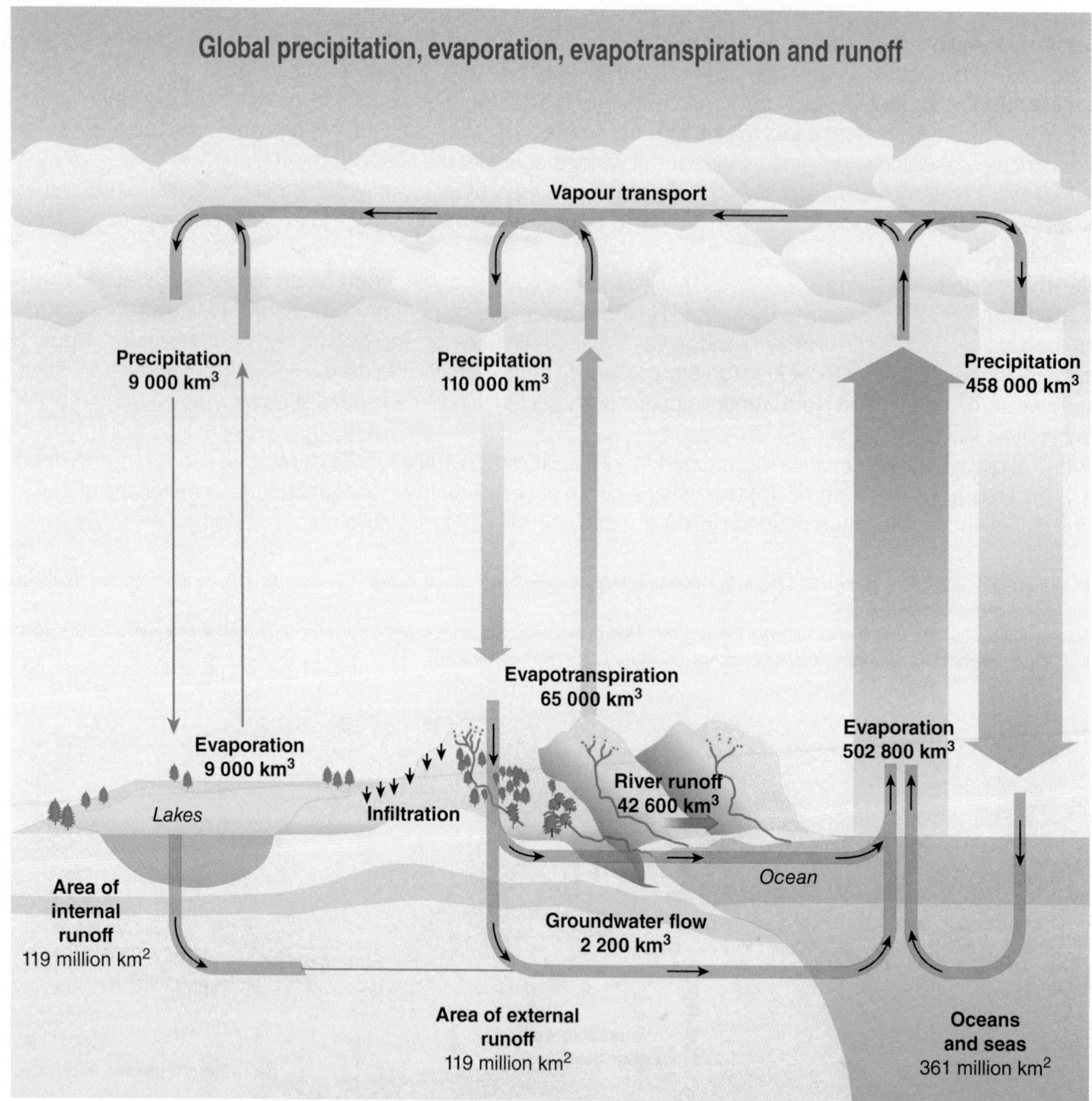

FIGURE 9.6 The world's water cycle. The width of the arrows is proportional to the volumes of transported water.

Reproduced from Vital Water Graphics: Freshwater Resources, by United Nations Environment Programme, © 2002 United Nations. Reproduced by permission of the publisher. Available at: http://www.unep.org/dewa/assessments/ecosystems/water/vitalwater/05.htm. Accessed June 5, 2017.

Freshwater Resources

FIGURE 9.7 provides estimates of world freshwater reserves. No more water exists on earth than there was 2,000 years ago, but the human population has grown enormously since ancient times.[5]

Here are some facts about water availability[5]: although about 70% of the earth's surface is covered by water, most of this water is unusable ocean water. Approximately 3% of all water is freshwater, of which the majority is unavailable for human use. This water is unavailable because nearly 75% of freshwater supplies are frozen in the polar ice caps and in glaciers. The remaining 1% of readily accessible water comes from surface freshwater; sources include lakes, rivers, and shallow underground aquifers. These readily accessible sources constitute the water supply that is renewed by the hydrological cycle.

Estimates suggest that only 0.01% of the world's total supply of water can be accessible for use by the human population.[5] This amount consists of about 12.5 to 14 billion cubic meters (441 to 490 billion cubic feet), or about 9,000 cubic meters (318,000 cubic feet) for each person on earth. (One cubic meter is approximately 1,000 liters [264 gallons].) Unfortunately, the water supply is not distributed evenly across the world. This unequal distribution is exacerbated by the fact that there may be annual cycles of drought and flooding.

FIGURE 9.8 shows the earth's distribution of freshwater resources represented by glaciers, surface water, and other reserves. Many of the geographic areas of the world that receive heavy rainfall are located away

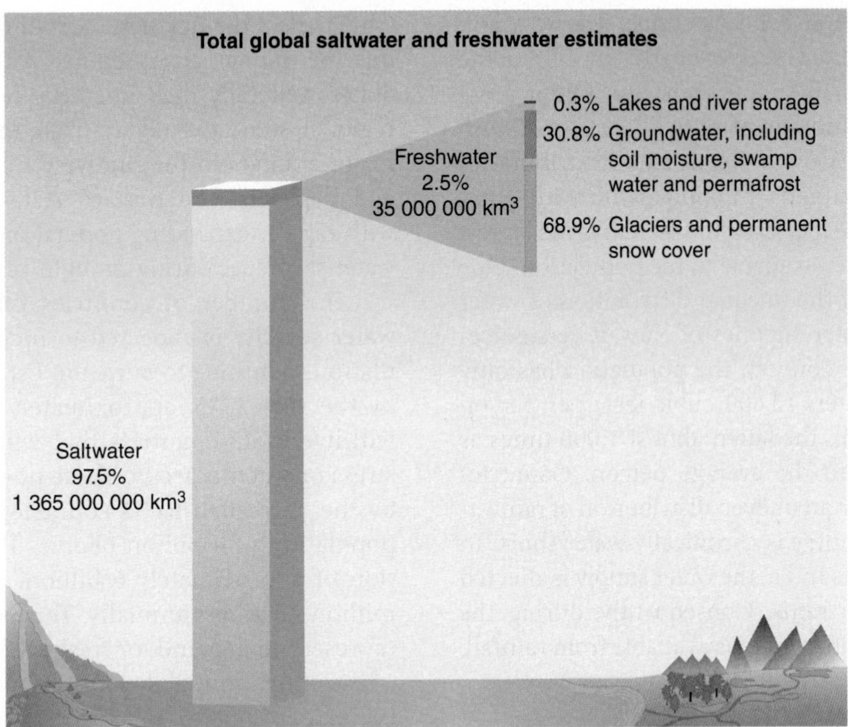

FIGURE 9.7 A world of salt.

Reproduced from Vital Water Graphics: Freshwater Resources, by United Nations Environment Programme, © 2002, United Nations. Reproduced by permission of the publisher. Available at: http://www.unep.org/dewa/assessments/ecosystems/water/vitalwater/01.htm. Accessed June 5, 2017.

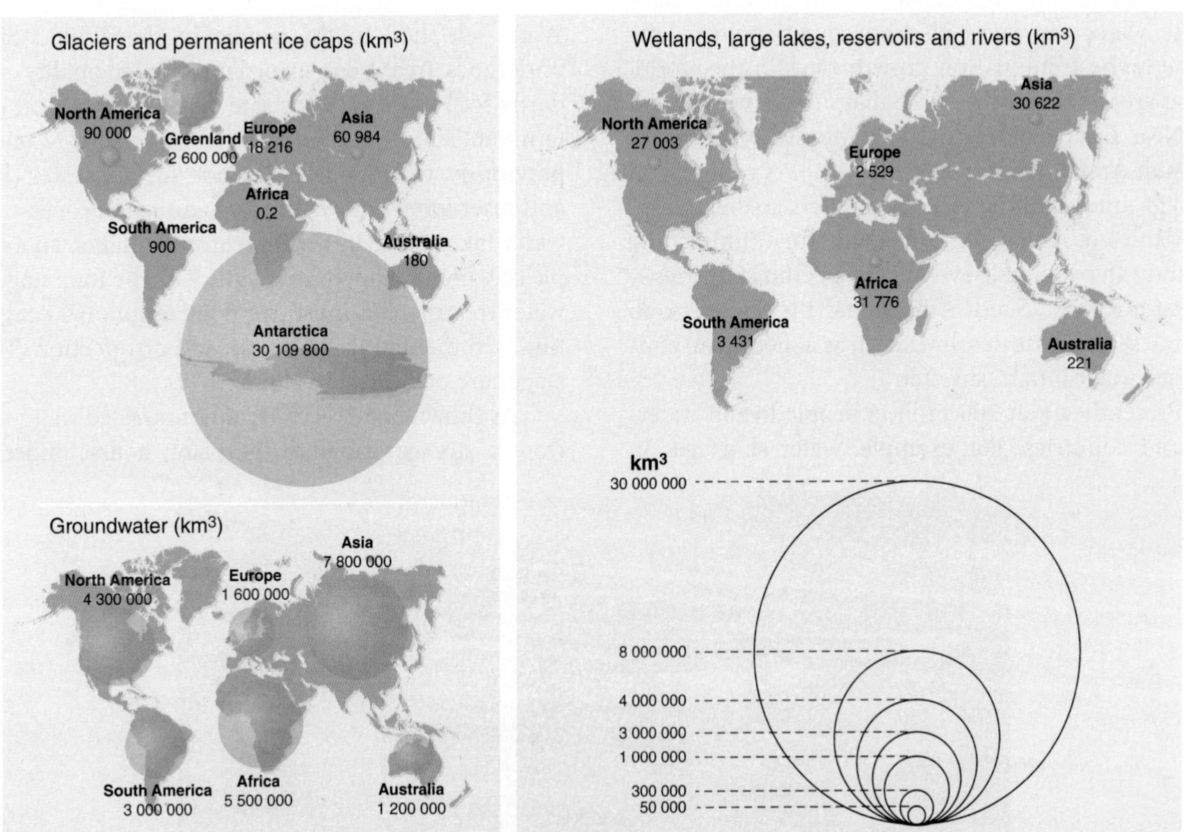

FIGURE 9.8 Global freshwater resources: quantity and distribution by region. (Estimates refer to standing volumes of freshwater.)

Reproduced from Vital Water Graphics: Freshwater Resources, by United Nations Environment Programme, © United Nations. Reprinted by permission of the publisher. Available at: http://www.unep.org/dewa/assessments/ecosystems/water/vitalwater/02.htm. Accessed June 5, 2017.

from large population concentrations. The Amazon basin in South America receives nearly one-fifth of the global average rainfall runoff each year. Other areas that receive high amounts of rainfall are Asia, South America, and the Congo River and its tributaries on the African continent. In comparison with other regions, the North American continent has the largest amount of freshwater available to the population. One extreme example of the unequal distribution of water resources is the water supplies of Kuwait versus Iceland. In the former country, the population has only about 75 cubic meters (2,600 cubic feet) per person of available water; in the latter, almost 1,000 times as much is available to the average person. Countries such as Mexico have an uneven distribution of rainwater; 90% of this country is chronically water short. In parts of Asia, such as India, the water supply is affected greatly by seasonal rains. Consequently, during the drier months very little water is available from rainfall.

Water Scarcity and Water Stress

In several areas of the world, the demand for water presently exceeds the available water supply as a result of water scarcity and water stress. (Refer to Exhibit 9.1 for definitions of water scarcity and water stress.) Ironically, two sections of the world that currently have severe water shortages also are experiencing some of the highest population growth rates in the world. These areas are Africa (sub-Saharan and north) and the Near East, regions that encompass 20 countries. In Saudi Arabia, for example, the supplies of so-called fossil groundwater that exist in aquifers are being rapidly depleted. As a means of supplanting diminishing groundwater sources, several Persian Gulf countries—Bahrain, Kuwait, Saudi Arabia, and the United Arab Emirates—now use desalinization as a method to convert sea water into freshwater.

Presently about 200 million people live in water-stressed countries. For example, water shortages in China affect the northern part of the country, including the capital city, Beijing. Although the United States generally has adequate amounts of water, regional shortages occur from time to time. Some major aquifers in the Midwest are being overutilized and depleted. The western states (e.g., California), with rapidly expanding populations, periodically face water shortages during drought years.

The number of countries that will experience water scarcity is expected to increase as their populations continue to burgeon. Estimates suggest that by the year 2025, approximately 2.8 billion people will live in 48 countries that will experience water stress or water scarcity. These numbers will increase by the year 2050 to 54 countries with a combined population of 4 billion people. The world's population of approximately 6 billion is adding about 80 million members annually. This population increase represents a demand for freshwater that is about that of the entire annual flow of Europe's Rhine River (64 billion cubic meters per year [2.3 trillion cubic feet]).

Treatment of Water for Residential Consumption

Water supplied to the public in the United States undergoes treatment in order to meet quality standards set by the EPA for safe levels of chemical contaminants and waterborne microorganisms. As noted previously, water sources include surface water (lakes and reservoirs) and water from aquifers. Processing of water takes place in water treatment plants, an example of which is shown in **FIGURE 9.9**. The four stages of water treatment in most plants are as follows: coagulation, sedimentation, filtration, and disinfection. These stages are presented in **FIGURE 9.10**.

As shown in **FIGURE 9.11**, after untreated water flows from a storage area into the plant, it first undergoes

FIGURE 9.9 Purification plant for drinking water.

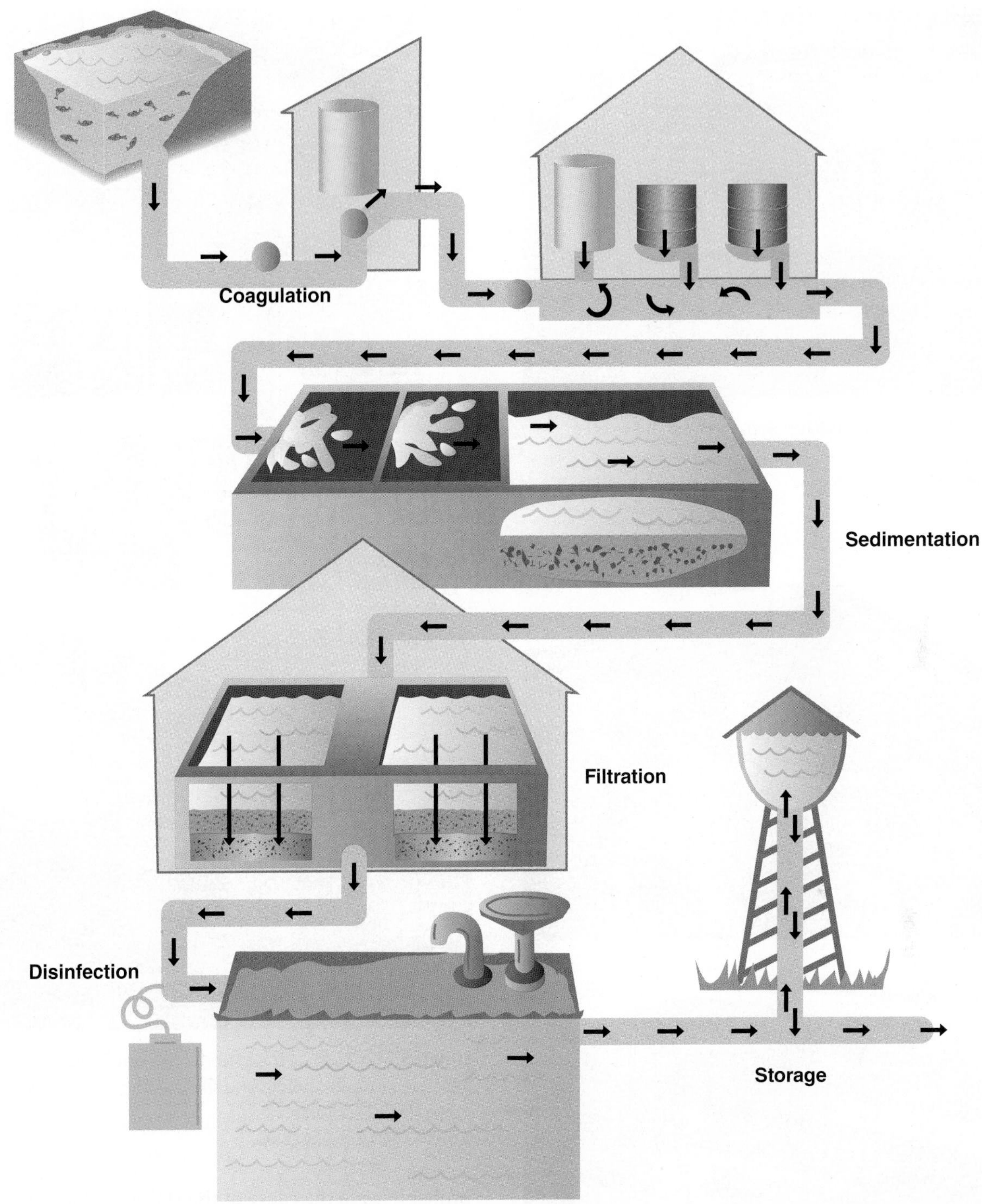

FIGURE 9.10 The stages of treatment at a water treatment plant.

Modified from Centers for Disease Control and Prevention. Drinking Water: Water Treatment. Available at: https://www.cdc.gov/healthywater/drinking/public/water_treatment.html

coagulation to remove suspended material. Aluminum sulfate is used as the coagulating agent. After the coagulating agent has been mixed with the water, the mixture is transferred to sedimentation tanks similar to the ones shown in **FIGURE 9.12**. The water then is filtered to remove smaller impurities. The filter is made of progressively finer layers of sand and a layer of activated charcoal. Subsequently, the water is treated with a disinfectant such as chlorine to destroy pathogens. (Refer to **FIGURE 9.13**.)

Fluoridation of Water

Some communities in the United States add fluoride to public drinking water in order to prevent tooth decay (dental caries). The rationale for this procedure came from the informal observations of dentists during the early 1900s, followed by epidemiologic studies.[7] Dr. Frederick McKay established a dental practice in 1901 in Colorado Springs, Colorado. (Refer to

FIGURE 9.11 Storage tank for aluminum sulfate used to coagulate solids in water.

Courtesy of Irvine Ranch Water District, Irvine, California

FIGURE 9.13 On the top is a model of a sand and charcoal filter used in water processing. On the bottom is a chlorinator room for water disinfection.

Courtesy of Irvine Ranch Water District, Irvine, California

FIGURE 9.12 Tanks used for additional skimming during secondary processing.

Courtesy of Irvine Ranch Water District, Irvine, California

FIGURE 9.14.) He soon observed that many of his patients' teeth displayed stains, which he referred to as "mottled enamel" or "Colorado brown stain." McKay concluded that high levels of fluoride in the local drinking water probably were the cause of the stains, and also seemed to confer resistance against tooth decay.

Subsequently, dental researchers including Dr. H. Trendley Dean (during the 1930s) further investigated links between fluoride in drinking water and absence of dental caries. Dean used the term **fluorosis** to describe the dental condition in which teeth have been discolored by fluoride. In epidemiologic studies, Dean demonstrated an inverse relationship between the prevalence of dental caries and levels of naturally occurring drinking water fluoride levels. The optimal fluoride level was set at 0.7 to 1.2 parts per million (ppm) of drinking water. This level was thought to minimize the occurrence of fluorosis and confer protection against dental caries.

FIGURE 9.14 Dr. Frederick S. McKay, one of the pioneers in water fluoridation.

Courtesy of CDC

Water fluoridation as a means of preventing dental caries was tested in prospective field trials in four pairs of cities (one intervention and one control in each pair) beginning in 1945. For example, the intervention city Grand Rapids, Michigan was paired with the control city Muskegon, Michigan. The field trials demonstrated a 50 to 70% reduction in the prevalence of dental caries with no apparent increase in dental fluorosis above what is seen in communities that have low levels of naturally occurring fluoride in their drinking water. Following these successful trials, many communities in the United States added fluoride to public drinking water.

Treatment of Water from Aquifers

Aquifers are a common source of potable water in many communities. For high-quality water from aquifers, minimal aeration, filtration, and disinfection are necessary. In some cases, water drawn from aquifers is free from microorganisms, but undesirable for human consumption because of impurities and coloration that impair the aesthetic qualities of this essential liquid. Such is the case of deep aquifers (present at 600 meters [2,000 feet]) found in some regions of the world (e.g., southern California). This water can be made acceptable through a process of filtration that uses ultrafine filters, as shown in **FIGURE 9-15**. The

A. Pump

B. Bank of filters

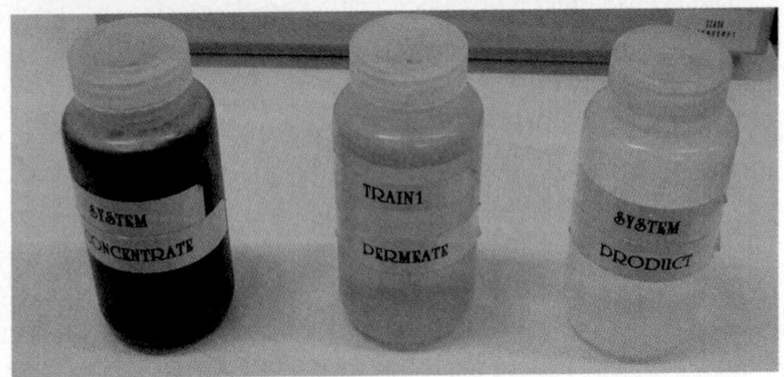

C. Samples of water before and after filtration

FIGURE 9.15 A, B, and C. Water filtration system for water from aquifer.

Courtesy of Irvine Ranch Water District, Irvine, California

figure shows pumps used to draw the water from the aquifer, banks of filters, and samples of water before and after filtration, in addition to the impurities that have been removed. In this example, the residual impurities removed through filtration are piped to the city's sanitary sewage processing plant.

Drinking Water Contamination

Many consumers in the United States are concerned about the safety of water supplied by municipal water plants. Some dramatic instances of waterborne diseases and the presence of lead and other heavy metals in water from the public water supply have been reported by the media. An instance of lead-contaminated drinking water occurred in Flint, Michigan, in 2014 when the city's water supply was changed from Lake Huron to the Flint River as a cost-saving measure. Despite these unusual instances of drinking water contamination, the public water supply generally is monitored carefully and must adhere to EPA safety standards. Nevertheless, consumers increasingly are adopting bottled water (shall we say "prestige water") as a substitute for tap water (refer to the following text box for a discussion of the relative merits of tap water versus bottled and vended water).

Potable water (drinking water) includes water from wells and runoff from the land's surface. Almost all water in its natural state is impure, because of common naturally occurring and anthropogenic sources of pollution. Naturally occurring sources of pollution arise from the diversity of aquatic animals and plants that inhabit the bodies of water used eventually for human consumption. In addition to microbial organisms that live in water, fish, aquatic animals, and wildlife produce wastes that contaminate the water. Soils in contact with the water also harbor microorganisms. Decaying tree leaves and branches contribute organic materials. Natural rock and soil formations may introduce radionuclides, nitrogen compounds, and heavy metals such as arsenic, cadmium, chromium, lead, and selenium, as well as other chemicals.

FIGURE 9.16 demonstrates how human-made pollutants may enter the urban water supply. Water that courses across the surface of the land may incorporate various contaminants, which include the following[8,9]:

- Chemicals and nutrients (e.g., fertilizers and nitrates from agricultural lands)
- Rubber, heavy metals, and sodium (from roads)
- Petroleum by-products and organic chemicals (from dry cleaners, service stations, and leaking underground storage tanks)
- Chemicals used in the home (solvents, paints, used motor oil, lead, and copper)
- Heavy metals and toxic chemicals (from factories)
- Microbial pathogens (from human and animal wastes)

Runoff from urban streets is a growing contributor to water pollution, especially after periods of heavy rainfall. Pet wastes that are washed into storm drains can represent a hazard to human and animal health. As a result, many cities are attempting to curtail this source of water pollution through local ordinances. (See **FIGURE 9.17**.)

WHICH IS SAFER: TAP WATER, BOTTLED WATER, OR VENDED WATER?

The growing public acceptance of bottled water may be the result of taste and health concerns. Also popular (particularly among Hispanic and Asian populations) is vended water. Here are some facts about the three types of water:

- Residents of communities that fluoridate water may be at increased risk of tooth decay should they drink bottled water exclusively, because bottled water usually is not fluoridated.
- Some bottled water may contain higher bacterial counts than tap water; all public drinking water sources are regulated by the EPA to ensure quality. The opened personal water bottles of school children have been reported to have fecal coliform contamination, possibly as a result of lapses in personal hygiene.
- Water purchased from water vending machines that are poorly maintained may be contaminated with coliform and other bacteria. Vended water comes from an approved source such as tap water; vending machines provide additional treatment to tap water by using carbon filtration and other methods.
- Bottled water, especially luxury brands, is much more expensive than tap water. Often the millions of used water bottles must be discarded in landfills, creating an adverse environmental impact.

Data from Lalumandier JA, Ayers LW. Fluoride and bacterial content of bottled water vs tap water. *Arch Fam Med*. 2000;9:246–250; Oliphant JA, Ryan MC, Chu A. Bacterial water quality in the personal water bottles of elementary students. *Can J Public Health*. 2002;95:366–367; Schillinger J, Du Vall Knorr S. Drinking-water quality and issues associated with water vending machines in the city of Los Angeles. *J Environ Health*. 2004;66(6):25–31; Postman A. The truth about tap. National Resources Defense Council. January 5, 2016. Available at: https://www.nrdc.org/stories/truth-about-tap?gclid=EAIaIQobChMIxan9-KWv1gIVEJFpCh3CswIyEAAYASAAEgJ5fvD_BwE. Accessed September 18, 2017.

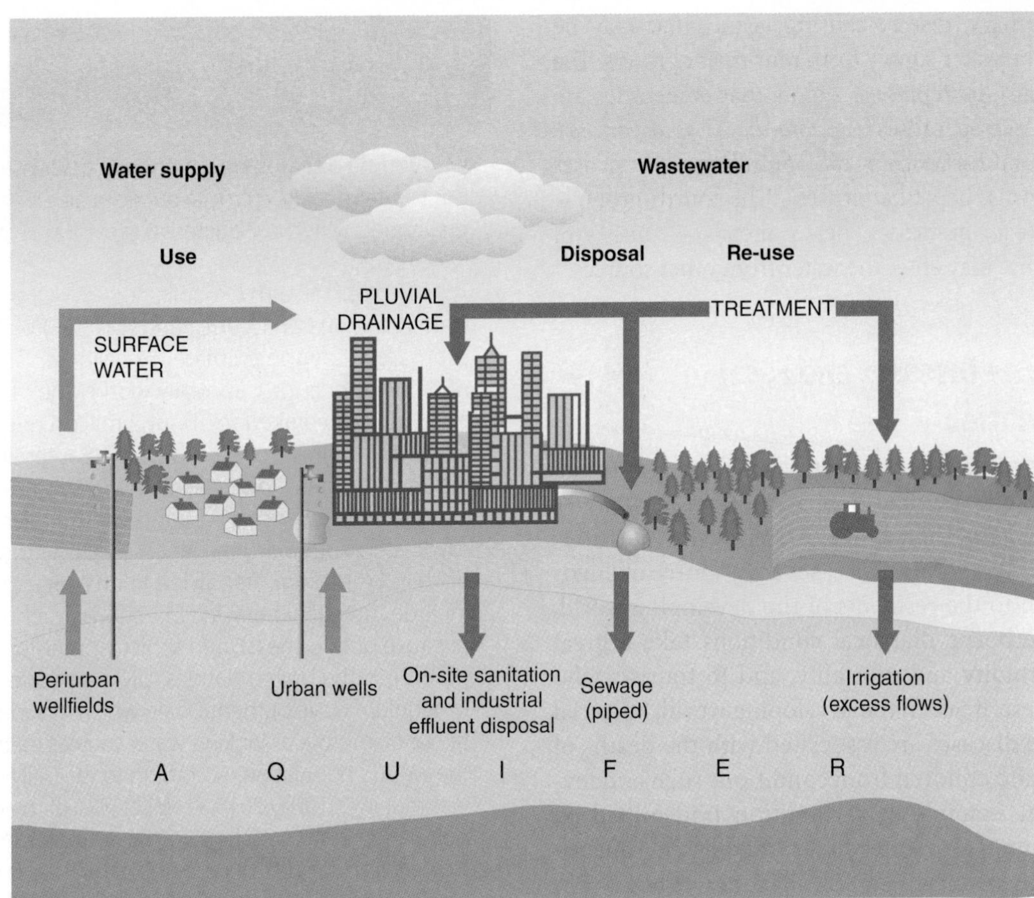

FIGURE 9.16 The urban water cycle: humanity's impact on groundwater.

Reproduced from Vital Water Graphics: Freshwater Resources, by United Nations Environment Programme, © 2002 United Nations. Reproduced by permission of the publisher. Available at: http://www.unep.org/dewa/assessments/ecosystems/water/vitalwater/a3.htm. Accessed June 5, 2017.

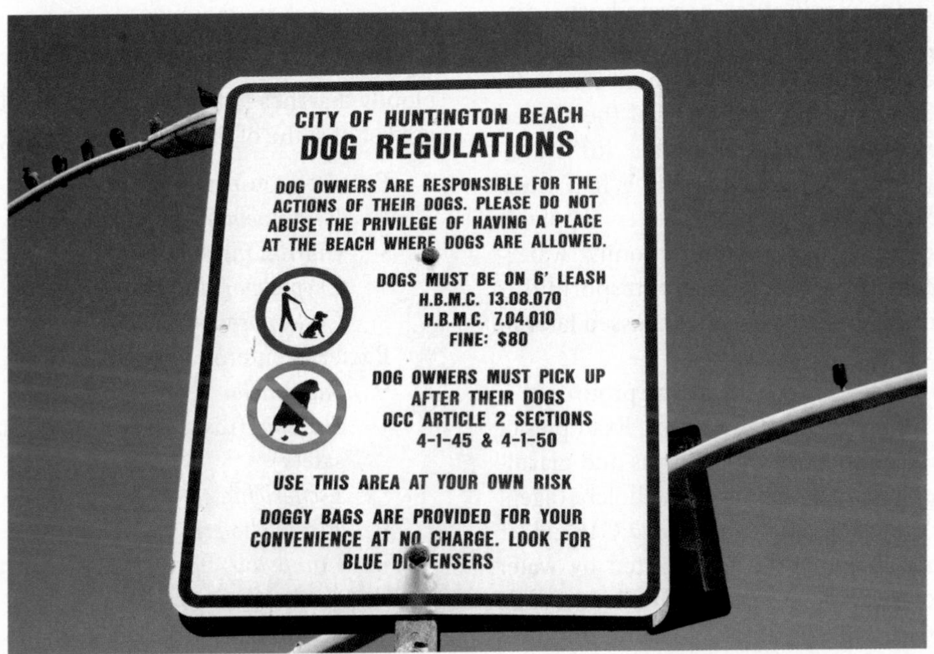

FIGURE 9.17 In order to prevent water and other pollution, this sign admonishes city residents to pick up after their pets.

In summary, disease-causing agents that may be present in the water supply form four major groups. The first three groups represent agents that sometimes are excreted in feces: parasites (e.g., *Giardia, Cryptosporidia*), bacteria (e.g., *Escherichia coli, Shigella*), and viruses (e.g., norovirus, hepatitis viruses). The fourth group—chemicals (e.g., pesticides, heavy metals)—consists of pollutants that may enter the water from other sources.

Waterborne Diseases and Fecal Contamination

Waterborne diseases are conditions that are "transmitted through the ingestion of contaminated water and water acts as the passive carrier of the infectious agent."[10(p371)] Waterborne diseases are a source of ongoing concern to the residents of the developing world, where waterborne diarrheal conditions take a great toll in morbidity and mortality, and to tourists who travel in these areas. In the developing world, bacterial waterborne diseases are associated with the deaths of newborns and children from conditions such as dehydration. An example of a condition transmitted by water is acute gastroenteritis, defined as an "inflammation of the stomach and or intestines or both."[11] In addition to gastroenteritis, many serious diseases can be spread through the water supply. To some extent, these diseases also affect developed countries including the United States. The text box titled *The Global Importance of Preventing Waterborne Diseases* presents facts about the significance of waterborne diseases caused by bacterial agents.

Despite reassurances by officials about its safety, the public water supply in some areas of the United States occasionally may be responsible for transmitting serious waterborne diseases. When such transmission occurs via the public water supply, it is usually a consequence of a malfunctioning water-processing system (e.g., the 1993 cryptosporidiosis outbreak in Milwaukee, Wisconsin, discussed later in the chapter).

Many pathogens—bacterial, viral, and protozoan—are responsible for waterborne infections. Examples of waterborne diseases are cryptosporidiosis and giardiasis (both caused by protozoan agents), cholera (agent *Vibrio cholerae*), and infection with *E. coli* O157:H7.[10] Other conditions that may be transmitted by water include infections caused by *Legionella pneumophila* and certain viruses (e.g., hepatitis A virus). Symptoms of waterborne diseases caused by bacteria include a form of severe diarrhea that can result in dehydration (as in cholera), fevers and abdominal distress of long duration (as in typhoid fever), and in some cases acute bloody diarrhea (as in *E. coli* O157:H7). The following is a list of some of the waterborne pathogens:

THE GLOBAL IMPORTANCE OF PREVENTING WATERBORNE DISEASES

Waterborne diseases are inextricably associated with poor water quality. A crucial public health measure for preventing waterborne illness is provision of improved water sources.

The World Health Organization (WHO) and the United Nations Children's Fund (UNICEF) in 2000 estimated that approximately 4 billion episodes of diarrhea occurred annually worldwide.[a] These episodes were linked to about 2 million deaths, which primarily impacted children. Bacterial enteropathogens (discussed in text) such as *Salmonella, Escherichia coli*, and *Vibrio cholerae* are associated with diarrheal illnesses. Other diseases from microbial agents carried in unsanitary water include intestinal parasitic infections, trachoma (a cause of blindness), and schistosomiasis.

In 2000 the United Nations (UN) established Millennium Development Goals, which included a target of improved drinking water sources for 88% of the world's population. According to the UN, this target was met in 2010; by 2015 a total of 91% of the global population benefitted from improved water sources.[b] However, about 80% of residents of rural areas did not have improved drinking water sources as of 2015.

[a] WHO/UNICEF Joint Monitoring Programme for Water Supply and Sanitation. *Global Water Supply and Sanitation*. Geneva, Switzerland: World Health Organization and United Nations Children's Fund; 2000.

[b] UNICEF, World Health Organization. *Progress on Sanitation and Drinking Water—2015 Update and MDG Assessment*. Geneva, Switzerland: UNICEF World Health Organization and United Nations Children's Fund; 2015.

- Enteric protozoal parasites
  - *Entamoeba histolytica*
  - *Giardia intestinalis*
  - *Cryptosporidium parvum*
  - *Cyclospora cayetanensis*
- Bacterial enteropathogens
  - *Salmonella*
  - *Shigella* (discussed in the chapter on food safety)
  - *Escherichia coli*
  - *Vibrio cholera*
  - *Campylobacter* (discussed in the chapter on food safety)
- Viral pathogens
  - Enteroviruses
  - Adenoviruses
  - Noroviruses (formerly called Norwalk-like viruses)

- Other agents
  - *Dracunculus medinensis*
  - *Legionella pneumophila*

Enteric Protozoal Parasites

Cryptosporidiosis. The infectious agent that causes cryptosporidiosis is called *Cryptosporidium parvum*, a protozoal organism. The symptoms of cryptosporidiosis include watery diarrhea, abdominal cramping, nausea, vomiting, and fever. The condition can be fatal among immunocompromised individuals.[12] **TABLE 9.1** presents a chronology of cryptosporidiosis in the United States. The first human case was diagnosed in 1976. Since then, several outbreaks have been linked to drinking water sources including public drinking water supplies.

The complex life cycle of the different species of *Cryptosporidium* is shown in **FIGURE 9.18**. (Note that *Cryptosporidium parvum* is the species of *Cryptosporidium* that causes infections in humans.) *Cryptosporidium* oocysts, which are the infectious component for cryptosporidiosis, are excreted in the stools of infected persons.[13] Water used for recreation and drinking may be contaminated with sewage that contains the oocysts of *C. parvum*. Following ingestion by other human hosts, the oocysts (see **FIGURE 9.19**) reproduce. Resistant to chlorine treatment, infective oocysts may survive treatment by water treatment plants, even modern installations.

The most notorious waterborne outbreak of cryptosporidiosis in the United States occurred in southern Milwaukee in late March and early April 1993. More than 400,000 people were estimated to be affected during this major outbreak.[14] This number actually may undercount the number of affected persons, as the case definition was limited to watery diarrhea.[12]

The outbreak was linked to the Milwaukee Water Works (MWW), the supplier of water for the city of Milwaukee and nine surrounding municipalities in Milwaukee County. Investigators believed that *Cryptosporidium* oocysts (see Figure 9-19) from untreated

TABLE 9.1 Cryptosporidiosis (Chronology)

1976	First human case diagnosed
1984	First well water outbreak
1987	First river water outbreak
1992	Multiple municipal water supply outbreaks
1993	Largest recorded waterborne outbreak in US history (Milwaukee, Wisconsin) Fresh-pressed apple cider outbreak (central Maine)
1994	First outbreak in community with state-of-the-art water treatment (Las Vegas, Nevada)

Modified and reproduced from CDC Public Health Image Library. ID #109. NCID Content Provider. Available at: http://phil.cdc.gov/Phil/details.asp. Accessed September 18, 2017.

Selected outbreaks after 1994	
1997	Children playing in a water sprinkler fountain at a zoo, Minnesota[a]
2007	A splash park, Idaho, 2007[b]
2013	Baker City, Oregon, from municipal water supply contamination[c]

[a]Centers for Disease Control and Prevention. Outbreak of cryptosporidiosis associated with a water sprinkler fountain—Minnesota, 1997. *MMWR*. 1998;47:856-860; [b]Centers for Disease Control and Prevention. Outbreak of cryptosporidiosis associated with a splash park—Idaho, 2007. *MMWR*. 2009(22);58:615-618; [c]DeSilva MB, Schafer S, Kendall Scott M, et al. Communitywide cryptosporidiosis outbreak associated with a surface water-supplied municipal water system—Baker City, Oregon, 2013. *Epidemiol Infect*. 2016;144(2):274–284.

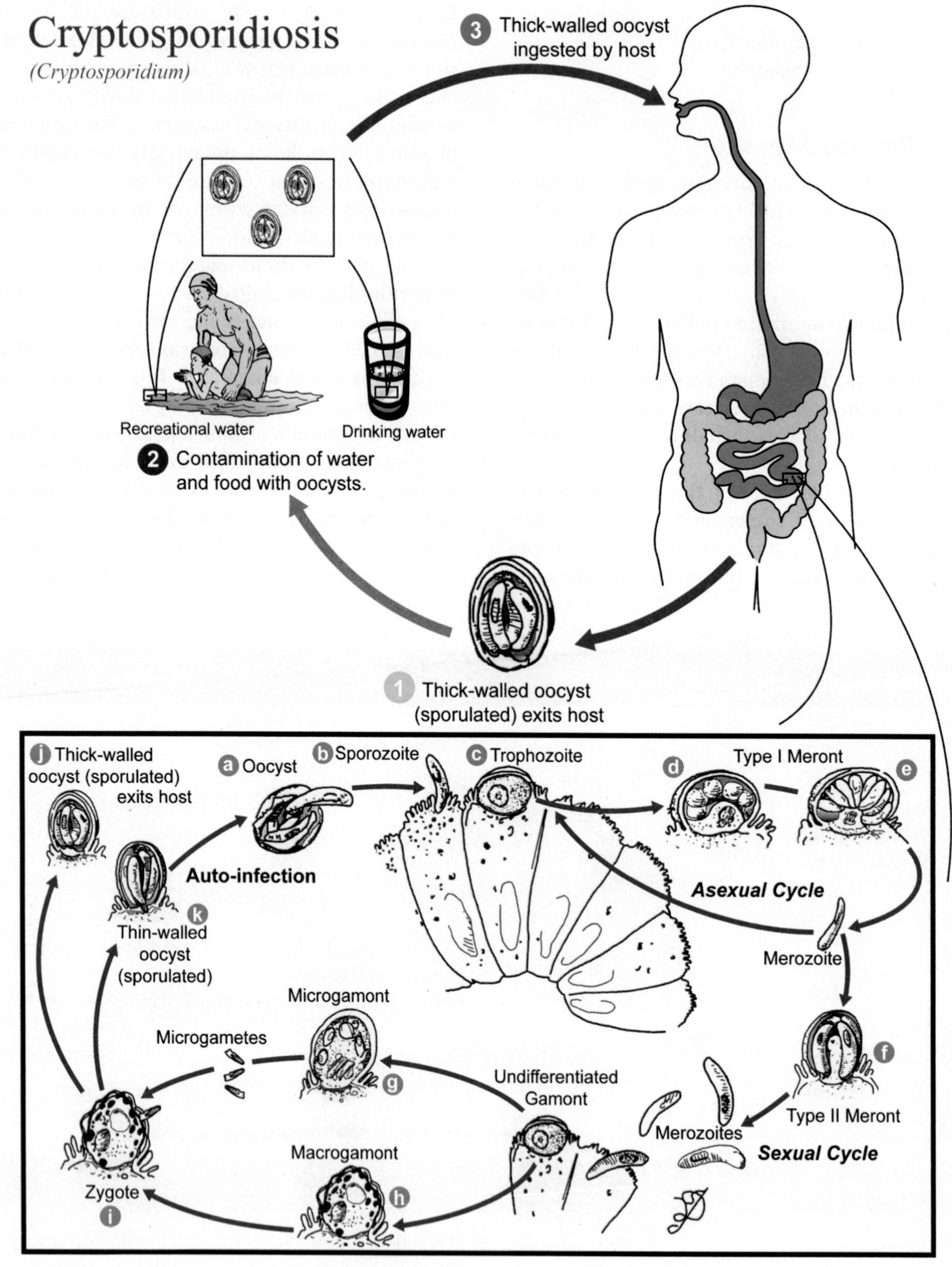

FIGURE 9.18 This illustration depicts the life cycle of different species of *Cryptosporidium*, the causal agents of cryptosporidiosis.

Courtesy of CDC/Alexander J. da Silva, PhD/Melanie Moser

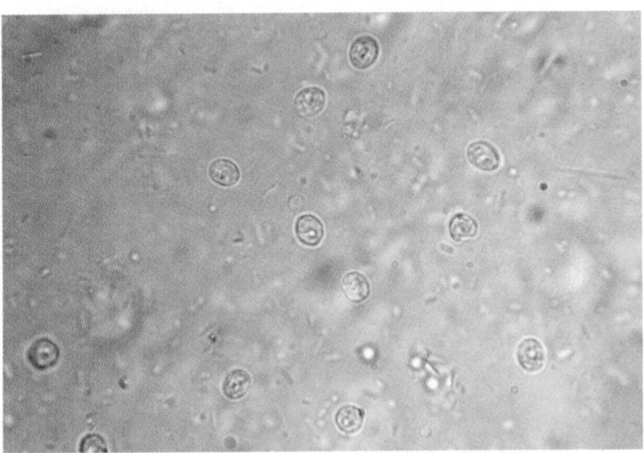

FIGURE 9.19 Oocysts of *Cryptosporidium parvum*. Oocysts are spheroidal objects, 4 to 6 microns (µm) in diameter.
Courtesy of CDC/ Dr. Peter Drotman

Lake Michigan water were drawn into the water treatment plant. The usual water treatment processes (e.g., coagulation, filtration, and chlorination) were not adequate to remove the oocysts.

Two major rivers flow into the Milwaukee harbor. Along the rivers are slaughterhouses and effluents of human sewage. During a springtime period of heavy precipitation, the rivers may have transported oocysts into Lake Michigan, a source of water for the MWW. Between approximately March 21 and April 5, a marked increase in the turbidity of treated water was noted. On April 7 the MWW issued an advisory to customers to boil their water; subsequently, the MWW closed the plant temporarily on April 9. This massive outbreak in Milwaukee suggests the need to continuously monitor the quality of drinking water (especially for turbidity), to conduct **public health surveillance** of diarrheal illnesses caused by water, and to recognize the hazards that *Cryptosporidium* outbreaks pose for the public.

Contaminated water has been linked to most known outbreaks of cryptosporidiosis in the United States.[15] A study conducted in the San Francisco Bay area evaluated the risk of transmission of endemic cryptosporidiosis by drinking water. The study did not support the hypothesis that drinking water is a risk factor for cryptosporidiosis among the immunocompetent population. Cases of cryptosporidiosis identified by a surveillance system found that travel to another country was a strong and significant risk factor for cryptosporidiosis.

Other sources of *Cryptosporidium* infections are public swimming pools and water parks used by large numbers of diapered children. These settings have been implicated in outbreaks of cryptosporidiosis. It is known that very low doses of *C. parvum* are required for infection. The oocysts are resistant to chlorine treatment; therefore, bathers should avoid drinking water used in public swimming pools. Parents of diapered children should be educated regarding the supervision of their children so that swimming pools do not become contaminated with fecal material.[16]

Amebiasis. Amebiasis is caused by the protozoal parasite *Entamoeba histolytica*, which produces cysts that are carried in human feces.[17] Transmission occurs via the ingestion of cysts that are contained in food and water that have been contaminated by feces. Both asymptomatic and symptomatic illnesses are associated with the organism. The latter include invasive intestinal amebiasis (e.g., dysentery, colitis, and appendicitis) and extraintestinal amebiasis (e.g., abscesses of the lungs and liver). When traveling to a less developed area of the world where sanitary conditions are wanting, tourists should avoid eating fresh fruits and vegetables they did not peel themselves, unpasteurized milk and other unpasteurized dairy products, and food sold by street vendors.[18]

Giardiasis. The agent that is responsible for giardiasis is *Giardia lamblia* (also known as *Giardia duodenalis*), a protozoal organism.[19] (Refer to **FIGURE 9.20**.) *G. lamblia* produces cysts that transmit the condition via contaminated food and water. The cysts have the ability to survive for long periods in cold water. The incubation period for giardiasis varies from 3 to 25 days, with approximately 7 to 10 days being typical. Symptoms of giardiasis include gastrointestinal effects such as acute or chronic diarrhea; some cases are asymptomatic.[13]

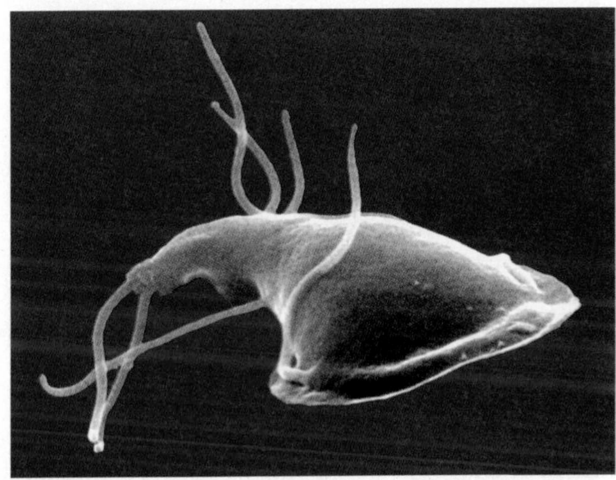

FIGURE 9.20 The protozoan parasite *Giardia lamblia*, causative agent for giardiasis.
Courtesy of CDC/Dr. Stan Erlandsen; Dr. Dennis Feely

Cyclosporiasis (an emerging parasitic disease). In common with other protozoal parasites discussed in this section, the causative organism *Cyclospora cayetanensis* is transmitted via ingestion of food and water that have been contaminated with cysts (oocysts) from the organism. Foodborne outbreaks have involved fresh produce (e.g., snow peas and raspberries). During May through July 2007, a cluster of 29 cases of cyclosporiasis, reported in British Columbia, Canada, was linked to imported basil.[20] The protozoal parasite causes watery diarrhea and other symptoms of abdominal distress in addition to low-grade fever.

Bacterial Enteropathogens

Examples of microorganisms in this category are *Salmonella*, *Escherichia coli* O157:H7, and *Vibrio cholerae*. *Salmonella* bacteria are associated with a number of foodborne (see the chapter on food safety) and waterborne diseases. The variant (serogroup) of *Salmonella* called *Salmonella enterica* serogroup Typhi (*S. enterica* Typhi) causes typhoid fever, which is transmitted by fecal-contaminated water and food (e.g., sold by street vendors). Typhoid fever is a very acute disease that causes fever, headache, and chills; in severe cases, it may produce confusion, delirium, and death.[21] According to the CDC about 400 typhoid fever cases occur in the United States each year. Globally, typhoid fever causes approximately 21 million cases of disease and 200,000 deaths annually. Historically, there have been many examples of waterborne outbreaks: during 1973, an outbreak of 188 proved or presumptive cases occurred at a migrant labor camp in Dade County, Florida.[22] Another outbreak was reported in 1985 in Haifa, Israel, and involved 77 confirmed cases.[23]

The bacterium *E. coli* may be transmitted through contact with contaminated lakes and swimming pools and via ingestion of contaminated water. The same organism also may be transmitted by contaminated food, as discussed in the chapter on food safety. The following text box describes a waterborne outbreak of *E. coli*.

The bacterium that causes cholera is known as *Vibrio cholerae*. The inhabitants of developed countries such as the United States (annual incidence 0 to 5 cases) have very low risk of contracting cholera, which primarily affects the developing world.[24] A continuing pandemic, lasting for four decades in Asia, Africa, and Latin America, has been a cause of epidemic diarrhea. The underlying factors responsible for cholera outbreaks are an inadequate infrastructure for processing water and migrations of large numbers of people to urban areas. Although cholera is a treatable condition, the disease has a case fatality rate of as high as 50% when untreated. The symptoms of cholera are vomiting and profuse watery diarrhea that can culminate in shock and death.

A WATERBORNE OUTBREAK OF *ESCHERICHIA COLI* O157:H7

In the summer of 1998, a large outbreak of *Escherichia coli* O157:H7 infections occurred in Alpine, Wyoming. [There were] 157 ill persons; stool from 71 (45%) yielded *E. coli* O157:H7. In two cohort studies, illness was significantly associated with drinking municipal water. . . . The unchlorinated water supply had microbiologic evidence of fecal organisms and the potential for chronic contamination with surface water. Among persons exposed to water, the attack rate was significantly lower in town residents than in visitors (23% vs. 50%, $p < 0.01$) and decreased with increasing age. The lower attack rate among exposed residents, especially adults, is consistent with the acquisition of partial immunity following long-term exposure. Serologic data, although limited, may support this finding. Contamination of small, unprotected water systems may be an increasing public health risk.

Reproduced from Olsen SJ, Miller G, Brever T, et al. A waterborne outbreak of *Escherichia coli* O157:H7 infections and hemolytic uremic syndrome: implications for rural water systems. *Emerg Infect Dis*. 2002;8:370–375.

Viral Pathogens

Over 100 kinds of viruses are found in human stools and pose a potential for transmission by water.[25] "These viruses are more resistant to environmental conditions and sewage treatment processes, including chlorination and ultraviolet (UV) radiation, than many of the sewage-associated bacteria."[25(p179)] Virus-associated conditions that may be spread through water are viral gastroenteritis and viral hepatitis. "The human enteric virus group, which includes Norwalk virus, rotavirus, hepatitis A virus, adenovirus, and enterovirus, is one of the leading causes of human illness."[25(p179)] The term *enteric* is defined as relating to the intestines. Disease outbreaks caused by enteric viruses have been associated with water and food that were contaminated by virus-laden human stools.

Two causes of viral gastroenteritis are noroviruses and adenoviruses. Infection with norovirus produces nausea, vomiting, and watery diarrhea with cramping.

Symptoms of norovirus-associated gastroenteritis usually continue for 24 to 72 hours. Most patients recover fully and do not have residual medical issues. Young children, elderly adults, and immunocompromised individuals can experience serious illness. Transmission routes for noroviruses include ingestion of contaminated food or water and close contact with those who are infected.[26]

The most common means for spreading adenoviruses are direct person-to-person contact, via the air from coughing and sneezing, and from contaminated surfaces. The stools of infected persons (e.g., on diapers) may carry some forms of adenoviruses. Among the diseases that adenoviruses cause are respiratory illnesses (most frequent outcome) and a range of conditions such as gastroenteritis and conjunctivitis. Less commonly, adenoviruses also are spread by waterborne transmission. Adenoviruses have been detected in swimming pools and small lakes.[27] High concentrations of these viruses have been detected near the mouths of rivers in southern California.[25]

Viral hepatitis A (caused by the hepatitis A virus) can be spread via person-to-person contact (e.g., the fecal-oral route) and by contaminated food (e.g., shellfish, fruits, and vegetables), water, and ice.[13,28,29] Countries with poor sanitation can be settings for the spread of hepatitis A. Infected persons who have deficient personal hygiene can spread the virus as well. Viral hepatitis E (a liver infection with the hepatitis E virus) is also transmitted by the fecal-oral route and is found in developing countries with inadequate sanitation and drinking-water contamination. The most frequent mode of transmission is via polluted drinking water. In order to prevent hepatitis E infection, travelers who visit developing countries with unsafe water supplies, for example, in Asia, the Middle East, Africa, and Central America need to avoid drinking tap water.

Due to the presence of infectious agents and toxins in sewage, workers who come into contact with sewage are at increased risk in comparison to the general population of contracting infectious diseases, such as hepatitis C (another form of viral hepatitis), caused by agents present in fecal matter.[30]

Other Agents

Guinea worm. Although there are several important waterborne diseases in this category, one that is noteworthy for the developing world is dracunculiasis (also called guinea worm disease).[31] As the result of an eradication campaign, the range for guinea worm is confined to rural sections of a small number of African countries. The causative organism is the nematode (a kind of roundworm) *Dracunculus medinensis*, which forms larvae that enter the water supply. The larvae infect small crustaceans in the water that are ingested in unfiltered water. In the abdominal cavity of the human host, the larvae mature into worms. Female worms that range in length from 3 to 6 feet (70 to 120 cm) then migrate to the surface of the skin of the feet where they produce painful blisters. The female worm then erupts from this lesion and, when the victim comes into contact with a body of water, releases larvae back into the water, thus perpetuating the cycle of infection. Refer to **FIGURE 9.21** for an illustration of the clinical features of dracunculiasis as well as additional information.

Legionellosis. The term *legionellosis* refers to illnesses caused by the Legionnaires' disease bacterium; the condition is classified as a waterborne disease.[13] The two distinct clinical forms of legionellosis are Legionnaires' disease and Pontiac fever (a milder form of legionellosis with pneumonia absent). Legionnaires' disease produces fever, cough, and pneumonia[13,32] and is associated with a case-fatality rate of about 15%. More than 25,000 Legionnaires' disease cases (and over 4,000 deaths) occur annually.[32]

The Legionnaires' disease bacterium known as *Legionella pneumophila* was first identified in 1977 after causing an outbreak of pneumonia that resulted in 34 deaths in 1976 at a meeting of the American Legion Convention—hence the name Legionnaires' disease. Microbiologists now recognize that more than 43 species of *Legionella* exist and that more than 20 of them are associated with human disease. However, *L. pneumophila* is regarded as the causative agent for most cases of legionellosis.[32]

Low levels of Legionnaires' disease bacteria are found in freshwater lakes, streams, and rivers. In the environment, some species of protozoa (e.g., some varieties of amoeba) serve as a host for *L. pneumophila*.[33] The organism can grow in domestic water systems, cooling systems, and whirlpool spas.[13] Transmission of the bacterium can occur when water that is rich in *L. pneumophila* becomes aerosolized and is inhaled or ingested in such a way that it enters the respiratory system. The United States Department of Labor notes that "[c]ooling towers, evaporative condensers, and fluid coolers use a fan to move air through a recirculated water system. This allows a considerable amount of water vapor and sometimes droplets to be introduced into the surroundings, despite the presence of drift eliminators designed to limit droplet release. This water may be in the ideal

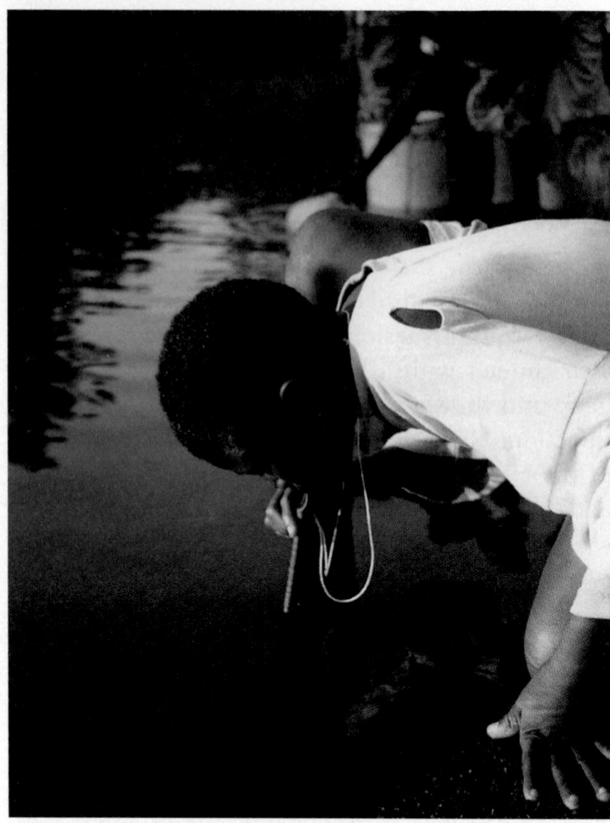

A. Portable pipe filters are a convenient method for preventing Guinea worm disease. A young boy uses a portable pipe filter to prevent transmission of Guinea worm disease.

The Carter Center/E. Staub

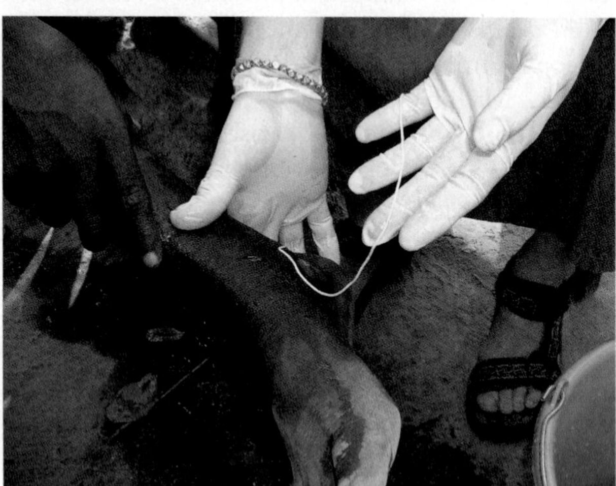

B. Traditional treatment of Guinea worm disease consists of wrapping the two- to three-foot-long (61- to 91-cemtimeter -long) worm around a small stick and extracting it: a slow, painful process that often takes weeks. The figure shows an emerging Guinea worm.

The Carter Center/E. Wolfe

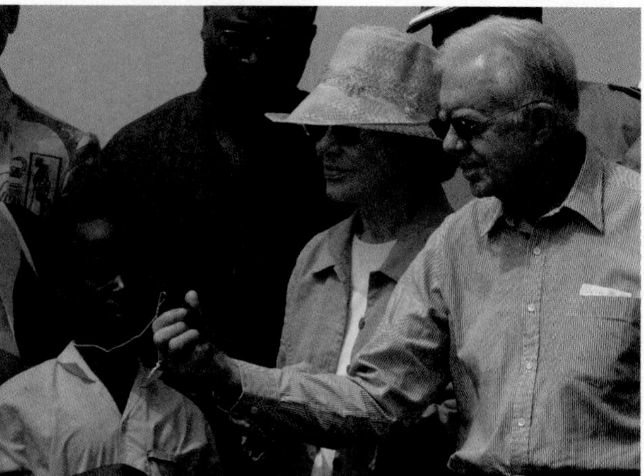

C. Former U.S. President Jimmy Carter and his wife, Rosalynn, examine a pipe filter in northern Ghana. Since 1986, The Carter Center has led the global campaign to eradicate Guinea worm disease.

The Carter Center/A. Poyo

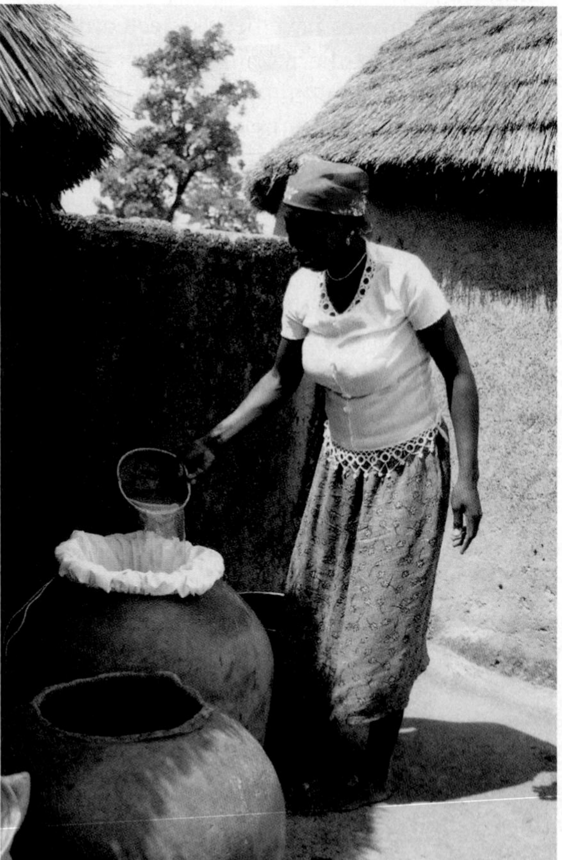

D. Guinea worm disease incapacitates its victims so adults cannot work and children cannot play or attend school. Filters catch the microscopic water fleas that contain infective larvae. This woman is using such a cloth filter.

The Carter Center/E. Staub

FIGURE 9.21 Guinea worm disease and the use of filters.

temperature range for Legionnaires' disease bacteria (LDB) growth, 20°–50°C (68°–122°F)."[34] **FIGURE 9.22** illustrates a cooling tower. Because the growth of *L. pneumophila* is possible in some types of cooling towers, proper maintenance of them is essential.

Chemicals in the Water Supply

As previously noted, chemicals enter the water supply from natural and anthropogenic sources. These substances may be present as contaminants in drinking water.

> A number of chemical contaminants have been identified in drinking water. These contaminants reach drinking water supplies from various sources, including municipal and industrial discharges, urban and rural run-off, natural geological formations, drinking water distribution materials and the drinking water treatment process. Chemical contaminants for which epidemiologic studies have reported associations include the following: aluminum, arsenic, disinfection by-products, fluoride, lead, pesticides and radon. Health effects reported have included various cancers, adverse reproductive outcomes, cardiovascular disease and neurological disease.[35(pS13)]

Pharmaceutical and personal care products (PPCPs) represent another source of water contamination. These chemicals may be either washed off or excreted from the body; examples of PPCPs that may end up in untreated sewage effluent, surface water, groundwater, and drinking water are analgesics, oral contraceptive agents, drugs for lowering cholesterol, and anticonvulsants.[36] In fact, the term *sewage epidemiology* describes the field of monitoring excreted drugs in the sewer system to assess the level of illicit drug use in the community. The chapters on toxic metals and elements and pesticides and other organic chemicals cover some of the hazards associated with metals, pesticides, and other chemicals. The following section will provide additional information about the effects of chemical contamination of drinking water.

Water Disinfection By-Products (DBPs)

The chemicals used to disinfect water include chlorine, chloramines, chlorine dioxide, and ozone. These chemicals are associated with by-products of chlorination called DBPs.

For example, chlorine is associated with trihalomethanes (THMs), which are among the most common and widely measured DBPs.[37,38] Researchers have probed the relationship between DBPs and several forms of cancer (e.g., bladder and rectal cancer) as well as adverse reproductive outcomes. Several epidemiologic studies have investigated the association between chlorination of drinking water and risk of birth defects; two examples of supportive studies were those conducted by Smith et al.[39] and Hwang and colleagues.[37]

FIGURE 9.22 Air conditioning cooling tower.

Smith et al. reported an association between exposure to THMs during pregnancy and lowered birth weight; the relationship differed by ethnicity of study participants.[39] The setting for the study was Bradford, a highly diverse city in northern England. Many of Bradford's residents experience social and economic deprivation. Research participants (n = 7,428) were babies from the "Born in Bradford" cohort as well as their mothers. A questionnaire was used to assess each mother's water consumption during pregnancy. Data regarding THM concentrations were provided by the local water company.

Hwang and colleagues examined data from a cross-section of 285,631 Norwegian births in 1993–1998.[37] The results of this cross-sectional study suggested that consumption of chlorinated waters that contained large amounts of natural organic materials was related to increased risk of birth defects.

Nevertheless, at this stage, little is known about the potential adverse health effects of exposure to DBPs.[38] Many researchers who have addressed this topic have been confronted with methodological difficulties, especially with respect to exposure assessment. Typically, studies rely on approximate methods of exposure assessment that do not account for individual variability in patterns of water consumption (e.g., the amount of water consumed, whether water is consumed at work as well as at home, and use of bottled and filtered water).[40]

Solvent-Contaminated Drinking Water

Industrial chemicals may infiltrate the underground aquifers used for public water supplies. An example is contamination of water by a leaking underground solvent tank used by a semiconductor plant in Santa Clara County, California.[41,42] Although the tank held many different chemicals, the chemical solvent trichloroethane (discussed in the chapter on pesticides and other organic chemicals) was present in the highest concentration. Unfortunately, the tank was located near a local water company's drinking water well. The leak was discovered in November 1981; by December 7 the levels of trichloroethane at the well were 1,700 parts per billion (ppb) and in mid-December reached 8,800 ppb. The level of 1,700 ppb exceeded by 8.5 times the level required by the state of California for remediation. Residents of an area affected by the contaminated water supply believed that an excessive number of spontaneous abortions and birth defects were occurring in their neighborhood. An investigation conducted in 1983 showed significant associations between exposure to the contaminated water and adverse pregnancy outcomes (congenital malformations and spontaneous abortions), but was unable to demonstrate a causal connection. A follow-up study suggested that the solvent leak was unlikely to have caused the increased numbers of adverse pregnancy outcomes.

Beach and Coastal Pollution

The final topic in this chapter is pollution of ocean water. The seemingly boundless world's ocean, which covers 70% of the planet, performs essential functions necessary for maintaining life on earth. According to the National Ocean Service (a branch of the National Oceanic and Atmospheric Administration [NOAA]), these actions include production of more than half of global oxygen and regulation of climate.[43] Other benefits include marine transportation, recreation, economic resources, food, and medicine. Alarmingly, sewage, untreated wastewater, and urban runoff endanger coastal areas. Plastics (e.g., plastic bags, disposable drink containers, discarded plastic medical devices, and microbeads in personal care products) wash up on our coasts and drift far out into the ocean where they form gargantuan masses that threaten sea creatures. Other troubling sources of ocean pollution are from drilling oil, petroleum releases from oil platforms, and oil spills from tankers.

Some estimates suggest that half of the world's coastal areas are endangered.[44] **FIGURE 9.23** shows a number of human activities that lead to coastal degradation. These actions include draining of coastal ecosystems, dredging, solid waste disposal, construction of dams for flood control, discharge of wastes from farms and industries, and logging activities.

The approximately 1 billion people who live near coastal areas cause great stress on coastal ecosystems. Growth rates of populations near coastal regions are estimated to be twice the rate of worldwide population growth.[45] As the population mushrooms, coastal areas are threatened by overdevelopment, poor planning, and economic expansion. Each day the world's coastal regions are the recipients of billions of gallons of treated and untreated wastewater. During heavy rains urban runoff into the oceans degrades the quality of ocean water by adding microbial agents, nutrients, and chemical toxins. The problem of urban runoff is compounded by developments that result in the loss of **wetlands**, which act as natural water-filtering mechanisms. Excessive amounts of nutrients that enter the oceans may cause harmful blooms of algae, resulting in reduced levels of oxygen in the water (anoxic conditions). An anoxic ocean environment can bring about fish kills and damage other forms of ocean life.

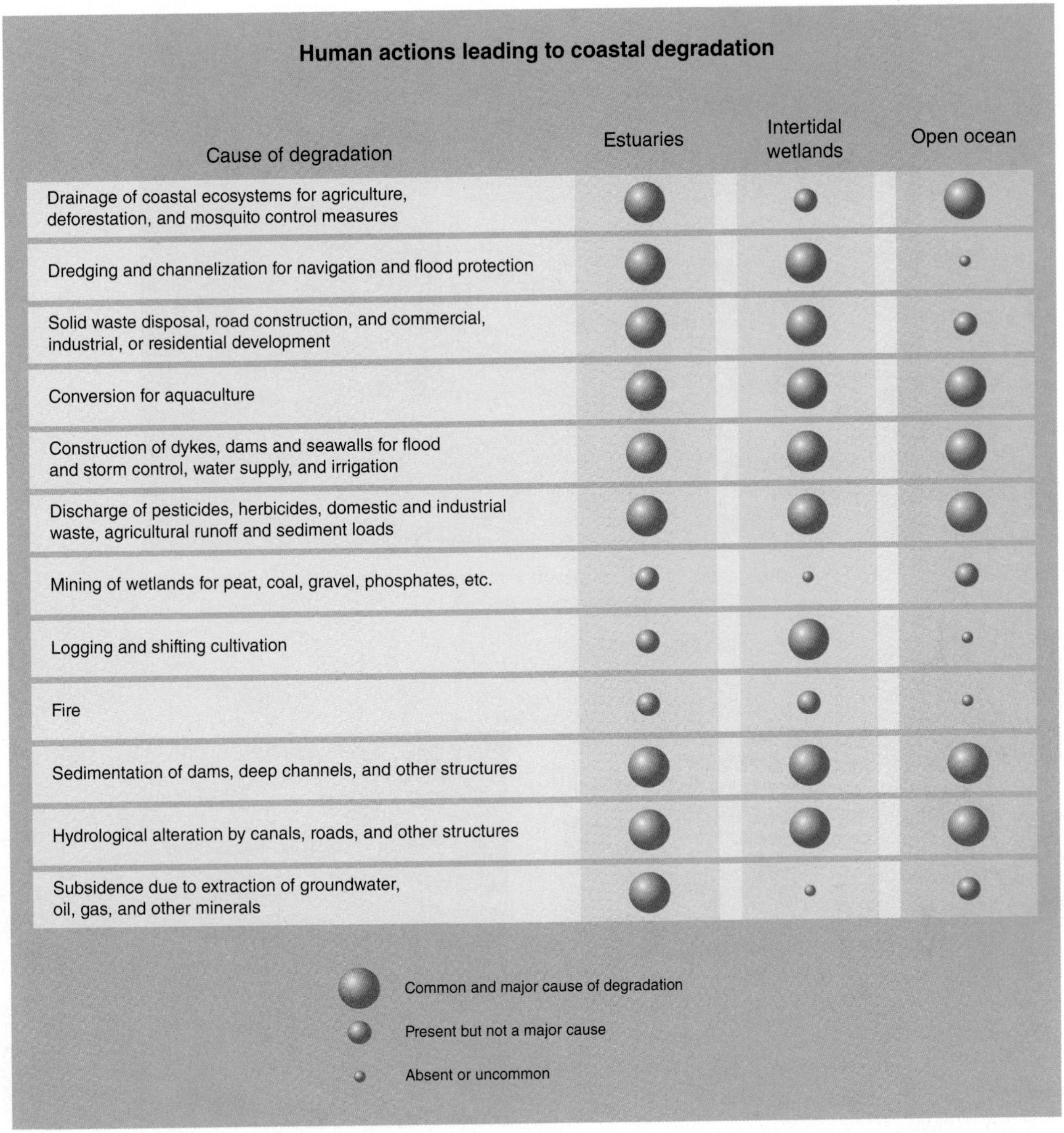

Human actions leading to coastal degradation

Cause of degradation	Estuaries	Intertidal wetlands	Open ocean
Drainage of coastal ecosystems for agriculture, deforestation, and mosquito control measures	●	•	●
Dredging and channelization for navigation and flood protection	●	●	·
Solid waste disposal, road construction, and commercial, industrial, or residential development	●	●	•
Conversion for aquaculture	●	●	●
Construction of dykes, dams and seawalls for flood and storm control, water supply, and irrigation	●	●	●
Discharge of pesticides, herbicides, domestic and industrial waste, agricultural runoff and sediment loads	●	●	●
Mining of wetlands for peat, coal, gravel, phosphates, etc.	•	·	•
Logging and shifting cultivation	•	●	·
Fire	•	•	·
Sedimentation of dams, deep channels, and other structures	●	●	●
Hydrological alteration by canals, roads, and other structures	●	●	●
Subsidence due to extraction of groundwater, oil, gas, and other minerals	●	·	•

● Common and major cause of degradation
• Present but not a major cause
· Absent or uncommon

FIGURE 9.23 Human actions leading to coastal degradation.

Reproduced from Vital Water Graphics: Coastal and Marine, by United Nations Environment Programme, © 2002 United Nations. Reproduced by permission of the publisher. Available at: http://www.unep.org/dewa/assessments/ecosystems/water/vitalwater/38.htm. Accessed June 5, 2017.

Urban runoff and sewage contamination of the ocean expose swimmers to waterborne diseases—gastrointestinal, respiratory, skin, and eye infections. As a result of ocean water contamination, beach closings may transpire some parts of the United States, especially after heavy rainfall that creates urban runoff. (See **FIGURE 9.24**.)

The EPA conducts an ongoing survey of beaches as authorized by the Beaches Environmental Assessment and Coastal Health Act (BEACH Act of 2000.)[46] The EPA surveyed a total of 3,762 beaches regarding advisories (meaning possible risks of swimming in the water) or closings that occurred during the 2012 swimming season and reported that about 40% had one or more advisories or closures. By far, the most frequent cause of beach closings was elevated bacteria levels that exceeded water quality standards. **FIGURE 9.25** shows the percentage of beaches with one of more notification actions.

Storm water runoff is a leading cause of ocean pollution. Garbage that is discarded carelessly on the street may end up in the ocean or on beaches, as **FIGURE 9.26** demonstrates. Following heavy rains, visitors to the

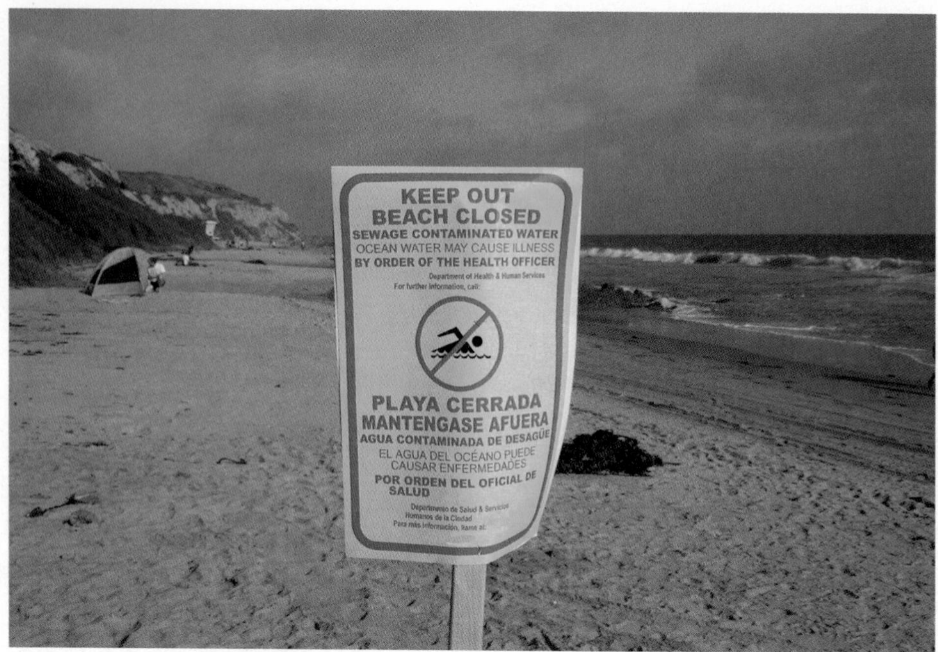

FIGURE 9.24 Beach closing sign. With increasing levels of pollution in some areas, beach closings have become more frequent.

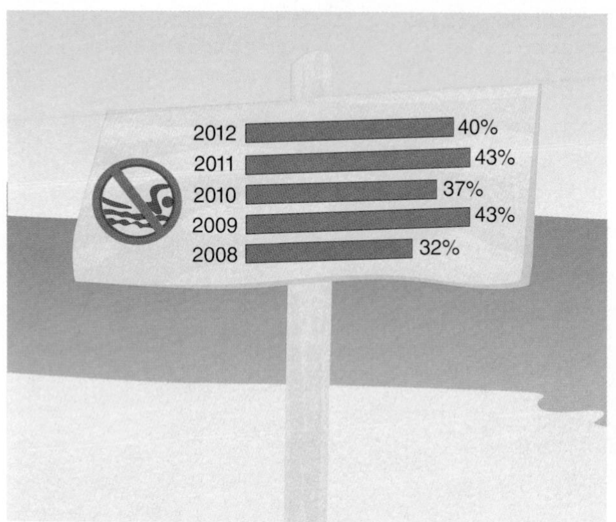

FIGURE 9.25 Percent of beaches with one or more notification actions.

Reproduced from U.S. Environmental Protection Agency. *EPA's BEACH Report*: 2012 Swimming Season. June 2013. EPA 820-F-13-014, p. 2

FIGURE 9.26 Ocean pollution evident in the harbor at the port town of Cobh, Ireland.

beach are likely to find massive quantities of nonbiodegradable plastics (e.g., Styrofoam cups, rings from beverage six packs, and plastic bags) that are littering the shoreline. In addition to being offensive, these materials endanger birds and fish. Motor vehicle oil and automotive chemicals that are dumped carelessly into storm drains also eventually pollute the ocean.

Petroleum Spills

Oil spills from tankers and offshore drilling platforms can have a devastating impact on the shoreline, aquatic life, mammals, and birds. One of the worst oil spills in US history was caused by the tanker *Exxon Valdez*, described in the following text box and shown in **FIGURE 9.27**.

Oil platforms similar to the one shown in **FIGURE 9.28** have been associated with accidental releases of crude oil. January 28, 1969 was the date of an extensive oil spill off the coast of the city of Santa Barbara, California.[47] The spill originated from an offshore drilling rig known as platform Alpha, operated by the Union Oil Company. This episode, which lasted for 11 days and allowed 3 million gallons (11 million liters) of oil to escape, caused

EXXON VALDEZ OIL SPILL, ALASKA

On March 24, 1989, the tanker *Exxon Valdez*, en route from Valdez, Alaska to Los Angeles, California, ran aground on Bligh Reef in Prince William Sound, Alaska. The vessel was traveling outside normal shipping lanes in an attempt to avoid ice. Within 6 hours of the grounding, the *Exxon Valdez* spilled approximately 10.9 million gallons [41.3 million liters] of its 53 million gallon (200 million liter) cargo of Prudhoe Bay Crude. Eight of the 11 tanks on board were damaged. The oil would eventually impact over 1,100 miles (1,770 kilometers) of noncontinuous coastline in Alaska, making the *Exxon Valdez* the largest oil spill to date [as of 1992] in US waters.

The response to the *Exxon Valdez* involved more personnel and equipment over a longer period of time than did any other spill in US history. Logistical problems in providing fuel, meals, berthing, response equipment, waste management, and other resources was one of the largest challenges to response management. At the height of the response, more than 11,000 personnel, 1,400 vessels, and 85 aircraft were involved in the cleanup....

Concern over oil-related wildlife mortality was intense during the spill. The grounding occurred at the beginning of the bird migration season. The US Fish and Wildlife Service estimated that mortalities directly related to the spill ranged from 350,000 to 390,000 birds, especially common and thick-billed murres, assorted sea ducks, bald eagles, and pigeon guillemots; 3,500 to 5,500 sea otters; and 200 harbor seals. In addition, killer whales may have been affected by the spill as their numbers in the area declined shortly after the spill. Of the 1,630 birds (over 36,000 dead birds were collected) and 357 sea otters that were trapped and treated by the International Bird Rescue Research Center (IBRRC)–run facilities (established in Homer, Kodiak, Seward, and Valdez in response to this spill), the total survival rate was 50.7% for birds, and 62% for sea otters. These survival rates are considered very good for oil impacted animals.

Unlike birds, sea otters had to be anesthetized to be washed, which increases the risk to the animal, and increases the cost of rehabilitation. The sea otter rehabilitation program was complex, with a total of 29 veterinarians, and 9 veterinarian technicians scheduled to provide 24-hour care. The resulting cost of the sea otter rehabilitation program was at least $51,000 per sea otter. The highest percentages of sea otter fatalities (60%) were recorded in the first 3 weeks of the spill.

Reproduced from National Oceanic and Atmospheric Administration. *Oil Spill Case Histories, 1967–1991, Summaries of Significant US and International Spills*. Report No. HMRAD 92-11. Seattle, Washington; 1992.

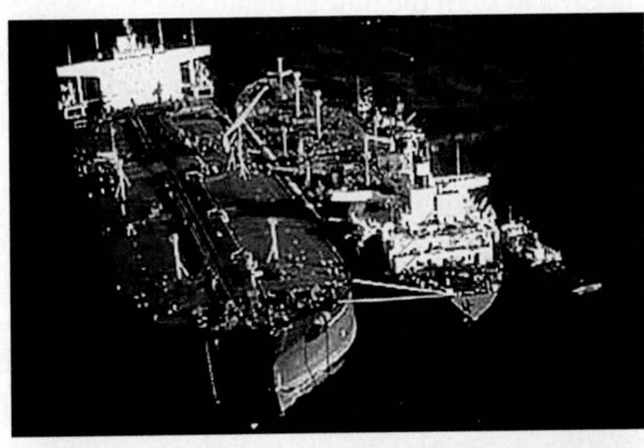

FIGURE 9.27 The *Exxon Valdez* oil spill. Upper left: Oil being removed (lightered) from the ship. Upper right: *Exxon Valdez* surrounded by a boom. Lower left: Biologist measuring depth of oil penetration. Lower right: Workers using high-pressure, hot-water washing to clean an oiled shoreline.

Reproduced from National Oceanic and Atmospheric Administration, NOAA's Ocean Service, Office of Response and Restoration. Image Galleries: *Exxon Valdez* Oil Spill. Available at: http://response.restoration.noaa.gov/gallery_gallery_photo.php? Accessed April 5, 2010.

FIGURE 9.28 Oil platform off the California coast.

an ecological catastrophe. The affected shoreline is an environmentally sensitive area noted for its pristine waters and abundant sea life. As a result of the spill, 3,600 ocean-feeding sea birds were killed. The carcasses of numerous poisoned seals and dolphins had to be removed from the beaches. This catastrophe was responsible for killing an untold number of fish and invertebrates and wreaked havoc on sensitive kelp forests.

In addition to the *Exxon Valdez* and Santa Barbara oil spills, there have been numerous other incidents. For example, the *Sea Empress* spilled approximately 72,000 tons (65,317 metric tons) of crude oil and 360 tons (326 metric tons) of heavy fuel oil when it ran aground at the entrance of Milford Haven harbor near Wales on February 15, 1996.[48] Nearby residents reported strong smells and complained of various symptoms that were associated with the spill. When contrasted with people living in control areas, residents in the exposed area indicated that they had higher rates of physical and psychological symptoms, which were consistent with the known toxicological effects of oil. The *Nakhodka* oil spill, which occurred on January 2, 1997, caused more than 6,000 tons (5,443 metric tons) of oil to be spilled into the Sea of Japan. Most of the oil drifted to the shoreline of the western portion of the island of Honshu, which is Japan's main island. Cleanup of the spill required the use of hand tools such as ladles and buckets because the area was inaccessible to machines. Local residents and those who were involved in the cleanup efforts experienced a number of acute symptoms of exposure, including low-back pain, headache, and effects on the eyes and skin.[49]

The process of extracting oil from petroleum fields has caused significant environmental damage in some areas of the world. The tropical forest areas of the Ecuadorian Amazon hold large oil reserves. As a result of drilling and extraction activities, billions of gallons of untreated wastes and oil have been dumped into the environment in this region. To prevent further damage to this ecologically sensitive area, controls will be needed to stem the flow of pollutants.[50]

On April 20, 2010, an explosion at the *Deepwater Horizon* oil platform and the resulting damage to the wellhead caused the largest marine oil spill in history up to that time. The catastrophe released 5 million barrels of oil into the Gulf of Mexico, from a location that is about 50 miles (80 kilometers) south of the Louisiana coastline. Information about the spill is provided in the following text box. **FIGURE 9.29** illustrates firefighting efforts at the site. A report of the effects of the oil spill is available from the Institute of Medicine.[51]

Conclusion

Water is an essential element of life on earth—a human being can survive for only about one week without water. As the global population expands and increasing industrialization creates ever-growing demands for water, supplies of this vital commodity have become endangered. Several areas of the world currently face chronic shortages and water stress; often these are the same areas that are experiencing skyrocketing population growth.

Surface water and groundwater are used as the sources of potable water in the majority of the countries of the world. These sources are vulnerable to contamination from human activities such as sewage

THE *DEEPWATER HORIZON* OIL SPILL, GULF OF MEXICO

Background: On the night of April 20, 2010, an explosion followed by a conflagration devastated the *Deepwater Horizon* oil platform, which eventually sank into the depths of the Gulf of Mexico. The drilling rig, owned and operated by Transocean Ltd. under contract with the British Petroleum Oil Company, was located in the Gulf approximately 50 miles (80 kilometers) southeast of the Mississippi River delta. The drilling rig extracted crude oil from a well that was about 5,000 feet (1,524 meters) below the surface of the water. When the platform sank, oil began gushing from the well. The extreme depth of the well challenged efforts to staunch the oil flow.

Dates of the spill: The spill commenced on April 20, 2010, with the explosion on the drilling rig, and ended on July 15, 2010, when the well was finally capped.

Significance of the spill: The *Deepwater Horizon* oil spill was history's largest accidental marine oil spill. By the time the wellhead had been capped, nearly 5 million barrels of oil are believed to have escaped from the well. Although some of the oil was recovered, an estimated 4.2 million barrels of oil were released into the Gulf of Mexico. Initial reports suggested that only about 1,000 to 5,000 barrels of oil per day were spewing from the well, a figure that greatly underestimated the scope of the disaster. Subsequent estimates increased periodically as the full scale of the catastrophe became known. The official tally eventually grew to up to 60,000 barrels per day.

This catastrophe eclipsed all previous marine oil spills that occurred in the United States and anywhere else in the world. The second largest marine release of oil in history also occurred in the Gulf during 1979 in the Bay of Campeche from the Mexican oil rig *Ixtoc 1*. Estimates placed the total from this spill at 3.3 million barrels.

Fate of workers onboard the rig: The platform was staffed by a crew of 126 at the time. A total of 115 crew members were able to evacuate. An additional 11 workers were assumed to have perished at sea and remained unaccounted for despite an extensive 3-day search by the US Coast Guard.

Immediate impact of the spill: Oil slicks from the gushing well covered nearly 30,000 square miles (78,000 square kilometers) of Gulf waters. As oil moved toward landfall, it eventually impacted the coastal areas of Louisiana and the neighboring states of Mississippi, Alabama, and Florida. Crude oil worked its way into the ecologically sensitive marshes at the mouth of the Mississippi River; a substantial die-off of wildlife occurred in Louisiana. The spill had immediate and devastating consequences for the local economy in oil-affected areas. In addition to discouraging tourism, a mainstay of the region, the catastrophe severely curtailed the fishing industry. Government authorities prohibited fishing at the mouth of the Mississippi River and nearby affected areas. An additional concern was the long-term effects of the thousands of gallons of dispersants applied to the floating crude oil upon marine life forms.

Partial data from Robertson C, Krauss C. Gulf spill is the largest of its kind, scientists say. *The New York Times*. August 2, 2010. Available at: http://www.nytimes.com/2010/08/03/us/03spill.html. Accessed September 17, 2017; and The Encyclopedia of Earth. Deepwater Horizon oil spill. Available at: http://editors.eol.org/eoearth/wiki/Deepwater_Horizon_oil_spill. Accessed September 17, 2017.

FIGURE 9.29 *Deepwater Horizon* fire.

Courtesy of US Coast Guard, Visual Information Gallery. *Deepwater Horizon fire*. Available at: https://www.defense.gov/Photos/Photo-Gallery/igphoto/2001161964/. Accessed June 6, 2017. The appearance of U.S. Department of Defense (DoD) visual information does not imply or constitute DoD endorsement.

disposal, large-scale farms, manufacturing facilities, and urban runoff. Standard water-processing methods are unable to remove many toxic chemicals and some pathological microbes. In the developing world, where adequate water processing methods are not in place, waterborne infections are a major cause of morbidity and mortality. The world's oceans, which seem to be vast and invulnerable to pollution, are rapidly degrading, especially in the coastal areas. The 21st century will need to become the era for implementing pollution abatement measures in order to protect a finite and increasingly endangered resource—water.

The cost of water to the consumer varies greatly across the globe and depends on local supply availability and demand. In some countries, bottled water accounts for 10% or more of drinking water supplies because water is unavailable or unsafe. Melting of the world's glaciers, which in some regions replenish vitally needed surface water, ultimately may cause further exacerbation of water shortages when the glaciers disappear completely.

Study Questions and Exercises

1. Define the following terms:
   a. Surface water
   b. Groundwater
   c. Hydrology
   d. Hydrological cycle
2. Describe the principal reserves for water. Which areas of the world have adequate supplies and which are facing a chronic shortage?
3. How is the majority of water used in the average household? What are some ways in which households might conserve water and prevent waste of water?
4. Why has bottled water garnered favor in recent years? To what extent is bottled water safer than tap water? What safeguards are in place to protect the quality of bottled water?
5. Explain the four stages of water treatment. How is it possible for waterborne pathogens such as *Cryptosporidium* to contaminate water that has been treated in modern treatment plants?
6. Describe three waterborne diseases and suggest methods for their prevention.
7. What historical observations led to the fluoridation of water? Take a position for or against the fluoridation of water.
8. What are some examples of toxic chemicals that may enter the public water supply? Describe some of the health effects that are attributed to toxic chemicals present in drinking water.
9. Defend or refute the hypothetical statement that the world's oceans, because of their vast size, are invulnerable to pollution from microbial agents and toxic chemicals.
10. Find an article in your local newspaper or other media source regarding local efforts by citizens to clean up the beach at a nearby lake or ocean. What steps can individuals take to prevent water pollution?
11. Arrange a visit to your local water plant to observe how water is processed for delivery to customers as finished water.
12. Based on your reading of this chapter and your own opinions, describe risk factors for oil spills from platforms and oil tankers.
13. Conduct a review of media reports of methods for water conservation. Suggest three methods whereby agriculture and industry might conserve water.
14. State three human actions associated with coastal degradation in highly populated areas. In your own opinion, what can be done to reduce the impact of these factors? (Refer to Figure 9-23.)
15. Define the term guinea worm disease. What procedures have led to a reduction in the frequency of this condition? (Refer to Figure 9-21.)

For Further Reading

Water: Our Thirsty World, A special issue of *National Geographic*, April. 2010.

References

1. US Environmental Protection Agency, Office of Water. Drinking water and health: what you need to know! EPA 816-K-99-001; 1999.
2. Tibbetts J. Water world 2000. *Environ Health Perspect.* 2000;108:A69–A73.
3. US Environmental Protection Agency. Safe Drinking Water Act. Drinking water facts and figures. EPA 816-F-04-036; 2004.
4. Holt MS. Sources of chemical contaminants and routes into the freshwater environment. *Food Chem Toxicol.* 2000;38 (1 suppl):S21–S27.
5. Hinrichsen D, Robey B, Upadhyay UD. Solutions for a water-short world. *Population Reports*. Baltimore, MD: Johns Hopkins University School of Public Health, Population Information Program; 1998. Series M, No. 14.
6. United Nations Environment Programme. Vital water graphics: freshwater resources. 2002. Available at: http://staging.unep.org/dewa/assessments/ecosystems/water/vitalwater/05.htm. Accessed June 5, 2017.
7. Centers for Disease Control and Prevention. Achievements in public health, 1900–1999: fluoridation of drinking water to prevent dental caries. *MMWR*. 1999;48:933–940.

8. Tong ST, Chen W. Modeling the relationship between land use and surface water quality. *J Environ Manage.* 2002;66:377–393.
9. US Environmental Protection Agency. *Water on Tap: What You Need to Know*. Washington, DC: EPA Office of Water; 2003.
10. Leclerc H, Schwartzbrod L, Dei-Cas E. Microbial agents associated with waterborne diseases. *Crit Rev Microbiol.* 2002;28:371–409.
11. Centers for Disease Control and Prevention. Norovirus. Available at: https://www.cdc.gov/norovirus/about/symptoms .html. Accessed June 24, 2017.
12. MacKenzie WR, Hoxie NJ, Proctor ME, et al. A massive outbreak in Milwaukee of *Cryptosporidium* infection transmitted through the public water supply. *N Engl J Med.* 1994;331:161–167.
13. Heymann DL, ed. *Control of Communicable Diseases Manual.* 20th ed. Washington, DC: American Public Health Association; 2015.
14. Dillingham RA, Lima AA, Guerrant RL. Cryptosporidiosis: epidemiology and impact. *Microbes Infect.* 2002;4:1059–1066.
15. Khalakdina A, Vugia DJ, Nadle J, et al. Is drinking water a risk factor for endemic cryptosporidiosis? A case-control study in the immunocompetent general population of the San Francisco Bay Area. *BMC Public Health.* 2003;3:11.
16. Carpenter C, Fayer R, Trout J, et al. Chlorine disinfection of recreational water for *Cryptosporidium parvum*. *Emerg Infect Dis.* 1999;5:579–584.
17. Centers for Disease Control and Prevention. Parasites—amebiasis. Causal agent. Available at: https://www.cdc.gov /parasites/amebiasis/pathogen.html. Accessed June 22, 2017.
18. Centers for Disease Control and Prevention. Parasites—amebiasis—*Entamoeba histolytica infection.* Available at: https://www.cdc.gov/parasites/amebiasis/general-info.html. Accessed June 22, 2017.
19. Centers for Disease Control and Prevention. Parasites and health: giardiasis. Available at: https://www.cdc.gov /parasites/giardia/pathogen.html. Accessed June 22, 2017.
20. Shah L, MacDougall L, Ellis A, et al. Challenges of investigating community outbreaks of cyclosporiasis, British Columbia, Canada. *Emerg Infect Dis.* 2009;15:1286–1288.
21. Centers for Disease Control and Prevention. Typhoid fever. For healthcare professionals. Available at: https://www.cdc .gov/typhoid-fever/health-professional.html. Accessed June 23, 2017.
22. Hoffman TA, Ruiz CJ, Counts GW, et al. Waterborne typhoid fever in Dade County, Florida. Clinical and therapeutic evaluation of 105 bacteremic patients. *Am J Med.* 1975;59:481–487.
23. Finkelstein R, Markel A, Putterman C, et al. Waterborne typhoid fever in Haifa, Israel: clinical, microbiologic and therapeutic aspects of a major outbreak. *Am J Med Sci.* 1988;296:27–32.
24. Centers for Disease Control and Prevention. Cholera—*Vibrio cholerae* infection: sources of infection & risk factors. Available at: https://www.cdc.gov/cholera/infection-sources .html. Accessed June 23, 2017.
25. Jiang S, Noble R, Chu W. Human adenoviruses and coliphages in urban runoff-impacted coastal waters of southern California. *Appl Environ Microbiol.* 2001;67:179–184.
26. Centers for Disease Control and Prevention. Norovirus. Clinical overview. Available at: http://www.cdc.gov/ncidod/dvrd/revb /gastro/norovirus-foodhandlers.htm. Accessed June 23, 2017.
27. Centers for Disease Control and Prevention. Adenoviruses. Available at: https://www.cdc.gov/adenovirus/index.html. Accessed June 23, 2017.
28. Centers for Disease Control and Prevention. Viral hepatitis. Hepatitis A questions and answers for the public. Available at: https://www.cdc.gov/hepatitis/hav/afaq.htm#whatHepA. Accessed June 23, 2017.
29. Centers for Disease Control and Prevention. Viral hepatitis. Hepatitis E FAQs for health professionals. Available at: https://www.cdc.gov/hepatitis/hev/hevfaq.htm#section2. Accessed June 23, 2017.
30. Brautbar N, Navizadeh N. Sewer workers: occupational risk for hepatitis C—report of two cases and review of literature. *Arch Environ Health.* 1999;54:328–330.
31. Centers for Disease Control and Prevention. DPDX. Dracunculiasis. Available at: https://www.cdc.gov/dpdx /dracunculiasis/index.html. Accessed June 23, 2017.
32. US Department of Labor. Occupational Safety and Health Administration. Legionnaires' disease. Available at: https:// www.osha.gov/dts/osta/otm/legionnaires/disease_rec.html. Accessed June 23, 2017.
33. Cirillo JD, Falkow S, Tompkins LS. Growth of *Legionella pneumophila* in *Acanthamoeba castellanii* enhances invasion. *Infect Immun.* 1994;62:3254–3261.
34. US Department of Labor. Occupational Safety and Health Administration. Legionnaires' disease: cooling towers, evaporative condensers, and fluid coolers. Available at: https://www.osha.gov/dts/osta/otm/legionnaires/cool_evap .html. Accessed June 23, 2017.
35. Calderon RL. The epidemiology of chemical contaminants of drinking water. *Food Chem Toxicol.* 2000;38(1 suppl):S13–S20.
36. Potera C. Drugged drinking water. *Environ Health Perspect.* 2000;108:A446.
37. Hwang B-F, Magnus P, Jaakkola JK. Risk of specific birth defects in relation to chlorination and the amount of natural organic matter in the water supply. *Am J Epidemiol.* 2002;156:374–382.
38. Nieuwenhuijsen MJ, Toledano MB, Eaton NE, et al. Chlorination disinfection byproducts in water and their association with adverse reproductive outcomes: a review. *Occup Environ Med* . 2000;57:73–85.
39. Smith RB, Edwards SC, Best N, et al. Birth weight, ethnicity, and exposure to trihalomethanes and haloacetuc acids in drinking water during pregnancy in the Born in Bradford cohort. *Environ Health Perspect* 2016;124:681–689.
40. Zender R, Bachand AM, Reif JS. Exposure to tap water during pregnancy. *J Expo Anal Environ Epidemiol.* 2001;11:224–230.
41. Deane M, Swan SH, Harris JA, et al. Adverse pregnancy outcomes in relation to water contamination, Santa Clara County, California, 1980–1981. *Am J Epidemiol.* 1989;129:894–904.
42. Wrensch M, Swan S, Lipscomb J, et al. Pregnancy outcomes in women potentially exposed to solvent-contaminated drinking water in San Jose, California. *Am J Epidemiol.* 1990;131:283–300.
43. National Oceanic and Atmospheric Administration (NOAA). National Ocean Service. Why should we care about the ocean? Available at: http://oceanservice.noaa.gov/facts /why-care-about-ocean.html. Accessed June 24, 2017.
44. United Nations Environment Programme. Vital water graphics: coastal and marine. 2002. Available at: http://staging. unep.org/dewa/assessments/ecosystems/water/vitalwater/38 .htm. Accessed June 5, 2017.

45. Hendrickson SE, Wong T, Allen P, et al. Marine swimming-related illness: implications for monitoring and environmental policy. *Environ Health Perspect*. 2001;109:645–650.
46. US Environmental Protection Agency. *EPA's BEACH Report: 2012 Swimming Season*. EPA 820-F-13-014; 2013.
47. Clarke KC, Hemphill JJ. The Santa Barbara oil spill: a retrospective. In: Danta D, ed. *Yearbook of the Association of Pacific Coast Geographers*. Honolulu, HI: University of Hawaii Press; 2002:157–162.
48. Lyons RA, Temple JM, Evans D, et al. Acute health effects of the *Sea Empress* oil spill. *J Epidemiol Community Health*. 1999;53:306–310.
49. Morita A, Kusaka Y, Deguchi Y, et al. Acute health problems among the people engaged in the cleanup of the *Nakhodka* oil spill. *Environ Res*. 1999;81:185–194.
50. San Sebastián M, Armstrong B, Stephens C. Outcomes of pregnancy among women living in the proximity of oil fields in the Amazon basin of Ecuador. *Int J Occup Environ Health*. 2002;8:312–319.
51. Institute of Medicine. *Assessing the Effects of the Gulf of Mexico Oil Spill on Human Health*. Washington, DC: The National Academies Press; 2010.

CHAPTER 10
Air Quality

LEARNING OBJECTIVES

By the end of this chapter the reader will be able to:

- Discuss the impacts of historically important air pollution episodes.
- List health effects associated with air pollution.
- Describe potential hazards linked to indoor air.
- Enumerate the typical components of urban ambient air pollution.
- Explain the concept of global warming.

Introduction

This chapter covers the sources and causes of air pollution, the components of air pollution (e.g., gases and particles), and some of the health and environmental effects that have been linked to air pollution, including greenhouse gases and global warming. A theme of the chapter is that clean air is intimately connected with the health of the earth and is a prerequisite for the well-being of humanity. While some may argue that controlling air pollution is too costly, this opinion has not been borne out by the evidence. For example, in the United States the Clean Air Act of 1990 sought to clear the nation's air of damaging pollutants. According to the US Environmental Protection Agency (EPA), by 2020 this act will be responsible for substantial reductions in adult and infant mortality from particle pollution and ozone and lowered morbidity from conditions such as chronic bronchitis, heart attacks, and asthma exacerbations. These benefits to quality of life will greatly exceed the costs of implementing clean air standards.[1]

Overview: Causes and Effects of Poor Air Quality

The causes of poor outdoor air quality and smog include the combustion of fossil fuels by motor vehicles, power plants, and industrial processes. In turn, such combustion releases harmful pollution into the air we breathe. The products of fossil fuel consumption are sulfur dioxide, particles, ground-level ozone, nitrogen oxides, carbon monoxide, and lead. Air pollution is particularly acute in low- and middle-income countries, where 98% of cities fail to meet World Health Organization (WHO) air quality standards.[2]

In the developed world many high-income nations have formulated and implemented air quality standards for safeguarding public health. Nevertheless, in high-income countries more than half of residents of live in cities that do not meet WHO air quality standards.[2] With respect to the United States, the American Lung Association notes that approximately one-half of Americans in 2016 were residents of counties that had unhealthful amounts of particles and ozone.[3] Refer to **FIGURE 10.1** for

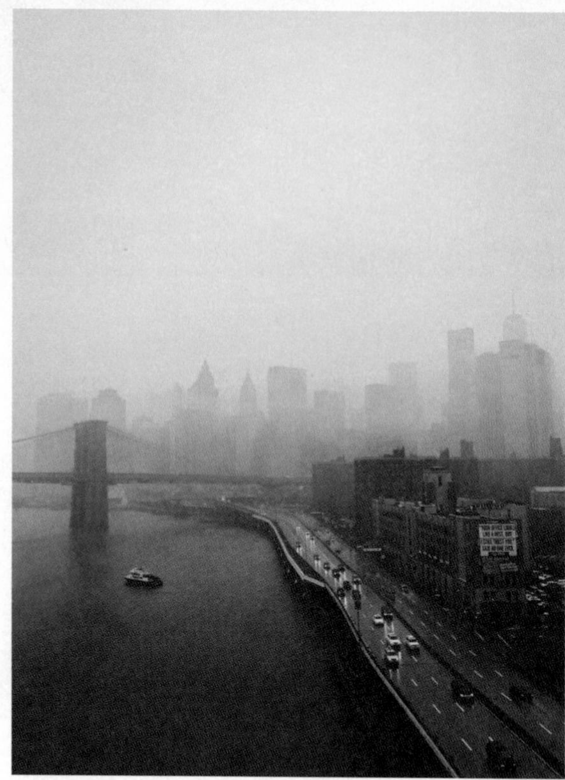

FIGURE 10.1 Smog in New York City.

an example of air quality in New York City on a smoggy day; air pollution obscures famous landmarks such as the Brooklyn Bridge. Over the past decades, air quality has improved considerably in most of the United States. However, much needs to be done to protect the gains that have been made and to improve the air in cities that do not have satisfactory air quality.

The potential damaging effects of air pollution are numerous: adverse human health effects (e.g., lung damage) and adverse environmental effects (e.g., acid rain and global warming). Most people who live in urban areas have experienced, at one time or another, the effects of outdoor air pollution. Symptoms can include burning eyes, aching lungs, difficulty breathing, wheezing, coughing, headache, and other symptoms such as an irritated throat and nose.[4,5] In recent years the United States and many other developed countries have made significant progress in improving air quality as a result of measures such as catalytic converters in automobiles, control of emissions from factories, and use of less-polluting fuels. Nevertheless, in the large cities of many developing nations, for example, Beijing, China, and Mexico City, Mexico, air quality has continued to degrade.

Often people refer to air pollution as **smog**. Originally, smog referred to smoke plus fog. Also, smog denotes

> ... A mixture of pollutants, principally ground-level ozone [discussed later in this chapter], produced by chemical reactions in the air involving smog-forming chemicals. [These chemicals can arise from both anthropogenic (due to human activities) and natural sources.] A major portion of smog formers comes from burning of petroleum-based fuels such as gasoline. Major smog occurrences are often linked to heavy motor vehicle traffic, sunshine, high temperatures, and calm winds or temperature inversion (weather condition in which warm air is trapped close to the ground instead of rising).[6]

The term **smog complex** refers to "eye irritation, irritation of the respiratory tract, chest pains, cough, shortness of breath, nausea, and headache" associated with smog.[4] In our homes, at work, and in other enclosed spaces, we may be exposed to secondhand cigarette smoke, allergens, potentially toxic molds, and hazardous chemicals.

High levels of air pollution can endanger our health and can kill people as well, especially when reactive gases are discharged into the atmosphere and when vulnerable persons are exposed. Health effects associated with air pollution can include some forms of cancer such as lung cancer and skin cancer (from possible depletion of the ozone layer), damage to vital tissues and organs such as the nervous system, and impairment of lung and breathing function. There have been numerous occasions of air pollution–caused mortality. For example, in the chapter on pesticides and other organic chemicals it was noted that during a release of toxic air pollutants in Bhopal, India, large numbers of people were killed by toxic fumes released into the environment. A second incident happened in London, England, where a 1952 great "smog disaster" killed several thousand people. Similar examples will be cited later in this chapter.

Air pollution damages the environment and causes property damage. It reduces visibility in national parks such as the Grand Canyon in the United States and can even interfere with aviation. Environmental damage linked to air pollution includes harm to forests, lakes and other bodies of water, wildlife, and buildings. Visit many large cities of the world and you will see that their structures often are blackened from many years of exposure to smog and smoke. These types of environmental damage are not geographically limited in scope; air pollution is a global problem. One example is shown in **FIGURE 10.2**, which portrays haze across northern India in a picture taken from a satellite above the earth. Satellites and other technologies make it possible to view similar clouds of air pollution over many parts of the earth.

FIGURE 10.2 Haze above northern India as seen from a satellite.

Reproduced from Rapid Response – LANCE – Terra/MODIS 2016/340 05:45 UTC Haze in northern India. Available at: https://lance.modaps.eosdis.nasa.gov/cgi-bin/imagery/single.cgi?image=India.A2016340.0545.1km.jpg. Accessed June 6, 2017.

FIGURE 10.3 Boy wearing a face mask for protection against air pollution.

The problem of urban air pollution is continuing to grow more acute in the developing world, where populations are expanding rapidly and fast-paced industrialization is coupled with increasing use of motor vehicles. In the developing areas, the use of fossil fuels has contributed to worsening air pollution. In some megacities, for example, Beijing, Delhi, Jakarta, and Mexico City, poor air quality threatens the health of these cities' inhabitants (see **FIGURE 10.3**).

Notorious Air Pollution Episodes in History

Major lethal air pollution episodes include those in the Meuse Valley in Belgium (1930); Donora, Pennsylvania (1948); and London, England (1952).[4]

Meuse Valley, Belgium

One of the earliest modern episodes of hazardous air pollution occurred during the first week of December 1930, in the Meuse Valley in Belgium. The Meuse Valley is located near Liege, Belgium, which at the time had a high concentration of steel industry operations (e.g., foundries, mills, coke ovens, and smelters). During an infamous episode of severe air pollution, sulfur dioxide, sulfuric acid mists, and fluoride gases rose to extremely high levels. This noxious mixture was thought to be associated with the deaths of more than 60 persons during the last 2 days of the incident. The resulting mortality was more than 10 times the normal mortality rate.[7] Most of the fatalities occurred among the elderly who had preexisting heart and lung diseases.[4]

Donora, Pennsylvania

During the time interval from approximately October 27 to October 30, 1948, an environmental air pollution disaster occurred in Donora, Pennsylvania, a small town located on the Monongahela River about 30 miles south of Pittsburgh. The disaster was associated with an air inversion (defined later in the chapter) in the valley in which the town was located. In this very severe episode of air pollution, fog combined with particulate matter and industrial and other contaminants. The sources of the contaminants were iron and steel mills, coal-fired home stoves, factories that burned coal, coke ovens, and metal works. This episode caused some form of illness among approximately half of the town's 14,000 residents; about 400 of them were hospitalized, and 20 died before the smog finally lifted at the end of October.[8] **FIGURES 10.4** and **10.5**, respectively, show how the town was obscured by air pollution on various days around the time of the smog episode and the locations of people who died.

Although the episode of late October affected a wide cross-section of age groups, those who were aged 55 years and older tended to be the most severely affected. During the incident, persons who were stricken reported respiratory symptoms (cough, sore throat, and difficulty breathing) and gastrointestinal symptoms (nausea and vomiting). Previous histories of heart disease and lung disease were contributing factors to the adverse outcomes of smog exposure.[4]

The Lethal London Fog of 1952

During December 5 through December 9, 1952, a very severe episode of air pollution settled on London, England. Traditionally, London had been known for its foggy climate coupled with smoke caused by the use of coal and other fossil fuels to heat homes, run power plants, and operate factories. At least 1 million coal stoves spewed forth choking sulfurous smoke. As a result, so-called "pea-souper" fogs were well known to the residents of the English metropolis. The consequence of the particularly lethal fog that occurred in December 1952 was a reported excess of 3,000 deaths above normal. Many environmental health experts consider the London fog of 1952 to be a landmark for the study of the health effects of air pollution and a catalyst that has led to research in this field.[7]

The foregoing examples (in the Meuse Valley, Donora, and London) illustrate the growing accumulation of evidence regarding associations between acute episodes of air pollution and increased mortality,

FIGURE 10.4 Views of Donora, Pennsylvania taken at about the time of the 1948 smog episode.

Modified and reproduced from US National Library of Medicine, National Institutes of Health, Images from the History of Medicine. Available at: https://collections.nlm.nih.gov. Accessed June 6, 2017.

FIGURE 10.5 A sketch of Donora made by Charles Shinn for the US Public Health Service in 1949. The circle indicates the locations of more than half of the deaths that occurred during the killer smog.

Data from Davis D. *When Smoke Ran Like Water: Tales of Environmental Deception and the Battle Against Pollution*. New York, NY: Basic Books; 2002. Sketch by Charles Shinn. Modified and reproduced from US National Library of Medicine, National Institutes of Health, Images from the History of Medicine. Available at: https://collections.nlm.nih.gov/catalog/nlm:nlmuid-101450758-img. Accessed June 6, 2107.

evidence that spurred air pollution research and galvanized public health officials.

Sources and Causes of Air Pollution

What is meant by the term **air pollution**? First consider perfectly clean air—which is not found in the natural environment, even in the most remote locations such as up in the mountains far away from human activities. In the laboratory, it is possible to produce clean air artificially by using filtration and other methods. After removing contaminants, we would find that the largest components (by weight) are nitrogen (76%), oxygen (23%), argon (1%), carbon dioxide (0.03%), plus a variety of other gases in lesser amounts and water vapor depending upon the relative humidity.[9]

The normal ambient air—air that we breathe—might be thought of as a kind of soup composed of particles and vapors that arise from natural and anthropogenic sources. Each liter of air may contain thousands of invisible suspended particles and also hundreds of invisible vapors; because these constituents are invisible, we are unaware of their presence.[9] Air pollution refers to the presence, in various degrees, of those substances (e.g., suspended particles and vapors) not found in perfectly clean air.

Numerous sources emit pollutants; these sources can be dichotomized into natural and anthropogenic sources. The former include volcanic ash, dusts, and organic materials carried by the wind. The latter are subdivided roughly into stationary sources (e.g., power plants and oil refineries) and mobile sources (e.g., on-road vehicles, off-road vehicles, and nonroad vehicles).

The types of pollutants may be classified as either primary or secondary. Primary air pollutants (discussed later in the chapter) are those emitted directly by sources of pollution. Examples are ozone, nitrogen dioxide, and carbon monoxide. Secondary air pollutants are those generated from atmospheric chemical reactions among primary pollutants.[9] An example would be the production of smog (discussed subsequently).

Naturally Occurring Air Pollution

Several natural events produce air pollutants, especially particulate matter. These phenomena include wind storms that spread dust clouds, salt evaporation along the earth's coasts, and production of materials that have a biologic origin (e.g., mold spores, pollen, and organic material from plants and animals).[10] In some areas of the world, raging forest fires emit huge amounts of smoke and ash into the atmosphere. The smoke trails can be seen by satellites hovering above the earth. In some cases, forest fires are truly natural events, such as when caused by lightning strikes. In other cases they are human-made events, for example, when caused by human carelessness, arson, or efforts to clear vegetation from the land.

Volcanic eruptions are another natural source of air pollution. Examples are the dust clouds produced by the eruptions of El Chichon near Mexico City; Galunggung in West Java, Indonesia; and Mount Pinatubo, in the Philippines. In the US state of Washington, the 1980 Mount Saint Helens eruption spewed forth tons of fine ash that descended upon a wide geographic area. (Refer to **FIGURE 10.6**.) Volcanic dusts often contain free crystalline silica, which is a toxic mineral.[11] The acute effects of exposure to volcanic dusts are irritation of the eyes and upper respiratory tract. Repeated exposure is thought to be a risk factor for the lung disease **pneumoconiosis**, especially when fine particles (less than 10 micrometers in diameter) are inhaled. However, a 5-year longitudinal follow-up study of loggers who were exposed to ash from Mount Saint Helens indicated that the risks of lung diseases such as chronic bronchitis and pneumoconiosis are minimal when exposure is initially high and then tapers off over time.[11]

Anthropogenic Sources of Air Pollution

As noted previously, anthropogenic sources of air pollution include stationary sources and mobile sources.

FIGURE 10.6 Eruption of Mount Saint Helens on May 18, 1980.

FIGURE 10.7 Electric power plant—example of a stationary source of air pollution.

Stationary Sources of Air Pollution

Stationary sources include electric generating plants (see **FIGURE 10.7**), factories and manufacturing complexes, oil refineries (see **FIGURE 10.8**), chemical plants, and incinerators.

FIGURE 10.8 Stationary source of air pollution: oil refinery.

Mobile Sources of Air Pollution

Mobile sources include on-road vehicles (e.g., cars, trucks, and buses), off-road vehicles (e.g., dune buggies and snowmobiles), and nonroad vehicles (e.g., airplanes, ships, and trains).[4,12] According to the US Environmental Protection Agency (EPA), motor vehicles produce approximately half of two major causes of smog—volatile organic compounds (VOCs) and nitrogen oxides (NO_x)—almost 75% of carbon monoxide, and more than half of emissions of toxic air pollutants.[5]

- *On-road vehicles*—The contribution of motor vehicles to air pollution is likely to continue as the number of drivers on the road increases, although technical improvements have reduced emissions from passenger vehicles. (Refer to **FIGURE 10.9** for a view of rush hour traffic on the Golden Gate bridge in northern California.) The number of

FIGURE 10.9 Rush hour on the Golden Gate bridge, California.

miles traveled each year on US highways increased by 178% between 1970 and 2005.[5] During the current century, vehicle miles increase by about 2% to 3% annually. In the United States, most drivers commute to work alone despite the availability of car pool lanes in many locations. Many geographically dispersed regions of the United States do not have convenient public transportation systems in place. This situation contrasts with developed European countries and Japan, two areas of the world that have excellent public transportation systems, which make the use of cars unnecessary or, at least, less frequent. Also contributing greatly to the air pollution problem are buses and trucks, which can cause more pollution (e.g., particle emissions) than automobiles.

- Modern automobiles produce much less pollution than automobiles did in the 1970s. Improvements include low-sulfur fuels, emission control technology, removal of lead from gasoline, and use of alternative fuels. Other innovations to reduce air pollution include electrically powered and hybrid vehicles.
- *Off-road vehicles*—Although their use is less frequent than that of on-road vehicles, off-road vehicles such as dune buggies and snowmobiles can add to air pollution in the areas where they are used. For example, they may be a cause of air pollution in national parks and wilderness areas where their use is permitted.
- *Nonroad vehicles*—Trains, airplanes, and ships also contribute to the burden of air pollution. Regarding the US transportation sector, the EPA has determined that large commercial jets are the third largest contributor of greenhouse gases.[13] These emissions from the jet engines of aircraft imperil human health and have been linked to climate change.

Components of Air Pollution

How do scientists classify air contaminants? According to Phalen, "One classification segregates air contaminants into the broad categories of infectious agents, allergens, chemical irritants, and chemical toxicants (biologic and nonbiologic). Another classification segregates air contaminants into particles, gases, and vapors. . . ."[9(p2)]

As noted, air pollution is composed of a number of components; these include solids and gases.[14] What we see as air pollution (for example, in Figure 10.13) is actually a mix of contaminants, particularly in the case of smog. Note also that there may be substances present that are not visible to the human eye. As a result of chemical reactions (described in the section on temperature inversion) that take place in the presence of certain atmospheric conditions, some of the contaminants become visible. Among the constituents of air pollution from motor vehicles (e.g., cars and trucks) are sulfur dioxide, particulate matter, greenhouse gases (e.g., CO_2), carbon monoxide, **hydrocarbons**, and nitrogen oxides.[15] Other polluting chemicals can include heavy metals and toxics such as benzene.

The term **criteria air pollutants** is used to describe "a group of very common air pollutants regulated by EPA on the basis of criteria (information on health and/or environmental effects of pollution)."[6[p45]] Criteria air pollutants are present everywhere in the United States.[16] Criteria air pollutants are ground level ozone, particulate matter, nitrogen oxides, carbon monoxide, sulfur dioxide, and lead.[16] In addition, several other pollutants are relevant to air quality: volatile organic compounds and smog (a mixture of pollutants). **TABLE 10.1** presents a summary of the names of criteria air pollutants, sources, health effects, and environmental effects. The following discussion presents a detailed examination of these pollutants, with the exception of lead and smog (defined previously). Lead, once a universal component of gasoline, has been phased out or is being phased out in most areas of the world; nevertheless, lead remains an important environmental contaminant that arises from many sources in addition to air pollution. Refer to the chapter on toxic metals and elements for coverage of lead.

Ozone is a gas that is one of the molecular forms of oxygen. The ordinary oxygen gas found in air is a molecule made up of two oxygen atoms bonded together (O_2). Ozone (O_3) consists of three oxygen atoms bonded together into a molecule.[6] Occurring in nature, ozone has the sharp smell associated with sparks from electrical equipment. Ozone is a main component of photochemical smog. Found near ground level, ozone in smog stems from a series of chemical reactions among the products of combustion of fossil fuels (including gasoline) in combination with various chemicals found in products such as solvents and paints.

Previously, in the chapter on ionizing and nonionizing radiation, we considered the role of ozone in protecting the earth from ultraviolet (UV) light as well as the influence of human activities on the environment that are hypothesized to have reduced the ozone layer. The health effects attributed to ground-level ozone include respiratory difficulties such as reduced lung function, exacerbation of allergic respiratory disease, local irritation of the eyes and respiratory tract, and reduction of the ability to fight off colds and related respiratory infections. Besides reducing visibility, environmental effects include damage to forests

TABLE 10.1 Overview of Criteria Air Pollutants

Name of Pollutant (Symbol)	Example of Source	Health Effects	Environmental Effects
Ozone (O_3) (ground-level ozone)	Variety of oxygen formed by chemical reaction of pollutants	Breathing impairment; chest pain, coughing, throat irritation	Affects vegetation (plants and trees) and ecosystems; main ingredient of smog.
Nitrogen dioxide (NO_2), nitrogen oxides (NO_x)	Combustion of fuels by cars, trucks, and vehicles, and by power plants	Irritates airways; aggravates respiratory diseases, e.g., asthma	Contributes to acid rain, which damages trees and lakes.
Carbon monoxide (CO)	Combustion of fossil fuels by cars, trucks, and other vehicles	Reduction in oxygen-carrying capacity of the blood	Very high levels not likely outdoors.
Particulate matter (PM_{10} and $PM_{2.5}$)	Many, e.g., fires, power plants, industries, vehicles	Airway irritation; lung and heart effects; premature death in people with heart of lung disease	Haze; increased acidity of lakes and streams; forest and crop damage; discolors buildings, furniture, and clothing.
Sulfur dioxide (SO_2)	Burning of fossil fuels	Breathing problems; lung damage; aggravates asthma	Contributes to acid rain; damage to trees and lakes; building damage.
Lead (Pb)	Leaded gasoline, paint, batteries; lead smelters	Can damage nervous system, immune system, and kidneys	Persists in environment; accumulates in soil, sediments; can harm plants and animals.

Data from US Environmental Protection Agency. Criteria air pollutants. Available at: https://www.epa.gov/criteria-air-pollutants. Accessed June 19, 2017.

and plants and oxidation effects that cause damage to certain kinds of fabric and rubber.

Nitrogen oxides (NO_x) refer to gases made up of a single molecule of nitrogen combined with varying numbers of molecules of oxygen. Nitrogen oxides are "produced from burning fuels, including gasoline and coal, and react with volatile organic compounds to form smog. Nitrogen oxides are also major components of acid rain."[6(p46)]

Although clean air is mostly nitrogen, some forms of nitrogen are considered to be major environmental pollutants.[17] Although the primary source of nitrogen pollution is fertilizer used in agriculture, approximately 25% of nitrogen generation (in the form of NO_x or NO_2—nitrogen dioxide) is the product of combustion of fossil fuels, for example, gasoline used in automobiles, as well as other fuels such as natural gas, diesel oil, and coal. Air pollution scientists attribute urban ozone, acid rain, and oxygen depletion of coastal waters to NO_x. Travelers to London and European cities can observe readily the damage to ancient marble-faced buildings, stone structures, and public monuments that has resulted from the acids in air pollution. With respect to human health, NO_2 and NO_x are potentially harmful to the respiratory system.

Carbon monoxide (CO) is defined as "a colorless, odorless, poisonous gas, produced by the incomplete burning of solid, liquid, and gaseous fuels. Appliances fueled with natural gas, liquified petroleum (LP gas), oil, kerosene, coal, or wood may produce CO. Burning charcoal produces CO and car exhaust contains CO."[6(p45)] Exposure to high levels of CO can result in death or serious health consequences. Indoor use of unventilated charcoal barbecues and inappropriate use of ovens (powered by natural gas) for space heating have been associated with fatalities. Smokers are exposed routinely to carbon monoxide from tobacco smoke, although this level of exposure is not sufficient to cause death. Carbon monoxide levels can become dangerous in poorly ventilated mines. On January 2,

2006, an explosion at the Sago coal mine in West Virginia was linked to the deaths of 12 of 13 miners who were trapped deep underground. The sole immediate survivor demonstrated symptoms of severe carbon monoxide poisoning for which treatment was rendered in a hyperbaric chamber. High levels of toxic carbon monoxide impeded rescue efforts at a Montcoal, West Virginia mine, where an explosion on April 5, 2010 killed 29 miners.

Inhaled carbon monoxide aggravates coronary heart disease, as well as circulatory, lung, and respiratory diseases, due to reduced oxygen-carrying capacity of the blood and increased demand on the heart and lungs. Carbon monoxide has more than 200 times more affinity for binding with hemoglobin in the blood than oxygen does. The combination of carbon monoxide and hemoglobin is called **carboxyhemoglobin**. High levels of carboxyhemoglobin interfere with the capacity of blood to transfer oxygen and carbon dioxide.[18] Patients who suffer from carbon monoxide poisoning can experience a range of symptoms, such as visual disturbances and impairment of mental and physical functioning.

Aerosol particles, also known as **particulate matter (PM)**, include dust, soot, and other finely divided solid and liquid materials that are suspended in and move with the air.[6] Particulate air pollution is a worldwide problem, producing—in some cases—what we observe as visible haze. The sources of particles include diesel exhaust from trucks and buses; smoke from incineration of garbage, wastes from crops, and slash burning; industrial activities; and effluents from wood-burning fireplaces. Particulate matter, which has been linked to lung damage, bronchitis, and early mortality, also causes environmental degradation through the deposition of soot on vehicles, clothing, and buildings. Such pollutants are known to irritate the eye, nose, and throat.

Particulate matter refers generically to a mixture of particles that can vary in size. (Refer to **FIGURE 10.10**). Let's consider as a reference point the thickness of a human hair (about 50-70 microns in diameter) and compare it with two particle sizes that are used as air pollution standards: PM_{10} and $PM_{2.5}$. Particles classified as PM_{10} are larger than 2.5 micrometers in diameter and range up to 10 micrometers in diameter. The $PM_{2.5}$ pollutants include the class of particles that are called ultrafine (diameter of 0.01 to 0.1 micrometers) and fine (diameter of 0.1 to 2.5 micrometers). Both are criteria air pollutants, as noted previously. The general health effects of particulate matter include respiratory system irritation, lung damage, and development of bronchitis. (More information on the health effects associated with PM is presented later in this chapter.)

In comparison with the larger particles, the invisible particles of diameter 2.5 micrometers ($PM_{2.5}$) or smaller have been of greater concern recently. $PM_{2.5}$ particles have the capability to bypass the body's normal defenses and can be inhaled deeply into the lungs where they are deposited; if they do not dissolve, the body's natural clearance mechanisms are unable to remove them efficiently. Some of the fine particles may contain liquid acid condensates and toxic heavy metals. Some researchers have claimed that $PM_{2.5}$ is associated with approximately 60,000 smog-related deaths annually in the United States;[19] not all agree that the figure is this large.

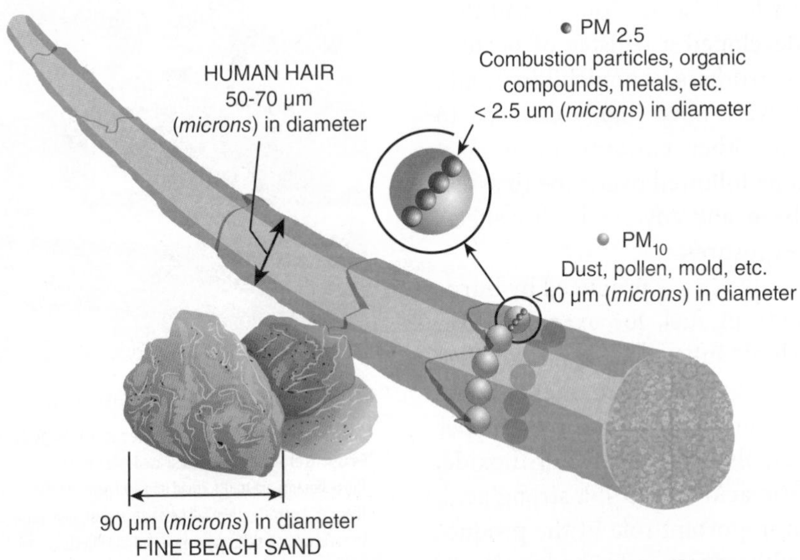

FIGURE 10.10 Size comparisons for PM particles.

Reproduced from US Environmental Protection Agency. Particulate matter (PM) basics. Available at: https://www.epa.gov/pm-pollution/particulate-matter-pm-basics#PM. Accessed June 19, 2017.

Regarding $PM_{2.5,}$ the EPA has set a standard of 12.0 micrograms of suspended particles per cubic meter of air (μg/m³). This is known as EPA's primary standard for public health protection from airborne particles. Given this standard, let's compare countries with respect to WHO's reports of levels of particulate pollution.[20] WHO reported generally acceptable levels of particulate pollution in the developed regions of the world. However, the same cannot be said of the developing regions.

Based on WHO's data from 2014, the United States, with a mean reading of 8.2 μg/m³, met the primary standard. In northern Europe, Scandinavian countries (e.g., Denmark, Norway, and Sweden) had levels that were below 12.0 μg/m³. In western Europe, some countries had levels at or close to the primary standard (e.g., France, Germany, and Switzerland). Belgium, Italy, and the Netherlands were examples of countries that slightly exceeded the standard. Some eastern European countries (e.g., Bulgaria, Bosnia and Herzegovina, Hungary, and Poland) exceeded the standard by approximately two to three times. Very high levels of particulate air pollution tended to occur in developing countries. For example, several Middle Eastern countries (e.g., Saudi Arabia and Qatar) demonstrated levels that were roughly nine times the standard. Other developing countries with high levels of particulate pollution were located in Africa and the rapidly industrializing countries of India and China.

Large amounts of PM were released following the collapse of the World Trade Center in New York City after the terrorist attacks on September 11, 2001. Examples of materials that could have been incorporated in the dusts include asbestos, fiberglass, and crystalline mineral components of concrete and wallboard. The Centers for Disease Control and Prevention (CDC) has developed a registry of persons (e.g., nearby residents, workers, schoolchildren, and first responders) who were most highly exposed to the dusts, smoke, and other airborne materials.[21] These individuals will be followed over time to assess whether there have been any adverse health effects associated with their exposures.

Sulfur dioxide (SO_2) is a gas produced by burning sulfur contaminants in fuel, for example, coal. Power plants that use high-sulfur coal or do not have effective emission controls are a source of SO_2. Some industrial processes, including production of paper and smelting of metals, also produce sulfur dioxide, which can form sulfuric acid (H_2SO_4), a strong acid. Thus, SO_2 can play an important role in the production of acid rain.[6] Health effects associated with sulfur dioxide include bronchoconstriction (as in asthma attacks) and production of excess mucus, which can produce coughing.

Volatile organic compounds (VOCs) refer to a class of chemicals that contain carbon (C), the basic chemical element found in living beings. Essentially all carbon-containing chemicals belong to the group that is called organic. Volatile chemicals evaporate and thus escape into the air easily.[6] Not classified as criteria air pollutants, VOCs are the product of fuel combustion (e.g., diesel fuel, gasoline, coal, and natural gas). Products used frequently in the workplace and at home release VOCs; examples are paints and lacquers, some types of glues, construction materials, and solvents such as benzene and toluene. One of the most significant sources of VOCs is automobiles. VOCs interact with other pollutants to form smog. The impact of VOCs on human health can be serious (e.g., cancers at various bodily sites). Some VOCs are known to be injurious to plants.

Acid Rain

The term **acid rain** occurs "...when sulfur dioxide (SO_2) and nitrogen oxides (NO_x) are emitted into the atmosphere and transported by wind and air currents. The SO_2 and NO_x react with water, oxygen and other chemicals to form sulfuric and nitric acids. These then mix with water and other materials before falling to the ground.[22] Installations such as electric utility plants emit SO_2 and NO_x. (Refer to **FIGURE 10.11**, which shows the acid rain pathway.) Eventually the

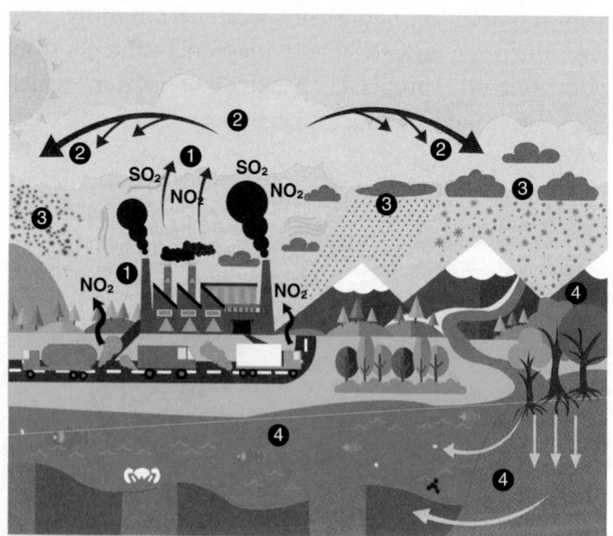

FIGURE 10.11 Acid rain pathway.

* Numbers shown in figure refer to the pathway for acid rain in our environment.

[1] Emissions of SO_2 and No_x are released into the air.

[2] The pollutants are transformed into acid particles that may be transported long distances.

[3] These acid particles then fall on earth as dust, rain, snow, and other materials.

[4] The acid rain particles may cause harmful effects on soil, forests, streams, and lakes.

Reproduced from US Environmental Protection Agency. What is acid rain? Available at: https://www.epa.gov/acidrain/what-acid-rain. Accessed June 6, 2017.

FIGURE 10.12 Deterioration of the exterior of a building in Dublin, Ireland. Air pollution, solar radiation, moisture, and cold–heat cycling are contributing causes.

acid rain settles on the earth, creating abnormally high levels of acidity that are potentially damaging to the environment, wildlife, and human health. For example, acid rain is believed to harm forests and certain species of fish, and to contribute to the deterioration of structures. (Refer to **FIGURE 10.12**.)

Temperature Inversion

The term **temperature inversion** refers to an atmospheric condition during which a warm layer of air stalls above a layer of cool air that is closer to the surface of the earth, as shown in **FIGURE 10.13**. A temperature inversion is the reverse of the usual situation in which the air closer to the surface of the earth is warmer than the air in the upper atmosphere. Solar radiation causes the earth's surface to become heated and warms the air near the surface. The usual temperature gradient (hot air below and cool above) allows convection of warm air from the earth's surface into the upper atmosphere, thus removing pollutants from the breathing zones of people. During a temperature inversion, pollutants (e.g., smog, smog-forming chemicals, and VOCs) can build up when they are trapped close to the earth's surface. Continuing release of smog-forming pollutants from motor vehicles and other sources during an inversion exacerbates air pollution. A temperature inversion layer as seen from an airplane is shown in **FIGURE 10.14**.

Temperature inversions contribute to the creation of smog, which is aggravated in cities such

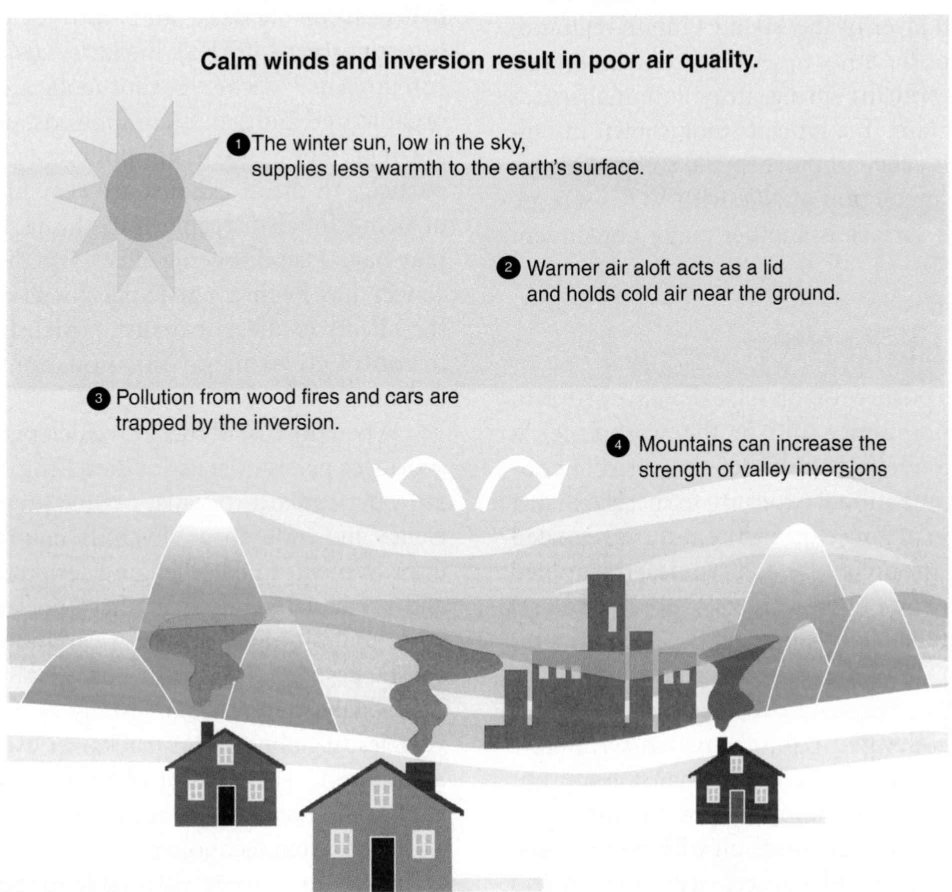

FIGURE 10.13 Diagram of temperature inversion.

Reproduced from National Weather Service. What are temperature inversions? Available at: http://www.wrh.noaa.gov/slc/climate/TemperatureInversions.php. Accessed June 7, 2017.

FIGURE 10.14 Inversion layer as seen from an airplane.

FIGURE 10.15 Air pollution in southern California on a spring day.

as Los Angeles that have unfavorable topographical characteristics and prevailing winds. In the Los Angeles Basin, under certain conditions, coastal on-shore winds push pollutants farther inland, away from their sources; these pollutants are stopped by nearby mountains, creating higher levels of air pollution than existed along the coast. This concentration of pollutants then may be trapped under an inversion layer; a series of chemical reactions among the pollutants produces smog. **FIGURE 10.15** shows smog caused by an inversion layer in the Inland Empire region of southern California. You can see that, even on this relatively clear day during spring, air pollution obscures a nearby mountain. The type of smog shown in Figure 10.15 also is called photochemical smog, which is caused by the interaction of NO_x with VOCs. Ozone near the earth's surface is another major constituent of smog.

Diesel Exhaust

Widespread exposure to diesel exhaust occurs in the general community and in the workplace. A meta-analysis reviewed studies of the possible etiologic role of occupational exposure to diesel exhaust in the induction of lung cancer. The authors reported that across the group of research reports, the pooled risk of lung cancer among workers exposed to diesel exhaust was statistically significantly elevated after controlling for smoking.[23] As a result of researchers' concerns that exposure to diesel exhaust may be linked with elevated risk of lung cancer, public health experts have named diesel exhaust as a probable carcinogen.[24] For example, epidemiologic evidence suggests that in comparison with nonexposed groups, two categories of workers (truck drivers and railroad crews) exposed directly to diesel exhaust have lung cancer incidence rates that are 20% to 40% higher.[25] The level of exposure to diesel exhaust among both types of transportation workers is roughly equivalent.[26] One study reported that some groups of miners were exposed to diesel exhaust levels above recommended standards.[24]

Diesel exhaust, a complex mixture of particles and gases, includes the element carbon, condensed hydrocarbon gases, and **polycyclic aromatic hydrocarbons (PAHs)**, the latter suspected of being carcinogens.[24] Other constituents are hundreds of organic and inorganic compounds, some of which are regarded as toxic air pollutants.[25] Most of the particles in diesel exhaust are very fine and capable of being inhaled deeply into the lungs where they may have a carcinogenic effect. For this reason, lung cancer has been a particular focus of research on the effects of diesel exhaust, which is also thought to contribute to airway inflammation, allergies, and asthma.

When the cost of fuel for vehicles is high and as oil shortages persist, the use of diesel engines will tend to grow in popularity because of their greater energy efficiency and endurance.[25] Even though there have been improvements in diesel engine design that reduce NO and particulate emissions, these newer engines continue to emit higher levels of carcinogenic and toxic substances than do conventional gasoline engines that have catalytic converters.[25] Many of the diesel-driven vehicles (e.g., buses and trucks) in use in the United States are 15 years old or older, indicating that these vehicles do not have the advantages of the latest emissions reduction technology.

Likely to be more vulnerable to the effects of diesel exhaust exposure than adults are children, whose defenses to ward off disease are not as well developed as those of adults. Children who ride in school buses powered by diesel engines are exposed to diesel exhaust. Residents (both children and adults) of urban areas may be exposed to diesel exhaust from motor vehicle traffic. For example, adolescent residents of upper Manhattan in New York City were reported to receive frequent exposures to diesel exhaust from vehicles that travel on city streets.[27] The role that these exposures play in exacerbation of lung disorders needs to be explored more fully.

Despite the supportive findings regarding the potentially carcinogenic role of diesel exhaust exposures, this topic of research remains highly controversial.[28] Some researchers remain skeptical about whether, in fact, a risk to human health exists from exposures to diesel exhaust. Researchers have questioned the validity of data and methods used in risk assessments of diesel exhaust exposure. For example, some of the studies of truck drivers have not controlled adequately for smoking status and the level of diesel exhaust exposure.[29]

Air Quality Standards

This section presents information on the Air Quality Index and National Ambient Air Quality Standards (NAAQS).

The Air Quality Index

The **Air Quality Index (AQI)** is used to provide the public with an indication of air quality in a local area:

> The AQI is an index for reporting daily air quality. It tells you how clean or polluted your air is, and what associated health effects might be a concern for you. The AQI focuses on health effects you may experience within a few hours or days after breathing polluted air. EPA calculates the AQI for five major air pollutants regulated by the Clean Air Act: ground-level ozone, particle pollution (also known as particulate matter), carbon monoxide, sulfur dioxide, and nitrogen dioxide. For each of these pollutants, EPA has established national air quality standards to protect public health. . . . .[30] (summarized in the following table).

Air Quality Index (AQI) Values	Levels of Health Concern	Colors
When the AQI is in this range:	*...Air quality conditions are:*	*...As symbolized by this color:*
0 to 50	Good	Green
51 to 100	Moderate	Yellow
101 to 150	Unhealthy for sensitive groups	Orange
151 to 200	Unhealthy	Red
201 to 300	Very unhealthy	Purple
301 to 500	Hazardous	Maroon

Reproduced from US Environmental Protection Agency. Air Quality Index (AQI) basics. Available at: https://airnow.gov/index.cfm?action=aqibasics.aqi. Accessed June 7, 2017.

Each category corresponds to a different level of health concern. The six levels of health concern and what they mean are:

- "Good" AQI is 0–50. Air quality is considered satisfactory, and air pollution poses little or no risk.
- "Moderate" AQI is 51–100. Air quality is acceptable; however, for some pollutants there may be a moderate health concern for a very small number of people. For example, people who are unusually sensitive to ozone may experience respiratory symptoms.
- "Unhealthy for Sensitive Groups" AQI is 101–150. Although general public is not likely to be affected at this AQI range, people with lung disease, older adults, and children are at a greater risk from exposure to ozone, and persons with heart and lung disease, older adults, and children are at greater risk from the presence of particles in the air.
- "Unhealthy" AQI is 151–200. Everyone may begin to experience some adverse health effects, and members of the sensitive groups may experience more serious effects.
- "Very Unhealthy" AQI is 201–300. This would trigger a health alert signifying that everyone may experience more serious health effects.
- "Hazardous" AQI is greater than 300. This would trigger a health warning. . . of emergency conditions. The entire population is more likely to be affected.[30]

National Ambient Air Quality Standards

Federal standards for air pollution are called the National Ambient Air Quality Standards (NAAQS). The EPA reviews the scientific literature at 5-year intervals and decides whether to revise each standard. Current standards (as of 2010) are presented in the following text box.

NATIONAL AMBIENT AIR QUALITY STANDARDS

The Clean Air Act, which was last amended in 1990, requires EPA to set National Ambient Air Quality Standards (40 CFR part 50) for pollutants considered harmful to public health and the environment. The Clean Air Act identifies two types of national ambient air quality standards.

Primary standards provide public health protection, including protecting the health of "sensitive" populations such as asthmatics, children, and the elderly. ***Secondary standards*** provide public welfare protection, including protection against decreased visibility and damage to animals, crops, vegetation, and buildings.

The EPA has set National Ambient Air Quality Standards for six principal pollutants, which are called "criteria" air pollutants. Periodically, the standards are reviewed and may be revised. The current standards are listed in **TABLE 10.2**. Units of measure for the standards are parts per million (ppm) by volume, parts per billion (ppb) by volume, and micrograms per cubic meter of air (μg/m^3).

TABLE 10.2 National Ambient Air Quality Standards for Criteria Air Pollutants

Pollutant		Primary/ Secondary	Averaging Time	Level	Form
Carbon monoxide (CO)		Primary	8 hours	9 ppm	Not to be exceeded more than once per year
			1 hour	35 ppm	
Lead (Pb)		Primary and secondary	Rolling 3-month average	0.15 μg/m^3	Not to be exceeded
Nitrogen dioxide (NO_2)		Primary	1 hour	100 ppb	98th percentile of 1-hour daily maximum concentrations, averaged over 3 years
		Primary and secondary	1 year	53 ppb	Annual mean
Ozone (O_3)		Primary and secondary	8 hours	0.070 ppm	Annual fourth-highest daily maximum 8-hour concentration, averaged over 3 years
Particule pollution (PM)	$PM_{2.5}$	Primary	1 year	12 μg/m^3	Annual mean, averaged over 3 years
		Secondary	1 year	15.0 μg/m^3	Annual mean, averaged over 3 years
		Primary and secondary	24 hours	35 μg/m^3	98th percentile, averaged over 3 years
	PM_{10}	Primary and secondary	24 hours	150 μg/m^3	Not to be exceeded more than once per year on average over 3 years
Sulfur dioxide (SO_2)		Primary	1 hour	75 ppb	99th percentile of 1-hour daily maximum concentrations, averaged over 3 years
		Secondary	3 hours	0.5 ppm	Not to be exceeded more than once per year

Modified and reproduced from US Environmental Protection Agency. Criteria air pollutants. National Ambient Air Quality Standards (NAAQS) table. Available at: http://www.epa.gov/criteria-air-pollutants/naaqs-table. Accessed June 18, 2017.

Reproduced from US Environmental Protection Agency. NAAQS table. Available at: https://www.epa.gov/criteria-air-pollutants/naaqs-table. Accessed February 24, 2017.

Health Effects of Air Pollution

The aesthetic impacts of air pollution include reduction in the quality of our lives by obscuring the natural environment, by damaging property, and, in some cases, by being malodorous. This section will focus on the human health effects of air pollution and will present information on methods for the measurement of air pollution in the laboratory.

Measurement of Air Pollution in Experimental Studies

Improvements in air quality in the United States have reduced the levels of pollutants in the air substantially. Consequently, increasingly sensitive techniques are needed to define the current health effects of air pollution. Examples of these sensitive techniques are shown in **FIGURE 10.16**, which illustrates scientific devices used in air pollution health effects studies, primarily with small animals. Air purifiers are used to create artificially purified air that has known levels of contaminants. Laser particle analyzers are used to quantify precisely the amount of particulate matter in a sample of air. Exposure chambers administer precise amounts of pollutants under controlled conditions to small mammals. Through the use of devices such as these, air pollution researchers are able to measure exposure to air pollution much more definitively than is possible through studies conducted in the ambient environment.

Human Health Effects

Among human populations, air pollution is associated with short-term (acute) and long-term effects. The former can include irritation of the eyes, nose, and throat;

A

B

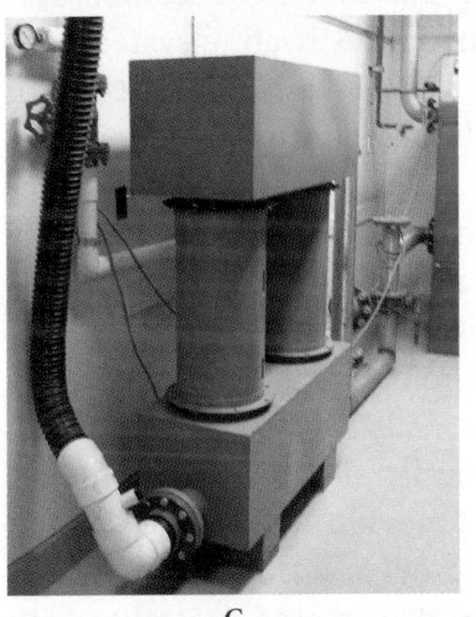

C

D

E

FIGURE 10.16 Devices employed in experimental studies of air pollution. The exposure chambers are used with small animals: A. Laser particle analyzer. B. Exposure chamber. C. Air purifier. D. Exposure chamber. E. Exposure chamber.

aching lungs, bronchitis and pneumonia, wheezing, coughing, nausea, and headaches. Examples of long-term effects are heart disease, chronic obstructive pulmonary disease (COPD), and lung cancer.

Population subgroups differ in their susceptibility to air pollution. Vulnerable groups who are at increased risk of being affected by air pollution include the elderly, persons afflicted with chronic diseases, and growing children. Persons who have low educational attainment or diabetes or who are black appear to be more susceptible to the adverse affects of air pollution and might be particularly likely to benefit from air pollution emissions controls.[31] On the other hand, another viewpoint is that introduction of emission controls could pose an economic burden for small businesses that employ minority groups that have low incomes.

As they grow and develop, children who live in some areas of the world are exposed to air pollution numerous times over the course of their lives. A meaningful question for researchers pertains to the effects of the cumulative impact of air pollution on children over long time periods.[32] These exposures that begin early in life may lead to future cases of asthma and serious lung diseases as well as chronic cough and respiratory irritation.

Three of the major health effects are airway sensitivity disorders (e.g., asthma), lung cancer, and heart attacks. In addition, heavy episodes of air pollution have been correlated with increased mortality rates. The specific effects of air pollution are related to total exposure—duration of exposure (for example, how long one lives in a smoggy environment) and the type and concentration of the pollutant. Other considerations in health effects are environmental and temporal factors—temperature, climate, and timing of exposure.[4]

Anatomy of the Lung and Impact of Air Pollution on the Body

This section will describe how components of air pollution enter the human lung and how they impact the body. (Refer to **FIGURE 10.17**.) During respiration, air is drawn in through the nose and mouth. The air then passes through the windpipe into the right and left lungs, which have three and two lobes, respectively. Each lobe is supplied with bronchial tubes that branch several times and terminate in bronchioles and alveoli. The lobes of the lungs and bronchi are shown in Figure 10.17A. Figures 10.17B and D illustrate the structure of the bronchial tubes, bronchioles, and alveoli; exchange of oxygen between inhaled air and the circulatory system takes place in the alveoli. Figure 10.17C (left side) shows the complete structure of a right lung that was taken from a dog.

Consider one of the major components of air pollution—smog. According to animal studies, smog has the capacity to damage the cellular structure of airways (e.g., windpipe and bronchial tubes) to the lungs. In response to ozone exposure, the exposed airways sometimes manifest swelling and inflammation. People who are exposed to smog may experience breathing difficulties and may develop acute or chronic cough.

Other conditions that are believed to be associated with the irritating effects of smog are chronic obstructive pulmonary disease and, among sensitive persons, exacerbation of asthma. Also suspected of being associated with smog and other forms of air pollution are some types of cancer.

Epidemiologic Studies of Air Pollution Health Effects

This section covers the following outcomes that have been studied in relation to air pollution: mortality, coronary heart disease, chronic obstructive pulmonary disease, asthma, and lung cancer.

Mortality

The World Health Organization estimated that 7 million people died from air pollution in 2012. From the global perspective, alarmingly high air pollution levels in cities located in developing countries substantially increase the risk of human mortality. In the United States, researchers have studied relationships between air pollution and mortality. Dockery et al. demonstrated statistically significant associations between air pollution and mortality in a prospective study of 8,111 adults who resided in six US cities.[33] In a study conducted among the predominantly Mormon residents of Utah Valley in central Utah, elevated concentrations of particles (PM_{10}) were associated with increased mortality, particularly from respiratory and cardiovascular causes.[34] Other epidemiologic studies conducted in large urban areas such as New York City have investigated associations between rising air pollution levels and mortality.

One type of epidemiologic study has examined the relationship between daily mortality and air pollution levels. The historically significant extreme air pollution episodes noted previously (e.g., the London fog in 1952) appeared to provide at least anecdotal evidence of the correlation of increased mortality with the times that air pollution levels rose to very high levels. Ito et al. examined London data for the period 1965 through 1972 for levels of air pollution classified

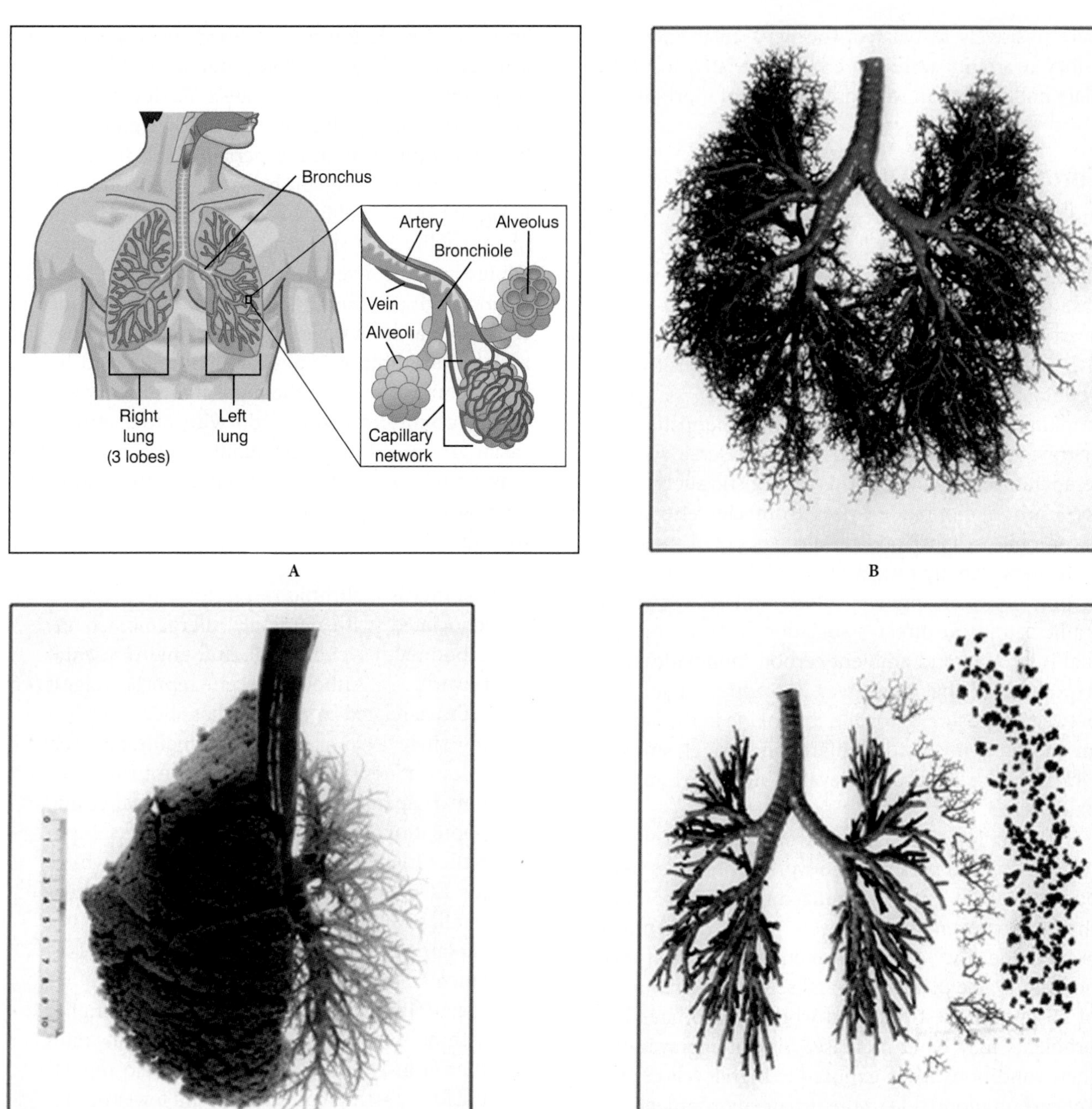

FIGURE 10.17 Lung anatomy: A. Diagram of the lungs. B. Human tracheobronchial tree. C. Right lung and left tracheobronchial tree. D. The bronchial tree dissected showing (respectively) bronchial tubes, bronchioles, and alveoli.

(A) Reproduced from Centers for Disease Control and Prevention. Basic information about lung cancer. Available at: http://www.cdc.gov/cancer/lung/basic_info. Accessed May 5, 2010. (B, C, and D) Dr. Robert Phalen, Air Pollution Laboratory, University of California, Irvine.

according to type of pollutant.[35] Statistically significant associations were found between air pollution variables and mortality.

Another study that reported a correlation between increases in total daily mortality and increased air pollution (SO_2 and smoke shade) was conducted in New York City.[36] Daily mortality was related to particulate air pollution in St. Louis, Missouri and counties in eastern Tennessee.[37] Researchers estimated the daily mortality rate associated with gaseous air pollution and particulate air pollution—inhalable particles (PM_{10}) and fine particles ($PM_{2.5}$). For each 100 micrograms per cubic meter increase in PM_{10}, the total mortality was reported to increase by 16% in St. Louis, Missouri, and by 17% in the eastern Tennessee counties. Weaker associations were reported for $PM_{2.5}$; associations between mortality and gaseous pollutants (e.g., SO_2, NO_2, and O_3) were not statistically significant. Another study conducted by Dominici et al. found that daily variations in PM_{10} and daily variations in mortality were positively associated. Cardiovascular and respiratory mortality were more closely related to particulate pollution than were other forms of mortality.[38]

Coronary Heart Disease

Individuals who have preexisting heart or lung disease may be at particular risk for the fatal or aggravating effects of air pollution that lead to coronary heart disease (CHD). People who are afflicted with ischemic heart disease combined with arrhythmias or congestive heart failure may have heightened sensitivity to pollutants from motor vehicle emissions.[39] Cigarette smoking and air pollution may act synergistically to aggravate lung diseases such as emphysema. Extreme temperatures may add to this synergistic effect.

Carbon monoxide arises from cigarette smoke, automobile exhaust, and certain types of occupational exposures. An investigation into the possible association between angina pectoris and heavy freeway traffic found no direct association between myocardial infarction and ambient carbon monoxide. It was hypothesized that there was an indirect association between exposure to carbon monoxide in the ambient air and acute myocardial infarction through smoking, which is associated with elevated blood carbon monoxide levels.[40]

The air of one large metropolitan city exposed pedestrians and persons who worked outdoors to carbon monoxide levels that ranged from 10 to 50 ppm; there were even higher levels in poorly ventilated areas of the city.[41] The recommended standard for carbon monoxide in occupational settings is a maximum of 50 ppm. Pedestrians and workers who have heart problems may be at increased risk of aggravation of their conditions when exposed to high levels of ambient carbon monoxide in the urban environment.

Chronic Obstructive Pulmonary Disease

A major investigation, the Tucson Epidemiological Study of Airways Obstructive Diseases, has tracked the etiology and natural history of COPD.[42] (The chapter on occupational health defines and provides more information on COPD.) Based on a multistage stratified cluster sample of white households in the Tucson, Arizona area, it has generated many important findings, including the effects of passive smoking on children.[43]

Asthma

One of the most frequent chronic conditions in the United States, **asthma** has shown an increasing prevalence since 1980.[44–46] This increase has happened even as air pollution has declined steadily. Currently asthma is regarded as a key public health problem that confronts the United States.[45] Residents of inner-city areas (notably children) seem to be more affected by the condition than are persons who live in other areas.[47] The growing problem of childhood asthma is significant to public health because of the burden of morbidity and suffering that this condition contributes to the lives of children and their parents. Interestingly, in the United States, the prevalence of childhood asthma varies by socioeconomic status and racial and ethnic group. Black children who live in low-income areas have the highest asthma prevalence and morbidity in comparison with other ethnic and racial groups.[48] Indoor air quality and genetic factors may be relevant to the occurrence of this significant respiratory disease. Some facts regarding asthma are the following:

> Asthma is a chronic respiratory illness often associated with familial, allergenic, socioeconomic, psychological, and environmental factors. . . . Although recent reports suggest asthma-related mortality has been declining since 1996, a disparity remains between rates for non-Hispanic whites and those for non-Hispanic blacks and other racial/ethnic populations. . . . Non-Hispanic blacks experience higher rates than non-Hispanic whites for ED [emergency department] visits, hospitalizations, and deaths; these trends are not explained entirely by higher asthma prevalence among non-Hispanic blacks. . . . Other racial/ethnic populations experience higher asthma mortality and hospitalization rates than non-Hispanic whites while also reporting lower asthma prevalence and fewer outpatient and ED visits.[46(p147)]

Symptoms of asthma—difficulty breathing and tightness of the chest—are a response to muscle contractions that constrict the trachea, lungs, and other parts of the airway. Excess mucus secretion also contributes to breathing difficulties. Asthma is believed to have a complex etiology that involves both genetic and environmental factors.[49] Among the triggers for asthma attacks are allergens, pollutants, and viral infections; examples of sources of allergens are vermin (i.e., cockroaches and rodents), dust mites, mold, and household pets (i.e., dogs and cats).[50] Other potential triggers are **environmental tobacco smoke**, cold air, exercise, stress, and aspirin.

Air pollution has been studied as a possible factor in asthma occurrence. Exploration of this association presents methodologic difficulties due to the facts that air pollution contains a complex mixture of substances and asthma has a complex, heterogeneous etiology.[51] Examples of components of air pollution studied in relation to asthma include particles, molds, nitrogen dioxide, ozone, and pollen.

Epidemiologic studies have reported an association between high concentrations of air pollution and asthma prevalence.[52] Among schoolchildren in Japan, incidence rates of asthma were associated with air pollution components such as particles (PM_{10}) and nitrogen dioxide.[52] A Los Angeles study of asthma among black children found that particles and molds were related to respiratory symptom occurrence.[53]

Lung Cancer

Although air pollution is a suspected cause of lung cancer, researchers have had difficulty in establishing a definitive association because of problems in measuring the effects of low-level exposures and problems in measuring air pollution itself. An example of research on air pollution's health effects is that conducted by Henderson et al.[54] Census tracts in Los Angeles were aggregated into 14 study areas that represented homogeneous air pollution profiles. The study reported a correlation between the geographic distribution of lung cancer cases and the general location of emission sources for hydrocarbons.

One of the components of air pollution that may be implicated in lung cancer mortality is fine particulate matter. Pope et al. conducted an analysis of a data set that contained information on risk factors, air pollution levels, and mortality from approximately 500,000 adults who resided in metropolitan areas throughout the United States.[55] These individuals were part of a large, ongoing, prospective study called the Cancer Prevention II study, which was conducted by the American Cancer Society. The investigators concluded that exposure to fine particulate matter over extended time periods was a risk factor for both lung cancer and cardiopulmonary mortality.

In large population studies, the effects of air pollution exposure may be impossible to disentangle from the effects of other exposures, for example, tobacco smoke. A retrospective study conducted in Erie County, New York found that air pollution alone was not associated with lung cancer.[56] However, the same study reported that the risk of lung cancer increased among heavy smokers who had extensive exposure to air pollution.

A Swedish study reported that air pollution from vehicle exhaust emissions may be important for increasing the risk of lung cancer.[57] According to the Swedish group, one of the components of vehicle emissions that may increase risk of lung cancer is NO_2.

Indoor Air Quality

Indoor air that is of poor quality can be a significant factor in the etiology of lung disease and exacerbation of existing conditions such as asthma and bronchitis. A report authored by four leading health and governmental organizations states:

> Studies from the United States and Europe show that persons in industrialized nations spend more than 90 percent of their time indoors. . . . For infants, the elderly, persons with chronic diseases, and most residents of any age, the proportion is probably higher. In addition, the concentrations of many pollutants indoors exceed those outdoors. The locations of highest concern are those involving prolonged, continuing exposure—that is, the home, school, and workplace.[58(p1)]

The indoor environment provides many opportunities for exposure to potentially irritating and harmful substances. Examples are aerosolized chemicals, fumes from gas appliances, components of building materials, and secondhand smoke from tobacco products. Adverse health outcomes that have been linked to indoor pollution include respiratory diseases such as asthma, Legionnaires' disease, the sick building syndrome, hypersensitivity pneumonitis, and multiple chemical sensitivity.[59]

Indoor Air Quality in the Home

Residential indoor air pollution poses risks for respiratory illness of various types.[60] For example, certain environmental characteristics tend to be found in the dwellings of inner-city children who have asthma; these characteristics include the presence of cockroaches, persistent dampness, wall-to-wall carpeting in children's bedrooms, exposure to environmental tobacco smoke, infestations with rodents, and pets that have fur.[47] Other factors that influence the quality of indoor air are effluents from gas stoves and the use of construction materials that contain formaldehyde.[60] Dust mites, molds, and bacteria, all of which are linked with asthma, also may be identified in indoor air.[61] Evidence suggests that children who become sensitized to allergenic indoor air pollutants such as molds and animal fur may incur an increased risk for the development of asthma.[61] Dwellings that have damp, odiferous interiors may be conducive to the growth of fungi, dust mites, and bacterial pathogens, all of which, in turn, tend to be related to respiratory symptoms.[62]

Indoor Use of Biomass Fuels

A frequent cause of indoor air pollution in developing countries is the use of unventilated indoor cooking stoves that burn biomass fuels (wood, animal dung, or cuttings from crops). (Refer to **FIGURE 10.18**.)

Approximately half the global population uses biomass fuels for cooking and heating; in some areas of the world (e.g., south Asia and sub-Saharan Africa), this figure reaches 80%. Often biomass fuels are combusted in crude stoves or pits that have been dug into the floor. Due to inefficient combustion and the lack of ventilation in household areas used for cooking, the house fills up with smoke. As a result, the persons who occupy the dwelling may be exposed to substantial amounts of pollutants: PM_{10}, carbon monoxide, nitrous oxides, sulfur compounds, and polycyclic aromatic hydrocarbons (PAHs). Some estimates indicate that the level of indoor air pollutants in such homes exceeds US EPA standards by several orders of magnitude. As shown in Figure 10.18, women who are responsible for cooking receive the largest doses of indoor air pollution, as do infants who may be strapped to the women's backs. Other household residents who may be at risk of the adverse effects of indoor air pollution are the elderly, who spend more time indoors than do other family members. A study conducted in India reported that elderly men and women who live in homes that use biomass fuels have a higher prevalence of asthma than elderly persons who live in homes that use cleaner fuels.[49]

FIGURE 10.18 Woman cooking with an indoor stove.

Photo by Nigel Bruce. Reproduced with permission from Van Hise Heart J. *The Air That They Breathe*. UC Berkeley Public Health, Berkeley, CA: School of Public Health; Spring 2003. 5.

Sick Building Syndrome and Building-Related Illness

Sick building syndrome (SBS) is another example of a condition ascribed to indoor air pollution.[63,64] SBS

> ...is used to describe situations in which building occupants experience acute health and comfort effects that appear to be linked to time spent in a building, but no specific illness or cause can be identified. The complaints may be localized in a particular room or zone, or may be widespread throughout the building. In contrast, the term "building related illness" (BRI) is used when symptoms of diagnosable illness are identified and can be attributed directly to airborne building contaminants.[65]

SBS is a temporary phenomenon that is relieved when affected persons are no longer inside the building. Symptoms of SBS can include headache, respiratory tract irritation, dry skin, and fatigue; the cause of the symptoms of SBS is unknown. In comparison, **building-related illness (BRI)** describes symptoms of diagnosable illnesses that can be linked to specific pollutants; examples are Legionnaires' disease and hypersensitivity pneumonitis. Possible causes of SBS are inadequate building ventilation, chemical contaminants, and biological contaminants.[65]

A condition that bears similarity to sick building syndrome is multiple chemical sensitivity. The onset of multiple chemical sensitivity is described as the development of sensitivity at about the same time to several chemicals, which are present in the ambient environment at low levels. The onset of the condition occurs in many cases following exposure to indoor air that has quality problems.[59] Also hypothesized as being related to low-quality indoor air is hypersensitivity pneumonitis, an illness "characterized by flu-like symptoms that include fever, chills, fatigue, cough, chest tightness, and shortness of breath."[59(p402)] Causal factors associated with the illness include exposure to molds, fungi, and bacteria found in indoor air.

Secondhand Smoke

FIGURE 10.19 demonstrates the lungs of two smokers—the lungs on the left are those of a smoker who had quit briefly before succumbing to a heart attack. The lungs shown on the right are those of a smoker who quit for several years. Note that the long-term quitter's lungs show horizontal striping, which corresponds to remaining deposits of carbon from cigarette smoke. The smoker who had quit for only a short period of time had lungs with more extensive deposits of carbon.

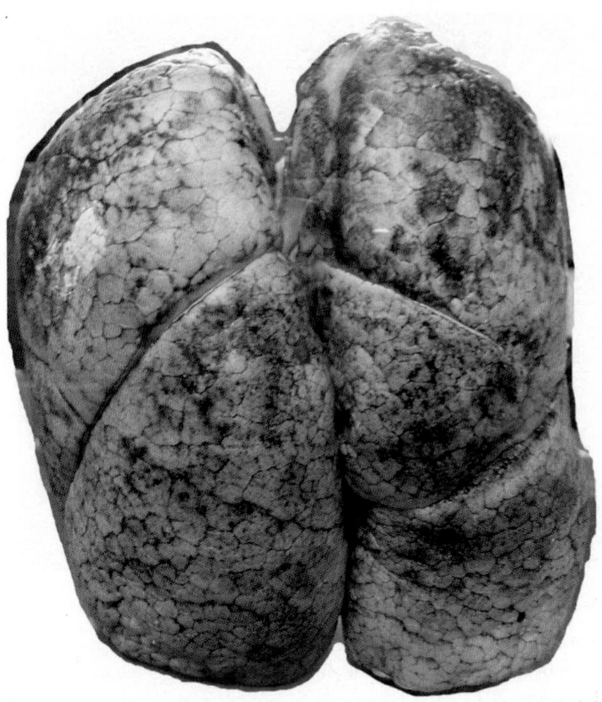

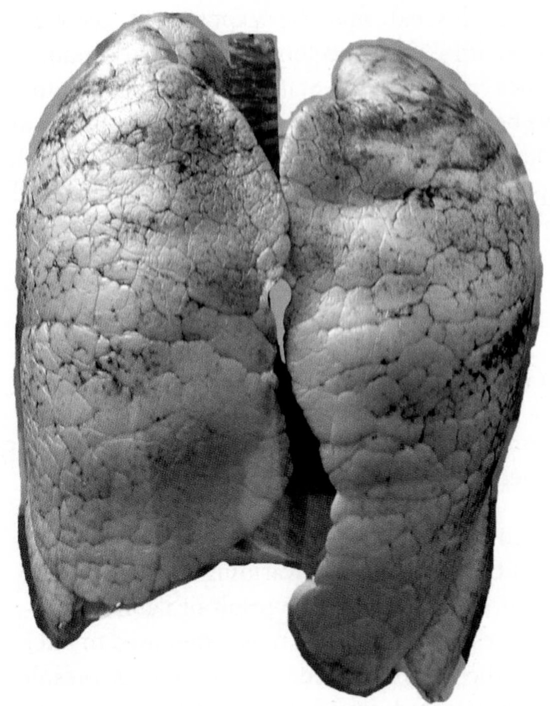

FIGURE 10.19 A. Dorsal view of the lungs of a male heavy smoker who quit 18 months prior to a fatal heart attack at 71 years of age. B. The lungs of a smoker who had quit smoking for a number of years.

Courtesy of Dr. Robert Phalen, Air Pollution Lab, University of California, Irvine.

The harmful effects of smoking upon smokers have been well known for some time. Some of these harmful effects are lung conditions such as emphysema and chronic obstructive pulmonary disease, heart disease, as well as lung cancer and cancers at other bodily sites. More recently the effects of exposure to environmental tobacco smoke (ETS) have become recognized more clearly. Exposure to ETS is also called "secondhand smoking," "passive smoking," or "involuntary smoking."[58]

The term **passive smoking**, also known as exposure to sidestream cigarette smoke, refers to the involuntary breathing of cigarette smoke by nonsmokers in an environment where there are cigarette smokers present. In restaurants, waiting rooms, international airliners, and other enclosed areas where there are cigarette smokers, nonsmokers may be unwillingly (and, perhaps, unwittingly) exposed to a potential health hazard. The effects of chronic exposure to cigarette smoke in the work environment were examined in a cross-sectional study of 5,210 cigarette smokers and nonsmokers. Nonsmokers who did not work in a smoking environment were compared with nonsmokers who worked in a smoking environment as well as with smokers. Exposure to smoke in the work environment among the nonsmokers was associated with a statistically significant reduction in pulmonary function test measurements in comparison with the nonsmokers in the smoke-free environment.[66] A 1992 report from the US Environmental Protection Agency concluded that environmental tobacco smoke is a human lung carcinogen responsible for approximately 3,000 lung cancer deaths annually among US nonsmokers.[67] The nonsmoking spouses of smokers are at greater risk of dying than are the nonsmoking spouses of nonsmokers. Eleven studies conducted in the United States have shown an average 19% increase in risk of death among nonsmoking spouses of smokers.[68] Among children, passive smoking is associated with bronchitis, pneumonia, fluid in the middle ear, asthma incidence, and aggravation of existing asthma.

Research on passive smoking presents several methodologic difficulties. Relatively small increases in risk of death from passive smoking are difficult to demonstrate, given the use of questionnaires to quantify smoking by spouses, the long- and short-term variability in exposures to cigarette smoke from sources other than the spouse (e.g., those at work, restaurants, and entertainment venues), and the long latency period between exposure to cigarette smoke and onset of disease. Additional research will be required to improve methods, such as the use of biologic markers, for assessing exposure to cigarette smoke.

Many states (e.g., California, Massachusetts) in the United States have banned smoking in locations such as restaurants where nonsmokers (both customers and employees) may be exposed to secondhand smoke. Smoking is no longer permitted on domestic airlines in the United States. Many foreign countries have begun to introduce smoking bans in public areas including restaurants; an example is Ireland. These steps will help to protect the public's health by reducing exposure to secondhand cigarette smoke.

Global Climate Change and Global Warming

According to one source, the term **global warming** is defined as follows:

> An increase in the near surface temperature of the Earth. Global warming has occurred in the distant past as the result of natural influences, but the term is most often used to refer to the warming predicted to occur as a result of increased emissions of greenhouse gases. Scientists generally agree that the Earth's surface has warmed by about 1 degree Fahrenheit [0.55 degrees Celsius] in the past 140 years.[69]

The use of fossil fuels, including coal and petroleum-based fuels, causes the release of gases, including **carbon dioxide (CO_2)**, methane, chlorofluorocarbon gases, and nitrous oxide. Chlorofluorocarbon gases used in air conditioners have been linked to the depletion of the ozone layer. Reduction of ozone levels in the upper atmosphere produces increased levels of ultraviolet radiation, normally absorbed by stratospheric ozone. The result could be an increased incidence of skin cancers and cataracts.

Gases that arise from natural sources and anthropogenic activities such as the burning of fossil fuels may accumulate in the atmosphere. Sometimes these gases are referred to as **greenhouse gases**, owing to the fact that in sufficient concentrations in the atmosphere, they may have the effect of trapping heat and causing the earth's temperature to rise. This principle is illustrated in **FIGURE 10.20**.

Have the earth's surface temperatures actually increased during the past century? According to the National Oceanic and Atmospheric Administration (NOAA), the answer is "yes," with some qualifications.[70] Since the end of the 19th century, temperatures at the surface of the earth have increased generally by about 0.74 degrees Celsius (1.3 degrees Fahrenheit), and for the past 50 years the linear trend of increase is about 0.13 degrees Celsius (0.23 degrees Fahrenheit), although some areas (e.g., the southeastern United States and sections of the north Atlantic) have cooled.

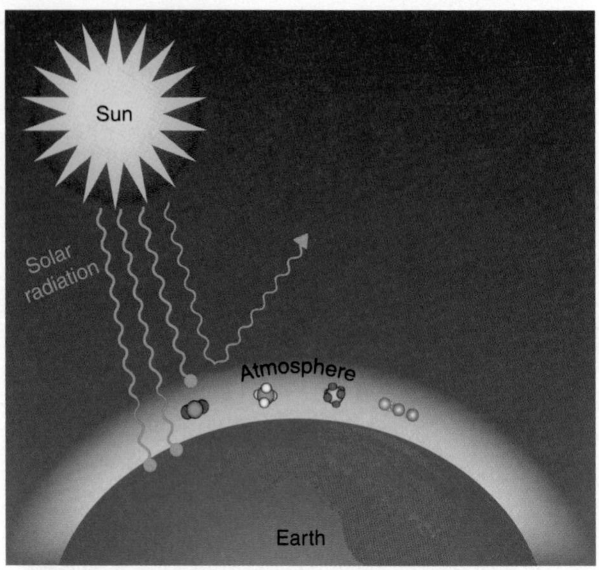

FIGURE 10.20 The greenhouse effect.

Reproduced from Deluisi B, National Oceanic & Atmospheric Administration. The greenhouse effect. Available at: https://www.esrl.noaa.gov/gmd/outreach/carbon_toolkit/basics.html. Accessed June 6, 2017.

Temperatures in North America and Eurasia have shown the greatest warming trend. **FIGURE 10.21** portrays global temperature anomalies during the period 1880 to 2016 (anomalies refer to departures from the long-term average temperature during the same time period). The trends shown by the data in Figure 10.21 suggest that global temperatures have shown an increase of 0.07 degrees Celsius (0.126 degrees Fahrenheit) per decade since 1880.

An international environmental group known as the Intergovernmental Panel on Climate Change (IPCC) has pointed to the likelihood of the role of human activities in causing global warming.[71] Increases in greenhouse gases have been linked with warming of the planet. With continuing emissions of greenhouse gases, models of future trends portend temperature increases of 0.5 to 8.6 degrees Fahrenheit (0.28 to 4.8 degrees Celsius) by the end of the 21st century.

A key issue is whether the increases in global temperature that have been observed during the past century represent natural temperature variability or are the effect of greenhouse gases. Analyses of temperature changes over the past millennium have used data from tree rings and ice cores. These analyses indicate that the recent large temperature increases during the late 20th century are most likely due to anthropogenic activities that have increased levels of greenhouse gases in the atmosphere.[72]

Deforestation

Contributing to the greenhouse effect is deforestation, which decreases the capacity of trees and the

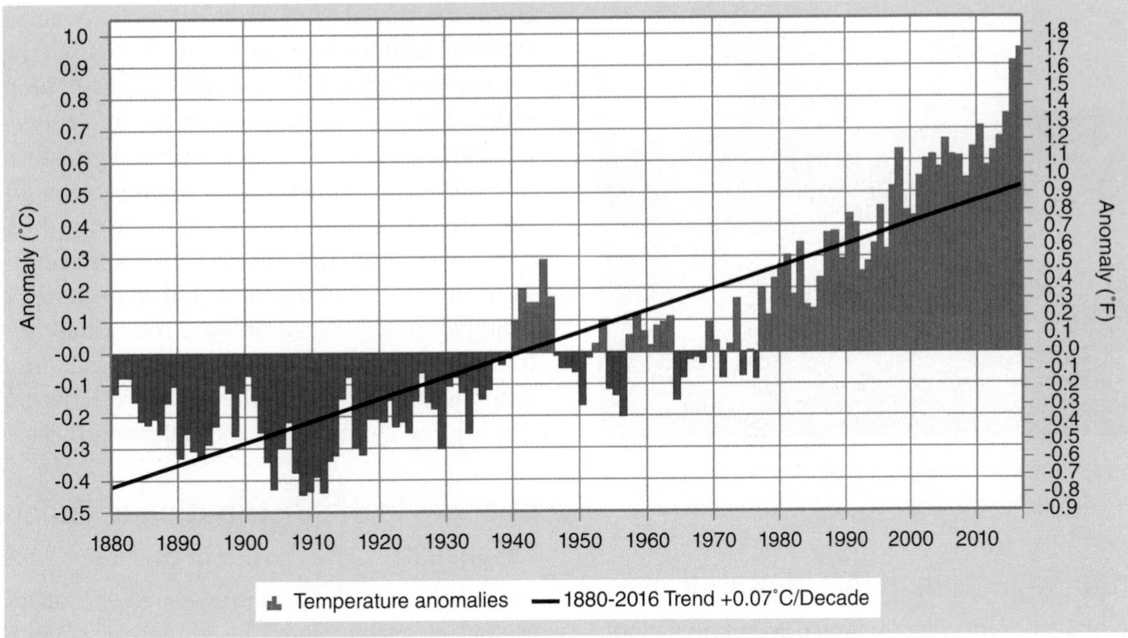

FIGURE 10.21 Global land and ocean temperature anomalies—January to December.

Reproduced from NOAA, National Centers for Environmental Information. Climate at a Glance: Global Time Series. Available at: http://www.ncdc.noaa.gov/cag/. Accessed August 17, 2017.

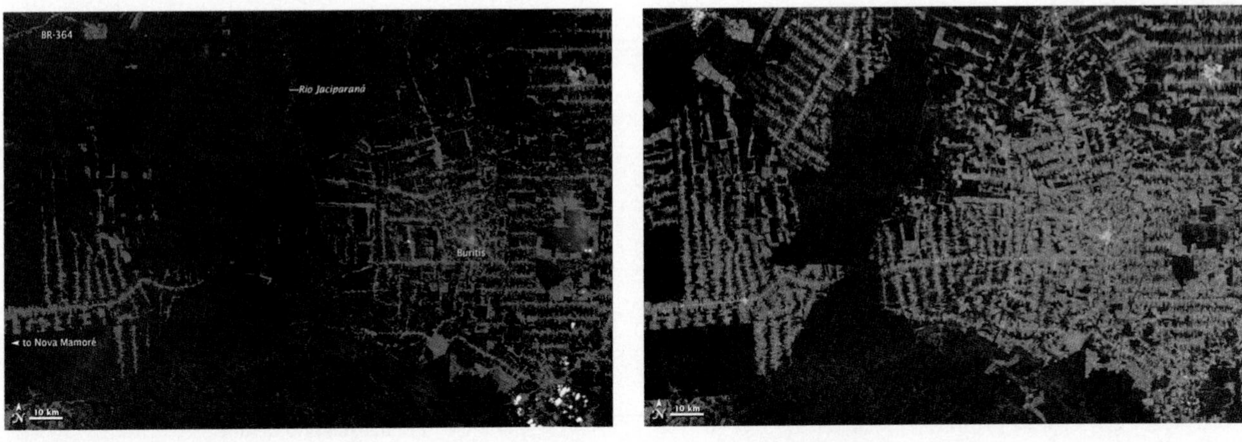

FIGURE 10.22 Brazilian deforestation of the state of Rondônia.

Reproduced from NASA, Earth Observatory. Amazon deforestation. Available at: https://earthobservatory.nasa.gov/Features/WorldOfChange/deforestation.php. Accessed June 7, 2017.

forest ecosystem to reduce the amount of CO_2 in the atmosphere. When deforestation occurs, biodiversity is reduced, resulting in the loss of plants that someday might yield valuable medicines and other products. Deforestation often is the result of burning huge swaths of land, further adding to air pollution.

Extensive areas of the world have been affected by deforestation including the Brazilian Amazon jungle, the largest tropical forest in the world (31% of the world's total). **FIGURE 10.22** provides earlier and later satellite images that show the progress of deforestation during a 12-year period.

Deforestation is not limited to the Amazon region; vast sections of the United States also have suffered from the widespread removal of trees. Forests, which take up about 33% of the land mass in the United States, contribute clean air, water, habitat for animals, recreational opportunities, and valuable lumber. One of the areas that has been affected is the US Pacific Northwest. In this region, logging of old-growth forests by the method of clear-cutting has caused habitat fragmentation, whereby continuously forested areas are segmented into smaller parcels that are separated by nonforested zones. The net result is the destruction of habitat needed for the survival of animals that must leave the protective cover of the forest and traverse nonforested segments in order to seek food and water. (Refer to **FIGURE 10.23** for an example of clear-cutting.)

FIGURE 10.23 Deforestation caused by clear-cutting practices in the United States.

Potential Impacts of Global Warming

The potential consequences of global warming include disturbances in the native habitats of plant and animal species. Another possible end result could be the production of an environment that is conducive to the growth of vector-borne diseases, growth of organisms in the ocean that cause foodborne seafood poisoning, and exacerbation of the effects of air pollution. Global warming, which has been linked to extreme climatic conditions such as heat waves, droughts, and monsoons, may cause disruption of the food supply and dwindling of food resources, especially in developing areas. Refer to **FIGURE 10.24** for an overview of the potential health effects associated with climate variability and change. Potential health effects shown in the figure range from temperature-related illnesses and deaths to vector- and rodent-borne diseases. Moderating influences involve nonclimatic factors such as population growth, changes in the standard of living, and improvements in health care and the public health infrastructure. Adaptation measures are steps that can be taken to lessen the possibility of adverse health outcomes linked to climate change. Examples of these measures are increasing the protection of populations via public health programs such as immunizations as well as the use of protective technologies (e.g., air conditioning and water treatment).[73]

Changes in the Distribution of Endemic Diseases

As global mean temperatures increase, changes in the ecological system favor the growth of some disease-causing agents (e.g., bacteria and fungi) and disease-carrying vectors. The increases in the prevalence and incidence of emerging and reemerging infectious diseases have been attributed to some degree to climate changes that are occurring in some regions of the world.[74] For example, as the temperature increases, the range of arthropods such as the *Aedes aegypti* mosquito can expand farther north. Normally, freezing temperatures during winter kill off this species of mosquito, which can carry viruses that cause diseases such as dengue fever and yellow fever. Also, extreme weather conditions such

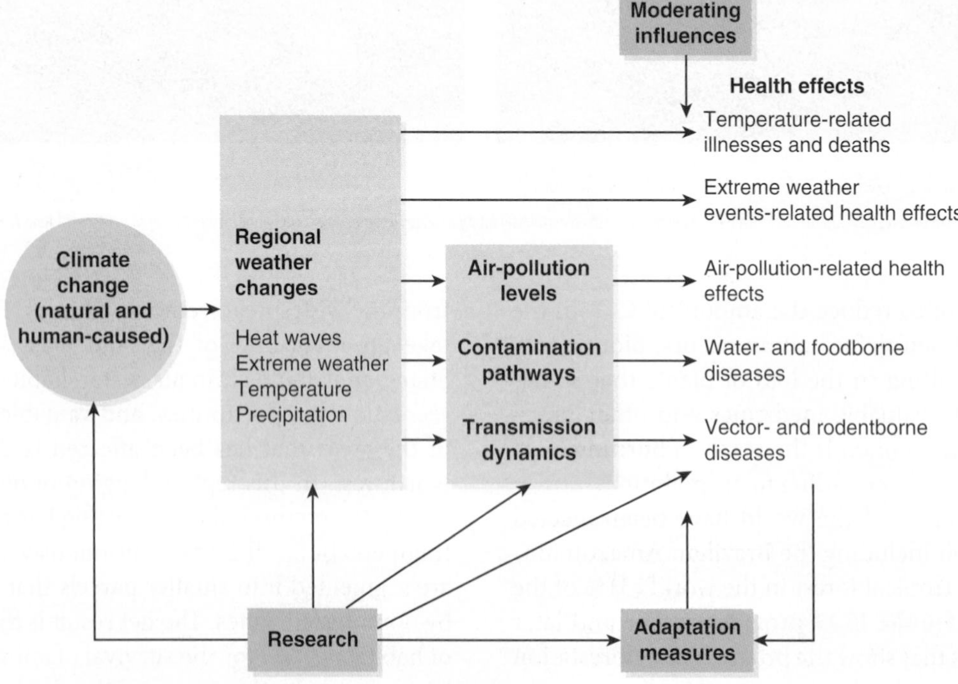

FIGURE 10.24 The relationship between potential health effects and variability and change in climate.

Modified and reproduced from Patz JA, McGeehin MA, Bernard SM, et al. The potential health impacts of climate variability and change for the United States: executive summary of the report of the health sector of the US National Assessment. *Environ Health Perspect*. 2000;108:368.

as heavy rainfall provide breeding grounds that contribute to the growth of mosquito populations.

Some Canadian experts believe that global warming might cause higher temperatures and longer summers that enable disease vectors to expand northward.[75] Another consequence could be that a warmer climate in Canada would permit the establishment of formerly exotic zoonoses (defined in the chapter on zoonotic and vector-borne diseases). As a result, Canadians would be at increased risk of contracting vector-borne diseases such as arboviral encephalitis, rickettsial diseases, and Lyme arthritis. Increased temperatures might contribute to the contamination of food sources derived from animals due to *Salmonella* and *E. coli* O157:H7. If summers become longer and winters milder, rodents and small mammal species that are reservoirs for zoonotic diseases might be able to breed and survive for longer time periods. Consequently, the human population would be at increased risk of zoonotic diseases such as hantavirus infection and plague.

Retreating Glaciers and Rising Oceans

Evidence of a changing global climate is the retreat of glaciers in high mountain areas worldwide. High-altitude areas on earth that are showing the retreat of perennial snow and glaciers are Glacier National Park in Montana (United States), Mount Kilimanjaro in Africa, and the Alps in Europe.[76] Regions such as these could be completely devoid of glaciers in the future. **FIGURE 10.25** demonstrates how the Columbia Glacier in Alaska has receded from 1986 to 2016.

Another apparent effect of global warming has been for the average height of the world's oceans to rise gradually as water stored in frozen sources melts. Refer to the following text box.

Particularly at risk of flooding due to a rise in the level of the sea are low-lying areas (especially low-lying coastal cities), deltas, and coral atolls.[77] Rising sea levels will cause intrusion of salt water and erosion of coastal areas. Countries that would be most affected are Vietnam, Egypt, Bangladesh, and low-lying island nations. According to a mid-range scenario in which the level of the ocean rises by 16 inches (40 centimeters) by the year 2080, as many as 200 million people would be affected by surges in the ocean level. An even greater increase in sea level by 1 meter (3.3 feet) would affect additional millions of people in China, Bangladesh, Egypt, and Indonesia.

IS SEA LEVEL RISING?

Global mean sea level has been rising at an average rate of 1.7 mm/year (plus or minus 0.5 mm) (0.08 in/year plus or minus 0.02 in) over the past 100 years, which is significantly larger than the rate averaged over the last several thousand years. Depending on which greenhouse gas increase scenario is used (high or low) projected sea-level rise is projected to be anywhere from 0.18 (low greenhouse gas increase) to 0.59 meters (0.59 feet to 1.9 feet) for the highest greenhouse gas increase scenario.

Reproduced from National Oceanic and Atmospheric Administration. Global Warming Frequently Asked Questions: Is sea level rising? Available at: https://www.ncdc.noaa.gov/monitoring-references/faq/global-warming.php. Accessed March 1, 2017.

Extreme Climatic Conditions

Evidence suggests that global warming is associated with extreme climatic conditions including heat waves and severe rain storms.

> Small changes in global mean temperatures can produce relatively large changes in the frequency of extreme temperatures. . . . Higher average ambient air temperatures are likely to induce more vigorous cycles of evaporation and precipitation. Indeed, a trend of increasing climate variability and extreme

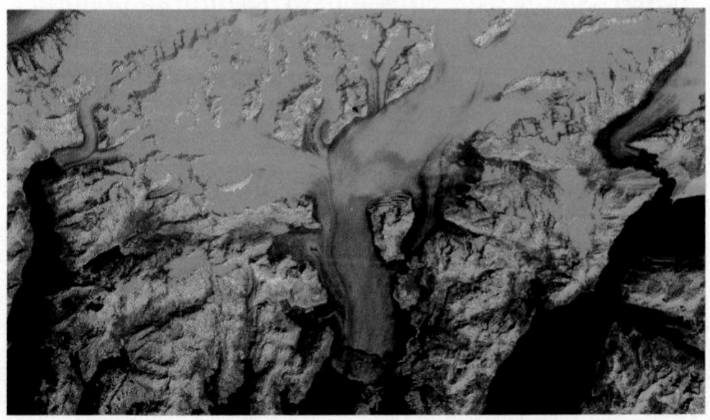

FIGURE 10.25 Columbia Glacier, Alaska—1986–2016.

Reproduced from NASA, Earth Observatory. Columbia Glacier, Alaska. Available at: https://earthobservatory.nasa.gov/Features/WorldOfChange/columbia_glacier.php. Accessed June 7, 2017. Images by Jesse Allen and Robert Simmon, using Landsat 4, 5, and 7 data from the USGS Global Visualization Viewer.

precipitation events has been observed over the past century, and recent models strongly correlate this trend with anthropogenic production of greenhouse gases.[78(p2283)]

From 1979 to 1995, the United States experienced a total of 6,615 heat-related deaths that were distributed as follows: heat due to weather conditions (2,792), heat due to human-made factors (327), and unspecified heat-related cause (3,496).[79] The major risk factors for heat-related death include excessive physical exertion during high temperatures, alcohol consumption, being overweight, using certain medications, and age. (Refer to **FIGURE 10.26**.)

Among significant episodes of heat-related mortality, two of particular note occurred: in Chicago, Illinois (US), and in France. The July 12 to July 16, 1995 heat wave in Chicago caused the daily high temperatures to rise to the range of 93°F to 104°F (33.9°C to 40.0°C); the city had a record-high temperature of 119°F (48.3°C) on July 13. The resulting high temperatures were responsible for 465 deaths that were certified as heat related during the period of the heat wave.[80]

Figure 10.26 presents more recent data on heat-related deaths in the United States, between 1979 and 2014. Heath-related deaths are a significant cause of

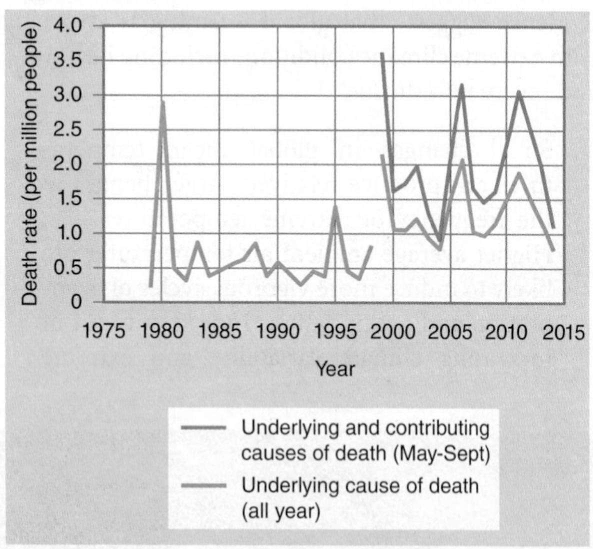

FIGURE 10.26 Deaths classified as "heat related" in the United States, 1979–2014.

Reproduced from US Environmental Protection Agency. Climate Change Indicators: Heat-Related Deaths. Available at: https://www.epa.gov/climate-indicators/climate-change-indicators-heat-related-deaths. Accessed June 7, 2017. Data sources: Centers for Disease Control and Prevention. CDC WONDER database: Compressed mortality file, underlying cause of death. Available at: http://wonder.cdc.gov/mortSQL.html. Accessed February 2016; Centers for Disease Control and Prevention. Indicator: Heat-related mortality. National Center for Health Statistics. Annual national totals provided by National Center for Environmental Health staff in June 2016. Available at: http://ephtracking.cdc.gov/showIndicatorPages.action.

mortality in this country. Such deaths may become common as global warming and temperatures become more extreme in the future. From 1979 to 2014, about 9,000 deaths from heat-related causes have occurred in the United States. A peak in deaths occurred in 2006, an extremely hot year in this country.

During August 2003, a blistering heat wave descended on France, causing a death toll of almost 15,000.[81] Many of the deaths occurred among the elderly after temperatures reached the low 100s Fahrenheit (low 40s Celsius) in a country where it is uncommon to use air conditioning. One account indicated that "morgues and funeral parlours coped with an overflow of victims. Refrigerated storerooms were set up; and temporary workers hired to collect bodies from private homes and hotels. Grave diggers worked overtime."[82(p411)]

Controlling Air Pollution and Global Warming

At the beginning the 21st century, the United States was the leading producer of greenhouse gases with 23% of the earth's emissions of such pollutants.[83] The major source of air pollution in the United States is combustion of fossil fuels, particularly by coal-fired electric generating plants and internal combustion engines. With only about 4% of the world's population, the United States is currently the second leading source of carbon dioxide emissions (producing 16% of the global total).[84] China has the distinction of being the world's leader in global-warming pollution and is responsible for approximately 28% of total emissions.

The world's developing nations are increasing their contribution of greenhouse gases as these countries add to their populations and increase their levels of industrialization. A number of steps have been proposed to reduce the emissions of harmful air pollution: technological controls, the climate protocols such as the Paris Agreement, and energy conservation.

Using Technology to Control Particulate Matter

Several mechanical devices that are used to reduce industrial emissions of particulate matter are scrubbers, filters, and electrostatic precipitators. Scrubbers are machines that transfer particles in gases to a collecting liquid. An electrostatic precipitator (see **FIGURE 10.27**) confers negative or positive electrical charges to particles

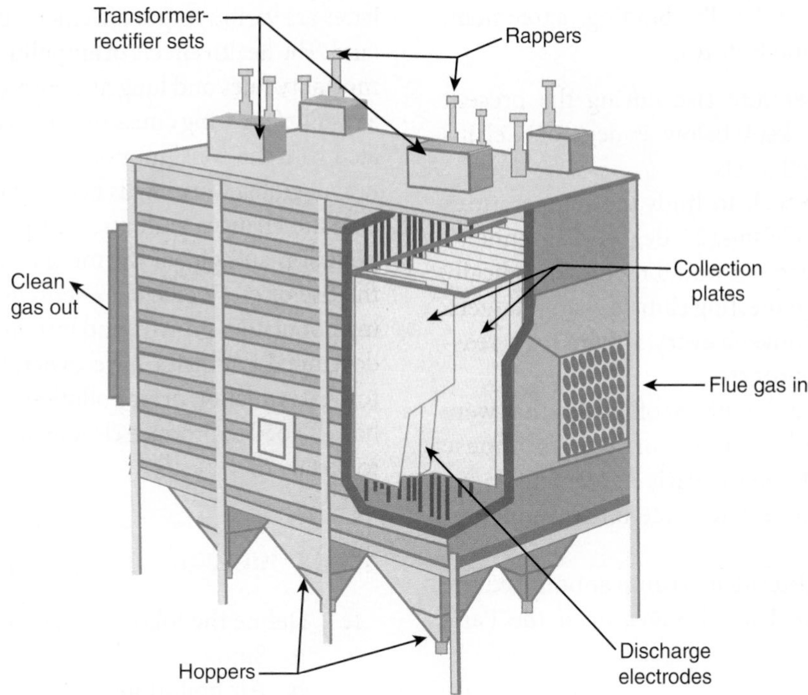

FIGURE 10.27 Conventional electrostatic precipitator.

Modified and reproduced from US Environmental Protection Agency. Basic Concepts in Environmental Sciences, Module 6: Air Pollutants and Control Techniques—Particulate Matter. Available at: http://www.epa.gov/apti/bces/module6/matter/control/control.htm#precipit. Accessed April 13, 2010.

that are present in gases and collects them on a surface that has the opposite charge.[85]

The Kyoto Protocol

The Kyoto Protocol was an international treaty that addressed climate change.[86,87] Initiated in Kyoto, Japan in December 1997, the provisions of the protocol were a legally binding commitment among the ratifying nations. The purpose of the protocol was to reduce emissions of greenhouse gases believed to be the cause of recent climate changes. The protocol required developed countries to reduce their emissions by targeted amounts. For example, the United States would be required to cut emissions by 7% and European countries (Switzerland, central and eastern Europe, and the European Union) by 8%. The Kyoto Protocol specified that it would come into effect 90 days after it was ratified by a sufficient number of industrialized countries that in combination produce at least 55% of the world's total CO_2 emissions. For some time, implementation of the Kyoto Protocol was stalled because an adequate number of countries, including the United States, had failed to ratify it. By February 2005, 141 nations including Russia had ratified the protocol, meaning that it could be implemented. The Kyoto Protocol went into force on February 16, 2005.[87] The United States, the world's largest producer of CO_2 at the time of the climate change pact, did not ratify the protocol; Australia was the only other developed country that has not signed the agreement.

In December 2009, almost 115 world leaders and more than 40,000 participants assembled in Copenhagen, Denmark to deliberate over climate change policy.[88,89] One of the products of the Copenhagen Climate Change Conference was an agreement known as the Copenhagen Accord, which articulated a political intention for reigning in carbon emissions and coping with climate change. Participants in the Copenhagen climate talks asserted that climate change was a major contemporary environmental challenge that should be addressed by curbing greenhouse gases. Delegates proposed that any future increase in global temperature by should be kept below 2 degrees Celsius (3.6 degrees Fahrenheit) above levels that prevailed during the preindustrial era. This goal was set to be reviewed in 2015. The agreement was not legally binding and was opposed by many countries, especially in the developing world. At the conclusion of the conference, talks became deadlocked. Nevertheless, this international conference ultimately may have had the positive effect of increasing the world's awareness of control of greenhouse gases and related issues.

Paris Agreement

The Paris climate conference in December 2015 culminated with the Paris Agreement.[90–93] A total of 195 countries adopted the legally binding agreement, which had the following features:

- The global temperature rise during the present century is to be kept below 2 degrees Celsius above preindustrial levels.
- Countries would seek to limit the temperatures rise to 1.5 degrees Celsius (2.7 degrees Fahrenheit).
- Parties to the agreement would report periodically on their progress in meeting climate change targets.
- Parties would reconvene every 5 years for a reassessment of their progress.
- Thirty days after, at least 55% of parties that were responsible for at least a total of 55% greenhouse gases officially ratified the Paris Agreement.
- The agreement came into force on November 4, 2016.
- On June 1, 2017, President Trump announced the withdrawal of the United States from the Paris Agreement.

Energy Conservation

From a global perspective, the Kyoto Protocol, the Copenhagen Accord, and similar agreements should be helpful in reducing the emission of greenhouse gases. In the United States, individual states such as California have assumed leadership in addressing climate change. Examples of steps that can be taken to reduce the emission of greenhouse gases and prevent global warming are the following:

- Increase the efficiency of older power plants.
- Develop more renewable and alternative energy sources (e.g., wind turbines and solar panels).
- Use energy-efficient designs in home construction and electrical appliances; try to reduce dependence on such appliances.
- Increase the fuel efficiency of motor vehicles as in the use of fully electric, hybrid gas-electric, and other high-mileage designs.
- Increase the use of public transportation; use bicycles for getting around in cities.

Conclusion

Clean air is an essential element for the survival of life on earth. Air that is polluted can endanger our health and cause damage to trees, wildlife, and property. Sources of air pollution can be subdivided into natural and anthropogenic sources. Natural pollution arises from volcanoes, dust storms, and fires. Anthropogenic sources arise from stationary and mobile sources. The former include factories, oil refineries, and incinerators; examples of the latter are various types of vehicles used on air, water, and land. The health effects of air pollution include increased mortality rates and lung and cardiovascular diseases. As a result of growing emissions of greenhouse gases associated with the consumption of fossil fuels, the phenomenon of global warming is believed to be causing extreme climatic changes and rises in the average sea level. Air pollution and global warming can be reduced through the use of more efficient energy technologies, by limiting population growth, and by conserving energy. Many developed countries have expended significant efforts toward reduction of air pollution. Although these efforts have helped to produce cleaner air, additional steps need to be taken.

Study Questions and Exercises

1. Define the following terms in your own words:
   a. Smog
   b. Air pollution
   c. Criteria air pollutants
   d. Normal ambient air
2. Give examples of how air pollution endangers human health. Include in your discussion a review of the short-term and long-term health effects of air pollution.
3. How does air pollution impact the environment? In what ways can the pollutants NO_x and SO_2 damage structures and monuments over long periods of time?
4. Define the following sources of air pollution and give examples:
   a. Stationary
   b. Mobile
   c. Natural
   d. Anthropogenic
5. Describe how ground-level ozone can be harmful to our health while ozone in the upper stratosphere plays an important role in protecting our health.
6. Explain the adverse health effects attributed to carbon monoxide. Which population groups are likely to be most vulnerable to the effects of carbon monoxide exposure?
7. Define what is meant by a temperature inversion. Explain how temperature inversions contribute to the creation of smog.
8. Explain why environmental health experts are concerned about the possible adverse consequences of exposure to diesel exhaust. Describe one major adverse health effect that has been attributed to diesel exhaust. Name the risk groups that are most affected by exposure to diesel exhaust.

9. Refer to the diagram that shows the anatomy of the human respiratory system (Figure 10-17, part A, page 257). Which parts of the respiratory system are most likely to be affected by fine particulate matter?
10. Define the term *Air Quality Index*. How does the Air Quality Index aid in protecting the health of the population?
11. Epidemiologic analyses have demonstrated a correlation between an increase in total daily mortality and an increase in air pollution. Give examples of historically significant fatal air pollution episodes that were characterized by extreme increases in air pollution and accompanying increases in mortality.
12. Explain the difference between sick building syndrome and building-related illness. Describe the sources of indoor air pollution that can affect human health adversely.
13. Define what is meant by greenhouse gases and describe how they contribute to the greenhouse effect.
14. Describe three methods for the control of greenhouse gases. Which method, if any, is likely to be most effective?
15. Define the term *global warming* and present arguments for and against the proposition that global warming has occurred during the past century. What environmental outcomes have been attributed to global warming?

For Further Reading

Davis, Devra. *When Smoke Ran Like Water*, 2002.
Gore, Al. *Earth in the Balance*, 1992.
Dawson, Kate Winkler. *Death in the Air*, 2017.

References

1. US Environmental Protection Agency (EPA). Benefits and costs of the Clean Air Act amendments of 1990. Available at: https://www.epa.gov/sites/production/files/2015-07/documents/factsheet.pdf. Accessed June 19, 2017.
2. World Health Organization (WHO). WHO global urban ambient air pollution database (update 2016). Available at: http://www.who.int/phe/health_topics/outdoorair/databases/cities/en/. Accessed June 19, 2017.
3. American Lung Association. *State of the Air 2016*. Chicago, IL: American Lung Association; 2016.
4. South Coast Air Quality Management District (AQMD). Smog and health. Available at: http://www.aqmd.gov/home/library/public-information/publications/smog-and-health-historical-info. Accessed June 16, 2017.
5. US Environmental Protection Agency. *The Plain English Guide to the Clean Air Act*. Research Triangle Park, NC: Office of Air Quality Planning and Standards; 2007.
6. US Environmental Protection Agency. National Risk Management Research Laboratory. *Environmental Curricula Handbook*. EPA 625/R-02/009. Appendix B: Glossary of Terms. Washington, DC: US Environmental Protection Agency; 2002.
7. Bell ML, Davis DL. Reassessment of the lethal London fog of 1952: novel indicators of acute and chronic consequences of acute exposure to air pollution. *Environ Health Perspect*. 2001;109(suppl. 3):389–394.
8. Helfand WH, Lazarus J, Theerman P. Donora, Pennsylvania: an environmental disaster of the 20th century. *Am J Public Health*. 2001;91:553–554.
9. Phalen RF. *The Particulate Air Pollution Controversy: A Case Study and Lessons Learned*. Boston, MA: Kluwer Academic Publishers; 2002.
10. Health Effects Institute. Understanding the health effects of components of the particulate matter mix: progress and next steps. Boston, MA: Health Effects Institute; 2002.
11. Centers for Disease Control and Prevention. Epidemiologic notes and reports cytotoxicity of volcanic ash: assessing the risk for pneumoconiosis. *MMWR*. 1986;35:265–267.
12. California Environmental Protection Agency, Air Resources Board. Glossary of air pollution terms. Available at: https://www.arb.ca.gov/html/gloss.htm. Accessed June 16, 2017.
13. US Environmental Protection Agency. EPA determines that aircraft emissions contribute to climate change endangering public health and the environment. Press release. Available at: https://www.epa.gov/newsreleases/epa-determines-aircraft-emissions-contribute-climate-change-endangering-public-health. Accessed June 20, 2017.
14. National Institutes of Health. MedlinePlus. Air pollution. Available at: http://www.medlineplus.gov/airpollution.html. Accessed June 19, 2017.
15. Union of Concerned Scientists. Cars, trucks, and air pollution. Available at: https://www.ucsusa.org/clean-vehicles/vehicles-air-pollution-and-human-health/cars-trucks-air-pollution#.WkP7pBqo6bM. Accessed December 27, 2017.
16. US Environmental Protection Agency. Criteria air pollutants. Available at: http://www.epa.gov/criteria-air-pollutants. Accessed February 24, 2017.
17. Kaiser J. The other global pollutant: nitrogen proves tough to curb. *Science*. 2001;294(5545):1268–1269.
18. Medical Dictionary. Carboxyhemoglobin. Available at: http://medical-dictionary. The free dictionary.com/carboxy hemoglobin. Accessed June 19, 2017.
19. [No authors listed]. Novel technologies measure ultrafine air pollution in the L.A. Basin. *J Environ Health*. 2004;66(8):51.
20. World Health Organization. Annual mean concentrations of fine particulate matter ($PM_{2.5}$) in urban areas (μg/m3). Available at: http://gamapserver.who.int/gho/interactive_charts/phe/oap_exposure/atlas.html. Accessed June 20, 2017.
21. Centers for Disease Control and Prevention. Potential exposures to airborne and settled surface dust in residential areas of lower Manhattan following the collapse of the World Trade Center—New York City, November 4–December 11, 2001. *MMWR*. 2003;52:131–136.
22. US Environmental Protection Agency. What is acid rain? Available at: https://www.epa.gov/acidrain/what-acid-rain. Accessed June 17, 2017.
23. Lipsett M, Campleman S. Occupational exposure to diesel exhaust and lung cancer: a meta-analysis. *Am J Public Health*. 1999;89:1009–1017.
24. Cohen HJ, Borak J, Hall T. Exposure of miners to diesel exhaust particulates in underground nonmetal mines. *AIHA J*. 2002;63:651–658.

25. Kagawa J. Health effects of diesel exhaust emissions—a mixture of air pollutants of worldwide concern. *Toxicol.* 2002;181–182:349–353.
26. Liukonen LR, Grogan JL, Myers W. Diesel particulate matter exposure to railroad train crews. *AIHA J.* 2002;63:610–616.
27. Northridge ME, Yankura J, Kinney PL, et al. Diesel exhaust exposure among adolescents in Harlem: a community-driven study. *Am J Public Health.* 1999;89:998–1002.
28. Stayner L. Protecting public health in the face of uncertain risks: the example of diesel exhaust [editorial]. *Am J Public Health.* 1999;89:991–993.
29. Steenland K. Lung cancer and diesel exhaust: a review. *Am J Ind Med.* 1986;10:177–189.
30. US Environmental Protection Agency. AIRNow: Air Quality Index (AQI)—A guide to air quality and your health. Available at: https://www.airnow.gov/index.cfm?action=aqibasics.aqi. Accessed June 17, 2017.
31. Levy JI, Greco SL, Spengler JD. The importance of population susceptibility for air pollution risk assessment: a case study of power plants near Washington, DC. *Environ Health Perspect.* 2002;110:1253–1260.
32. Health and Clean Air. Saving the children. *Health & Clean Air Newsletter.* Spring 2003. Available at: http://www.healthandcleanair.org/newsletters/spring2003.html. Accessed June 17, 2017.
33. Dockery DW, Pope CA III, Xu X, et al. An association between air pollution and mortality in six U.S. cities. *New Engl J Med.* 1993;329:1753–1759.
34. Pope CA III. Adverse health effects of air pollutants in a nonsmoking population. *Toxicol.* 1996;111:149–155.
35. Ito K, Thurston GD, Hayes C, et al. Associations of London, England, daily mortality with particulate matter, sulfur dioxide, and acidic aerosol pollution. *Arch Environ Health.* 1993;48:213–220.
36. Schimmel H, Greenberg L. A study of the relation of pollution to mortality: New York City, 1963–1968. *J Air Pollut Control Assoc.* 1972;22:607–616.
37. Dockery DW, Schwartz J, Spengler JD. Air pollution and daily mortality: associations with particulates and acid aerosols. *Environ Res.* 1992;59:362–373.
38. Dominici F, McDermott A, Zeger SL, et al. National maps of the effects of particulate matter on mortality: exploring geographical variation. *Environ Health Perspect.* 2003;111:39–43.
39. Mann JK, Tager IB, Lurmann F, et al. Air pollution and hospital admissions for ischemic heart disease in persons with congestive heart failure or arrhythmia. *Environ Health Perspect.* 2002;110:1247–1252.
40. Kuller LH, Radford EP, Swift DP, et al. Carbon monoxide and heart attacks. *Arch Environ Health.* 1975;30:477–482.
41. Wright GR, Jewizyk S, Onrot J, et al. Carbon monoxide in the urban atmosphere. *Arch Environ Health.* 1975;30:123–129.
42. Lebowitz MD, Holberg CJ, Knudson RJ, Burrows B. Longitudinal study of pulmonary function development in childhood, adolescence, and early adulthood. *Am Rev Respir Dis.* 1987;136:69–75.
43. Lebowitz MD. The relationship of socio-economic factors to the prevalence of obstructive lung diseases and other chronic conditions. *J Chronic Dis.* 1977;30:599–611.
44. Mannino DM, Homa DM, Pertowski CA, et al. Surveillance for asthma—United States, 1960–1995. *MMWR Surveill Summ.* 1998;47:1–28.
45. Mannino DM, Homa DM, Akinbami LJ, et al. Surveillance for asthma—United States, 1980–1999. *MMWR Surveill Summ.* 2002;51:1–13.
46. Rhodes L, Bailey CM, Moorman JE, et al. Asthma prevalence and control characteristics by race/ethnicity—United States, 2002. *MMWR.* 2004;53:145–148.
47. Crain EF, Walter M, O'Connor GT, et al. Home and allergic characteristics of children with asthma in seven U.S. urban communities and design of an environmental intervention: the Inner-City Asthma Study. *Environ Health Perspect.* 2002;110:939–945.
48. Klinnert MD, Price MR, Liu AH, et al. Unraveling the ecology of risks for early childhood asthma among ethnically diverse families in the Southwest. *Am J Public Health.* 2002;92:792–798.
49. Mishra V. Effect of indoor air pollution from biomass combustion on prevalence of asthma in the elderly. *Environ Health Perspect.* 2003;111:71–77.
50. National Institutes of Health. Of air and asthma: air pollution's effects. Bethesda, MD: National Institutes of Health (NIH); 2008.
51. Delfino RJ. Epidemiologic evidence for asthma and exposure to air toxics: linkages between occupational, indoor, and community air pollution research. *Environ Health Perspect.* 2002;110(suppl 4):573–589.
52. Shima M, Nitta Y, Ando M, et al. Effects of air pollution on the prevalence and incidence of asthma in children. *Arch Environ Health.* 2002;57:529–535.
53. Ostro B, Lipsett M, Mann J, et al. Air pollution and exacerbation of asthma in African-American children in Los Angeles. *Epidemiol.* 2001;12:200–208.
54. Henderson BE, Gordon RJ, Menck H, et al. Lung cancer and air pollution in south-central Los Angeles County. *Am J Epidemiol.* 1975;101:477–488.
55. Pope CA III, Burnett RT, Thun MJ, et al. Lung cancer, cardiopulmonary mortality, and long-term exposure to fine particulate air pollution. *JAMA.* 2002;287:1132–1141.
56. Vena JE. Air pollution as a risk factor in lung cancer. *Am J Epidemiol.* 1982;116:42–56.
57. Nyberg F, Gustavsson P, Jarup L, et al. Urban air pollution and lung cancer in Stockholm. *Epidemiol.* 2000;11:487–495.
58. The American Lung Association (ALA), the Environmental Protection Agency (EPA), the Consumer Product Safety Commission (CPSC), and the American Medical Association (AMA). *Indoor Air Pollution: An Introduction for Health Professionals.* Publication 1994-523-217/81322. Washington, DC: US Government Printing Office; 1994.
59. Oliver LC, Shackleton BW. The indoor air we breathe: a public health problem of the '90s. *Public Health Rep.* 1998;113:398–409.
60. Lebowitz MD, Holberg CJ, Boyer B, Hayes C. Respiratory symptoms and peak flow associated with indoor and outdoor air pollutants in the southwest. *J Air Pollut Control Assoc.* 1985;35:1154–1158.
61. Vojta PJ, Friedman W, Marker DA, et al. First national survey of lead and allergens in housing: survey design and methods for the allergen and endotoxin components. *Environ Health Perspect.* 2002;110:527–532.
62. Engvall K, Norrby C, Norbäck D. Asthma symptoms in relation to building dampness and odour in older multifamily houses in Stockholm. *Int J Tuberc Lung Dis.* 2001;5:468–477.
63. Jaakkola JJK, Tuomaala P, Seppänen O. Air recirculation and sick building syndrome: a blinded crossover trial. *Am J Public Health.* 1994;84:422–428.

64. Mendell MJ, Fine L. Building ventilation and symptoms—where do we go from here? *Am J Public Health.* 1994;84:346–348.
65. US Environmental Protection Agency. Indoor air facts no. 4 (revised): sick building syndrome. Available at: https://www.epa.gov/sites/production/files/2014-08/documents/sick_building_factsheet.pdf. Accessed June 17, 2017.
66. White JR, Froeb HF. Small-airways dysfunction in nonsmokers chronically exposed to tobacco smoke. *New Engl J Med.* 1980;302:720–723.
67. US Environmental Protection Agency. *Respiratory Health Effects of Passive Smoking: Lung Cancer and Other Disorders.* EPA publication 600/6-90/006F. Washington, DC: EPA; 1992.
68. Boyle P. The hazards of passive—and active—smoking. *N Engl J Med.* 1993;328:1708–1709.
69. National Aeronautics and Space Administration (NASA). Earth Observatory glossary. Available at: https://earthobservatory.nasa.gov/Glossary/?mode=alpha&seg=f&segend=h. Accessed June 17, 2017.
70. National Oceanic and Atmospheric Administration (NOAA). Global warming. Available at: https://www.ncdc.noaa.gov/monitoring-references/faq/global-warming.php. Accessed June 20, 2017.
71. US Environmental Protection Agency. *Inventory of U.S. Greenhouse Gas Emissions and Sinks: 1990-2015.* EPA 430-P-17-001; 2017.
72. Crowley TJ. Causes of climate change over the past 1000 years. *Science.* 2000;289:270–277.
73. Patz JA, McGeehin MA, Bernard SM, et al. The potential health impacts of climate variability and change for the United States: executive summary of the report of the health sector of the U.S. National Assessment. *Environ Health Perspect.* 2000;108:367–376.
74. McMichael AJ, Patz J, Kovats RS.Impacts of global environmental change on future health and health care in tropical countries. *Br Med Bull.* 1998;54:475–488.
75. Charron DF. Potential impacts of global warming and climate change on the epidemiology of zoonotic diseases in Canada. *Can J Public Health.* 2002;93:334–335.
76. Yohe E. Sizing up the Earth's glaciers. 2004. NASA Earth Observatory. Available at: http://earthobservatory.nasa.gov/Features/GLIMS/. Accessed June 17, 2017.
77. Patz JA, Kovats RS. Hotspots in climate change and human health. *BMJ.* 2002;325:1094–1098.
78. Patz JA, Khaliq M. Global climate change and health: challenges for future practitioners. *JAMA.* 2002;287:2283–2284.
79. Centers for Disease Control and Prevention. Heat-related mortality—United States, 1997. *MMWR.* 1998;47:473–476.
80. Centers for Disease Control and Prevention. Heat-related mortality—Chicago, July 1995. *MMWR.* 1995;44:577–579.
81. CBSNEWS.com. France ups heat toll. September 25, 2003. Available at: http://www.cbsnews.com/stories/2003/08/29/world/main570810.shtml. Accessed June 17, 2017.
82. Dorozynski A. Heat wave triggers political conflict as French death rates rise. *BMJ.* 2003;327:411.
83. A wake-up call for environmental health. *Lancet.* 2003;362:587.
84. Natural Resources Defense Council. Global warming 101. Available at: https://www.nrdc.org/stories/global-warming-101. Accessed June 17, 2017.
85. US Environmental Protection Agency, Air Pollution Training Institute (APTI). Basic concepts in environmental sciences. Module 6: Air pollutants and control techniques. Particulate matter: control techniques. Available at: https://www.apti-learn.net/lms/content/epa/courses/re_100/index.htm?cid=2&userID=43049&bookmarks=. Accessed June 21, 2017.
86. ThoughtCo. What is the Kyoto Protocol? Available at: https://www.thoughtco.com/what-is-the-kyoto-protocol-1204061. Accessed June 17, 2017.
87. United Nations Framework Convention on Climate Change (UNFCCC). Kyoto Protocol Reference Manual. Bonn, Germany: UNFCCC; 2008.
88. United Nations Framework Convention on Climate Change (UNFCCC). Copenhagen Climate Change Conference—December 2009. Available at: http://unfccc.int/meetings/copenhagen_dec_2009/meeting/6295txt.php. Accessed June 22, 2017.
89. *The Guardian.* Climate change talks yield small chance of global treaty. Available at: http://www.guardian.co.uk/environment/2010/apr/11/climate-change-talks-deal-treaty. Accessed June 17, 2017.
90. United Nations Framework Convention on Climate Change (UNFCCC). The Paris Agreement. Available at: http://unfccc.int/paris_agreement/items/9485.phphttp://unfccc.int/paris_agreement/items/9485.php. Accessed June 22, 2017.
91. United Nations Framework Convention on Climate Change (UNFCCC). The Paris Agreement—status of ratification. Available at: http://unfccc.int/paris_agreement/items/9444.php. Accessed June 22, 2017.
92. European Commission. Paris Agreement. Available at: https://ec.europa.eu/clima/policies/international/negotiations/paris_en. Accessed June 22, 2017.
93. Shear MD. Trump will withdraw U.S. from Paris climate agreement. *The New York Times.* June 1, 2017. Available at: https://www.nytimes.com/20 17/06/01 /climate/trump-paris-climate-agreement.html. Accessed June 22, 2017.

© Jean-Luc Rivard/EyeEm/Getty Images

CHAPTER 11

Food Safety

LEARNING OBJECTIVES

By the end of this chapter the reader will be able to:

- State three measures for preventing foodborne illness.
- Discuss 10 microbiological agents that are implicated in foodborne illness.
- Explain the significance of foodborne illness for the world's population.
- List five categories of contaminants in the food supply.
- Describe one major regulation for protecting the safety of food from carcinogens.

Introduction

The focus of this chapter is on food safety, including foodborne diseases, foodborne infections, and foodborne outbreaks. We will learn that some of the causes of acute foodborne illness are microbial agents and toxic chemicals such as pesticides and heavy metals. In addition, other contaminants in foods include food additives, antibiotics used to promote growth in animals used for food, and low levels of potential carcinogens. Sources of food contamination are pollutants in water used to process foods, chemicals used by the agricultural industry, and even pollutants found in the air. You will learn about the potential for foodborne contaminants to affect human health adversely. The role that such contaminants play in causing ill-defined gastrointestinal and other diseases is suspect, yet unclear. Many cases of such illnesses probably go unreported, so their scope is unknown. A pressing question is whether, and to what extent, human exposure to low levels of toxic contaminants in foods increases the risk of adverse health effects or is entirely benign.

A later topic in this chapter is foodborne disease prevention, which can be achieved through the safe handling of foods, safety inspections, and enforcement of government regulations. One of the major purviews of local and federal government agencies is to ensure the quality of foodstuffs. This goal is supported by cadres of government-employed environmental health workers who inspect the food supply and investigate foodborne illness outbreaks. The chapter will conclude with a review of new trends in production of foods and possible implications for human health.

The Global Burden of Foodborne Illness

From the global perspective, foodborne illness is a major cause of morbidity (and occasionally mortality). What is meant by *foodborne illness* (**foodborne disease**)? This term denotes illness caused by foods that have been consumed. Often such illnesses are inaccurately called "food poisoning." The major causes of foodborne illnesses include toxins formed by bacterial growth in the intestines or in food itself and by infections with bacteria and other microbes following the ingestion of contaminated food.[1] The former category of illnesses is called **foodborne intoxications** and the latter **foodborne infections**. A **foodborne outbreak** indicates "the occurrence of a similar illness among two or more people which an investigation linked to consumption of a common meal or food items, except for botulism (one case is an outbreak)."[2(p4)] Foodborne illness can be both acute and long term. Illnesses transmitted by foods can cause adverse birth outcomes, chronic illnesses, and disabilities (e.g., miscarriages, neurologic sequelae from meningitis, and kidney failure).

The problem of foodborne illness is well known to the general public, due to the media's frequent coverage of outbreaks. For example, the media (both in print and online) carry frequent accounts of foodborne illness outbreaks that happen in restaurants in your community; even the major restaurant chains are not immune to such incidents. Another example comes from media reports of dramatic foodborne illness outbreaks on cruise ships; often these outbreaks devastate the vacations of passengers who must terminate their cruises prematurely. Other foodborne outbreaks are associated with foreign travel; at one time or another, many travelers to foreign countries have experienced illnesses caused by unsanitary or improperly prepared meals. Travelers to exotic locales may acquire foodborne pathogens—uncommon in their usual place of residence—that challenge healthcare providers who are unfamiliar with the resulting diseases.[3]

Within the past few years, the incidence of foodborne illnesses has increased in industrialized nations.[4] This increase has been attributed to changes in agricultural and food processing methods, globalization in food distribution, and other social and behavioral changes among the human population.

According to the Centers for Disease Control and Prevention (CDC), foodborne illness continues to be a public health problem of great importance for the United States.[5] Some estimates suggest that foodborne illnesses (reported and unreported) affect almost one-quarter of the population each year in the United States. Other estimates indicate that foodborne illnesses cause 9,000 deaths annually with an economic cost of $5 billion.[6] In recent years, the incidence of foodborne illness in the United States has tended to remain stable over time, with recent declines in some conditions (e.g., *Escherichia coli* [*E. coli*] O157 infection) and the need to sustain these declines.

Surveillance of Foodborne Illness

In the United States, the CDC maintains responsibility at the federal level for surveillance of foodborne illness. For many types of foodborne diseases, only a small proportion may be reported by so-called passive surveillance systems, which rely on the reporting of cases of foodborne illness by clinical laboratories to state health departments and ultimately to the CDC.[7] In contrast, the CDC Foodborne Diseases Active Surveillance Network (FoodNet) is an active system whereby public health officials maintain frequent direct contact with clinical laboratory directors to identify new cases of foodborne illness. The FoodNet program is the main foodborne disease part of the CDC's Emerging Infections Program (EIP). Foodborne diseases are monitored in several US sites, which are part of the EIP. Examples of the foodborne diseases that are monitored are those caused by parasites such as *Cyclospora* and *Cryptosporidium* (see the chapter on water quality) and bacterial agents such as *Campylobacter*, *E. coli* O157:H7, *Listeria monocytogenes*, *Salmonella*, *Shigella*, *Vibrio*, and *Yersinia enterocolitica*. These conditions are described later in the chapter.

The CDC has developed a model, called the burden of illness pyramid, for describing how reporting of foodborne disease takes place. **FIGURE 11.1** illustrates the burden of illness pyramid.

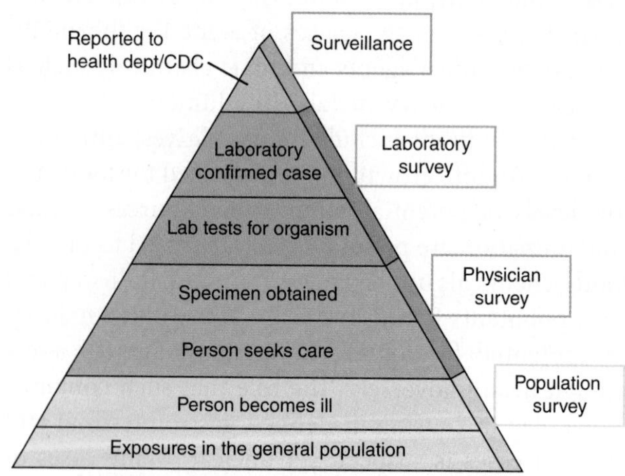

FIGURE 11.1 FoodNet surveillance—burden of illness pyramid.

Reproduced from Centers for Disease Control and Prevention. FoodNet. Foodborne Diseases Active Surveillance Network. Available at: https://www.cdc.gov/foodnet/lessons_learned.pdf. Accessed May 8, 2017.

[The pyramid] illustrates steps that must occur for an episode of illness in the population to be registered in surveillance. Starting from the bottom of the pyramid:

- Some members of the general population are exposed to an organism.
- Some of these exposed persons become ill.
- Some of these ill people seek medical care.
- A specimen is obtained from some persons and submitted to a clinical laboratory.
- A laboratory tests some of these specimens for a given pathogen.
- The laboratory identifies the causative organism in some of these tested specimens and thereby confirms the case. The laboratory-confirmed case is reported to a local or state health department.

FoodNet conducts laboratory surveys, physician surveys, and population surveys to collect information about each of these steps.[8]

TABLE 11.1 tallies the number of outbreaks that occurred in 2014 by cause; during 2014 a total of 864 outbreaks and 13,264 outbreak-associated illnesses were reported.

FIGURE 11.2 shows data on annual reports of combined foodborne disease outbreaks from 1998 to 2008. During this time period the CDC received reports of a total of 13,405 foodborne disease outbreaks. For outbreaks with known etiology, the most common causes were viruses, bacteria, chemical and toxic agents, and parasites. These data are

TABLE 11.1 Number of Reported Foodborne Disease Outbreaks and Outbreak-Associated Illnesses, by Etiology, United States, 2014*

Etiology	No. of Outbreaks	No. of Illnesses
Bacterial	317	5,858
Chemical and toxin	52	171
Parasitic	8	46
Viral	288	5,390
Single etiology	665	11,465
Multiple etiologies	12	479
Unknown etiology	187	1,302
Total 2014	**864**	**13,264**

* Totals include confirmed and suspected etiology.

Data from Centers for Disease Control and Prevention. *Surveillance for foodborne disease outbreaks, United States, 2014: Annual Report*. Atlanta, Georgia: US Department of Health and Human Services, CDC; 2016: 5.

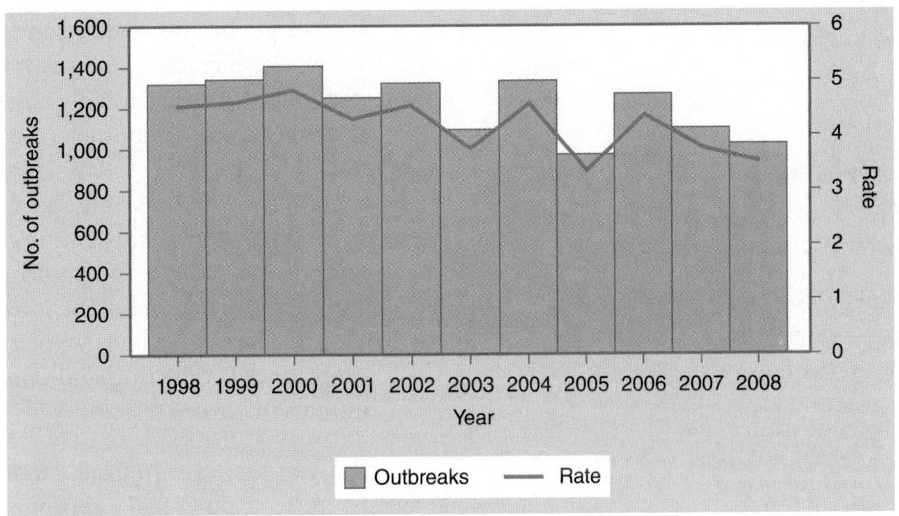

* N = 13,405.
† Per 1 million population.

FIGURE 11.2 Number and rate of foodborne disease outbreaks, by year—Foodborne Disease Surveillance System, United States, 1998–2008.

Reproduced from Centers for Disease Control and Prevention. Surveillance for Foodborne-Disease Outbreaks—United States, 1998-2008. Surveillance Summaries, June 13, 2013. *MMWR*. 2013;62(No. SS-2):10.

helpful in suggesting continued efforts that will be required to reduce further the incidence of some types of foodborne illness.

Categories of Food Hazards

The "big three" categories of food hazards are biological, physical, and chemical.[9] In addition, nutritional hazards are those associated with the presence of nutrients and other food constituents in excessive or deficient amounts that lead to disease. **TABLE 11.2** provides a list of contaminants that may be present in food. Physical hazards encompass foreign objects in food such as stones, glass, metal (e.g., bullets), and pieces of wood. Food hazards may be introduced when foods are harvested, processed, shipped, or stored.

Common Microbial Agents of Foodborne Illness

The media and the CDC publish frequent reports of disease outbreaks caused by microbial agents.

TABLE 11.2 Contaminants that May Be Present in Food

Pathogenic microbial agents

- Examples: bacteria, worms, protozoa, fungi, viruses, prions, and (in certain cases) toxins from such organisms

Chemically related foodborne hazards

- Marine toxins, mushroom toxins
- Heavy metals
- Pesticides, herbicides, and fungicides
- Preservatives and additives

Residues of medicines administered to food animals

- Antibiotics, growth-promoting hormones

Foreign objects and other physical contaminants

- Natural components: bones, shells, seeds
- Other: glass and metal fragments, stones

Radioactive materials

Materials used in packaging

- Residues of plastics
- Waxes

Miscellaneous contaminants

- Debris from insects (insect parts, ova) and from rodents (fecal material, fur)
- Cleaning agents used in the food processing environment

Some microbial pathogens (e.g., *Salmonella*) are more common agents of foodborne illness than are others (e.g., *Clostridium botulinum*). In addition, a type of microbial agent referred to as an emerging foodborne disease pathogen increasingly is causing foodborne infections. Emerging foodborne disease pathogens are a subset of agents from the general category of emerging infectious disease agents, which cause emerging infections. These infections, which include cholera, Rift Valley fever, and Lyme disease, "have newly appeared in the population, or have existed but are rapidly increasing in incidence or geographic range. . . ."[10(p7)] Emerging foodborne disease pathogens include *E. coli* O157:H7, *Campylobacter jejuni*, and *Listeria monocytogenes*. As they increase in frequency and scope, emerging foodborne diseases are a major concern of environmental health workers across the globe.

The chapter on zoonotic and vector-borne diseases focused mainly on zoonotic emerging infectious diseases. Some of these same emerging infectious disease pathogens represent hazards to the food supply. The rise in newly recognized pathogens has altered the epidemiology of foodborne illness by increasing the occurrence and sequaelae of emerging foodborne diseases.[3] Some of the microbial agents that cause foodborne illness have adapted to changes in food production, causing their reemergence and the development of new pathogens.[11] Another factor that has contributed to the rise of these pathogens is globalization of the food supply, as explained in an editorial in the *New England Journal of Medicine*:

> The surge in livestock production around the world, including most parts of Asia, has resulted in unprecedentedly large populations of closely confined animals, particularly pigs and chickens. . . . Today, large hog-raising operations, with tens of thousands of animals on a single farm, are common throughout parts of Asia, North and South America, and Europe. . . . A number of infectious agents tend to be associated with an increasing intensity of production and concentration of animals in limited spaces.[12(p1280)]

The editorial also indicates that these infectious agents "pose a potentially serious threat to human health."[12(p1280)] Other potential hazards arise from fruits and vegetables that originate in developing countries and are consumed in developed countries. In addition, the processes (e.g., cooking, treatment, and pasteurization) designed to deactivate pathogens "can and do fail."[12(p1281)]

TABLE 11.3 Abbreviated Listing of Foodborne Pathogens and Diseases

Type of Organism	Name of Disease
Bacteria: Pathogenic Bacteria	
Salmonella	Salmonellosis
Clostridium botulinum	Botulism
Clostridium perfringens	*Clostridium perfringens* food intoxication
Staphylococcus aureus	Staphylococcal food intoxication
Campylobacter jejuni	Vibrionic enteritis
Shigella	Shigellosis
Listeria monocytogenes	Listeriosis
Bacteria: E. coli *Diarrheal Diseases*	
Escherichia coli O157:H7 enterohemorrhagic (EHEC) serotype	*E. coli* O157:H7 infection
Parasites: Worms	
Trichinella spiralis	Trichinellosis, trichinosis
Taenia solium, Taenia saginata	Taeniasis, cysticercosis (*T. solium*)
Viruses	
Norovirus (formerly called Norwalk virus group)	Epidemic viral gastroenteropathy
Other Pathogenic Agents	
Prions	Bovine spongiform encephalopathy (BSE), commonly referred to as mad cow disease

TABLE 11.3 provides an abbreviated list of microbial agents responsible for foodborne illness. The agents form three general classes (bacteria, parasites, and viruses) plus a fourth class called "other." Some also are classified as agents of emerging foodborne disease. Excluded from this section are parasitic protozoal organisms, which were discussed in the chapter on water quality. Microbes vary greatly in the dose that is required to produce human illness.[11] In some instances the ingestion of only a single virus particle can cause disease; in others only a few organisms (viruses, bacteria, or parasites) need be consumed.

Bacterial Agents

The pathogens *Salmonella*, *Clostridium botulinum*, *Staphylococcus aureus*, and *Clostridium perfringens* are bacteria—unicellular organisms that have characteristic shapes, such as rod shaped (in some instances a slender, curved rod), spherical (cocci), or spiral. Bacteria can be moving (motile)—propelled by flagella or projections from cells—or nonmoving (nonmotile).

Bacteria may be classified as *gram positive* or *gram negative*, terms that refer to Gram's method, which is a staining technique for classifying bacteria. Using this methodology, technicians stain bacteria with a dye known as gentian violet; then they expose the bacteria to Gram's solution (a mixture of iodine, potassium iodide, and water) and other chemicals. After washing, bacteria that retain the gentian violet are gram positive and those that do not retain it are gram negative.

Foodborne infections should be differentiated from foodborne intoxications. Foodborne infections are induced by infectious agents such as some bacteria (e.g., *Salmonella*) that cause foodborne illness directly; symptoms of such infection are variable and may include nausea, diarrhea, vomiting, headaches, and abdominal pain. Foodborne intoxications result from other agents that do not cause infections directly but produce spores or toxins as they multiply in the food; these toxins then can affect the nervous system (neurotoxins) or the digestive system. An example of such a bacterial agent is *Staphylococcus aureus*.

Salmonella

Nontyphoidal *Salmonella* bacteria include approximately 2,500 serotypes. They can cause human illness when transmitted by a variety of sources, e.g., breakfast cereals.[13] Used to identify subspecies of bacteria, the process of serotyping involves determining whether bacterial isolates react with the blood serum from animals that have formed antibodies against specific types of bacteria. Bacteria that react with specific sera are called members of that serotype.

The causative agents for the foodborne infection salmonellosis, *Salmonella* are described as rod-shaped, motile, gram-negative, and non-sporeforming bacteria. **FIGURE 11.3** shows an image of *Salmonella* bacteria. (Note that some serotypes of *Salmonella* are nonmotile.)

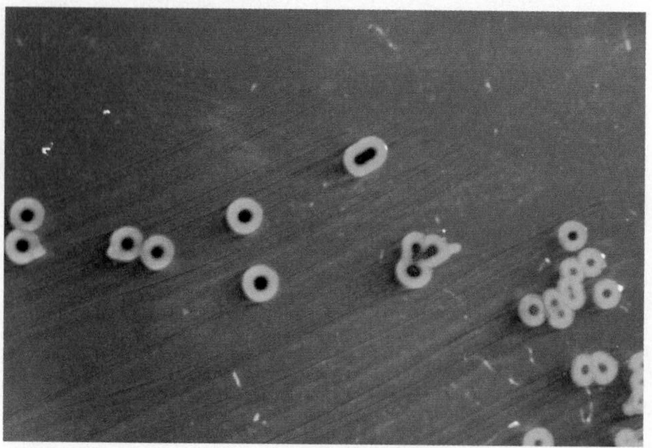

FIGURE 11.3 Picture of *Salmonella* Typhimurium bacteria.
Courtesy of CDC/ Dr. Mike Miller

Foodborne illnesses associated with *Salmonella* infections cause about 380 deaths and 19,000 hospitalizations each year.[14] The two most common serotypes in the United States are *Salmonella* serotype Enteritidis and *Salmonella* serotype Typhimurium. In 2011, a total of 30% of laboratory-confirmed cases of salmonellosis were caused by these two serotypes.[15] Four serotypes (Enteritidis, Typhimurium, Newport, and Javiana) accounted for 48% of isolates. A variety of *Salmonella* called *S. typhi* causes typhoid fever. (This disease, which can be transmitted by contaminated food and water, was discussed in the chapter on water quality.)

Salmonellosis is ranked among the most frequent types of foodborne illness in the United States. Approximately 1 million cases of domestically acquired nontyphoidal salmonellosis (reported and nonreported) are estimated to occur each year in the United States; about 1,800 domestic-origin cases of typhoid fever (a severe type of *Salmonella* infection) arise in the United States annually.[16] A total of 45,828 cases of laboratory-confirmed salmonellosis were reported in the United States in 2011.[15] Refer to **FIGURE 11.4** for the incidence rate of laboratory-confirmed human *Salmonella* infection reported to the CDC. The upper panel shows all serotypes of *Salmonella* compared with serotypes Typhimurium and Enteritidis, the two most common serotypes of *Salmonella* in the United States since 1970. A spike in *Salmonella* incidence in 1985 was from an outbreak related to pasteurized milk. The lower panel compares all serotypes of Salmonella with more than 1,000 infections reported to CDC in 2011.

Salmonella are known to occur widely in wild and domestic animal reservoirs.[1] Animals that are used for human consumption—poultry, swine, cattle—may harbor *Salmonella*. Pet animals such as cats, dogs, and turtles also can be reservoirs for *Salmonella*. Some animals and birds are chronic carriers of the bacteria. *Salmonella* bacteria may be transferred to environmental surfaces at work and at home (e.g., the kitchen) from raw meats, poultry, and seafood; from animal feces; and from contaminated water and soil. Symptoms and characteristics of *Salmonella* infections (nontyphoidal) are shown in the text box.

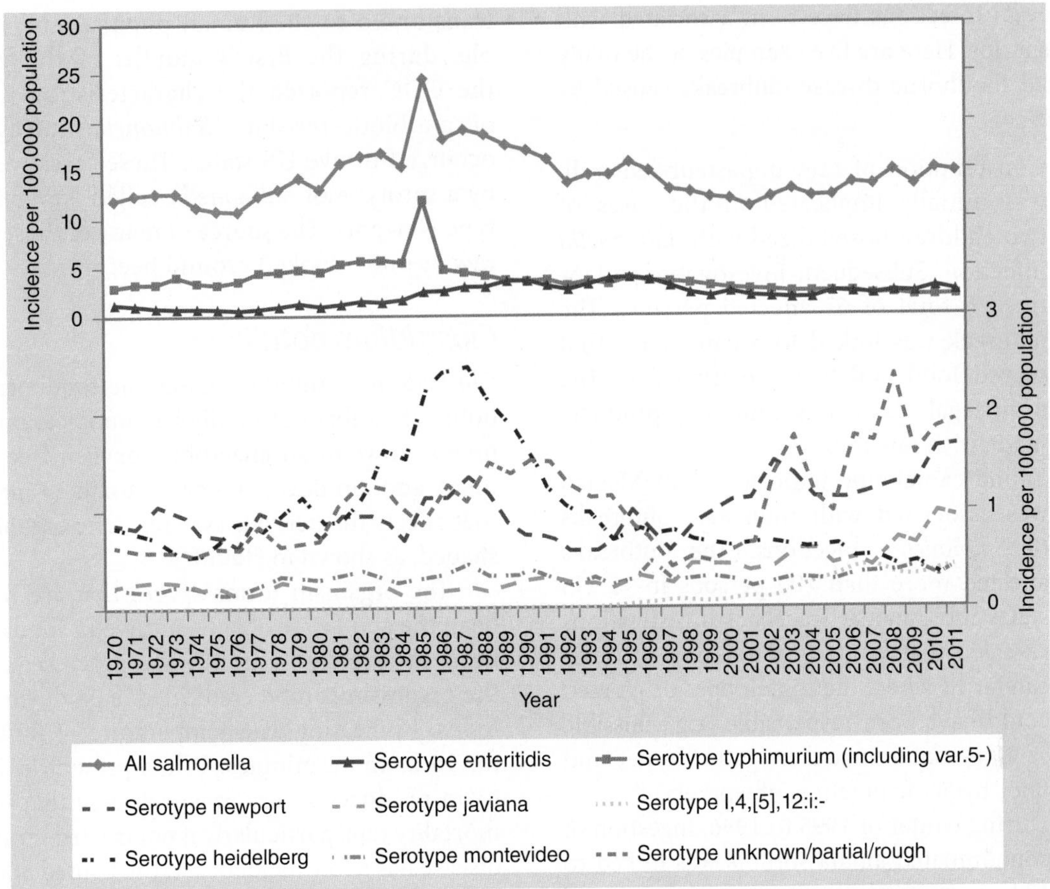

FIGURE 11.4 Incidence rate of laboratory-confirmed human *Salmonella* infection reported to CDC, United States, 1970–2011.*

* All serotypes compared with serotypes Typhimurium and Enteritidis in upper panel and all serotypes compared with serotypes with more than 1,000 infections reported to CDC in 2011 in lower panel.

Reproduced from Centers for Disease Control and Prevention. National Enteric Disease Surveillance. *Salmonella* Annual Report, 2011. Available at: https://www.cdc.gov/ncezid/dfwed/pdfs/salmonella-annual-report-2011-508c.pdf. Accessed May 8, 2017.

NONTYPHOIDAL SALMONELLOSIS (*SALMONELLA* SPP.)

- **Onset and symptoms:** Onset time is 6 to 72 hours after exposure. Symptoms include nausea, vomiting, abdominal cramps, diarrhea, fever, and headache.
- **Infective dose:** As low as one cell, depending on age and health of host and strain differences among members of the genus.
- **Duration of symptoms:** Symptoms generally last 4 to 7 days, with acute symptoms usually lasting 1 to 2 days or longer, depending on host factors, the dose ingested, and strain characteristics.
- **Diagnosis of human illness:** Serological identification of culture isolated from stool.
- **Associated food sources:** A few examples of foods that have been linked to *Salmonella* illness include meats, poultry, eggs, milk and dairy products, fish, shrimp, spices, yeast, coconut, sauces, freshly prepared salad dressings made with unpasteurized eggs, cake mixes, cream-filled desserts and toppings that contain raw egg, dried gelatin, peanut butter, cocoa, produce (fruits and vegetables such as tomatoes, peppers, and cantaloupes), and chocolate. Various *Salmonella* species have long been isolated from the outside of egg shells.
- **Target populations:** Anyone, of any age, may become infected with *Salmonella*. Particularly vulnerable are people with weak immune systems, such as the very young and the elderly, people with human immunodeficiency virus (HIV) or chronic illnesses, and people on some medications; for example, chemotherapy for cancer or the immunosuppressive drugs used to treat some types of arthritis. People with HIV are estimated to have salmonellosis at least 20 times more than does the general population and tend to have recurrent episodes.

Modified and reproduced from US Food and Drug Administration. *Bad Bug Book, Foodborne Pathogenic Microorganisms and Natural Toxins*. Second Edition. *Salmonella* species. 2012.

Salmonella infections have been associated with diverse scenarios. Here are five examples of the many documented foodborne disease outbreaks caused by *Salmonella*:

1. Consumption of raw, unpasteurized milk was initially implicated in the cases of two children hospitalized with *Salmonella* infection. Subsequent investigations identified a total of 62 affected persons. The outbreak was linked to a dairy farm that served food and had a petting zoo. The farm legally sold raw milk and products made with raw milk.[17]
2. Eating cantaloupe imported from Mexico was associated with multistate outbreaks of *Salmonella* infections. The outbreaks affected more than 100 persons in 12 US states and Canada. Traceback investigations revealed that sanitary conditions at Mexican farms where the cantaloupes originated could have been responsible (e.g., possible sewage-contaminated irrigation water and poor hygienic practices of workers).[18]
3. During winter of 1995 to 1996, ingestion of contaminated alfalfa sprouts was linked to 133 cases of salmonellosis in Oregon and British Columbia. Investigations suggested that the source of the *Salmonella* bacteria was contaminated alfalfa seeds.[19]
4. A multistate outbreak of *Salmonella* serotype Agona infections affected more than 200 persons in 11 states during April through May 1998. The outbreak was linked to Toasted Oats cereal that had been contaminated with this serotype of *Salmonella* bacteria.[13]
5. During 1985, the CDC became aware of the largest number of culture-confirmed cases of *Salmonella* ever reported for a single outbreak—5,770 by April 16, 1985.[20] This large number of cases was responsible for the increase shown in Figure 11.4. The cause of the incident was pasteurized milk, which ordinarily would be free from *Salmonella*. Apparently, the pasteurization process was inadequate, or the milk became contaminated after pasteurization.

Several strains of *Salmonella* have developed antibiotic resistance, making their treatment more difficult. The development of antimicrobial-resistant *Salmonella* has been tied to the administration of antibiotics to animals consumed for food; these resistant forms of *Salmonella* are capable of being transmitted to the human population.[21] For example, during the first 4 months of the year 2002, the CDC reported the characteristics of 47 cases of antibiotic-resistant *Salmonella* infections that occurred in five US states. These cases were caused by a serotype of *Salmonella* called *Salmonella* serotype Newport. The source of many of these cases was raw or undercooked ground beef.[22]

Clostridium botulinum

Clostridium botulinum causes the foodborne disease botulism, a form of foodborne intoxication. *C. botulinum* grows in an anaerobic (oxygen-free) environment and produces a potent toxin (a neurotoxin) that affects the nervous system.[23] The organism is rod shaped, as shown in **FIGURE 11.5**.

The organism forms spores that are resistant to heat; the spores are able to survive in foods that have been incorrectly or minimally processed. However, the organism can be eradicated, e.g., during canning foods, by heating at a temperature of 80°C (176°F) for 10 or more minutes. When present in foods, the toxin produces a very severe disease that has a high mortality rate, particularly if not treated promptly. The disease has a low incidence in the United States; about 10 to 30 outbreaks occur each year. The form of the disease called infant botulism has been linked to consumption of honey by infants. Other facts regarding botulism are listed in the following text box.

EXHIBIT 11.1 presents two case studies of botulism that were reported by the CDC.

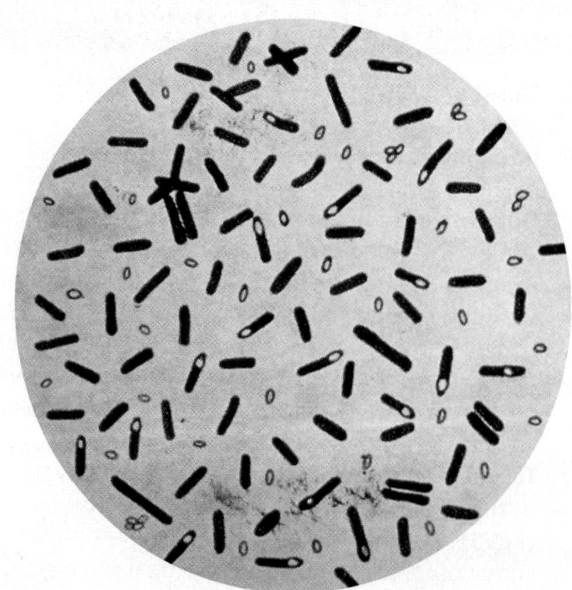

FIGURE 11.5 A photomicrograph of *Clostridium botulinum* bacteria.

Courtesy of CDC

CLOSTRIDIUM BOTULINUM (AMONG ADULTS)

- **Onset and symptoms**: Onset of symptoms in foodborne botulism is usually 18 to 36 hours after ingesting food containing the toxin, although times have varied from 4 hours to 8 days. Early signs of intoxication consist of marked lassitude, weakness, and vertigo, usually followed by double vision and progressive difficulty in speaking and swallowing. Difficulty in breathing, weakness of other muscles, abdominal distention, and constipation may also be common symptoms. Tiny amounts of the toxin can cause paralysis, including paralysis of the breathing muscles.
- **Associated food sources**: The types of foods involved in botulism vary according to food preservation and cooking practices. Most often, illnesses are due to home-canned foods that weren't processed or cooked properly. Any food conducive to outgrowth and toxin production can be associated with botulism. This can occur when food processing allows spore survival and the food is not subsequently heated before consumption, to eliminate any live cells. Almost any type of food that is not very acidic (pH above 4.6) can support growth and toxin production by *C. botulinum*. A variety of foods, such as canned corn, peppers, green beans, soups, beets, asparagus, mushrooms, ripe olives, spinach, tuna fish, chicken and chicken livers and liver pate, luncheon meats, ham, sausage, stuffed eggplant, lobster, and smoked and salted fish have been associated with botulinum toxin.

Modified and reproduced from US Food and Drug Administration. *Bad Bug Book, Foodborne Pathogenic Microorganisms and Natural Toxins*. Second Edition. *Clostridium botulinum*. 2012.

EXHIBIT 11.1

Case Study: Botulism

1. On June 30, 1994, a 47-year-old resident of Oklahoma was admitted to an Arkansas hospital with subacute onset of progressive dizziness, blurred vision, slurred speech, difficulty swallowing, and nausea. . . . He developed respiratory compromise and required mechanical ventilation. The patient was hospitalized for 49 days, including 42 days on mechanical ventilation, before being discharged. The patient had reported that, during the 24 hours before onset of symptoms, he had eaten home-canned green beans and a stew containing roast beef and potatoes. Although analysis of the leftover green beans was negative for botulism toxin, type A toxin was detected in the stew. The stew had been cooked, covered with a heavy lid, and left on the stove for 3 days before being eaten without reheating.[24(pp200–201)]
2. On November 23, 1997, a previously healthy 68-year-old man became nauseated, vomited, and complained of abdominal pain. During the next 2 days, he developed diplopia, dysarthria, and respiratory impairment, necessitating hospitalization and mechanical ventilation. Possible botulism was diagnosed, and . . . antibotulinum toxin was administered. . . . A food history revealed no exposures to home-canned products; however, the patient had eaten pickled eggs that he had prepared 7 days before onset of illness; gastrointestinal symptoms began 12 hours after ingestion. The patient recovered after prolonged supportive care. The pickled eggs were prepared using a recipe that consisted of hard-boiled eggs, commercially prepared beets and hot peppers, and vinegar. The intact hard-boiled eggs were peeled and punctured with toothpicks then combined with the other ingredients in a glass jar that closed with a metal screw-on lid. The mixture was stored at room temperature and occasionally was exposed to sunlight. Cultures revealed *Clostridium botulinum* type B, and type B toxin was detected in samples of the pickled egg mixture. . . .[25(pp778–779)]

Modified from Centers for Disease Control and Prevention. Foodborne botulism from eating home-pickled eggs—Illinois, 1997. *MMWR*. 2000;49:778–780; and Centers for Disease Control and Prevention. Foodborne botulism—Oklahoma, 1994. *MMWR*. 1995;44:200–202.

Clostridium perfringens

Clostridium perfringens causes perfringens food poisoning, a common source of foodborne illness in the United States.[26] The CDC estimates that almost 1,000,000 cases occur each year and is the second most frequent bacterial cause of foodborne illness following *Salmonella. C. perfringens* is anaerobic, rod shaped, and forms spores. The organism occurs commonly in the environment (e.g., in soil and sediments), especially in those areas contaminated with feces; *C. perfringens* is a frequent resident of the intestines of humans and animals. The following text box contains some facts regarding *C. perfringens*.[26]

Here is an example of an outbreak: In 1993, *C. perfringens* caused illness outbreaks after diners consumed St. Patrick's Day meals of corned beef in Ohio and Virginia. The Ohio incident sickened at least 156 persons. The Virginia outbreak caused 115 persons to become ill. In both situations, the causative

GASTROINTESTINAL FORM OF *CLOSTRIDIUM PERFRINGENS* (C. PERFRINGENS)

- **Onset and symptoms:** Symptoms occur about 16 hours after consumption of foods containing large numbers [of cells or spores of] *C. perfringens* capable of producing the enterotoxin [a toxin that causes food poisoning]. Common characteristics include watery diarrhea and mild abdominal cramps. The disease generally lasts 12 to 24 hours. In the elderly or infants, symptoms may last 1 to 2 weeks.
- **Associated food sources:** In most instances, the actual cause of poisoning by this organism is temperature abuse of cooked foods. Small numbers of the organism often are present after the food is cooked, due to germination of its spores, which can survive high heat and can multiply rapidly as a result of a fast doubling time. Therefore, during cool-down (109–113°F) (42.8°C–45.0°C) and storage of prepared foods, this organism can reach levels that cause food poisoning much more quickly than can other bacteria. Meats (especially beef and poultry), meat-containing products (e.g., gravies and stews), and Mexican foods are important vehicles.

Modified and reproduced from US Food and Drug Administration. *Bad Bug Book, Foodborne Pathogenic Microorganisms and Natural Toxins*. Second Edition. *Clostridium perfringens*. 2012.

factor was *C. perfringens* intoxication, apparently from consuming meat that had been kept at improper holding temperatures, which allowed the spores and bacteria to proliferate.[27]

Staphylococcus aureus

Staphylococcus aureus is a spherically shaped bacterium that causes a foodborne intoxication with rapid onset.[28] **FIGURE 11.6** illustrates a grape-like cluster of *S. aureus* bacteria. During growth in foods, *S. aureus* elaborates a toxin that is extremely resistant to cooking at high temperatures. The organism can thrive in an environment that has heavy concentrations of salt or sugar (ham is a frequent vehicle for foodborne *S. aureus* poisoning); most other bacteria are unable to tolerate such an environment. Frequently, *S. aureus*–associated foodborne outbreaks are linked to foods that have not been stored in a safe temperature range (below 45°F [7.2°C] and above 140°F [60°C]).

Some of the additional characteristics of staphylococcal food poisoning are listed in the following text box.

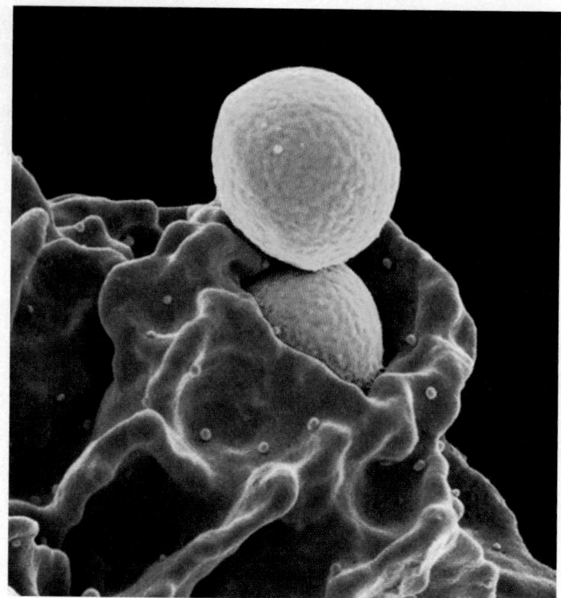

FIGURE 11.6 Scanning electron micrograph of two methicillin-resistant *Staphylococcus aureus* (MRSA) bacteria.

Courtesy of National Institute of Allergy and Infectious Diseases (NIAID)

STAPHYLOCOCCUS AUREUS

- **Onset and symptoms**: The onset of symptoms usually is rapid (1 to 7 hours) and in many cases acute, depending on individual susceptibility to the toxin, amount of toxin ingested, and general health. When ingested, the enterotoxin may rapidly produce symptoms, which commonly include nausea, abdominal cramping, vomiting, and diarrhea.
- **Associated food sources:** Foods frequently implicated in staphylococcal food poisoning include meat and meat products; poultry and egg products; salads, such as egg, tuna, chicken, potato, and macaroni; bakery products, such as cream-filled pastries, cream pies, and chocolate éclairs; sandwich fillings; and milk and dairy products. Foods that require considerable handling during preparation and are kept slightly above proper refrigeration temperatures for an extended period after preparation are frequently involved in staphylococcal food poisoning.
- **Reservoir:** Staphylococci are widely distributed in the environment. They can be found in the air, dust, sewage, water, milk, and food, or on food equipment, environmental surfaces, humans, and animals. Staphylococci are present in the nasal passages and throats and on the hair and skin of 50% or more of healthy individuals.

Modified and reproduced from US Food and Drug Administration. *Bad Bug Book, Foodborne Pathogenic Microorganisms and Natural Toxins*. Second Edition. *Staphylococcus aureus*. 2012.

EXHIBIT 11.2

Case Study: *S. aureus* Food Outbreak Associated with Ham

On September 27, 1997, a community hospital in northeastern Florida notified the St. Johns County Health Department about several persons who were treated in the emergency department because of gastrointestinal illnesses suspected of being associated with a common meal. Self-administered questionnaires were distributed to the 125 attendees to document food histories, illnesses, and symptoms. A case was defined as nausea and/or vomiting in a person who attended the party or consumed food served at the party and who became ill within 8 hours after eating. Leftover food was collected and submitted for laboratory analysis. Food preparers were interviewed about the purchase and preparation of food served at the party.

Of the approximately 125 persons who attended the party, 98 completed and returned questionnaires. Of these, 31 persons attended the event but ate nothing, and none of them became ill; they were excluded from further analysis. A total of 18 (19%) persons had illnesses meeting the case definition, including 17 party attendees and one person who ate food brought home from the party. Ill persons reported nausea (94%), vomiting (89%), diarrhea (72%), weakness (67%), sweating (61%), chills (44%), fatigue (39%), myalgia (28%), headache (11%), and fever (11%). Onset of illness occurred at a mean of 3.4 hours after eating (range: 1–7 hours); symptoms lasted a median of 24 hours (range: 2–72 hours). Seven persons sought medical treatment, and two of those were hospitalized overnight.

Illness was strongly associated with eating ham. Of the 18 ill persons, 17 (94%) had eaten ham. The ill person who had not attended the party had eaten only leftover ham. None of the other foods served at the party were significantly associated with illness. One sample of leftover cooked ham and one sample of leftover rice pilaf were analyzed and were positive for staphylococcal enterotoxin type A. Samples of stool or vomitus were not obtained from any ill persons, and cultures from nares or skin were not obtained from the food preparers.

On September 25, a food preparer had purchased a 16-pound precooked packaged ham, baked it at home at 400°F [204°C] for 1.5 hours, and transported it to her workplace, a large institutional kitchen, where she sliced the ham while it was hot on a commercial slicer. The food preparer reported having no cuts, sores, or infected wounds on her hands. She reported that she routinely cleaned the slicer in place rather than dismantling it and cleaning it according to recommended procedures and that she did not use an approved sanitizer. All 16 pounds of sliced ham had been placed in a 14-inch by 12-inch by 3-inch plastic container that was covered with foil and stored in a walk-in cooler for 6 hours, then transported back to the preparer's home and refrigerated overnight. The ham was served cold at the party the next day. The rice pilaf was prepared the day of the party by a different person.[29(pp1189–1190)]

Reproduced from Centers for Disease Control and Prevention. Outbreak of Staphylococcal food poisoning associated with precooked ham—Florida, 1997. *MMWR*. 1997;46:1189–1191.

The CDC reported the occurrence of an *S. aureus*–associated foodborne outbreak in 1997. Refer to the case study in **EXHIBIT 11.2**.

Escherichia coli O157:H7 (*E. coli* O157:H7)

E. coli O157:H7 is referred to as an enterohemorrhagic (EHEC) strain of *E. coli*. Some *E.coli* bacteria elaborate a toxin (Shiga toxin) that sometimes causes severe symptoms. The acute disease caused by *E. coli* O157:H7 is called hemorrhagic colitis, because one of the dramatic symptoms of the illness produced by the toxin from *E. coli* O157:H7 is bloody diarrhea. **FIGURE 11.7** shows an electron micrograph of *E. coli* bacteria.

Nonpathogenic strains of *E. coli* reside harmlessly in the intestinal tracts of animals and humans and are part of the normal bacterial flora. Apparently, they retard the growth of pathogenic bacteria and synthesize vitamins. However, some strains of *E. coli* are pathogenic, an example being the EHEC strain *E. coli* O157:H7.

The agent *E. coli* O157:H7 gained notoriety when it was associated with consumption of undercooked hamburger meat at a fast-food restaurant.[3] The organism,

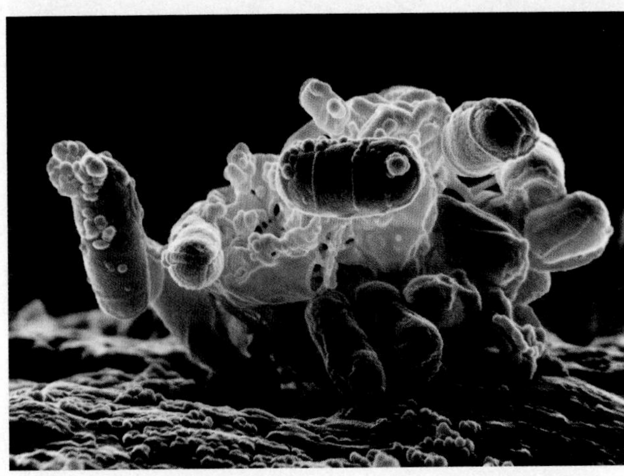

FIGURE 11.7 A cluster of *E. coli* bacteria.

Reproduced from US Department of Agriculture, Agricultural Research Service. News & Events, Image Gallery, Image Number K11077-2. Photo by Eric Erbe, colorization by Christopher Pooley. Available at: https://www.ars.usda.gov/oc/images/photos/mar04/k11077-2/. Accessed May 10, 2017.

found in the intestines of healthy cattle, may invade meat during the slaughter process.[30] Outbreaks have occurred among visitors to petting farms.[31] In the United States, this organism—which causes both bloody and nonbloody diarrhea—has been linked to approximately 73,000 cases and 61 deaths annually.[32] Young children and immunocompromised individuals tend to be more susceptible to complications from infection with *E. coli* O157:H7 than other groups.[33] An example of a complication is hemolytic uremic syndrome (HUS), which produces such adverse consequences as acute kidney failure, end-stage renal disease, and hypertension. Refer to **TABLE 11.4** for a chronology of the *E. coli* O157:H7 pathogen, and refer to the following text box for the characteristics of *E. coli* O157:H7–associated foodborne illness.

TABLE 11.4 *E. coli* O157:H7 (Chronology)

1982	First recognized as a pathogen
1985	Associated with hemolytic uremic syndrome
1990	Outbreak from drinking water
1991	Outbreak from apple cider
1993	Multistate outbreak from fast-food hamburgers
1995	Outbreak from fresh produce
1996	Multistate outbreak from unpasteurized apple juice
	Outbreak in Japan
Selected outbreaks since 2000	
2006	Fresh spinach
2010	Shredded romaine lettuce from one processing facility
2014	Ground beef suspected of being produced by a packing company
2015	Rotisserie chicken salad sold by a warehouse store
2016	Alfalfa sprouts produced by a Wisconsin farm

Modified and reproduced from Centers for Disease Control and Prevention. Public Health Image Library, ID #107. Available at: http://phil.cdc.gov/Phil/details.asp. Accessed August 20, 2017; and Centers for Disease Control and Prevention. Reports of Selected *E. coli* Outbreak Investigations Available at: https://www.cdc.gov/ecoli/outbreaks.html. Accessed August 20, 2017.

ENTEROHEMORRHAGIC *ESCHERICHIA COLI* (EHEC)—*E. COLI* O157:H7

- **Onset and symptoms**: Symptoms usually begin 3 to 4 days after exposure, but the time may range from 1 to 9 days. [Symptoms include] severe cramping (abdominal pain), nausea or vomiting, and diarrhea that initially is watery, but becomes grossly bloody. In uncomplicated cases, duration of symptoms is 2 to 9 days, with an average of 8 days.
- **Associated food sources**: Raw or undercooked or ground beef and beef products are the vehicles most often implicated in O157:H7 outbreaks. Produce, including bagged lettuce, spinach, and alfalfa sprouts, increasing is being implicated in O157:H7 infections.

Modified and reproduced from US Food and Drug Administration. *Bad Bug Book, Foodborne Pathogenic Microorganisms and Natural Toxins*. Second Edition. Enterohemorrhagic *Escherichia coli* (EHEC). 2012.

Among the foodborne illness outbreaks associated with *E. coli* O157:H7 is an incident that affected the patrons of a hamburger chain and impacted four western states. Refer to **EXHIBIT 11.3**.

Shigella

Shigella bacteria, which are highly infectious, are the cause of shigellosis (bacillary dysentery); an example of this agent is *Shigella sonnei*. The bacteria can be

SHIGELLA SPECIES

- **Onset and symptoms:** May include abdominal pain, cramps, diarrhea, fever, vomiting, and blood, pus, or mucus in stools. The onset time for shigellosis ranges from 8 to 50 hours. In otherwise healthy people, the disease is usually self-limiting, although some strains are associated with fatality rates as high as 10% to 15%. Severe cases… tend to occur primarily in immunocompromised or elderly people and young children…
- **Associated food sources:** Most cases shigellosis are caused by ingestion of fecally contaminated food or water. *Shigella* is commonly transmitted by foods consumed raw; for example, lettuce, or as nonprocessed ingredients, such as those in a five-layer bean dip. Salads (potato, tuna, shrimp, macaroni, and chicken), milk and dairy products, and poultry also are among the foods that have been associated with shigellosis.

Modified and reproduced from US Food and Drug Administration. *Bad Bug Book, Foodborne Pathogenic Microorganisms and Natural Toxins*. Second Edition. *Shigella* species. 2012.

EXHIBIT 11.3

Case Study: Multistate *E. coli* O157:H7 Outbreaks

From November 15, 1992 through February 28, 1993, more than 500 laboratory-confirmed infections with *E. coli* O157:H7 and four associated deaths occurred in four states—Washington, Idaho, California, and Nevada.

On January 13, 1993, a physician reported to the Washington Department of Health a cluster of children with hemolytic uremic syndrome (HUS) and an increase in emergency room visits for bloody diarrhea. During January 16–17, a case-control study comparing 16 of the first cases of bloody diarrhea or postdiarrheal HUS identified with age- and neighborhood-matched controls implicated eating at chain A restaurants during the week before symptom onset. On January 18, a multistate recall of unused hamburger patties from chain A restaurants was initiated.

As a result of publicity and case-finding efforts, during January–February 1993, 602 patients with bloody diarrhea or HUS were reported to the state health department. A total of 477 persons had illnesses meeting the case definition of culture-confirmed *E. coli* O157:H7 infection or postdiarrheal HUS. Of the 477 persons, 52 (11%) had close contact with a person with confirmed *E. coli* O157:H7 infection during the week preceding onset of symptoms. Of the remaining 425 persons, 372 (88%) reported eating in a chain A restaurant during the 9 days preceding onset of symptoms. Of the 338 patients who recalled what they ate in a chain A restaurant, 312 (92%) reported eating a regular-sized hamburger patty. Onsets of illness peaked from January 17 through January 20. Of the 477 patients, 144 (30%) were hospitalized; 30 developed HUS, and three died. The median age of patients was 7.5 years (range: 0–74 years). [During approximately the same time period, additional cases occurred in Idaho, California, and Nevada. These cases also were linked to the chain A restaurants.]

During the outbreak, chain A restaurants in Washington state linked with cases primarily were serving regular-sized hamburger patties produced on November 19, 1992; some of the same meat was used in "jumbo" patties produced on November 20, 1992. The outbreak strain of *E. coli* O157:H7 was isolated from 11 lots of patties produced on those two dates; these lots had been distributed to restaurants in all states where illness occurred. Approximately 272,672 (20%) of the implicated patties were recovered by the recall.

A meat traceback by a CDC team identified five slaughter plants in the United States and one in Canada as the likely sources of carcasses used in the contaminated lots of meat and identified potential control points for reducing the likelihood of contamination. The animals slaughtered in domestic slaughter plants were traced to farms and auctions in six western states. No one slaughter plant or farm was identified as the source.

Modified and reproduced from Centers for Disease Control and Prevention. Update: multistate outbreak of *Escherichia coli* O157:H7 infections from hamburgers—western United States, 1992–1993. *MMWR*. 1993;42:258–259, 261.

transmitted via the fecal-oral route through contaminated water and foods that are handled in an unsanitary manner. In the United States, an estimated 130,000 cases of shigellosis occur each year; approximately 31% of these cases is estimated to be caused by food.[34] Other facts regarding *Shigella* are presented in the following text box.

Campylobacter jejuni

Campylobacter jejuni is responsible for an illness known as campylobacteriosis. Among the bacterial causes of foodborne infections, *C. jejuni* is among the most commonly reported agents. About 2.1 to 2.4 million cases of campylobacteriosis occur each year in the United States.[35] Shown in **FIGURE 11.8** are data on the relative rates of laboratory-confirmed infections with *Campylobacter*, Shiga toxin–producing *Escherichia coli* (STEC) O157, *Listeria*, *Salmonella*, and *Vibrio* reported by the CDC's FoodNet program.

Some of the characteristics of foodborne campylobacteriosis are listed in the text box titled "*Campylobacter jejuni*."

Among the *C. jejuni*–associated outbreaks reported by the CDC was the occurrence of 75 cases of illness in December 2002 related to consumption of unpasteurized milk in Wisconsin.[36] The unpasteurized milk was distributed by a retail dairy farm store. Unpasteurized milk can serve as a vehicle for *C. jejuni* and several other pathogens. The CDC noted that the facility that supplied milk to patients was a Grade A organic dairy farm with 36 dairy cows. The farm also had a retail store in which milk and other food products were available. In addition, farm operators provided unpasteurized milk samples at community events and to persons who toured the farm, including children from childcare facilities. Because unpasteurized milk cannot be sold legally to consumers in Wisconsin, the dairy distributed unpasteurized milk through a cow-leasing

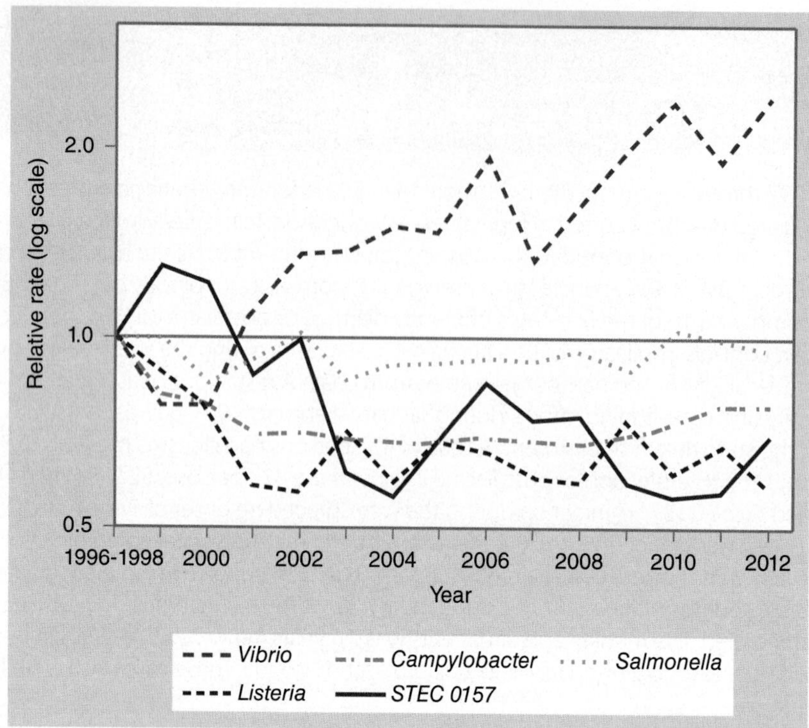

FIGURE 11.8 Relative rates of laboratory-confirmed infections with *Campylobacter*, STEC* O157, *Listeria*, *Salmonella*, and *Vibrio* compared with 1996–1998 rates, by year — Foodborne Diseases Active Surveillance Network, United States, 1996–2012.†

* Shiga toxin–producing *Escherichia coli*.

† The position of each line indicates the relative change in the incidence of that pathogen compared with 1996–1998. The actual incidences of these infections cannot be determined from this figure.

Reproduced from Centers for Disease Control and Prevention. Incidence and Trends of Infection with Pathogens Transmitted Commonly Through Food — Foodborne Diseases Active Surveillance Network, 10 U.S. Sites, 1996–2012. *MMWR*. 2013;62:285.

CAMPYLOBACTER JEJUNI

- **Onset and symptoms:** The disease caused by *C. jejuni* infections is called campylobacteriosis. The incubation period, from time of exposure to onset of symptoms, generally is 2–5 days. Fever, diarrhea, abdominal cramps, and vomiting are the major symptoms. Most cases of campylobacteriosis are self-limiting. The disease typically lasts from 2 to 10 days.
- **Associated food sources:** Major food sources linked to *C. jejuni* infections include improperly handled or undercooked poultry products, unpasteurized ("raw") milk and cheeses made from unpasteurized milk, and contaminated water. *Campylobacter* infection in humans has been linked to handling and eating raw or undercooked meat and poultry, whether fresh or frozen.

Modified and reproduced from US Food and Drug Administration. *Bad Bug Book, Foodborne Pathogenic Microorganisms and Natural Toxins*. Second Edition. *Campylobacter jejuni*. 2012.

program. Customers paid an initial fee to lease part of a cow. Farm operators milked the cows and stored the milk from all leased cows together in a bulk tank. Either customers picked up milk at the farm or farm operators had it delivered. On December 8, investigators obtained a milk sample from the farm's bulk milk tank, and cultures of the milk samples grew *C. jejuni* with a pulsed-field gel electrophoresis (PFGE) pattern that matched the outbreak strain. Farm operators were ordered to divert all milk to a processor for pasteurization.[36(p548)]

Listeria monocytogenes

The illness listeriosis is caused by the bacterium *Listeria monocytogenes*, which has been found in some domestic mammals and several species of birds. The bacterium occurs widely in plant-related materials such as vegetables that may have been contaminated with sewage used as a fertilizer.[37] These vegetables may be consumed raw or after minimal processing. In the United States, *L. monocytogenes* is associated with about 1,600 cases of illness and 260 fatalities per year.[38] High-risk groups include pregnant women, their newborn infants, elderly persons, and immunocompromised individuals. The characteristics of listeriosis are listed in the following text box.

Sporadic outbreaks of listeriosis are reported from time to time. A 2002 outbreak in 8 northeastern US states was responsible for 46 confirmed cases,

LISTERIA MONOCYTOGENES

- **Onset and symptoms:** *L. monocytogenes* can cause two forms of disease. One can range from mild to intense symptoms of nausea, vomiting, aches, fever, and, sometimes, diarrhea, and usually goes away by itself. The other, more deadly form occurs when the infection spreads through the bloodstream to the nervous system (including the brain), resulting in meningitis and other potentially fatal problems. Gastroenteritis caused by *L. monocytogenes* has a relatively short incubation period, from a few hours to 2 or 3 days. The severe, invasive form of the illness can have a very long incubation period, estimated to vary from 3 days to 3 months.
- **Associated food sources:** Many foods have been associated with *L. monocytogenes*. Examples include raw milk, inadequately pasteurized milk, chocolate milk, cheeses (particularly soft cheeses), ice cream, raw vegetables, raw poultry and meats (all types), fermented raw-meat sausages, hot dogs and deli meats, and raw and smoked fish and other seafood. *L. monocytogenes* can grow in refrigerated temperatures, which makes this organism a particular problem for the food industry.

Modified and reproduced from US Food and Drug Administration. *Bad Bug Book, Foodborne Pathogenic Microorganisms and Natural Toxins*. Second Edition. *Listeria monocytogenes*. 2012.

7 fatalities, and 3 stillbirths or miscarriages.[39] The multistate outbreak was associated with eating sliceable turkey deli meat.

Worms

Trichinella

Trichinosis is a foodborne disease associated with eating meat that contains a nematode (also called a roundworm) from the genus *Trichinella*.[40] The classic agent of trichinosis is *Trichinella spiralis*. Pigs (including feral pigs), black bears, and cougars may harbor *Trichinella* larva.[41] The agent can be found in many carnivorous and omnivorous animals worldwide. The disease may be transmitted from animals such as these to humans when meat that contains the cysts of *T. spiralis* has not been cooked adequately. For example, outbreaks have been associated with eating pork from local farms and from consumption of grizzly bear meat.[42]

Formerly trichinosis was common in the United States, because garbage that contained raw meat was used routinely as feed for swine. Trichinosis outbreaks have been prevented following the elimination of this practice. Consumers can protect themselves further by cooking pork and wild game products adequately; also, freezing kills the worms of *T. spiralis*.

Infection with *T. spiralis* can be asymptomatic, as in the instance of light infections, or can include gastrointestinal effects in more severe infections. After about 1 week, when the organism migrates into muscle tissues, symptoms can range from fever and facial edema to life-threatening effects upon the heart and central nervous system.

The life cycle of *Trichinella* is shown in **FIGURE 11.9**. Following ingestion of meat that contains cysts, the larvae are released due to the action of stomach acids on the cysts. The larvae subsequently may invade the small intestine and develop into adult worms. The female adult worms release larvae that can migrate into muscle tissue and become encysted. The human host is a dead-end host, meaning that the disease cannot be spread to other humans.

Tapeworms

Taeniasis is a parasitic disease caused by tapeworms.[43] One form is caused by the beef tapeworm (*Taenia saginata*) and the other by the pork tapeworm (*Taenia solium*). These organisms may induce human illness following the consumption of raw or undercooked infected beef or pork. Usually the symptoms of taeniasis are limited to mild abdominal distress; one of the main symptoms is the passage of the proglottids (the section of the worm that contains eggs) of *T. saginata* and *T. solium* in stools. *T. solium* can cause a serious condition termed cysticercosis, in which the organism migrates to muscle or brain tissue and forms cysts. The worms of *T. saginata*, which reside in human intestines, can be quite long. (See **FIGURE 11.10**.)

Viral Agents

Hepatitis A Virus

The chapter on water quality indicated that hepatitis A virus (HAV) can be transmitted by both contaminated food and water. Foods that are associated commonly with HAV outbreaks include shellfish and salads.[44] A common mode for contamination of foods is by HAV-infected workers in food processing plants and restaurants. Approximately 1,600 cases of hepatitis A reported in the United States annually are believed to be foodborne.

One instance of a large hepatitis A outbreak occurred from February to March 1997; this incident caused 213 cases in Michigan and 29 cases in Maine,

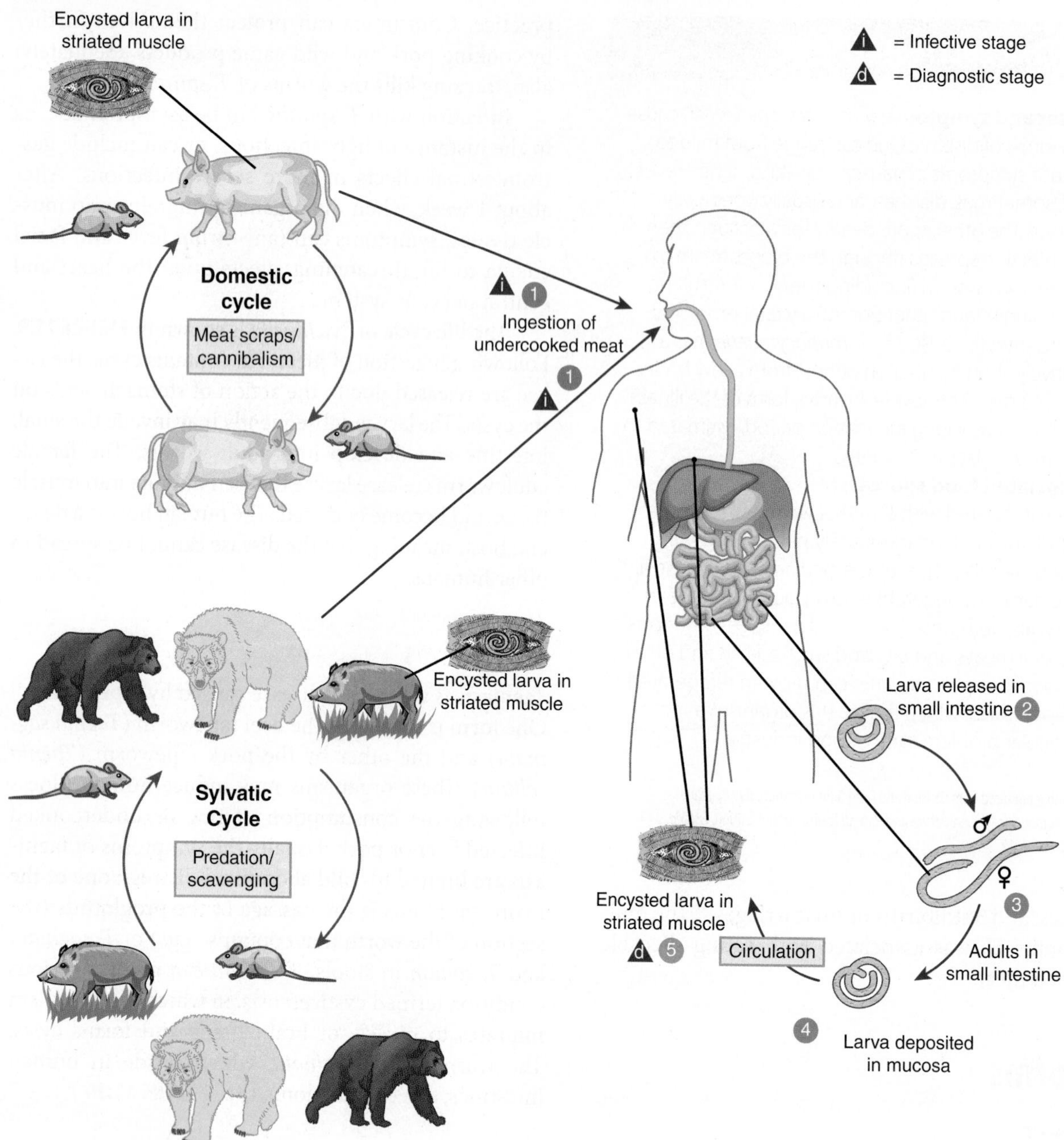

FIGURE 11.9 Life cycle of *Trichinella*.

Reproduced from Centers for Disease Control and Prevention. Trichinellosis. Available at: https://www.cdc.gov/dpdx/trichinellosis/index.html. Accessed May 11, 2017.

and seemed to be linked to sporadic cases in several other states.[45] The source of the outbreak was frozen strawberries fed to schoolchildren in schools that had purchased the contaminated product from the same commercial processor. Other examples of hepatitis A outbreaks include those linked to sandwich shops and to national hamburger chains.

Hepatitis A outbreaks that involved almost 1,000 cases occurred in Tennessee, North Carolina, Georgia, and Pennsylvania from September to November 2003.[46] The outbreak in Pennsylvania was associated with a single restaurant and was responsible for more than 600 infected persons and 3 deaths. This outbreak was the largest episode of foodborne hepatitis A that had ever been reported in the United States. The Pennsylvania outbreak, as well as those in the other three states, was associated with green onions that had been imported from Mexico. This widespread occurrence of hepatitis A demonstrates the need to cook green onions thoroughly instead of serving them raw.

Norovirus

The chapter on water quality presented the topic of noroviruses, particularly with respect to waterborne norovirus infections. This virus is transmitted easily

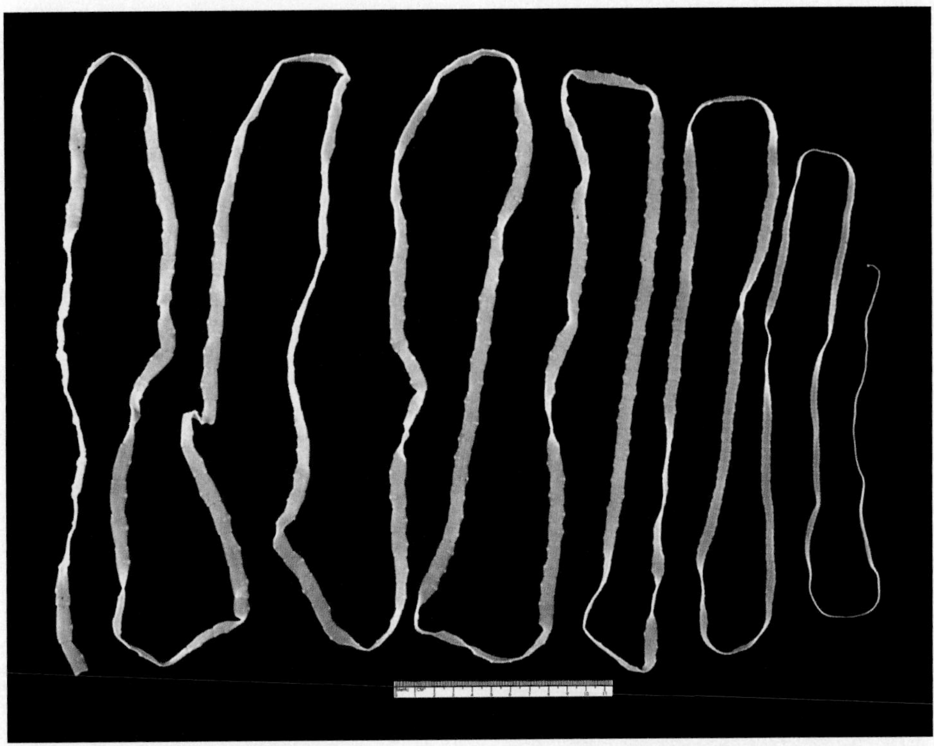

FIGURE 11.10 Adult tapeworm (*Taenia saginata*).

Courtesy of CDC

FIGURE 11.11 Cruise ship. The closed environment of a cruise ship may permit the ready transmission of viral and other agents that cause gastrointestinal illness.

within closed environments such as cruise ships.[47] (See **FIGURE 11.11**.) Some cases may be brought on board by passengers who fall ill just before embarking on a cruise. It is also possible that crew members and shipboard environmental contamination can act as disease reservoirs.[48]

Outbreaks of gastrointestinal illness have been responsible for sickening hundreds of passengers on cruise ships.[49] The culprit for many outbreaks is believed to be a virus similar to the Norwalk virus, which is also called the winter vomiting virus. During 1 month in fall 2002, the toll was estimated to be 1,400 infected passengers and crew members on 10 ships. One outbreak that occurred on the *Amsterdam* (operated by Holland America) afflicted 163 passengers and 18 crew members out of 1,905 primarily elderly persons who went on the cruise. Illness symptoms consisted of vomiting and diarrhea as well as associated dehydration and abdominal cramps. These symptoms lasted from 1 to 2 days.

Just as the occurrence of gastrointestinal illness has continued to be reported throughout the United States, the CDC expects to continue to see increases in the number of reported gastrointestinal illness cases on some cruises.[50] Annually, about 6.8 million passengers in North America took cruises during the late 1990s. During the 6-year period from 1993 to 1998, the number of passengers who embarked on cruises from the United States increased by 50%.[51] The number of North American cruise passengers increased by 11% from 2003 to 2004 (a record year with nearly 9 million passengers) and by 9% between 2004 and 2005.

US government regulations state:

> Cruise vessels sailing to U.S. ports are required to notify the CDC of all reported gastrointestinal illnesses that have been reported to the ships' medical staff. This report must be filed 24 hours prior to arrival at a U.S. port, from

a foreign port. If the number of ill passengers or crewmembers reaches 2%, the vessel is required to file a special report. The CDC continues to closely monitor illness reports on a daily basis. An "outbreak" of gastrointestinal illness is defined as having 3% or more of either passengers or crew reported with a gastrointestinal illness.[50]

The CDC operates the Vessel Sanitation Program (VSP), which aims to keep the risks of diarrheal disease outbreaks among passengers at a minimum. A component of the VSP program is twice-yearly inspections of cruise ships that enter US waters; these inspections are not announced in advance.[50] Refer to the text box for an overview of outbreaks that occurred on cruise ships from 2008 through 2014 plus an example of an outbreak on a cruise ship.

ACUTE GASTROENTERITIS ON CRUISE SHIPS, 2008–2014

From 1990 to 2004, the reported rates of diarrheal disease (three or more loose stools or a greater than normal amount in a 24-hour period) on cruise ships decreased 2.4%, from 29.2 cases per 100,000 travel days to 28.5 cases. In 2001, the Vessel Sanitation Program and the cruise industry expanded the diarrheal illness case definition to include acute gastroenteritis (diarrhea, or vomiting that is associated with loose stools, bloody stools, abdominal cramps, headache, muscle aches, or fever). The most common causative agent has been norovirus.

From 2008 to 2014, the rate of acute gastroenteritis on cruise ships decreased among passengers from 27.2 cases per 100,000 travel days in 2008 to 22.3 in 2014, while the rate among crew members was essentially unchanged. The rate among both passengers and crew members was higher in 2012 compared with the preceding and following years, likely because of the emergence of a new norovirus strain. Among 73,599,005 passengers on cruise ships during 2008–2014, a total of 129,678 (0.18%) cases of acute gastroenteritis were reported during outbreak and nonoutbreak voyages; among 28,281,361 crew members, 43,132 (0.15%) cases were reported. Only a small proportion of those cases were part of a norovirus outbreak.

Cases of acute gastroenteritis illness on cruise ships are relatively infrequent. Norovirus, the most common causative agent of outbreaks, accounted for 14,911 cases among passengers and crew members during 2008–2014, 0.01% of the estimated number of norovirus cases in the United States during the study period. To further reduce acute gastroenteritis on cruise ships, travelers should practice good hand hygiene, especially after using the toilet and before touching the face or eating; persons experiencing diarrhea or vomiting should promptly report their illness for proper assessment, treatment, and monitoring.[a]

Example of an outbreak: cruise ship *Oosterdam*

Voyage dates: November 3 through November 18, 2016

Number of passengers reported ill/total passengers onboard: 90/1,843 (4.88%)

Number of crew reported ill/total number of crew onboard: 22/796 (2.76%)

Predominant symptoms: vomiting, diarrhea

Causative agent: norovirus (suspected)[b]

Reprinted and adapted from the following sources:

[a]Modified from Freeland AL, Vaughan GH, Banerjee SN. Acute gastroenteritis on cruise ships—United States, 2008-2014. *MMWR*. 2016;65(1):1-5.

[b]Reproduced from Centers for Disease Control and Prevention. Vessel Sanitation Program. Investigation update on the Oosterdam. Available at: https://www.cdc.gov/nceh/vsp/surv/outbreak/2016/nov18_oosterdam.htm. Accessed February 28, 2017.

Other Agents

Prions

Prions are hypothesized to be a modification of a normal protein; the modified protein is called a prion protein, which is a pathogenic agent that can damage the central nervous system of cattle. Infections with prions are regarded as the cause of **bovine spongiform encephalopathy (BSE)**, a neurologic disease in cattle.[52] The course of BSE (also known as mad cow disease) is progressive, ultimately fatal, and is potentially transmissible among cattle. The CDC reported that the first infections with BSE among cows probably began in the 1970s. A BSE epizootic in the United Kingdom peaked in January 1993. At the conclusion of 2010, the number of BSE cases in cattle exceeded 184,500 in the UK. Since the outbreak in the UK, cases also have been reported in 20 European countries as well as Japan, Israel, Canada, and the United States.

In North America (Canada and the United States), a total of 24 cases of BSE were reported as of February 2015. The majority of these cases (20) occurred in Canada. Authorities detected a case of BSE in Canada on February 12, 2015.

On December 23, 2003, the first known case of BSE was reported in the United States. An account of that case is presented in **EXHIBIT 11.4**. As of February 2015, a total of four cases (one from a cow born in Canada) were reported in the United States.

EXHIBIT 11.4

Case Study: Bovine Spongiform Encephalopathy in a Dairy Cow—Washington State, 2003

On December 23, 2003, the US Department of Agriculture (USDA) made a preliminary diagnosis of bovine spongiform encephalopathy (BSE) in a single "downer" (i.e., nonambulatory disabled) dairy cow in Washington state. . . . The BSE-positive cow was aged 6.5 years when it was slaughtered on December 9. Before slaughter, the cow was nonambulatory; its condition was attributed to complications from calving. The animal was examined by a USDA Food Safety and Inspection Service (FSIS) veterinary medical officer both before and after slaughter. After examination, the carcass was released for use as food for human consumption. Tissues (e.g., brain, spinal cord, and small intestine) considered to be at high risk for the transmission of the BSE agent were removed from the cow during slaughter and sent for inedible rendering (often used for nonruminant animal feed). Because the cow was nonambulatory at slaughter, brain tissue samples were taken by USDA's Animal and Plant Health Inspection Service (APHIS) as part of its targeted surveillance for BSE. On December 23, a presumptive diagnosis of BSE was made, and the herd to which this cow belonged was placed under a state hold order.

On December 24, FSIS recalled beef from cattle slaughtered in the same plant on the same day as the BSE-positive cow. Some of the beef subject to the recall had been shipped to several establishments, which processed it further. Meat products manufactured from the recalled meat were distributed primarily to locations in Oregon and Washington, with smaller quantities distributed to locations in California, Idaho, Montana, and Nevada. FSIS continues to verify the distribution and control of all recalled products.

APHIS, in collaboration with the Canadian Food Inspection Agency (CFIA), traced the birth of the BSE-positive cow to a farm in Alberta, Canada. On January 6, USDA and CFIA announced that DNA evidence had confirmed this traceback to Canada with a high degree of certainty. This line of investigation indicates that the BSE-positive cow was one of 82 animals from a Canadian herd cleared for shipment to the United States; 81 of the cattle listed on the Canadian animal health certificate entered the United States on September 4, 2001 through Oroville, Washington.

Modified and reproduced from Centers for Disease Control and Prevention. Bovine spongiform encephalopathy in a dairy cow—Washington State, 2003. *MMWR*. 2004;52:1280–1281.

BSE appears to be spread among cattle by feeds that contain rendered cattle products. **FIGURE 11.12** shows a cow that has BSE and brain tissue affected by BSE. A crucial issue is whether the agent that causes BSE in cattle can be transmitted to humans. **Variant Creutzfeldt-Jakob disease (vCJD)** is one of the four human prion diseases. According to *Control of Communicable Diseases Manual*, 20th edition, "Variant CJD . . . has been linked causally with bovine spongiform encephalopathy (BSE), a prion disease of cattle."[1(p485)] Further, "[t]he mechanism of transmission of BSE from cattle to humans has not been established, but the favored hypothesis is that humans are infected through dietary consumption of the BSE agent."[1(p487)]

Chemically Related Foodborne Hazards

Referring back to Table 11.2, you will see that in addition to microbial agents, other types of agents represent foodborne hazards. These include toxins, heavy metals, pesticides, herbicides, and fungicides; veterinary medicines in meat; and additives used for various purposes. Rocourt et al. point out that "Exposure to chemicals in food can result in acute and chronic toxic effects ranging from mild and reversible to serious and life threatening. These effects may include cancer, birth defects and damage to the nervous system, the reproductive system and the immune system."[2(p2)] **TABLE 11.5** presents selected examples of chemical hazards as well as their associated incubation periods and clinical symptoms.

Toxins

Naturally occurring toxins, such as those from seafood and mushrooms, may be associated with foodborne illness. Other naturally occurring toxins (not covered in detail in this textbook) include oxalic acid found in rhubarb and other plant leaves, nitrites and nitrates found in beets and other vegetables, and aflatoxin (a type of mycotoxin) produced by molds that grow on peanuts and some grains.

Toxins Associated with Marine Environments and Seafood

One source of toxins in the marine environment is from a harmful algal bloom (HAB).[53] These toxins are produced by two kinds of marine phytoplankton: dinoflagellates (minute sea plants) and other tiny organisms

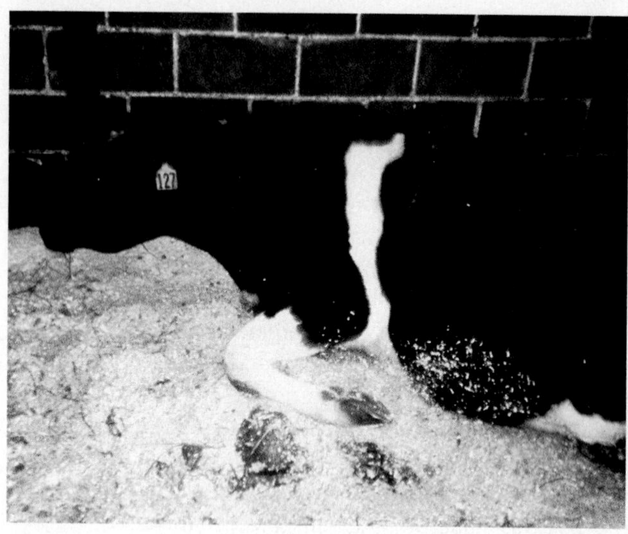

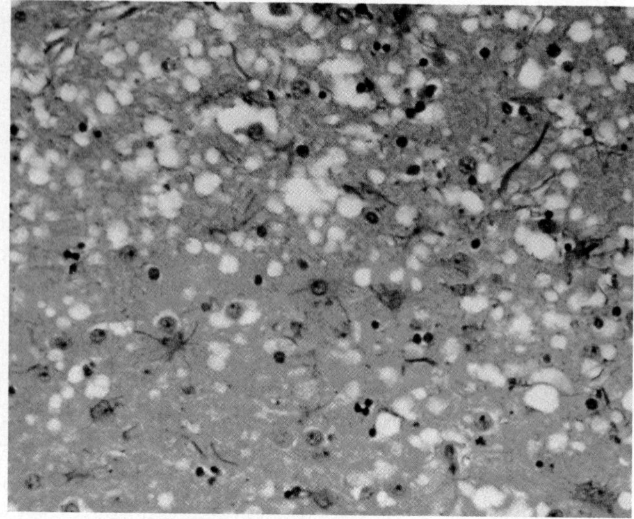

FIGURE 11.12 A. BSE-affected cow, showing incoordination and difficulty standing. B. Brain tissue from a BSE-affected cow with vacuoles, or microscopic holes.

Photo credit (A): Courtesy of Dr. Art Davis and U.S. Dept. of Agriculture - Animal and Plant Health Inspection Service, APHIS. (B): Courtesy of Dr. Al Jenny and U.S. Dept. of Agriculture - Animal and Plant Health Inspection Service, APHIS.

TABLE 11.5 Foodborne Chemical Hazards

Etiologic Agent	Incubation Period	Clinical Syndrome
1. Marine toxins		
a. Ciguatoxin	1–48 hrs, usually 2–8 hrs	Usually gastrointestinal symptoms followed by neurologic symptoms (including paresthesia of lips, tongue, throat, or extremities) and reversal of hot and cold sensation
b. Scombroid toxin (histamine)	1 min–3 hrs, usually < 1 hr	Flushing, dizziness, burning of mouth and throat, headache, gastrointestinal symptoms, urticaria, and generalized pruritus
c. Paralytic or neurotoxic shellfish poison	30 mins–3 hrs	Paresthesia of lips, mouth or face, and extremities; intestinal symptoms or weakness, including respiratory difficulty
d. Puffer fish, tetrodotoxin	10 mins–3 hrs, usually 10–45 mins	Paresthesia of lips, tongue, face, or extremities, often following numbness; loss of proprioception or "floating" sensations
2. Heavy metals a. Antimony b. Cadmium c. Copper d. Iron e. Tin f. Zinc	5 mins–8 hrs, usually < 1 hr	Vomiting, often metallic taste
3. Monosodium glutamate (MSG)	3 mins–2 hrs, usually < 1 hr	Burning sensation in chest, neck, abdomen, or extremities; sensation of lightness and pressure over face or heavy feeling in chest

4. Mushroom toxins		
a. Shorter-acting toxins:	≤ 2 hrs	Usually vomiting and diarrhea, other symptoms differ with toxin as follows:
Muscimol		Confusion, visual disturbance
Muscarine		Salivation, diaphoresis
Psilocybin		Hallucinations
Coprinus artrementaris		Disulfiram-like reaction
Ibotenic acid		Confusion, visual disturbance
b. Longer-acting toxin (e.g., *Amanita* spp.)	6–24 hrs	Diarrhea and abdominal cramps for 24 hrs followed by hepatic and renal failure

Modified and reproduced from Bean NH, Goulding JS, Lao C, et al. Surveillance for foodborne disease outbreaks—United States, 1988–1992. In: CDC Surveillance Summaries, *MMWR*. 1996;45(SS-5):63–65.

called marine diatoms. An example of a toxin from an HAB is called domoic acid. (See **FIGURE 11.13**.) During an HAB, toxins can accumulate in seafood, which may not show any obvious signs such as an off taste or smell even though contaminated. Examples of poisoning from marine toxins are paralytic shellfish poisoning (PSP) and ciguatera fish poisoning.

Scombroid toxin and puffer fish toxin are additional categories of toxins from seafood. Scombroid toxin is elaborated by bacterial deterioration of certain species of fish (e.g., tuna and mackerel). Puffer fish toxin is present in puffer fish, a delicacy for sushi lovers in Japan. Highly trained chefs avoid the poisonous organs in puffer fish so that sushi from puffer fish is safe to consume. Some forms of poisoning with marine toxins are mild and resolve after a few hours, whereas others (e.g., severe paralytic shellfish poisoning and poisoning from puffer fish) can be life threatening or fatal.

Mushroom Toxins

Mushroom toxins are produced naturally by several types of mushrooms, referred to commonly as toadstools. The symptoms and effects of mushroom poisoning vary depending upon the amount consumed; some can cause fatal injuries and irreversible organ damage.[54] Persons affected most often are mushroom hunters who misidentify wild mushrooms and persons from foreign countries who mistake US mushroom varieties for wild mushrooms that may be safe in their native country. Most of the mushroom toxins that are linked to human poisoning cannot be deactivated by cooking, canning, freezing, or any other kind of processing; the only sure way to avoid mushroom poisoning is by avoiding consumption of toxic mushrooms.

Heavy Metals

The chapter on toxic metals and elements covered the health effects of heavy metals (e.g., antimony, cadmium, copper, iron, tin, and zinc). When present in high concentrations, heavy metals can cause vomiting that occurs a few minutes to several hours (but in most cases in less than 1 hour) after ingestion. In addition to metal poisoning caused by high concentrations of heavy metals, there are many possible routes of lower-level exposures to foodborne metals:

- One source of metal poisoning is metals that have leached into foods (especially acidic foods) from metal containers and cooking vessels.
- In California, an instance of lead poisoning occurred when children consumed a variety of lollipops manufactured in Mexico. The candy's wrapper contained extremely high lead concentrations.[55]
- Poultry consumers were exposed to arsenic, which was used in the past as an approved food supplement for poultry to control intestinal parasites. The use of the supplement, e.g., roxarsone, has been discontinued.[56]
- As noted previously, certain species of fish—shark, swordfish, king mackerel, and tilefish—have levels of mercury that may be high enough to be harmful to developing human fetuses.

Pesticides in Foods

Each year, insects, fungi, and other unwanted pests cause extensive damage to vegetables and fruits, with untold economic losses. Applied to crops to reduce the impact of insects and other hazards, pesticides may leave residues that remain on food crops. These residues should be removed by washing and peeling fruits and vegetables carefully. Dietary sources are one of the principal means of exposure of the population to persistent organic chemicals, which include the family of organochlorine pesticides and polychlorinated biphenyls.[57] As noted in the chapter on pesticides and other organic chemicals, the health effects of low-level exposures to pesticides are generally unknown or have not been ascertained definitively at this time.

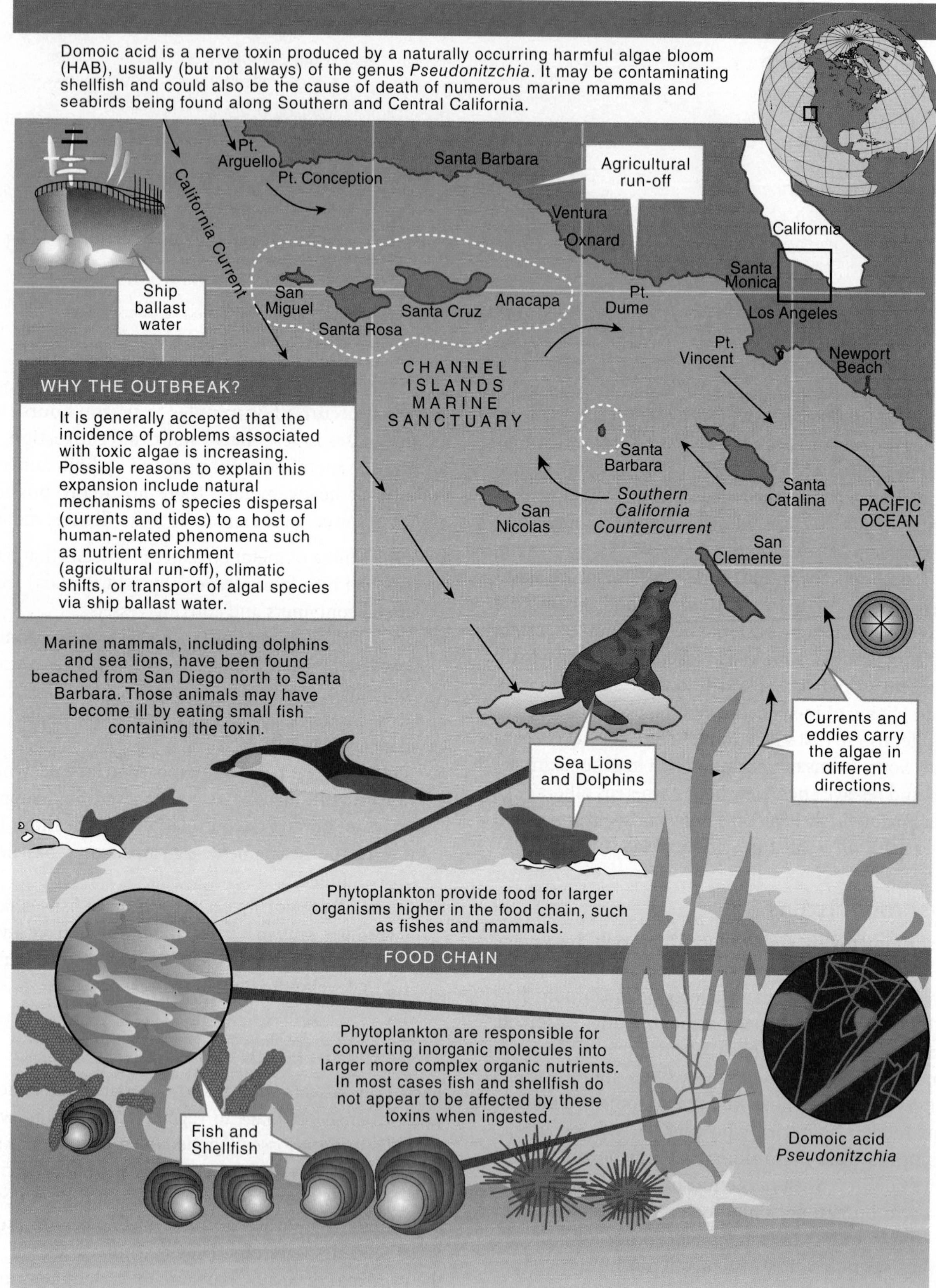

FIGURE 11.13 Domoic acid.

Reproduced courtesy of Channel Islands National Marine Sanctuary. Domoic Acid.

There are several examples of foods that contain low levels of pesticides. In a study of pesticides in baby foods, a public interest group found that 53% of a random sample of jarred fruits and vegetables had traces of at least one pesticide.[58] Nevertheless, the levels found were below federal safety standards for human consumption. However, these levels may not account for the possible special vulnerability of infants to potentially toxic chemicals, coupled with multiple low-level exposures from several different sources, for example, from foods and breast milk.

In comparison with low-level foodborne pesticide exposures, the effects of high-level exposures often are acute. Foodborne pesticides can sicken humans or be potentially toxic if consumed in large amounts. An example is an occurrence of foodborne aldicarb poisoning during early July 1985. Three adults developed symptoms of poisoning following the consumption of a solid green watermelon purchased in Oakland, California.[59] Victims reported the rapid development of symptoms, which included nausea, vomiting, heavy sweating, and reduced heart rate. Around the same time, several persons in Oregon also developed illnesses subsequent to eating striped watermelons. Oregon health officials announced that the pesticide aldicarb had been detected in other watermelons suspected of causing illnesses similar to those already reported in Oregon and California. The California Department of Health Services then issued a statewide embargo on the sale of watermelons and advised residents not to eat them. During the remainder of July, many additional similar illness reports were received. To investigate the cause of the outbreak, officials tested 250 watermelons for aldicarb—only 4% tested positive, possibly due to the lack of sensitivity of the laboratory tests that were used.

This episode was the largest foodborne pesticide illness outbreak that had ever been recorded in North America. California reported 692 probable cases and other states and Canada reported an additional 483 cases. The pesticide aldicarb is regarded as appropriate for use on citrus fruits and potato crops and is not registered for use on melons. This episode demonstrated that aldicarb can cause very severe illnesses if used inappropriately on food crops.

A second example of foodborne pesticide poisoning was a report of endrin poisoning. (Refer to **EXHIBIT 11.5**.)

Antimicrobials in Meat

Antimicrobials that are similar or identical to those used by humans are employed for disease prevention and growth enhancement among animals (cattle, swine, poultry, and fish).[60] Four examples of the numerous common antimicrobials that have been approved in the United States for administration to food animals are amoxicillin, penicillin, bacitracin, and erythromycin. In addition, sulfonamide and several other antimicrobials sometimes are administered to farm-raised fish. In a procedure known as **metaphylaxis**, antimicrobials are given to an entire collection of animals, for example, chickens on a poultry farm or fish raised in pens, for the purpose of curing disease among the sick animals and preventing disease among the remainder. Regarding the use of antimicrobials for growth enhancement of animals (with the exception of farmed fish), antimicrobials increase the efficiency of feeds by reducing the amount of feeds that animals require.

Residues of antimicrobial agents may persist in meat that is destined for human consumption. To reduce the levels of potentially harmful residues in meats, administration of antimicrobials must be discontinued for prescribed periods before feed animals are slaughtered and shipped to market. According to Greenlees,

> "Unlike chemical residues, there is an additional concern for the use of antimicrobial animal drugs of an indirect toxicity to the human consumer, characterized by the loss of an effective therapeutic intervention, resulting in adverse health effects. This potential impact on human health has resulted in the concern for microorganisms that are resistant to . . . antimicrobial drugs. . . ."[61(p133)]

At present, because of concerns about potentially undesirable consequences of antimicrobial use, some countries restrict their administration in animal feeds. Improved conditions for raising food animals and the use of vaccines can reduce the need for microbial agents.

Biotechnology and Genetically Modified Foods

Biotechnology and genetically modified (GM) foods hold promise for addressing the need for more food supplies as the world's population increases and, at the same time, available land and water for agricultural uses diminish.[62,63] Examples of GM foods are those that result from genes that have been manipulated (e.g., through recombinant DNA technologies). These technologies may lead to increases in agricultural productivity and food availability for a burgeoning world

EXHIBIT 11.5

Case Study: Pesticide Poisoning with Endrin

In mid-March 1988, three family members in Orange County, California became dizzy and nauseated within 1 hour of eating taquitos, a snack consisting of a corn tortilla wrapped around a meat filling. Two of the three subsequently had multiple grand mal seizures. The taquitos, a commercial product sold frozen in sealed plastic bags of 48, had been purchased 5 days earlier.

After receiving the reports of illness, the County of Orange Health Care Agency (COHCA) requested that the product be removed from the shelves of the store where the implicated bag was purchased. Several remaining taquitos from the implicated bag were tested by the US Department of Agriculture (USDA), Food Safety and Inspection Service, and found to contain endrin, a pesticide known to cause seizures. Samples of taquitos removed from the store and tested in USDA laboratories were negative for pesticides. The USDA reviewed the operations of the plant where the taquitos were produced, but no evidence of the pesticide was found. However, 90 cases of taquitos were destroyed by the plant owner as a precautionary measure. The USDA concluded that the poisonings were an isolated incident and closed the case April 20, 1988.

Subsequently, in September 1988, COHCA was informed of a 17-year-old boy who, in mid-March, had four seizures 30 minutes after eating taquitos purchased from the same store. After the seizures in March, he had been diagnosed as epileptic and begun on long-term anticonvulsants. At a hearing to determine the 17-year-old's continued eligibility for a driver's license, the hearing officer remarked that he had presided at a similar case (that of the father of the index family) the week before. (In California, seizures and loss of consciousness are reportable conditions for the purpose of determining eligibility for a driver's license.) This new information implicated a second bag of taquitos and indicated that other illnesses may have been misdiagnosed, resulting in serious medical and social consequences. Therefore, the investigation was reopened by the COHCA and the State of California. A state-issued press release and a mailing by the store to over 40,000 customers generated 100 calls to the local health department. As a result of this publicity, [investigators identified] two additional persons who had suffered seizures less than or equal to 12 hours after eating taquitos. All five seizure patients had eaten taquitos purchased from the same discount store within a 5-day period in March. Families other than the index family had no remaining taquitos.

California Department of Food and Agriculture laboratories confirmed the presence of endrin in leftover taquitos from the index family. Endrin was present in the tortillas but not in the meat filling. The store, the manufacturing plant, and the manufacturer's suppliers were thoroughly inspected, but no source of endrin was found. Because of the limited nature of the outbreak and failure to find evidence of contamination in the plant inspections, the California Department of Health Services suspects deliberate tampering as the cause of the outbreak.

Reproduced from Centers for Disease Control and Prevention. Endrin poisoning associated with taquito ingestion—California. *MMWR*. 1989;38:345–346.

population. The advantages of GM foods are purported to be the following:

- Develop genetically engineered herbicide-resistant plants that survive application of herbicides used to destroy weeds.
- Increase the amount of arable land by developing plants that are able to grow in poor-quality soils.
- Increase resistance of plants to attacks from insect pests; such resistance would reduce the need for toxic insecticides and thus benefit the environment.
- Improve the nutritional value of food crops.
- Increase the nutritional value of meat, for example, by reducing its fat content.
- Increase the resistance of food animals to disease, thereby reducing costs and protecting the food supply from animal pathogens.

Despite the advantages of GM foods, environmental groups have expressed concerns over their potential negative consequences.[64–66] Some of the concerns raised are the following:

- Use of GM plants may reduce genetic diversity; GM plants used for food crops might be more susceptible to eradication than existing plants in the event of a disaster such as the appearance of a new plant disease.
- GM plants and animals could escape into the environment, supplanting natural species. For example, farm-raised GM salmon might escape their holding pens and compete more successfully for food and reproduction success than wild salmon.
- Concern about the health effects of genetically modified foods (e.g., allergic reactions and carcinogenesis) has led some countries to ban their use.
- Pollen from GM plants could cross-pollinate with non-GM plants, causing loss of biodiversity.

- The number of antibiotic-resistant strains of bacteria might increase via gene transfer from GM foods.
- Genes from GM plants could migrate into non-GM plants (called outcrossing) with undetermined effects on food safety and security.

Although many scientific studies have verified the safety of GM foods, their applications require more research, especially regarding long-term consequences. In the United States, the National Research Council is studying the safety of animal biotechnology as it pertains to food production.[65] The disadvantages of GM foods need to be balanced against their potential advantages. A paper delivered at the National Academy of Sciences stated that:

> ...biotechnology adds value across the system from crop to farmer, customer and consumer. Biotechnology can, and is, enhancing the quality of food in addition to improving the quantity of food. Biotechnology can improve the sustainability of production systems by requiring fewer inputs to control pests and better protect the quality of water and land mass around us. Biotechnology can add health and vitality to humans. As we look at food production in a more holistic way, biotechnology will be an important component of that holistic system.[67(p5971)]

Food Additives

The term **food additives** is defined as "substances that become part of a food product when added (intentionally or unintentionally) during the processing or production of that food."[68] There are three categories of food additives: intentional, incidental, and malicious.

- *Intentional (direct) additives* are added to foods in order to improve their quality.
- *Incidental (indirect) additives* may be present in foods as a result of unintentional contamination during packaging, storage, and handling. Examples are pesticide residues on fruits, substances from packaging, or antimicrobial agents present in meats; foods may contain parts of insects that were not removed during processing or that have infested a food item.
- *Malicious additives* include substances such as poisons that saboteurs introduce into foods for various reasons (e.g., poisoning of taquitos from the pesticide endrin).

Examples of types of additives and their purposes are shown in **TABLE 11.6**. Additives include emulsifiers, stabilizers, and thickeners; vitamins and minerals; preservatives; and many other substances. Some of the types and functions of additives shown in the table are explained in the following section.

Preservatives

Preservatives that processors add to foods act as antimicrobial agents and antioxidants. Among the antimicrobial preservatives are nitrates (e.g., sodium nitrate and potassium nitrate), **BHA (butylated hydroxyl anisole)** and **BHT (butylated hydroxytoluene)**, disodium EDTA (ethylenediaminetetraacetic acid), sulfites, and propionates. The function of preservatives is to arrest the proliferation of microbes such as bacteria, yeasts, and molds in foods. For example, the preservative sodium nitrite is used on meats for color retention and retardation of the growth of spores that can produce the botulinum toxin. Antioxidants prevent fatty foods from acquiring a rancid taste; they also prevent some foods from turning brown and reduce the loss of vitamins by other foods.

One concern about the use of nitrates and nitrites is that they may interact with amines (chemicals related to ammonia) to produce substances known as nitrosamines, which have been implicated as suspected carcinogens. To minimize the consumer's exposure to nitrosamines, manufacturers are required to keep sodium nitrate at safe levels that have been set by the US Food and Drug Administration (FDA).

BHA and BHT are used as antioxidants for foods that contain fats and oils; for example, some cereals and other dry foods contain BHT. Sulfites are employed mainly as antioxidants that maintain the color of fruits and vegetables. Because they destroy the vitamin thiamin (vitamin B1), sulfites may not be added to foods such as flour that are used to supply this nutrient. Some individuals are allergic to sulfites and must avoid them in their diets; since 1986, the FDA has banned the use of sulfites in fruits and vegetables that are destined to be consumed raw, e.g., in salad bars. One type of sulfite is sodium sulfite. At present, sodium sulfite "is generally recognized as safe when used with good manufacturing practice, except that it is not used in meats; in food recognized as a source of vitamin B1; on fruits and vegetables intended to be served raw to consumers or sold raw to consumers, or to be presented to consumers as fresh."[69] Propionates (e.g., calcium propionate) are used as antimicrobial agents to retard spoilage of foods. Refer to the FDA website for a brochure titled *Food Ingredients and Colors*.[70]

Several concerns about the safety of food additives exist among the general public and health professionals. Some persons have allergic reactions to

TABLE 11.6 Some of the Types of Food Additives

Type of Additive	Purpose	Examples
Emulsifiers, stabilizers, and thickeners	Impart a consistent texture to products; prevent separation of foods such as salad dressings.	Soy lecithin, gelatin, and carrageenan
Anticaking agents	Enable products such as table salt to flow freely.	Calcium silicate
Nutrients	Enrichment (replacement of nutrients lost during processing) and fortification (adding to the nutritional value of foods).	Many types of vitamins, e.g., folic acid, beta-carotene, vitamin D; many types of minerals, e.g., iron, iodine
Preservatives (antimicrobials) Preservatives (antioxidants)	Retard spoiling by microbial agents. Prevent fats and oils from becoming rancid; prevent cut fresh fruits from turning brown.	Ascorbic acid, citric acid, sodium nitrate, calcium propionate, BHA, and BHT
Leavening agents	Cause bread and baked goods to rise during baking.	Baking soda
Flavor enhancers (synthetic and natural)	Enhance flavor of foods.	MSG (monosodium glutamate)
Sweeteners (synthetic and natural)	Add sweetness with or without the extra calories.	Aspartame, sucrose
Color additives (artificial and natural)	Impart color to foods.	FD&C Yellow No. 6, caramel color
Fat replacers	Impart texture and creamy "mouth feel" to foods.	Olestra, cellulose gel

Data from International Food Information Council (IFIC) Foundation and US Food and Drug Administration. *Food Ingredients and Colors*. Washington, DC: International Food Information Council (IFIC) Foundation and US Food and Drug Administration. November 2004; revised April 2010. Available at: https://www.fda.gov/food/ingredientspackaginglabeling/foodadditivesingredients/ucm094211.htm. Accessed September 10, 2017.

some additives (e.g., MSG, sulfites, and the artificial sweetener aspartame). Sodium-containing additives (e.g., sodium bicarbonate) may add unwanted sodium to the diets of persons afflicted with hypertension. Finally, other additives may be known or potential carcinogens; for example, nitrosamine compounds are known carcinogens. The artificial sweetener saccharin has been shown to cause cancer in laboratory animals; however, the carcinogenic potency of saccharin for humans is extremely low compared with many other carcinogens. BHA and BHT may be a cause of tumors in laboratory animals. The safety of food additives is monitored by the FDA; this topic is covered in the next section.

Regulation of Food Safety

In the United States, a body of governmental regulations is designed to assure the quality of food products and to protect the public from possible harm resulting from additives and chemicals in food. The US Food and Drug Administration (FDA) is the federal agency charged with regulation and enforcement of food safety. The US Department of Agriculture (USDA) regulates the safety of meat and poultry products. Food safety regulations that are in force at present are the result of an incremental process of change: among the significant US food safety regulations that have been legislated or that have evolved over time are the following[71]:

- 1906 Food and Drugs Act
- The Federal Food, Drug, and Cosmetic Act of 1938
- Miller Pesticide Amendments, 1954
- Food Additives Amendment, 1958
- Color Additive Amendments, 1960
- Animal Drug Amendments, 1968
- Food Quality Protection Act (FQPA), 1996

Food, Drug, and Cosmetic Act of 1938

The 1906 Food and Drugs Act prohibited the sale of foods considered "misbranded" or "adulterated."[71] The broadened Federal Food, Drug, and Cosmetic Act, formulated in 1938, introduced several important standards for food, drugs, cosmetics, and medical devices:

> The new law brought cosmetics and medical devices under control, and it required that drugs be labeled with adequate directions for safe use. Moreover, it mandated pre-market approval of all new drugs, such that a manufacturer would have to prove to FDA that a drug were safe before it could be sold. It irrefutably prohibited false therapeutic claims for drugs, although a separate law granted the Federal Trade Commission jurisdiction over drug advertising. The act also corrected abuses in food packaging and quality, and it mandated legally enforceable food standards. Tolerances for certain poisonous substances were addressed. The law formally authorized factory inspections, and it added injunctions to the enforcement tools at the agency's disposal.[72]

The 1938 Act was amended during the period of 1954 to 1968 to take into account the safety of pesticide residues and food additives. A congressional select committee, established in 1952 and chaired by Representative James Delaney of New York, was responsible for initiating many of the amendments designed to regulate pesticides in foods and food additives.

Miller Pesticide Amendments

Adopted in 1954, the Miller Pesticide Amendments to the Food, Drug, and Cosmetic Act of 1938 governed the levels of pesticides that were permitted in foods. Pesticide residues were required to fall within tolerances set by the FDA. The 1938 act and the 1954 amendments remain in force, but were modified by additional standards for the regulation of pesticides in foods that were adopted subsequently, as described in the following text.

Food Additives Amendment, 1958

The Food Additives Amendment was implemented in 1958 in recognition of the public's growing concern over chemicals that were being used increasingly in foods. The 1958 amendment, which applies exclusively to the use of food additives as well as chemicals that could come into contact with food (e.g., from food packaging), defined what was meant by the terms *food additive* and *unsafe food additive*. Under the amendment, food manufacturers were required, by means of a premarket review, to demonstrate the safety of food additives to the FDA.[73] As part of the Food Additives Amendment, the Delaney Clause regulated the permissible levels of additives in foods. Specifically, the clause prohibited the use of food additives, including pesticides, that had been determined to cause cancer in human beings or animals.

Many substances that already had been in use over a long period of time and had established safety records were exempted from review. These substances fell under the Generally Recognized as Safe (GRAS) rule; examples of such additives are the many types of salts, flavoring agents, and preservatives that were believed to be safe. Food processors that intended to use other additives not on the GRAS list were required to demonstrate their safety before they could be marketed.

Adopted in 1960, the Color Additive Amendments contained provisions for the premarket approval of colors that were designated for use in foods, drugs, and cosmetics. Finally, the Animal Drug Amendments of 1968 govern the safety and efficacy of drugs and feed additives that are administered to food animals and control the levels of residues of such drugs in meats destined for human consumption.

Food Quality Protection Act (FQPA), 1996

One of the consequences of the Delaney Clause was that it provoked the wrath of food growers and processors by prohibiting the application of any pesticide that might appear in processed foods, should that pesticide carry even a minimal risk of causing cancer. The 1996 FQPA repealed the Delaney Clause and adopted a strict standard for chemicals applied to raw and processed food. The standard was that a substance could not cause a lifetime incidence of more than one cancer case per 1 million exposed persons.[74] According to the EPA:

> The Food Quality Protection Act (FQPA) was passed unanimously by Congress and then signed into law by President Clinton on August 3, 1996. The FQPA amended the

> Federal Insecticide, Fungicide, and Rodenticide Act (FIFRA) and the Federal Food, Drug, and Cosmetic Act (FFDCA) and thus fundamentally changed EPA's regulation of pesticides.... The FQPA requires that EPA... make a safety finding when setting tolerances, i.e., that the pesticide can be used with 'a reasonable certainty of no harm.'"[75]

Overall, FQPA won the favor of environmental groups as well as food processors because it placed into action more effective standards to protect children and to test chemicals for their teratogenic effects.

Agencies That Regulate Food Safety

FIGURE 11.14 illustrates agencies that regulate food safety at the international level as well as the national, state, and local levels in the United States. For the sake of example, the state of California has been chosen to illustrate the state level. However, the illustration would apply to most of the states in the country. In a similar vein, examples of local jurisdictions chosen are located in southern California. The choice of these health agencies is arbitrary; similar illustrations could be selected from other local jurisdictions across the United States.

The following are detailed accounts of the functions of each of the agencies shown in Figure 11.14:

- *International:* "The Food and Agriculture Organization of the United Nations (UN FAO) was founded in 1945 with a mandate to raise levels of nutrition and standards of living, to improve agricultural productivity, and to better the condition of rural populations. Today, FAO is the largest specialized agency in the United Nations system and the lead agency for agriculture, forestry, fisheries and rural development. An intergovernmental organization, FAO has 180 member countries

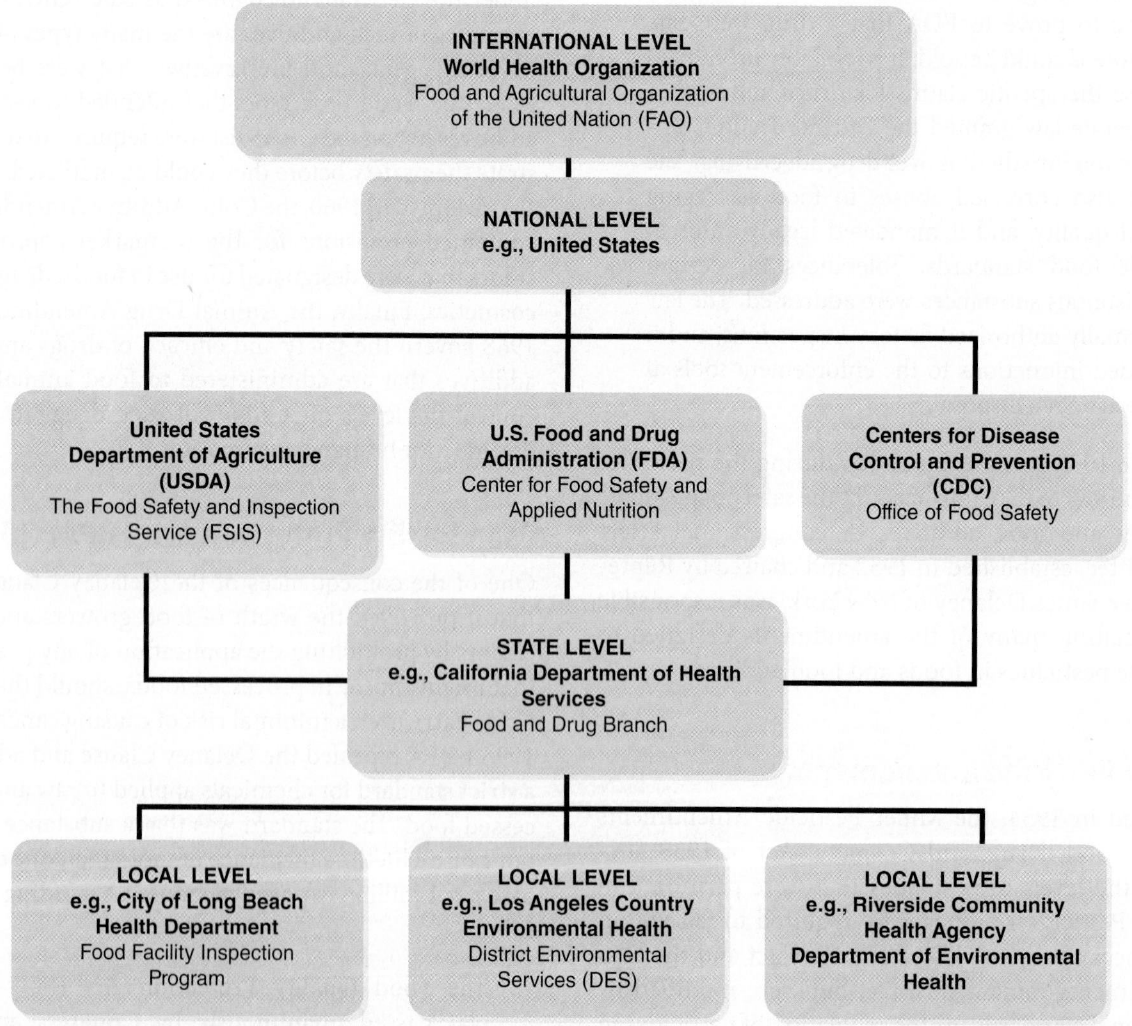

FIGURE 11.14 Key players in food safety and regulation from the perspective of the international, national, state, and local levels.

Courtesy of the author, with the assistance of L. Francisco.

plus one member organization, the European Community. Since its inception, FAO has worked to alleviate poverty and hunger by promoting agricultural development, improved nutrition and the pursuit of food security—defined as the access of all people at all times to the food they need for an active and healthy life."[76]

- *National:* "The Food Safety and Inspection Service (FSIS) is the public health regulatory agency in the U.S. Department of Agriculture responsible for ensuring that the nation's commercial supply of meat, poultry, and egg products is safe, wholesome, and correctly labeled and packaged."[77]

FIGURE 11.15 shows poultry that has been condemned by the USDA.

- *National:* The Center for Food Safety and Applied Nutrition (CFSAN) is an arm of the US Food and Drug Administration (FDA). "The Center regulates $417 billion worth of domestic food, $49 billion worth of imported foods, and over $60 billion worth of cosmetics sold across state lines. . . . FDA's responsibility in the area of cosmetics covers all domestic and imported products. . . . This regulation takes place from the products' point of U.S. entry or processing to their point of sale. . . . FDA's responsibility in the food area generally covers all domestic and imported food . . . [with the following exceptions:]"[78]
    - Meat, poultry, and frozen, dried and liquid eggs; these are "under the authority of the U.S. Department of Agriculture (USDA) Food Safety and Inspection Service (FSIS). . . ."[78]
    - Labeling of alcoholic beverages (above 7% alcohol) and tobacco; these "are regulated by the U.S. Department of the Treasury's Alcohol and Tobacco Tax and Trade Bureau (TTB). . . ."[78]
    - Establishment of tolerances for pesticide residues in foods and setting of requirements for drinking water; these areas are the responsibility of the US Environmental Protection Agency (EPA).
- *National:* At the federal level, the Centers for Disease Control and Prevention (CDC) collaborates with other agencies for promotion of food safety. Two of these are the US Food and Drug Administration (FDA) and the US Department of Agriculture (USDA) Food Safety and Inspection Service. The CDC also forms partnerships with local and state health departments and the food industry. The CDC monitors foodborne disease outbreaks and provides leadership in their investigation. The CDC maintains several nationwide foodborne disease surveillance systems, for example, the Foodborne Diseases Active Surveillance Network (FoodNet) in operation since 1996 and the Foodborne Disease Outbreak Surveillance System (FDOSS). Another function of CDC is to develop state-of-the-art techniques and systems for rapid identification of certain foodborne viruses and DNA fingerprinting of bacteria that may be transmitted by foods; this information helps participating public health officials identify contaminated foods that are implicated in outbreaks. CDC manages PulseNet, a DNA fingerprinting network.[79]
- *State:* The mission of the California Department of Health Services, Food and Drug Branch "is to protect and improve the health of all California residents by assuring that foods, drugs, medical devices, cosmetics and certain other consumer products are safe and are not adulterated, misbranded nor falsely advertised; and that drugs and medical devices are effective."[80]
- *Local:* The City of Long Beach Health and Human Services Food Facility Inspection Program oversees the safety of the regulation of food facilities only within the City of Long Beach. "The goal of the Consumer Protection Program is to ensure that the food sold and served in the City of Long Beach is safe, wholesome, properly labeled and advertised, and produced under sanitary

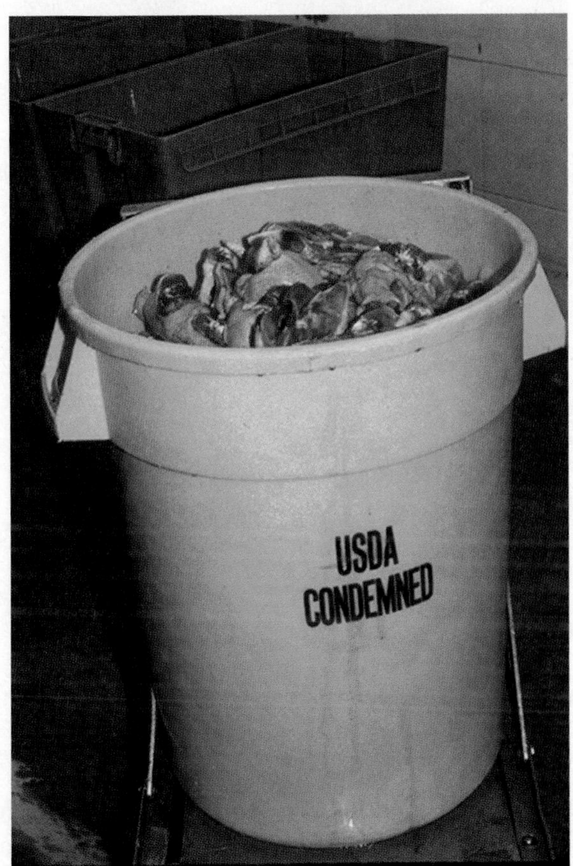

FIGURE 11.15 Condemned poultry.
Reproduced courtesy of United States Department of Agriculture.

conditions. . . . The Food Inspection Program . . . enforces state and local food safety regulations in over 2,000 food facilities in the city."[81]

- *Local:* Los Angeles County Department of Public Health. Environmental Health: District Surveillance and Enforcement Branch "... is comprised of twenty-nine District Offices and three Complaint Investigation Units. We are responsible for inspection and enforcement of public health laws at restaurants and markets, licensed housing and associated swimming pools at apartments and condominiums, temporary food facilities at community events such as fairs or carnivals, pet food stores, food warehouses, theaters, and self-service laundries. The Branch also handles general sanitation complaints such as refuse, sewage, and green pools, at private or commercial properties."[82]
- *Local:* The mission of the Riverside County Community Health Agency, Department of Environmental Health "is to enhance the quality of life in Riverside County through implementation of environmental health programs that protect public health and safety as well as the environment."[83]

Foodborne Disease Prevention

In the United States, two main factors are related to foodborne disease outbreaks: improper temperatures for keeping foods and poor personal hygiene of food handlers.[84] Less developed areas of the world may be deficient in one or both of these elements and in addition not have a reliable water supply. Consider the following two graphic examples of how foodborne illness may erupt in less developed regions of the world. **FIGURE 11.16** shows food preparation techniques found in some parts of the world; in these examples, food is being prepared out in the open, on a street. Such obvious unsanitary conditions for the storing and preparation of food place consumers at risk of illness outbreaks.

TABLE 11.7 lists risk factors identified by the CDC as being associated with foodborne illnesses, in order of their importance. Note that the percentages total more than 100% because more than one factor at a time can be present as a cause of foodborne illness. Routine inspections of restaurants and food-handling facilities focus on this set of risk factors, some of which are discussed in the following sections.

FIGURE 11.16 A. Woman cooking food on the sidewalk in Xian, China. B. A Peruvian street vendor selling a corn-based drink he made at home using a single glass that he fills with a dipper.

(Part A) Courtesy of CDC/ Dr. Edwin P. Ewing, Jr. (Part B) Courtesy of CDC

Safe Cooling and Cold Holding Temperatures

In order to prevent microorganisms from multiplying in foods, they must be kept in the safe temperature range: above 140°F (60°C) for hot foods and below 40°F (4°C) for foods that are kept in cold storage. The range between 40°F and 140°F (4°C and 60°C) is the danger zone, the temperature range in which microbes that can cause foodborne illness produce toxins or multiply. Refer to **FIGURE 11.17**.

TABLE 11.7 Top Risk Factors for Foodborne Illness

Percentage*	Risk Factor
63%	Inadequate cooling and cold holding temperatures
29%	Preparing food ahead of planned service
27%	Inadequate hot holding temperatures
26%	Poor personal hygiene/infected persons
25%	Inadequate reheating
9%	Inadequate cleaning of equipment
7%	Use of leftovers
6%	Cross-contamination
5%	Inadequate cooking or heat processing
4%	Containers adding toxic chemicals
2%	Contaminated raw ingredients
2%	Intentional chemical additives
1%	Incidental chemical additives
1%	Unsafe sources

* Percentages refer to foodborne illness outbreaks in which this risk factor is present.
Modified and reproduced from County of Los Angeles Public Health, Environmental Health.

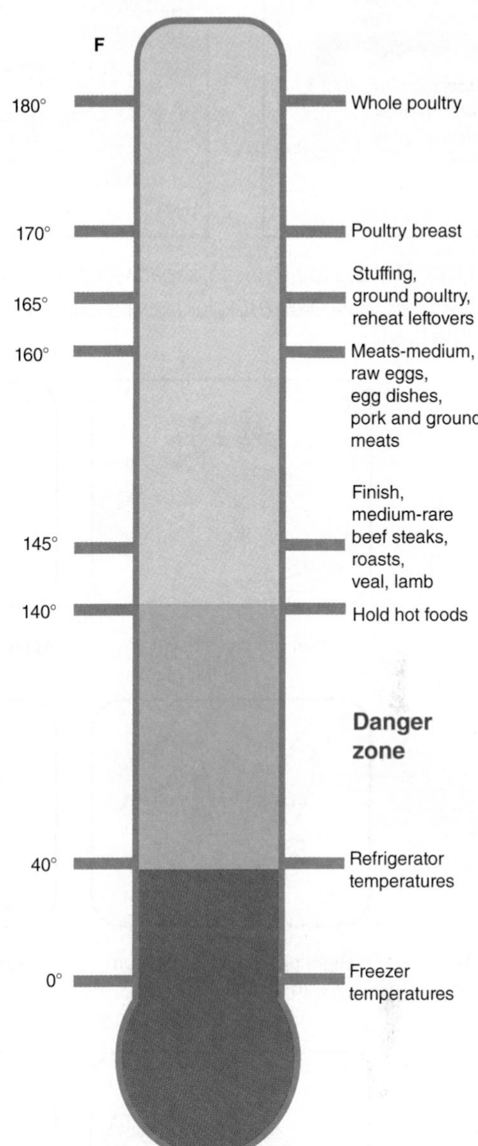

FIGURE 11.17 Apply the heat chart to determine safe temperature levels for preserving, storing, and cooking foods.

Modified and reproduced from US Food and Drug Administration, Center for Food Safety and Applied Nutrition. Food safety for moms to be: Educator tools—Apply the heat. Available at: http://www.fda.gov/Food/ResourcesForYou/HealthEducators/ucm109104.htm. Accessed May 11, 2017.

Preventing Growth of Foodborne Pathogens

FIGURE 11.18 illustrates four steps to prevent the growth of foodborne pathogens in food. These steps include thorough hand washing, prevention of cross-contamination, cooking at sufficiently high temperatures, and storing cooked and vulnerable foods at proper temperatures in a refrigerator. An example of cross-contamination of foods is the use of a cutting board to cut raw poultry and subsequent use of the same cutting board to cut bread without washing the cutting board before reuse.

The recommended method of safe hand washing is illustrated further in **FIGURE 11.19**. For more information, refer to the steps shown in the figure. Warm water, soap, friction, and adequate duration of washing are essential for cleaning one's hands.

Irradiation

As noted previously, foodborne illness continues to cause numerous cases of disease and deaths in the United States despite regulations and procedures that are in place to protect the public by monitoring the safety of the food supply. The CDC reports that "[f]ood irradiation reduces or gets rid of pathogens,

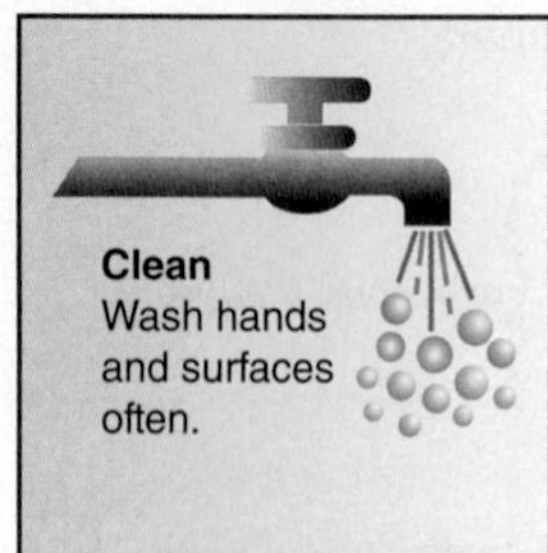

Clean
Wash hands and surfaces often.

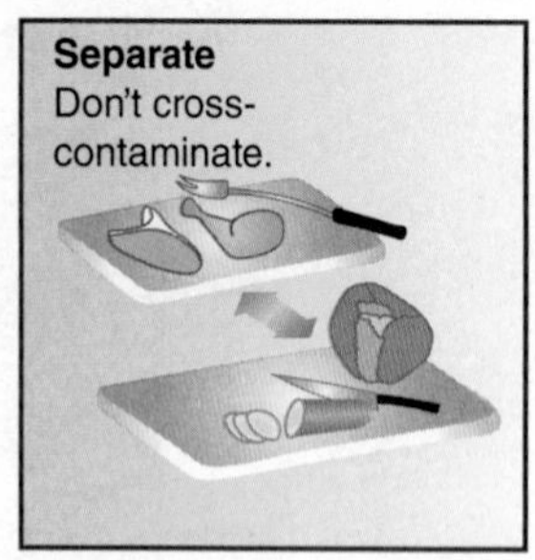

Cook
Cook to proper temperatures.

FIGURE 11.18 The big four—clean, separate, cook, and chill.

Reproduced from United States Department of Agriculture, Food Safety and Inspection Service. *The Food Safety Educator*. 2001;6(2). Available at: http://www.fsis.usda.gov/OA/educator/educator6-2a.htm. Accessed May 11, 2017.

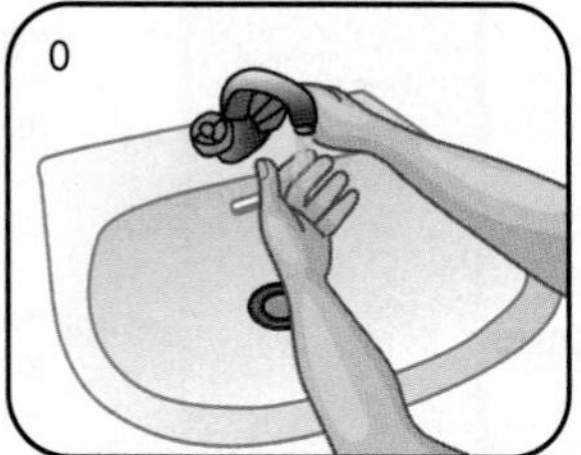

Wet hands with water

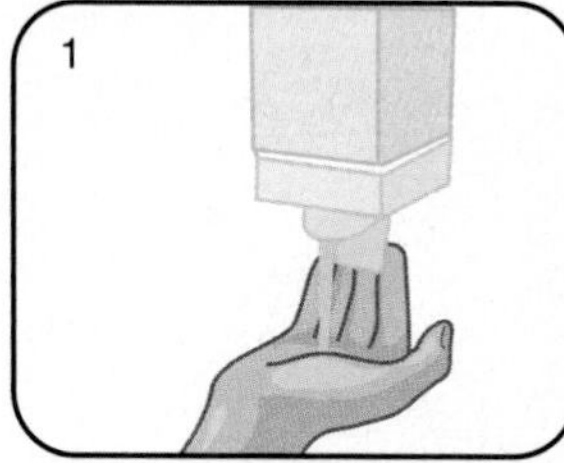

Apply enough soap to cover all hand surfaces.

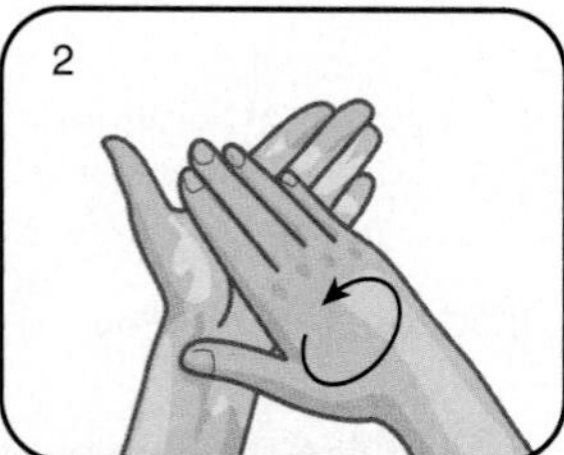

Rub hands palm to palm

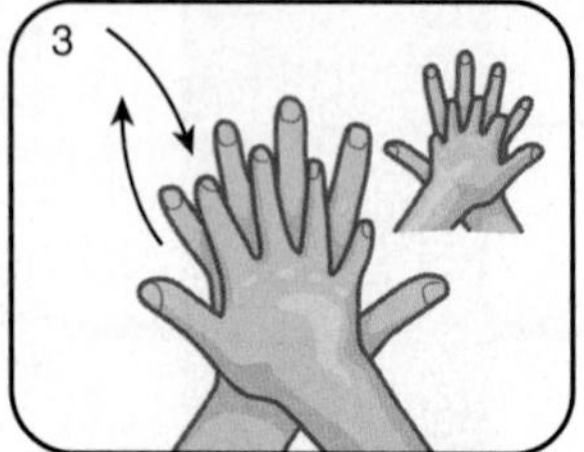

Right palm over left dorsum with interlaced fingers and vice versa

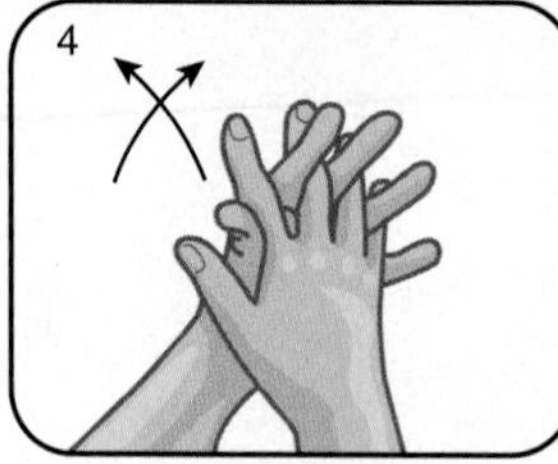

Palm to palm with fingers interlaced

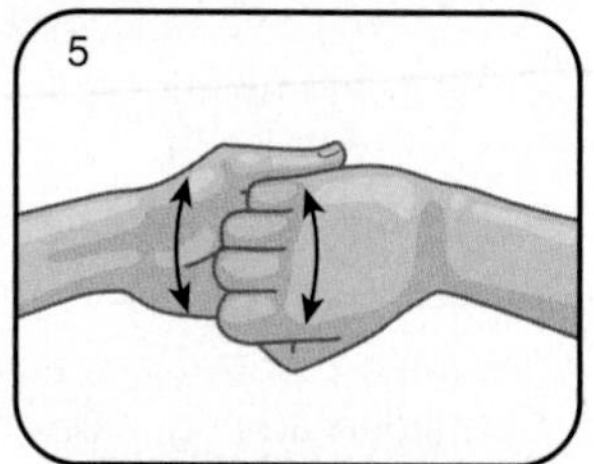

Backs of fingers to opposing palms with fingers interlocked

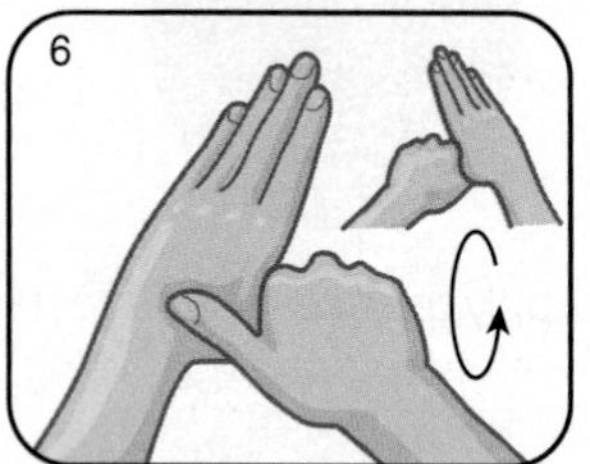

Rotational rubbing of left thumb clasped in right palm and vice versa

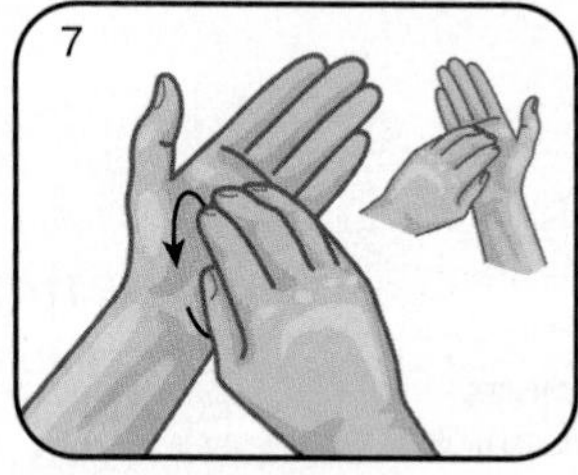

Rotational rubbing, backwards and forwards with clasped fingers of right hand in left palm and vice versa.

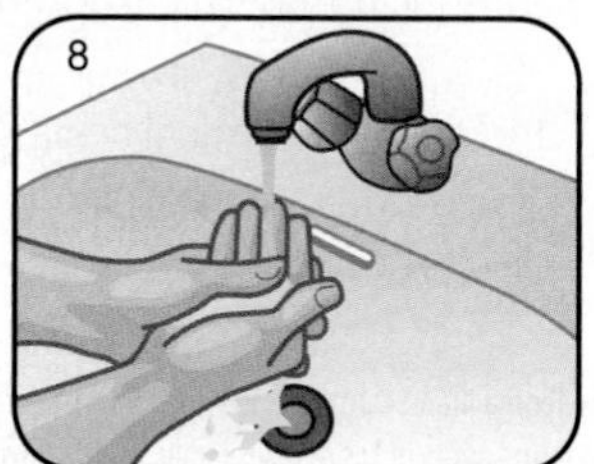

Rinse hands with water

Dry throughly with a single use towel

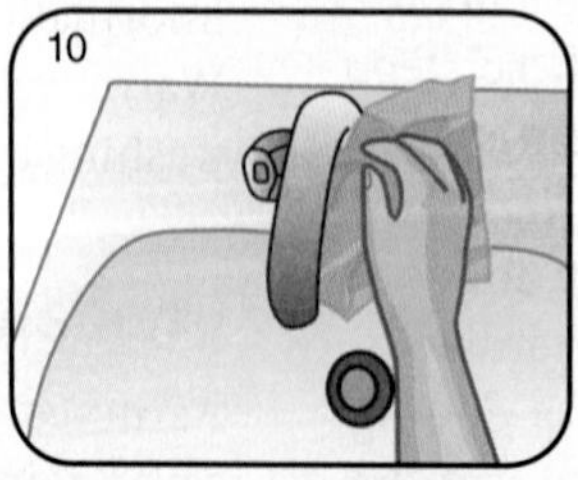

Use towel to turn off faucet

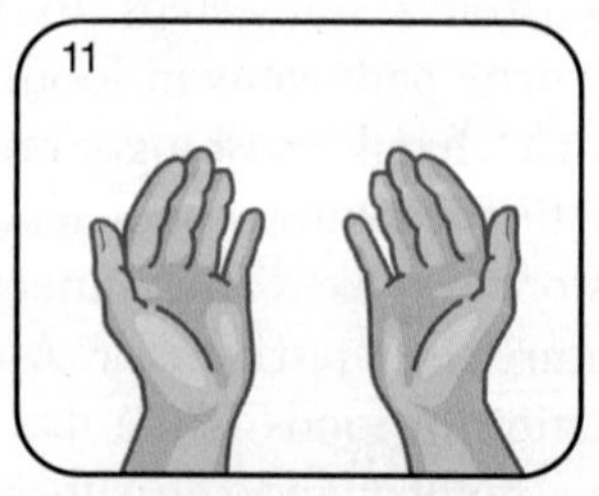

...and your hands are safe

FIGURE 11.19 Proper hand washing technique.

Reproduced from World Health Organization. Clean hands protect against infection. Available at: http://www.who.int/gpsc/clean_hands_protection/en/. 2018, Copyright 2018.

such as bacteria and molds that spoil food and cause food poisoning and other illness. For example, irradiation can kill *E. coli*, *Campylobacter* and *Salmonella* bacteria. These bacteria make millions of people sick and send thousands of people to the hospital each year. Animal feed also can contain *Salmonella*. Irradiation can prevent the spread of these bacteria to livestock."[85]

During the process of irradiation, energy from ionizing radiation passes through foods and destroys bacteria and other pathogens; the procedure does not cause food to become radioactive. Exposing foods to radiation does not cause their temperatures to increase as with other methods of food preservation. Hence, irradiation of foods is referred to as cold pasteurization.[86]

The sources of ionizing radiation used in cold pasteurization include gamma radiation from cobalt-60 and cesium-137, electron beam radiation, and X-rays. The latter two sources do not require a radioactive source and can be switched on and off by the operator. A low dose of radiation (< 0.1 kiloGrays) is effective at killing insect pests and parasitic organisms. Medium doses in the range of 1.5 kiloGrays to 4.5 kiloGrays are lethal to most bacteria. High doses up to 45 kiloGrays are needed to kill bacterial spores and some viruses.[87] These doses are needed for immediate eradication of the respective organisms.

Irradiation of food has been certified as a safe method for reducing the risk of foodborne disease.[86] Irradiation also reduces spoilage and increases the shelf life of foods. Use of ionizing radiation for food sterilization has been subjected to evaluation over a period of 50 years. More than 40 countries have approved the irradiation of food, and the procedure has received the endorsements of many health- and food-related organizations.

In the case of many foods, irradiation produces little change in the nutritional value and taste and other sensory qualities of the product. Foods that retain their quality include meat, poultry, seafood, soft fruit and some produce, and grains.[87,88,] Leafy vegetables and sprouts wilt following irradiation; meats that have high fat contents may develop slightly unpleasant odors. Cold pasteurization cannot be regarded as a panacea for foodborne illness. Following irradiation, pathogens can be reintroduced into foods during later stages of processing.

Food irradiation has been slow to gain implementation, perhaps because of resistance from the public and possible higher costs.[87,88] Surveys indicate that about half the US population is ready to accept irradiated foods; this percentage may increase once consumers learn of the benefits of irradiation in reducing the risk of foodborne illness. One irradiated product—ground beef—has been available in supermarkets since 2000. The FDA requires that irradiated foods are identified by the Radura symbol shown in **FIGURE 11.20**.

FIGURE 11.20 Radura logo, an international symbol used for irradiated food.

Reproduced from U.S. Food and Drug Administration. Food irradiation: what you need to know. Available at: https://www.fda.gov/food/resourcesforyou/consumers/ucm261680.htm. Accessed November 22, 2017.

Hazard Analysis of Critical Control Points

Hazard analysis of critical control points (HACCP) is a system for reducing the risk of foodborne illness through application of the seven principles shown in the text box.[89] Using HACCP, foods are monitored from the time of harvest to the time of consumption. The methodology was adopted in the 1950s to assure the safety of food used in the US space program. In the 1990s, the FDA established HACCP for the processing of seafood. During the late 1990s, the US Department of Agriculture, which has regulatory oversight of meat and poultry, initiated HACCP for plants that process these products. HACCP was applied to the juice industry beginning in 2002. The seven principles of HACCP are described in the text box below.

Investigating and Reporting a Foodborne Outbreak

Tracking the cause of a foodborne disease outbreak requires a great deal of detective work and provides many challenges to local public health departments, which often are charged with the responsibility for safety of foods sold in restaurants, grocery stores, and other establishments. The following list provides examples of steps that may be required in the investigation of foodborne illness following a suspected outbreak:

- Identify as many of the cases involved as possible. This stage may require searching for cases (case finding).
- Develop hypotheses regarding the cause of the outbreak from interviews with affected persons.
- Test hypotheses regarding the outbreak.
    - Perform a descriptive epidemiologic analysis: describe person, time, and place variables regarding affected and nonaffected persons who were involved in the outbreak. For affected

SEVEN PRINCIPLES OF HAZARD ANALYSIS OF CRITICAL CONTROL POINTS

1. **Perform a hazard analysis.** The first principle is about understanding the operation and determining what food safety hazards are likely to occur. (Examples of hazards are microbes, chemicals, and foreign objects.) The manager needs to understand how the people, equipment, methods, and foods all affect each other. The processes and procedures used to prepare the food are also considered. This usually involves defining the operational steps (receiving, storage, preparation, cooking, etc.) that occur as food enters and moves through the operation. Additionally, this step involves determining the control measures that can be used to eliminate, prevent, or reduce food safety hazards. Control measures include such activities as implementation of employee health policies to restrict or exclude ill employees and proper hand washing.
2. **Decide on the critical control points (CCPs).** Once the control measures in principle #1 are determined, it is necessary to identify which of the control measures are absolutely essential to ensuring safe food. An operational step where control can be applied and is essential for ensuring that a food safety hazard is eliminated, prevented, or reduced to an acceptable level is a critical control point (CCP).
3. **Determine the critical limits.** Each CCP must have boundaries that define safety. Critical limits are the parameters that must be achieved to control a food safety hazard. For example, when cooking pork chops, the *Food Code* sets the critical limit at 145°F (63°C) for 15 seconds. When critical limits are not met, the food may not be safe. Critical limits are measurable and observable.
4. **Establish procedures to monitor CCPs.** Once CCPs and critical limits have been determined, someone needs to keep track of the CCPs as the food flows through the operation. Monitoring involves making direct observations or measurements to see that the CCPs are kept under control by adhering to the established critical limits.
5. **Establish corrective actions.** While monitoring CCPs, occasionally the process or procedure will fail to meet the established critical limits. This step establishes a plan for what happens when a critical limit has not been met at a CCP. The operator decides what the actions will be, communicates those actions to the employees, and trains them in making the right decisions. This preventive approach is the heart of HACCP. Problems will arise, but you need to find them and correct them before they cause illness or injury.
6. **Establish verification procedures.** This principle is about making sure that the system is scientifically sound to effectively control the hazards. In addition, this step ensures that the system is operating according to what is specified in the plan. Designated individuals like the manager periodically make observations of employees' monitoring activities, calibrate equipment and temperature measuring devices, review records/actions, and discuss procedures with the employees. . . .
7. **Establish a record keeping system.** There are certain written records or kinds of documentation that are needed in order to verify that the system is working. These records will normally involve the HACCP plan itself and any monitoring, corrective action, or calibration records produced in the operation of the HACCP system.

Modified and reproduced from US Food and Drug Administration, Center for Food Safety and Applied Nutrition. *Managing Food Safety: A Manual for the Voluntary Use of HACCP Principles for Operators of Food Service and Retail Establishments*. OMB Control No. 0910-0578. College Park, MD: US Food and Drug Administration, Center for Food Safety and Applied Nutrition; 2006:9–11.

persons, describe their symptoms, including time of onset.

- List the food items served and compare the occurrence of symptoms between those who ate and those who did not eat each particular food listed.
- Create a map of the cases.
- Collect samples of blood, stools, and vomitus for diagnosis of causative agent.
- Collect samples of implicated foods, if possible, and subject to analysis.

- Take corrective actions, as necessary, such as recalling suspected foods and requiring improvements of facilities where food was produced.[90]

Employment Roles in Food Safety

Many employment roles, particularly in the government sector, relate to food safety. Restaurant inspectors are responsible for overseeing the safety of foods served in local restaurants. Epidemiologists investigate outbreaks of foodborne illness in their communities. Microbiologists are crucial to the identification of microbes that may be present in foods. The US federal government employs inspectors who are involved with the processing of food animals. The following paragraphs describe the role of food inspectors:

> Entry-level food inspectors in private commercial slaughtering plants provide the first line of defense against diseased and adulterated meat and poultry. They are responsible for much of the day-to-day in-plant inspection of animals before and after slaughter.
>
> One career path for a food inspector is to the consumer safety inspector position. Consumer safety inspectors work in one or more privately owned meat, poultry, and egg

processing plants. They ensure the plant is operating within its written plans for HACCP, sanitation, and processing.

In addition, they conduct regulatory oversight activities inside the plants in matters relating to other areas of consumer protection, e.g., misbranding.

Another career path for a food inspector is to the import inspector position. Import inspectors are stationed at ports and other points of entry to the United States. They make sure that products imported from other countries are as safe as those produced domestically.[91]

Conclusion

This chapter covered food safety, including the causes and prevention of foodborne illness and regulations to protect the safety of the food supply at the international, national, state, and local levels. Foodborne illness is a major cause of morbidity and mortality in the developing world and also is a growing source of concern for developed countries. Estimates indicate that foodborne illness impacts one-quarter to one-third of the US population each year; also, foodborne illness can be a vexing problem for the closed environments of places like cruise ships.

Foodborne hazards include biological, chemical, and physical contaminants. Of growing concern for the safety of the food supply are emerging foodborne disease pathogens. Other potential threats to the food supply are residues of pesticides, medicines administered to food animals, genetically modified foods, and additives in foods. In the United States, numerous statutes regulate the quality of the food supply. These statutes are monitored and enforced by regulatory agencies such as the FDA and the USDA as well as by state and local authorities. As a consequence of regulatory and enforcement actions, the food supply of the United States is regarded as one of world's safest.

Study Questions and Exercises

1. Define the following terms:
   a. Foodborne disease
   b. Foodborne outbreak
   c. Passive versus active surveillance systems
   d. Food additive
   e. Radura logo
2. Describe the categories of hazards that may be present in foods and suggest methods for their control or prevention.
3. Name two emerging foodborne disease pathogens. Select one of them and describe an outbreak that involved the pathogen. Describe how foodborne disease outbreaks caused by the pathogen can be prevented.
4. Give examples of pesticides that have caused food-associated poisoning. To what extent can these poisonings be prevented?
5. State the names of two microbial agents of foodborne disease that occur among food animals. Give examples of how these agents may be transmitted by foods and suggest methods for prevention of their occurrence among animals used for food.
6. What foodborne pathogen has been associated with stillbirths? What foods have been suspected of carrying the pathogen? Suggest methods of prevention of foodborne illness from this cause.
7. Name the agent responsible for trichinosis. What foods may be responsible for passing trichinosis to human hosts? To what extent do you believe trichinosis currently poses a hazard to the US food supply? Discuss methods for prevention of trichinosis.
8. What types of microbial agents have been linked to outbreaks of illnesses on cruise ships? What can be done to prevent such outbreaks?
9. Name three acts and/or amendments that regulate food safety in the United States. Describe their purposes and the types of foods and components of foods to which they apply.
10. Give examples of international and national agencies charged with the enforcement of food safety regulations. Contact your state and local health departments and write a brief description of their activities with respect to enforcement of food safety.
11. Discuss the risk factors for foodborne illness. Describe practical methods for the prevention of foodborne illness and indicate how you apply them in your own home or business.
12. What are the procedures that a local health department might use for investigating an outbreak of foodborne illness? (Refer to Exhibit 11.2). What factors could have been responsible for the *S. aureus* outbreak described in the exhibit?

For Further Reading

Royte, Elizabeth. Waste not, want not. National Geographic, March 2016, 30-55.

References

1. Heymann DL, ed. *Control of Communicable Diseases Manual*. 20th ed. Washington, DC: American Public Health Association; 2015.
2. Rocourt J, Moy G, Vierk K, et al. *The Present State of Foodborne Diseases in OECD Countries*. Geneva, Switzerland: World Health Organization; 2003.
3. Altekruse SF, Cohen ML, Swerdlow DL. Emerging foodborne diseases. *Emerg Infect Dis*. 1997;3:285–293.
4. McMeekin TA, Brown J, Krist K, et al. Quantitative microbiology: a basis for food safety. *Emerg Infect Dis*. 1997;3:541–549.
5. Centers for Disease Control and Prevention. Preliminary FoodNet data on the incidence of infection with pathogens transmitted commonly through food—10 states, 2009. *MMWR*. 2010;59:418–422.
6. Gist GL. Emerging foodborne diseases and NEHA's response. *J Environ Health*. 1999;61:4,61.
7. Centers for Disease Control and Prevention. FoodNet surveillance—what is FoodNet? Available at: https://www.cdc.gov/foodnet/surveillance_pages/whatisfoodnet.htm. Accessed August 19, 2017.
8. Centers for Disease Control and Prevention. FoodNet surveillance. Active laboratory surveillance. Burden of illness. Available at: https://www.cdc.gov/foodnet/surveillance.html. Accessed August 19, 2017.
9. University of Nebraska—Lincoln. Institute of Agriculture and Natural Resources. UNL Food. Physical hazards. http://food.unl.edu/physical-hazards. Accessed May 8, 2017.
10. Morse SS. Factors in the emergence of infectious diseases. *Emerg Infect Dis*. 1995;1:7–15
11. Woteki CE, Kineman BD. Challenges and approaches to reducing foodborne illness. *Annu Rev Nutr*. 2003;23:315–344.
12. Osterholm MT. Emerging infections—another warning [editorial]. *New Engl J Med*. 2000;342:1280–1281.
13. Centers for Disease Control and Prevention. Multistate outbreak of *Salmonella* serotype Agona infections linked to toasted oats cereal—United States, April–May, 1998. *MMWR*. 1998;47:462–464.
14. Centers for Disease Control and Prevention. Salmonella. Available at: https://www.cdc.gov/salmonella/index.html. Accessed December 28, 2017.
15. Centers for Disease Control and Prevention. National Enteric Disease Surveillance. *Salmonella* annual report, 2011. Available at: https://www.cdc.gov/ncezid/dfwed/pdfs/salmonella-annual-report-2011-508c.pdf. Accessed May 8, 2017.
16. US Food and Drug Administration (FDA). *Bad Bug Book, Foodborne Pathogenic Microorganisms and Natural Toxins*. 2nd ed. *Salmonella* species. 2012.
17. Centers for Disease Control and Prevention. Multistate outbreak of *Salmonella* serotype Typhimurium infections associated with drinking unpasteurized milk—Illinois, Indiana, Ohio, and Tennessee, 2002–2003. *MMWR*. 2003;52:613–615.
18. Centers for Disease Control and Prevention. Multistate outbreaks of *Salmonella* serotype Poona infections associated with eating cantaloupe from Mexico—United States and Canada, 2000–2002. *MMWR*. 2002;51:1044–1047.
19. Van Beneden CA, Keene WE, Strang RA, et al. Multinational outbreak of *Salmonella enterica* serotype Newport infections due to contaminated alfalfa sprouts. *JAMA*. 1999;281:158–162.
20. Centers for Disease Control and Prevention. Epidemiologic notes and reports update: milk-borne salmonellosis—Illinois. *MMWR*. 1985;34:215–216.
21. White DG, Zhao S, Sudler R, et al. The isolation of antibiotic-resistant *Salmonella* from retail ground meats. *New Engl J Med*. 2001;345:1147–1154.
22. Centers for Disease Control and Prevention. Outbreak of multidrug-resistant *Salmonella* Newport—United States, January–April 2002. *MMWR*. 2002;51:545–548.
23. US Food and Drug Administration. *Bad Bug Book, Foodborne Pathogenic Microorganisms and Natural Toxins*. 2nd ed. *Clostridium botulinum*. 2012.
24. Centers for Disease Control and Prevention. Foodborne botulism—Oklahoma, 1994. *MMWR*. 1995;44:200–202.
25. Centers for Disease Control and Prevention. Foodborne botulism from eating home-pickled eggs—Illinois, 1997. *MMWR*. 2000;49:778–780.
26. US Food and Drug Administration. *Bad Bug Book, Foodborne Pathogenic Microorganisms and Natural Toxins*. 2nd ed. *Clostridium perfringens*. 2012.
27. Centers for Disease Control and Prevention. Clostridium perfringens gastroenteritis associated with corned beef served at St. Patrick's Day meals—Ohio and Virginia, 1993. *MMWR*. 1994;43:137–138, 143–144.
28. US Food and Drug Administration. *Bad Bug Book, Foodborne Pathogenic Microorganisms and Natural Toxins*. 2nd ed. *Staphylococcus aureus*. 2012.
29. Centers for Disease Control and Prevention. Outbreak of staphylococcal food poisoning associated with precooked ham—Florida, 1997. *MMWR*. 1997;46:1189–1191.
30. Centers for Disease Control and Prevention. Update: multistate outbreak of *Escherichia coli* O157:H7 infections from hamburgers—western United States, 1992–1993. *MMWR*. 1993;42:258–263.
31. Crump JA, Sulka AC, Langer AJ, et al. An outbreak of *Escherichia coli* O157:H7 infections among visitors to a dairy farm. *New Engl J Med*. 2002;347:555–560.
32. Mead PS, Slutsker L, Dietz V, et al. Food-related illness and death in the United States. *Emerg Infect Dis*. 1999;5:607–625.
33. US Food and Drug Administration. *Bad Bug Book, Foodborne Pathogenic Microorganisms and Natural Toxins*. 2nd ed. Enterohemorrhagic *Escherichia coli* (EHEC). 2012.
34. US Food and Drug Administration. *Bad Bug Book, Foodborne Pathogenic Microorganisms and Natural Toxins*. 2nd ed. *Shigella* species. 2012.
35. Altekruse SF, Stern NJ, Fields PI, et al. *Campylobacter jejuni*—an emerging foodborne pathogen. *Emerg Infect Dis*. 1999;5:28–35.
36. Centers for Disease Control and Prevention. Outbreak of *Campylobacter jejuni* infections associated with drinking unpasteurized milk procured through a cow-leasing program—Wisconsin, 2001. *MMWR*. 2002;51:548–549.
37. Beuchat LR. *Listeria monocytogenes*: incidence on vegetables. *Food Control*. 1996;7:223–228.
38. Centers for Disease Control and Prevention. Listeria (listeriosis). Available at: https://www.cdc.gov/listeria/index.html. Accessed September 7, 2017.
39. Centers for Disease Control and Prevention. Public health dispatch: outbreak of listeriosis—northeastern United States, 2002. *MMWR*. 2002;51:950–951.
40. Centers for Disease Control and Prevention. Trichinellosis. Available at: https://www.cdc.gov/dpdx/trichinellosis/index.html. Accessed May 11, 2017.
41. Centers for Disease Control and Prevention. Parasites—trichinellosis. Available at: https://www.cdc.gov/dpdx/trichinellosis/index.html. Accessed May 11, 2017.

42. Centers for Disease Control and Prevention. Trichinosis—Maine, Alaska. *MMWR*. 1986;35:33–35.
43. Centers for Disease Control and Prevention. DPDx. Taeniasis. Available at: https://www.cdc.gov/dpdx/taeniasis/index.html. Accessed May 8, 2017.
44. US Food and Drug Administration. *Bad Bug Book, Foodborne Pathogenic Microorganisms and Natural Toxins*. 2nd ed. Hepatitis A virus. 2012.
45. Hutin YJF, Pool V, Cramer EH, et al. A multistate, foodborne outbreak of hepatitis A. *New Engl J Med*. 1999;340:595–602.
46. Acheson DWK, Fiore AE. Preventing foodborne disease—what clinicians can do. *New Engl J Med*. 2004;350:437–440.
47. Centers for Disease Control and Prevention. Outbreaks of gastroenteritis associated with noroviruses on cruise ships—United States, 2002. *MMWR*. 2002;51:1112–1115.
48. Larkin M. Passengers implicated in gastroenteritis outbreaks on cruise ships. *Lancet*. 2002;360:2052.
49. Charatan F. Viral gastroenteritis sickens hundreds on cruise ships. *BMJ*. 2002;325:1192.
50. Centers for Disease Control and Prevention. Vessel Sanitation Program. Investigation update on the *Sundream, Olympia Voyager, Carnival Spirit,* and *Sun Princess*. Investigation update—February 4, 2003. Available at: https://www.cdc.gov/nceh/vsp/surv/outbreak/2003/update_feb4.htm. Accessed September 7, 2017.
51. Cramer EH, Gu DX, Durbin RE, et al. Diarrheal disease on cruise ships, 1990–2000: the impact of environmental health programs. *Am J Prev Med*. 2003;24:227–233.
52. Centers for Disease Control and Prevention. BSE (bovine spongiform encephalopathy, or mad cow disease). Available at: https://www.cdc.gov/prions/bse/. Accessed May 11, 2017.
53. Centers for Disease Control and Prevention. Marine environments. Available at: https://www.cdc.gov/habs/illness-symptoms-marine.html. Accessed September 8, 2017.
54. US Food and Drug Administration. *Bad Bug Book, Foodborne Pathogenic Microorganisms and Natural Toxins*. 2nd ed. Mushroom toxins: amanitin, gyromitrin, orellanine, muscarine, ibotenic acid, muscimol, psilocybin, coprine. 2012.
55. US Food and Drug Administration. Supporting document for recommended maximum level for lead in candy likely to be consumed frequently by small children. Available at: https://www.fda.gov/Food/FoodborneIllnessContaminants/Metals/ucm172050.htm. Accessed May 8, 2017.
56. Nigra AE, Nachman KE, Love DC, et al. Poultry consumption and arsenic exposure in the U.S. population. *Environ Health Perspect*. 2017; 125:370–377.
57. Fattore E, Fanelli R, La Vecchia C. Persistent organic pollutants in food: public health implications. *J Epidemiol Community Health*. 2002;56:831–832.
58. Pesticides in baby food. *Environ Health Perspect*. 1995; 103:1082–1083.
59. Centers for Disease Control and Prevention. Epidemiologic notes and reports aldicarb food poisoning from contaminated melons—California. *MMWR*. 1986;35:254–258.
60. McEwen SA, Fedorka-Cray PJ. Antimicrobial use and resistance in animals. *Clin Infect Dis*. 2002;34(suppl 3):S93–S106.
61. Greenlees KJ. Animal drug human food safety toxicology and antimicrobial resistance—the square peg. *Int J Toxicol*. 2003;22:131–134.
62. Council on Scientific Affairs, American Medical Association. Biotechnology and the American agricultural industry. *JAMA*. 1991;265:1429–1436.
63. Louis HE, Chassy BM, Ochanda JO. Genetically modified food crops and their contribution to human nutrition and food quality. *Trends Food Sci Technol*. 2003;14:191–209.
64. Jones L. Science, medicine and the future: genetically modified foods. *BMJ*. 1999;318:581–584.
65. EH Update: NRC reports on concerns about animal biotechnology. *J Environ Health*. 2003;65(6):42.
66. World Health Organization. Frequently asked questions on genetically modified foods. Available at: http://www.who.int/foodsafety/areas_work/food-technology/faq-genetically-modified-food/en/. Accessed January 1, 2018.
67. Kishore GM, Shewmaker C. Biotechnology: enhancing human nutrition in developing and developed worlds. *Proc Natl Acad Sci USA*. 1999;96:5968–5972.
68. MedlinePlus. Medical encyclopedia: food additives. Available at: http://medlineplus.gov/ency/article/002435.htm. Accessed May 11, 2017.
69. US Food and Drug Administration. CFR—Code of Federal Regulations Title 21—Food and Drugs. Available at: http://www.accessdata.fda.gov/scripts/cdrh/cfdocs/cfcfr/CFRSearch.cfm?fr=182.3798. Accessed May 8, 2017.
70. International Food Information Council (IFIC) Foundation and US Food and Drug Administration. *Food Ingredients and Colors*. Washington, DC: International Food Information Council (IFIC) Foundation and US Food and Drug Administration; 2010:2–3. Available at: https://www.fda.gov/downloads/Food/IngredientsPackagingLabeling/ucm094249.pdf. Accessed May 12, 2017.
71. Merrill RA. Food safety regulation: reforming the Delaney Clause. *Annu Rev Public Health*. 1997;18:313–340.
72. US Food and Drug Administration. FDA History—Part II: The 1938 Food, Drug, and Cosmetic Act. Available at: http://www.fda.gov/AboutFDA/WhatWeDo/History/Origin/ucm054826.htm. Accessed May 11, 2017.
73. US Department of Health and Human Services, Food and Drug Administration. Proposed rules: substances generally recognized as safe. *Federal Register*. 1997;62(74):18937–18964.
74. Lee G. In food safety changes, victories for many. *The Washington Post*. July 28, 1996:A4.
75. US Environmental Protection Agency. Laws & regulations. Summary of the Food Quality Protection Act. Available at: https://www.epa.gov/laws-regulations/summary-food-quality-protection-act. Accessed May 8, 2017.
76. United Nations Food and Agriculture Organization. World markets for organic fruit and vegetables—Chapter 4: assistance to developing countries in organic agriculture. 2001. Available at: http://www.fao.org/docrep/004/y1669e/y1669e0o.htm. Accessed April 27, 2017.
77. United States Department of Agriculture, Food Safety and Inspection Service. About FSIS. Available at: https://www.fsis.usda.gov/wps/portal/informational/aboutfsis. Accessed May 8, 2017.
78. US Food and Drug Administration, Center for Food Safety and Applied Nutrition. CFSAN—What we do. Available at: https://www.fda.gov/aboutfda/centersoffices/officeoffoods/cfsan/whatwedo/default.htm. Accessed May 8, 2017.
79. Centers for Disease Control and Prevention. CDC and food safety. March 2016. Available at: https://www.cdc.gov/foodsafety/pdfs/cdc-and-food-safety.pdf. Accessed May 8, 2017.
80. California Department of Public Health, Food and Drug Branch. Available at: https://www.cdph.ca.gov/Programs/CEH/DFDCS/Pages/FoodandDrugBranch.aspx. Accessed May 8, 2017.

81. City of Long Beach, Health and Human Services. Consumer protection food facility inspection program. Available at: http://www.longbeach.gov/health/services/directory/food-program/. Accessed May 9, 2017.
82. Los Angeles County Department of Public Health. Environmental Health: District Surveillance and Enforcement Branch. Available at: http://publichealth.lacounty.gov/eh/DSE/aboutDSE.htm. Accessed May 9, 2017.
83. Riverside County Community Health Agency.Department of Environmental Health. Available at: http://www.rivcoeh.org/About-Us/who. Accessed May 9, 2017.
84. Collins JE. Impact of changing consumer lifestyles on the emergence/reemergence of foodborne pathogens. *Emerg Infect Dis*. 1997;3:471–479.
85. US Environmental Protection Agency. Food irradiation. Available at: https://www3.epa.gov/radtown/food-irradiation.html. Accessed September 9, 2017.
86. Wood OB, Bruhn CM. Position of the American Dietetic Association: food irradiation. *J Am Diet Assoc*. 2000;100:246–253.
87. Tauxe RV. Food safety and irradiation: protecting the public from foodborne infections. *Emerg Infect Dis*. 2001;7 (3 suppl):516–521.
88. Osterholm MT, Potter ME. Irradiation pasteurization of solid foods: taking food safety to the next level. *Emerg Infect Dis*. 1997;3:575–577.
89. US Food and Drug Administration, Center for Food Safety and Applied Nutrition. *Managing Food Safety: A Manual for the Voluntary Use of HACCP Principles for Operators of Food Service and Retail Establishments*. OMB Control No. 0910-0578. College Park, MD: US Food and Drug Administration, Center for Food Safety and Applied Nutrition; 2006:9–11.
90. Centers for Disease Control and Prevention. Multistate and nationwide foodborne outbreak investigations: a set-by-step guide. Available at: http://www.cdc.gov/foodsafety/outbreaks/investigating-outbreaks/investigations/index.html. Accessed May 9, 2017.
91. United States Department of Agriculture, Food Safety and Inspection Service. Career opportunities: food inspector & consumer safety inspector positions. Available at: https://www.fsis.usda.gov/wps/portal/fsis/topics/careers/opportunities-and-types-of-jobs/food-inspector-and-consumer-safety-inspector. Accessed May 9, 2017.

CHAPTER 12

Solid and Liquid Wastes

LEARNING OBJECTIVES

By the end of this chapter the reader will be able to:

- Describe trends in the production of solid waste.
- Summarize methods for source reduction of solid waste.
- Discuss the role of landfills for disposing of solid waste.
- Describe methods for primary, secondary, and tertiary sewage treatment.
- Discuss hazards of poorly designed solid waste disposal sites and improperly processed sewage.

Introduction

This chapter covers two crucial topics for modern society—disposal of solid wastes and treatment and processing of liquid wastes. The first part will examine the history of solid waste disposal, challenges caused by the growing volume of waste, and advantages and disadvantages of various methods for waste disposal. A related theme will be a review of proposed solutions for reduction in solid waste (e.g., source reduction, recycling, and composting). The growing environmental impact of the disposal of hazardous materials and wastes will be reviewed. Finally, the second part of the chapter will cover the topic of disposal of liquid wastes, including sewage and animal wastes.

Definitions of Waste

The term **waste** is used to describe materials that are perceived to be of negative value.[1] Why should citizens be concerned about waste disposal? One reason is that acute problems can occur if wastes are not disposed of properly. A major concern for waste disposal is the pollution that can result from the process of disposal.[1] The term **hazardous waste** refers to "[b]y-products of society that can pose a substantial or potential hazard to human health or the environment when improperly managed. [Hazardous waste] [p]ossesses at least one of four characteristics (ignitability, corrosivity, reactivity, or toxicity), or appears on special EPA lists."[2] The term **pollution** "can be defined as the introduction into the natural environment by humans of substances, materials, or energy that cause hazards to human health, harm to living resources and ecological systems, damage to structures and amenities or that interfere with the legitimate uses of the environment."[1(p72)]

Continuing increases in the volume of waste are exacerbating the difficult problem of waste disposal; in some regions of the world, including parts of the United States, space reserved for dump sites is being rapidly depleted. Consequently, during the 21st century much work will need to be done by the global society to cope with its growing output of solid wastes. In addition, a particularly acute problem is the fact that the burgeoning populations of less developed countries are introducing large amounts of liquid wastes directly into waterways and oceans without adequate processing. The presence of such wastes endangers already limited drinking water supplies as well as the viability of aquatic environments.

Historical Background—Solid Waste Disposal

The concept of waste management is relatively recent; until the 20th century, the individual was responsible for discarding wastes. Often wastes were discarded carelessly into the environment. The efficient processes that have been realized in the United States and other developed countries were not available early in the 20th century. Since that time, the waste management industry in the United States has evolved to the point where many individuals discard waste without thinking about its ultimate fate. Although some people resent contemporary legislative proposals directed toward abatement of wastes and penalties that are levied against those persons and industries that discharge wastes illegally, almost all people dislike pollution. It is ironic that at the same time, some of the same people who dislike pollution do not recognize that they contribute to the pollution problem through their own consumption practices.[1] For example, foods are wasted and recyclable items are thrown into the trash pile.

New York City, Early 20th Century

The following passage provides a graphic description of the status of solid waste disposal in New York City during the late 19th and early 20th century:

> Until the years following the Civil War, New York City had no municipal health department, no sanitation department, no public water system that could provide pure water to a population that was literally doubling in size every 10 to 15 years. The city was powered, quite literally, by horses—well over 100,000 of them—each depositing between 20 and 25 pounds of manure on the city's streets each day. Dead horses, pigs, goats, and other animals littered the streets and mixed with the overflow of outdoor privies located behind the tenements, shacks, and shanties that housed the city's poor. In 1880, 15,000 horse carcasses were removed from the city's streets; by 1912, over 20,000 horses, cattle, mules, and donkeys were being removed.[3(p14)]

The early methods for waste disposal in New York City were quite primitive and sometimes consisted of dumping garbage into rivers. The sketches and cartoons in **FIGURE 12.1** depict unsanitary conditions in New York City during the mid- to late-19th century.

The Growing Solid Waste Problem

Trash or garbage also is called **municipal solid waste (MSW)**. In 2014, the United States—residents, businesses, and institutions—produced approximately 258 million tons (234 million metric tons) of MSW (before recycling). This amount translated to approximately 4.44 pounds (2.0 kilograms) of waste per person per day, which was almost twice the amount for 1960.[4] However, increasing amounts of MSW are being recycled or composted each year. In 2014, 89 million tons (81 million metric tons) were recycled; an additional 33 million tons (30 million metric tons) were burned for energy generation. Refer to **FIGURE 12.2** for trends in MSW generation.

Finding sites for the disposal of MSW is becoming a challenging problem as several areas of the United States, e.g., Fresh Kills landfill on Staten Island in New York City, have closed their landfills. MSW from the city is being transported to other states.[5] However, although many sites are being closed, at the same time more and more MSW is being recycled and larger regional landfills have been constructed reducing the need for additional landfills. As of 2009, about 1,900 MSW landfill sites were in use in the United States.[6]

A consequence of shutting down landfills is increases in disposal fees known as tipping fees, which increased from about $20 in 1982 to about $50 per ton in 2014. However, despite the closure of landfill sites in many states, lack of available disposal sites presently does not seem to be a problem at the national level.[4] Availability of disposal sites varies greatly according to the specific region of the country.

The problem of waste disposal is not confined to developed countries. The less developed areas of the world also have shown an increase in the amount of solid waste production. Although there are abundant examples, take the case of China. Urban residents of China have experienced a rise in their living standard that has caused a rapid increase in the annual amount of domestic garbage. Now the average city dweller in China produces about 2.5 pounds (about 1.2 kilograms) of garbage per day.[7] This growing waste problem is paralleled in other parts of the developing world.

FIGURE 12.1 A dumping ground (1866) and street scenes in New York City (1881) demonstrate unsanitary conditions.

Modified and reproduced from US National Library of Medicine, National Institutes of Health, Images from the History of Medicine. Image numbers: upper left (Dumping ground), 101436260; upper right (Happy Thought), 101435654; lower left (Hygea), 10435656; lower left (Street scene), 101435658. Available at: https://collections.nlm.nih.gov/. Accessed August 20, 2017.

Components of the Municipal Solid Waste Stream

Solid waste includes the subset of materials referred to as municipal solid waste plus other types of solid waste. The components of the municipal solid waste stream include packaging, food waste, paper, electronics, appliances, and organic materials such as yard trimmings from landscaping.[4] Usually not included as part of MSW are other types of solid waste such as materials from construction sites, sewage sludge, and some forms of industrial wastes that are not hazardous. Refer to **TABLE 12.1** for a list of materials that typically make up solid waste.

FIGURE 12.3 shows the relative proportion of materials discarded in the municipal solid waste stream. The figure illustrates that paper makes up the largest percentage of solid waste, about 27% in 2014.

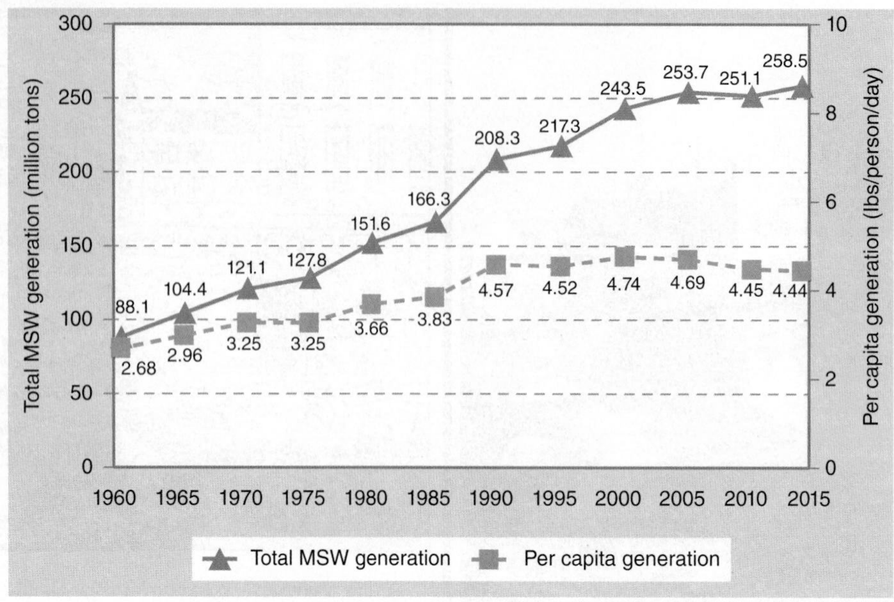

FIGURE 12.2 Municipal solid waste generation rates, 1960 to 2014.

Modified and reproduced from US Environmental Protection Agency, Office of Land and Emergency Management (5306P). Advancing Sustainable Materials Management: 2014 Fact Sheet. EPA530-R-17-01. Washington, DC: US Environmental Protection Agency; 2016, p.2.

TABLE 12.1 What Is Solid Waste?

Garbage, also known as municipal solid waste (e.g., milk cartons and coffee grounds); see Figure 12.3 for percentages of specific materials generated.
Refuse (e.g., metal scrap, wall board, and empty containers).
Sludges from waste treatment plants, water supply treatment plants, or pollution control facilities (e.g., scrubber slags).
Industrial wastes (e.g., manufacturing process wastewaters and nonwastewater sludges and solids).
Other discarded materials, including solid, semisolid, liquid, or contained gaseous materials resulting from industrial, commercial, mining, agricultural, and community activities (e.g., boiler slag).

Reproduced from US Environmental Protection Agency. Resource Conservation and Recovery Act (RCRA) Overview. Available at: https://www.epa.gov/rcra/resource-conservation-and-recovery-act-rcra-overview. Accessed February 20, 2017.

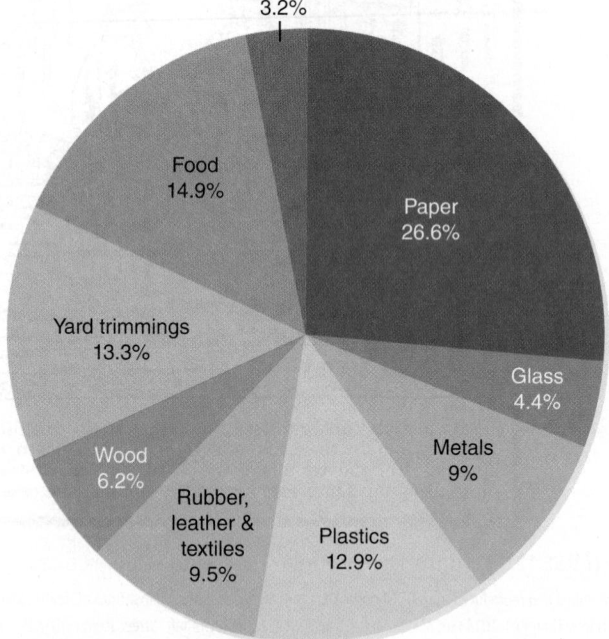

FIGURE 12.3 Total municipal solid waste generation (by material), 2014—258 million tons (234 million metric tons).

Modified and reproduced from US Environmental Protection Agency, Office of Land and Emergency Management (5306P). Advancing Sustainable Materials Management: 2014 Fact Sheet. EPA530-R-17-01. Washington, DC: US Environmental Protection Agency; 2016, p.7.

Solid Waste Management

Municipal solid waste disposal is an industrial enterprise that has four main dimensions: recycling, landfilling, composting, and combustion.[8] Refer to **FIGURE 12.4**; note that when MSW decomposes in landfills or is combusted, a by-product can be the production of useful energy. This energy may be produced from methane gas that seeps from disposal sites or by combustion.

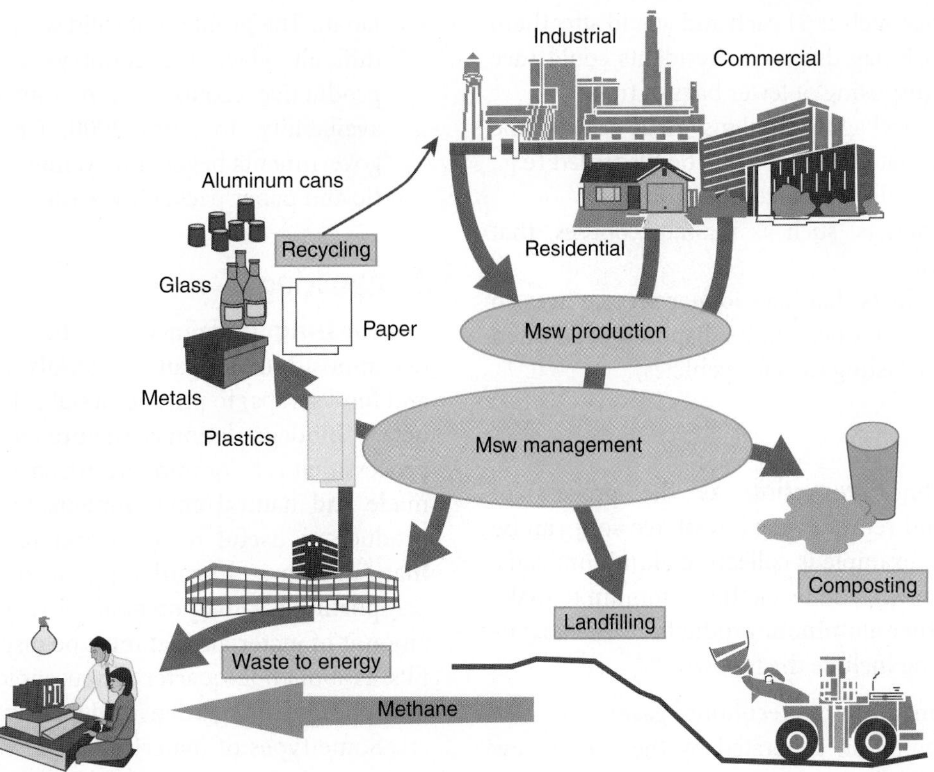

FIGURE 12.4 Major components of municipal solid waste management.

Reproduced from US Department of Energy, Energy Information Administration, Office of Coal, Nuclear, Electric and Alternate Fuels.

The US Environmental Protection Agency (EPA) has developed a hierarchy for the management of MSW.[9] Procedures at the top of the hierarchy are favored over those at the bottom. The hierarchy is as follows:

- Source reduction and reuse
- Recycling/composting
- Energy recovery
- Treatment and disposal

Source Reduction

The term **source reduction** refers to "reducing waste at the source, and is the most environmentally preferred strategy."[9] Through source reduction, the volume of solid waste that must be deposited in landfills is limited. Two important components of source reduction are *waste reduction* and *waste recycling*. Waste reduction aims to reduce the amount of waste produced at the source. Waste recycling refers to reuse of materials in the waste.[10] **FIGURE 12.5** shows some of the programs that may be used to achieve source reduction.

FIGURE 12.5 Programs for recycling wastes.

Modified and reproduced from US Environmental Protection Agency. Recycle City, Dumptown Game. Available at: https://www3.epa.gov/recyclecity/gameintro.htm. Accessed June 7, 2017.

The following are examples of source reduction methods that prevent waste from entering disposal channels:

- Financial incentives to homeowners, for example, PAYT, an acronym that refers to "pay as you throw." Municipalities that use the PAYT system require that the amount that residents pay for trash pickup is proportional to the amount of trash that they generate.[11] To illustrate, in a hypothetical PAYT system, residents would purchase stickers for a set price such as $1 each and would affix them to each trash bag disposed. Residents could save money by disposing of fewer bags of trash.

- Improved packaging designs that reduce the amount of materials that must be discarded (e.g., the use of smaller packages for products).
- Design products, such as refillable bottles, that can be reused.
- Design products that have longer service lives so that they will not need to be disposed of so often (e.g., longer-lasting tires for vehicles).

Recycling

The EPA defines **recycling** as the process of "[c]ollecting and reprocessing a resource so it can be used again. An example is collecting aluminum cans, melting them down, and using the aluminum to make new cans or other aluminum products."[12] The advantages of recycling include the following:[13]

- Reduces emissions of greenhouse gases.
- Prevents pollution generated by the use of new materials.
- Decreases the amount of materials shipped to landfills, thereby reducing the need for new landfills.
- Preserves natural resources.
- Opens up new manufacturing employment opportunities and increases US competitiveness.
- Saves energy.

As shown in **FIGURE 12.6A**, the rate of recycling of MSW in the United States has increased from 6.4% (5.6 million tons [5.1 million metric tons]) in 1960 to 34.6% (89.4 million tons [81.1 million metric tons]) in 2014; automobile batteries (lead-acid batteries) are the products that are recycled most frequently (about 99% of auto batteries). The materials that are most amenable to recycling include automobile batteries, paper, and landscaping materials. Although they may be placed in landfills, wastes from construction sites, sludge from wastewater treatment, and industrial wastes that are nonhazardous usually are not termed MSW.[14] **FIGURE 12.6B** presents the recycling rates for various materials in 1960 through 2014.

Examples of accomplishments achieved through recycling are the following:

- Germany: A very effective program for waste management and recycling of waste is in place. The country recycled 65% of its municipal waste in 2013.[15] The German city of Freiburg has developed a noteworthy system for waste management.[10]
- Slovenia, Austria, Belgium: These three European countries had recycling rates that were 55% or higher in 2013.[15]
- Japan: The problem of solid waste disposal is very difficult, given the country's dense population, productive economy, and limited landfill space availability. In April 2000, Japanese municipal governments began a recycling program for plastic and paper packaging waste.[10]

Composting

Composting is defined as "the aerobic biological decomposition of organic materials (e.g., leaves, grass, and food scraps) to produce a stable humus-like product. . . . Biodegradation is a natural, ongoing biological process that is a common occurrence in both human-made and natural environments."[16(p7–8)] Composting produces a useful material that resembles soil and that can be used in gardening. According to the EPA, composting has the potential for reducing greatly the amount of materials that must be disposed of in landfills, as about one-quarter of household wastes consists of clippings from gardens and food waste.

Some types of materials are acceptable for composting, whereas others are not appropriate. According to the EPA, some food products such as meats may yield poor quality compost or be a magnet for vermin. If you anticipate constructing your own compost bin, refer to **TABLE 12.2** for a list (although not all-inclusive) of acceptable and unacceptable materials to include.

Composting breaks down organic materials through the action of physical and chemical processes. If you dig through your own backyard compost pile, you may notice that the temperature of the materials in the compost pile can become quite high. In compost piles that have been maintained properly, optimal temperatures can reach 32°–60°C (90°–140°F) as a result of microbial action.[14] These high temperatures are believed to kill off many harmful components such as pathogenic bacteria, weed seeds, and insect larvae. Nevertheless, a criticism of composting is that it may be insufficient to eliminate pathogenic agents present in the materials that are being composted.[1]

Various types of microorganisms are very important in composting. Other organisms such as invertebrates (e.g., beetles, sowbugs, and earthworms) aid in decomposing the materials when these living organisms have adequate oxygen and water (provided by well-designed compost bins); however, they play a lesser role in composting than do microorganisms.[14] A specific type of composting, known as vermicomposting (composting by a special kind of earthworm), puts red wigglers to work consuming food scraps and uses their castings as a rich garden fertilizer.

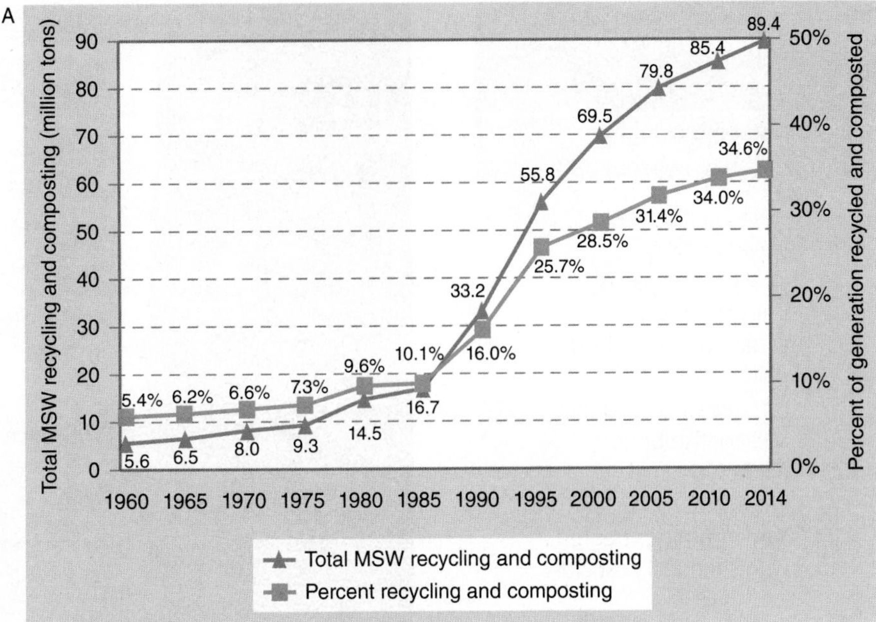

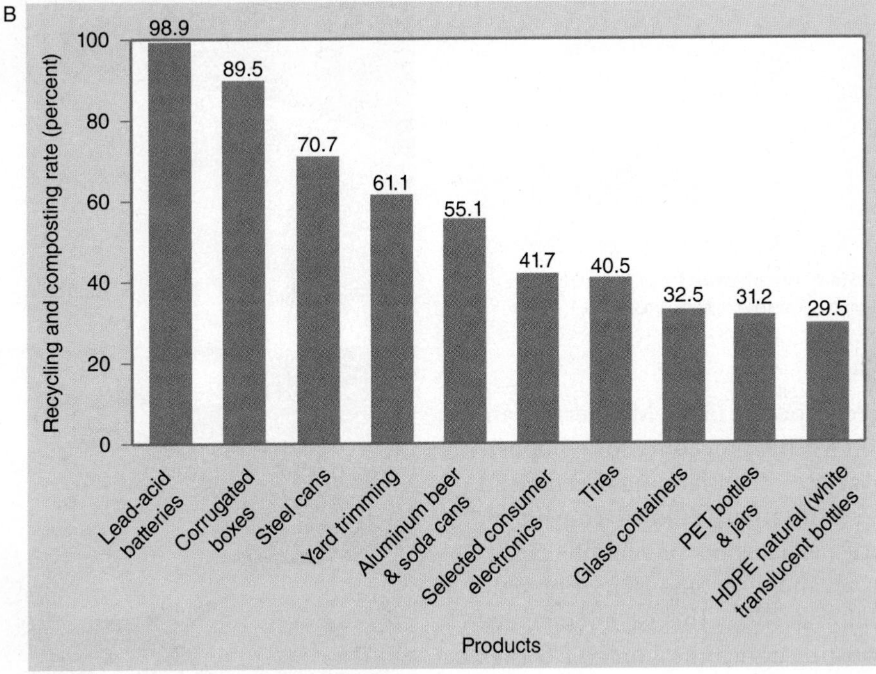

FIGURE 12.6 A. Municipal solid waste recycling rates, 1960–2014. B. Recycling rates of selected products, 2014 (not including combustion with energy recovery).

Modified and reproduced from US Environmental Protection Agency, Office of Land and Emergency Management (5306P). Advancing Sustainable Materials Management: 2014 Fact Sheet. EPA530-R-17-01. Washington, DC: US Environmental Protection Agency; 2016, p. 3, 5.

Five cities in the western United States have emerged as leaders in composting, recycling, and reduction of municipal waste. Four of them are in California (San Francisco, Los Angeles, San Jose, and San Diego); the remaining city is Portland, Oregon. Each city has been able to divert 68% or more of wastes from disposal sites. San Francisco, the leading city for recycling in the United States, has achieved an 80% reduction of wastes going to landfills. The City by the Bay has a mandatory composting program for compostable materials such as food. A California law passed in 1989 has established a 75% recycling goal by 2020.[17] **FIGURE 12.7** shows a citizen at work in a compost program and an example of a compost bin.

TABLE 12.2 Materials to Include and Exclude in Compost

Materials to Include	Materials to Exclude
Cardboard	Black walnut tree leaves or twigs
Paper	Coal or charcoal ash
Coffee grounds and filters	Dairy products
Cotton rags	Diseased or insect-ridden plants
Dryer and vacuum cleaner lint	Fats, grease, lard, or oils
Eggshells	Meat or fish bones and scraps
Fireplace ashes	Pet wastes
Fruits and vegetables	Yard trimmings treated with chemical pesticides
Grass clippings	
Hair and fur	
Hay and straw	
Houseplants	
Leaves	
Nut shells	
Sawdust	
Shredded newspaper	
Tea bags	
Wood chips	
Wool rags	
Yard trimmings	

Modified and reproduced from US Environmental Protection Agency. Composting at home. Available at: https://www.epa.gov/recycle/composting-home. Accessed June 15, 2017.

A

B

FIGURE 12.7 A. Example of a composting bin. B. Citizen putting organic materials into a composting bin.

Part A: © Harperdrewart/Dreamstime.com. Part B: © Gale Verhague/Dreamstime.com

The island of Crete, located in the Mediterranean, has developed a noteworthy procedure for composting organic solid wastes.[18] The island has been faced with increasing desertification (gradual conversion of land into desert), exhaustion of landfill capacity for waste disposal, and a chronic lack of organic material for agricultural soils. The island operates programs for composting agricultural organic wastes (e.g., olive leaves, branches, and pressed grape skins), manure, and sewage sludge. The resulting compost will have potential applications for agriculture and landscaping.

Landfills

Materials that cannot be recycled or composted need to be deposited in landfills. A landfill is composed of four major parts: a bottom liner, a system for collecting leachates, a cover, and an appropriate location that minimizes the contamination of groundwater by materials released from the site.[19] The term **leachate** refers to "[w]ater that collects contaminants as it trickles through wastes, pesticides or fertilizers. Leaching may occur in farming areas, feedlots, and landfills, and may result in hazardous substances entering surface water, ground water, or soil."[20] The typical design of a landfill is as follows: the bottom is lined with a dense layer of clay and sealed with thick plastic sheeting to contain leaks of hazardous materials.[21] A flexible membrane liner is designed to hold in toxic chemicals

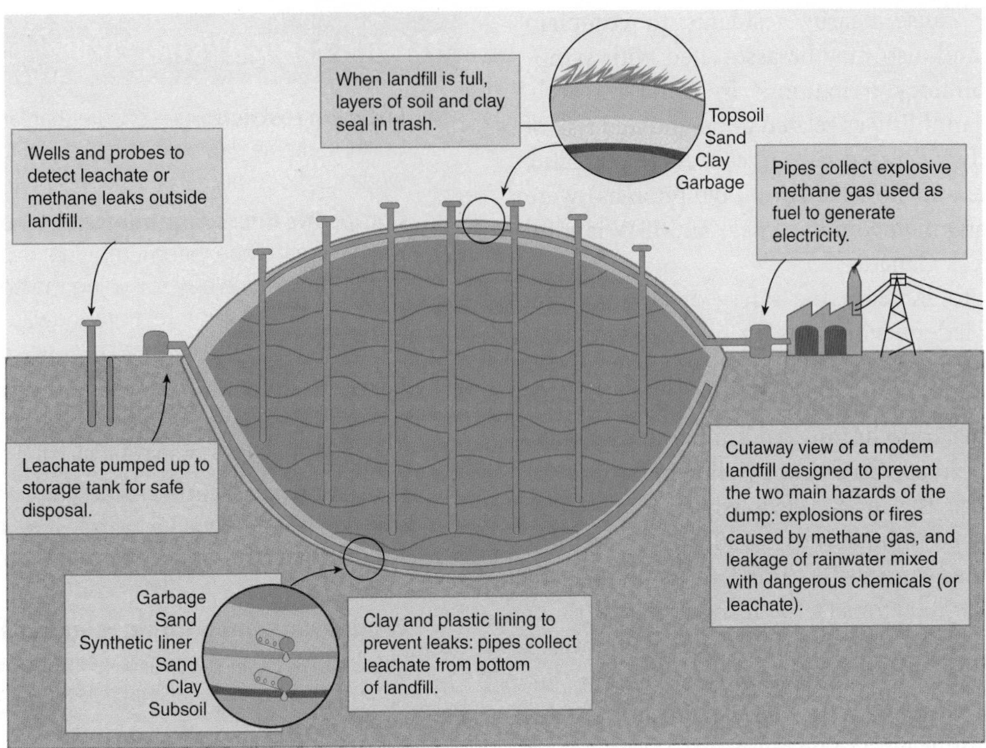

FIGURE 12.8 What is a municipal solid waste landfill?

Modified and reproduced from US Environmental Protection Agency. Municipal Solid Waste Landfills. Available at: https://www.epa.gov/landfills/municipal-solid-waste-landfills. Accessed June 8, 2017.

that might contaminate groundwater. A leachate sump collects leachates, which then can be subjected to further treatment. Garbage is piled up in rows that are 10 to 20 feet (3 to 6 meters) high; bulldozers and rollers aid in compacting the garbage; at the end of the day, the newly added garbage is covered with soil and other materials. After the garbage is covered, anaerobic bacteria aid in the decomposition of organic materials and produce methane gas. **FIGURE 12.8** shows the profile of a landfill.

An example of an operating landfill is shown in **FIGURE 12.9**. The bulldozer is in the process of crushing the garbage and covering it with soil. This procedure protects the site from seagulls and other scavengers that feed on garbage and can disseminate microbial pathogens into the environment. Newer designs incorporate a method for collecting leachates and disposing of them in an environmentally safe manner. Hazardous wastes such as pesticides and biological materials from hospitals should not be disposed of in MSW waste sites unless those sites are designed to accept such materials, which otherwise need to be processed separately.

Dangers associated with landfills, especially the older designs that were not well sealed against leakage, include air pollution and groundwater contamination.[22] Considerable amounts of leachate may be generated from landfill sites, even in arid countries,

FIGURE 12.9 A municipal solid waste landfill site.

such as Kuwait.[23] Hazardous wastes in landfills may contain toxic heavy metals and old batteries that have been discarded.[24] These materials may contaminate the environment by leaking out. Solvents and cleaning agents are among the other chemicals that may enter groundwater from leachates. MSW landfills generate methane, volatile organic compounds (VOCs), and other gases.[25] Methane vented from landfills poses a fire hazard and is a greenhouse gas.

Some VOCs are hazardous because of their potential carcinogenic properties. Emissions of VOCs from landfills may cause nearby residents to complain about odors and also may be associated with symptoms of respiratory irritation.[25] In some research, proximity to landfills was related to heightened risk of cancer. Slightly increased risks of cancers (pancreatic, liver, prostate, and non-Hodgkin's lymphomas) were reported among men who lived near an MSW landfill site in Montreal, Canada.[26]

In order to increase the safety of landfills, the EPA adopted federal standards for their construction in 1988.[22] The newer designs that have been implemented reduce leaching of chemicals from landfill sites, minimize air pollution, and limit access by scavengers that could spread disease. Following improved safety standards, the small, previously existing 10,000 landfills were combined into 3,500 larger landfills, which are called **megafills**. These megafills take in from 5,000 to 10,000 tons (4,536 to 9,072 metric tons) of trash per day and serve regional needs for waste disposal; some megafills, such as those located in Virginia, serve the disposal needs of several nearby states. The larger landfills tend to be more cost effective than incinerators for disposal of solid waste; consequently, landfills now are used for the disposal of most of the solid waste produced in the United States. Federal standards for the design of landfills are shown in the following text box.

Incineration

Another method for disposal of solid waste is incineration, which can be used to generate energy while at the same time reducing the volume and weight of waste. Incineration for disposal of MSW takes place in facilities such as the one shown in **FIGURE 12.10**. This plant has the capacity to burn about 2,250 tons (2,041 metric tons) of MSW per day at temperatures of 2,500 degrees Fahrenheit (1,371 degrees Celsius). No attempt is made to separate the trash into components; at this high temperature glass and aluminum in the trash melt. Metals from the residues of combustion can be recycled into scrap metal; remaining ash is deposited in landfills. The plant has sufficient electric-generating capacity for nearly 40,000 homes.[27]

Although the method of incineration of solid wastes can produce valuable energy, some facilities generate hazardous emissions.[1] The substances contained in these emissions may be hazardous to human health and to the environment. Examples of toxic materials that may be emitted from incinerators include polychlorinated biphenyls (PCBs), cadmium, and heavy metals, as well as other potentially carcinogenic substances. These materials may

FEDERAL LANDFILL STANDARDS

- **Location restrictions**—ensure that landfills are built in suitable geological areas away from faults, wetlands, flood plains, or other restricted areas.
- **Composite liners requirements**—include a flexible membrane (geomembrane) overlaying two feet of compacted clay soil lining the bottom and sides of the landfill, protect groundwater and the underlying soil from leachate releases.
- **Leachate collection and removal systems**—sit on top of the composite liner and remove leachate from the landfill for treatment and disposal.
- **Operating practices**—include compacting and covering waste frequently with several inches of soil to help reduce odor; control litter, insects, and rodents; and protect public health.
- **Groundwater monitoring requirements**—require testing groundwater wells to determine whether waste materials have escaped from the landfill.
- **Closure and postclosure care requirements**—include covering landfills and providing long-term care of closed landfills.
- **Corrective action provisions**—control and clean up landfill releases and achieve groundwater protection standards.
- **Financial assurance**—provides funding for environmental protection during and after landfill closure (i.e., closure and postclosure care).

Modified from US Environmental Protection Agency. Municipal solid waste: landfills. Available at: https://www.epa.gov/landfills/municipal-solid-waste-landfills. Accessed March 1, 2017.

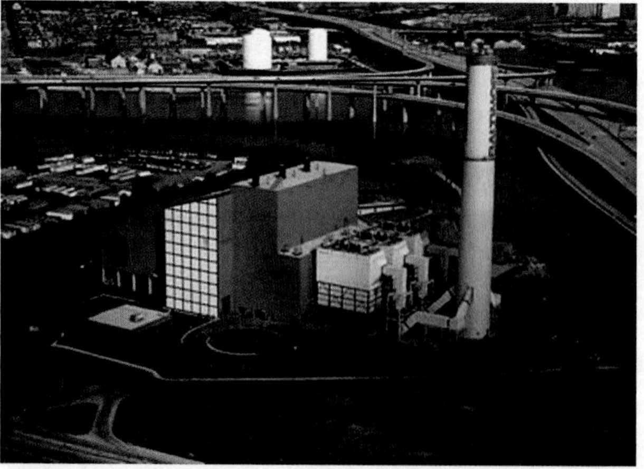

FIGURE 12.10 Plant for conversion of trash into energy in Baltimore, Maryland.

Reproduced from Energy Information Administration, Energy Kids. Baltimore RESCO waste-to-energy plant. Available at: https://www.eia.gov/Kids/energy.cfm?page=RESCOE_Plant. Accessed June 8, 2017.

cause air pollution or be deposited on the land.[28] In a waste-to-energy plant, pollutants in combustion gases are removed by air pollution control technologies;[29] examples are scrubbers and filters. High temperatures deactivate pathogens and some harmful chemicals.

Disposal of Hazardous Materials and Wastes

Hazardous wastes are disposed of legally in the United States by discarding them on the surface of the land, storing them in slurry ponds, dumping them into landfills or into the ocean, and incinerating them.[30] Some waste disposal sites used before the 1970s were designed poorly; untreated wastes were dumped into open pits or deposited onto land. Other wastes were placed in illegal dumping sites via a practice called "midnight dumping." Examples of contaminated lands are urban abandoned properties, old industrial sites, and mines that are no longer in use. Most contaminated land came about from antiquated practices.[31]

Hazardous wastes originate from the following sources (this list is not exhaustive):

- *Hazardous materials used in the home*—leftover pesticides, cleaning products, automotive products, painting supplies, and other toxic, reactive, flammable, and nonflammable products.[32] Almost every home contains products that incorporate hazardous substances.[33]
- *Medical waste*—in addition to general waste, medical waste may include chemicals, infectious agents, and radioactive materials.[34]
- *Industrial hazardous waste*—examples include toxic chemicals, solvents, heavy metals from plating operations, and residues from the manufacture of pesticides.
- *Radioactive waste*—materials from spent nuclear fuel and tailings from uranium processing.
- *Mining wastes and extraction wastes*—toxic chemicals left over from mining operations include acids and heavy metals.

Scope of the Hazardous Waste Problem

Worldwide, an enormous quantity of hazardous wastes are produced each year—more that 400 million tons (363 million metric tons) according to some estimates.[35] These toxic substances are being produced in increasing amounts everywhere in the world.[36] For example, India is concerned about the fact that industrial and hazardous wastes are being deposited on the land in an uncontrolled fashion, causing increasing levels of pollution.[37]

The developed world produces most of the toxic wastes. In order to dispose of hazardous materials, industrialized countries have adopted the practice of shipping them to developing countries that welcome cash payment for accepting the shipments. Of concern for the environment is that many developing nations do not have adequate facilities for safe disposal of hazardous wastes, which may endanger the health of the local population and the ecosystem. Several notorious episodes of clandestine shipments of toxic wastes have gained public attention:

> Among these were a number of so-called "toxic cargo" incidents involving ships loaded with hazardous waste that were forced to travel from one international port to another in search of a final disposal site. For example, a vessel named the *Khian Sea* departed from Philadelphia, Pennsylvania, in August 1986 loaded with 14,000 tons of municipal incinerator ash. After dumping 4,000 tons of the ash on a beach in Haiti . . . , the ship plied the waters of five continents for 27 months looking for a country that would accept the remainder of its cargo, most of which is now suspected of having been dumped in the Indian Ocean.[37(pA411)]

In 1989, a treaty titled the Basel Convention on the Control of Transboundary Movements of Hazardous Wastes and Their Disposal came into effect under the sponsorship of the United Nations. The purpose of the treaty is to regulate international transport of hazardous wastes and promote the disposal of such wastes in a safe manner. Of continuing concern is the international trade of hazardous recyclable materials such as lead-acid batteries, which have great commercial value and are being shipped to developing countries. The receiving nations, in many cases, do not possess safe methods for extracting materials from recyclables.

Uncontrolled Hazardous Waste Sites in the United States

Toxic materials contained in uncontrolled hazardous waste sites raise major concerns for public health and the environment in the United States.[38] At least 40,000 of these sites have been reported to federal agencies; as of March 2010, about 1,300 were on the National Priorities List (discussed in

FIGURE 12.11 A. Toxic waste dump, Superfund site. B. Workers in protective clothing to guard against contamination from hazardous waste.

Courtesy of CDC

the chapter on toxic metals and elements). Superfund legislation (described in the chapter on environmental policy and regulation) mandates the cleanup of hazardous waste sites in the United States. Stakeholders from many sectors of the US population have expressed their concerns about hazardous substances that may be released into the environment from uncontrolled hazardous waste sites. The environmental impacts of releases of toxic materials from hazardous waste sites include potential adverse human health effects, the high costs of cleanup, reductions in property values, and various types of potential ecological damage.[38] **FIGURE 12.11** illustrates a Superfund dump site for hazardous materials. Also shown in the figure are hazardous waste workers dressed in protective clothing.

Environmental Migration of Hazardous Wastes

Toxic chemicals present in hazardous wastes may enter the drinking water supply and, through biomagnification, may enter the food chain. According to one expert,

> Solid wastes deposited in terrestrial environments are subject to leaching by surface and ground waters. Leachates may then be transported to other surface waters and drinking water aquifers through hydrologic transport. Leachates also interact with natural organic matter, clays, and microorganisms in soils and sediments. These interactions may render chemical constituents in leachates more or less mobile, possibly change chemical and physical forms, and alter their biologic activity. Oceanic waste disposal practices result in migration through diffusion and ocean currents. . . . Sediments serve as major sources and sinks of chemical contaminants. Food chain transport in both aquatic and terrestrial environments results in the movement of hazardous chemicals from lower to higher positions in the food web.[30(p295)]

The foregoing pathways for transport of hazardous chemicals are shown in **FIGURE 12.12**.

Love Canal

The Love Canal toxic waste site discussed in this section illustrates some of the concepts presented in Figure 12.12. (Although it is not an ideal example of all the concepts presented in the figure, the Love Canal incident represents a situation in which hazardous wastes entered the water, air, and soil.)

The Love Canal incident is significant to the field of environmental health because it is identified with hazardous chemical exposures and their possible harmful influences on human health.[39] The Love Canal, located near Niagara Falls, New York, was excavated in the 1890s for a proposed hydroelectric power project but never was completed. The Occidental Chemical Corporation, formerly known as Hooker Chemicals and Plastics, used the site beginning in 1942 for the disposal of approximately 20,000 tons (18,144 metric tons) of toxic waste sealed in metal drums. The chemical wastes included dioxin, pesticides, and other organic compounds.

When the disposal of toxic chemicals ended 10 years later in 1952, the Love Canal was covered and subsequently deeded to the Niagara Falls Board of Education. The new land that was created then became an area for construction of schools and homes. From the 1960s to the late 1970s, the presence of numerous toxic chemicals was observed near the former landfill site. Consequently, families who resided on the site

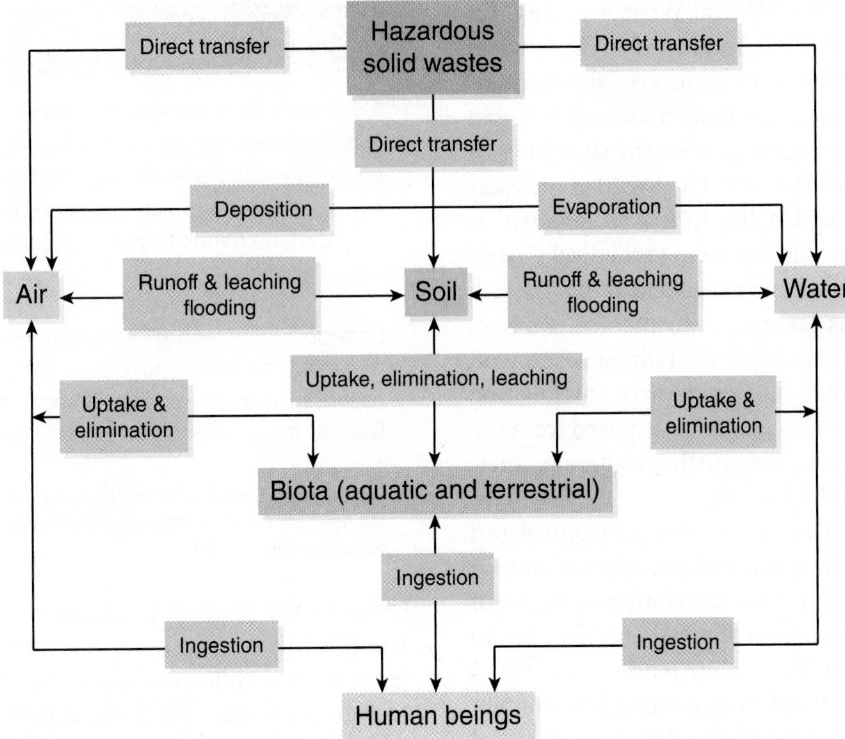

FIGURE 12.12 Major physical and biological transport pathways of hazardous chemicals derived from solid waste disposal.

Reproduced from Van Hook RI. Transport and transportation pathways of hazardous chemicals from solid waste disposal. *Environ Health Perspect*. 1978;27:296.

were evacuated, and officials declared a national emergency. Residents of the immediate area complained of adverse health outcomes such as miscarriages, birth defects, cancer, and urinary tract diseases.

One of the outcomes of the Love Canal incident was the creation of the Superfund, administered by the EPA. According to Superfund regulations, responsible parties must assume liability for cleanup of environmental hazards that they cause. Both the Occidental Chemical Corporation and the city of Niagara Falls paid more than $20 million to the former residents of the Love Canal site. More than $200 million was paid to New York State and the federal government for cleanup of the site. In the 1990s the site was declared free from toxic waste and the original neighborhood was repopulated.[40–42]

Medical Waste

Medical waste is defined as "a subset of wastes generated in healthcare facilities, such as hospitals, physicians' offices, dental practices, blood banks, and veterinary hospitals/clinics as well as medical research facilities and laboratories."[43] Toward the end of the 20th century, the United States produced more than 3.5 million tons (3.2 million metric tons) of medical waste annually.[34] Some examples of medical waste are shown in **TABLE 12.3**.

TABLE 12.3 Examples of Healthcare Waste

■ Infectious waste: waste contaminated with blood and other bodily fluids (e.g., discarded diagnostic samples, bandages, and disposable medical devices)
■ Pathological waste: e.g., human tissues, contaminated animal carcasses
■ Sharps: e.g., syringes, scalpels
■ Chemicals: e.g., solvents, mercury from broken thermometers
■ Pharmaceuticals: e.g., drugs, vaccines
■ Genotoxic waste: e.g., highly hazardous cancer treatment drugs
■ Radioactive waste: e.g., materials contaminated by radionuclides from diagnostic material and radiotherapeutic materials
■ Nonhazardous general waste

Modified from World Health Organization. Healthcare waste. Fact sheet No. 253, updated November 2015. Available at: http://www.who.int/mediacentre/factsheets/fs253/en/. Accessed June 13, 2017.

Congress enacted the Medical Waste Tracking Act (MWTA) of 1988, which expired in 1991. The law was created in response to the public's concerns about potentially hazardous medical wastes such as used syringes and needles that were washing up on the shorelines of several Eastern states. The MWTA provided a 2-year time period during which the EPA was directed to develop regulations for management of medical wastes. When the MWTA ended, the EPA decided that the risk of medical waste was far greater to medical workers than to the general public. Since 1991, the states instead of the EPA have assumed the primary role of regulating medical wastes by applying guidance acquired from the MWTA. Individual states have further elaborated their own programs for medical wastes.[43]

According to the Centers for Disease Control and Prevention (CDC), the levels and amounts of microbial agents in medical wastes are similar to those of household waste.[44] Consequently, it is permissible to deposit medical waste in sanitary landfills, provided that workers are prevented from coming into contact with the waste during disposal. In addition to disposal in landfills, medical waste may be decontaminated by chemicals, incinerated, irradiated, or subjected to high temperatures (autoclaving).[44]

As noted, the greatest danger from medical wastes is to those who are occupationally exposed at the point where they are produced; in comparison, the general population has a much lower risk of disease.[43] One of the hazards associated with medical wastes is potential injury by *sharps* (a term that refers to sharp objects such as used needles), which can transmit infectious agents via accidental needle sticks. Rules for disposal of sharps vary according to the locality where they are used. Some methods for safe disposal of sharps include special collection sites, household hazardous waste collection programs, and services for gathering residential special waste.[45]

Safe Disposal of Hazardous Wastes

Hazardous wastes may be held temporarily in storage at waste transfer stations. At the site where they are produced, hazardous wastes are transferred by trucks or rail tank cars to waste storage facilities, which use containers, tanks, or containment buildings. Some communities offer programs for the collection of household hazardous wastes so that they can be disposed of in an environmentally friendly way. These programs help to reduce the amount of hazardous wastes that are dumped down the drain or into landfills that are not designed for the disposal of toxic materials.

Use of land disposal units is the most frequently employed method for disposal of hazardous wastes (refer to the preceding text box).[46] Hazardous waste landfills must be designed with heavy liners and systems for the collection of leachates in order to protect groundwater. Typically, liquid wastes are placed in hazardous waste injection wells, into which technicians inject wastes deep underground at extremely high pressures. Liquid wastes in injection wells are also regulated by the Safe Drinking Water Act, because such wastes may impact groundwater.

LAND DISPOSAL OF TOXIC WASTES

"Land disposal means the placement in or on the land,... [Examples include] placement in a landfill, surface impoundment, waste pile, injection well, land treatment facility, salt dome formation, salt bed formation, underground mine or cave, or placement in a concrete vault, or bunker intended for disposal purposes."

Congress has mandated the treatment of hazardous wastes in an acceptable manner prior to their disposal; uncontrolled disposal of untreated hazardous wastes is forbidden because of potential hazards to our health and environment.

Reproduced from US Environmental Protection Agency. Land disposal restrictions for hazardous waste. Available at: https://epa.gov/hw/land-disposal-restrictions-hazardous-waste. Accessed June 13, 2017.

Sewage Processing and Disposal

The remainder of this chapter covers methods for processing and disposal of sewage. Sewage denotes "[t]he waste and wastewater produced by residential and commercial sources and discharged into sewers."[47] In order to prevent the spread of disease and damage to the environment, municipalities in developed countries pipe sewage to sewage treatment plants, which process sewage by microbial digestion, filtration, aeration, and addition of chemicals."[47]

The methods for sewage disposal evolved over time from the advances of the ancient world to declines in use during the Middle Ages to the innovations of the modern era. One of the major achievements of the ancient Romans was transportation of water in large volumes through the use of aqueducts.[48] A network of aqueducts carried fresh water into Rome. The Romans developed technologically advanced sewer systems that in many respects were superior to sewer disposal practices used during the Middle Ages. The Romans constructed a system of underground tunnels for transporting sewage to the Tiber River.

The Middle Ages were a time of regression for the sanitary disposal of sewage. In medieval cities, the populace casually discarded human waste into the streets, where it coursed through open gutters. Human and animal wastes were not isolated from drinking water supplies, allowing the wastes to seep into drinking water wells. Excrement provided sustenance for rodents and increased the number of ticks and fleas that could spread the plague. Waterborne diseases such as typhoid fever and cholera could be spread easily. The result was that pestilence became rampant in crowded medieval cities, leading to the deaths of hundreds of thousands of people. The connection between human wastes and disease was unknown at the time.

During the 19th century, with the recognition of the connection between sewage-polluted water and cholera, methods for the treatment of sewage began to improve. For example, in London sewers were constructed so that human excrement could be separated from drinking water supplies. In the mid-1800s, New York City installed a system of sewers that drained into nearby rivers. During the 20th century, sewage was processed to remove sludge, which (along with industrial waste) was dumped into the ocean about 100 miles (160 kilometers) offshore; this practice was banned after 1992.[49]

Modern technology for sewage treatment removes solids, deactivates microbes, and produces wastewater that can be returned safely to waterways or in some cases can be reused or recycled. Sewage treatment involves preliminary processing plus primary, secondary, and tertiary phases; these steps are described in **FIGURE 12.13**.

FIGURE 12.14A–D illustrates primary, secondary, and tertiary sewage processing. The primary stage aims to remove large materials, which can be composted or shipped to landfills. Secondary processing promotes microbial digestion of organic material that remains in the sewage; microorganisms that are present naturally in sewage or that may be added to enhance microbial action aid in the digestion of the liquor during aeration. **FIGURE 12.15** presents a view of microorganisms found in sewage, as seen through a microscope. Following this stage, the sewage sometimes is discharged into waterways.

Tertiary (high-level) processing is directed at removal of remaining solids and microorganisms from the liquid portion of sewage. Various methods exist for high-level processing including filtration through sand and charcoal filters and deactivation of microorganisms (disinfection) by using chlorine or ultraviolet (UV) radiation. **FIGURE 12.16A** demonstrates the progression of sewage from raw effluent to finished product. **FIGURE 12.16B** shows the wastewater from sewage after it has been filtered and disinfected.

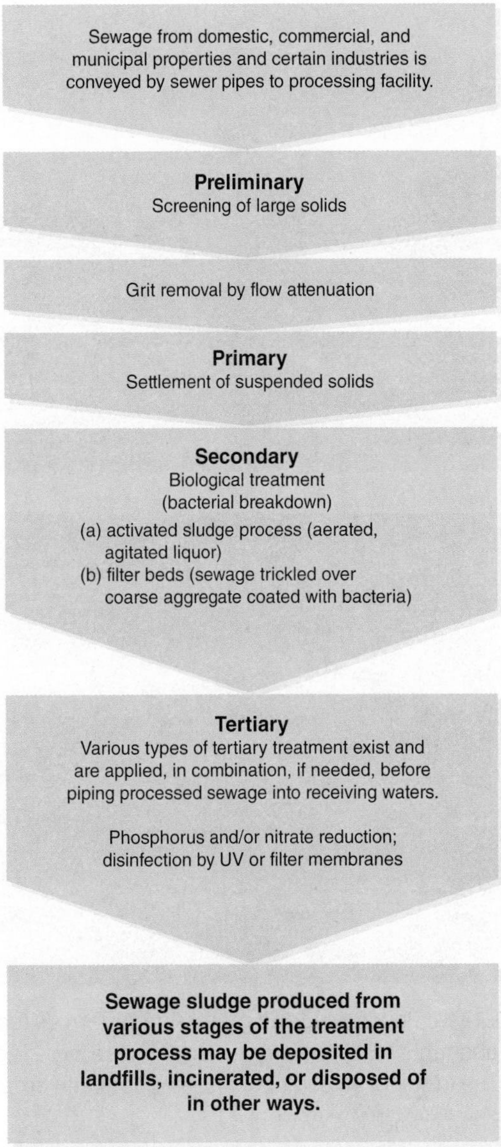

Sewage from domestic, commercial, and municipal properties and certain industries is conveyed by sewer pipes to processing facility.

Preliminary
Screening of large solids

Grit removal by flow attenuation

Primary
Settlement of suspended solids

Secondary
Biological treatment
(bacterial breakdown)
(a) activated sludge process (aerated, agitated liquor)
(b) filter beds (sewage trickled over coarse aggregate coated with bacteria)

Tertiary
Various types of tertiary treatment exist and are applied, in combination, if needed, before piping processed sewage into receiving waters.

Phosphorus and/or nitrate reduction; disinfection by UV or filter membranes

Sewage sludge produced from various stages of the treatment process may be deposited in landfills, incinerated, or disposed of in other ways.

FIGURE 12.13 Diagram of sewage treatment process cycle.

Modified and reproduced with permission from Department for Environment, Food and Rural Affairs (DEFRA). *Sewage treatment in the UK: UK Implementation of the EC Urban Waste Water Treatment Directive.* London: Department for Environment, Food and Rural Affairs (DEFRA); 2002:10. © Crown copyright 2002

The finished wastewater may be recycled; for example, the Irvine Ranch Water District (IRWD) in southern California recycles processed wastewater for watering landscapes and makes it available for flushing toilets in high-rise buildings.

Another issue is clean-up of water originating from urban run off. The water for the marsh shown in **FIGURE 12.17** is all urban runoff that is diverted by IRWD from the San Diego Creek (Irvine) running alongside marsh ponds. The water is diverted to the ponds and then cleaned via the natural processes of the marsh ponds before being sent back into the San Diego Creek, cleaner than when it was originally diverted.

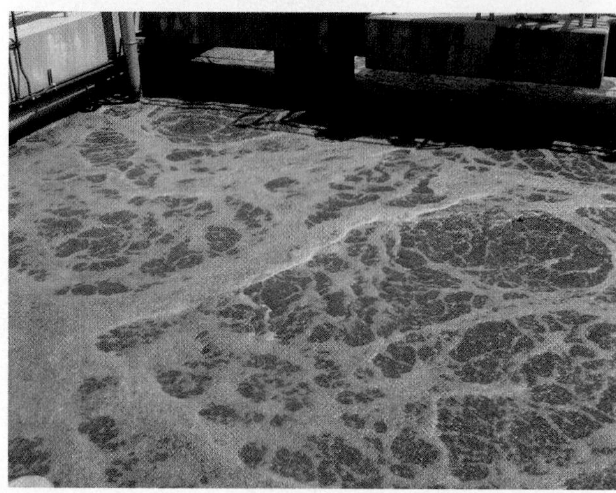

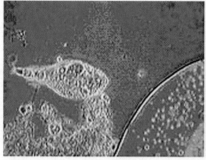

FIGURE 12.14 Primary sewage processing system. A. (upper left) Primary sewage processing tank. B. (upper right) Array of pipes and tanks for secondary sewage processing. C. (lower left) Aeration of sewage during secondary processing. D. (lower right) The inside of an aeration tank that has been emptied for maintenance.

Courtesy of the Irvine Ranch Water District, Irvine, California

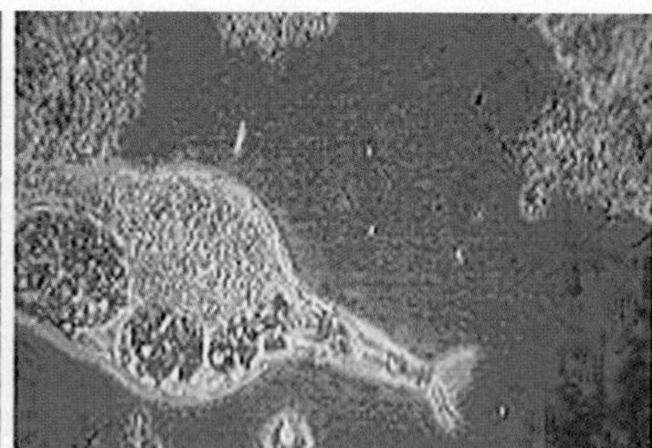

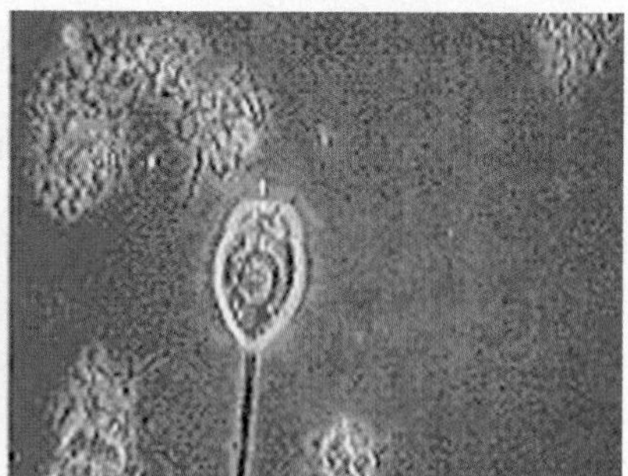

FIGURE 12.15 Microorganisms found in sewage liquor.

Courtesy of the Irvine Ranch Water District, Irvine, California

This use of recycled water from sewage saves millions of gallons of fresh water each year. Most jurisdictions in the United States require that wastewater receive at least secondary treatment. Water that has received only primary treatment is not recommended for any use and generally needs secondary or tertiary treatment for common purposes such as landscape irrigation. Some municipal water districts augment their supply of potable water by using recycled water that has been subjected to tertiary treatment to recharge aquifers and augment surface waters. **FIGURE 12.18** provides examples of the appropriate uses of recycled water that correspond to the level of treatment of wastewater.

FIGURE 12.16 A. The stages in sewage processing from raw sewage on the left to the finished product on the right. B. The finished product.

Courtesy of the Irvine Ranch Water District, Irvine, California

FIGURE 12.17 Water from urban runoff used to create a pond in an arid region and then cleaned via natural processes.

Courtesy of the Irvine Ranch Water District, Irvine, California

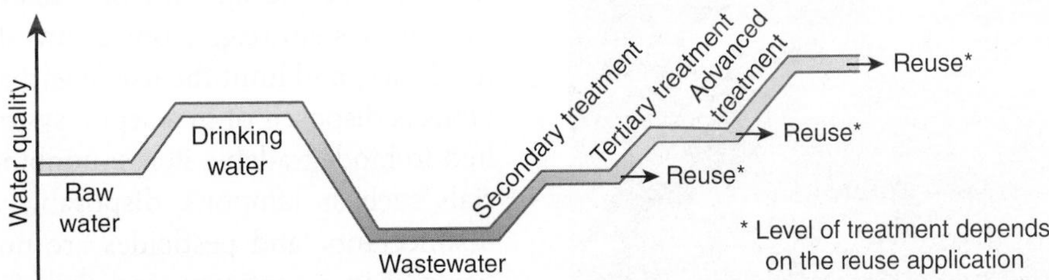

FIGURE 12.18 Treatment technologies are available to achieve any desired level of water quality for recycled water.

Modified from US Environmental Protection Agency. Office of Wastewater Management. *Guidelines for Water Reuse*. EPA/600/R-12/618. Washington, D.C.;2012;p1–7.

Other Methods for Sewage Disposal

Composting Toilets

Composting toilets convert human excrement into a soil-like material that must be buried or hauled away for disposal.[50] They require little water to operate and contain and deactivate pathogens in human wastes. The composting of waste occurs as the result of aerobic bacterial action. There are several brands, including the Clivus Multrum (trade name) system shown in **FIGURE 12.19**.

Septic System

A **septic system** is a type of underground "well" that receives human sanitary wastes. A septic system includes "a septic tank and a subsurface fluid distribution system."[51] It is "used to dispose of human sanitary waste or effluent from dwellings, businesses, community centers or other places where people congregrate."[51] Most septic systems consist of a septic tank and drainfield.[52] **TABLE 12.4** describes how a septic system works. **FIGURE 12.20** illustrates a septic system.

Septic systems are a common onsite method for disposal of sewage, especially in rural areas. A home septic system is a passive method for waste (home sewage) treatment that uses both anaerobic and aerobic stages. Anaerobic action treats waste material as it sits in the septic tank; solids collect on the bottom of the tank. Effluents from the tank flow to an absorption field, where aerobic microorganisms in the soil aid in sanitizing liquid wastes. Most home septic systems are designed for a 20- to 30-year life, after which they eventually fail because the drainfield becomes clogged and the soil loses its ability to assimilate wastes. The life of a septic system can be maximized through proper maintenance (e.g., periodic inspection and pumping out of the septic tank). Also, homeowners should try to minimize the amount of wastewater that is sent into the system (e.g., from dishwashers and washing machines) and limit the use of garbage disposal units. Objects disposed of in a septic system should be limited to biodegradable items; nonbiodegradable materials such as tampons, disposable diapers, solvents, disinfectants, and pesticides are not appropriate for disposal in a septic system. Failed septic systems as well as the presence of numerous septic systems in a densely populated area can contaminate groundwater, causing a potentially serious health hazard.

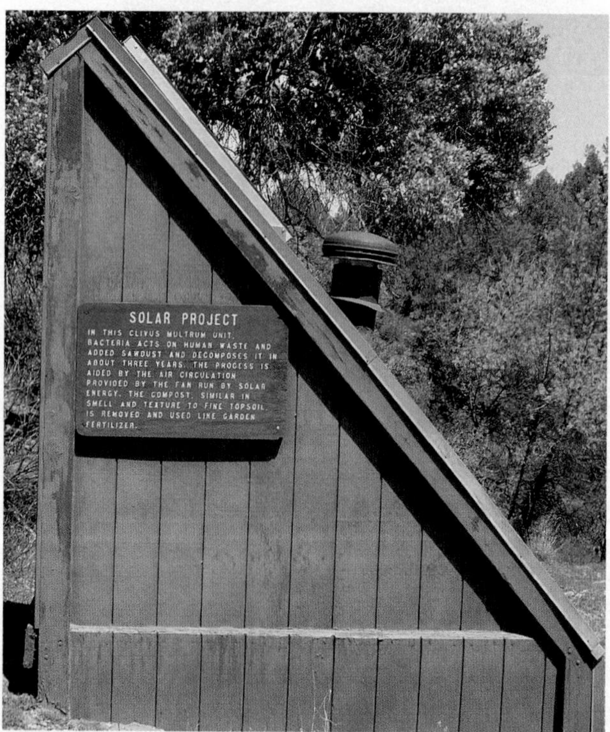

FIGURE 12.19 Experimental unit that uses solar energy to aid in composting of human wastes.

TABLE 12.4 How a Typical Septic System Works

1. All water runs out of your house from one main drainage pipe into a septic tank.
2. The septic tank is a buried, watertight container usually made of concrete, fiberglass, or polyethylene. Its job is to hold the wastewater long enough to allow solids to settle down to the bottom forming sludge, while the oil and grease floats to the top as scum. Compartments and a T-shaped outlet prevent the sludge and scum from leaving the tank and traveling into the drainfield area.
3. The liquid wastewater (effluent) then exits the tank into the drainfield.
4. The drainfield is a shallow, covered excavation made in unsaturated soil. Pretreated wastewater is discharged through piping onto porous surfaces that allow wastewater to filter though the soil. The soil accepts, treats, and disperses... wastewater as it percolates through the soil, ultimately discharging to groundwater. If the drainfield is overloaded with too much liquid, it will flood, causing sewage to flow to the ground surface or create backups in toilets and sinks.
5. Finally, the wastewater percolates into the soil, naturally removing harmful coliform bacteria, viruses, and nutrients. Coliform bacteria is a group of bacteria predominantly inhabiting the intestines of humans or other warm-blooded animals. It is an indicator of human fecal contamination.

Reproduced from US Environmental Protection Agency. How your septic system works. Available at: https://www.epa.gov/septic/how-your-septic-system-works. Accessed June 15, 2017.

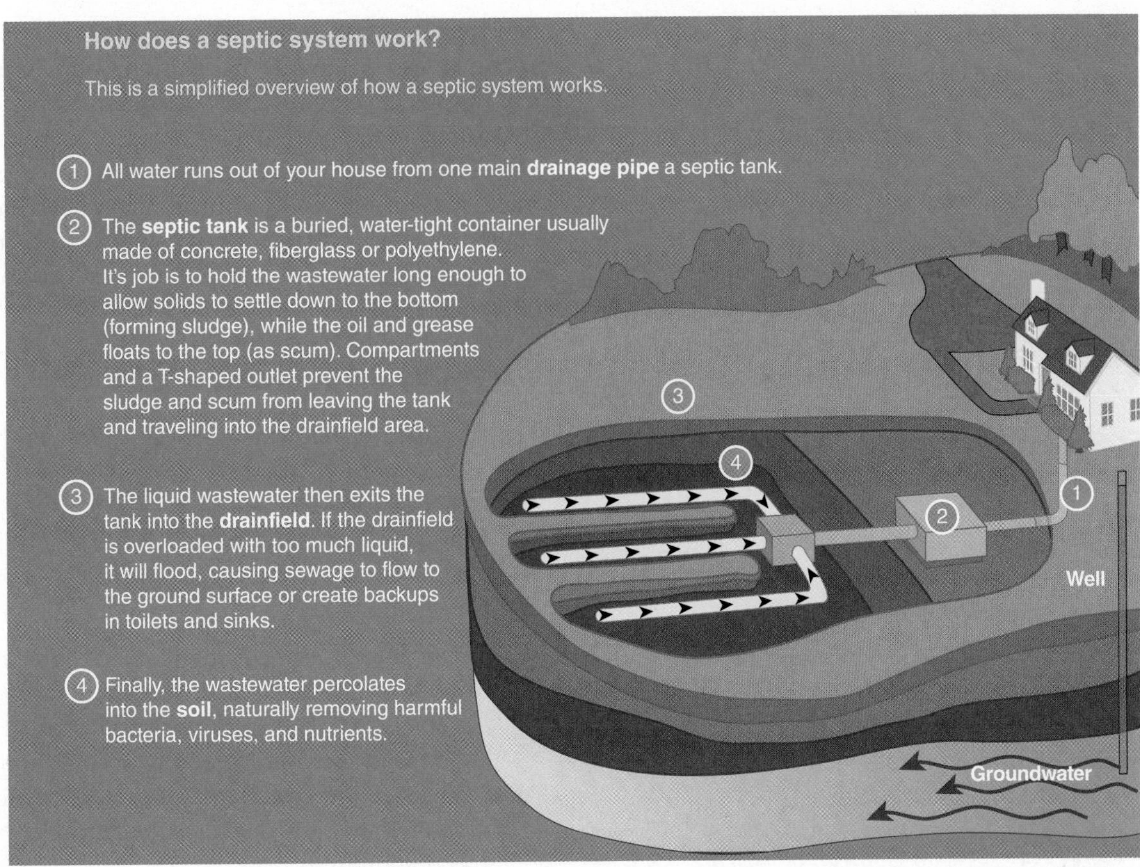
How does a septic system work?
This is a simplified overview of how a septic system works.
1 All water runs out of your house from one main drainage pipe a septic tank.
2 The septic tank is a buried, water-tight container usually made of concrete, fiberglass or polyethylene. It's job is to hold the wastewater long enough to allow solids to settle down to the bottom (forming sludge), while the oil and grease floats to the top (as scum). Compartments and a T-shaped outlet prevent the sludge and scum from leaving the tank and traveling into the drainfield area.
3 The liquid wastewater then exits the tank into the drainfield. If the drainfield is overloaded with too much liquid, it will flood, causing sewage to flow to the ground surface or create backups in toilets and sinks.
4 Finally, the wastewater percolates into the soil, naturally removing harmful bacteria, viruses, and nutrients.
Well
Groundwater

FIGURE 12.20 Home septic system.

[1]Drainage pipe
[2]Septic tank
[3]Drainfield
[4]Soil.

Modified and reproduced from US Environmental Protection Agency. *Do Your Part—Be SepticSmart*. EPA-832-B-12-005.

Animal Wastes

Modern intensive methods of raising livestock crowd large numbers of animals into confined areas in order to maximize efficiency. (Refer to **FIGURE 12.21**.) These methods produce substantial quantities of wastes: manure, urine, carcasses of dead animals, and reproductive tissues.[53] For example, the figure shows cattle awaiting slaughter. Often, during this time, huge piles of manure accumulate in the feed lot, where the cattle must be in contact with their own wastes.

Animal wastes are associated with the following components: nutrients, organic materials, microorganisms, residues of animal medicines, and potentially toxic gases. These materials pose a number of environmental hazards including the buildup of nutrients on the land, surface contamination, and water pollution due to runoff. Animal manure can be used as a source of fertilizer to improve agricultural productivity; however, application of excessive amounts is an environmentally undesirable practice that can result in the runoff of contaminants that later may infiltrate the water supply.[54]

In addition, workers in contact with livestock and other farm animals are exposed to pathogenic microorganisms contained in feces.[53] Occupational activities involve cleaning of animal holding areas, disposal of wastes, and equipment maintenance. Workers need to handle animals directly during birthing, treatment, and slaughter and may be exposed to diseased tissues or inhale or ingest pathogenic materials. Examples of microbes present in manure are *Salmonella*, *Escherichia coli*, protozoa, and zoonotic agents. A Danish study found that wastes from pig farms were contaminated with *Salmonella*; slurry from pig wastes that was disposed of on agricultural soil continued to show contamination with *Salmonella* for up to 2 weeks after it was deposited on the land.[55]

Antimicrobial agents are given to livestock routinely for therapeutic purposes and for prevention of illness (e.g., spread of disease) and at low doses to promote growth.[53,54] Use of antibiotics and other microbial agents is thought to contribute to the development of resistant strains of bacteria, which may have negative impacts on the health of people in contact with the

FIGURE 12.21 Cattle as far as the eye can see in a feed lot: an example of intensive methods for processing cattle.

animals. An estimated 75% of antimicrobial agents that are destined for animals find their way back into the environment via excreta.[53] When the antimicrobials enter the water supply, they may foster the production of antibiotic-resistant organisms that later can become a hazard to human health.

Conclusion

This chapter covered two topics related to waste disposal: solid waste and sewage. Before the 20th century, the disposal of waste was handled casually. Often solid waste was dumped onto the land, and liquid waste from human and animal excrement flowed through open sewers without much regard for the environmental consequences of such disposal methods. With increasing industrialization and growing populations worldwide, solid and liquid wastes threaten the environment and require safe methods for their disposal. Of particular concern are hazardous wastes, which can cause dangerous water pollution, contamination of the land, and toxic air pollution. As a result of intensive methods for raising food animals, excrement from animals represents a growing environmental hazard.

Progress has been made in source reduction and recycling municipal solid waste and industrial waste. Incineration and use of specially designed land disposal units that prevent toxic chemicals from reaching groundwater are two common methods for solid waste disposal. The procedures for sewage disposal include primary, secondary, and tertiary processing. Through high-level processing, wastewater can be made suitable for recycling. Although many industrialized countries have disposal programs in place, developing countries face growing hazards associated with increasing amounts of pollution from sewage and solid waste; developing countries often lack the infrastructure and finances to build waste processing facilities. Through the assistance of international agencies, increased funding and technical know-how could be made available to developing countries in order to help them improve their infrastructures for treatment of wastes.

Study Questions and Exercises

1. Define the following terms in your own words:
   a. Waste
   b. Pollution
   c. Municipal solid waste
   d. Waste recycling
   e. Composting
2. Describe some of the methods of source reduction for solid waste. Visit your local waste management facility; what methods for reduction of solid waste are available in your community?
3. What are the advantages of each of the following methods of waste management?
   a. Landfilling
   b. Composting
   c. Recycling
   d. Incineration
4. Why has recycling become a thriving industry for less developed countries? What hazards can be associated with waste recycling by developing countries?
5. Describe the design of a landfill disposal site. What microbial and physical processes take place in a landfill? How do modern landfills protect the environment from toxic chemicals?
6. What health hazards are associated with uncontrolled and older waste sites in the United States? Describe the processes through which hazardous solid wastes can affect human beings.
7. How do current methods for treating sewage differ from those that were used in the Middle Ages? Historically speaking, what factors spurred the development of today's sanitary sewage systems?
8. Describe the stages for processing sewage. At what stage of processing is it permissible in the United States to dispose of wastewater from sewage into waterways?
9. What health-related considerations relate to the use of recycled water? What level of processing is required in order for wastewater to be recycled? Describe some of the uses for recycled water.

10. Describe alternatives to sewer systems for disposal of human sewage. To what extent could these alternatives replace municipal sewer systems? What are their advantages and disadvantages?
11. Discuss the reasons why the Love Canal incident was significant to the field of environmental health.

For Further Reading

Humes, Edward. *Garbology: Our Dirty Love Affair with Trash*, 2013.

References

1. Hamer G. Solid waste treatment and disposal: effects on public health and environmental safety. *Biotechnol Adv.* 2003;22:71–79.
2. US Environmental Protection Agency. Terms and acronyms. Hazardous waste. Available at: https://iaspub.epa.gov/sor_internet/registry/termreg/searchandretrieve/termsandacronyms/search.do. Accessed June 14, 2017.
3. Rosner D. Introduction, "Hives of sickness and vice." In: Rosner D, ed. *"Hives of Sickness": Public Health and Epidemics in New York City*. New Brunswick, NJ: Rutgers University Press; 1995.
4. US Environmental Protection Agency, Office of Land and Emergency Management (5306P). Advancing sustainable materials management: 2014 fact sheet. EPA530-R-17-01. Washington, DC: US Environmental Protection Agency; 2016.
5. Zimlich R. Regional landfill capacity problems do not equate to a national shortage. Waste 360. Available at: http://www.waste360.com/operations/regional-landfill-capacity-problems-do-not-equate-national-shortage. Accessed August 31, 2017.
6. US Environmental Protection Agency. Municipal solid waste landfills. Available at: https://www.epa.gov/landfills/municipal-solid-waste-landfills. Accessed September 1, 2017.
7. Jiang Y, Kang M, Liu Z, et al. Urban garbage disposal and management in China. *J Environ Sci.* 2003;15:531–540.
8. Carlin J. Municipal solid waste profile. In: US Department of Energy, Energy Information Administration, Office of Coal, Nuclear, Electric and Alternate Fuels. *Renewable Energy Annual 1996*. Washington, DC: US Department of Energy; 1996.
9. US Environmental Protection Agency. Sustainable materials management: non-hazardous materials and waste management hierarchy. Available at: https://www.epa.gov/smm/sustainable-materials-management-non-hazardous-materials-and-waste-management-hierarchy. Accessed September 1, 2017.
10. Uela K, Koizumi H. Reducing household waste: Japan learns from Germany. *Environ.* 2001;43:20–32.
11. Canterbury JL. Pay-as-you-throw: a growing MSW management success story. *Resource Recycling*. October 1997;16–22.
12. US Environmental Protection Agency. Terms and acronyms: recycling. Available at: https://iaspub.epa.gov/sor_internet/registry/termreg/searchandretrieve/termsandacronyms/search.do. Accessed June 14, 2017.
13. US Environmental Protection Agency. Wastes—resource conservation—reduce, reuse, recycle: recycling basics. Available at: https://www.epa.gov/recycle/recycling-basics. Accessed June 12, 2017.
14. US Environmental Protection Agency, Office of Solid Waste. Municipal solid waste in the United States: 2007 facts and figures. EPA 530-R-08-010. Washington, DC: US Environmental Protection Agency; 2008.
15. McCarthy N. The countries winning the recycling race [infographic]. *Forbes*. March 4, 2016. Available at: https://www.forbes.com/sites/niallmccarthy/2016/03/04/the-countries-winning-the-recycling-race-infographic/#1e2c866a2b3d. Accessed September 3, 2017.
16. US Environmental Protection Agency. Decision maker's guide to solid waste management, Volume II. EPA 530-R-95-023. Washington, DC: US Environmental Protection Agency; 1995.
17. Clarke C. 5 cities that are recycling superstars. *TakePart*. September 17, 2014. Available at: http://www.takepart.com/article/2014/09/17/5-cities-are-recycling-superstars/. Accessed June 14, 2017.
18. Manios T. The composting potential of different organic solid wastes: experience from the island of Crete. *Environ Int.* 2004;29:1079–1089.
19. Environmental Research Foundation. The basics of landfills. Available at: http://www.ejnet.org/landfills/. Accessed June 13, 2017.
20. US Environmental Protection Agency. Terms of environment: Glossary: leachate. https://iaspub.epa.gov/sor_internet/registry/termreg/searchandretrieve/termsandacronyms/search.do. Accessed June 15, 2017.
21. Utah State University. Intermountain Herbarium. Fun facts about fungi. Waste not, want not: fungi as decomposers. Available at: http://www.herbarium.usu.edu/fungi/funfacts/Decay.htm. Accessed June 13, 2107.
22. Taylor D. Talking trash: the economic and environmental issues of landfills. *Environ Health Perspect.* 1999;107:A404–A409.
23. Al-Yaqout AF, Hamoda MF. Evaluation of landfill leachate in arid climate—a case study. *Environ Int.* 2003;29:593–600.
24. National Institute of Environmental Health Sciences. Hazardous material/waste. Available at: https://www.niehs.nih.gov/health/topics/agents/haz-waste/index.cfm. Accessed June 13, 2017.
25. Deloraine A, Zmirou D, Tillier C, et al. Case-control assessment of the short-term health effects of an industrial toxic waste landfill. *Environ Res.* 1995;68:124–132.
26. Goldberg MS, Siemiatyck J, DeWar R, et al. Risks of developing cancer relative to living near a municipal solid waste landfill site in Montreal, Quebec, Canada. *Arch Environ Health.* 1999;54:291–296.
27. Energy Information Administration. Energy Kids: a report from energy ant—my trip to the RESCO trash-to-energy plant in Baltimore, Maryland. Available at: https://www.eia .gov/Kids/energy.cfm?page=RESCOE_Plant. Accessed June 15, 2017.
28. Gonzalez CA, Kogevinas M, Gadea E, et al. Biomonitoring study of people living near or working at a municipal solid-waste incinerator before and after two years of operation. *Arch Environ Health.* 2000;55:259–267.
29. US Energy Information Agency. Waste-to-energy (municipal solid waste). Available at: https://www.eia.gov/energyexplained/index.cfm/data/index.cfm?page=biomass_waste_to_energy. Accessed June 15, 2017.
30. Van Hook RI. Transport and transportation pathways of hazardous chemicals from solid waste disposal. *Environ Health Perspect.* 1978;27:295–308.

31. US Environmental Protection Agency. Chapter 3: Land. In: *EPA's 2008 Report on the Environment*. Washington, DC: National Center for Environmental Assessment. EPA/600/R-07/045F; 2008. Available from the National Technical Information Service, Springfield, VA.
32. US Environmental Protection Agency. Household hazardous waste (HHW). Available at: https://www.epa.gov/hw/household-hazardous-waste-hhw. Accessed June 13, 2017.
33. US Department of Homeland Security, Federal Emergency Management Agency (FEMA). Are you ready? Household chemical emergencies. Available at: https://www.fema.gov/pdf/areyouready/areyouready_full.pdf. Accessed June 13, 2017.
34. Lee BK, Ellenbecker MJ, Moure-Ersaso R. Alternatives for treatment and disposal cost reduction of regulated medical wastes. *Waste Manag*. 2004;24:143–151.
35. Schmidt CW. Trading trash: why the US won't sign on to the Basel Convention. *Environ Health Perspect*. 1999;107:A411–A413.
36. Gupta JP, Babu BS. A new hazardous waste index. *J Hazard Mater*. 1999;A67:1–7.
37. Vijay R, Sihorwala TA. Identification and leaching characteristics of sludge generated from metal pickling and electroplating industries by toxicity characteristics leaching procedure (TCLP). *Environ Monit Assess*. 2003;84:193–202.
38. Johnson BL, DeRosa CT. The toxicologic hazard of Superfund hazardous waste sites. *Rev Environ Health*. 1997;12:235–251.
39. Gibbs LM. Learning from Love Canal: a 20th anniversary retrospective. (1998). The Envirolink Network. Welcome to EnviroArts: Orion Online. Available at: http://arts.envirolink.org/arts_and_activism/LoisGibbs.html. Accessed June 13, 2017.
40. Tuchman G. Despite toxic history, residents return to Love Canal. CNN.com; August 7, 1998. Available at: http://www.cnn.com/US/9808/07/love.canal. Accessed June 13, 2017.
41. Questia: Encyclopedia: Love Canal. *The Columbia Encyclopedia*, 6th ed. New York, NY: The Columbia University Press; © 2017) Available at: https://www.questia.com/read/1E1-Lovecana/love-canal. Accessed June 13, 2017.
42. US Environmental Protection Agency. Superfund site: Love Canal, Niagara Falls, NY. Available at: https://cumulis.epa.gov/supercpad/cursites/csitinfo.cfm?id=0201290. Accessed June 13, 2107.
43. US Environmental Protection Agency. Medical waste. Available at: https://www.epa.gov/rcra/medical-waste. Accessed June 13, 2017.
44. Centers for Disease Control and Prevention. Perspectives in disease prevention and health promotion summary of the Agency for Toxic Substances and Disease Registry Report to Congress: the public health implications of medical waste. *MMWR*. 1990;39:822–824.
45. US Food and Drug Administration. Best way to get rid of used needles and other sharps. Available at: https://www.fda.gov/MedicalDevices/ProductsandMedicalProcedures/HomeHealthandConsumer/ConsumerProducts/Sharps/ucm263240.htm. Accessed June 16, 2017.
46. US Environmental Protection Agency. Wastes: land disposal. Available at: https://archive.epa.gov/epawaste/hazard/tsd/td/web/html/disposal.html. Accessed June 16, 2017.
47. US Environmental Protection Agency. Terms and acronyms: sewage. Available at: https://iaspub.epa.gov/sor_internet/registry/termreg/searchandretrieve/termsandacronyms/search.do. Accessed June 14, 2017.
48. Smith N. Roman hydraulic technology. *Scientific American*. May 1978:154–161.
49. US Environmental Protection Agency. History: Reilly in New York to mark end of sewage sludge dumping (EPA press release); June 30, 1992. Available at: https://archive.epa.gov/epa/aboutepa/reilly-new-york-mark-end-sewage-sludge-dumping.html. Accessed June 14, 2107.
50. US Environmental Protection Agency. Office of Water. Water efficiency technology fact sheet—composting toilets. EPA 832-F-99-066. Washington, DC: US Environmental Protection Agency; 1999.
51. US Environmental Protection Agency. Region 5: Office of the Regional Administrator. Glossary of underground injection control terms: septic system. Available at: https://ofmpub.epa.gov/sor_internet/registry/termreg/searchandretrieve/glossariesandkeywordlists/search.do?details=&glossaryName=Underground%20Injection%20Control. Accessed June 14, 2107.
52. US Environmental Protection Agency. How your septic system works. Available at: https://www.epa.gov/septic/how-your-septic-system-works. Accessed June 15, 2017.
53. Cole DJ, Hill VR, Humenik FJ, et al. Health, safety, and environmental concerns of farm animal waste. *Occup Med*. 1999;14:423–448.
54. Campagnolo ER, Johnson KR, Karpati A, et al. Antimicrobial residues in animal waste and water resources proximal to large-scale swine and poultry feeding operations. *Sci Total Environ*. 2002;299:89–95.
55. Baloda SB, Christensen L, Trajcevska S. Persistence of a *Salmonella enterica* serovar Typhimurium DT12 clone in a piggery and in agricultural soil amended with *Salmonella*-contaminated slurry. *Appl Environ Microbiol*. 2001;67:2859–2862.

© Jean-Luc Rivard/EyeEm/Getty Images

CHAPTER 13
Occupational Health

LEARNING OBJECTIVES

By the end of this chapter the reader will be able to:

- Compare two historically important occupational health incidents.
- Discuss the significance of the occupational environment for health.
- List physical and psychological effects of occupational hazards.
- State five categories of health impacts of occupational hazards.
- Illustrate three methods for the prevention of occupationally related disease.

Introduction

This chapter explores hazardous agents found in the occupational environment, reviews their potential health effects, and identifies procedures for prevention of occupational illnesses and injuries. You will learn that the occupational environment poses numerous health hazards to workers employed in a variety of positions. Among the potential dangers encountered in the workplace are high noise levels, fumes and dusts, toxic chemicals, ionizing and non-ionizing radiation, high temperatures, biohazards, and psychological and social hazards including stress. In some instances, exposure to workplace hazards is associated with traumatic injuries and deaths as well as a range of conditions such as hearing loss, respiratory diseases, cancer, and adverse birth outcomes. The list of possible hazards that occur in the workplace is extensive and growing with the introduction of new industrial processes such as nanotechnologies used in high tech industries. The expansion of industrialization in the developing world increases possible hazards to workers who may not have the same level of safeguards that exist in developed countries. Nowadays, many individuals telecommute or work at home; the result is an expansion of the work environment to the home setting, with transfer of employment insecurity and job stresses to one's home environment.[1]

Occupational Health Concepts

The field of occupational medicine specializes in detection and prevention of diseases that arise from the work environment.[1] **Occupational disease** is defined as those health outcomes that are "caused or influenced by exposure to general conditions or specific hazards encountered in the work environment."[1(p97)]

Although the scope of occupational hazards and their effects is vast, too little research has been conducted on the health consequences of these hazards. A classic 1972 statement (which remains valid today) on the status of the occupational environment expressed the following poignant sentiment regarding workplace hazards and the health of workers:

> We are dealing here with the tip of a treacherous ecological iceberg. Few exact studies have been made to measure the full dimensions of occupational illness, occupational pollution, [and] occupational exposures. It is, unhappily, mostly guess work. We do know, for instance, how certain occupations lead to a high rate of specific kinds of cancer. But there are only scant and fragmentary epidemiological studies of the health effects of chemical or noise pollution on whole groups of workers. Most of the data consists of broad hints of a far flung, unfathomed problem yet to be accurately measured.[2(p4)]

Background and History

The history of occupational health contains many notorious examples of how workers' health was impacted by unsafe working conditions. Recognition of occupational risks from mining occurred during Greek and Roman times. Subsequent to the writings of Galen (CE 129–200), through the Dark Ages and the medieval period until the time of Ramazzini (early 1700s), the issue of occupational health was accorded little notice in Europe. However, exceptions during the Renaissance included Agricola's (1494–1555) discussion of the health effects of mining and smelting of gold and silver and the work of Paracelsus (discussed in the chapter on environmental toxicology). From about the 1700s onward, recognition grew with respect to the contribution of occupationally related exposures to adverse health conditions. Many examples of explorations of the impacts of unsafe and hazardous working environments on the health of workers can be cited. One body of writing focused on the effects of miners' exposures to toxic metals and other hazards. Among the persons who made significant contributions to occupational health are those listed in **TABLE 13.1**. Note that there was a void in occupational health information during the Dark Ages in Europe; however, during this time scholars in the Middle East and Asia continued to make contributions to the field. A notable Middle Eastern scholar of occupational health was Abu Bakr Muhammad ibn Zakariya

TABLE 13.1 Abridged List of Noteworthy Figures in the History of Occupational Health

Name	Dates (Birth–Death)	Contribution
Hippocrates	460–377 BCE (estimated)	Discussed hazards of metal working and lead.
Pliny the Elder	CE 23–70	Described hazards of dust.
Galen	129–200 (estimated)	Described hazards to miners.
Rhazes	850–923 (estimated)	Used occupational classifications in medical case descriptions.
Paracelsus	1493–1541	Wrote book on occupational diseases.
Agricola	1494–1555	Described hazards of mining and producing gold and silver.
Bernardino Ramazzini	1633–1714	"Father of occupational medicine."
Sir Percival Pott	1714–1788	Identified scrotal cancer among chimney sweeps.
Dr. Alice Hamilton	1869–1970	Publicized dangerous occupational conditions (e.g., phossy jaw).
Irving Selikoff	1915–1992	Identified health effects of asbestos exposure.

al-Razi (about 850–923). Also known as Rhazes, this Persian scholar used occupational classifications (e.g., goldsmith) in describing medical cases.

Paracelsus (1493–1541) became one of the founders of the field of toxicology, a discipline that is used to examine the toxic effects of chemicals found in environmental venues such as the workplace (see the chapter on environmental toxicology). In 1556, *De Re Metallica*, a book that described the environmental and occupational hazards of mining, was published. The work is attributed to Georgius Agricola, who lived in Germany.

Bernardino Ramazzini, whose likeness is shown in **FIGURE 13.1**, has been called the founder of the field of occupational medicine.[3] He is credited with creating elaborate descriptions of the manifestations of occupational diseases among many different types of workers.[1] His descriptions covered a plethora of occupations from miners to cleaners of privies to fabric workers. The father of occupational medicine is also considered to be a pioneer in the field of ergonomics, by pointing out the hazards associated with postures assumed in various occupations. Ramazzini authored *De Morbis Artificum Diatriba (Diseases of Workers)*, published in 1700. His book highlighted the risks posed by hazardous chemicals, dusts, and metals used in the workplace.

Sir Percival Pott tracked the occurrence of scrotal cancer among chimney sweeps in London during the late 1700s. As noted previously in the chapter on

ISTITUTO NAZIONALE MEDICO FARMACOLOGICO - ROMA

RITRATTI DI MEDICI E NATURALISTI ITALIANI DAL SECOLO XV° AL XVIII°

BERNARDINO RAMAZZINI

(dal busto in marmo esistente nell' Università di Modena)

FIGURE 13.1 Bernardino Ramazzini, founder of occupational medicine.

Courtesy of National Library of Medicine.

environmental epidemiology, Pott is regarded as the first person to describe an environmental cause of cancer.

Dr. Alice Hamilton's description of phossy jaw (phosphorus necrosis of the jaw) was among her contributions to occupational health. **FIGURE 13.2** shows Hamilton's picture. Phossy jaw was an affliction that occurred among workers who manufactured matches during the late 1800s. The condition, which affected the victims' jawbones, was accompanied by severe pain and abscesses that drained fetid-smelling pus. Over time, as the disease process unfolded, the victim gradually became disfigured; surgery, the only means available to save the life of the patient, led to removal of the jaw bone and further disfigurement.

Phossy jaw was caused by exposure to white phosphorus. Affected workers were those who dipped matchsticks into white phosphorus paste. Many of the workers were children who labored in vapor-filled, poorly ventilated rooms. The condition, which developed slowly over a period of years, produced debilitation, neurologic disturbances, and lung hemorrhages. Remarkably, phossy jaw was a completely avoidable condition, because red phosphorus worked as well in matches as white phosphorus but was much safer. During the early 20th century, many countries passed laws that prohibited the use of white phosphorus in matches. **EXHIBIT 13.1** provides a graphic description of the condition.

In addition to phossy jaw, described in the foregoing section, other conditions are historically important to the field of occupational health. Some of these work-related diseases are listed in **TABLE 13.2**.

FIGURE 13.2 Alice Hamilton, MD

EXHIBIT 13.1

Case Study: Phossy Jaw

Physician Alice Hamilton, who crusaded against dangerous industrial working conditions during the late 1800s, wrote the following description of a condition caused by exposure to white phosphorus.

Phossy jaw is a very distressing form of industrial disease. It comes from breathing the fumes of white or yellow phosphorus, which gives off fumes at room temperature, or from putting into the mouth food or gum or fingers smeared with phosphorus. Even drinking from a glass which has stood on the workbench is dangerous. The phosphorus penetrates into a defective tooth and down through the roots to the jawbone, killing the tissue cells which then become the prey of suppurative germs from the mouth, and abscesses form. The jaw swells and the pain is intense, for the suppuration is held in by the tight covering of the bone and cannot escape, except through a surgical operation or through a fistula boring to the surface. Sometimes the abscess forms in the upper jaw and works up into the orbit, causing the loss of an eye. In severe cases one lower jawbone may have to be removed, or an upper jawbone—perhaps both. There are cases on record of men and women who had to live all the rest of their days on liquid food. The scars and contractures left after recovery were terribly disfiguring, and led some women to commit suicide. Here was an industrial disease which could be clearly demonstrated to the most skeptical. Miss Addams told me that when she was in London in the 1880s she went to a mass meeting of protest against phossy jaw and on the platform were a number of pitiful cases, showing their scars and deformities.[4(p116)]

Reproduced from Hamilton A. *Exploring the Dangerous Trades: The Autobiography of Alice Hamilton, M.D.* Nabu Press; 2011.

Mad Hatter's Disease

The term *mad hatter's disease* stems from Lewis Carroll's 1865 book *Alice in Wonderland*, in which Carroll portrayed the Mad Hatter as a zany character. The phrase *mad as a hatter* had a grounding in reality. At one time, mercury-containing solutions were applied during the process of converting fur into the felt that was used to manufacture hats. Millinery workers were exposed to mercury fumes in poorly ventilated rooms during the course of their careers. After a period of time, some of the individuals developed "mad hatter's disease," a condition also known as mercurial erethism. This neuropsychiatric syndrome is characterized by

TABLE 13.2 Classic Occupational Diseases Found in Historical Literature

Name of Disease	Definition/Etiology
Miners' asthma	The common name for pneumoconiosis among miners exposed to dusts such as coal dust.
Potters' rot, miners' phthisis	Silicosis, respiratory disease from inhalation of silica dust.
Brass-founders' ague	A type of metal fume fever caused by inhalation of fumes from welding brass. This self-limiting condition is associated with fever and other symptoms that resolve after 24 hours.
Filecutters' paralysis	Paralysis of the hands caused by lead exposure.
Painters' colic	Abdominal pain associated with anemia caused by exposure to white lead in paint.
Bakers' itch	A skin reaction (eczema) caused by contact with the components of baked goods (e.g., sugar).
Mule spinners' cancer, also known as mule spinners' disease	The mule was a textile spinning machine; the disease referred to scrotal cancer that occurred among male cotton textile workers who were exposed to mineral oils over long time periods as they used the mule.
Hatters' shakes; the mad hatter's disease (described in text)	Mercury poisoning among millinery workers.
Caisson disease (decompression sickness)	A disease caused by decompression when workers emerged from caissons, which were used to construct the anchoring piers for the Brooklyn Bridge in New York City and for similar projects. The laborers were exposed to air under very high pressures for extended time periods.
Phossy jaw (described in text)	Phosphorus necrosis of the jaw.

symptoms that include irritability, personality change, loss of self-confidence, depression, memory loss, and reduced ability to concentrate. **FIGURE 13.3** shows a hat maker at work in a poorly ventilated environment in which workers could be exposed to fumes of elemental mercury.

Historically Significant Occupational Accidents

A report published in the *Morbidity and Mortality Weekly Report* observes: "At the beginning of this century [the 20th century], workers in the United States faced remarkably high health and safety risks on the job."[5(p461)] Many workers remained on the job for 16-hour periods during 6- and 7-day workweeks.[6] Two major incidents—the Triangle Shirtwaist Company fire and the Gauley Bridge disaster—illustrated the deplorable conditions that workers often were forced to endure. These poor environmental conditions included contact with hazardous machinery, crowding, lack of ventilation, poor lighting, and infrequent protection against hazards. These two events as well as others led to reforms that resulted eventually in the improvement of working conditions.

Triangle Shirtwaist Company Fire

March 25, 1911 marks the date of New York City's worst factory fire, which occurred at a 10-story structure formerly named the Asch building. The disaster, also one of the worst industrial accidents in American history, claimed the lives of 146 women within the brief time span of 15 minutes. Several hundred women labored on the top three floors used by the Triangle Shirtwaist

FIGURE 13.3 Man working in a Connecticut hat-making plant and exposed to steam laden with mercury (ca. 1937).
Courtesy of CDC/ Barbara Jenkins, NIOSH; U.S. Public Health Service

Factory. Doors were locked to prevent the women from leaving their sewing machines; fire escapes were nonfunctional. As a result, when a fire erupted about 4:30 in the afternoon, many of the women (especially on the ninth floor) perished from the fire when they were unable to escape or were killed when they jumped from windows or attempted to slide down elevator cables.[7,8]

Gauley Bridge Disaster

Beginning about 1931, workers began a tunneling project near the small town of Gauley Bridge, located in West Virginia. During tunneling operations, workers were exposed to high levels of silica dust from which they did not have adequate protection. Estimates indicated that as many as 1,500 workers contracted the lung disease silicosis and 1,000 died from this cause.[9] (More information on silicosis is provided later in this chapter.)

Significance of the Occupational Environment for Health

Currently, fatal and nonfatal occupational illnesses and injuries exact a significant social and economic toll in the United States. The National Institute for Occupational Safety and Health (NIOSH) reported that on a typical day during the 1990s, 11,000 employees endured disabling injuries; occupationally related diseases caused the deaths of 130 individuals.[10] The direct costs of all occupational injuries and illnesses in 2002 were about $45.8 billion and the indirect costs as much as $229 billion.[10]

Workers' Memorial Day, observed on April 28 of each year, honors

> "...those workers who have died on the job, to acknowledge the grievous suffering experienced by families and communities, and to recommit ourselves to the fight for safe and healthful workplaces for all workers. It is also the day the Occupational Safety and Health Act (OSHA) was established in 1971. Under the Occupational Safety and Health Act of 1970, employers are responsible for providing safe and healthful workplaces for their workers. OSHA's role is to ensure these conditions for America's working men and women by setting and enforcing standards, and providing training, education and assistance. Every year, events are held across the country to remember workers who have died on the job

and honor them by continuing to fight for improved worker safety."[11]

During 2015, approximately 2.9 million nonfatal work-related illnesses and injuries were reported in private industry.[12] Of this total, the vast majority (95.2% or 2.8 million cases) were injuries. (Refer to **FIGURE 13.4**.) The remaining 4.8% were illnesses, a category that included poisonings, respiratory conditions, and skin diseases. **FIGURE 13.5**, which is of historical interest, demonstrates time trends in reported occupational illnesses from 1972 to 2001. The numbers peaked between 1992 and 1996 and subsequently declined. During 2002, the number of newly reported nonfatal

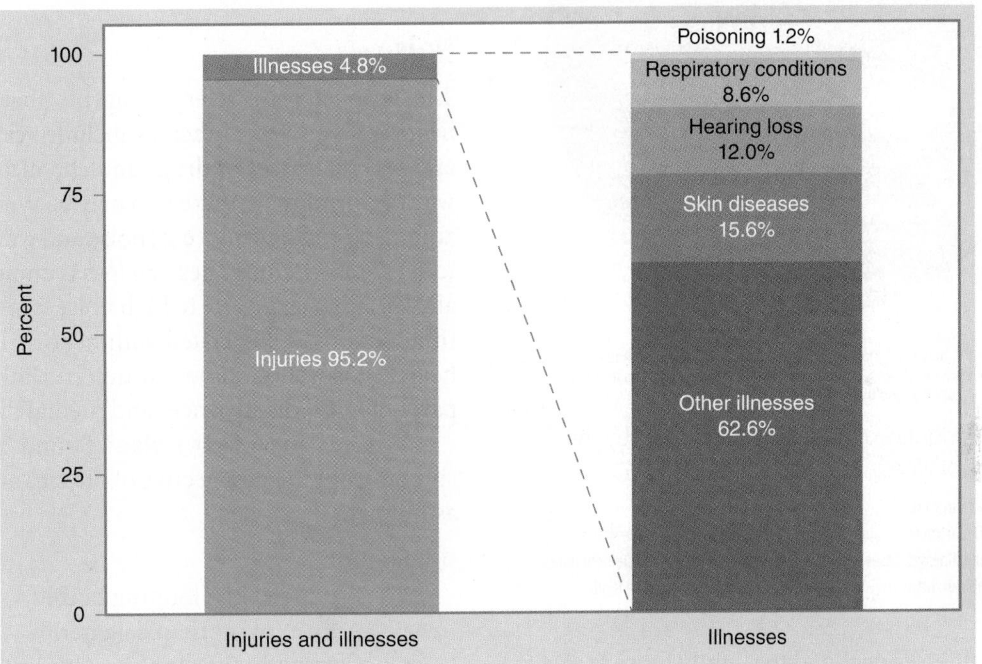

FIGURE 13.4 Distribution of nonfatal occupational injury and illness cases (n = 2,905,900) by category of illness, private industry, 2015.

Reproduced from U.S. Bureau of Labor Statistics, U.S. Department of Labor. 2015 Survey of occupational injuries and illnesses. Summary estimates charts package, October 27, 2016 . Available at: https://www.bls.gov/iif/oshwc/osh/os/osch0057.pdf

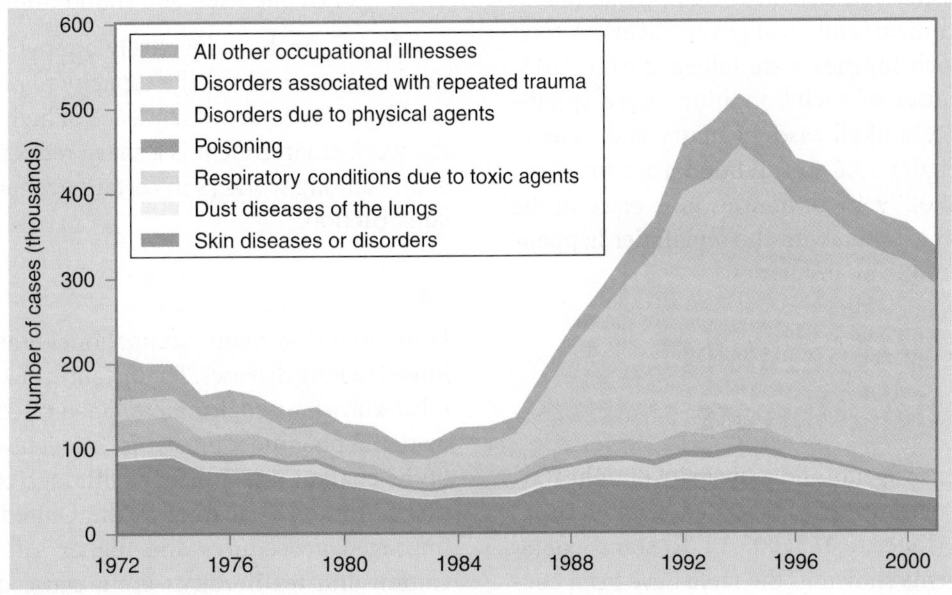

FIGURE 13.5 Number of illness cases in private industry by type of illness, 1972–2001.

Reproduced from Sestito JP, Lunsford RA, Hamilton AC, Rosa RR, eds. *Worker Health Chartbook 2004*. NIOSH Publication Number 2004-146. Cincinnati, OH: Department of Health and Human Services, Centers for Disease Control and Prevention, National Institute for Occupational Safety and Health; 2004:19.

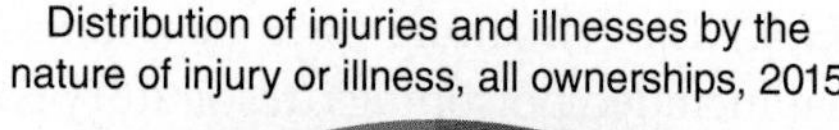

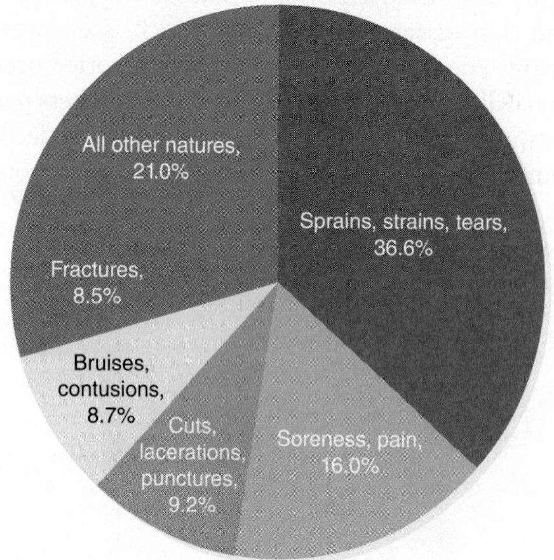

FIGURE 13.6 Distribution of injuries and illnesses by the nature of injury or illness, all ownerships, 2015.*

* Cases with days away from work.

Reproduced from U.S. Department of Labor, Bureau of Labor Statistics. 2015 Nonfatal Occupational Injuries and Illnesses: Cases with days away from work. Case and Demographics. November 2016. Available at: https://www.bls.gov/iif/oshwc/osh/case/osch0058.pdf

injuries and illnesses declined to slightly more than 4.7 million cases. The number of occupational injuries and illnesses has shown a declining trend since 2002.

FIGURE 13.6 presents the relative distribution of occupational injuries and illnesses by nature, 2015. The data are for cases that required days away from work for all business ownerships (e.g., private industry, state government, and local government). A total of 1,153,490 such injuries were tallied during 2015. The leading causes of such conditions were sprains and strains (36.6% of all cases of injury and illness). Refer to the Figure 13.6 for additional information. In 2015, a total of 79.4% of injuries took place in the service-providing sector, with the remainder happening in goods-producing industries.

Overview of Agents of Occupational Disease

A large number of agents and a diversity of unhealthful workplace settings are associated with occupational disease. (Refer to **TABLE 13.3**.) To some extent, many of the agents shown in the table have been covered in previous chapters. Nevertheless, a review and discussion of the relevance of these disease-causing factors to the work environment is in order.

As indicated by Table 13.3, agent factors associated with occupational diseases and injuries include noise, dusts, toxic heavy metals and their fumes, carbon monoxide, chemicals, ionizing radiation, microbial agents, lifting heavy weights, repetitive motion, accidents, and work-related stress. The following paragraphs describe some of these agent factors and their health effects.

Noise

The term *ototoxic* refers to agents that can produce hearing loss. Ototoxic agents include very loud sounds and several classes of drugs and chemicals used in the work environment; there also may be synergistic effects among ototoxic agents (e.g., noise and workplace chemicals). Noise-induced hearing loss is cumulative; generally speaking, exposure to high noise levels for extended time periods is associated with hearing loss and other hearing disorders. Some research has linked noise with psychological disturbances and mental disorders.

What is meant by **noise**? Sound and noise are similar from the perspective of physics and are defined as follows:

- Sound[13]
  - Produced by vibrating bodies.
  - A form of mechanical energy.
  - Carried in the air as longitudinal waves.
  - Impinges upon the auditory system and produces physiological responses.
- Noise[14]
  - A type of sound that is unwanted.
  - Can be made up of pure tones, a narrow band of frequencies, and sound impulses.

Noise can be subjectively annoying, can disrupt performance, and has the capacity to produce adverse health effects. One consequence of high noise levels in the work environment is a sharp reduction in hearing acuity among workers in certain occupations, such as construction.

Dusts

Dusts found in many occupational environments are linked to lung diseases. For instance, black lung disease (also known as coal miners' pneumoconiosis or coal workers' pneumoconiosis) is an occupational hazard of the coal mining industry (although the threat from this condition is abating in the United States due to improved procedures and higher safety standards). Cotton and textile dusts are a cause of brown lung disease. In addition to pneumoconiosis, exposure to dusts may induce emphysema, lung cancer, and other pulmonary conditions.

TABLE 13.3 Categories of Occupationally Associated Diseases or Injuries, Agent Factors, and Associated Work Settings

Disease or Injury	Agent Factors	Work Settings
Hearing loss	Noise	Transport, manufacturing, and construction
Lung diseases: asthma, chronic obstructive pulmonary disease, pneumoconiosis	Dusts, e.g., coal, cotton, and textile dusts	Mining, textile manufacturing, and construction
Cancer and lead poisoning	Toxic heavy metals (e.g., lead) and their fumes	Metal working, plumbing, and ceramics manufacture
Carbon monoxide poisoning	Carbon monoxide	Mining, smelting, and indoor use of gasoline-powered engines
Allergic and irritant dermatitis	Chemicals	Manufacturing, numerous industries
Fertility and pregnancy abnormalities	Chemicals, ionizing radiation	Manufacturing, hospitals
Infectious diseases	Microbial agents, e.g., bacteria and viruses	Health care, public utilities, and agriculture
Low-back disorders	Lifting heavy weights	Manufacturing, health care
Musculoskeletal disorders of the upper extremities	Repetitive motion	Manufacturing, office work
Traumatic injuries	Workplace accidents	All occupations potentially at risk
Adverse mental health outcomes	Work-related stress	All occupations potentially at risk

Toxic Heavy Metals and Their Fumes

Some metals used in the work environment have potentially toxic or injurious properties, as noted previously in the chapter on toxic metals and elements. The toxic heavy metals (arsenic, lead, mercury, cadmium, chromium, and nickel) are potential hazards to human health. During the processing and milling of heavy metals, employees are at risk of breathing fumes and dusts that contain toxic levels of these metals. Consider the widespread occupational exposures to lead: workers who have potential for such exposures are plumbers, radiator repair workers, painters, ceramic workers, and metal workers. (Refer to the following text box.) Another group of lead-exposed individuals are those who patronize or work in firing ranges; such workers include coaches and law enforcement personnel who regularly practice shooting guns.

Carbon Monoxide

Carbon monoxide (CO), which is odorless and can pose an undetected threat, is one of the hazardous toxic gases found in the work environment. CO causes death by depriving the body of oxygen; CO combines more readily than oxygen with the hemoglobin in the blood, thus accounting for CO's hazardous nature. The product formed by the combination of CO and hemoglobin is known as carboxyhemoglobin (COHb). The oxygen-carrying capacity of blood decreases in proportion to the amount of COHb that is present. In comparison to healthy individuals, workers who have preexisting diseases that affect the circulation and oxygen-carrying capacity of the blood are at increased risk of the effects of CO poisoning. Persons who should avoid CO exposure are those who have conditions such as lung disease, heart disease, and anemia. Cigarette smoking is known to increase blood COHb levels.

Among the many sources of CO in the occupational environment are small gasoline-powered engines (e.g., those used in lawn mowers, pressure washers, and generators). Care should be taken to provide adequate ventilation when using these devices. Other sources of CO are kilns, furnaces, forges, space heaters, and fires and explosions. Some mines can have high levels of CO.

Chemicals

Literally thousands of chemical substances are used in the modern workplace. For example, the list of substances associated with the manufacture of plastics and paints includes epoxy resins, acrylates, and formaldehyde. Other widely used industrial chemicals are colorants, dyes, disinfectants, pharmaceuticals, preservatives, and solvents.[15] Chemical exposures have been associated with dermatoses, liver disease, and many other conditions. As noted previously, agricultural workers are exposed routinely to potentially toxic pesticides.

Ionizing Radiation

Sources of ionizing radiation exist in many occupational settings. One example is found in healthcare facilities, such as nursing homes, specialized clinics, dentistry suites, and hospitals. Commonly, healthcare personnel might encounter ionizing radiation from sources such as radionuclides and X-ray machines. Other occupational venues that involve the use of ionizing radiation include research facilities, nuclear generating plants, and plants for the production of nuclear weapons. The chapter on ionizing and nonionizing radiation discussed hazards associated with ionizing radiation.

Microbial Agents

Microbial agents (sometimes referred to as biologic agents) of disease are a source of health risks for workers in many occupational categories, for example, hospital employees, workers exposed to sewage, and agricultural workers. Persons on the front line of patient contact—physicians and nurses—interact

OCCUPATIONAL LEAD EXPOSURES AT A SHIPYARD—DOUGLAS COUNTY, WISCONSIN, 2016

On March 28, 2016, the Minnesota Poison Control System was consulted by an emergency department provider regarding clinical management of a shipyard worker with a blood lead level (BLL) >60 μg/dL; the National Institute for Occupational Safety and Health defines elevated BLLs as ≥5 μg/dL. The Minnesota Poison Control System notified the Minnesota Department of Health (MDH). Concurrently, the Wisconsin Department of Health Services (WDHS) received laboratory reports concerning two workers from the same shipyard with BLLs >40 μg/dL. These three workers had been retrofitting the engine room of a 690-foot vessel since January 4, 2016.

Work was suspended during March 29–April 4 in the vessel's engine room, the presumptive primary source of lead exposure. On March 29, the shipyard partnered with a local occupational health clinic to provide testing for workers. Employees and their household members were also tested by general practitioners and local laboratories. The shipyard hired sanitation crews for lead clean-up and abatement and provided personal protective equipment for its employees. On April 1, WDHS and MDH issued advisories to alert regional healthcare organizations, local public health agencies, and tribal health departments to the situation and launched a joint investigation on April 4. Subsequently, WDHS activated its Incident Command System and worked with MDH to compile a list of potentially exposed workers. By August 31, a total of 357 workers who might have been employed at the shipyard during December 2015–March 2016 had been identified.

During April–July 2016, WDHS and MDH attempted telephone interviews with workers.... information [was gathered] regarding employment history, work tasks, personal exposure prevention, symptoms commonly associated with lead exposures, and take-home contamination prevention and household composition...

As of August 31, a total of 233 (65.3%) of 357 workers received at least one BLL test and 185 (51.8%) completed interviews. Among 233 tested workers..., 171 (73.4%) had BLLs ≥5 μg/dL, 151 (64.8%) had BLLs ≥10 μg/dL, 33 (14.2%) had BLLs ≥40 μg/dL, and two (0.9%) had BLLs ≥60 μg/dL. Among 341 household members identified through worker interviews, 46 (13.5%) received a BLL test; none had an elevated BLL. Not all exposed workers and household members were tested for lead, and not every BLL test result might have been reported to WDHS or MDH....

The Occupational Safety and Health Administration (OSHA) enforcement investigation began on February 10, 2016 because of lead exposure hazards and revealed that shipyard workers were exposed to lead at ≤20 times the reduced permissible exposure limit of 40 μg/m^3.

This investigation highlights timely laboratory-based BLL reporting and efficient interstate collaboration. Moreover, it emphasizes the importance of implementing proper engineering controls and periodic BLL monitoring of employees exposed to lead and providing correct personal protective equipment for workers in the shipbuilding industry.

Modified and reproduced from Weiss D, Yendell SJ, Baertlein, LA, et al. Occupational lead exposures at a shipyard—Douglas County, Wisconsin, 2016. *MMWR*. 2017;66:34.

directly with patients who may be affected by communicable diseases. Microbial agents originate from a number of sources including human and animal blood and bodily fluids, feces from humans and animals (e.g., manure, droppings from rodents), persons who are carrying infectious diseases, infected primates and other animals, and disease-carrying insects. For example, of particular concern are the following agents and associated diseases:

- Bloodborne pathogens
  - Hepatitis B virus (HBV)
  - Hepatitis C virus (HCV)
  - Human immunodeficiency virus (HIV)
- Bacteria
  - Tuberculosis
  - Anthrax
- Viruses
  - Influenza
  - Other communicable respiratory diseases

Work-Related Stresses

Beyond the category of physical and chemical hazards are the psychological and somatic effects of work-related stresses. Chronic stress has been implicated in a range of somatic conditions (e.g., coronary heart disease) and mental disorders including depression. The term *going postal* refers to employees of the post office and other companies who react to stressful conditions of their environment, particularly the work environment, by committing violent acts. This topic is described in more detail later in the chapter.

Specific Occupationally Associated Diseases and Conditions

Referring back to Table 13.3, you will observe examples of the conditions that have been associated with occupational hazards. NIOSH developed the National Occupational Research Agenda (NORA), which was first publicized in 1996 and identified research priorities to create a safer work environment during the 21st century. NORA targeted 21 research priorities in the broad areas of disease and injury, the work environment and workforce, and research tools and approaches. For example, the area of disease and injury targeted for special focus the following topics:

- Allergic and irritant dermatitis
- Asthma and chronic obstructive pulmonary disease
- Fertility and pregnancy abnormalities
- Hearing loss
- Infectious diseases
- Low-back disorders
- Musculoskeletal disorders
- Traumatic injuries[16]

Allergic and Irritant Dermatitis (Contact Dermatitis)

Because of direct exposure to the occupational environment, the skin is one of the most common sites of contact with chemicals used in the work setting.[17] Consequently, the skin is vulnerable to numerous injuries and dermatologic conditions; the full extent of job-related dermatoses is unknown, because many skin diseases that result from occupational sources are not reported. Workers at highest risk of occupational skin diseases are employed in manufacturing, construction, food production, and activities such as metal plating and engine service.[18] Referring back to Figure 13.4, observe that in 2015 occupationally related skin diseases or disorders in private industry accounted for 15.6% of occupational illnesses. **FIGURE 13.7** shows a case of dermatitis, which has affected the subject's forearms, wrists, and hands.

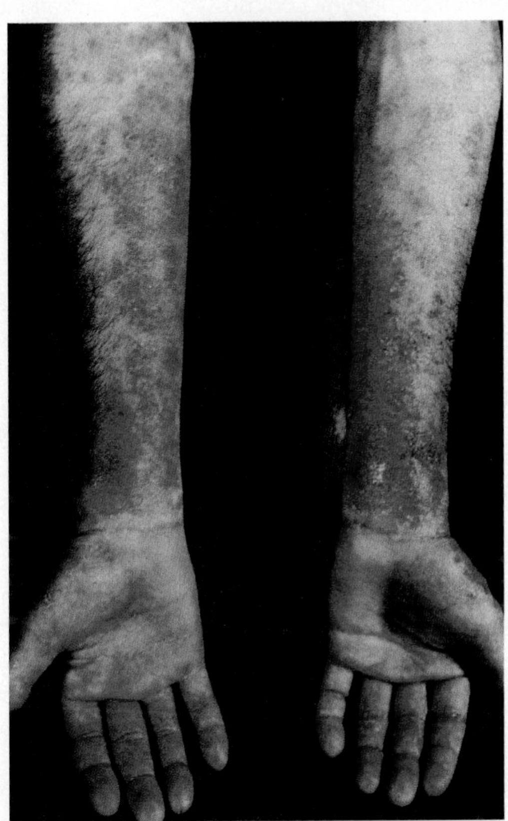

FIGURE 13.7 Occupational contact dermatitis.

Reproduced from Key MM, Henschel AF, Butler J, et al., eds. *Occupational Diseases: A Guide to Their Recognition*. (Rev. ed.) Washington, DC: National Institute for Occupational Safety and Health; 1977:78.

Occupationally Associated Respiratory Diseases

A number of significant respiratory diseases are linked to occupational exposures. Examples of these conditions, listed in **TABLE 13.4**, include asbestosis, coal workers' pneumoconiosis, silicosis, byssinosis, mesothelioma, lung cancer, and other noteworthy diseases that affect the respiratory system. Many of the work-related respiratory diseases are chronic conditions that have long latency periods. Some of the respiratory diseases are indirect or underlying, and not a direct cause of mortality; hence the frequency of their contributions as causes of morbidity and mortality is likely to be underestimated. Hazards arise not only from visible dust particles, but also from unseen dust particles. A discussion of some of the major conditions follows.

Chronic Obstructive Pulmonary Disease (COPD) and Asthma

As reported in statistics for the leading causes of death, COPD is called "chronic lower respiratory disease (CLRD)" and includes chronic and unspecified bronchitis, emphysema, asthma, and other chronic lower respiratory diseases.[19] In 2014, the category of CLRD (COPD) was the third leading cause of mortality in the United States.[19] COPD causes significant reduction in work time, economic losses, morbidity, and, ultimately, increases in the risk of mortality. Here are some additional facts regarding diseases in this classification[16].

- About 15–19% of cases of COPD and asthma can be linked to occupational exposures.
- Asthma has become the most frequently diagnosed occupational respiratory disease in occupational medicine clinics.

TABLE 13.4 Occupationally Associated Respiratory Diseases

Name of Lung Disease	Symptoms or Condition	Exposure
Asbestosis	Progressive scarring of lung tissue	Asbestos fibers
Coal workers' pneumoconiosis	Breathing difficulties	Coal dust
Silicosis	Presence of scar tissue in lungs	Crystalline silica from various sources
Byssinosis (brown lung disease)	Obstruction of small airways; impaired lung function	Dusts from hemp, flax, and cotton
Malignant mesothelioma	Uncommon cancer of chest lining	Asbestos
Hypersensitivity pneumonitis	Inflammation of air sacs of lungs; loss of normal breathing function	Organic dusts, fungus spores, and bird droppings
Asthma	Symptoms of asthma (e.g., wheezing, difficulty breathing)	Allergens in workplace (e.g., pollen, animal dander)
Chronic obstructive pulmonary disease	Chronic cough, loss of lung function	Air pollution and contamination
Respiratory tuberculosis	Symptoms of tuberculosis (e.g., night sweats, persistent cough)	Tuberculosis bacillus
Occupationally associated lung cancer	Cancer of the lung, trachea, and bronchus	Workplace carcinogens (e.g., beryllium, radon)
Disorders due to smoking in the workplace	Lung cancer, chronic bronchitis, and cough	Sidestream tobacco smoke

- Occupational exposures may exacerbate preexisting asthma.
- COPD is related to workplace exposure to dusts.

Pneumoconiosis

The term *pneumoconiosis* is defined as "the accumulation of dust in the lungs and the tissue reactions to its presence. For the purpose of this definition, 'dust' is meant to be an aerosol composed of solid inanimate particles."[20] Pneumoconiosis can result in fibrosis of the lungs (development of fibrous tissues) and lung nodules. The following list provides examples of substances that, when inhaled, may cause pneumoconiosis:

- Asbestos (associated with asbestosis and a lung condition known as mesothelioma, defined below)
- Cotton dust (byssinosis)
- Silica-containing dusts (silicosis)
- Coal dust (black lung disease, coal workers' pneumoconiosis)

FIGURE 13.8 reports trends in mortality from pneumoconioses. After 1972, the number of deaths from this cause declined, shrinking to 2,745 cases in 1999. Formerly, coal workers' pneumoconiosis was a leading cause of lung-associated mortality; the chart shows an increasing trend in mortality from asbestosis.

Asbestosis

Asbestosis refers to a type of pulmonary fibrosis associated with exposure to asbestos, the name assigned to six different fibrous minerals found in nature. Asbestosis results from inhalation of large amounts of asbestos fibers over long time periods. The condition occurs mainly among workers exposed to asbestos and is uncommon among the larger population. Asbestosis is related to declines in pulmonary function and increased risk of lung cancer, especially among smokers. Asbestos-associated mesothelioma is a rare form of cancer that invades the chest lining. From 1968 to 1999, over 18,000 deaths were caused by asbestosis. (Refer to Figure 13.8.)

The epidemiologist Irving Selikoff identified hazards of asbestos exposure during the 1950s. Although their use has declined greatly, asbestos-containing products were incorporated for many years into materials for fireproofing, insulation, and motor vehicle brakes. Asbestos has many useful attributes such as fire resistance and insulating ability. In 1973, the EPA banned the spraying of materials that contained

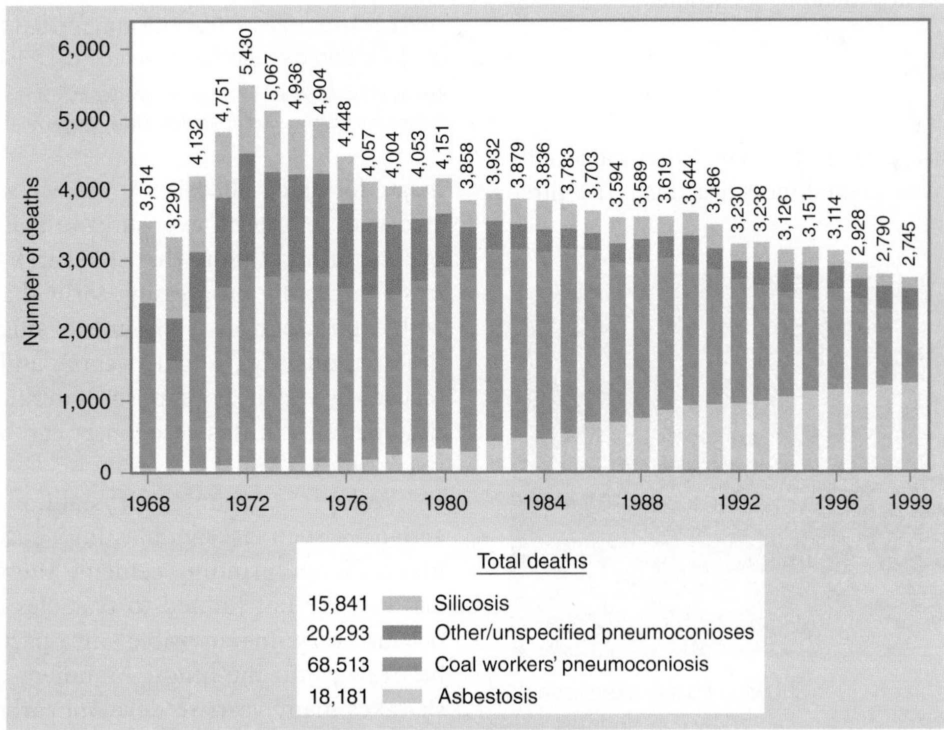

FIGURE 13.8 Number of deaths of US residents aged 15 or older with pneumoconiosis recorded as an underlying or contributing cause on the death certificate, 1968–1999.

Reproduced from Sestito JP, Lunsford RA, Hamilton AC, Rosa RR, eds. *Worker Health Chartbook 2004*. NIOSH Publication Number 2004-146. Cincinnati, OH: Department of Health and Human Services, Centers for Disease Control and Prevention, National Institute for Occupational Safety and Health; 2004.

asbestos on buildings; in 1989, the agency issued a regulation that would prohibit most other uses of asbestos. The EPA's 1989 regulation was overturned in 1991 by a court action that continued to permit many uses of asbestos.[21] Currently asbestos is banned from six categories of products including certain types of construction materials, certain paper goods, and all new applications of asbestos. The majority of car manufacturers have adopted substitutes for asbestos-containing brake linings.

Coal Workers' Pneumoconiosis (Black Lung Disease)

Coal workers' pneumoconiosis (CWP) is a potentially fatal lung disease that is associated with breathing coal dust, which then becomes deposited in the lungs as shown in **FIGURE 13.9**. Over time, the lungs of persons who are afflicted with CWP lose their elasticity, causing respiration to become increasingly difficult.

In the United States, the federal government regulates dust levels in coal mines. Following the enactment of the Federal Coal Mine and Safety Act of 1969, average dust levels declined sharply in coal mines. The 1969 law provided black lung disability benefits to coal miners who had incurred disabilities caused by mine-related dust exposure. **FIGURE 13.10** shows a poster that proclaims the availability of such benefits.

Silicosis

The lung disease *silicosis* is a condition caused by inhaling respirable crystalline silica dust. The three major forms of silicosis are *chronic silicosis* (associated with long-term exposures to silica at low levels over a period of more than 20 years), *accelerated silicosis* (a rapidly progressing form that is associated with exposure to large amounts of silica during short time periods, e.g., 5 to 15 years), and *acute silicosis* (caused by very heavy exposures to silica during short time periods; in this case, onset can be as soon as a year's time).[22]

Workers encounter crystalline silica during activities such as sandblasting, cutting tiles and masonry, and grinding cement. The dangers of silica dust are not limited to particles of visible size, because very fine, invisible silica particles also may be drawn into the lungs. As noted, chronic silicosis takes many years to develop; early in the course of the disease, few symptoms may be present or, if observable, may consist of coughing, shortness of breath, and weight loss. Silicosis causes the lungs and

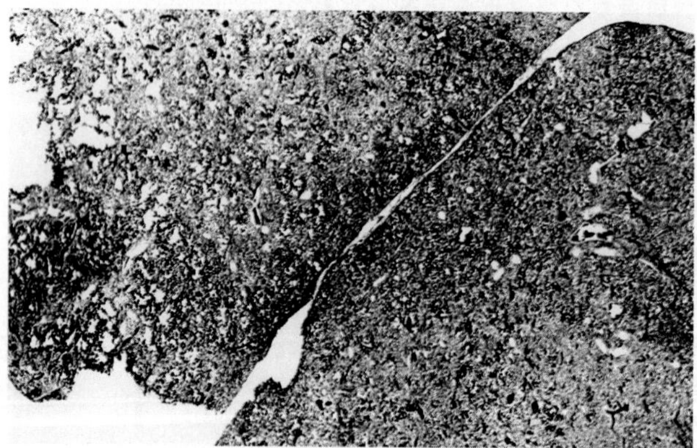

FIGURE 13.9 Section of a lung showing coal workers' pneumoconiosis. The dark areas are caused by coal deposits.

Reproduced from Key MM, Henschel AF, Butler J, et al., eds. *Occupational Diseases: A Guide to Their Recognition*. (Rev. ed.) Washington, DC: National Institute for Occupational Safety and Health; 1977:102.

FIGURE 13.10 This 1970 poster announced the availability of black lung benefits for disabled coal miners.

Reproduced from US Social Security Administration. Social Security history: black lung benefits. Available at: http://www.ssa.gov/history/blung.html. Accessed June 10, 2017.

chest lymph nodes to swell. Silicosis may result in pulmonary fibrosis, loss of lung function, and connective tissue disease. Persons afflicted with silicosis are at high risk of tuberculosis. Additional complications include lung cancer and respiratory failure. **FIGURE 13.11** shows a worker who is not protected from silica dust. The inset illustrates a close-up of silica dust.

Fertility and Pregnancy Abnormalities

Fertility and pregnancy abnormalities include dysfunctions that affect fetuses and infants; examples are birth defects, prematurity, low birth weight, spontaneous abortions, and developmental disabilities. There are many opportunities for workers to be exposed to workplace hazards that could be associated with these abnormalities. A special exposure group includes women of childbearing age and pregnant women. Here are some observations regarding workplace hazards and fertility and pregnancy abnormalities:

- Thousands of chemicals are used in the work setting, with more being added each year. A mere 4,000 of these chemicals have been tested for adverse reproductive effects.[16]
- Of particular concern are exposure to anti-cancer drugs, which may affect human reproduction.[16]
- Some hazards to human reproduction are known—lead, solvents, and ionizing radiation.

The etiologies of most birth defects and developmental disabilities have not been ascertained. With the exception of some known teratogens, the role of occupational exposures in these outcomes requires additional research.

Hearing Loss Caused by Noise

Hearing loss caused by noise in the work environment occurs very commonly and takes a major toll, particularly among construction workers such as carpenters as well as among employees in manufacturing. Here are some facts regarding hearing loss in the workplace[23]:

- Work-related noise is associated with hearing impairment. Data from the 2011–2012 National Health and Examination Survey indicated that

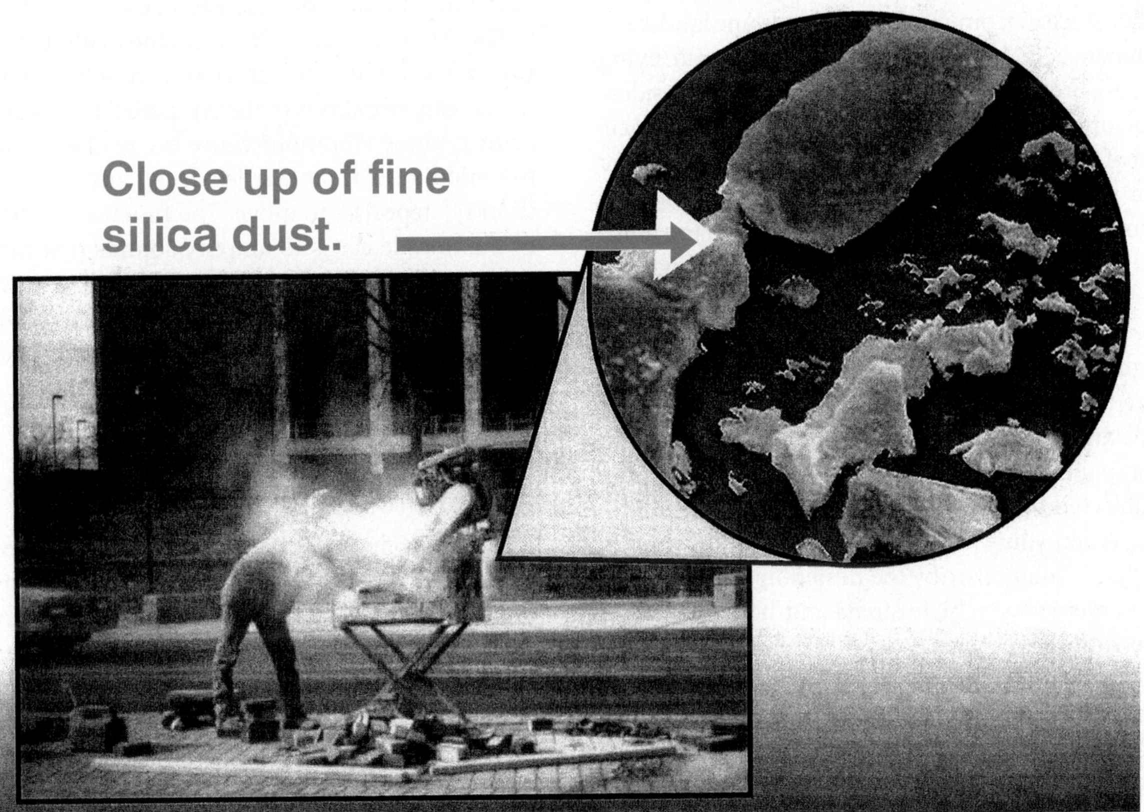
Close up of fine silica dust.

FIGURE 13.11 The worker shown is sawing bricks while unprotected from breathing silica dust.

Reproduced from CDC, National Institute for Occupational Safety and Health (NIOSH). Silicosis: Learn the Facts. DHHS (NIOSH) Publication No. 2004-108. Washington, DC: NIOSH; 2004:6.

about 33% of individuals exposed to noise at work had auditory impairment in comparison with 20% of those not exposed.[24]

- High levels of potentially dangerous job-related noise impact up to 22 million workers each year in the United States.[23]
- Each year workers' compensation for disability from hearing loss costs an estimated $242 million.[23]

Noise, a term that stems from the Latin word "nausea," is a feature of modern life that at best is an annoyance and at worst can affect our health adversely, both physically and mentally. According to a professor of otolaryngology, "Noise is the most ubiquitous global pollutant, which permeates all aspects of life throughout the inhabited world."[25(p3)] Although one of the primary direct consequences of excessive noise is hearing loss, other health consequences may include cardiovascular effects (e.g., exacerbation of high blood pressure), increases in stress levels, and sleep deprivation.[26] Noise has been recognized as harmful for more than 2,500 years: as early as 600 BCE, the Sybarites of Greece banned metal work that involved noise-producing hammering within city limits.[14]

Noise is a pervasive environmental problem, whether it arises from motor vehicles, home landscape maintenance, construction projects, jet aircraft flying overhead, hovering helicopters, or even an inconsiderate person playing music too loudly in a public place. Those of us (the majority in any urban area) who live near a major highway (e.g., an interstate highway in the United States, the *autobahn* in Germany, the *autostrada* in Italy, or any other heavily traveled highway in the world) are aware of the roar of trucks and other motor vehicles. Homeowners who are unfortunate enough to live under the flight path of an airport must endure the ear-splitting sounds produced by jets taking off, landing, and flying overhead. City dwellers who rely on public transportation—as in New York City—are subjected to the head-splitting screech of subway trains. When you walk around in most cities, you are impacted involuntarily by the deafening sounds emitted by emergency vehicle sirens and honking horns. In the suburbs, especially on weekends, one cannot fail to notice the racket produced by power lawnmowers, leaf blowers, trimmers, and similar machines. It is difficult to escape the din of modern society; even in remote national parks, noise from motor vehicles, snowmobiles, and aircraft intrudes.

As many as 30 million persons in the United States (11% of the population), and a similar percentage of Europeans, face the risk of hearing loss due to excessive noise.[25] Estimates suggest that about 12% of the population of the developing world is confronted with potential hearing loss due to high noise levels. The residents of megacities in Asia must cope with the incessant cacophony from motor vehicle engines, blowing of horns, and crowded shopping areas.

Even when noise levels are not sufficient to damage hearing, they may interfere with everyday life and impact the quality of our lives adversely. Some reactions to noise are annoyance, sleep disturbance, and reduction in cognitive performance. Think of the difficulty of trying to conduct classes in an environment such as a university located in a large, noisy city or under the flight path of an airport. Noise in the work environment may interfere with workers' alertness and communication and induce stress over time. Some experts believe that prolonged exposure to noise may result in psychological reactions that have adverse consequences for people's immune system status and physical well-being.[27]

Physiology of Hearing

The human ear translates the energy from sound waves into neurological impulses that are heard as sound through the actions of the ear canal, eardrum, bones of the middle ear, cochlea, and cochlear nerve. (Refer to **FIGURE 13.12**, which illustrates the major parts of the ear.) After sounds enter and pass through the ear canal, they create vibrations in the tympanic membrane (eardrum). These vibrations cause the ossicles—bones in the middle ear (hammer [maleus], anvil [incus], and stirrup [stapes])—to move. The function of the bones in the middle ear is to amplify sounds that move the oval window of the cochlea. In turn, this movement causes fluid in the cochlea to move and to stimulate hairs, which then stimulate cells that send neurological impulses along the cochlear nerve to the brain. These impulses are perceived as sound.

The Nature of Sound and Hearing

Sound is produced by oscillating waves of various frequencies. The term *Hertz (Hz)* denotes the number of cycles per second associated with the oscillation of a given sound wave, with low Hz numbers characterizing low tones and high Hz numbers characterizing high tones. The approximate range of human hearing is from a low of 20 Hz to 20,000 Hz on the upper end of the frequency scale. The usual sounds we hear in everyday life range from about 60 Hz to about 6,000 Hz.

Audiologists measure hearing level (HL) with an audiometer, an instrument that generates test tones of

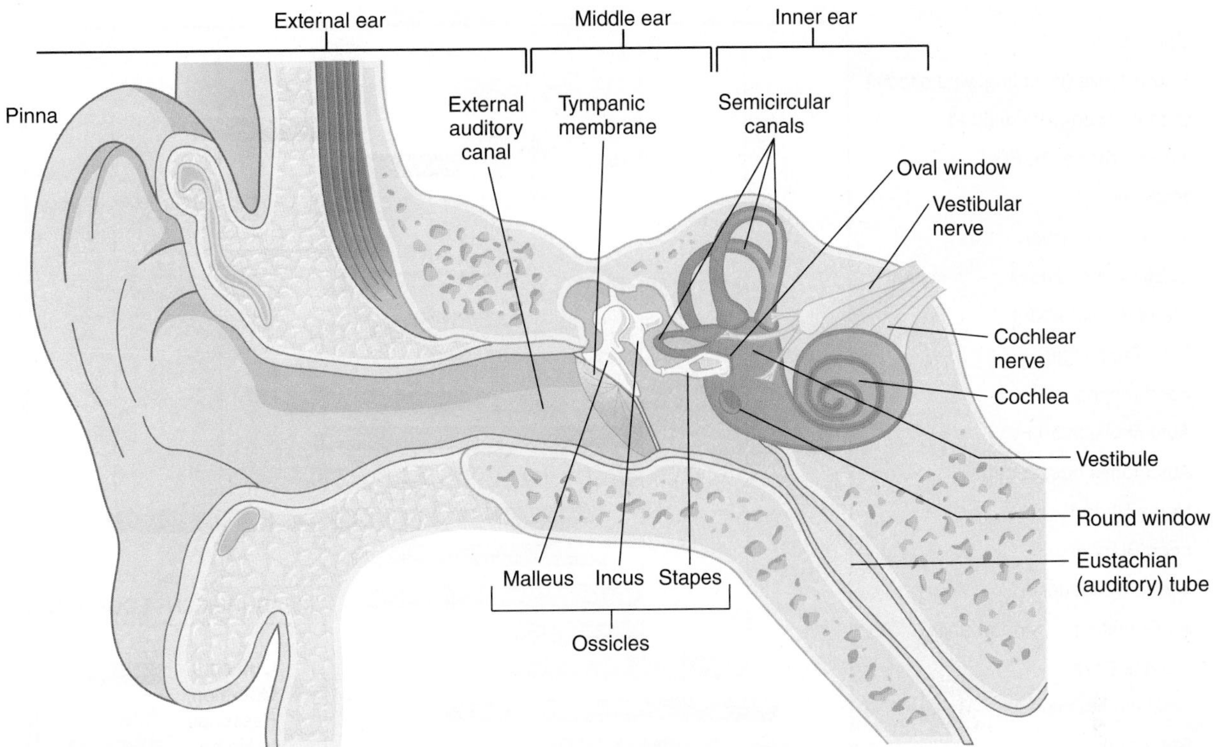

FIGURE 13.12 The outer, middle, and inner ear.

Modified and reproduced from NASA. The effects of space flight on the human vestibular system. Available at: https://www.nasa.gov/audience/forstudents/9-12/features/F_Human_Vestibular_System_in_Space.html. Accessed June 10, 2017.

known intensities. In a type of hearing test known as air conduction testing, hearing is measured by using pure tones generated by an audiometer; the hearing threshold refers to the lowest level of a tone to which a subject responds correctly 50% of the time. A patient's hearing level is defined as the difference between the threshold level that the tested person can hear and the average level that the general population can hear.

The sound pressure level (SPL) is a measure of the intensity of sound. The SPL is reported on a logarithmic scale that uses **decibels (dBs)**, the standard measure of the strength of a sound. While the issue of sound measurement is complex, the dB scale refers to 10 times the log of a ratio to the base 10. To illustrate, the logs of 10, 100, and 1,000 to the base 10 are 1, 2, and 3, respectively. Thus, 10 dB is 10 times the log of 10; a level of 30 dB is 10 times the log of 1,000; and a level of 60 dB is 10 times the log of 1,000,000—and so forth, for various points on the scale. Thus, an increase of 10 dB (e.g., from 20 dB to 30 dB) represents a 10-fold increase in sound intensity and not a linear jump of 10 units. Human beings can perceive differences between sound intensities from the lowest levels to the highest levels, which vary by a factor of more than 1 trillion. A logarithmic scale is advantageous for characterizing the large variability in the range of sounds that the human ear can perceive.

The dB scale is useful for describing SPLs of common sounds, hearing levels (HLs), and the sound levels that are associated with hearing loss:

- The threshold of hearing is 0 dB/HL.
- Sounds at 85 dB/SPL can produce slight hearing loss.
- Brief exposure to sounds at 100 dB/SPL can cause severe, permanent hearing loss.

FIGURE 13.13 lists the approximate SPLs in dB of common sounds: for example, refrigerator (45–68 dB), subway (80–115 dB), and chain saw (105–115 dB). The SPLs of some other sounds not shown in the chart are breathing (10 dB), quiet sounds in an office (50–60 dB), average traffic noise (80 dB), shotgun firing (130 dB), close proximity to a jet engine (140 dB), and threshold for pain (about 125 dB).[28]

Noise in the Workplace

Noise is a significant feature of most work environments, ranging from "quiet" offices to active construction sites. Two examples of employment categories affected greatly by noise are construction and health care.

Construction Industry. Sources of noise in the construction industry include bulldozers, heavy trucks,

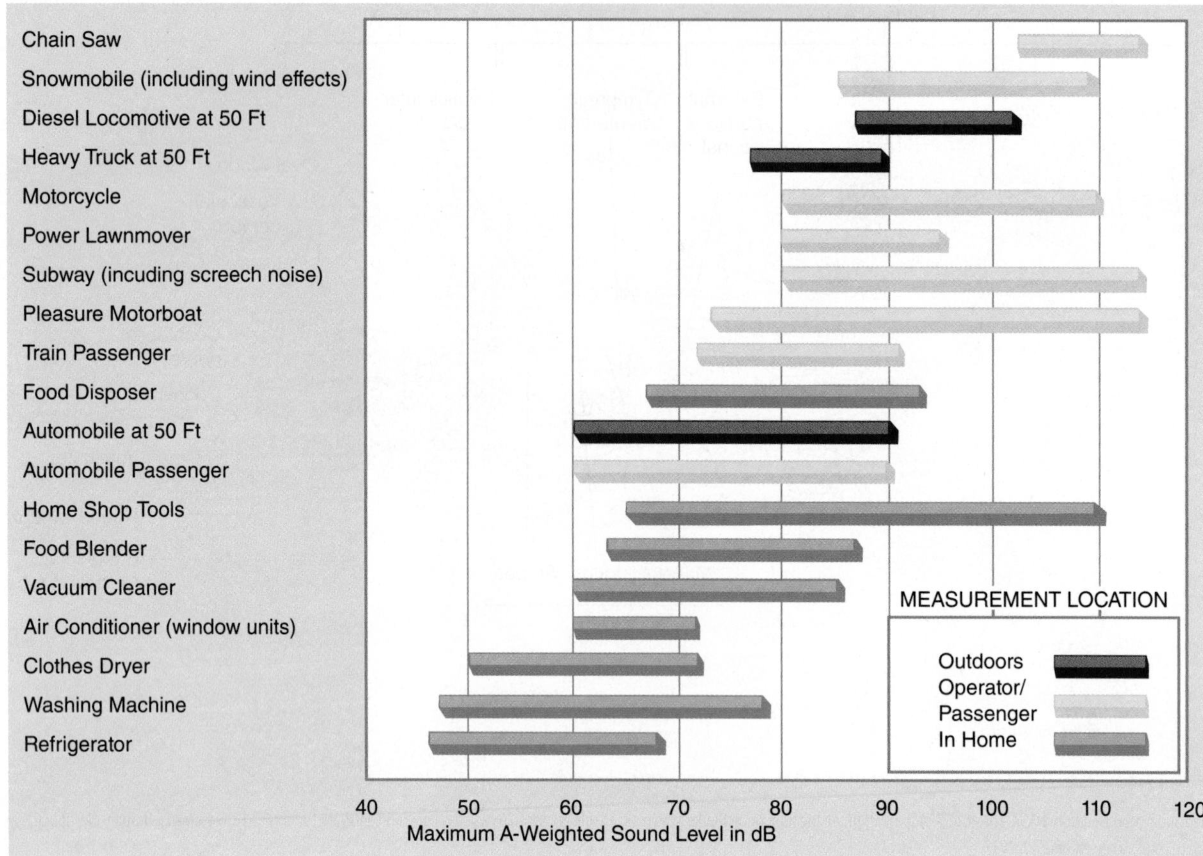

FIGURE 13.13 Typical range of common sounds.

Reproduced from US Environmental Protection Agency. Protective Noise Levels: Condensed Version of EPA Levels Document. EPA 550/9-79-100. Washington, DC: US EPA, Office of Noise Abatement & Control; November 1978:5.

and loading machines. Other noisy equipment includes jackhammers, nail guns, and drills. As a result, construction workers of almost all types are exposed to high noise levels.[29] **FIGURE 13.14** shows the sound levels associated with tools used by carpenters. For example, a chain saw produces sounds that reach the 110 dB level; other power tools used in carpentry may exceed this level.

As a consequence of exposure to high noise levels, many older carpenters have diminished hearing acuity. Also noteworthy is the fact that by age 25 carpenters typically have the hearing acuity of persons who are 50 years old and who have had no noise exposure. (Refer to **FIGURE 13.15**.)

Healthcare Industry. Noise levels in hospital operating rooms are reported to reach 118 dB; sources include high-speed bone-cutting drills and suctioning devices.[30] Operating room physicians, nurses, and other personnel are exposed to high noise levels from these sources. In dental offices, dentists and hygienists are exposed to excessive noise from air compressors, ultrasonic instrument cleaners, high-speed dental drills, and scalers.[31]

Permissible Noise Levels in the Workplace. The NIOSH recommended exposure limit (REL) for noise exposure in the workplace during an 8-hour shift is 85 dBA (using a type of weighting on a sound meter called "A-weighting" and meaning that exposure is an 8-hour time-weighted average).[32] The REL remains as of the publication of the third edition of this textbook (2018). Employers are required to post warning signs in work areas where noise levels exceed 85 dBA. For higher noise levels, exposure times must be limited, as shown in **FIGURE 13.16**.

Infectious Diseases

Several modes of transmission of infectious diseases in occupational settings are possible, as noted in the following examples:

- Healthcare workers are at increased risk of exposure to the hepatitis virus (e.g., hepatitis B and C) and the human immunodeficiency virus (HIV), particularly through accidental needle sticks. Other sources of exposure include biohazards resulting from direct patient contact as well as contact with blood and blood products.

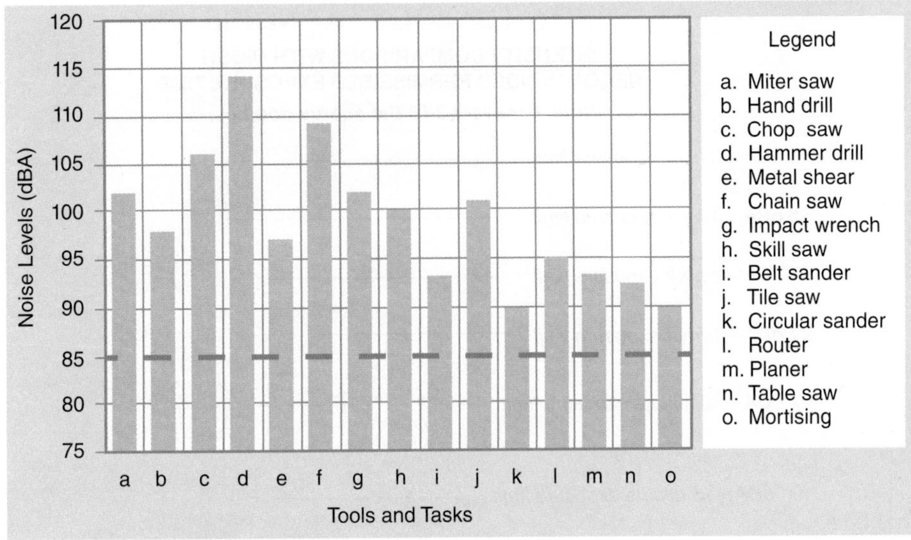

FIGURE 13.14 Carpenters' noise exposures from tools used for various tasks.

Reproduced from CDC, National Institute for Occupational Safety and Health (NIOSH). Noise and Hearing Loss Prevention. Available at: https://www.cdc.gov/niosh/topics/noise/chart-carpenters.html. Accessed June 10, 2017.

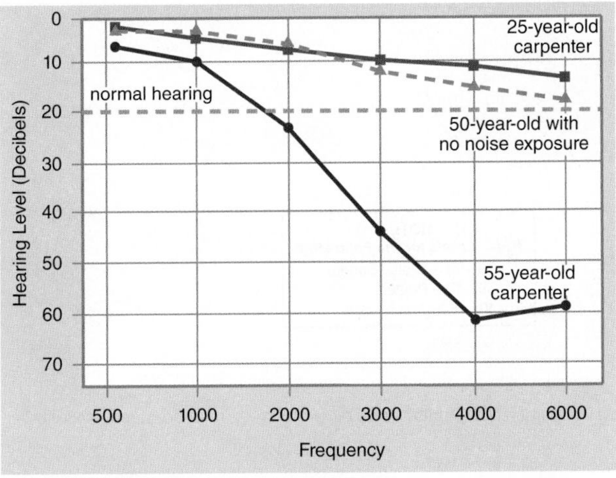

FIGURE 13.15 The average 25-year-old carpenter has the ears of a 50-year-old person who has not been exposed to noise.

Reproduced from CDC, National Institute for Occupational Safety and Health (NIOSH). Noise and Hearing Loss Prevention. Available at: https://www.cdc.gov/niosh/topics/noise/chart-50yrold.html. Accessed June 10, 2017.

- Public utility workers may be exposed to the hazards of disease carried in raw sewage.
- Agricultural workers are at increased risk of exposure to zoonotic diseases through contact with animals and disease agents contained in the soil.
- Social service workers and corrections personnel may be exposed to tuberculosis.
- Clinical laboratory specialists may be exposed to infectious agents from specimens including bloodborne pathogens, viruses, and bacterial agents (e.g., *Yersinia pestis* and *Salmonella typhi*, among others).
- Mortuary workers are exposed to a range of microbial agents as human bodies are embalmed or exhumed from burial sites.
- Adult film industry workers are at risk from a range of bloodborne and sexually transmitted infections.

Musculoskeletal Disorders

The term **musculoskeletal disorder (MSD)** refers to "an injury or disorder of the muscles, nerves, tendons, joints, cartilage, or spinal discs."[33] These conditions occur widely in the work environment. For example, a total of 356,910 cases of musculoskeletal disorders were reported in 2015 among workers in private industry, state government, and local government; MSDs accounted for 31% of all nonfatal occupational injury and illness cases requiring days away from work.[34] A total of 317,440 cases of MSDs were reported in 2008. Of this total 74% were caused by sprains, strains, and tears. Refer to **FIGURE 13.17**.

Carpal Tunnel Syndrome

Carpal tunnel syndrome (CTS) is a condition associated with compression or squeezing of the median nerve, which runs from the forearm to the palm of the hand through the carpal tunnel of the wrist. CTS may produce symptoms of numbness or pain in the wrist and hand.[35] The condition may necessitate long periods of recuperation and absence from work. CTS is related to repetitive work activities with the hand and wrist; one of the work domains that has been described as having a high frequency of CTS is the meat packing industry.[36] The use of computer keyboards for extended time periods also has been identified as a risk factor for CTS.

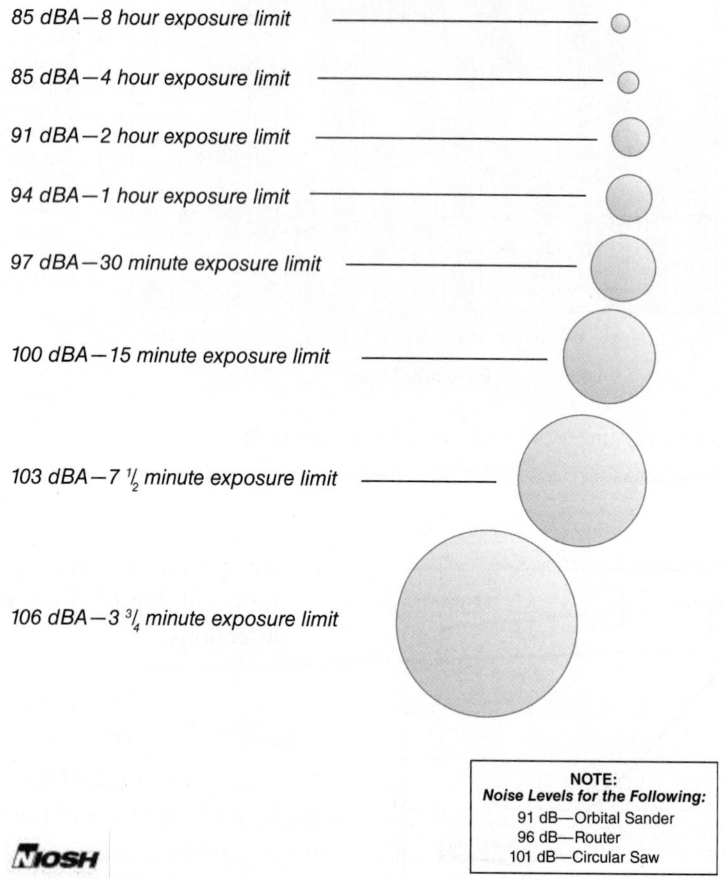

FIGURE 13.16 Permissible noise levels in the US workplace.

Reproduced from CDC, National Institute for Occupational Safety and Health (NIOSH). Noise and Hearing Loss Prevention. Available at: https://www.cdc.gov/niosh/topics/noise/chart-lookatnoise.html. Accessed June 10, 2017.

Back Injuries

Back injuries are an extremely common form of work-related morbidity. In 2015, the Bureau of Labor Statistics (BLS) reported a total of 200,500 back-related injuries or illnesses among private, state, and local government employers. Such injuries resulted in a median of 8 days away from work. Back injuries accounted for about 17% of all the occupational injuries and illnesses that required time away from work in 2015.[34] Factors that contribute to job-associated back pain and injuries include performing repetitive movements that affect the back, remaining sedentary for long periods, and putting too much strain on the back, (e.g., by heavy lifting).[37] One of the occupations with a high risk of back injuries is the nursing profession, in which back injuries may occur when nurses or aides are required to lift patients. Approximately 38% of nurses are reported to have sustained back injuries.[38] Such injuries may be caused by lifting over barriers or lifting from an awkward position. Other workers with a high frequency of back problems include laborers, carpenters, and drivers.

Work-Related Injuries and Fatalities

Work-related injuries and fatalities have continued to be an issue of major consequence for society worldwide, in terms of both human suffering and economic costs. In the United States, workplace injuries exact billions of dollars each year in medical expenses, foregone wages, reduced productivity, and other costs.

What is meant by an injury? Injuries are defined as the result of energy applied to the human body in such a manner that the energy exceeds the body's physiological tolerance. Refer to the chapter on injuries with a focus on unintentional injuries and deaths for more information on this definition and the topic of injuries in general. The present section focuses primarily on fatal occupational injuries.

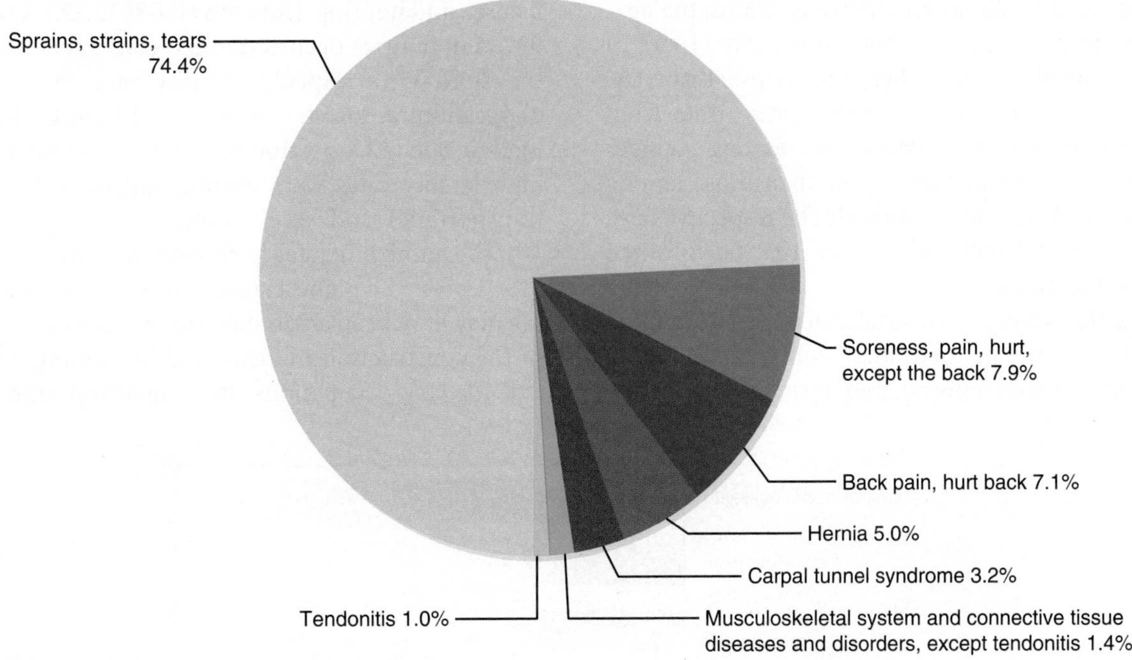

FIGURE 13.17 Distribution of musculoskeletal disorders by nature of injury or illness, 2008.

Reproduced from US Department of Labor, Bureau of Labor Statistics. Injuries, Illnesses, and Fatalities. Case and demographic characteristics for work-related injuries and illnesses involving days away from work. Available at: http://www.bls.gov/iif/oshcdnew.htm. Accessed June 10, 2017.

Work-related traumatic injuries are serious injuries from causes such as vehicle collisions, workplace violence, and falls. In 2007, a total of about 5,400 persons, or approximately 15 people per day, died in the United States from traumatic workplace injuries. The total number of workplace-related deaths in 2015 was 4,836.[39] Consequently, because of the large toll that they take, fatal injuries in the workplace are a very significant public health concern. The National Institute for Occupational Safety and Health (NIOSH) states:

> Over 93,000 workers were fatally injured while working in the United States from 1980 through 1995. Each day an average of 16 people died—simply by doing their jobs. These deaths result from exposure to many different hazards on the job. Leading causes of traumatic occupational fatalities include motor vehicles, homicides, machines, falls, electrocutions, and falling objects. . . . Workers are at risk of fatal injury in many different ways. Workers who operate motor vehicles or machines risk injury due to overturns or collisions. Taxicab drivers risk being killed during robbery attempts while construction workers risk fatal falls while working from heights. Electrical linemen risk electrocution while repairing power lines and loggers risk being struck by a falling tree during tree harvesting operations. These scenarios highlight some of the risks that are a daily presence in many of the industries and occupations in the United States. These jobs are frequently noted as being the jobs with the highest fatal injury rates. However, the potential for work-related injury exists in any job where injury risks are present and not controlled.[40(pxi)]

Sources of information on occupational fatalities include NIOSH and the BLS in the US Department of Labor. NIOSH is the federal agency charged with conducting research on occupational illnesses and injuries. NIOSH gathers information on fatal occupational injuries among workers aged 16 years and older.[40] This information is derived from cases of work-related injury reported on death certificates from 52 vital statistics–reporting units that encompass the 50 US states, New York City, and the District of Columbia. The National Traumatic Occupational Fatalities (NTOF) Surveillance System was implemented by NIOSH in the 1980s to conduct a census of death certificate–based information in order to perform descriptive and analytic epidemiologic analyses of causes of job-related mortality. A second source of occupational fatality information is the BLS Census of Fatal Occupational Injuries (CFOI), which collects information from various state and federal data sources.[41]

Here are some facts regarding the 4,836 fatal occupational injuries reported in 2015. (Refer to **FIGURE 13.18**.) Most fatalities in 2015 (42%) concerned transportation incidents and highway deaths; the next most frequent category—about one-sixth (16%) of fatalities—involved falls, slips, and trips. The CDC reported age disparities in fatality rates. Data from 2007 revealed that the fatality rates among younger (aged 15–24) workers were lower than those among older (aged 25 and older) workers. The respective rates were 3.6 per 100,000 and 4.4 per 100,000 full-time equivalent workers.[42]

In 2015, several industrial categories accounted for the largest number of deaths—construction; transportation and warehousing; and agriculture, forestry, fishing, and hunting. (Refer to **FIGURE 13.19**.) A total of 937 fatal injuries occurred in the construction industry alone. With respect to fatality rates, the category of agriculture, forestry, fishing, and hunting had the highest rate (22.8 per 100,000). Other industries with high fatality rates were mining and the category of transportation and warehousing.

The number of fatal work-related injuries reported in 2015 was the highest since 2008.[39] In part this finding may have been attributable to increases in fatalities in the construction industry and in farming, fishing, and forestry occupations. In comparison with 2014,

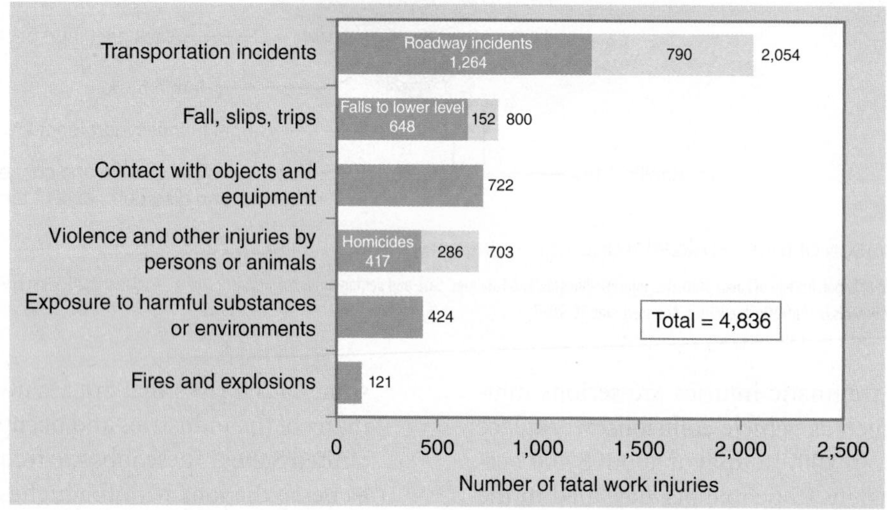

FIGURE 13.18 Fatal occupational injuries by major event, 2015.

Reproduced from U.S. Department of Labor, Bureau of Labor Statistics, 2016. Fatal occupational injuries in 2016. Available at: https://www.bls.gov/iif/oshwc/cfoi/cfch0014.pdf. Accessed August 21, 2017.

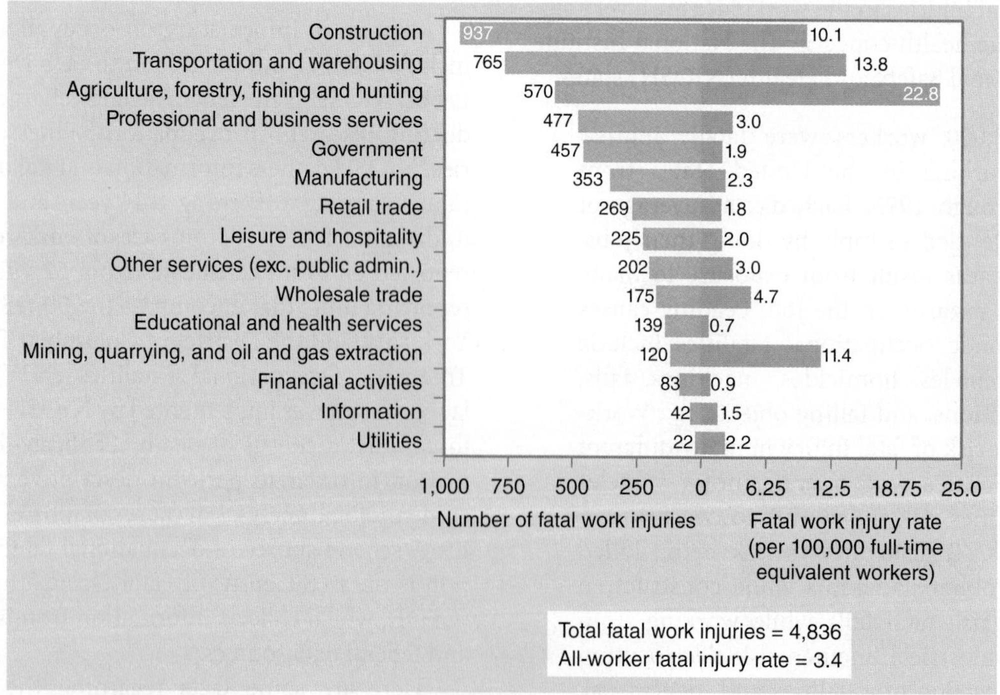

FIGURE 13.19 Number and rate of fatal work injuries by industry sector, 2015.

U.S. Bureau of Labor Statistics, Current Population Survey, Census of Fatal Occupational Injuries, 2016. Available at: https://www.bls.gov/iif/oshwc/cfoi/cfch0015.pdf

declines in fatalities were reported for workplace suicides; however, homicides increased.

Job Stress and Associated Conditions

Job stress can affect the health of workers adversely and results in significant time away from work. The term **job stress** is "defined as the harmful physical and emotional responses that occur when the requirements of the job do not match the capabilities, resources, or needs of the worker. Job stress can lead to poor health and even injury."[43(p6)] The BLS reported 5,659 cases of anxiety, stress, and neurotic disorders that involved absence from work in 2001 (most recently available data).[41] As shown in **FIGURE 13.20**, the rates of anxiety, stress, and neurotic disorders declined from 0.8 per 10,000 full-time workers to 0.6 per 10,000 workers (about 25%) between 1992 and 2001.

More recent data from 2016 indicate that self-reported job stress (feeling tense and stressed out at work) is prevalent in the American workplace. Job stress varies according to the generational membership of an employee. About 45% of millennials report being under stress at work in comparison with slightly more than 20% of boomers. Job stress is associated with adverse physical health effects (e.g., experiencing physical symptoms) and adverse mental health consequences (e.g., depression). These physical and mental health outcomes vary according to generations, being most pronounced among millennials. (See **FIGURE 13.21**.)

One of the sources of job stress arises from violent events that transpire in the workplace or that are related to the performance of one's occupation. *Critical incident stress* refers to intense reactions that arise

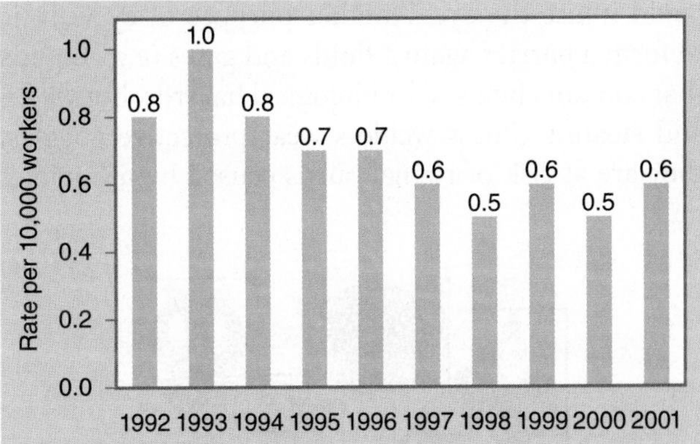

FIGURE 13.20 Annual rates of anxiety, stress, and neurotic disorder cases involving days away from work in private industry, 1992–2001.

Reproduced from Sestito JP, Lunsford RA, Hamilton AC, Rosa RR, eds. *Worker Health Chartbook 2004*, Chapter 2: Fatal and Nonfatal Injuries, and Selected Illnesses and Conditions. NIOSH Publication Number 2004-146. Cincinnati, OH: Department of Health and Human Services, Centers for Disease Control and Prevention, National Institute for Occupational Safety and Health; 2004:35.

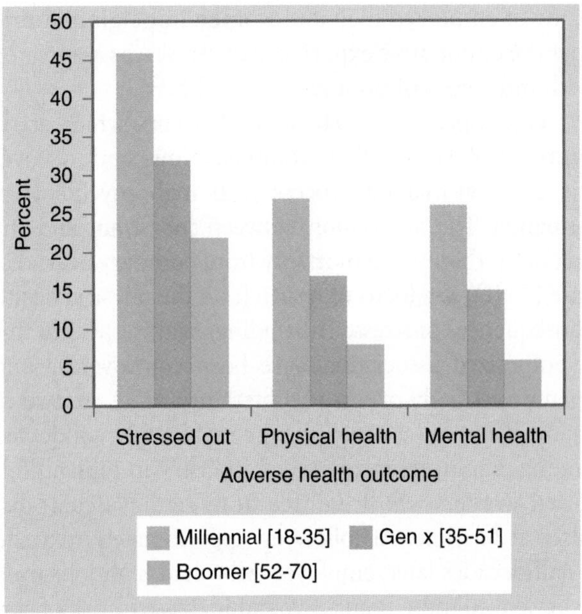

FIGURE 13.21 Generational effects of job stress.

Definitions:
Stress—feeling tense and stressed out at work.
Physical health—experiencing physical symptoms (e.g., dizziness) during workday.
Mental health—depression or other mental health issues limited goal achievement.
Data from American Psychological Association. 2016 Work and Well-being Survey. Available at: http://www.apaexcellence.org/assets/general/2016-work-and-wellbeing-survey-results.pdf?_ga=2.257129099.1693186252.1497117907-309235544.1497117907. Accessed June 10, 2017

from witnessing violent behavior or from dealing with traumatic incidents such as those experienced by paramedics or personnel who are first responders to disasters. Critical incident stress management (CISM) is a method of helping persons cope with the aftermath of the experience of a traumatic event. One form of CISM involves debriefing the individual.[44]

Sources of job stress include:

- Work overload—long hours, high-pressure deadlines
- Job dissatisfaction
- Job insecurity due to fear of layoff, changes in the nature of work, global competition, and deregulation of labor
- Workers' lack of control over their environment and conditions of employment; e.g., employment insecurity and shift work are associated with work-related stress.[45]
- Assembly line work that involves repetitive tasks
- Dealing with members of the public who may be abusive
- Inadequate compensation and lack of benefits such as health insurance

Job stress has been researched extensively as a possible contributor to both impaired mental health status and poor physical health status. In one example that addressed mental health, a large, 1-year, prospective study of employed men and women in Belgium found that the cumulative experience of stress was associated with indicators of poor mental health.[46]

One aspect of job stress is job strain, which arises from employment that combines low control over the decision-making process with high psychological demands. The association between job strain and the risk of morbidity and mortality from coronary heart disease (CHD) is a focus of research on the physical health consequences of stress. The findings with respect to this hypothesized association have been controversial and span a vast body of research literature. Here are two of many examples: a prospective cohort study conducted among employees in the metal industry in Finland followed workers who were free from cardiovascular diseases at baseline. At follow-up approximately two and a half decades later, employees who had high job strain experienced more than a twofold increase in risk of cardiovascular mortality.[47] In contrast, a prospective study conducted among subjects from the Framingham Offspring Study did not find that job strain was a significant risk factor for CHD or CHD mortality in men and women.[48]

Prevention of Occupational Disease

The best method for control of occupational hazards is primary prevention, which includes the following[49]:

- Engineering controls
- Modification of work practices
- Administrative controls

Examples of engineering controls are physical modifications of the work environment to reduce hazards, such as use of quieter machinery, installation of protective guards, and improvement of building ventilation to remove dusts and vapors. Examples of modification of work practices include use of safety education and training programs in order to change work procedures so that they are safer than the usual methods. Examples of administrative controls are organization of work shifts and rotation of employees to minimize exposure to hazards. Several other methods for protecting workers from occupational hazards are use of personal protective equipment, public health surveillance, and establishment and enforcement of exposure limits.

Personal Protective Equipment

A procedure for reducing exposure to workplace hazards, and the one that is the last line of defense, is the use of personal protective equipment. The Occupational Safety and Health Administration (OSHA) defines **personal protective equipment (PPE)** as apparatuses "designed to protect employees from serious workplace injuries or illnesses resulting from contact with chemical, radiological, physical, electrical, mechanical, or other workplace hazards. Besides face shields, safety glasses, hard hats, and safety shoes, PPE includes a variety of devices and garments such as goggles, coveralls, gloves, vests, earplugs, and respirators."[49]

Devices to Protect against Airborne Hazards

The two basic types of PPE for protecting workers against airborne hazards are contaminant removers and clean air suppliers. Contaminant removers filter airborne particles or remove airborne chemicals and gases. Clean air suppliers are either attached to a line that brings in clean air from a remote source or are self-contained breathing apparatuses that issue clean air from tanks.[50] **FIGURE 13.22** illustrates some of the types of PPE for protection against airborne hazards.

Devices to Protect Hearing

Examples of devices used to protect hearing are ear plugs and ear muffs, as shown in **FIGURE 13.23**.

Protective Eyeware

Protective eyeware includes goggles, face shields, safety glasses, and respirators that cover the full face. Eyeware protects against solid objects such as flying chips, metal fragments, and other small objects that could injure the eye. Another purpose of eyeware is to form a barrier against fluids and gases (e.g., liquids that contain chemical or biological hazards, hot water, and steam). Unless welders wear protective goggles, they are at risk of retinal burns caused by ultraviolet

FIGURE 13.22 Respirators for protection against gases and chemicals.

Reproduced from US Department of Labor, Occupational Safety and Health Administration (OSHA). Woodworking eTool, Finishing/Chemicals, Personal Protective Equipment. Available at: http://www.osha.gov/SLTC/etools/woodworking/finishchems_ppe.html. Accessed June 8, 2017.

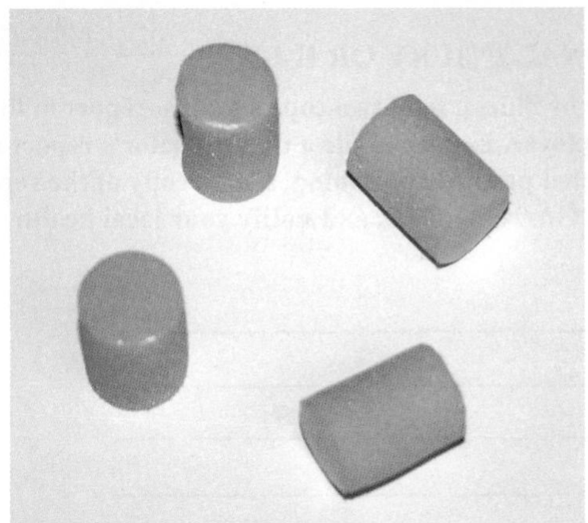

FIGURE 13.23 Hearing protection devices. Earplugs (top) and ear muffs (bottom).

radiation. **FIGURE 13.24** demonstrates a worker who is wearing a protective shield.

Public Health Surveillance

Public health surveillance is defined as "the ongoing systematic collection, analysis, and interpretation of health data essential to the planning, implementation, and evaluation of public health practices, closely integrated with the timely dissemination of these data to those who need to know."[51(p2)] Surveillance systems include the collection of information about occupational injuries and illnesses and maintenance of databases on exposures (exposure databases) to occupational hazards such as noise, industrial chemicals, and ionizing radiation. Exposure surveillance systems and databases can be instrumental in advancing the health and safety of the workplace.[52] These systems can provide the foundation for epidemiologic studies of adverse health outcomes in the workplace.

FIGURE 13.24 Worker who is wearing protective goggles and a face shield.

Reproduced from US Department of Labor, Occupational Safety and Health Administration. Shipyard Employment eTool, Ship Repair, PPE Selection, Eye and Face Protection. Available at: http://www.osha.gov/SLTC/etools/shipyard/standard/ppe/general_ppe/eye_face.html. Accessed June 9, 2017.

Several US federal agencies collaborate in surveillance programs; these agencies include NIOSH, the National Center for Health Statistics (NCHS), BLS, OSHA, and the Mine Safety and Health Administration (MSHA).[51] Examples of surveillance programs operated by NIOSH are:

- National Surveillance System of Pneumoconiosis Mortality (NSSPM)
- National Traumatic Occupational Fatalities (NTOF) Surveillance System
- State-based Sentinel Event Notification System for Occupational Risks (SENSOR)

A noteworthy example of state-level surveillance of occupational diseases and injuries is the Occupational Health Surveillance and Evaluation Program (OHSEP) operated by the State of California's Department of Health Services.[53] OHSEP examines the occurrence of conditions such as work-related asthma, pesticide-related illnesses, and fatal occupational injuries through data obtained from reporting systems. The Doctor's First Report of Occupational Injury or Illness is one of the reporting forms used by OHSEP. (Refer to **FIGURE 13.25**.)

Exposure Limits

In most situations, workplace exposures to hazardous agents are higher than the exposures of the general public. In order to protect workers from these higher exposure levels to hazardous agents, organizations and government agencies have established guidelines and regulations for limitation of exposures. These

Print Form

STATE OF CALIFORNIA

DOCTOR'S FIRST REPORT OF OCCUPATIONAL INJURY OR ILLNESS

Reset Form

Within 5 days of your initial examination, for every occupational injury or illness, send two copies of this report to the employer's workers' compensation insurance carrier or the insured employer. Failure to file a timely doctor's report may result in assessment of a civil penalty. In the case of diagnosed or suspected pesticide poisoning, send a copy of the report to Department of Industrial Relations, P.O. Box 420603, San Francisco, CA 94142-0603, and notify your local health officer by telephone within 24 hours.

1. Insurer Name and Address

2. Employer Name

3. Address No. and Street | City | Zip Code

4. Nature of business (e.g. food manufacturing, building construction, retailer of women's clothes.)

5. Patient Name (first Name, middle initial, last name) | 6. Sex | 7. Date of Birth

8. Address No. and Street | City | Zip Code | 9.Phone Number

10. Occupation (Specific job title) | 11. Social Security Number | 12. Address No.& Street Where Inj. Occurred

City Where Injury Occ. | County | 13. Date and hour of injury or onset of illness

14. Date last worked | 15. Date and hour of 1st exam or treatment | 16. Have you or your office previously rendered treatment

Patient please complete this portion, if able to do so. Otherwise, doctor please complete immediately, inability or failure of a patient to complete this portion shall not affect his/her rights to workers' compensation under the California Labor Code.

17. **Describe how the accident or exposure happened.** (Give specific object, machinery or chemical. Use reverse side if more space is required.)

18. **SUBJECTIVE COMPLAINTS**

19. **Objective Findings**

A. Physical Examination

B. X-ray and laboratory results (State if none or pending.)

Form 5021 (Rev. 5) 10/2015 Sheet 1 of 3

FIGURE 13.25 Doctor's first report of occupational injury or illness.

STATE OF CALIFORNIA **DOCTOR'S FIRST REPORT OF OCCUPATIONAL INJURY OR ILLNESS**

20. **DIAGNOSES** (if occupational illness specify etiologic agent and duration of exposure.) Chemical or toxic compounds involved? [ ]

1. ______________________ ICD-10 ________
2. ______________________ ICD-10 ________
3. ______________________ ICD-10 ________
4. ______________________ ICD-10 ________
5. ______________________ ICD-10 ________
6. ______________________ ICD-10 ________
7. ______________________ ICD-10 ________
8. ______________________ ICD-10 ________
9. ______________________ ICD-10 ________
10. ______________________ ICD-10 ________
11. ______________________ ICD-10 ________
12. ______________________ ICD-10 ________

21. Are your findings and diagnosis consistent with patient's account of injury or onset of illness? [ ] If "no," please explain below:

22. Is there any other current condition that will impede or delay patient's recovery? [ ] If "yes," please explain below:

23. **TREATMENT RENDERED** (Use reverse side if more space is required.)

24. If further treatment required, specify treatment plan/estimated duration.

25. If hospitalized as inpatient, give hospital name and location

Date admitted [ ] Estimated length of stay [ ]

26. WORK STATUS - Is patient able to perform usual work? ☐ Yes ☐ No

If "no", date when patient can return to Regular work [ ] Modified work [ ]

Specify restrictions

Form 5021 (Rev. 5) 10/2015

Sheet 2 of 3

FIGURE 13.25 (Continued)

STATE OF CALIFORNIA

DOCTOR'S FIRST REPORT OF OCCUPATIONAL INJURY OR ILLNESS

Physician Signature: *(original signature, do not stamp)*

I declare under penalty of perjury that this report is true and correct to the best of my knowledge and that I have not violated Labor Code section 139.3.

Physician signature ______________________ Cal. License Number: ______________

Executed at: ______________________ Date *(mm/dd/yyyy)*: ______________

Physician Name ______________________ Specialty: ______________

Physician address: ______________________ Phone Number ______________

Any person who makes or causes to be made any knowingly fraudulent material statement or material representation for the purpose of obtaining or denying workers' compensation benefits or payments is guilty of a felony.

PRIVACY NOTICE: The Administrative Director is authorized to maintain the records of the Division of Workers' Compensation (DWC). (Cal. Lab. Code § 126.) The Information Practices Act of 1977 and the Federal Privacy Act require the Administrative Director to provide this notice to individuals who submit information to the DWC pertaining to a workers' compensation claim. (Cal. Civ. Code § 1798.17; Public Law 93-579.)

The principal purpose for requesting information from injured workers, dependents, lien claimants, physician, employers or their representatives is to administer the California workers' compensation system. Each form shows which fields are required to be completed for DWC to process the form. If a required field in a form is incomplete or unreadable, the DWC may return the form to the individual for correction or may reject the form. Providing a social security number is required on this form pursuant to Labor Code § 6409. If you do not provide your security number, the DWC may return the form to you for correction or reject the form. If you do not have a social security number, indicate this in the space provided for the injured worker's social security number. As permitted by law, social security numbers are used to help properly identify injured workers and to conduct statistical research as allowed under the Labor Code.

As authorized by law, information furnished on this form may be given to: you, upon request; the public, pursuant to the Public Records Act; a governmental entity, when required by state or federal law; to any person, pursuant to a subpoena or court order pursuant to any other exception in Civil Code § 1798.24.

An individual has a right of access to records containing his/her personal information that are maintained by the Administrative Director. An individual may also amend, correct, or dispute information in such personal records. (Cal. Civ. Code §§ 1798.34-1798.3.) You may request a copy of the DWC's policies and procedures for inspection of records at the address below. Copies of the procedures and all records are ten cents ($0.10) per page, payable in advance. (Cal. Civ. Code § 1798.33.) Requests should be sent to: Division of Workers' Compensation- Medical Unit, P.O. Box 71010, Oakland, CA 94612. Tel: (510) 286-3700 or (800) 794.6900. Fax: (510) 622-3467.

Form 5021 (Rev. 5) 10/2015 Sheet 3 of 3

FIGURE 13.25 (Continued)

standards make use of the terms **threshold limit values (TLVs)**, *biological exposure indices (BEIs)*, *permissible exposure limits (PELs)*, and *maximum allowable concentrations (MACs)*. The definitions of the foregoing terms are as follows:

- *Threshold limit value (TLV)*—"Refers to airborne concentrations of substances and represents conditions under which it is believed that nearly all workers may be unaffected."[54] Guidelines for TLVs are published by the American Conference of Governmental Industrial Hygienists (ACGIH) for use by trained industrial hygienists to assist them in decision-making regarding safe exposure levels to chemicals and physical agents in the workplace.[55]
- *Biological exposure indices (BEIs)*—"A measure of the amount of chemical absorbed into the body."[54] The ACGIH also publishes BEIs annually.
- *Permissible exposure limits (PELs)*—"An OSHA standard . . . [that] is subject specific and has the supporting documentation for promulgation and enforcement of occupational health regulations. [PELs denote] [a]n allowable exposure level in the workplace air [that is] averaged over an eight-hour shift."[54] This type of average is called a time-weighted average (TWA). PELs refer to airborne concentrations of substances, although they may refer also to dermatologic exposures.[56]
- *Maximum allowable concentrations (MAC)*—"define permissible levels of exposure to chemicals."[54]

Conclusion

The scope of the occupational environment is diverse and poses hazards to workers who are employed in many different types of employment settings. The field of occupational health has a lengthy history, beginning with the first humans who engaged in occupations such as mining and smelting metals. During the Industrial Revolution, the 19th century, and the early 20th century, workers often were required to work long hours under extremely hazardous conditions. Work-related diseases such as black lung disease and silicosis affected a large proportion of miners. With the introduction of labor laws and industrial reforms, the contribution of the occupational environment to adverse health conditions among workers has diminished. Nevertheless, occupational illnesses and injuries are a substantial cause of morbidity and mortality in the United States. Primary prevention programs—those directed at eliminating adverse health outcomes before they occur—should be a goal of occupational physicians and nurses, policymakers, employers, and industrial hygienists. Through the use of surveillance systems for monitoring occupational illnesses, introduction of environmental controls, and supplying of personal protective equipment, employers can achieve the realistic goal of minimizing the occurrence of adverse health outcomes that are associated with work.

Study Questions and Exercises

1. Define the following terms:
   a. Pneumoconiosis
   b. Byssinosis
   c. Mesothelioma
   d. Black lung disease
   e. Ototoxic
   f. Decibel (dB)
   g. Trauma
   h. Mule spinners' cancer
2. List five persons who were significant to the history of occupational health and state their contributions.
3. The following list describes occupations that involve exposures to hazardous substances (e.g., metals or chemicals). Indicate the hazards associated with each occupation. Which of these occupations, if any, no longer exist?
   a. Chimney sweep
   b. Welder
   c. Farm worker
   d. Miner
   e. Attendant at a firing range
   f. Pottery maker
4. What are two of the most common occupational illnesses (as opposed to injuries) found in private industry? What are the causal factors for these conditions?
5. The following is a list of agents of occupationally associated diseases and injuries. Describe the work settings in which they may be encountered and the types of adverse health outcomes they produce:
   a. Microbial agents
   b. Toxic chemicals and metals
   c. Repetitive motion
   d. Noise
   e. Work-related stress
6. In 2015, what was the leading cause of fatal occupational injuries? What were the second and third leading causes of fatal occupational

injuries? Which five industry sectors tended to have the highest rates of fatal occupational injuries? (Hint: refer to Figure 13.18 and Figure 13.19).

7. For the construction industry, a total of 975 fatal injuries were reported in 2008, whereas the number for agriculture, forestry, fishing, and hunting was 672. The rate of fatal injuries was higher for the category that included agriculture, forestry, fishing, and hunting than for the category of construction. Explain how the rate could be lower for construction than for agriculture, forestry, fishing, and hunting.
8. In your own words, describe methods for preventing occupational fatalities.
9. What are the five leading causes of nonfatal occupational injuries? What can be done to prevent such injuries?
10. The number of occupational illnesses has declined since 1994. What factors have led to this decline? What steps can be taken to reduce occupational hazards and thereby improve the health of workers?
11. Even though workers in professional occupations usually are not exposed directly to hazardous agents, they are prone to occupationally associated illnesses.
    a. What types of hazards predominate in professional occupations?
    b. What illnesses are associated with these hazards?
    c. What interventions would you propose to mitigate these hazards?

For Further Reading

Occupational Health and Safety for the 21st Century, Robert H. Friis, Burlington, MA: Jones & Bartlett Learning, 2016.

References

1. Gochfeld M. Chronologic history of occupational medicine. *J Occup Environ Med.* 2005;47:96–114.
2. Wallick F. *The American Worker: An Endangered Species.* New York, NY: Ballantine Books; 1972.
3. Franco G. Ramazzini and workers' health. *Lancet.* 1999;354:858–861.
4. Hamilton A. *Exploring the Dangerous Trades: The Autobiography of Alice Hamilton, M.D.* Boston, MA: Little, Brown and Company; 1943.
5. Centers for Disease Control and Prevention. Improvements in workplace safety—United States, 1900–1999. *MMWR.* 1999;48:461–469.
6. Rosner D, Markowitz G. Labor Day and the war on workers. *Am J Public Health.* 1999;89:1319–1321.
7. US National Park Service. Triangle Shirtwaist factory building. Available at: http://www.nps.gov/history/nr/travel/pwwmh/ny30.htm. Accessed June 10, 2017.
8. Linder D. The Triangle Shirtwaist Factory Fire Trial. University of Missouri–Kansas City School of Law; 2002. Available at: http://law2.umkc.edu/faculty/projects/ftrials/triangle/triangleaccount.html. Accessed June 10, 2017.
9. Kuschner WG. Introduction to the symposium. *Postgraduate Medicine.* 2003;113(4):69. Available at: http://www.tandfonline.com/doi/abs/10.3810/pgm.2003.04.1397. Accessed June 10, 2017.
10. US Department of Health and Human Services, Public Health Service. *Healthy People 2010.* Progress review: occupational safety and health. February 18, 2004.
11. US Department of Labor. Occupational Safety and Health Administration (OSHA). Workers' Memorial Day, April 28, 2017. Available at: https://www.osha.gov/workersmemorialday/. Accessed August 23, 2017.
12. US Department of Labor, Bureau of Labor Statistics. *News Release: Workplace Injuries and Illnesses—2015.* Washington, DC: US Department of Labor, Bureau of Labor Statistics; October 27, 2016. USDL-16-2056.
13. Berglund B, Lindvall T, eds. *Community Noise.* Archives of the Center for Sensory Research. 1995;2(1):1–195.
14. Kam PCA, Kam AC, Thompson JF. Noise pollution in the anaesthetic and intensive care environment. *Anaesthesia.* 1994;49:982–986.
15. European Agency for Safety and Health at Work. Fact sheet 40—Skin sensitisers. Available at: https://osha.europa.eu/sites/default/files/publications/documents/en/publications/factsheets/40/Factsheet_40_-_Skin_sensitisers.pdf. Accessed June 10, 2017.
16. Centers for Disease Control and Prevention, National Institute for Occupational Safety and Health (NIOSH). The Team Document. Ten years of leadership advancing the National Occupational Research Agenda. Washington, DC: NIOSH; 2006.
17. Centers for Disease Control and Prevention. Current trends leading work-related diseases and injuries. *MMWR.* 1986;35:561–563.
18. Peate WF. Occupational skin disease. *Am Fam Physician.* 2002;66:1025–1032, 1039–1040.
19. Kochanek KD, Murphy SL, Xu JQ, Tejada-Vera B. Deaths: Final data for 2014. *National Vital Statistics Reports.* 2016;65(4). Hyattsville, MD: National Center for Health Statistics.
20. Stellman JM, ed. *Encyclopaedia of Occupational Health and Safety.* 4th ed. Geneva, Switzerland: International Labour Organization; 1998.
21. US Environmental Protection Agency. Asbestos: asbestos ban and phase out. Available at: https://www.epa.gov/asbestos/asbestos-ban-and-phase-out-federal-register-notices. Accessed June 11, 2017.
22. MedlinePlus. Silicosis. Available at: https://medlineplus.gov/ency/article/000134.htm. Accessed June 11, 2017.
23. US Department of Labor. OSHA Occupational noise exposure. Available at: https://www.osha.gov/SLTC/noisehearingcon-servation/. Accessed June 12, 2017.
24. Centers for Disease Control and Prevention. Vital signs: noise-induced hearing loss among adults—United States 2011-2012. *MMWR.* 2017;66(5):139–144.
25. Alberti PW. Noise, the most ubiquitous pollutant. *Noise & Health.* 1998;1:3–5.

26. Noise Pollution Clearinghouse (NPC). About noise, noise pollution, and the Clearinghouse. Available at: http://www.nonoise.org/aboutno.htm. Accessed June 11, 2017.
27. Prasher D. Issue of environmental noise and annoyance [editorial]. *Noise & Health*. 1999;3:1.
28. National Institute on Deafness and Other Communication Disorders (NIDCD). I love what I hear! Common sounds. Available at: https://www.nidcd.nih.gov/health/i-love-what-i-hear-common-sounds. Accessed June 11, 2017.
29. Schneider S, Johanning E, Bélard J-L, et al. Noise, vibration, and heat and cold. *Occup Med*. 1995;10:363–383.
30. Ray CD, Levinson R. Noise pollution in the operating room: a hazard to surgeons, personnel, and patients. *J Spinal Disord*. 1992;5:485–488.
31. Merrell HB, Claggett K. Noise pollution and hearing loss in the dental office. *Dent Assist J*. 1992;61(3):6–9.
32. Centers for Disease Control and Prevention, National Institute for Occupational Safety and Health (NIOSH). *Criteria for a Recommended Standard: Occupational Noise Exposure, Revised Criteria 1998*. DHHS (NIOSH) Publication No. 98-126. Cincinnati, Ohio: US Department of Health and Human Services; 1998.
33. Centers for Disease Control and Prevention. National Institute for Occupational Safety and Health (NIOSH). Musculoskeletal disorders. Available at: https://www.cdc.gov/niosh/programs/msd/risks.html. Accessed August 24, 2017.
34. US Department of Labor. Bureau of Labor Statistics. Nonfatal occupational injuries and illnesses requiring days away from work, 2015. USDL-16-2130.
35. National Institute of Neurological Disorders and Stroke. Carpal tunnel syndrome fact sheet. Available at: https://www.ninds.nih.gov/Disorders/Patient-Caregiver-Education/Fact-Sheets/Carpal-Tunnel-Syndrome-Fact-Sheet. Accessed June 11, 2017.
36. Isolani L, Bonfiglioli R, Raffi GB. Different case definitions to describe the prevalence of occupational carpal tunnel syndrome in meat industry workers. *Int Arch Occup Environ Health*. 2002;75:229–234.
37. Mayo Clinic. Back pain at work: preventing pain and injury. Available at: https://www.mayoclinic.org/healthy-lifestyle/adult-health/in-depth/back-pain/art-20044526. Accessed December 8, 2017.
38. Stetler CB, Burns M, Sander-Buscemi K, et al. Use of evidence for prevention of work-related musculoskeletal injuries. *Orthop Nurs*. 2003;22(1):32–41.
39. US Department of Labor. Bureau of Labor Statistics. National census of fatal occupational injuries in 2015. News Release. USDL-16-2304.
40. Marsh SM, Layne LA. *Fatal Injuries to Civilian Workers in the United States, 1980–1995 (National and State Profiles)*. DHHS (NIOSH) Publication No. 2001-129S. CDC, National Institute for Occupational Safety and Health; 2001.
41. Sestito JP, Lunsford RA, Hamilton AC, Rosa RR, eds. *Worker Health Chartbook 2004*. NIOSH Publication Number 2004-146. Cincinnati, OH: US Department of Health and Human Services, Centers for Disease Control and Prevention, National Institute for Occupational Safety and Health (NIOSH); 2004.
42. Centers for Disease Control and Prevention. Occupational injuries and deaths among younger workers—United States, 1998–2007. *MMWR*. 2010;59:449–455.
43. Centers for Disease Control and Prevention, National Institute for Occupational Safety and Health (NIOSH). *Stress . . . at Work*. DHHS (NIOSH) Publication No. 99-101. Cincinnati, OH: National Institute for Occupational Safety and Health; 1999.
44. MedicineNet.com. Definition of critical incident stress management. Available at: http://www.medterms.com/script/main/art.asp?articlekey=26281. Accessed June 11, 2017.
45. Elovaino M, Kuusio H, Aalto AM, et al. Insecurity and shiftwork as characteristics of negative work environment: psychosocial and behavioural mediators. *J Adv Nurs*. 2010;66:1080–1091.
46. Godin I, Kittel F, Coppieters Y, et al. A prospective study of cumulative job stress in relation to mental health. *BMC Public Health*. 2005;5:67.
47. Kivimaki M, Leino-Arjas P, Luukkonen R, et al. Work stress and risk of cardiovascular mortality: prospective cohort study of industrial employees. *BMJ*. 2002;325:857.
48. Eaker ED, Sullivan LM, Kelly-Hayes M, et al. Does job strain increase the risk for coronary heart disease or death in men and women? The Framingham Offspring Study. *Am J Epidemiol*. 2004;159:950–958.
49. US Department of Labor, Occupational Safety and Health Administration (OSHA). OSHA fact sheet: personal protective equipment. Washington, DC: Occupational Safety and Health Administration; 2006.
50. Centers for Disease Control and Prevention, National Institute for Occupational Safety and Health (NIOSH), National Personal Protective Technology Laboratory. Respirators. Available at: http://www.cdc.gov/niosh/npptl/topics/respirators. Accessed June 11, 2017.
51. Centers for Disease Control and Prevention, National Institute for Occupational Safety and Health (NIOSH). *Tracking Occupational Injuries, Illnesses, and Hazards: The NIOSH Surveillance Strategic Plan*. DHHS (NIOSH) Publication No. 2001-118. Cincinnati, Ohio: NIOSH; 2001.
52. LaMontagne AD, Herrick RF, Van Dyke MV, et al. Exposure databases and exposure surveillance: promise and practice. *AIHA J*. 2002;63:205–212.
53. California Department of Public Health, Occupational Health Branch. Occupational Health Surveillance and Evaluation Program (OHSEP). Available at: https://archive.cdph.ca.gov/programs/ohsep/Pages/Default.aspx. Accessed June 12, 2017.
54. Extension Toxicology Network (EXTOXNET). Toxicology information brief: standards. Available at: http://pmep.cce.cornell.edu/profiles/extoxnet/TIB/standards.html. Accessed June 12, 2017.
55. American Conference of Governmental Industrial Hygienists (ACGIH). Policy statement on the uses of TLVs® and BEIs.® Available at: http://www.acgih.org/tlv-bei-guidelines/biological-exposure-indices-introduction. Accessed June 12, 2017.
56. US Department of Labor, Occupational Safety and Health Administration (OSHA). OSHA permissible exposure limits (PELs). Available at: https://www.osha.gov/SLTC/hazardoustoxicsubstances/. Accessed June 12, 2017.

CHAPTER 14

Injuries With a Focus on Unintentional Injuries and Deaths

LEARNING OBJECTIVES

By the end of this chapter the reader will be able to:

- Define the term *intentionality of injury.*
- Describe environmental factors associated with injuries.
- State time trends in mortality from injuries in the United States.
- List types of injuries associated with particular age groups, e.g., children, adults, and the elderly.
- Illustrate a theoretical framework used for injury prevention.

Introduction

Because of their association with environmental factors, injuries are a noteworthy topic for an environmental health textbook. This chapter delimits the scope of injuries (both intentional and unintentional), describes how environmental factors contribute to such injuries, and suggests methods for prevention of injuries. The field of injury studies covers a vast domain, e.g., car crashes, falls, poisonings, shootings, recreational injuries, and industrial injuries. The present chapter will focus primarily on unintentional injuries and consider in detail three major areas: motor vehicle injuries and transport-related injuries, life stages and unintentional injuries, and methods of injury prevention. The chapter will also describe methodologies used in injury research, identify some of the most important types of injuries, and discuss the economic impact of injuries.

The Significance of Injuries

Injuries are a significant phenomenon worldwide because everyone is at risk. From the worldwide perspective, injuries of all types (both intentional and unintentional) account for an estimated 5.8 million deaths annually[1]; unintentional injuries cause more than 3.5 million deaths, 6% of all deaths, and two-thirds of all injury deaths during a typical year.[2] With respect to the United States, unintentional injuries are the fourth leading cause of mortality in the population as a whole and the leading cause of death for the population aged 1 to 44 years (2014 data).[3] These dramatic contributions to mortality (as well as to morbidity) worldwide and domestically often are preventable and can be linked with environmental influences. The death toll from injuries is projected to increase substantially by the year 2020.[4]

Beyond the direct harm to people, injuries burden the healthcare system and rehabilitation facilities. The adverse effects caused by injuries include the immediate impact not only upon the affected individual but also upon that person's family members and social support system. Nonfatal injuries can lead to permanent disability and drastically affect one's independence—not to mention the pain and suffering that may be experienced by the injured person.[5] Injuries may create stress and severe emotional responses among family members and exact a substantial economic cost.

Noteworthy is the fact that injuries manifest characteristic social and demographic distributions, disproportionately affecting the poor in all countries of the world.[6] A total of 90% of unintentional injury deaths transpire in low- and middle-income countries.[2] Some environmental conditions found in many less-developed countries increase the likelihood of unintentional injuries; these aspects of the environment include poorly maintained sidewalks, lack of paved streets, unavailability of pedestrian zones in cities, public transportation facilities in disrepair, and unsafe working conditions in factories.

At the international level, persons between the ages of 5 and 44 years are at special risk of injuries of all types; among this age group injury-related causes account for 6 of the 10 leading causes of death. About half of injury-related mortality occurs in the group aged 15 to 44 years; mortality from injuries among men is double that among women.[4] Injuries account for approximately one-third of deaths among children aged 1 to 14 years in Europe, although death rates differ greatly between eastern and western European countries.[7] For example, the main contributors to injury deaths among children in the former Soviet Union countries of eastern Europe are drowning, poisoning, fires, and falls. Traffic-related injuries are the leading cause of injury-related mortality among children and young people in the northwestern part of Europe.

In the United States, all forms of injuries accounted for 199,752 deaths during 2014. (Refer to **FIGURE 14.1**.) These were distributed according to the five general categories shown on the left of Figure 14.1. These categories, defined according to International Classification of Diseases (ICD) codes, are unintentional injuries, suicide, homicide, undetermined, and legal intervention/war (shown as legal intervention in the figure). The three leading causes of death from injuries were unintentional injuries, suicide, and homicide. More findings regarding fatalities from unintentional injuries are discussed later in the chapter. The aggregate categories shown in the figure can be subdivided into finer subcategories and rearranged into other classifications. For example, the general category of suicide can also fall under specific mechanisms and intent of death (e.g., firearm suicide, suicide by drowning, and suicide by poisoning). Note that the

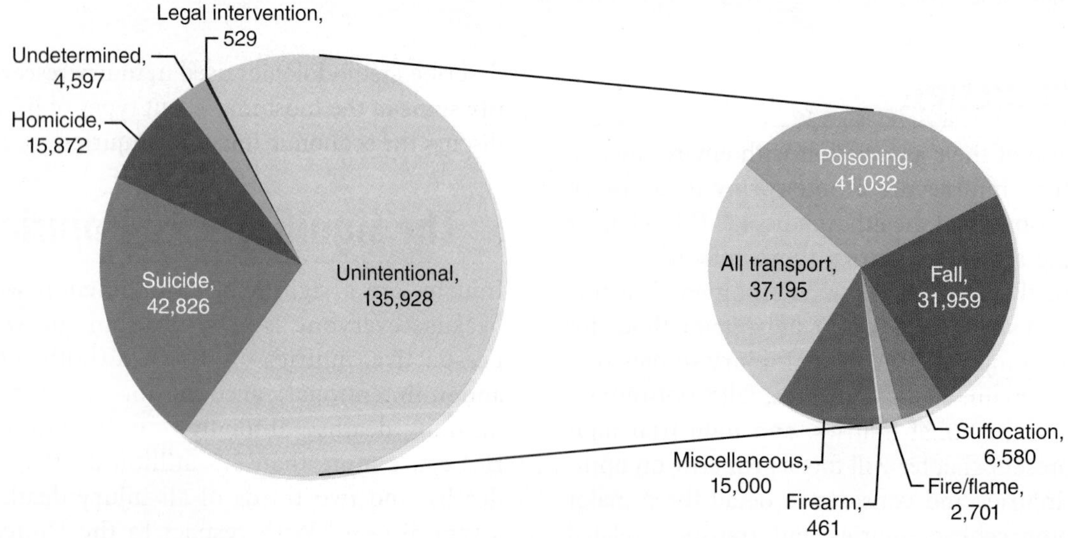

FIGURE 14.1 Number of injury deaths in the United States, 2014; all injury deaths (n = 199,752) and unintentional injury deaths (n = 135,928).

Data from Kochanek KD, Murphy SL, Xu JQ, Tejada-Vera B. Deaths: Final data for 2014. *National vital statistics reports*. Hyattsville, MD: National Center for Health Statistics. 2016;65(4):87–88.

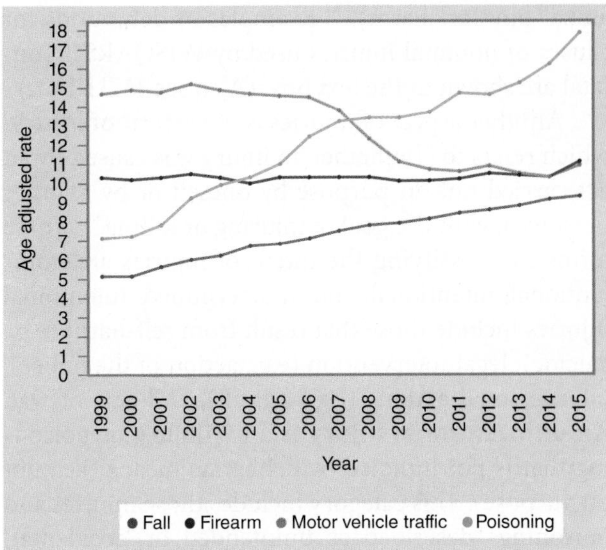

FIGURE 14.2 Age-adjusted death rates for four leading causes of injury death, by year*—United States, 1999–2015.

* Year and injury mechanism and all other leading causes.

Data from Centers for Disease Control and Prevention, National Center for Health Statistics. Underlying Cause of Death 1999-2015 on CDC WONDER Online Database, released December, 2016. Data are from the Multiple Causes of Death Files, 1999-2015, as compiled from data provided by the 57 vital statistics jurisdictions through the Vital Statistics Cooperative Program. Available at: http://wonder.cdc.gov/ucd-icd10.html. Accessed May 13, 2017.

right circle in Figure 14.1 provides data regarding unintentional injuries.

Time trends in age-adjusted death rates for the four leading causes of injury death from 1999 to 2015 are presented in **FIGURE 14.2**. According to the Centers for Disease Control and Prevention (CDC), "[W]ithin external causes of injury death, unintentional poisoning was the leading mechanism of injury mortality in 2014, followed by unintentional motor vehicle traffic–related injuries."[3(p2)] Data for time trends for individual years show that beginning in 2008, unintentional poisoning replaced motor vehicle traffic–related injuries as the leading mechanism of injury mortality. This outcome was a reversal of a trend that had existed from 1999 to 2007. (See Figure 14.2).

Note that in comparison with Figure 14.1, the data in Figure 14.2 combine information on several causes of a specific category of injury death as described in the following list. The four leading causes of injury deaths shown in Figure 14.2 encompass the following subcategories:

- Poisoning deaths include unintentional poisoning deaths, suicide, homicide, undetermined, and legal intervention or war (2014 data: number = 51,966; age-adjusted rate = 16.2 per 100,000).[3]
- Motor vehicle traffic deaths include fatalities to occupants of motor vehicles, motorcyclists, pedal cyclists, pedestrians, other, and unspecified (2014 data: number = 33,736; age-adjusted rate = 10.3 per 100,000).[3]
- Firearm deaths include unintentional firearm deaths, suicide, homicide, undetermined, and legal intervention or war (2014 data: number = 33,594; age-adjusted rate = 10.3 per 100,000).[3]
- Falls (2014 data: number=33,018; age-adjusted rate = 9.1 per 100,000).[3]

How Injuries are Classified

Specialists in the field of injury research have devoted considerable attention to methods for describing and classifying injuries. Development of a standardized framework for classification of injuries facilitates making comparisons internationally and domestically within the United States. For example, the rates for various types of injuries can be compared more easily from one year to the next as well as across states and regions. Accurate and reliable information helps policy makers develop injury prevention and intervention programs.

One of the methods for classifying injuries uses the *International Classification of Diseases* (ICD) injury matrices, which organize injury data into helpful groupings in order to make international and national comparisons.[8] The ICD classifies injuries according to two dimensions: *external cause* (e.g., car crash) and *nature of the injury* (e.g., fracture).[9] Within each of these two dimensions are two axes, which are collapsed into a single code. For example, the first dimension, external cause, is presented as a single code that combines the mechanism of injury (e.g., firearm) and intent of injury (e.g., intentional). The second dimension is the nature of the injury, also a single code for a two-factor variable that aggregates the nature of the injury itself (e.g., open wound) with the region of the body affected (e.g., chest).

The term **injury** refers to "the physical damage [to the person] that results when a human body is suddenly or briefly subjected to intolerable levels of energy. It can be a bodily lesion resulting from acute exposure to energy in amounts that exceed the threshold of physiological tolerance, or it can be an impairment of function resulting from a lack of one or more vital elements (i.e., air, water, and warmth), as in drowning, strangulation, or freezing. The time between exposure to the energy and the appearance of an injury is short."[6(p5)] Examples of energy that can cause injuries are mechanical force, radiant energy from light or shock waves, extremes of temperature, electrical energy, and chemical energy. Injuries that do not result in death—nonfatal injuries—are defined as "bodily harm resulting from severe exposure to an external force or substance (mechanical, thermal, electrical, chemical, or radiant) or a submersion. This bodily harm can be unintentional or violence-related."[10]

The **cause (mechanism) of injury** denotes "the way in which the person sustained the injury; how the person was injured; or the process by which the injury occurred. . . . The underlying cause is what starts the chain of events that leads to an injury. The direct cause is what produces the actual physical harm. The underlying and direct causes can be the same or different. For example, if a person cuts his or her finger with a knife, the cut is both the underlying and direct cause. However, if a child falls and hits his or her head on a coffee table, the fall is the underlying cause (the action that starts the injury event), and the contact with the table is the direct cause (the action that causes the actual physical harm)."[10] Examples of definitions for causes of nonfatal injuries used by WISQARS™ Nonfatal are shown in the text box. (Also see **TABLE 14.1**.)

Another aspect of injuries is the **intent of injury**, which refers to "[w]hether an injury was caused by an act carried out on purpose by oneself or by another person(s), with the goal of injuring or killing."[10] Three terms for classifying the intent of injuries are unintentional, intentional, and undetermined. Intentional injuries include those that result from self-harm (e.g., suicide), legal intervention (e.g., action of the police), interpersonal actions (e.g., assault), and acts of war. An **unintentional injury** is an "[i]njury or poisoning that is not inflicted by deliberate means (i.e., not on purpose). This category includes those injuries and poisonings described as unintended or 'accidental,' regardless of whether the injury was inflicted by oneself or by another person."[10]

An unintentional injury is the consequence of an unanticipated event that results in harm or death.

TABLE 14.1 Data Sources and Types of Information Available on Injuries

Data Source	Sponsor	Population Coverage	Information Available
Behavioral Risk Factor Surveillance System	CDC	All of United States (persons aged 18 years old and older)	Risk behaviors for leading causes of injury and death; injury-related data (e.g., occurrence of falls, use of seat belts).
Central nervous system surveillance: traumatic brain injury (TBI)	CDC	Participating US states	Extent of injury among persons who experience TBIs. Risk factors for TBI.
National Electronic Injury Surveillance System—All Injury Program	US Consumer Product Safety Commission	US hospitals	Originally for injuries associated with consumer products. Now includes all nonfatal injuries treated in hospital emergency departments.
National Hospital Discharge Survey	CDC—National Center for Health Statistics	US national sample of about 500 hospitals	Injuries among persons who survive and are discharged from inpatient hospital care.
National Vital Statistics System	CDC—National Center for Health Statistics	All of United States	Deaths causes by injuries and violence.
Web-based Injury Statistics Query and Reporting System (WISQARS)	CDC	All of United States	Injury morbidity and mortality data. Fatal and nonfatal unintentional and violent injuries. Permits online user queries via interactive database.

Data from National Center for Injury Prevention and Control. *CDC Injury Fact Book*. Atlanta, GA: Centers for Disease Control and Prevention; 2006:16.

EXAMPLE CAUSES OF NONFATAL INJURIES

Cut/pierce/stab: Injury resulting from an incision, slash, perforation, or puncture by a pointed or sharp instrument, weapon, or object. This category does not include injury from being struck by or against a blunt object (such as the side of a night stand) or bite wounds; these injuries fall in the category "struck by/against."

Inhalation/ingestion/suffocation: Inhalation, aspiration, or ingestion of food or other object that blocks the airway or causes suffocation; intentional or accidental mechanical suffocation due to hanging, strangulation, lack of air in a closed place, plastic bag, or falling earth. This category does not include injury resulting from a foreign body that does not block the airway.

Natural/environmental: Injury resulting from exposure to adverse natural and environmental conditions (such as severe heat, severe cold, lightning, sunstroke, large storms, and natural disasters) as well as lack of food or water.

Poisoning: Ingestion, inhalation, absorption through the skin, or injection of so much of a drug, toxin (biologic or nonbiologic), or other chemical that a harmful effect results, such as drug overdoses. This category does not include harmful effects from normal therapeutic drugs (i.e., unexpected adverse effects to a drug administered correctly to treat a condition) or bacterial illnesses.

Struck by/against or crushed: Injury resulting from being struck by (hit) or crushed by a human, animal, or inanimate object or force other than a vehicle or machinery; injury caused by striking (hitting) against a human, animal, or inanimate object or force other than a vehicle or machinery.

Transportation-related causes: Injury involving modes of transportation, such as cars, motorcycles, bicycles, and trains. This category is divided into five subcategories according to the person injured: motor vehicle occupant, motorcyclist, pedal cyclist, pedestrian, and other transport. This category also involves another factor—whether the injury occurred in traffic (on a public road or highway).

Pedal cyclist: Injury to a pedal cycle rider from a collision, loss of control, crash, or some other event involving a moving vehicle or pedestrian. This category includes riders of unicycles, bicycles, tricycles, and mountain bikes. This category does not include injuries unrelated to transport (moving), such as repairing a bicycle.

Modified and reproduced from Centers for Disease Control and Prevention, Injury Center. Definitions for WISQARS™ Nonfatal. Available at https://www.cdc.gov/injury/wisqars/nonfatal_help/definitions.html. Accessed March 1, 2017.

The use of the term unintentional injury is preferred to accident; the latter implies a random event that cannot be prevented. Most unintentional injuries are highly preventable; for example, laws that require seat belts and air bags in cars have contributed to a decline in motor vehicle driver and passenger deaths. Other preventive measures include safety messages directed to the public via media campaigns, prohibition of consumption of alcoholic beverages in public parks, school-based safety programs for school children, and licensing of firearms.

In practice—for example, in some government reports—the terms unintentional injury and accident often are used interchangeably, perhaps because of the public's familiarity with the latter term and its use in common parlance. An **accident** is defined as "an unanticipated event—commonly leading to INJURY or other harm—in traffic, the workplace, or a domestic, recreational, or other setting. The primary event in a sequence that leads ultimately to injury if that event is genuinely not predictable. Epidemiological studies have demonstrated that the risk of accidents is often predictable and that accidents are preventable."[11] Because the word *accident* is not in favor as a scientific term, the National Center for Health Statistics has added the term "unintentional injuries" in parentheses next to the category of accidents. The National Highway Traffic Safety Administration (NHTSA) favors the use of the term car *crash* to *accident.*

The World Health Organization states, "Injuries and violence have been neglected from the global health agenda for many years, despite being predictable and largely preventable. Evidence from many countries shows that dramatic successes in preventing injuries and violence can be achieved through concerted efforts that involve, but are not limited to, the health sector."[1]

Injury epidemiology is a branch of epidemiology that studies the distribution and determinants of injuries in the population. The results of epidemiologic investigations are applied to the prevention and control of injuries. For example, descriptive epidemiologic studies aid in the development of analytic research into the risk factors for and causes of injuries. From this information, policies and procedures to prevent injuries are facilitated. Currently, such research programs may be coupled with ongoing injury surveillance programs and databases that store information about the occurrence of injuries.

The chapter on environmental epidemiology covered the use of natural experiments, one of the study designs employed in analytic epidemiology. When new laws to increase safety are adopted (e.g., improvement of automobile safety through the mandatory use of passenger safety belts), they are similar to natural experiments, which can be evaluated by the use of epidemiologic methods.

Sources of Injury Data

As noted previously, high quality (reliable and valid) data are necessary for developing descriptive and analytic research on the occurrence of and risk factors for unintentional injuries. See Table 14.1 for examples of data sources; as shown in the table, one method for collecting information about injuries is the use of **surveillance systems**. Public health surveillance refers to the systematic and continuous gathering of information about the occurrence of diseases and other health phenomena. As part of the surveillance process, personnel analyze and interpret the data they have collected and distribute the data and associated findings to planners, health workers, and members of the community. Injury surveillance systems specialize in collecting information about injuries, for example, occupational injuries, motor vehicle injuries, risk behaviors, traumatic brain injuries, and injuries to consumers.

In addition to listing information regarding surveillance data, Table 14.1 gives other examples of data sources that may be accessed for information about injuries. The National Hospital Discharge Survey compiles information about patients who receive hospital care for injuries. The National Vital Statistics System collects and reports data on injuries listed on death certificates. The *National Vital Statistics Reports* published by the CDC is an example of a report that displays information on the frequency and distribution of intentional and unintentional injuries. Many of the data sources shown in the table can be accessed via the Internet.

Unintentional Injuries

The term *unintentional injury* was defined in an earlier section. Five of the categories of unintentional injuries are:

- Injuries to children—e.g., children who live in substandard housing units are at increased risk of injuries, as are youngsters who play in unsupervised playgrounds.
- Motor vehicle injuries—factors associated are human errors, driving under the influence, and poorly designed highways. Although progress has been made in recent years, unsafe vehicles continue to contribute to the toll of injuries on public roadways; consider the case of an engineering defect that is alleged to have caused one brand of motor vehicle to accelerate out of control. Another example is unsafe airbags that injure passengers when they explode during a crash.
- Firearm injuries—failure to educate the public about safe methods for storage and handling of weapons contributes to the number of unintentional firearm deaths.
- Falls—inadequately illuminated stairways and slippery surfaces contribute to falls, a major cause of morbidity among the elderly.
- Workplace injuries—the work setting, a venue for the use of dangerous machinery and exposure to toxic chemicals, is a domain for numerous injuries to employees. This topic was covered previously in the chapter on occupational health.

In the United States, unintentional injuries were the fourth leading cause of mortality in 2014. During 2014, a total of almost 135,928 deaths from unintentional injuries (5% of the total deaths) were recorded.[3] Figure 14.1 indicates that of the 135,928 deaths due to unintentional injuries, the three leading causes of death were poisonings, transport-related fatalities, and falls. The category of transport injuries included motor vehicle traffic injuries (e.g., vehicle occupants, motorcyclists, and pedestrians), other land-transport injuries, and injuries that occurred on water and in the air and space. The crude and adjusted death rates for unintentional injuries were 42.6 and 40.5 per 100,000, respectively.

During 2013, a total of 28.1 million visits were made to hospital emergency departments for unintentional injuries.[12] The three most frequent types of visits were for falls, motor vehicle traffic events, and "struck against or struck accidentally by objects or persons." **FIGURE 14.3** displays the percentage distribution of visits to emergency departments for unintentional injuries.

The top 10 causes of unintentional injuries during 2015 among age groups from under age 1 to age 65 and older are presented in **FIGURE 14.4**. You can see that unintentional poisonings dominate as the leading cause of injury deaths in all age groups combined and in 4 of the 10 age groups shown in the chart. Among infants (younger than 12 months old), suffocation is the leading cause of unintentional injury, followed by homicide (unspecified and specified) and motor vehicle traffic injuries. With respect to children and young adults (persons aged 5 years to 24 years), motor vehicle traffic injuries lead the causes of injury death; also, unintentional drowning, homicide/firearm, and fires and burns are frequent causes of mortality from injuries. In age groups 35 to 44, 45 to 54, and 55 to 64, unintentional poisonings are at the top of the list of causes of injury deaths. Unintentional falls are the leading cause of injury death among persons 65 years of age and older. Refer to Figure 14.4 for information about other major causes of injury death.

Economic Impacts of Unintentional Injuries

The economic impacts of unintentional injuries include direct medical costs for treatment and indirect costs such as those due to:

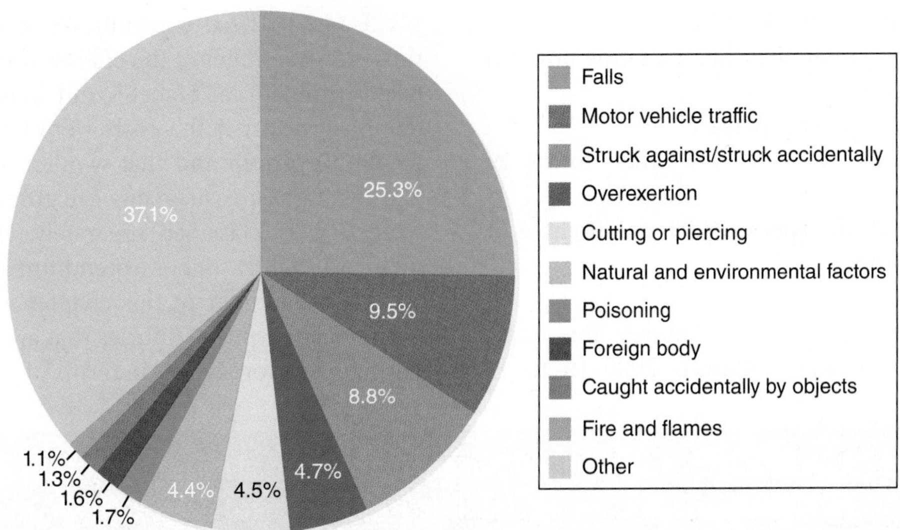

FIGURE 14.3 Percentage distribution of visits to emergency departments for 10 leading unintentional injuries, United States, 2013.

Data from Rui P, Kang K, Albert M. National Hospital Ambulatory Medical Care Survey: 2013 Emergency Department Summary Tables. Available from: http://www.cdc.gov/nchs/data/ahcd/nhamcs_emergency/2013_ed_web_tables.pdf. Accessed May 16, 2017.

	Age Groups										
Rank	<1	1–4	5–9	10–14	15–24	25–34	35–44	45–54	55–64	65+	Total
1	Unintentional suffocation 1,125	Unintentional drowning 390	Unintentional MV traffic 351	Unintentional MV traffic 412	Unintentional MV traffic 6,787	Unintentional poisoning 11,231	Unintentional poisoning 10,580	Unintentional poisoning 11,670	Unintentional poisoning 7,732	Unintentional fall 28,486	Unintentional poisoning 47,478
2	Homicide unspecified 135	Unintentional MV traffic 352	Unintentional drowning 129	Suicide suffocation 234	Homicide firearm 4,140	Unintentional MV traffic 6,327	Unintentional MV traffic 4,686	Unintentional MV traffic 5,329	Unintentional MV traffic 5,008	Unintentional MV traffic 6,860	Unintentional MV traffic 36,161
3	Homicide other spec., classifiable 69	Homicide unspecified 153	Unintentional fire/burn 72	Suicide firearm 139	Unintentional poisoning 3,920	Homicide firearm 3,996	Suicide firearm 2,952	Suicide firearm 3,882	Suicide firearm 3,951	Suicide firearm 5,511	Unintentional fall 33,381
4	Unintentional MV traffic 64	Unintentional suffocation 131	Homicide firearm 69	Homicide firearm 121	Suicide firearm 2,461	Suicide firearm 3,118	Suicide suffocation 2,219	Suicide suffocation 2,333	Unintentional fall 2,504	Unintentional unspecified 5,204	Suicide firearm 22,018
5	Unintentional suffocation 50	Unintentional fire/burn 100	Unintentional other land transport 32	Unintentional drowning 87	Suicide suffocation 2,119	Suicide suffocation 2,504	Homicide firearm 2,197	Suicide poisoning 1,835	Suicide poisoning 1,593	Unintentional suffocation 3,837	Homicide firearm 12,979
6	Unintentional drowning 30	Unintentional pedestrian, other 75	Unintentional suffocation 31	Unintentional other Land transport 51	Unintentional drowning 504	Suicide poisoning 769	Suicide poisoning 1,181	Homicide firearm 1,299	Suicide suffocation 1,535	Unintentional poisoning 2,198	Suicide suffocation 11,855
7	Homicide suffocation 24	Homicide other spec., classifiable 73	Unintentional natural/ environment 24	Unintentional fire/burn 41	Suicide poisoning 409	Undetermined poisoning 624	Undetermined poisoning 699	Unintentional fall 1,298	Unintentional suffocation 777	Adverse effects 1,721	Unintentional unspecified 6,930
8	Unintentional fire/burn 22	Homicide firearm 50	Unintentional pedestrian other 20	Unintentional poisoning 36	Homicide cut/pierce 312	Unintentional drowning 445	Unintentional fall 492	Undetermined poisoning 828	Unintentional unspecified 696	Unintentional fire/burn 1,171	Unintentional unspecified 6,914
9	Undetermined unspecified 21	Unintentional suffocation 31	Unintentional poisoning 17	Unintentional suffocation 26	Undetermined poisoning 234	Homicide cut/pierce 399	Unintentional drowning 374	Unintentional suffocation 469	Homicide firearm 681	Suicide poisoning 1,005	Suicide poisoning 6,816
10	Four tied 12	Unintentional fall 30	Unintentional struck by or against 17	Suicide poisoning 23	Unintentional fall 217	Unintentional fall 324	Homicide cut/pierce 291	Unintentional drowning 450	Two tied: undet. poisoning unit. fire/burn 565	Suicide suffocation 908	Unintentional drowning 3,602

FIGURE 14.4 Ten leading causes of injury death by age group highlighting unintentional injury deaths, United States—2015.

Reproduced from Centers for Disease Control and Prevention. Ten Leading Causes of Death and Injury. Available at: https://www.cdc.gov/injury/images/lc-charts/leading_causes_of_injury_deaths_unintentional_injury_2015_1050w760h.gif. Accessed May 13, 2017.

- Lost productivity at work and home
- Charges for rehabilitative therapies (for physical and mental health)
- Coverage of caregiver expenses
- Personal injury lawsuits and other forms of litigation
- Other nonmedical expenses such as purchase of wheelchair ramps.[5]

Estimates of medical expenditures for treatment of injuries were $117 billion in 2000. (Refer to **TABLE 14.2**.) These expenditures are comparable to those for treatment of the effects of such major public health problems as diabetes and smoking.[5] Table 14.2 demonstrates that the costs were highest among the 45–64 age group and that women in this age group had the highest costs of any age group when men and women were classified separately. However, overall, men had slightly higher expenditures than women.

The remainder of this chapter will cover several of the major types of injuries that are among the most burdensome for society in terms of economic costs as

TABLE 14.2 Percentage and Number[a] of Persons Reporting Treatment for an Injury and Percentage and Amount of Medical Expenditures Attributable to Injuries, by Selected Characteristics—United States, 2000

	Persons Reporting Treatment for an Injury		Medical Expenditures Attributable to Injuries		Injury-Attributable Expenditures ($)	
Characteristic[b]	**(%)**	**No.[c]**	**%**	**MEPS[d]**	**NHA[e]**	**Per capita, NHA**
Total	16.3	44.7	10.3	64.7	117.2	427
Sex						
Male	17.3	23.1	12.5	33.2	59.8	448
Female	15.4	21.6	9.2	31.8	57.4	409
Age group (yrs)						
10	11.9	4.8	7.8	3.1	5.7	141
0–19	17.9	7.2	16.6	7.4	13.4	333
0–29	15.8	5.7	6.8	2.7	5	137
0–44	17.8	11.3	12.2	14.6	26.5	417
5–64	16.7	10.2	10.6	20.9	37.9	621
65	16.7	5.5	8.7	16	29	881
Sex by age group						
Male						
10	13.7	2.8	9.4	1.6	3	145
0–19	20.7	4.3	26.3	5.1	9.2	445
0–29	18	3.2	7.9	1	1.7	98

5–64	17.9	5.7	12.5	12.9	23.3	732
65	18.5	3.5	6.7	7.1	12.9	680
0–44	20	6.2	15.3	7.5	13.6	438
5–64	15.5	4.6	7.9	7.6	13.7	463
65	14.2	2	11.8	9.6	17.4	1,233
Female						
10	9.9	1.9	6.7	1.3	2.3	118
0–19	15	3	11.9	3	5.4	272
0–29	13.6	2.5	4.8	1.4	2.5	139
0–44	15.7	5.1	9.4	6.7	12.1	373

[a] In millions.
[b] On the basis of Medical Expenditure Panel Survey (MEPS) estimates.
[c] Results were weighted to be nationally representative.
[d] In billions. MEPS estimate of US medical expenditures in 2000 is restricted to the civilian, noninstitutionalized population.
[e] In billions. National Health Accounts (NHA) estimates include the US-based military and institutionalized populations and are calculated by multiplying the NHA estimate of US medical expenditures in 2000 by the percentage of medical expenditures attributable to injuries estimated by MEPS.
Modified from National Center for Injury Prevention and Control. *CDC Injury Fact Book*. Atlanta, GA: Centers for Disease Control and Prevention; 2006:3.

well as the extent of morbidity and mortality that they produce. Among the topics that we will explore are:

- Motor vehicle injuries and other types of transport injuries. Related topics are pedestrian injuries, child passenger safety, teenage and older adult drivers, risk factors for injuries, and methods for prevention of such injuries.
- Unintentional injuries among children and young adults. Examples are recreational and sports-related injuries.
- Common injuries such as falls among elderly persons
- A framework for injury prevention

Motor Vehicle Injuries

When one encounters the scene of a car crash, often the event is frightening and evokes intense emotions. (Refer to **FIGURE 14.5**.) Unfortunately, motor vehicle crashes are an all too common event on America's highways. Motor-vehicle traffic injuries were the second leading cause of injury mortality in 2014. During that year, the number of motor vehicle–related traffic

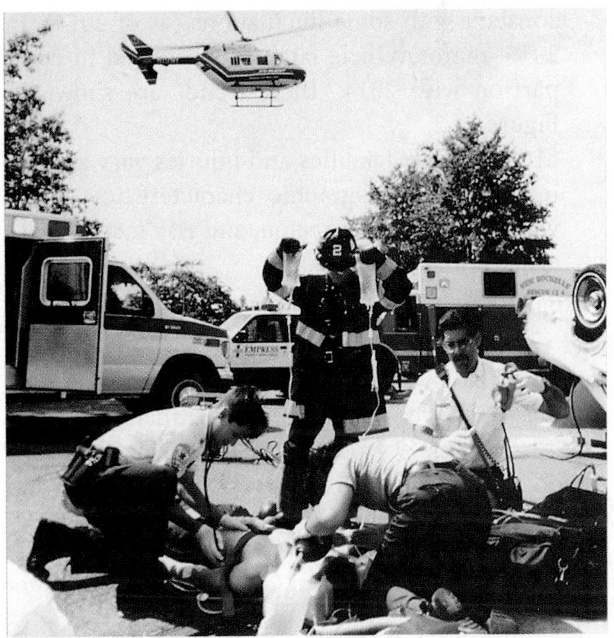

FIGURE 14.5 Emergency treatment at the scene of a car crash.

Reproduced from National Highway Traffic Safety Administration. NHTSA Image Library, Img 138 H. Available at: http://www.nhtsa.dot.gov/nhtsa/ImageLibrary/display.cfm. Accessed May 20, 2010.

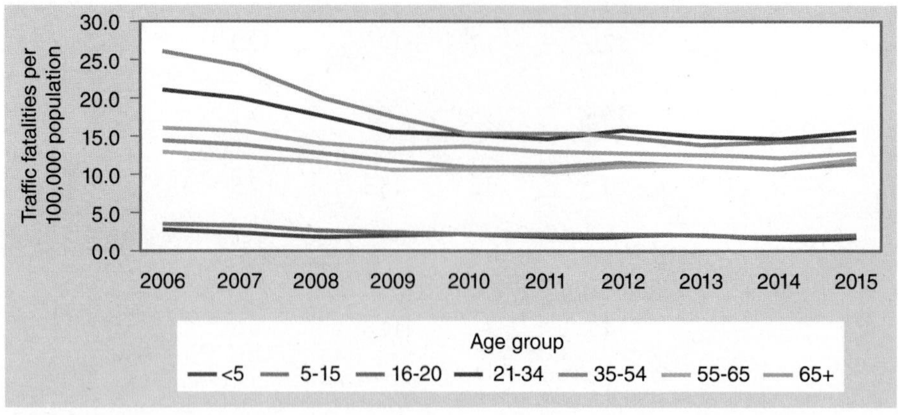

FIGURE 14.6 Motor vehicle traffic fatality rates by age group, 2006–2015.

Reproduced from National Center for Statistics and Analysis. 2015 older population fact sheet. (Traffic Safety facts. Report No. DOT HS 812 372). Washington, DC: National Highway Traffic Safety Administration. Published February 2017. Accessed May 13, 2017.

deaths totaled 33,736.[3] (Refer to Figure 14.2.) The following time trends pertain to motor vehicle deaths.

- From the late 1960s to 1992, death rates showed a 43% decline.[13]
- The annual death rates from 1999 to 2005 did not vary substantially; the rates ranged from 15.2 per 100,000 to 15.7 per 100,000. The rates tended to be constant despite the continuing introduction of safety measures and laws to prevent motor vehicle injuries.[13]
- Beginning around 2006, the occurrence of motor vehicle fatalities declined until about 2010; since then the trend in deaths has tended to remain constant with some fluctuations (as of 2014). In 2015, motor vehicle fatalities increased in comparison with 2014. These trends are shown in Figure 14.2.
- Motor vehicle fatalities and injuries vary according to the demographic characteristics of the victims, geographic region, and risk factors associated with crashes.[13]

Time trends in death rates for motor vehicle fatalities vary by sex and age of victim. The annual death rate for motor vehicle fatalities during 1999 to 2005 for males (21.7 per 100,000) was more than double the rate for females (9.4 per 100,000). During this same period, persons aged 15 to 24 years and 75 years and older had the highest rates, 26.8 per 100,000 and 25.9 per 100,000, respectively.[13]

FIGURE 14.6 presents data on time trends in motor vehicle fatalities by age group from 2006 to 2015.[14] As shown in the figure, the age groups 16 to 20 and 21 to 34 years of age had the highest death rates from motor vehicle fatalities during this time period. Among young drivers (15 to 20 years of age), the number of fatalities declined by 43% between 2006 and 2015—from 7,493 to 4,308 cases.[15] The rate of traffic fatalities among persons aged 65 years of age and older also reflected a declining trend; the fatality rate declined from 16.3 per 100,000 in 2006 to 12.9 per 100,000 in 2015.[14] **FIGURE 14.7** demonstrates that the excess of male over female motor vehicle fatalities was consistent across age groups (e.g., age group 16 to 20 years through age group 65 years and older).

Motor vehicle death rates vary by geographic region of the country, reaching the highest level in the South and the lowest in the Northeast in 2005. The respective rates by region were South (19.5 per 100,000), Midwest (14.7 per 100,000), West (14.2 per 100,000), and Northeast (9.8 per 100,000).[13] The rates for this cause also vary by race and ethnicity. American Indians/Alaska Natives had the highest rate (30.6 per 100,000) among four racial classifications.[16] (Refer to **FIGURE 14.8**.)

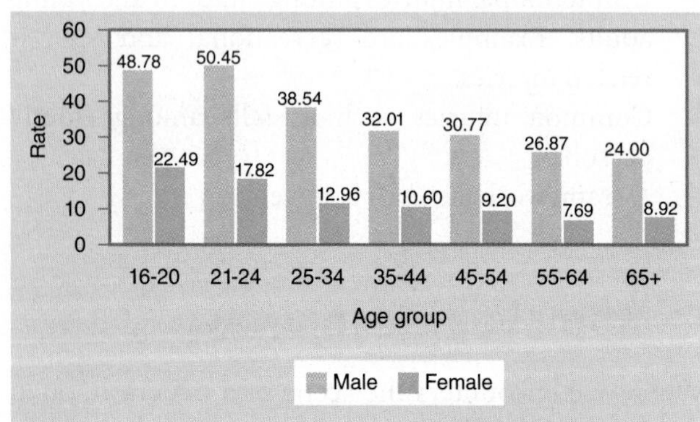

FIGURE 14.7 Driver involvement rates in fatal crashes by age and gender per 100,000 licensed drivers, 2015.

Reproduced from National Center for Statistics and Analysis. 2015 older population fact sheet. (Traffic Safety facts. Report No. DOT HS 812 372). Washington, DC: National Highway Traffic Safety Administration. Published February 2017. Accessed May 13, 2017.

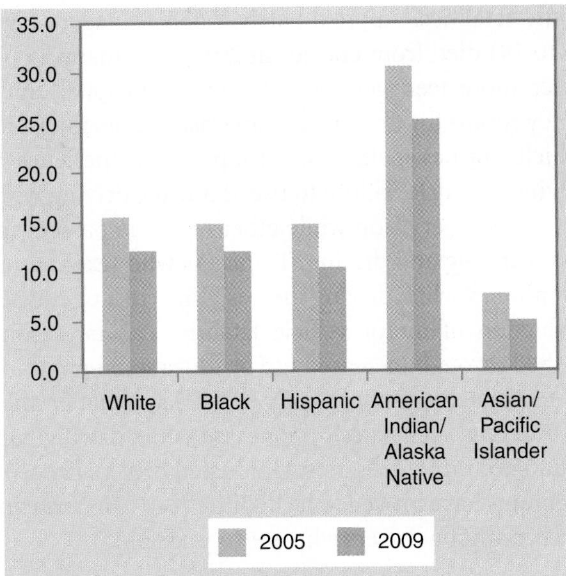

FIGURE 14.8 Age-adjusted rates of motor vehicle-related deaths, by race*/ethnicity, sex, and year—National Vital Statistics System, United States, 2005 and 2009.

* Persons of Hispanic ethnicity might be of any race or combination of races.

Data from West BA, Naumann RB. Motor vehicle-related deaths, United States, 2005 and 2009. *MMWR*. 2013;62:177.

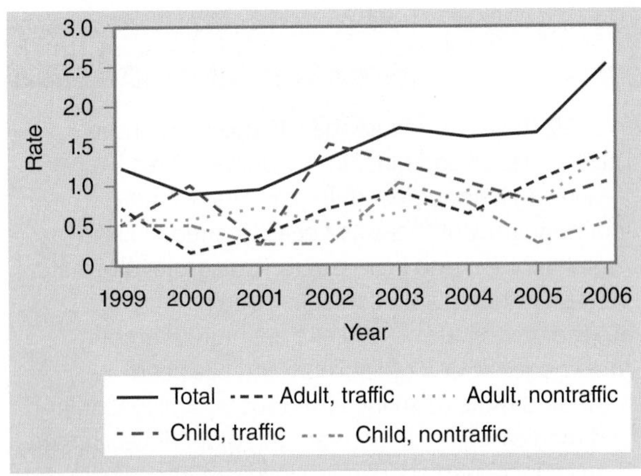

* Per 100,000 population.

† Adult is defined as a person aged ≥18 years; child is defined as a person aged <18 years.

§ Based on *International Classification of Deaths, 10th Revision*. Traffic crash defined as occurring on a public highway; nontraffic crash occurs entirely in any place other than a public highway.

¶ Includes only decedents who were west Virginia residents involved in crashes that occurred in West Virginia (N = 215).

FIGURE 14.9 Death rate* attributed to fatal all-terrain vehicle crashes, by age group† and crash classification§—West Virginia¶, 1999–2006.

Reproduced from Centers for Disease Control and Prevention. All-terrain vehicle fatalities—West Virginia, 1999-2006. *MMWR*. 2008;57:314.

Other Types of Transport-Related Injuries

Another type of transport-related injury is associated with the use of all-terrain vehicles (ATVs). A motorized ATV has low-pressure tires that enable it to travel off road. According to data for the 1990s, West Virginia had death rates from ATV crashes that were about eight times higher than the national average.[17] The state enacted several laws to reduce ATV fatalities; examples were reducing the distance that is permitted to travel on paved roads, reducing the speed of the vehicle, and requiring helmet use. Nevertheless, between 1999 and 2006, fatal ATV crashes increased by about 14% per year. (Refer to **FIGURE 14.9**.) During this period, 250 people died from ATV crashes in West Virginia. Factors related to ATV fatalities were lower socioeconomic status, being single or divorced, and having lower levels of education.

Pedestrians and Motor Vehicles

During 2015, 5,376 deaths to pedestrians resulted from traffic crashes. A large proportion of the deaths occurred among child pedestrians. In comparison with injuries to automobile passengers who are involved in crashes, injured pedestrians have a much higher likelihood of death.[18] Environmental factors are strongly implicated in pedestrian injuries. In illustration, roads in many US communities are devoid of sidewalks and shoulders, forcing pedestrians to walk close to vehicles. Convenient and safe pedestrian crossings may not be available: in the interest of promoting the flow of traffic, authorities may place crosswalks far apart, creating the temptation to jaywalk. Pedestrian fatalities can be prevented through the thoughtful design of roads and sidewalks. Other measures for enhancing pedestrian safety include:

- Increasing the awareness of the public, pedestrians, and drivers regarding pedestrian safety
- Developing safe-walking programs for children
- Modifying city streets to maximize pedestrian safety and encourage walking as a form of exercise
- Creating warning signs for motorists as well as speed bumps in areas frequented by children and other pedestrians

Older Adult Drivers

In 2015, the number of older drivers (aged 65 years and older) had increased by 33% over the previous decade (2006) to 40.1 million licensed drivers—about 18% of all licensed drivers.[14] Beginning at ages 70 to 75, the frequency of involvement in fatal crashes starts to increase and reaches the highest levels among drivers who are ages 85 and older.[19] Methods for the prevention of motor vehicle crashes among this age group include: (1) maintaining a regimen of regular exercise to increase strength and flexibility; (2) reviewing personal use of medications

OLDER ADULT DRIVERS: WHO IS MOST AT RISK?

In 2015, there were more than 40 million licensed drivers ages 65 and older in the United States.

Driving helps older adults stay mobile and independent. But the risk of being injured or killed in motor vehicle crash increases as you age. Involvement in fatal crashes, per mile traveled, begins increasing among drivers ages 70–74 and are highest among drivers ages 85 and older. This trend has been attributed more to an increased susceptibility to injury and medical complications among older drivers rather than an increased risk of crash involvement.

- Across all age groups, males have substantially higher death rates than females.
- Age-related declines in vision and cognitive functioning (ability to reason and remember), as well as physical changes, may affect some older adults' driving abilities.

Modified and reproduced from Centers for Disease Control and Prevention. Older adult drivers. Available at: https://www.cdc.gov/motorvehiclesafety/older_adult_drivers/. Accessed February 26, 2017.

that may impair driving; (3) obtaining annual vision checks; (4) creating a safe environment within the car's interior by eliminating distractions, e.g., turning down the radio and not using a cell phone; and (5) completing a driver's training course for senior drivers. Other facts related to older drivers are shown in the text box.

Teen Drivers

Figure 14.6 demonstrates that during 2006 to 2015, persons aged 16 to 20 (along with those who are 21 to 34 years of age) had the highest rates of motor vehicle traffic fatalities. Approximately 2,333 teenagers (ages 16 to 19) died from crashes in 2015.[20] As many as 100 times more teenagers were treated in hospital emergency rooms for this cause. Some risk factors for motor vehicle crashes in this age group are inexperience in driving a vehicle, failure to use seat belts, driving without adult supervision while other teens are passengers, and drinking and driving. Teenagers who text and use cell phones while driving increase their risk of crashes. Reduction of motor vehicle fatalities can be accomplished through increased use of educational programs for teenage drivers and their parents. Laws that prohibit distractions such as cell phone use while driving contribute to improved safety. Graduated drivers licensing programs have proved to be highly effective in reducing crashes among drivers who are 16 years old.[20]

Summary of Risk Factors for Motor Vehicle Injuries

Several risk factors associated with car crashes were noted previously. These include:

- Drinking and driving: alcohol-impaired driving crashes were implicated in almost one-third of all traffic-related deaths in the United States in 2014.[21] Particularly at risk were children and child passengers in a vehicle driven by an alcohol-impaired driver. Elevated blood alcohol concentrations are associated with fatal motorcycle crashes; those who have had prior drunk driving convictions have an increased risk of fatal crashes.
- In addition to alcohol consumption, impaired driving can be caused by use of prescription medications and substance abuse. Refer to **FIGURE 14.10**.
- Failure to "buckle up."

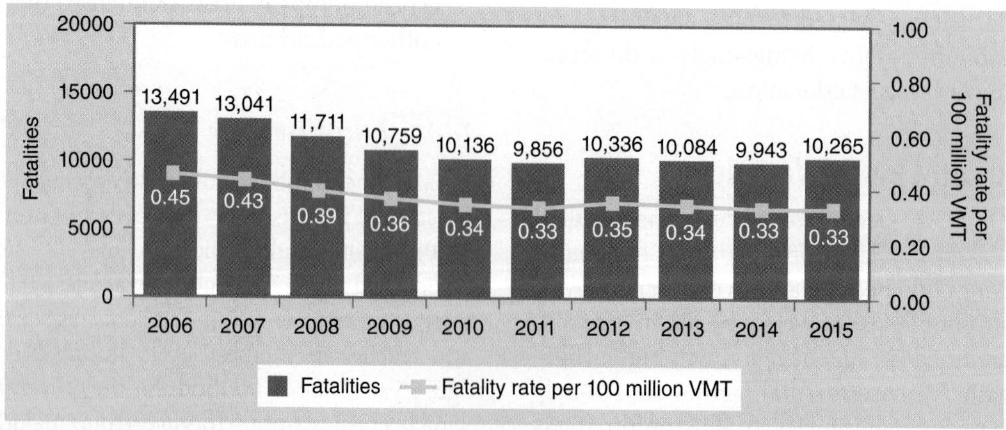

FIGURE 14.10 Fatalities and fatality rate per 100 million vehicle miles of travel (VMT) in alcohol-impaired-driving crashes, 2006–2015.

Reproduced from National Center for Statistics and Analysis. *Alcohol-impaired driving: 2015 data*. (Traffic Safety Facts. DOT HS 812 350). Washington, DC: National Highway Traffic Safety Administration. 2016.

In recent years, many US states have instituted laws that mandate the use of automobile safety belts. Primary laws (referring to primary enforcement of safety-belt laws) permit law enforcement personnel to stop drivers and passengers who are not wearing belts and issue citation for noncompliance.[22] Secondary laws allow police officers to issue a citation for noncompliance only when motorists have been stopped for some other violation. Evidence suggests that primary laws are more effective than secondary laws for encouraging motorists to wear safety belts and reducing fatalities caused by automobiles. Seat belt use tends to increase during both daytime and nighttime driving hours among jurisdictions that transition from secondary seat belt laws to primary laws.[23]

Child Safety Seats

Child safety seats, essential for protecting infants and young children during a car crash or sudden stop, should be placed in the rear seat of a passenger vehicle and never in the front seat because of the risk of injury from air bags. When transporting infants who are younger than 1 year old and weigh less than 20 pounds (9 kilograms), the seats should face toward the rear of the vehicle. **FIGURE 14.11** shows an infant who has been secured correctly in the seat.

FIGURE 14.11 Infant positioned correctly in a rear-facing infant–only car seat.

Courtesy of CDC/ American Academy of Pediatrics/Annemarie Poyo.

Unintentional Injuries Among Children

In the United States, unintentional injuries are the top source of morbidity and mortality among children.[24] Despite the fact that the statistics regarding injuries disclose this alarming fact, injuries among children tend to be an under-recognized public health issue. During the years 2000–2005, the death rate for unintentional injuries among children was 15.0 per 100,000. **FIGURE 14.12** shows unintentional injury death rates among children aged 0-19 years by age group and sex for 2006 through 2015. The figure reveals that males younger than 1 year of age followed by males aged 15 to 19 years had the highest mortality rates of the groups shown. The rates for males exceeded those for females across all age groups.

For all forms of injury (intentional and unintentional), substantial disparities exist among childhood injury death rates by race across age groups. The CDC reported that "the disparity between rates among blacks and whites was greatest in infancy, when the rate for blacks was 2.5 times that for whites. Rates

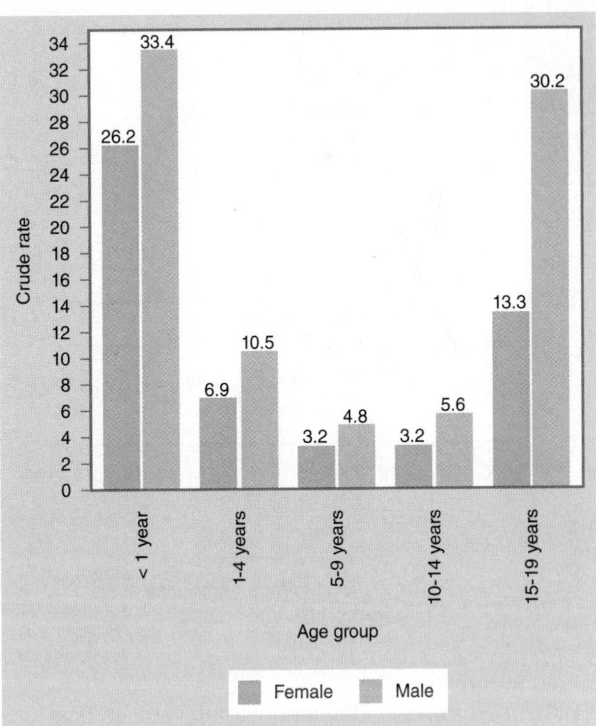

FIGURE 14.12 Crude unintentional injury death rate per 100,000 among children aged 0–19 years, by age group and sex, 2006–2015.

Data from Centers for Disease Control and Prevention, National Center for Health Statistics. Underlying Cause of Death 1999-2015 on CDC WONDER Online Database, released December, 2016. Data are from the Multiple Cause of Death Files, 1999-2015, as compiled from data provided by the 57 vital statistics jurisdictions through the Vital Statistics Cooperative Program. Available at http://wonder.cdc.gov/ucd-icd10.html. Accessed May 16, 2017.

among AI/AN (American Indian/Alaska Native) children were consistently two times those of white children for nearly every age from 0 to 19 years. Overall, for Hispanic children, injury death rates were comparable with or just below those for white children; in contrast, rates for A/PI (Asian/Pacific Islander) children were significantly lower than those for white children."[25(p4)] Refer to **FIGURE 14.13** for information on trends in unintentional injury death rates between 2006 and 2015; the figure demonstrates that these disparities among racial and ethnic groups have tended to persist over time.

Among children aged 0 to 19 years, the most frequent cause of unintentional injury death during 1999 through 2015 was transportation related: 7.3 per 100,000. These data were aggregated for 1999 through 2015. (See **FIGURE 14.14**.) Transportation-related causes include fatalities that occur among occupants of a motor vehicle, pedal cyclists, pedestrians, and users of other forms of transportation. Other major causes of deaths from unintentional injuries include drowning, falls, fires or burns, poisoning, and suffocation. According to the CDC, other deaths and injuries among children result from child maltreatment (abuse and neglect), child passenger safety, fireworks-related injuries, playground injuries, poisonings, residential fire-related injuries, suicide, traumatic brain injury, water safety, young drivers, and youth violence.[26] See the text box "Key Facts about Unintentional Injury Deaths among Children, United States—2000–2006" for additional information.

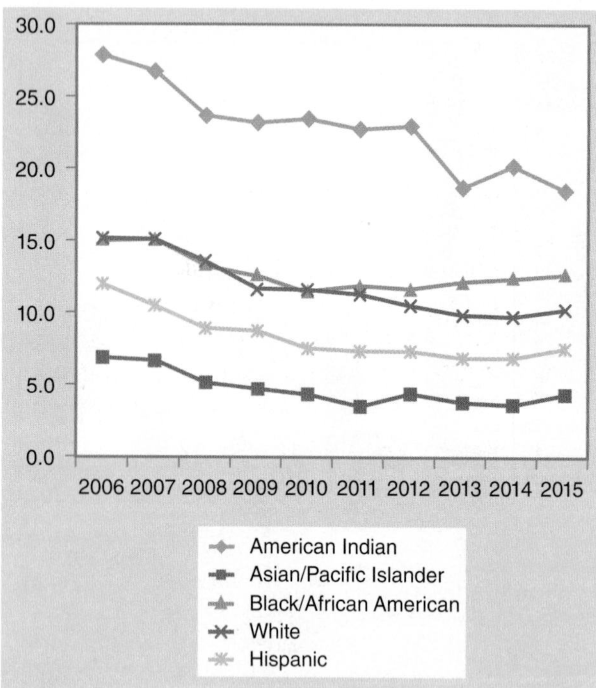

FIGURE 14.13 Trends in unintentional injury death rates* among children aged 0–19 years, by race/ethnicity† and year—United States, 2006–2015.

* Crude rate per 100,000.

† Hispanics can be of any race.

Data from Centers for Disease Control and Prevention, National Center for Health Statistics. Underlying Cause of Death 1999-2015 on CDC WONDER Online Database, released December, 2016. Data are from the Multiple Cause of Death Files, 1999-2015, as compiled from data provided by the 57 vital statistics jurisdictions through the Vital Statistics Cooperative Program. Available at http://wonder.cdc.govjucd-icd10.html. Accessed May 18, 2017.

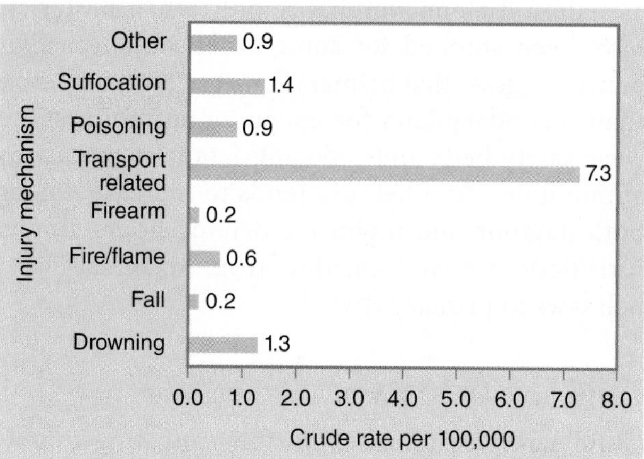

FIGURE 14.14 Unintentional injury death rates among children aged 0–19 years, by cause, United States, 1999–2015.

Data from Centers for Disease Control and Prevention, National Center for Health Statistics. Underlying Cause of Death 1999-2015 on CDC WONDER Online Database, released December, 2016. Data are from the Multiple Cause of Death Files, 1999-2015, as compiled from data provided by the 57 vital statistics jurisdictions through the Vital Statistics Cooperative Program. Available at http://wonder.cdc.govjucd-icd10.html. Accessed May 18, 2017.

KEY FACTS ABOUT UNINTENTIONAL INJURY DEATHS AMONG CHILDREN, UNITED STATES—2000–2006

- On average, 12,175 children aged 0 to 19 years died each year in the US from an unintentional injury.
- Males had higher injury death rates than females.
- Injuries due to transportation were the leading cause of death for children.
- The leading causes of injury death differed by age group.
- Risk for injury death varied by race. Injury death rates were highest for American Indian and Alaska Natives and were lowest for Asian or Pacific Islanders. Overall death rates for whites and blacks were approximately the same.
- Injury death rates varied by state depending upon the cause of death.
- For injury causes with an overall low burden, death rates greatly varied by age.

Modified and reproduced from Borse NN, Gilchrist J, Dellinger AM, et al. *CDC Childhood Injury Report: Patterns of Unintentional Injuries among 0–19 Year Olds in the United States, 2000–2006*. Atlanta, GA: Centers for Disease Control and Prevention, National Center for Injury Prevention and Control; 2008:3–4.

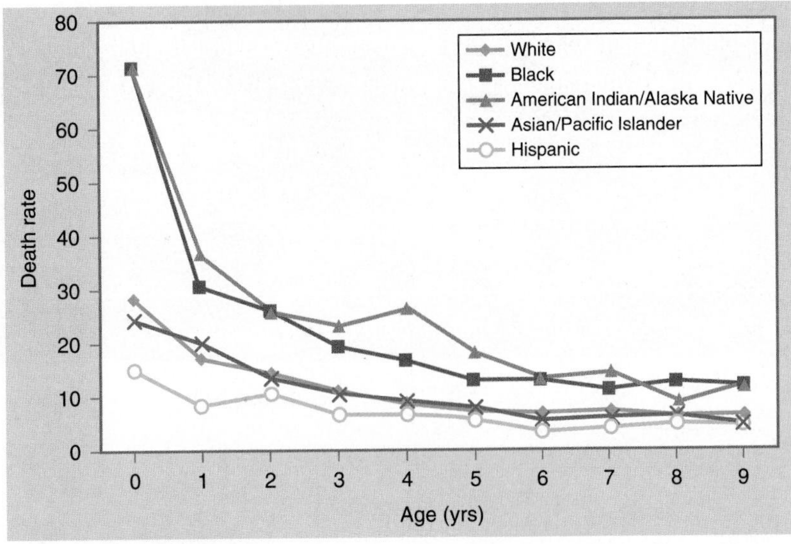

*Per 100,000 population.
†Hispanics are not included in any of the racial categories.

FIGURE 14.15 Rate* of injury death among children aged 0–9 years, by race/ethnicity†—United States, 1999–2002.

Reproduced from Bernard SJ, Paulozzi LJ, Wallace LJD. Fatal injuries among children by race and ethnicity—United States, 1999-2002. In: CDC Surveillance Summaries, May 18, 2007. Centers for Disease Control and Prevention. *MMWR*. 2007;56(No. SS-5):14.

The rate of injury death for all types of injuries among children aged 0 to 9 years by race/ethnicity in the United States (1999–2002) is shown in **FIGURE 14.15**. The highest death rates from injuries among children occur during the first 5 years of life and then stabilize by age 9. These data suggest the need for targeted interventions for reducing the toll of unintentional injuries among children, particularly during the early years of life. For example, such interventions should take into account the fact that infants and very young children are curious about their environment and like to explore, thus increasing their risk for poisoning and drowning. Older children are susceptible to bicycle and playground injuries because of immature cognitive development and limited physical coordination.

Playground Injuries

Playground injuries are a significant cause of nonfatal injuries and traumatic brain injuries among children. Annually in the United States, between 2001 and 2009, more than 200,000 visits to emergency departments were for playground injuries; about 8% were for traumatic brain injuries (TBIs). Refer to **TABLE 14.3** for more information. Environmental factors such as safe design of playground equipment and surfaces reduce the possibility of injuries. The surfaces shown in **FIGURE 14.16** are made of either soft sand or thick, padded rubber; these surfaces protect children in the event of falls. Playground sets should not have sharp edges and should be constructed securely so that they will not collapse. Playground equipment designed for home use may be flimsy and made of wood that can splinter or that contains toxic preservatives. Some older playground installations used lead-based paint, which children could ingest.

Sports- and Recreation-Related Injuries

In the United States, almost 40 million children and adolescents take part in organized sports; about 170 million adults engage in physical activity not connected with work. Participation in these activities incurs the risk of TBI, which can cause long-lasting adverse health effects such as behavioral changes and memory loss. A TBI is a type of brain injury that may occur from an external force such as a blow to the head. Many individuals present to hospital emergency departments for treatment of sports- and recreation-related injuries; a significant proportion of these injuries require hospitalization.

As the CDC reported:

> During 2001–2005, an estimated 207,830 patients with sports and recreation [SR]-related TBIs were treated in US hospital EDs [emergency departments] each year, accounting for 5.1% of all SR-related ED visits. . . . Overall, males accounted for approximately 70.5% of SR-related TBI ED visits. The highest rates of SR-related TBI ED visits for both males and females occurred among those aged 10–14 years, followed by those

TABLE 14.3 Estimated Annual Number of Emergency Department Visits for All Nonfatal Injuries and Nonfatal TBIs Related to Sports and Recreation Activities among Persons Aged ≤19 Years, by Type of Activity*

Activity	TBIs		All Visits for Sports and Recreation-Related Injuries		% of All Visits that Were TBIs
	No.	95% CI (±)	No.	95% CI (±)	
Activities that account for 10,000 more of annual emergency department visits					
Bicycling	26,212	(6,809)	323,571	(48,566)	8.1
Football	25,376	(4,845)	351,562	(47,448)	7.2
Playground	16,706	(5,198)	210,979	(37,050)	7.9
Basketball	13,987	(3,077)	375,601	(47,607)	3.7
Soccer	10,436	(3,736)	135,988	(39,167)	7.7
Activities that account for 10% or more of all visits that were TBIs					
All-terrain vehicle riding	6,337	(3,481)	59,533	(14,061)	10.6
Horseback riding	3,638	(1,266)	23,842	(5,169)	15.3
Tobogganing/sledding	2,377	(948)	23,306	(8,383)	10.2
Golf§§	1,887	(609)	17,078	(3,510)	11.0
Ice skating	1,673	(631)	14,608	(4,241)	11.4

* National Electronic Injury Surveillance System–All Injury Program, United States, 2001–2009.
CI = confidence interval.
§§ Includes injuries related to golf carts.
Modified from Centers for Disease Control and Prevention. Nonfatal Traumatic Brain Injuries Related to Sports and Recreation Activities among Persons Aged ≤19 Years—United States, 2001–2009. *MMWR*. 2011;60:1340.

FIGURE 14.16 Example of an outdoor playground set with safe surfaces.

aged 15–19 years (**FIGURE 14.17**). Activities associated with the greatest number of TBI-related ED visits included bicycling, football, playground activities, basketball, and riding all-terrain vehicles (ATVs). Activities for which TBI accounted for greater than 7.5% of ED visits for that activity included horseback riding (11.7%), ice skating (10.4%), riding ATVs (8.4%), tobogganing/sledding (8.3%), and bicycling (7.7%).[27(p733)]

Table 14.3 summarizes data on the estimated annual number of hospitalizations for nonfatal injuries and nonfatal TBIs in the United States related to sports and recreational activities among persons younger than 19 during 2001 through 2009. The five

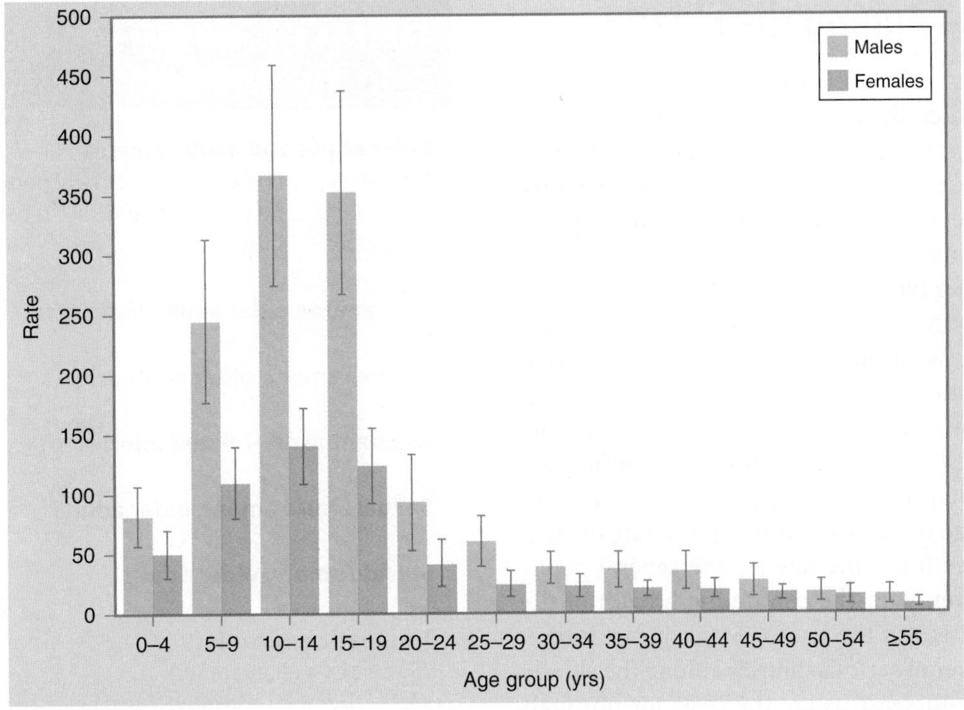

*Per 100,000 population.
†95% confidence interval.

FIGURE 14.17 Estimated annual rate* of nonfatal sports- and recreation-related traumatic brain injuries treated in emergency departments, by age group and sex—National Electronic Injury Surveillance System-All Injury Program, United States, 2001–2005.†

Reproduced from Centers for Disease Control and Prevention. Nonfatal traumatic brain injuries from sports and recreation activities—United States, 2001-2005. *MMWR.* 2007;56:735.

activities most frequently associated with visits to the emergency room for all injuries related to sports and recreational activities were bicycling, football, use of playgrounds, basketball, and soccer. The activity with the highest proportion (15.5%) of emergency department visits for TBIs was horseback riding.

Participation in collegiate sports, e.g., volleyball, can be a cause of injuries. The National Collegiate Athletic Association collected injury surveillance data for participation in women's volleyball from 1988–1989 through 2003–2004. Results indicated 2,216 injuries reported from 50,000 games and 4,725 injuries from 90,000 practices. The majority of reported injuries affected the lower extremities; ankle injuries were the most frequently reported type of injury.[28]

Prevention of Unintentional Injuries among Children

Among the major types of children's unintentional injuries are those related to motor vehicles (e.g., injuries to child passengers and young drivers), burns (e.g., from residential fires and improper handling of fireworks), playground mishaps, poisonings, and drowning. Environmental factors bear an important relationship to unintentional injuries among children, for example, at home and in recreational areas.[29] The impact of these influences is particularly acute for children who come from low-income and minority families. Environmental factors that contribute to unintentional injuries include:

- Aging, substandard housing facilities that increase risk of falls and other injuries
- Unavailability of safe playgrounds in combination with a lack of parental supervision while children are playing
- Failure to install home protective devices such as smoke alarms, locks on storage cabinets, and electrical outlet covers

Among the measures for preventing injuries are installing fences around pools in order to prevent children's unsupervised access, improving the design of playground equipment and surfaces, implementing municipal building codes that require smoke alarms and fire safety standards for construction, securing water heaters to prevent them from toppling over, packaging pharmaceuticals in child-safe containers, and mandating the use of car safety seats by small children.

Injuries Among the Elderly

Many forms of injuries (including those caused by violence) threaten the health and well-being of older adults—persons 65 years of age and older. Elder abuse is a potential cause of injuries among older persons. This phenomenon may take the form of physical abuse, which results in physical injuries, as well as sexual and other types of abuse including neglect and exploitation.[30] Individuals in the senior group have an increased risk of numerous types of injuries that may cause deaths or disability. In comparison with all other age groups, persons over the age of 64 years have the highest risk of fatal injuries and injuries that lead to a hospital stay.[31] Among individuals who have attained the age of 75 years or older, the rate of such injuries is three times the rate for the general population. Moreover, with the continuing growth of the elderly population in the United States in the future, the total number of serious injuries among the elderly is projected to increase as are the costs for hospitalizations and trauma care. During the first decade of the 21st century, morbidity and mortality from several types of fatal and nonfatal injuries have shown an increasing trend among the 65 years and older age group. For example, mortality from falls, motorcycle crashes, machinery, poisoning, and drowning increased.[32]

As shown in Figure 14.4, the leading causes of injury death among older people include falls (described in the following section), motor vehicle injuries, firearm deaths, fires and burns, and poisoning. **TABLE 14.4** provides a list of frequent causes of injuries among older persons. For some elderly individuals, a concomitant of the aging process is the loss of independence, which is the consequence of affliction with debilitating chronic diseases and other conditions. Consequently, some elderly persons are vulnerable to injury from maltreatment by care givers and other forms of abuse, as shown in the table.

TABLE 14.4 Types of Injury and Violence that Pose the Greatest Threat to Older Adults in the United States

Elder abuse and maltreatment With the aging process some elderly persons become more dependent on others and, as a result, more vulnerable to abuse.
Falls among older adults (Refer to text.)
Injuries among older adult drivers
Residential fire-related injuries
Sexual abuse among older adults
Suicide among older adults
Traumatic brain injury (especially among persons 75 years of age and older)

Data from Centers for Disease Control and Prevention. Injury center: injuries among older adults. Available at: http://www.cdc.gov/NCIPC/olderadults.htm. Accessed May 19, 2010.

Injuries and Fatalities that Result from Falling

According to the CDC, falls are the leading cause of fatal and nonfatal injuries for persons aged 65 years and older.[33] Approximately 17,000 persons in this age group died from falls during 2014. Data from the 2014 Behavioral Risk Factor Surveillance System Survey revealed that almost 29% of elderly persons fell at least one time during the previous 12 months of that year. Older respondents were more likely to report falling than younger individuals; for example, the percentage of falling among persons 65 to 74 years of age was 26.7% and increased to 29.8% and 36.5%, respectively, among individuals aged 75 to 84 years and 85 years and older.

Race and ethnicity, health status, and sex were related to falling. With respect to race and ethnicity, the percentage of falls was higher among whites (29.6%) and American Indian/Alaska natives (34.2%) than among blacks (23.2%) and Asian/Pacific Islanders (19.8%). Regarding health status, persons who had poor health status were more than two times as likely to report falling as those who had excellent health status (47.3% versus 19.2%). The percentage of falls (30.3%) was higher among women than among men (25.5%); the percentage of injuries from falls was also higher among women (12.6%) than among men (8.3%). As shown in **FIGURE 14.18**, the rates of fatal falls were higher among men than among women, a phenomenon that has persisted over time. For both sexes, the rates of fall-related death tended to increase between 2000 and 2013.

Falling is associated with hip fractures, which often are linked to death soon afterward. As shown in **FIGURE 14.19**, the rate of injury hospitalizations for hip fractures per 1,000 population is approximately 10 times higher among persons 85 and older than among persons 65 to 74.

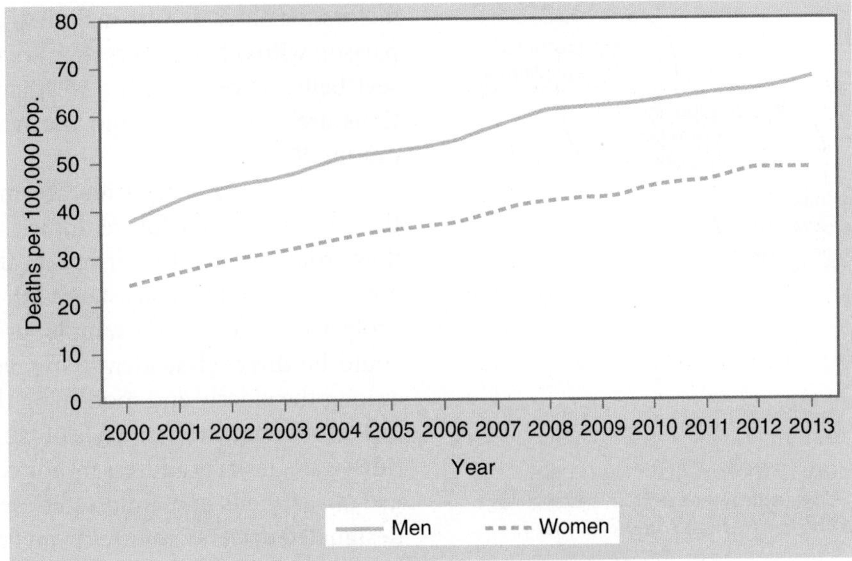

* Rates are age-adjusted using the 2000 U.S. standard population.
† Deaths from unintentional falls are identified using International Classification of Diseases, Tenth Revision (ICD-10) underlying cause of death codes W00-W19. There were 10,273 deaths in 2000 and 25,464 in 2013 from unintentional falls among adults aged ≥65.

FIGURE 14.18 Death rates* from unintentional falls† among adults aged ≥65 years, by sex—United States, 2000–2013.

Reproduced from Centers for Disease Control and Prevention. QuickStats: Death rates* from unintentional falls among adults aged ≥65 years, by sex–United States, 2000-2013. *MMWR.* 2015;64(16):450.

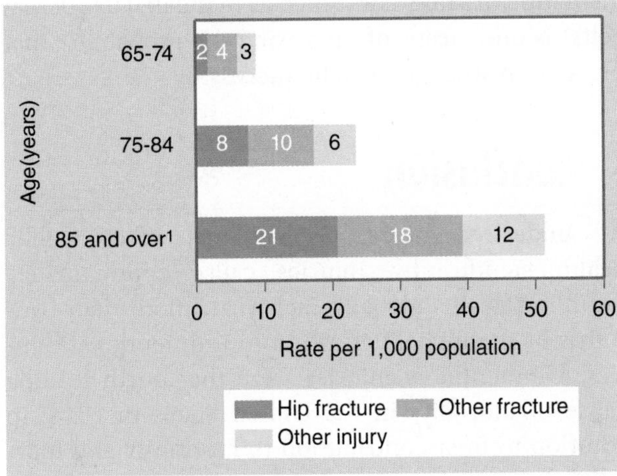

FIGURE 14.19 Injury hospitalizations, by age (older population) and selected diagnoses: United States, 2010.

Reproduced from Levant S, Chari K, DeFrances CJ. Hospitalizations for patients aged 85 and over in the United States, 2000–2010. NCHS data brief, no 182. Hyattsville, MD: National Center for Health Statistics. 2015, p. 4.

Prevention of Falls

Among the elderly, factors associated with falls are dizziness, impaired balance, and reduced visual acuity. Also, many elderly persons may be afflicted with symptoms associated with chronic illnesses and the sequelae of strokes. Community-based assessments of risk factors for falling linked with coordinated interventions are one possible method for the prevention of falls. An example was a clinical trial that consisted of visits to the homes of independent-living elderly persons who had fallen; during these visits researchers assessed risk factors and household hazards that might lead to falling; referrals were made for appropriate interventions in the community as well as exercise programs that provided strength and balance training.[34]

Another program investigated risk factors associated with falling among elderly women who had experienced strokes and who lived at home.[35] Among the most important risk factors for falling in this group were problems with balance while dressing. The findings suggested that interventions are needed to improve balance among female stroke victims when they are engaging in complex activities of daily living such as dressing; clinicians also should take into account the special risks and needs of stroke patients with respect to falling.

Schema for Injury Prevention

This chapter has demonstrated that injuries, both intentional and unintentional, rank among the leading causes of morbidity and mortality worldwide and in the United States. As a result, injuries are a significant public health problem that warrants preventive interventions at the population level. Prevention of injuries requires a systematic approach such as that depicted

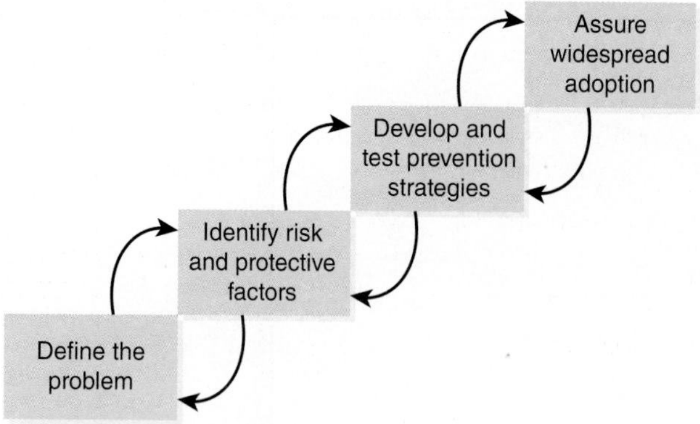

FIGURE 14.20 The public health approach to injury prevention.

Reproduced from Centers for Disease Control and Prevention, National Center for Injury Prevention and Control. The public health approach to violence prevention. Available at https://www.cdc.gov/violenceprevention/pdf/PH_App_Violence-a.pdf. Accessed May 15, 2017.

in **FIGURE 14.20**. The CDC identifies four components for the development of a strategy for the prevention of unintentional injuries.[36]

Problem definition refers to gathering data and analyzing datasets for patterns and trends. In this phase, injury specialists define subgroups of the population that have the highest prevalence of specific types of injuries. Descriptive epidemiologic studies form one of the keystones of problem identification. To illustrate, many types of injuries differ according to demographic variables and show disparities according to race and socioeconomic level. One example discussed previously is the higher rate of motor vehicle fatalities among men in comparison with women. Another example is the increased risk of falling among older persons in comparison with younger persons.

Risk factors are variables that increase the probability of the occurrence of an injury. Personal behavior and environmental components contribute to the risk of unintentional injuries. Examples of behavioral risk factors for motor vehicle fatalities are alcohol consumption, failure to wear seat belts, and driving while distracted. Examples of environmental risk factors pertain to both the home setting and the physical environment outside the home. At home, unsafe environmental conditions increase the risk of adverse outcomes such as falling, poisoning, and injuries to children; outside the home, environmental conditions are implicated in many types of injuries, including harm to pedestrians, playground mishaps, and sports-related injuries.

Protective factors are those that reduce the probability of an injury. Child safety seats and air bags protect the occupants of automobiles from severe or fatal injuries during a collision. Examples of protective factors for older drivers include the fact that in comparison with other drivers they are more likely to wear seat belts, drive when the weather and other conditions are safest, and avoid the use of alcohol when driving.[14]

Prevention strategies are methods to prevent the occurrence of future injuries. These strategies flow from empirical information that has been gathered and should be consistent with identified risk and protective factors. For example, prevention strategies could be directed at identifying risk factors for falls among the elderly and modifying the environment in order to reduce the incidence of falling. Other prevention strategies may address motor vehicle injuries (e.g., use of seat belts and child safety seats, assignment of designated drivers), injuries from fires and burns (e.g., installation of smoke alarms, wearing nonflammable clothing), drowning (barriers around swimming pools), and injuries from firearms (storage of weapons and ammunition in locked vaults).[37]

Widespread adoption means that successful prevention strategies are disseminated widely in order to encourage their adoption on a broad scale. Implementation of safety laws (such as mandated use of seat belts) is one means of increasing the probability that prevention strategies will be successful.

Conclusion

An under-recognized problem that affects public health significantly, injuries cause approximately 5 million deaths globally each year; more than two-thirds of these fatalities are from unintentional injuries. Unintentional injuries were the fourth leading cause of mortality in the United States in 2014. In addition to their contribution to morbidity and mortality, injuries can have devastating social and economic consequences; these include impacts upon the individual, family members, and society in terms of lost productivity, direct costs of health care, and related economic burdens.

The term *unintentional injury* is preferred to *accident*, which suggests that such events are random and cannot be prevented; in fact much progress has been achieved through the systematic examination of the distribution and causes of injuries. Injuries have differential distributions according to age, sex, race, and other variables. For example, males from all age groups are much more likely than females to die from injuries. Among the total US population, the top four categories of deaths from unintentional injuries are unintentional poisonings, motor vehicle fatalities, firearm-related deaths, and falls. Among persons aged

65 years and older, falls are the leading cause of fatal injuries. The public health approach to injury prevention involves four components that have implications for identification of environmental risk factors and development of intervention strategies, which are then disseminated widely.

Study Questions and Exercises

1. Define the following terms:
   a. Injury
   b. Causes of injuries
   c. Intent of injury
   d. Intentional injury
   e. Unintentional injury
   f. Injury epidemiology
2. Differentiate between the terms *unintentional injury* and *accident*. Why is the use of *unintentional injury* preferred?
3. Why do you need to be concerned about injuries? Discuss the significance of unintentional injuries for morbidity and mortality globally and in the United States. How do unintentional injuries rank in comparison with other causes of mortality? Describe the potential economic, social, and emotional costs of unintentional injuries.
4. Describe how the causes of unintentional injury death differ according to age group. Using your own ideas, discuss possibilities for prevention.
5. Describe variation in motor vehicle–related fatalities by demographic characteristics such as age, sex, and race. How can such fatalities be prevented among high-risk groups such as young male drivers?
6. Why do you think that death rates from motor vehicle injuries have not declined very much in recent years, despite the continuing introduction of safety measures and new laws to prevent motor vehicle injuries?
7. Why is the rate of fatalities so high among ATV drivers? What can be done to reduce the number of ATV-associated injuries?
8. Describe what is meant by disparities in the occurrence of unintentional injuries. Give some examples from the global and domestic (US) perspectives. How do you account for these disparities? Be sure to link them to environmental factors in your response.
9. What are the five most common sports and recreation activities associated with nonfatal injuries among all ages in the United States? How does the frequency of all injuries from sports and recreation activities compare with the frequency of traumatic brain injuries caused by sports and recreation activities? Using your own ideas, suggest methods for the prevention of the traumatic brain injuries linked with sports and recreation activities.
10. Describe the patterns and causes of unintentional injuries among children and the elderly. What can be done to protect vulnerable populations such as children and the elderly from unintentional injuries?
11. Invite a trauma specialist to your classroom and ask this individual to discuss the types of injuries seen in the hospital trauma center.
12. Arrange a debate in your classroom to discuss the causes and consequences of unintentional injuries. Assume that little can be done to prevent such events because they are random occurrences. Ask one group of students to present the *pro* side of this assumption and another group to present the *con* side of the assumption.
13. Conduct a walk-through of your home and identify potential injury hazards. Record these in a list and discuss them in class. Suggest how to eliminate these hazards.

For Further Reading

Sleet DA, Dellinger AM. Traffic injury prevention and environmental public health. In Friis, RH, ed. *The Praeger Handbook of Environmental Health*, Volume 4. Santa Barbara, CA: ABC-CLIO, LLC, 2012, 91-112.

References

1. World Health Organization. Injuries and violence: the facts. Geneva, Switzerland: World Health Organization; 2010.
2. Norton R, Hyder AA, Bishar, Peden M. Unintentional injuries. In Jamison DT, Breman JG, Measham AR, et al. (eds.) *Disease Control Priorities in Developing Countries.* Washington, DC: The International Bank for Reconstruction and Development/The World Bank; 2006.
3. Kochanek KD, Murphy SL, Xu J, et al. Deaths: final data for 2014. National Vital Statistics Reports. 2016;65(4). Hyattsville, MD: National Center for Health Statistics.
4. Peden M, McGee K, Sharma G. *The Injury Chart Book: A Graphical Overview of the Global Burden of Injuries.* Geneva, Switzerland: World Health Organization; 2002.
5. National Center for Injury Prevention and Control. *CDC Injury Fact Book.* Atlanta, GA: Centers for Disease Control and Prevention; 2006.
6. Holder Y, Peden M, Krug E, et al. (eds). *Injury Surveillance Guidelines.* Geneva, Switzerland: World Health Organization; 2001.
7. World Health Organization (WHO)/Europe. Children at risk! Main health effects of exposure to environmental risk factors. Fact Sheet EURO/05/03. Copenhagen, Denmark: WHO–Regional Office for Europe; November 27, 2003.

8. Centers for Disease Control and Prevention. Injury data and resources: ICD injury matrices. Available at: https://www.cdc.gov/nchs/injury/injury_matrices.htm. Accessed May 12, 2017.
9. Bergen G, Chen LH, Warner M, et al. *Injury in the United States: 2007 Chartbook*. Hyattsville, MD: National Center for Health Statistics; 2008.
10. Centers for Disease Control and Prevention, Injury Center. Definitions for WISQARS Nonfatal. Available at: https://www.cdc.gov/ncipc/wisqars/nonfatal/definitions.htm. Accessed May 12, 2017.
11. Porta M, ed. *A Dictionary of Epidemiology*. 6th ed. New York, NY: Oxford University Press; 2014.
12. Rui P, Kang K, Albert M. National Hospital Ambulatory Medical Care Survey: 2013 emergency department summary tables. Available from: http://www.cdc.gov/nchs/data/ahcd/nhamcs_emergency/2013_ed_web_tables.pdf. Accessed May 16, 2017.
13. Centers for Disease Control and Prevention. Motor vehicle-related death rates—United States, 1999–2005. *MMWR*. 2009; 58:161–165.
14. National Center for Statistics and Analysis. 2015 older population fact sheet. (Traffic Safety Facts. Report No. DOT HS 812 372). Washington, DC: National Highway Traffic Safety Administration (NHTSA); February 2017.
15. National Center for Statistics and Analysis. Young drivers: 2015 data. (Traffic Safety Facts. Report No. DOT HS 812 363). Washington, DC: National Highway Traffic Safety Administration (NHTSA); February 2017.
16. West BA, Naumann RB. Motor vehicle-related deaths, United States, 2005 and 2009. *MMWR*. 2013;62:177.
17. Centers for Disease Control and Prevention. All-terrain vehicle fatalities—West Virginia, 1999–2006. *MMWR*. 2008;57: 312–315.
18. Shieber RA, Vegega ME, eds. Executive summary: reducing childhood pedestrian injuries. *Injury Prevention*. 2002;8 (suppl I):i3–i8.
19. Centers for Disease Control and Prevention. Older adult drivers. Available at: https://www.cdc.gov/motorvehiclesafety/older_adult_drivers/. Accessed May 13, 2017.
20. Centers for Disease Control and Prevention. Teen drivers: get the facts. Available at: https://www.cdc.gov/motorvehiclesafety/teen_drivers/teendrivers_factsheet.html. Accessed May 13, 2017.
21. Centers for Disease Control and Prevention. Impaired driving. Available at: https://www.cdc.gov/MotorVehicleSafety/Impaired_Driving/impaired-drv_factsheet.html. Accessed May 15, 2017.
22. Centers for Disease Control and Prevention. Impact of primary laws on adult use of safety belts—United States, 2002. *MMWR*. 2004;53:257–260.
23. Masten S. Traffic safety facts: research note. The effects of changing to primary enforcement on daytime and nighttime seat belt use. Washington, DC: National Highway Transportation Safety Administration; 2007. DOT HS 810 743.
24. Borse NN, Gilchrist J, Dellinger AM, et al. *CDC Childhood Injury Report: Patterns of Unintentional Injuries among 0–19 Year Olds in the United States, 2000–2006*. Atlanta, GA: Centers for Disease Control and Prevention, National Center for Injury Prevention and Control; 2008.
25. Bernard SJ, Paulozzi LJ, Wallace LJD. Fatal injuries among children by race and ethnicity—United States, 1999–2002. *CDC Surveillance Summaries*. Centers for Disease Control and Prevention. *MMWR*. May 17, 2007;56(No. SS-5):1–16.
26. Centers for Disease Control and Prevention. Injury Center: Injuries among children and adolescents. Available at: http://www.cdc.gov/NCIPC/factsheets/children.htm. Accessed May 21, 2010.
27. Centers for Disease Control and Prevention. Nonfatal traumatic brain injuries from sports and recreation activities—United States, 2001–2005. *MMWR*. 2007;56:733–737.
28. Agel J, Palmieri-Smith RM, Dick R, et al. Descriptive epidemiology of collegiate women's volleyball injuries: National Collegiate Athletic Association Injury Surveillance System, 1988–1989 through 2003–2004. *Journal of Athletic Training*. 2007;42:295–302.
29. Grad J, Kunseler E. Policies to reduce unintentional injuries from falls, drowning, poisoning, fires and choking in children and adolescents. Copenhagen, Denmark: World Health Organization–Regional Office for Europe; 2007.
30. Centers for Disease Control and Prevention. Elder abuse. Available at: http://www.cdc.gov/violenceprevention/elderabuse/definitions.html. Accessed May 20, 2017.
31. MacKenzie EJ. Epidemiology of injuries: current trends and future challenges. *Epidemiologic Reviews*. 2000;22:112–119.
32. Hu G, Baker SP. Recent increases in fatal and non-fatal injury among people aged 65 years and over in the USA. *Inj Prev*. 2010;16:26–30.
33. Centers for Disease Control and Prevention. Falls and fall injuries among adults aged ≥ 65 years—United States, 2014. *MMWR*. 2016;65:994–998.
34. Elley CR, Robertson MC, Kerse NM, et al. Falls Assessment Clinical Trial (FACT): design, interventions, recruitment strategies and participant characteristics. *BMC Public Health*. 2007;7:185.
35. Lamb SE, Ferrucci L, Volapto S, et al. Risk factors for falling in home-dwelling older women with stroke: the Women's Health and Aging Study. *Stroke*. 2003;34:494–501.
36. Centers for Disease Control and Prevention. National Center for Injury Prevention and Control. The public health approach to violence prevention. Available at https://www.cdc.gov/violenceprevention/pdf/PH_App_Violence-a.pdf. Accessed May 15, 2017.
37. World Health Organization (WHO). Facts about injuries: preventing global injuries. Available at: http://www.who.int/violence_injury_prevention/resources/publications/en/injury_factsheet.pdf. Accessed August 14, 2017.

© Jean-Luc Rivard/EyeEm/Getty Images

Glossary

Note: Some definitions are quoted from other sources; refer to text for citations.

accident An unanticipated event commonly leading to injury or other harm; this term is not appropriate for use in scientific discourse.

acid rain The precipitation of acidic compounds formed when components of air pollution (e.g., SO_2 and NO_x) interact with other components in the air such as water, oxygen, and oxidants.

additive The combination of two chemicals produces an effect that is equal to their individual effects added together.

aerosol particles Particulate matter, including dust, soot, and other finely divided solid and liquid materials, which is suspended in and move with the air.

agenda setting Setting priorities, deciding at what time to deal with an environmental problem, and determining who will deal with the problem.

agent A factor, such as a microorganism, chemical substance, or form of radiation, whose presence, excessive presence, or (in deficiency diseases) relative absence is essential for the occurrence of a disease.

air pollution Air pollution refers to the presence, in various degrees, of substances not found in perfectly clean air.

Air Quality Index (AQI) An index for reporting daily air quality, which tells the public how clean or polluted the air is.

alpha particle Alpha particles (α) are positively charged and made up of two protons and two neutrons from the atom's nucleus. Alpha particles come from the decay of the heaviest radioactive elements, such as uranium, radium, and polonium.

analytic epidemiology Examines causal (etiologic) hypotheses regarding the association between an exposure and health condition(s).

antagonism Occurs when two chemicals administered together interfere with each other's actions or one interferes with the action of the other.

aquifer A layer or section of earth or rock that contains freshwater, known as groundwater.

arboviral diseases A group of viral diseases that are acquired most frequently when blood-feeding arthropod vectors—including ticks, sand flies, biting midges, and mosquitoes—infect a human host.

aromatic compound An organic molecule that contains a benzene ring; for example, benzene and toluene.

asthma A chronic respiratory illness characterized by difficulty breathing and tightness of the chest in response to muscle contractions that constrict the trachea, lungs, and other parts of the airway.

beta particle Beta particles (β) are small, fast-moving particles with a negative electrical charge that are emitted from an atom's nucleus during radioactive decay.

bias Deviation of results or inferences from the truth, or processes leading to such deviation. Any trend in the collection, analysis, interpretation, publication, or review of data that can lead to conclusions that are systematically different from the truth.

bioaccumulation Occurs when trace amounts of substances (such as heavy metals) become more concentrated and potentially more harmful as they move up the food chain; sometimes called biomagnification.

biodiversity Generally refers to the different types and variability of animal and plant species and the ecosystems in which they live.

biomagnification See *bioaccumulation*.

bovine spongiform encephalopathy (BSE) A neurological disease in cattle hypothesized to be caused by prions; also known as mad cow disease.

breast cancer 1, early onset (BRCA1) First breast cancer gene identified.

building-related illness (BRI) An identified diagnosable illness attributed directly to airborne building contaminants.

built environment Urban areas and structures (e.g., roads, parks, and buildings) constructed by human beings, as opposed to undeveloped, rural areas.

burden of disease The impact of disease in a population. An approach to the analysis of health problems, including loss of healthy years of life; one measure of the burden of disease is disability-adjusted life years (DALYs).

butylated hydroxyl anisole (BHA) Antioxidant used for foods that contain fats and oils.

butylated hydroxytoluene (BHT) Antioxidant used for foods that contain fats and oils.

carbon dioxide (CO_2) A by-product of combustion of fossil fuels, carbon dioxide is a greenhouse gas.

carbon monoxide (CO) A colorless, odorless, poisonous gas, produced by incomplete burning of carbon-based fuels, including gasoline, oil, and wood.

carboxyhemoglobin A combination of carbon monoxide and hemoglobin, which at high levels limits the oxygen-carrying capacity of blood.

carcinogen A chemical (or substance) that causes or is suspected of causing cancer, a disease associated with unregulated proliferation of cells in the body.

carrying capacity The maximum number of individuals that can be supported sustainably by a given environment.

case–control study A study in which subjects who participate are defined on the basis of the presence (cases) or absence (controls) of outcome of interest. Association between exposure and outcome is measured by the odds ratio (OR).

case series study Study in which information about individual patients who share a disease in common is gathered over time.

cause (mechanism) of injury The way in which a person sustains an injury.

chemical allergy An immunologically mediated adverse reaction to a chemical resulting from previous sensitization to that chemical or to a structurally similar one.

chronic obstructive pulmonary disease (COPD) A group of conditions subsumed under the category of "chronic lower respiratory diseases"; includes chronic and unspecified bronchitis, emphysema, asthma, and other chronic lower respiratory diseases.

coalitive interaction Several agents that individually have no known toxic effects interact to produce a toxic effect.

cohort study A type of study in which subjects are classified according to their exposure to a factor of interest and then are observed over time to document the occurrence of new cases (incidence) of disease or other health events. Relative risk (RR) is the measure of association.

colony collapse disorder (CCD) Occurs when worker bees abandon their hive, leaving behind the queen bee and young bees, resulting in the eventual death of the bee hive.

common units A system of measurement of radiation; still used often in the United States, it has been superseded by the SI system.

completed fertility rate The number of children a woman has given girth to when she completes childbearing; see also *total fertility rate.*

composting The controlled decomposition (breaking down through the action of physical and chemical processes) of organic materials, such as leaves, grass, and food scraps, by microorganisms.

composting toilet Unit that converts human excrement into a soil-like material that should be buried or hauled away for disposal.

confounding A situation in which a measure of the effect of an exposure on risk is distorted because of the association of the exposure with other factor(s) that influence(s) the outcome under study. The other factor or factors were not controlled for in the study design.

criteria air pollutants A group of very common air pollutants regulated by the U.S. Environmental Protection Agency (EPA), including ozone, nitrogen dioxide, carbon monoxide, sulfur dioxide, particulate matter, and lead.

cross-sectional study A study that examines the relationship between diseases (or other health-related characteristics) and other variables of interest as they exist in a defined population at one particular time.

Curie (Ci) A common unit of measure used to describe the amount of radioactivity in a sample of material.

decibel (dB) A logarithmic scale designed to measure sound pressure levels.

demographic transition Alterations over time in a population's fertility rate, mortality rate, and composition.

descriptive epidemiology The depiction of the occurrence of disease in populations according to classification by person (e.g., sex, age, race), place, and time variables.

dioxin(s) Dioxin usually refers to the chemical TCDD (2,3,7,8-tetrachlorodibenzo-p-dioxin), but dioxins refer to a family of dioxin-like compounds that are unintentional by-products of certain industrial, nonindustrial, and natural processes, usually involving combustion.

disability-adjusted life expectancy The number of years of healthy life expected in a particular population.

disability-adjusted life years (DALYs) Adjustment of life expectancy to allow for long-term disability as estimated from official statistics; a DALY lost is a measure of the burden of disease on a defined population.

dose The amount of a substance administered at one time.

dose-response relationship The relationship of observed outcomes (responses) in a population to varying levels of a protective or harmful agent such as a form of medication or an environmental contaminant.

ecologic study A study in which the units of analysis are populations or groups of people, rather than individuals; also called *ecological study.*

ecological model Proposes that the determinants of health (environmental, biological, and behavioral) interact and are interlinked over the life course of individuals.

ecosystem A dynamic complex of plant, animal, and microorganism communities and the nonliving environment interacting as a functional unit.

emerging zoonoses Zoonotic diseases that are caused by either apparently new agents or by known agents that occur in locales or species that previously did not appear to be affected by these known agents.

environment The complex of physical, chemical, and biotic factors (such as climate, soil, and living things) that acts upon an organism or an ecological community and ultimately determines its form and survival.

environmental epidemiology The study of diseases and health conditions that occur in the population and are linked to environmental factors.

environmental health Comprises those aspects of human health, including quality of life, that are determined by physical, chemical, biological, social, and psychosocial factors in the environment. It also refers to the theory and practice of assessing, correcting, controlling, and preventing those factors in the environment that potentially can affect adversely the health of present and future generations.

environmental impact assessment (EIA) A process for reviewing the potential impact of anthropogenic (human-related) activities (e.g., a proposed construction project) with respect to their general environmental consequences.

environmental justice The equal treatment of all people in society irrespective of their racial background, country of origin, and socioeconomic status, especially concerning environmental laws and policies.

environmental objectives Statements of policy intended to be assessed using information from a monitoring program.

environmental policy A statement by an organization (public or private) of its intentions and principles in relation to its overall environmental performance.

environmental risk transition Changes in environmental risks that happen as a consequence of economic development in the less developed regions of the world.

environmental sustainability The concept that resources should not be depleted faster than they can be regenerated; the concept also specifies that there should be no permanent change to the natural environment.

environmental tobacco smoke (ETS) Secondhand smoke; also known as sidestream smoke.

environmental toxicology The study of how ecological systems—their structure, dynamics, and function—are affected by pollutants.

epidemiologic transition A shift in the pattern of morbidity and mortality from causes related primarily to infectious and communicable diseases to causes associated with chronic, degenerative diseases.

epidemiologic triangle Fundamental model of disease causality that includes three major categories of factors: agent, host, and environment.

epidemiology The study of the distribution and determinants of health and diseases, morbidity, injuries, disability, and mortality in populations.

expectation of life See *life expectancy*.

essential metals Metals (e.g., copper, zinc, manganese, and iron) necessary for human nutrition in trace amounts, but that can be toxic if ingested in excessive amounts.

exposure Proximity and/or contact with a source of a disease agent in such a manner that effective transmission of the agent or harmful effects of the agent may occur.

extraterrestrial radiation Primary cosmic rays that originate from outer space (e.g., the earth's galaxy and the sun) and interact with the earth's atmosphere to produce secondary cosmic rays.

extremely low frequency radiation (ELF) A form of nonionizing radiation in the range of 50 to 60 hertz that originates from electric power poles, wiring in the walls of buildings, and some electrical appliances.

finished water Water leaving a treatment plant and ready to be used by consumers after being collected, treated, and, usually, filtered.

fluorosis A dental condition in which teeth have been discolored by fluoride.

food additives Substances that become part of a food product when added (intentionally or unintentionally) during the processing or production of that food.

food insecurity A situation in which supplies of wholesome foods are uncertain or may have limited availability.

foodborne diseases Illnesses acquired by consumption of contaminated food; they are frequently and inaccurately referred to as food poisoning.

foodborne infections Foodborne diseases induced by infectious agents such as some bacteria (e.g., *Salmonella*) that cause foodborne illness directly.

foodborne intoxications Foodborne diseases incited by agents that do not cause infections directly but produce spores or toxins as they multiply in food; these toxins then can affect the nervous system (neurotoxins) or the digestive system.

foodborne outbreak The occurrence among two or more people of cases of a similar illness, which an investigation has linked to consumption of a common meal or food items. In the instance of botulism, one case constitutes an outbreak.

freshwater lake A lake usually located at a high latitude containing freshwater as opposed to saltwater.

gamma rays Consisting of photons, these high-energy rays are part of the electromagnetic spectrum and are the most highly penetrating type of radiation.

glaciers and icecaps Frozen freshwater located in high elevations and near the earth's poles.

global warming An increase in the near-surface temperature of the earth, in part due to increased emissions of greenhouse gases and deforestation.

greenhouse gases Gases that arise from natural sources and anthropogenic activities such as the burning of fossil fuels and that may accumulate in the atmosphere. They are referred to as greenhouse gases because in sufficient concentrations in the atmosphere, they may have the effect of trapping heat and causing the earth's temperature to rise.

groundwater Any water that is stored naturally deep underground in aquifers or that flows through rock or soil, supplying springs and wells; this water is less susceptible to contamination than surface water.

hazard The inherent capability of an agent or a situation to have an adverse effect; a factor or exposure that may adversely affect health.

hazard analysis of critical control points (HACCP) A system for reducing the risk of foodborne illness in which foods are monitored from the time of harvest to the time of consumption.

hazardous waste By-products of society that can pose a substantial or potential hazard to human health or the environment when improperly managed. It possesses at least one of four characteristics (ignitability, corrosivity, reactivity, or toxicity), or appears on special EPA lists.

health impact assessment (HIA) A method for describing and estimating the effects that a proposed project or policy may have on the health of a population.

healthy worker effect The observation that employed populations tend to have lower mortality rates than the general population.

heavy metal A metal that has a high atomic weight with a specific gravity that exceeds the specific gravity of water by five or more times. Examples include arsenic, lead, and mercury.

hormesis The belief that some nonnutritional toxic substances also may impart beneficial or stimulatory effects at low doses, but at higher doses they produce adverse effects.

host A person or other living animal, including birds and arthropods, that affords subsistence or lodgment to an infectious agent under natural conditions.

hydrocarbon A substance that consists of only carbon and hydrogen atoms. Benzene is an example.

hydrological cycle The natural cycle by which water evaporates from oceans and other water bodies, accumulates as water vapor in clouds, and returns to oceans and other water bodies as precipitation.

incidence The occurrence of new disease or mortality within a defined period of observation (e.g., a week, month, year, or other time period) in a specified population.

incidence rate The rate at which new events occur in a population.

injury Physical damage to a body that results when it is suddenly or briefly subjected to intolerable levels of energy.

injury epidemiology A branch of epidemiology that studies the distribution and determinants of injuries in the population.

insecticide A pesticide compound specifically used to kill or prevent the growth of insects.

intent of injury Whether an injury was caused by an act carried out on purpose or by accident.

intervention study An investigation involving intentional change in some aspect of the status of the subjects. An example of such a change is the introduction of a preventive or therapeutic regimen.

ionizing radiation Radiation that has so much energy it can knock electrons out of atoms, a process known as ionization. Ionizing radiation can affect the atoms of living things, so it poses a health risk by damaging tissue and DNA in genes. Ionizing radiation comes from radioactive elements, cosmic particles from outer space, and x-ray machines.

isotope Each of two or more forms of the same element that contain equal numbers of protons but different numbers of neutrons in their nuclei, and hence differ in atomic mass but not in chemical properties; in particular, a radioactive form of an element.

itai-itai disease A condition attributed to ingestion of cadmium and characterized by excruciating pain across the entire body.

job stress The harmful physical and emotional responses that occur when the requirements of the job do not match the capabilities, resources, or needs of the worker.

latency The time period between initial exposure to a factor and a measurable response.

leachate A liquid produced as water percolates through wastes, collecting contaminants.

lethal dose 50 (LD_{50}) The dosage (mg/kg body weight) that causes death in 50% of exposed animals.

life expectancy The average number of years an individual is expected to live if current mortality rates continue to apply. Life expectancy at birth is the average number of years a newborn baby can be expected to live if current mortality trends continue.

medical waste A subset of wastes generated in healthcare facilities, such as hospitals, physicians' offices, dental practices, blood banks, and veterinary hospitals/clinics, as well as medical research facilities and laboratories.

megacities Urbanized areas that have 10 million or more inhabitants.

megafills Landfills that take in 5,000–10,000 tons (4,536 to 9,072 metric tons) of trash per day and serve regional needs for waste disposal.

melanosis A dermatologic condition that can cause the darkening of the skin of the entire body (diffuse melanosis) or spotted pigmentation of the skin or tissues inside the mouth (spotted melanosis); associated with chronic low-level arsenic exposure.

metaphylaxis A procedure in which antimicrobials are given to an entire collection of animals, for example, chickens on a poultry farm or fish raised in pens, for the purpose of curing disease among the sick animals and preventing disease among the remainder.

methyl isocyanate (MIC) An intermediate chemical used for the manufacture of carbamate pesticides.

microwave radiation A form of nonionizing radiation produced by radar; a major use is to heat foods. Microwave radiation is associated with ocular effects (lens opacities such as cataracts and clouding of the lenses of the eyes) at high levels.

Minamata disease A neurologic condition associated with ingestion of mercury; characterized by numbness of the extremities, deafness, poor vision, and drowsiness.

municipal solid waste (MSW) Trash or garbage; includes packaging, furniture, clothing, bottles, food waste, papers, batteries, and organic materials such as grass clippings from landscaping.

musculoskeletal disorder (MSD) An injury or disorder of the muscles, nerves, tendons, joints, cartilage, or spinal discs.

National Priorities List (NPL) The EPA's list of the most serious uncontrolled or abandoned hazardous waste sites identified for possible long-term remedial action under Superfund legislation.

natural experiments Naturally occurring circumstances in which subsets of the population have different levels of exposure to a supposed causal factor, in a situation resembling an actual experiment where human subjects would be randomly allocated to groups.

nitrogen oxides (NO_x) Gases made up of a single molecule of nitrogen combined with varying numbers of molecules of oxygen; produced from burning fuels, which react with volatile organic compounds to form smog and with water to form acid rain.

noise A type of sound that is unwanted; can be made up of pure tones, a narrow band of frequencies, and sound impulses.

nonionizing radiation Radiation that has enough energy to move atoms in a molecule around or cause them to vibrate, but not enough to remove electrons. Examples are sound waves, visible light, and microwaves.

nonrenewable water Water in aquifers and other natural reservoirs that is not recharged by the hydrological cycle or is recharged so slowly that significant withdrawal for human use causes depletion.

nuclide Any species of atom that exists for a measurable length of time. A nuclide can be distinguished by its atomic mass, atomic number, and energy state.

occupational disease Those health outcomes that are caused or influenced by exposure to general conditions or specific hazards encountered in the work environment.

odds ratio (OR) The measure of association between exposure and outcome used in case–control studies. The OR refers to the ratio of the odds in favor of exposure among the disease group (the cases) to the odds in favor of exposure among the no-disease group (the controls).

organic chemical Generally speaking, a substance that contains carbon atoms, although this definition is not always true, as in the case of the gas carbon dioxide (CO_2). (CO_2 contains carbon atoms but is not an organic chemical.) Organic chemicals are found in all living organisms, although many also can be synthesized.

organochlorines A wide range of chemicals that contain carbon, chlorine, and, sometimes, several other elements; these chemicals include many herbicides, insecticides, and fungicides as well as industrial chemicals such as polychlorinated biphenyls (PCBs).

organophosphate (OP) pesticides Pesticides that act on the nervous systems of insects and higher organisms by disrupting the transmission of nerve impulses.

ozone (O_3) A gas that consists of three oxygen atoms bonded together into a molecule; it is a main component of ground-level, photochemical smog.

particulate matter (PM) Refers generically to a mixture of airborne particles that can vary in size; it is one of the constituents of air pollution.

passive smoking The involuntary breathing of cigarette smoke by nonsmokers in an environment where there are cigarette smokers present; also known as exposure to sidestream cigarette smoke or environmental tobacco smoke (ETS).

persistent organic pollutants (POPs) Carbon-containing chemical compounds that, to varying degrees, resist photochemical, biological, and chemical degradation.

personal protective equipment (PPE) Apparatuses designed to protect employees from serious workplace injuries or illnesses resulting from contact with chemical, radiological, physical, electrical, mechanical, or other workplace hazards.

pesticide Any substance or mixture of substances intended for preventing, destroying, repelling, or mitigating pests. Also, any substance or mixture intended for use as a plant regulator, defoliant, or desiccant.

pests Unwanted organisms such as insects, rodents, and weeds.

photoallergy A type of dermatitis that is activated by sunlight, which sensitizes the skin to allergens.

phototoxicity A severe reaction to sunlight that occurs sometimes in conjunction with drugs (e.g., antibiotics and diuretics) that a person may be taking.

phthalates A class of plasticizer that is used to increase the flexibility of plastics; also used as additives in cosmetics, aspirin, and insecticides.

plasticizer A substance that is added during the manufacture of a plastic to make it more pliable and softer.

pneumoconiosis The accumulation of dust in the lungs and the tissue reactions to its presence. For the purpose of this definition, "dust" is meant to be an aerosol composed of solid inanimate particles.

point prevalence All cases of a disease that exist at a particular point in time relative to a specific population from which the cases are derived.

poison Any agent capable of producing a deleterious response in a biological system.

policy cycle The distinct phases involved in the policy-making process; comprises five stages: (1) problem definition, formulation, and reformulation; (2) agenda setting; (3) policy establishment (i.e., adoption and legislation); (4) policy implementation; and (5) policy assessment.

polluter-pays principle States that the polluter should bear the expenses of carrying out pollution prevention and control measures.

pollution The introduction into the natural environment by humans of substances, materials, or energy that cause hazards to human health, harm to living resources and ecological systems, and damage to structures and amenities, or that interfere with the legitimate uses of the environment.

polychlorinated biphenyls (PCBs) Classified as dioxin-like chemicals, these substances comprise an extensive class of organochlorines.

polycyclic aromatic hydrocarbons (PAHs) A group of over 100 different chemicals that are formed during the incomplete burning of coal, oil and gas, garbage, or other organic substances such as tobacco or charbroiled meat.

population at risk Those members of the population who are capable of developing the disease or condition being studied.

population dynamics The ever-changing interrelationships among the set of variables that influence the demographic makeup of populations as well as the variables that influence the growth and decline of population sizes.

potentiation One chemical that is not toxic causes another chemical to become more toxic.

precautionary principle States that preventive, anticipatory measures should be taken when an activity raises threats of harm to the environment, wildlife, or human health, even if some cause-and-effect relationships are not fully established.

preservatives Additives to foods that act as antimicrobial agents and antioxidants, retard spoiling by microbial agents, prevent fats and oils from becoming rancid, and prevent cut fresh fruits from turning brown.

prevalence The number of existing cases of a disease, health condition, or deaths in a population at some designated time.

prevention strategies Methods to prevent the occurrence of future injuries.

problem definition The process of gathering data and analyzing datasets for patterns and trends in order to prevent injuries.

protective factors Variables that reduce the probability of an injury.

public health surveillance The ongoing systematic collection, analysis, and interpretation of health data essential to the planning, implementation, and evaluation of public health practices, closely integrated with the timely dissemination of these data to those who need to know.

radiation Propagation of energy through space, or some other medium, in the form of electromagnetic waves or particles. Radiation also includes neutrons, which are part of the nucleus of an atom and do not have an electrical charge.

radioactive decay Reduction in activity of a quantity of radioactive material by disintegration of its atoms. Elements that undergo radioactive decay are said to be radioactive.

radioactivity The spontaneous emission of radiation from the nucleus of an unstable atom. As a result of this emission, the radioactive atom is converted, or decays, into an atom of a different element that might or might not be radioactive.

radiofrequency (RF) radiation Occurs when an electric charge alternates (oscillates back and forth) very rapidly in a radio antenna to produce radio waves, which move at the speed of light.

radioisotope A radioactive isotope. A common term for a radionuclide.

radiological dispersal device (RDD) Combines a conventional explosive, such as dynamite, with radioactive material; commonly known as a dirty bomb.

radionuclide A radioactive nuclide. An unstable isotope of an element that decays or disintegrates spontaneously, emitting radiation.

recycling The process of minimizing waste generation by recovering and reprocessing usable products that might otherwise become waste (e.g., recycling of aluminum cans, paper, and bottles).

relative risk (RR) The ratio of the incidence rate of a disease or health outcome in an exposed group to the incidence rate of the disease or condition in a nonexposed group; the measure of association between exposure and outcome used in cohort studies.

rem (roentgen equivalent in man) A measure of radiation dose related to biological effect (i.e., the equivalent dose or effective dose). A rem is a measure of dose deposited in body tissue, averaged over the mass of the tissue of interest. One rem is approximately the dose from any radiation corresponding to one roentgen of gamma radiation.

renewable water Freshwater that is continuously replenished by the hydrological cycle for withdrawal within reasonable time limits; includes water in rivers, lakes, or reservoirs that fill from precipitation or runoff.

risk The likelihood of experiencing an adverse effect.

risk assessment The process of determining risks to health attributable to environmental or other hazards; involves (1) hazard assessment, (2) dose–response assessment, (3) exposure assessment, and (4) risk characterization.

risk management The adoption of steps to eliminate identified risks or lower them to acceptable levels (often as determined by a government agency that has taken into account input from the public).

runoff Water originating as precipitation on land that then runs off the land into rivers, streams, and lakes, eventually reaching the oceans, inland seas, or aquifers, unless it evaporates first.

selection bias Bias in the estimated association or effect of an exposure on the outcome that arises from the procedures used to select individuals into the study or the analysis.

septic system A "well" that is used to emplace sanitary waste below the surface and typically comprises a tank and subsurface fluid distribution system or disposal system.

sewage The wastewater generated by people in homes and businesses.

SI (System International) units International System of Units, which has been implemented more recently than the common units system for measurement of radiation.

sick building syndrome (SBS) A situation in which building occupants experience acute health and comfort effects that appear to be linked to time spent in a building, but no specific illness or cause can be identified.

sister chromatid exchange (SCE) A reciprocal exchange of DNA between a pair of DNA molecules.

smog A mixture of pollutants, principally ground-level ozone, produced by chemical reactions in the air among certain chemicals (e.g., volatile organic chemicals).

smog complex Symptoms associated with smog, including eye irritation, irritation of the respiratory tract, chest pains, cough, shortness of breath, nausea, and headache.

social environment Influences upon the individual that arise from societal and cultural factors.

solvent A liquid substance capable of dissolving other substances.

source reduction Reducing waste at the source; the most environmentally preferred strategy.

source water The untreated and unfiltered water in rivers, streams, lakes, and aquifers from which water utilities draw water to be treated, filtered, and tested to produce drinking water.

specific gravity The relative density of a solid or liquid, usually when measured at a temperature of 20°C (68°F), compared with the maximum density of water (at 4°C [39.2°F]).

spectrum of toxic dose The toxicity or hazards that are related to exposure to a particular chemical.

sulfur dioxide (SO_2) Gas produced by burning sulfur contaminants in fuel (such as coal).

Superfund A federal fund provided through the Comprehensive Environmental Response, Compensation, and Liability Act (CERCLA) to clean up uncontrolled or abandoned hazardous-waste sites, as well as accidents, spills, and other emergency releases of pollutants and contaminants into the environment.

surface water Water from lakes, streams, rivers, and surface springs.

surveillance system The systematic gathering of information about the occurrence of diseases, injuries, and other health phenomena.

synergism As applied to toxicology, indicates that the combined effect of exposures to two or more chemicals is greater than the sum of their individual effects.

temperature inversion An atmospheric condition during which a layer of warm air stalls above a layer of cool air that is closer to the surface of the earth.

threshold As applied to toxicology, the lowest dose of a chemical at which a particular response may occur.

threshold limit value (TLV) Refers to airborne concentrations of substances and represents conditions under which it is believed that nearly all workers may be unaffected.

total fertility rate (TFR) The number of children a woman has given girth to when she completes childbearing; see also *completed fertility rate.*

toxic agent A material or factor that can be harmful to biological systems.

toxic substance A material that has toxic properties.

toxicants Toxic substances that are human-made or result from human (anthropogenic) activity.

toxicity The degree to which something is poisonous; denotes the amount of a substance that can produce a deleterious effect.

toxicologist Scientist who investigates in living organisms the adverse effects of chemicals (including their cellular, biochemical, and molecular mechanisms of action) and assesses the probability of the occurrence of such effects.

toxicology The study of the adverse effects of chemicals on living organisms.

toxin Usually refers to a toxic substance made by living organisms including reptiles, insects, plants, and microorganisms.

trace metals See *essential metals.*

trichloroethylene (TCE) A grease-dissolving solvent that evaporates readily; added to some cleaning agents, paint removers, and adhesives.

ultraviolet radiation (UVR) A form of nonionizing radiation in the optical range. Sources of UVR include the sun, welders' arcs, lamps used for tanning beds, and some flood lamps used in photography.

unintentional injury An injury or poisoning that is not inflicted on purpose. This category includes those injuries and poisonings described as unintended or "accidental," regardless of whether the injury was inflicted by oneself or by another person.

variant Creutzfeldt-Jakob disease (vCJD) One of the four main human prion diseases; it has been linked causally with bovine spongiform encephalopathy (BSE), a prion disease of cattle. The favored hypothesis is that humans are infected through dietary consumption of the BSE agent (from cattle).

vector An insect or any living carrier that transports an infectious agent from an infected individual or its wastes to a susceptible individual or its food or immediate surroundings.

vector-borne infection A type of infection transmitted by a vector; the several classes of vector-borne infections are determined by the interaction among the infectious agent, the human host, and the vector.

volatile organic compounds (VOCs) Organic compounds (often in paints and solvents) that easily become vapors or gases. Along with carbon, they contain one or more elements such as hydrogen, oxygen, fluorine, chlorine, bromine, sulfur, and nitrogen.

waste In the context of solid or liquid waste disposal, any material perceived to be of negative value.

water reservoir An artificial lake, produced by constructing physical barriers across flowing rivers, which allows the water to pool.

water scarcity An annual supply of renewable freshwater less than 1,000 cubic meters (35,000 cubic feet) per person.

water stress An annual supply of renewable freshwater between 1,000 and 1,700 cubic meters (35,000 to 60,000 cubic feet) per person.

water withdrawal Removal of freshwater for human use from any natural source or reservoir, such as a lake, river, or aquifer.

waterborne diseases Conditions that are transmitted through the ingestion of contaminated water, which acts as the passive carrier of the infectious agent.

wetlands Swamps, bogs, marshes, mires, lagoons, and floodplains.

widespread adoption Dissemination of prevention strategies on a broad scale.

xenobiotics Chemical substances that are foreign to the biological system; they include naturally occurring compounds, drugs, environmental agents, carcinogens, and insecticides.

X-rays Penetrating electromagnetic radiation whose wavelengths are shorter than those of visible light.

zoonosis An infection or infectious disease transmissible under natural conditions from vertebrate animals to humans.

Index

Exhibits, figures, and tables are indicated by exh, f, and t following the page numbers.

A

C

D

E

F

I

J

K

L

O

Q

R

S

T

U

V

W

X

Y

Z